Microbial Ecology

FUNDAMENTALS AND APPLICATIONS

Third Edition

Of Related Interest from the Benajmin/Cummings Series in the Life Sciences

General Biology

N. A. Campbell
Biology, third edition (forthcoming in 1993)

J. Morgan and E. B. Carter
Investigating Biology: A Laboratory Manual
(forthcoming in 1993)

Plant Biology

M. G. Barbour, J. H. Burk, and W. D. Pitts
Terrestrial Plant Ecology, second edition (1987)

J. Mauseth
Plant Anatomy (1988)

L. Taiz and E. Zeiger
Plant Physiology (1991)

Biochemistry and Cell Biology

W. M. Becker and D. W. Deamer
The World of the Cell, second edition (1991)

C. Mathews and K. H. van Holde
Biochemistry (1990)

G. L. Sackheim
Chemistry for Biology Students, fourth edition (1990)

W. B. Wood, J. H. Wilson, R. M. Benbow, and L. E. Hood
Biochemistry: A Problems Approach, second edition (1981)

Molecular Biology and Genetics

E. J. Ayala and J. A. Kiger, Jr.
Modern Genetics, second edition (1984)

L. E. Hood, I. L. Weissman, W. B. Wood, and J. H. Wilson
Immunology, second edition (1984)

R. Schief
Genetics and Molecular Biology (1986)

J. D. Watson, N. H. Hopkins, J. W. Roberts, J. A. Steitz,
and A. M. Weiner
Molecular Biology of the Gene, fourth edition (1987)

G. Zubay
Genetics (1987)

Microbiology

I. E. Alcamo
Fundamentals of Microbiology, third edition (1991)

J. Cappuccino and N. Sherman
Microbiology: A Laboratory Manual, third edition (1992)

T. R. Johnson and C. L. Case
Laboratory Experiments in Microbiology, brief edition,
third edition (1992)

G. J. Tortora, B. R. Funke, and C. L. Case
Microbiology: An Introduction, fourth edition (1992)

Evolution, Ecology, and Behavior

D. D. Chiras
Environmental Science, third edition (1991)

R. J. Lederer
Ecology and Field Biology (1984)

M. Lerman
Marine Biology: Environment, Diversity, and Ecology
(1986)

E. Minkoff
Evolutionary Biology (1983)

R. Trivers
Social Evolution (1985)

Animal Biology

H. E. Evans
Insect Biology: A Textbook of Entomology (1984)

E. N. Marieb
Essentials of Human Anatomy and Physiology, third
edition (1991)

E. N. Marieb
Human Anatomy and Physiology, second edition (1992)

E. N. Marieb and J. Mallatt
Human Anatomy (1992)

L. G. Mitchell, J. A. Mutchmor, and W. D. Dolphin
Zoology (1988)

A. P. Spence
Basic Human Anatomy, third edition (1991)

Microbial Ecology

FUNDAMENTALS AND APPLICATIONS

Third Edition

Ronald M. Atlas   *&*   **Richard Bartha**
University of Louisville   Rutgers University

THE BENJAMIN/CUMMINGS PUBLISHING COMPANY, INC.
Redwood City, California • Menlo Park, California • Reading, Massachusetts
New York • Don Mills, Ontario • Wokingham, U.K. • Amsterdam • Bonn • Sydney
Singapore • Tokyo • Madrid • San Juan

Senior Editor: Edith Beard Brady
Associate Aquisitions Editor: Lisa Donohoe
Editorial Assistant: Sissy Lemon
Editorial Consultant: Dr. Lawrence Parks
Production Editor: Gail Carrigan
Technical Consultant: Tony Jonick
Text Designer: Richard Kharibian & Associates
Cover Designer: Yvo Riezebos

Cover Photo: H. W. Jannasch and C. O. Wirsen, *Appl. Environ. Microbiol.,* 41:528–538 1981. Courtesy of Holger W. Jannasch, Woods Hole Oceanographic Institution, Woods Hole, MA.

Library of Congress Cataloging-in-Publication Data

Atlas, Ronald M., 1946–
    Microbial ecology: fundamentals and applications / Ronald M.
    Atlas & Richard Bartha.—3rd ed.
            p.   cm.
    Includes bibliographical references and index.
    ISBN 0-8053-0653-6
    1. Microbial ecology. I. Bartha, Richard. II. Title.
QR100.A87 1992   576´.15—dc20   92–14652
                                                      CIP

    4 5 6 7 8 9 10-CRW-999897969594

The Benjamin/Cummings Publishing Company, Inc.
390 Bridge Parkway, Redwood City, CA 94065

Preface

The field of microbial ecology has experienced dramatic growth during the past twenty-five years. While prior to 1960 this area of specialization was virtually unknown, an impressive body of literature has since accumulated in this field. The rapid development of microbial ecology as a scientific discipline was undoubtedly promoted by societal interest in environmental quality. Many of today's environmental problems, as well as their potential solutions, are intimately interwoven with the microbial component of the global ecosystem. Numerous practical implications add relevance and excitement to this new field; the volume and quality of the work performed in it is steadily increasing. We hope that the third edition of *Microbial Ecology: Fundamentals and Applications,* will continue to stimulate student interest in this exciting field.

The discipline of microbial ecology started to emerge in the early 1960s. It became firmly established as an area of specialization during the late 1970s with increased focus on the environment. Since then, studies in microbial ecology, using new methodologies and novel approaches, have been undertaken to help us understand the role of microorganisms in global ecology and environmental quality. The l990s have seen the introduction of molecular biology into the field of microbial ecology, with the development of new techniques for applying molecular methods to environmental studies. Recombinant DNA technology has raised the specter of using genetically engineered microorganisms for environmental applications, including pest control and removal of pollutants. Practical biotechnological applications to environmental remediation have occurred through the new field of bioremediation.

Microbial ecology cuts across traditional academic subjects and is truly an interdisciplinary science. Students who attend courses in microbial ecology have very diverse backgrounds and career goals. A challenge in writing this book has been to accommodate the varied backgrounds of students taking a course in microbial ecology. The first edition of this textbook in 1981 attempted to provide a comprehensive and flexible teaching tool designed for one-semester courses on the advanced undergraduate and graduate level. In the second edition, published in 1987, we raised the level of presentation and increased the usefulness of the book as a reference source for investigators in the field of microbial ecology. In the second edition, we also provided extensive reference lists for all chapters, and we have updated and expanded these references in this third edition. We have also added a wealth of new material to this third edition, including information on molecular and other new methods now used in microbial ecology. Like previous editions, the third edition of *Microbial Ecology: Fundamentals and Applications* contains both basic ecological and applied environmental information.

In the third edition we have removed the chapter on microbial taxonomy and included this information as an appendix; this material is generally covered in introductory courses. Chapter 1 contains a short historical introduction and discusses the field of microbial ecology in relation to other scientific disciplines. Chapter 2 discusses microbial evolution and the diversity of microorganisms, including the various physiological types of microorganism. Chapter 3 describes the interactions between microorganisms, and chapters 4 and 5 explore the interactions of microorganisms with plants and with animals, respectively. Chapter 6 is devoted to microbial communities and includes a discussion of successional processes. Chapter 7 discusses the quantitative measurement of numbers, biomass, and activity of microorganisms; chapter 8 examines the influence and measurement of environmental determinants. Chapter 9 presents air, water, and soil as microbial habitats and describes the typical composition of their communities. Chapters 10 and 11 contain an expanded discussion of the biogeochemical cycling activities performed by microbial communities. Chapters 12 through 15 deal with applied aspects of microbial ecology evident in biodeterioration control, sanitation, soil conservation, pollution control, resource recovery, and biological control. The appendices, provided primarily as resource material rather than for formal classroom teaching, discuss statistical concepts that are frequently used in microbial ecology research and the taxonomy of microorganisms. We have also added a glossary to aid the reader in finding new terms that are used in the field of microbial ecology.

Throughout the textbook, we discuss not only the state of our understanding of microbial ecology but also the methodology employed in obtaining this knowledge. The emergence of the field of microbial ecology has gone hand in hand with technical developments, and the field will continue to expand as new experimental tools become available. Experimental approaches need to be emphasized because most introductory laboratory courses in microbiology stress pure culture procedures and do little to prepare the student for the investigation of the dynamic interactions of organisms and their environment. We have liberally illustrated general statements with specific figures and tables taken from journal articles; the source articles are cited in the legends and are included among the references. Along with the literature references within the text, they provide access points to the relevant literature on specific topics.

We wish to acknowledge our colleagues, particularly Drs. J. Staley, J. C. Meeks, W. Mitsch, G. Cobbs, M. Finstein, D. Eveleigh, and D. Pramer who reviewed various sections of the first edition of this work; Drs. B. Olson, M. Klug, and C. Remsen, who reviewed the complete revised manuscript for the second edition; and Drs. M. Franklin, M. H. Franklin, J. Harner, M. Sadowsky, S. Schmidt, and S. Jennett, who made valuable suggestions for the third edition. Also we want to thank our many colleagues who generously took the time to respond to our inquiries regarding the strengths and weaknesses of the previous editions and who have informally made suggestions to us for improvements. Their suggestions proved most helpful in improving the final product. The production of the third edition was greatly aided by the efforts of Lisa Donohoe and Gail Carrigan of Benjamin/Cummings and by Tony Jonick, who helped us with the computer layout of the book. We are indebted to the many individuals, societies, and companies who generously provided permission to reprint illustrative material. We wish to thank Larry Parks especially for his tireless efforts in preparing the manuscript and revising the illustrations. We are again grateful to our families for their support during the writing of this book.

Louisville, Kentucky   R. M. A.

New Brunswick, New Jersey   R. B.

June 1992

Contents

Preface   v

PART ONE
Ecology and Evolution   1

1
Microbial Ecology: Historical Development   3

The Scope of Microbial Ecology   3
Historical Overview   4
   The Beginnings of Microbiology   4
   The Pure Culture Period   6
   Microbial Ecology into the Twentieth Century   8
Relation of Microbial Ecology to General Ecology   11
Sources of Information for the Microbial Ecologist   12
Opportunities for the Microbial Ecologist   17
Summary   18
References & Suggested Readings   18

2
Microbial Evolution   21

The Origins of Life   21
   Chemical Evolution   23
   Cellular Evolution   24
   Evolution of Organelles   28

Genetic Basis for Evolution   29
Evolution of Physiological Diversity   31
Summary   32
References & Suggested Readings   32

PART TWO
Population Interactions   35

3
Interactions among Microbial Populations   37

Interactions within a Single Microbial Population   37
   Positive Interactions   38
   Negative Interactions   40
Interactions between Diverse Microbial Populations   41
   Neutralism   41
   Commensalism   43
   Synergism (Protocooperation)   45
   Mutualism (Symbiosis)   49
   Competition   53
   Amensalism (Antagonism)   56
   Parasitism   57
   Predation   60
Summary   64
References & Suggested Readings   65

4
Interactions between Microorganisms and Plants   69

Interactions with Plant Roots   69
   The Rhizosphere   69
   Mycorrhizae   74
Symbiotic Nitrogen Fixation in Nodules   77
Interactions with Aerial Plant Structures   83
Microbial Diseases of Plants—Plant Pathogens   86
   Viral Diseases of Plants   89
   Bacterial Diseases of Plants   90
   Fungal Diseases of Plants   94
Summary   96
References & Suggested Readings   97

5
Microbial Interactions with Animals   103

Microbial Contributions to Animal Nutrition   103
   Predation on Microorganisms by Animals   103
   Cultivation of Microorganisms by Animals for Food and Nutrition   105
   Associations of Chemolithotrophs and Deep-Sea Hydrothermal Vent Animals   111
   Digestion within the Rumen   112
   Symbiotic Associations with Photosynthetic Microorganisms   115
Fungal Predation on Animals   117
   Nematode- and Rotifer-Trapping Fungi   117
   Fungal–Scale Insect Associations   118
Symbiotic Light Production   119
Ecological Aspects of Animal Diseases   120
Summary   124
References & Suggested Readings   125

6
Microbial Communities and Ecosystems   130

Development of Microbial Communities   130

Population Selection within Communities: r and K Strategies   131
Succession within Microbial Communities   131
Genetic Exchange in Microbial Communities   138
Structure of Microbial Communities   140
   Diversity and Stability of Microbial Communities   140
   Species Diversity Indices   141
   Genetic Diversity Indices   144
Ecosystems   145
   Experimental Ecosystem Models   145
   Mathematical Models   148
Microbial Communities in Nature   151
   Microbes within Macro-communities   151
   Structure and Function of Some Microbial Communities   154
Summary   156
References & Suggested Readings   156

PART THREE
Quantitative and Habitat Ecology   163

7
Measurement of Microbial Numbers, Biomass, and Activities   165

Sample Collection   166
   Soil Samples   166
   Water Samples   167
   Sediment Samples   169
   Air Samples   169
   Biological Samples   170
Sample Processing   171
Detection of Microbial Populations   172
   Phenotypic Detection   172
   Immunological Detection   173
   Gene Probe Detection   174
Determination of Microbial Numbers   178
   Direct Count Procedures   179
   Viable Count Procedures   186
Determination of Microbial Biomass   189

Biochemical Assays   190

Physiological Approaches to Biomass
Determination   193

Measurement of Microbial Metabolism   194

Heterotrophic Potential   194

Growth Rate Based upon Nucleotide
Incorporation   196

Productivity and Decomposition   196

Specific Enzyme Assays   198

Summary   199

References & Suggested Readings   199

8

Effects of Abiotic Factors and Environmental Extremes on Microorganisms   212

Abiotic Limitations to Microbial Growth   212

Liebig's Law of the Minimum   212

Shelford's Law of Tolerance   212

Environmental Determinants   214

Temperature   215

Radiation   220

Pressure   223

Salinity   227

Water Activity   229

Movement   231

Hydrogen Ion Concentration   232

Redox Potential   234

Magnetic Force   236

Organic Compounds   236

Inorganic Compounds   237

Summary   240

References & Suggested Readings   241

9

Microorganisms in Their Natural Habitats: Air, Water, and Soil Microbiology   246

The Habitat and Its Microbial Inhabitants   246

Atmo-ecosphere   247

Characteristics and Stratification of the
Atmosphere   247

The Atmosphere as Habitat and Medium for
Microbial Dispersal   248

Microorganisms in the Atmo-ecosphere   253

Hydro-ecosphere   255

Freshwater Habitats   255

Composition and Activity of Freshwater Microbial
Communities   260

Marine Habitats   263

Estuaries   264

Characteristics and Stratification of the
Ocean   265

Composition and Activity of Marine Microbial
Communities   269

Litho-ecosphere   271

Rocks   271

Soils   272

Summary   281

References & Suggested Readings   281

**PART FOUR
Biogeochemical Cycling   287**

10

Biogeochemical Cycling: Carbon, Hydrogen, and Oxygen   289

Biogeochemical Cycling   289

Reservoirs and Transfer Rates   291

The Carbon Cycle   292

Carbon Transfer through Food Webs   294

Carbon Cycling within Habitats   298

Methanogenesis and Methylotrophy   299

Acetogenesis   300

Carbon Monoxide Cycling   300

Limitations to Microbial Carbon Cycling   301

Microbial Degradation of Polysaccharides   301

Microbial Degradation of Lignin   304

Biodegradation and Heterotrophic Production
in Aquatic Environments   306

The Hydrogen Cycle   306
The Oxygen Cycle   308
Summary   310
References & Suggested Readings   310

11

Biogeochemical Cycling: Nitrogen, Sulfur, Phosphorus, Iron, and Other Elements   314

The Nitrogen Cycle   314
    Fixation of Molecular Nitrogen   316
    Ammonification   319
    Nitrification   320
    Nitrate Reduction and Denitrification   322
The Sulfur Cycle   323
    Oxidative Sulfur Transformations   324
    Reductive Sulfur Transformations   328
    Some Practical Implications of the
        Sulfur Cycle   330
The Phosphorus Cycle   332
The Iron Cycle   334
The Manganese Cycle   335
Calcium Cycling   336
Silicon Cycling   337
Interrelations Between the Cycling of Individual
Elements   338
Summary   341
References & Suggested Readings   341

PART FIVE
Biotechnological Aspects of Microbial Ecology   347

12

Ecological Aspects of Biodeterioration Control: Soil, Waste, and Water Management   349

Control of Biodeterioration   349
    Fouling Biofilms   352
Management of Agricultural Soils   354

Treatment of Solid Waste   355
    Landfills   356
    Composting   357
Treatment of Liquid Wastes   360
    Biological Oxygen Demand   360
    Aerobic Treatments   363
    Anaerobic Treatments   369
    Tertiary Treatments   372
    Disinfection   375
Treatment and Safety of Water Supplies   375
    Water Quality Testing   376
Summary   380
References & Suggested Readings   380

13

Microbial Interactions with Xenobiotic and Inorganic Pollutants   383

Persistence and Biomagnification of Xenobiotic
Molecules   384
    Recalcitrant Halocarbons   386
    Polychlorinated Biphenyls and Dioxins   388
    Synthetic Polymers   391
    Alkyl Benzyl Sulfonates   392
    Petroleum Hydrocarbons   393
    Pesticides   399
Microbial Interactions with Some Inorganic
Pollutants   404
    Acid Mine Drainage   404
    Microbial Conversions of Nitrate   405
    Microbial Methylations   407
    Microbial Accumulation of Heavy Metals and
        Radionuclides   410
Summary   411
References & Suggested Readings   412

14

Biodegradability Testing and Monitoring the Bioremediation of Xenobiotic Pollutants   417

Biodegradability and Ecological Side Effect
Testing   417

Testing for Biodegradability and
Biomagnification   417

Testing for Effects on Nontarget
Microorganisms   420

Biosensor Detection of Pollutants   420

Reporter Genes   421

Immunoassay Biosensors   421

Bioremediation   421

Bioremediation Efficacy Testing   422

Side Effects Testing   423

Approaches to Bioremediation   424

Environmental Modification for
Bioremediation   424

Microbial Seeding and Bioengineering
Approaches to the Bioremediation of
Pollutants   426

Bioremediation of Various Ecosystems   428

Bioremediation of Contaminated Soils and
Aquifers   428

Bioremediation of Marine Oil Pollutants   430

Bioremediation of Air Pollutants   431

Summary   433

References & Suggested Readings   433

15

Microorganisms in Mineral and Energy Recovery and Fuel and Biomass Production   439

Recovery of Metals   439

Microbial Assimilation of Metals   442

Copper Bioleaching   442

Uranium Bioleaching   444

Recovery of Petroleum   444

Production of Fuels   446

Ethanol   446

Methane   449

Hydrocarbons Other than Methane   450

Hydrogen   450

Production of Microbial Biomass   451

Summary   454

References & Suggested Readings   455

16

Ecological Control of Pests and Disease-causing Populations   460

Modification of Host Populations   460

Immunization   461

Modification of Reservoirs of Pathogens   463

Modification of Vector Populations   464

Microbial Amensalism and Parasitism to Control
Microbial Pathogens   464

Antifungal Amensalism and Parasitism   464

Antibacterial Amensalism and Parasitism   466

Microbial Pathogens and Predators for Controlling
Pest Populations of Plants and Animals   467

Microbial Control of Insect Pests   467

Microbial Control of Other Animal Pests   474

Microbial Control of Weeds and Cyanobacterial
Blooms   475

Genetic Engineering in Biological Control   475

Frost Protection   476

Bacillus thuringiensis Pesticides   477

Other Applications   477

Additional Practical Considerations   477

Summary   478

References & Suggested Readings   478

PART SIX
Appendices   483

Appendix 1

Statistics in Microbial Ecology   485

Data Collection and Experimental Design   485

Independent and Dependent Variables   485

Bias   486

Estimation of Central Tendency   487

Measurements of Variability   488

Range, Variance, and Standard Deviation   488

Hypothesis Testing   489

Significance and Hypothesis Testing   489

Parametric Tests of Significance   490

Nonparametric Tests of Significance   495
Correlation and Regression   496
Cluster Analysis   498
Factor Analysis   501
Computers—the Pragmatic Approach to
Statistics   503
References & Suggested Readings   504

Appendix 2 ⎯⎯⎯⎯⎯
Survey of Microorganisms   506

Acellular Microorganisms   506
Prions   506
Viroids   506
Viruses   506

Prokaryotes   508
Archaebacteria (Archaea)   508
Eubacteria (Bacteria)   511
Eukaryotes (Eucarya)   520
Fungi   520
Algae   524
Protozoa   526
References & Suggested Readings   528

Glossary   533

Index   547

PART ONE

Ecology and Evolution

1

Microbial Ecology: Historical Development

THE SCOPE OF MICROBIAL ECOLOGY

The unprecedented spurt of technological and economic growth in industry that followed the conclusion of World War II—and the accompanying mood of boundless expectations—gradually gave way in the 1960s to feelings of alarm over population explosion, deterioration of the environment, and rapid depletion of nonrenewable natural resources. Humankind, having acquired almost limitless powers to subdue and exploit Earth, appeared able neither to control its own population size nor to manage the limited resources of "Spaceship Earth" in a wise and sustainable fashion. Groups of scientists and economists drew up grim but credible scenarios of impending disaster while attempting to project current population, consumption, and pollution trends into the near future. Proposed remedies stressed population control, limits to technological and economic growth, pollution abatement, and increased reliance on renewable resources for energy and raw materials. Such ideas were presented in books understand-able to large segments of society and had notable effects on societal attitudes and on legislative action. Changes in societal attitudes led to various forms of antipollution legislation, to conservation and consumer activism, and to the formation of the United States Environmental Protection Agency (EPA) in 1970 and other environmental regulatory bodies around the world. Various legislative actions on air pollution, water pollution, and strip mining were enacted to conserve natural resources and to protect human health against deteriorating environmental quality.

The desire of a responsive segment of society to live in harmony with nature, rather than to disrupt it, changed the once rather esoteric concept of ecology into a common household word and stimulated a broad interest in this branch of biology. The term ecology is derived from the Greek words *oikos* (household or dwelling) and *logos* (law). Thus, ecology is "the law of the household" or, by its contemporary definition, the science that explores the interrelationships between organisms and their living (biotic) and nonliving (abiotic) environments. The term ecology was first defined and used in this sense by German biologist Ernst Häeckel (1866). Microbial ecology is the science that specifically examines the relationships between microorganisms and their biotic and abiotic environments. The term microbial ecology came into frequent use only in the early 1960s. The current popularity of microbial ecology and the rapid development of this field in the last thirty years relate to a general awareness of ecology and the role of microorganisms in maintaining a high level of environmental quality.

Microbial ecology emerged as an energetic and dynamic branch of science in this enlightened atmosphere because it was recognized that microorganisms occupy a key position in the orderly flow of materials and energy through the global ecosystem by virtue of their metabolic abilities to transform organic and inorganic substances. The environmental persistence of various synthetic chemicals and plastics, biomagnification of pollutants, eutrophication, acid mine drainage, nitrate pollution of well water, methylation of mercury, depletion of ozone by nitrous oxide in the atmosphere, and a plethora of other environmental problems reflect unfavorable and unintended interactions of human activities with the microbial component of the global ecosystem. At the same time, microorganisms are crucial to solving some of our pressing environmental and economic problems, such as how to dispose properly of various liquid and solid wastes; relieve nitrogen fertilizer shortage; recover metals from low-grade ores; biologically control pests; and produce food, feed, and fuel from by-products and waste materials. These and many other practical implications make microbial ecology a highly relevant and exciting subject for study.

HISTORICAL OVERVIEW

Although widespread use of the term microbial ecology is rather recent, ecologically oriented research on microbes was performed as soon as their existence was realized. Much of this research was identified as soil or aquatic microbiology, but general microbiology, too, contributed to the understanding of the microbial activities crucial to the balance of nature. We need to recognize, however, that some microbiological approaches often considered fundamental to the study of microbes, such as the isolation of pure cultures and their cultivation on synthetic media, are not conducive to ecological observations of the type that the general ecologist can readily make on plants and animals in their natural environments. Studying pure cultures of individual microorganisms eliminates the biological interactions that are the essence of ecological relationships. It is not surprising that some important observations relating to microbial ecology either predated the pure culture technique or were made later by investigators who unwittingly or of necessity dealt with mixed microbial populations rather than with pure cultures.

The Beginnings of Microbiology

Although microbes were probably the first living organisms on Earth, it was not until the mid-seventeenth century that the advent of the microscope permitted their observation. During this period, Robert Hooke, an English experimental philosopher, described his microscopic observations of fungi and protozoa (Hooke 1665). Antonie van Leeuwenhoek in the 1680s viewed the multitude and diversity of the previously hidden world of microbes, including such small forms as yeasts and some bacteria (Dobell 1932). Leeuwenhoek, an amateur scientist and microscope maker in Delft, Holland, used simple but powerful lenses to tediously observe the otherwise invisible microbes (Figure 1.1). His observations, including detailed and recognizable drawings, were recorded in a series of letters he sent between 1674 and 1723 to the Royal Society in London. Through the *Proceedings of the Royal Society,* his discoveries, which otherwise might have been lost, were rapidly disseminated. Leeuwenhoek's reports included descriptions of microbes in rainwater (microbes in their natural habitats) and of the effects of pepper on microbes (environmental influences on microbes), thus providing not only the earliest descriptions of bacteria but also the earliest studies on microbial ecology.

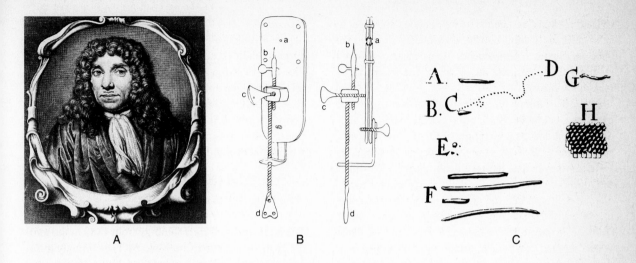

Figure 1.1
(A) Antonie van Leeuwenhoek (1632–1723) who developed simple but powerful microscopes and published the first sketches of microscopic observations of bacteria. (Source: National Library of Medicine.) (B) Early microscopes of Antonie van Leeuwenhoek were simple, consisting of (a) a spherical glass lens held by two metal plates; (b) a specimen holder; and (c) and (d) screws for positioning and focusing the specimen. (Source: C. Dobell 1932. Reprinted by permission of Russell and Russell.) (C) Drawings of microorganisms by Antonie van Leeuwenhoek show several recognizable common bacterial shapes: (a) and (f) rods of varying size; (b) and (e) cocci; (c) and (d) the path of a short motile rod; (g) spiral; (h) cluster of cocci. (Courtesy of Royal Society of London.)

After Leeuwenhoek, little additional progress in the development of the science of microbiology took place until the turn of the eighteenth century, when the Italian naturalist Lazzaro Spallanzani, debating the then-prevalent view of spontaneous generation, produced experimental evidence showing that putrefaction of organic substrates is caused by minute organisms that do not arise spontaneously but multiply by cell division. Spallanzani showed that heating destroys these organisms and that sealing the containers after heating prevents spoilage indefinitely. Heating, however, did not always prevent putrefaction. The heat resistance of bacterial endospores, unknown at that time, contributed to inconsistent and confusing results.

Despite the growing experimental evidence against their views, advocates of the spontaneous generation theory, who claimed that microorganisms arise spontaneously and are the result of the putrefaction of organic substances, were not convinced and tried to explain Spallanzani's experiments by the partial removal of air during the heat treatment. In retrospect, the idea may seem rather strange to us, yet in the first half of the nineteenth century even such eminent scientists as the German chemist Justus von Liebig (1839) firmly believed that yeasts and other microorganisms were the products rather than the causative agents of fermentations.

Using slightly different experimental protocols, Charles Cagniard-Latour (1838), Theodor Schwann (1837), and Friedrich Küntzing (1837) sought to disprove the theory of spontaneous generation, independently reaching the conclusion that living plantlike microorganisms (yeasts) caused alcoholic fermentations. Experiments by Schwann clearly implicated airborne microorganisms as agents of subsequent spoilage in heat-sterilized media, but the controversy dragged on until the classic experiments of Louis Pasteur silenced even the most obstinate advocates of the spontaneous generation theory.

Louis Pasteur, in a beautifully logical series of experiments (1860–1862), utterly demolished the theory of spontaneous generation (Pasteur 1861), reportedly stating, "No more shall the theory of spontaneous generation ever rear its ugly head again." By direct microscopic observation, as well as by cultural methods, he demonstrated the presence of microorganisms in the air and their role in initiating fermentation or spoilage in presterilized media. Pasteur placed yeast water, sugared yeast water, urine, sugar beet juice, and pepper water into ordinary flasks. He then reshaped the necks of the flasks so that there were several curves in each neck, resembling a swan's neck. By using curved necks, Pasteur could leave the flasks open to the air, thus overcoming a major criticism of previous experiments aimed at disproving spontaneous generation where air, an essential life force, had been excluded. Pasteur boiled the liquids until they steamed through the necks. Dust and microbes settled out in the curved necks of the flasks, and thus, while exposed to air, the broth did not become contaminated with microorganisms. Contrary to the opinion of those who believed in spontaneous generation, no change or alteration appeared in the liquid. These flasks, still sterile, were later sealed and may be seen at the Pasteur Institute in Paris.

In his efforts to disprove spontaneous generation, Pasteur found a powerful ally in the English physicist John Tyndall (1820–1893), who demonstrated that optically clear (particle-free) air does not cause spoilage of sterile media and solved the nagging problem of heat-resistant bacterial endospores by discontinuous heat treatments (Tyndall 1877). This technique resulted in germination of endospores and their destruction upon a subsequent heating, thus accomplishing the reliable sterilization of all media at 100°C.

One year before Tyndall devised his sterilization method by intermittent heating, Ferdinand Julius Cohn (1876), professor of botany at the University of Breslau, Germany, described the endospores of *Bacillus subtilis* and their heat-resistant properties. He began efforts to establish the taxonomy of bacteria and was the first scientist to recognize and encourage the work of Robert Koch on the cause of anthrax. He recognized that microorganisms have consistent and characteristic morphologies, a view not shared by many uncritical investigators who, observing a succession of forms in their impure cultures, believed them to be one and the same organism.

The Pure Culture Period

The mainstream of microbiology from 1880 to the middle of the twentieth century was overwhelmingly influenced by the towering figures of Louis Pasteur and Robert Koch (Figure 1.2). A large portion of Pasteur's and Koch's works were oriented toward the most pressing practical problems of their time. Pasteur was trained as a chemist and was prompted to begin his microbiological studies because of the problems of wine producers, brewers, distillers, and vinegar makers. His studies on fermentations between 1857 and 1876 clearly established microorganisms as the causative agents of the various fermentation processes. Both Pasteur and Koch also devoted much of their efforts to examining the role of microbes as causative agents of infectious diseases and to the prevention of such diseases through immunization and sanitation measures. The ability to determine the etiology (cause) of disease depended on the development of appropriate techniques and methodological approaches.

Koch and his coworkers developed simple methods for the isolation and maintenance of pure microbial cultures on solid media (Koch 1883). Previous attempts to obtain pure cultures by dilution or on slices of boiled potato had been cumbersome, uncertain, and not generally applicable. Gelatin and later agar, which Koch utilized to solidify liquid media of any desired composition, were eminently suitable for use in the identification of disease-causing agents. The use of solidified agar media allowed microbiologists to separate out a microbial type from a heterogenous community and study its activities, removed from complex interactions, in an exacting and reproducible

A   B

Figure 1.2
(A) Louis Pasteur (1822–1895), noted for his many accomplishments in microbiology. He demolished the theory of spontaneous generation, demonstrated the microbial origin of various fermentations and diseases, and contributed to the development of immunization by use of attenuated microorganisms. (Source: National Library of Medicine.) (B) Robert Koch (1843–1910), who made major contributions to the development of pure culture technique and the field of medical microbiology. (Source: National Library of Medicine.)

manner. The importance of this advance was monumental and transformed microbiology from a chancy art into an exact science. Many of the pure culture techniques devised by Koch are being used today, a hundred years later, without any substantial modification.

Koch's pure culture techniques established a trend in microbiology that has had a profound influence on the approach used by microbiologists: First isolate a pure culture and then see what it does. The terse statement by Oscar Brefeld (1881), "Work with impure cultures yields nothing but nonsense and *Penicillium glaucum*," summarizes this attitude, and few microbiologists would

argue with its wisdom even today. Unfortunately, this approach provides little opportunity for the observation of interactions between microorganisms and their biotic and abiotic surroundings; thus it has limited value in microbial ecology. We should remember that the inhibition zone around a chance *Penicillium* contaminant on a *Staphylococcus* culture, perceived and correctly interpreted by Sir Alexander Fleming (1929), sparked the line of investigations that eventually yielded penicillin and other lifesaving antibiotics. Microbial ecology depends upon the development of methods for understanding just such interactions.

Microbial Ecology into the Twentieth Century

The pure culture orientation of microbiologists undoubtedly contributed to the relative neglect of ecological research on microorganisms, but ecological research also had to overcome tremendous methodological difficulties. Small size, scarce morphological detail, and low behavioral differentiation of microorganisms are formidable obstacles to conventional ecological observation and research.

Nevertheless, work by microbiologists in the nineteenth century began to establish the role of microbial metabolic activities in global ecological processes. Pasteur's work clearly established the role of microorganisms in the biodegradation of organic substances. He anticipated but failed to prove transformations of inorganic substances by microorganisms in converting ammonium to nitrate (the nitrification process), but proof for such transformations was forthcoming from other investigators. In 1839, Nicolas-Théodore de Saussure reported his observation on the capacity of soil to oxidize hydrogen gas. Since this activity was eliminated by heating, by adding a 25% sodium chloride solution, or by adding a 1% sulfuric acid solution, he correctly concluded that the oxidation was microbial, though hydrogen-oxidizing bacteria were isolated only during the first decade of the twentieth century. Jean Jacques Theophile Schloesing and A. Müntz (1879) reported that ammonium in sewage was oxidized to nitrate during passage through a sand column. The nitrification activity of the column was eliminated by treatment with chloroform vapors but was restored by inoculation with a soil suspension. Again, microbial nitrifying activity was the only logical explanation for this experiment, which also pointed to the presence of nitrifying organisms in soil.

All the above early information was obtained without the aid of the pure culture technique, but soon thereafter Sergei Winogradsky (Figure 1.3), a Russian scientist working at various European universities and later in his life at the Pasteur Institute in Paris, successfully isolated nitrifiers (1890). Winogradsky, during his long and fruitful career, also described the microbial oxidation of hydrogen sulfide and sulfur (1887) and the oxidation of ferrous iron (1888) and correctly developed the concept of microbial chemoautotrophy—though one of his principal models, *Beggiatoa*, could not be grown as a chemoautotroph in pure culture until recently. He also described the anaerobic nitrogen-fixing bacteria and contributed to the studies of reduction of nitrate and symbiotic nitrogen fixation. He originated the nutritional classification of soil microorganisms into the autochthonous (humus-utilizing) and zymogenous (opportunistic) groups and is regarded by many as the founder of soil microbiology (Waksman 1953).

Equally important was the work of the Dutch microbiologist Martinus Beijerinck (Figure 1.4) at the University of Delft. In 1905, half a century before the

Figure 1.3

Sergei Winogradsky (1856–1953), who established the concept of microbial chemoautotrophy and made major contributions to the development of enrichment culture technique and to soil microbiology. (Source: Waksman Institute of Microbiology, Rutgers University.)

Figure 1.4

Martinus Beijerinck (1851–1931), known for his major contributions to the knowledge of nitrogen and sulfur cycling by microorganisms. His works included studies on symbiotic nitrogen fixation. (Source: Waksman Institute of Microbiology, Rutgers University.)

emergence of microbial ecology as an area of specialization, Beijerinck stated, "The way I approach microbiology . . . can be concisely stated as the study of microbial ecology, i.e., of the relation between environmental conditions and the special forms of life corresponding to them" (Van Iterson et al. 1940). Beijerinck isolated the agents of symbiotic (1888) and nonsymbiotic aerobic (1901) nitrogen fixation and also isolated sulfate reducers. He recognized the virtual ubiquity of most microbial forms and the selective influence of the environment that favors the development of certain types of microorganisms. Based on his principle that "everything is everywhere, the environment selects," Beijerinck, with contributions from Winogradsky, developed the immensely useful and adaptable technique of enrichment culture.

Tailoring culture conditions to favor microbes with a particular metabolic ability usually leads to the rapid enrichment and isolation of the desired organism even if its original numbers were very low in the sample (Van Iterson et al. 1940). With additional reports on nitrate reduction (Deherain 1897), methanogenesis and methanothrophy (Söhngen 1906), and the isolation of hydrogen bacteria (Kaserer 1906), the outlines of microbial cycling processes were clear (Doetsch 1960; Brock 1961; Doelle 1974).

The fundamental discoveries described here pertaining to the role of microorganisms in cycling of materials were accomplished by relatively few microbiologists who had an interest in unusual metabolic patterns. Further progress in our understanding of microbial metabolism as related to cycling processes and the unifying metabolic features among micro- and macroorganisms emerged from the work of Albert Jan Kluyver, Cornelius Bernardus van Niel, and Roger Stanier; these microbiologists were notable for the breadth, comparativeness, and ecological implications of their work in microbial physiology. Kluyver (Figure 1.5) succeeded Beijerinck as the leading microbiologist at the University of Delft, the hometown of Leeuwenhoek. He took a broad interest in microbial physiology and with his students and coworkers studied a variety of oxidative, fermentative, and chemoautotrophic microorganisms, including unusual metabolic types. His greatest contribution was perhaps his comparative approach, which stressed the unifying metabolic features within the diverse microbial world (Kamp et al. 1959).

A student of Kluyver's, C. B. van Niel (Figure 1.6), came from the same famous school of Dutch microbiologists. After settling in the United States early in his career (1929), he was instrumental in transplanting his school's tradition of comparative and ecologically oriented microbial physiology to his new country. Among his many contributions, his investigations on bacterial phototrophy have special importance. He pointed out the striking analogy of hydrogen sulfide (H_2S) and water (H_2O) in the photoprocesses of photosynthetic sulfur bacteria and green plants, respectively (Pfennig 1987). Roger Stanier, a student of van Niel's, continued the tradition of unraveling the role of microbial metabolism in ecological processes; his work with *Pseudomonas,* for example, showed the tremendous versatility of aerobic microorganisms in degrading complex organic compounds. Van Niel taught a world-famous laboratory course that emphasized the study of microorganisms from nature. The annual summer microbial ecology course at Woods Hole Oceanographic Institute, Woods Hole, Massachusetts, follows the approach established by van Niel. Robert Hungate, who took van Niel's course, expanded the study of microbial metabolism in the environment further by developing methods for culturing strict anaerobes; he also pioneered studies on the complex interactions of microbial populations by studying the microorganisms in the rumen of cows and the guts of termites (Hungate 1979). As president of the American Society for Microbiology, Hungate in 1970 appointed an ASM Committee on Environmental Microbiology. That same year the International Association of Microbiological Societies established an International Commission on Microbial Ecology and appointed Martin Alexander as chairperson.

Thus, in 1970, the importance of the role microorganisms play in nature received recognition, and the coherent discipline of microbial ecology was formally recognized. An important milestone in the development of microbial ecology was the International Symposium on Modern Methods in Microbial Ecology in Uppsala, Sweden (1972) sponsored by the International Commission on Microbial Ecology. Since 1977, international symposia on microbial ecology have been convened every three years (I., Dunedin, New Zealand, 1977; II., Coventry, England, 1980; III., East Lansing, Michigan, 1983; IV., Ljubljana, Yugoslavia, 1986; V., Kyoto, Japan, 1989; VI., Barcelona, Spain, 1992). At these symposia, hundreds of microbial ecologists

Figure 1.6
Cornelius Bernardus van Niel (1897–1985), who furthered the study of microorganisms in nature and established a model course for the study of microbial ecology. (Source: Hopkins Marine Station, Stanford University)

Figure 1.5
Albert Jan Kluyver (1888–1956), who emphasized the unifying features of microbial metabolism. (Source: Waksman Institute of Microbiology, Rutgers University.)

gather to exchange the latest results of their studies and to discuss their growing understanding of the roles microorganisms play in nature.

RELATION OF MICROBIAL ECOLOGY TO GENERAL ECOLOGY

While in the first half of the twentieth century microbiologists seem to have had little interest to spare for basic ecological theories, during that same time period general ecologists treated the ecological processes of decomposition, detritus formation, and mineral nutrient cycling with equal disdain. These processes were taken for granted and, in contrast to primary production, were seldom scrutinized or quantified. Considering that nutrient regeneration (recycling) can be one of the key constraints of primary production, this neglect is somewhat surprising. But even if general ecologists had been interested in quantifying microbial cycling processes, little in their training prepared them for this task. The microbiologists of that period could have offered little assistance. The principal approach of general ecology was field observation; for microbiologists, it was laboratory experimentation on pure cultures.

Because of their small size and short generation time, microorganisms have potential advantages for studies in population dynamics, but observational difficulties discouraged their use in such investigations. Notable exceptions are the classic experiments of George Francis Gause (1934) on populations of the ciliate protozoa *Paramecium caudatum* and *Didinium nasutum,* the latter preying on the former species. These and similar experiments with *Schizosaccharomyces pombe* (yeast), which is preyed upon by *Paramecium bursaria* (ciliate protozoan), revealed the same out-of-phase cyclic oscillations that are characteristic for some predator-prey interactions in nature. At that time, these ecologically oriented, mixed culture experiments elicited little response among microbiologists, while among zoologists and ecologists they sparked a great deal of additional research and lively debate.

Communication between classical and microbial ecologists in the early twentieth century was minimal or nonexistent; the two disciplines had no common language or approach. Some microbiological methods were, however, developed in this period that allowed microbes to be observed in ways that were compatible with classical ecological approaches. Among the few techniques developed for the *in situ* observation of microorganisms in the first half of this century, the contact slide (Cholodny 1930; Rossi et al. 1936) deserves mentioning. In this method, microscope slides buried in soil or sediment provide a surface not unlike that of soil particles for microbial growth. Careful removal and microscopic observation of these slides provides an opportunity to observe microorganisms as they grow and interact in their natural environment.

The early studies established the basis for later research in microbial ecology. Stimulated by a general interest in the environment and ecology, as well as by the availability of radiotracer, microanalytical and electron microscopic techniques, a rapid upswing in the pace and range of ecologically oriented microbiological research occurred in the early 1960s. Much of this research was prompted by the noticeable and troublesome presence in the environment of synthetic pollutants that were not, or not rapidly enough, degraded by microorganisms, and by eutrophication problems caused by detergent fillers and fertilizer runoff.

Microorganisms had been considered to be infallible biological incinerators. The recognition by Martin Alexander that some compounds produced by humans are recalcitrant, that is, totally resistant to microbial attack, led to increased research on the biodegradability of pollutants (Alexander 1981; Leisinger et al. 1982). In the late 1960s, Ronald Atlas and Richard Bartha began studies on the biodegradation of petroleum pollutants in the sea, pioneering the field of bioremediation, which uses the metabolic activity of microorganisms to remove environmental contaminants (Atlas and Bartha 1992). Additional concern arose about the effects of radiation and pollutants on microbial cycling activities in soil and natural waters. The biomagnification of DDT, PCBs, mercury, and other pollutants called attention to food web interrelations and to the key position of microbial forms as primary producers in aquatic food webs. The nitrogen fertilizer shortage that developed in the wake of the energy crisis of the early 1970s resulted in renewed and intense interest in symbiotic and nonsymbiotic nitrogen–fixation processes. The energy crisis also called for research on fuel generation from renewable resources such as organic wastes and plant biomass, with microorganisms as obvious candidates to effect such conversions.

A great stimulus to microbial ecology was space exploration. The possibility of soft-landing instruments on remote planets prompted the development of methods and instrument packages for detecting microbial life. The testing of these methods and instruments, in turn, stimulated interest in the microbial communities of extremely harsh terrestrial environments, such as the cold, dry valleys of Antarctica (Cameron 1971). Microorganisms living in very high-temperature environments, such as hot springs, were intensively studied by Thomas Brock (Brock 1978). Holger Jannasch further expanded our knowledge of life in extreme environments by his studies on the microorganisms living in proximity to deep-sea thermal vents (Jannasch and Mottl 1985). The progress in methodology and in our understanding of the structure and function of microbial ecosystems that has taken place in the last thirty years makes up a large portion of the material covered in this book.

SOURCES OF INFORMATION FOR THE MICROBIAL ECOLOGIST

Long before the time was ripe for microbial ecology to emerge as a distinct area of specialization, a considerable amount of fundamental knowledge on microbe-environment interactions accumulated in several microbial subdisciplines. As described earlier, the microbiology of soil, with its important agricultural implications, attracted the attention and talent of such distinguished investigators as Sergei Winogradsky and Martinus Beijerinck. Plant pathology, with its

related practical implications, also received early attention. In strong analogy to the progress in medical microbiology, the pure culture technique led rapidly to the isolation and characterization of various fungi and bacteria that cause diseases of plants. Beijerinck also made a crucial contribution to this field with his investigation of plant diseases caused by submicroscopic, filterable principles. Continuation of Beijerinck's unfinished work by others led to the startling discovery of a crystallizable plant pathogenic principle: the tobacco mosaic virus.

Freshwater microbiology, with its strong ties to the practical concerns of safe drinking water and acceptable sewage disposal practices, became an early concern to microbiologists and sanitary engineers. A great deal of information on total microbial and coliform counts in fresh water was amassed and utilized primarily as an indication of water quality. Nevertheless, the survival of nonindigenous coliforms in natural waters and the factors influencing such survival again led to general ecological insights. Marine microbiology, largely devoid of similar sanitary concerns, developed substantially later and largely due to the lifelong work of Claude ZoBell at the Scripps Institute of Oceanography in La Jolla, California. Not only did ZoBell and his coworkers collect an impressive body of data on the occurrence and function of microorganisms in marine waters and sediments, but ZoBell also wrote the first comprehensive text on marine microbiology (1946) and is regarded by many to be the founder of this important field. The ecological implications of his work are many and of foremost importance.

Humans have utilized microorganisms in the processing of food since prehistoric times and, conversely, have striven to prevent the microbial spoilage of stored food by drying, salting, and other methods. In the early years, of course, this was done without any understanding of the microbial nature of these processes. The insights of the nineteenth century into the microbial nature of spoilage and fermentation sparked the rapid development of the subdiscipline of food microbiology. Their applied nature notwithstanding, studies on fermentations and spoilage processes yielded a great deal of fundamental ecological information on the influence of environmental parameters on microorganisms, on survival under adverse conditions, and on microbial succession.

Clearly, a great deal of knowledge on microbial ecology was amassed in related fields prior to the time when microbial ecology emerged as a separate subdiscipline. The 1960s were apparently the right time to consolidate this knowledge and to identify it as a distinct area of specialization. A general ecological awareness, a recognition that many ecological problems by nature cut across arbitrary environmental boundaries and affect the global ecosystem rather than only the soil or the aquatic environment, may have speeded the evolution of microbial ecology. In all other respects, however, the birth of microbial ecology followed closely the process by which other fields of specialization have evolved.

Illustrative of the increased interest and research volume in microbial ecology during the last twenty-five years are Tables 1.1 and 1.2. To our knowledge, no book or periodical appeared prior to 1957 with *Microbial Ecology* or something comparable as its title, but impressive numbers of such books and periodicals have appeared thereafter. The tables briefly summarize the orientation of these books and periodicals.

Publications of interest to microbial ecologists are not limited to those included among the listed periodicals. Many are scattered in a great multitude of general science and specialty journals. To list all relevant journals would be tedious, if not impossible. Instead, Table 1.3 lists subject areas with some representative examples. The microbial ecologist studies interactions among various microorganisms, interactions of microorganisms with their abiotic environment, and interactions of microorganisms with plants and animals.

Obviously, the study of such interactions cuts across the lines of several established disciplines. To be effective, the microbial ecologist requires a broad background in the physicochemical and biological sciences. Specifically, the microbiology student planning to specialize in this area should take care to acquire a good foundation in physics, chemistry, general biology, molecular biology, and microbiology, including its biochemical aspects. Molecular biological methods have become increasingly important in all fields of biology, including microbial ecology. In addition to these basics, it is desirable for a microbial ecologist to have some degree of familiarity with all or at least some of the subjects of botany (especially phytoplankton and plant physiology) and zoology (especially zooplankton and animal physiology), and with aspects of soil science, limnology, oceanography, geochemistry, and climatology that are relevant as environmental determinants of microbial life. A sufficient background in statistics is essential for designing appropriate sampling schedules and for evaluating the significance of acquired data.

Table 1.1
Periodicals on microbial ecology

Journal	Description
Advances in Microbial Ecology. 1977–. Plenum Press, New York.	Review series on basic and applied aspects of microbial ecology.
Applied and Environmental Microbiology. 1976–. American Society for Microbiology, Washington, D.C.	Formerly *Applied Microbiology,* the title was modified to recognize and publish papers on applied and general microbial ecology.
FEMS Microbial Ecology. 1985–. Federation of European Microbiological Societies, Elsevier Science Publishers, Amsterdam, The Netherlands.	Scope similar to *Microbial Ecology.* Created as a publication outlet for European scientists working in this field.
Microbial Ecology. 1974–. Springer-Verlag, New York.	Research reports on all aspects of microbial ecology.

Table 1.2

Text and reference books on microbial ecology

Book	Description
Aaronson, A. 1970. *Experimental Microbial Ecology*. Academic Press, New York.	– A useful compendium of isolation techniques for various types of microorganisms.
Alexander, M. 1971. *Microbial Ecology*. John Wiley and Sons, New York.	– Text written in essay style. Useful mainly at the graduate level.
Alexander, M. 1977. *Introduction to Soil Microbiology*. John Wiley and Sons, New York.	– Text describes the ecology of soil microorganisms.
Alexander, M. (ed.). 1984. *Biological Nitrogen Fixation: Ecology, Terminology and Physiology*. Plenum Press, New York.	– A series of papers on the conversion of molecular nitrogen by nitrogen-fixing bacteria.
Blakeman, J. P. (ed.). 1981. *Microbial Ecology of the Phylloplane*. Academic Press, London.	– Collected papers on microorganisms living on the surfaces of above-ground plant structures.
Brock, T. D. 1966. *Principles of Microbial Ecology*. Prentice-Hall, Englewood Cliffs, N. J.	– First textbook on microbial ecology. Advanced undergraduate level. Out of print.
Brock, T. D. 1978. *Thermophilic Microorganisms and Life at High Temperatures*. Springer-Verlag, New York.	– A comprehensive work on the microorganisms that live at very high temperatures. Written by an authority in the field.
Bull, A. T., and J. H. Slater (eds.). 1982. *Microbial Interactions and Communities*. Academic Press, London.	– A collection of articles describing the basis for interactions between microbial populations.
Burns, R. G., and J. H. Slater (eds.). 1982. *Experimental Microbial Ecology*. Blackwell Scientific Publications, Oxford, England.	– An excellent collection of methodological approaches used in microbial ecology, indicating the applicability and limitations of each.
Campbell, R. E. 1983. *Microbial Ecology*. Blackwell Scientific Publications, Oxford, England.	– Introductory text stressing the role of microorganisms in biogeochemical cycles and food webs.
Codd, G. A. 1984. *Aspects of Microbial Metabolism and Ecology*. Academic Press, London.	– A series of papers on the ecological aspects of microbial metabolism and energetics.
Colwell, R. R., and R. Morita (eds.). 1974. *Effect of the Ocean Environment on Microbial Activities*. University Park Press, Baltimore.	– A collection of articles on marine microorganisms.
Colwell, R. R., R. K. Sizemore, J. F. Cooney, J. D. Nelson, Jr., R. Y. Morita, S. D. Van Valkenburg, and R. T. Wright. 1975. *Marine and Estuarine Microbiology Laboratory Manual*. University Park Press, Baltimore.	– A good selection of exercises on microbial ecology of saltwater environments.
Doetsch, R. N., and T. M. Cook. 1973. *Introduction to Bacteria and their Ecobiology*. University Park Press, Baltimore.	– Undergraduate level text with strong physiological emphasis.
Edmonds, P. 1978. *Microbiology—An Environmental Perspective*. Macmillan, New York.	– Introductory microbiology text with ecological-environmental emphasis.
Edwards, C. (ed.). 1990. *Microbiology of Extreme Environments*. McGraw-Hill, New York.	– Edited book with fine chapters on adaptations of thermophiles, acidophiles, alkalophiles, oligotrophs, halophiles, and metal-tolerant microorganisms.
Ehrlich, H. L. 1981. *Geomicrobiology*. Marcel Dekker, New York.	– Review of the actions of microorganisms in geochemical transformations.

Table 1.2

Text and reference books on microbial ecology (Continued)

Book	Description
Fenchel, T., and T. H. Blackburn. 1979. *Bacteria and Mineral Cycling*. Academic Press, London.	– Review of the activities of microorganisms that transform inorganic substances and establish biogeochemical cycles.
Fletcher, M., and G. D. Floodgate (eds.). 1985. *Bacteria in their Natural Environments*. Academic Press, London.	– A series of papers on the ecological adaptations of microorganisms and their critical metabolic activities in the environment.
Gaudy, A., and E. Gaudy. 1980. *Microbiology for Environmental Science Engineers*. McGraw-Hill, New York.	– General review of the environmental aspects of microbiology with emphasis on the practical engineering implications.
Gorden, R. W. 1972. *Field and Laboratory Microbial Ecology*. W. C. Brown, Dubuque, Iowa.	– Laboratory manual with a good selection of exercises, but lack of detail demands a lot of initiative by the instructor.
Gould, G. W., and J. E. L. Corry (eds.). 1980. *Microbial Growth and Survival in Extremes of Environment*. Academic Press, New York.	– A series of papers on the adaptations of microorganisms that permit growth and survival under extreme conditions.
Goyal, S. M., C. P. Gerba, and G. Bitton (eds.). 1987. *Phage Ecology*. John Wiley and Sons, New York.	– A concise compilation of the literature on the distribution and behavior of phage in the environment.
Gray, T. R. G., and D. Parkinson (eds.). 1968. *The Ecology of Soil Bacteria*. University of Toronto Press, Toronto, Canada.	– A collection of articles on the ecology of soil microorganisms.
Gregory, P. H. 1973. *The Microbiology of the Atmosphere*. John Wiley and Sons, New York.	– An excellent work about the microorganisms of the air.
Hattori, T., Y. Ishida, Y. Maruyama, R. Y. Morita, and A. Uchida (eds.). 1989. *Recent Advances in Microbial Ecology*. Japan Scientific Societies Press, Tokyo, Japan.	– Proceedings of the Fifth International Symposium on Microbial Ecology held in 1989 in Kyoto, Japan.
Klug, M. J., and C. A. Reddy. 1984. *Current Perspectives in Microbial Ecology*. American Society for Microbiology, Washington, D.C.	– Proceedings of the Third International Symposium on Microbial Ecology held in 1983 in East Lansing, Michigan.
Krumbein, C. W. E. (ed.). 1983. *Microbial Geochemistry*. Blackwell Scientific Publications, Oxford, England.	– A series of papers on the biogeochemical activities of microorganisms.
Kuznetzov, S. I. 1970. *The Microflora of Lakes and Its Geochemical Activity*. University of Texas Press, Austin, Tex.	– Classic Russian work on the microbiology of freshwater lakes.
Laskin, A. H., and H. Lechevalier (eds.). 1974. *Microbial Ecology*. CRC Press, Cleveland, Ohio.	– A collection of reviews on biodegradation of pesticides and on population dynamics in soil and mixed cultures.
Levin, M. A., R. J. Seidler, and M. Rogul (eds.). 1992. *Microbial Ecology: Principles, Methods, and Applications*. McGraw-Hill, New York.	– An extensive collection of papers describing modern approaches to the study of microbial ecology.
Loutit, M., and J. A. R. Miles (eds.). 1979. *Microbial Ecology*. Springer-Verlag, Berlin.	– Proceedings of the First International Symposium on Microbial Ecology held in 1977 in Dunedin, New Zealand. Papers cover a wide range of subjects.

Table 1.2

Text and reference books on microbial ecology (Continued)

Book	Description
Lynch, J. M., and N. J. Poole (eds.). 1979. *Microbial Ecology—A Conceptual Approach*. Blackwell Scientific Publications, Oxford, England.	– Good text for advanced students covering the major areas studied by microbial ecologists.
Marshall, K. C. 1976. *Interfaces in Microbial Ecology*. Harvard University Press, Cambridge, Mass.	– Important work on the occurrence and activities of microorganisms at interfaces, such as between air and water.
Megusar, F., and M. Gantar. 1986. *Perspectives in Microbial Ecology*. Slovene Society for Microbiology, Ljubljana, Yugoslavia.	– Proceedings of the Fourth International Symposium on Microbial Ecology held in 1986 in Ljubljana, Yugoslavia
Mitchell, R. 1974. *Introduction to Environmental Microbiology*. Prentice-Hall, Englewood Cliffs, N. J.	– Elementary microbiology text relevant to the environmental sciences, such as sanitary engineering.
Postgate, J. 1982. *Fundamentals of Nitrogen Fixation*. Cambridge University Press, New York.	– A concise discussion of microbial fixation of molecular nitrogen.
Postgate, J. 1984. *The Sulphate-Reducing Bacteria*. Cambridge University Press, New York.	– A thorough discussion of the bacteria that reduce sulfate.
Rheinheimer, G. 1986. *Aquatic Microbiology*. John Wiley and Sons, New York.	– Excellent work describing the ecology of microorganisms in freshwater habitats.
Richards, B. N. 1987. *The Microbiology of Terrestrial Ecosystems*. Longman/Wiley, New York.	– Undergraduate-graduate-level text emphasizing systems approach and plant-microbe interactions.
Rosswall, T. (ed.). 1972. *Modern Methods in the Study of Microbial Ecology*. Ecological Research Committee NFR Swedish National Science Reseach Council, Uppsala, Sweden.	– Proceedings of the first international meeting on microbial ecology. Useful collection of method descriptions.
Sieburth, J. M. 1979. *Sea Microbes*. Oxford University Press, New York.	– Magnificent collection of micrographs showing microorganisms in marine environments.
Slater, J. H., R. Whittenbury, and J. W. T. Wimpenny. 1983. *Microbes in their Natural Environment*. Society for General Microbiology, Symposium Series #34. Cambridge University Press, Cambridge, England.	– Proceedings of a symposium held at the University of Warwick in 1983. Articles emphasize metabolic activities of microorganisms in natural habitats.
Stolp, H. 1988. *Microbial Ecology: Organisms, Habitats, Activities*. Cambridge University Press, Cambridge, England.	– Concise text at undergraduate level aimed for microbiology and nonmicrobiology majors.
Williams, R. E. O., and C. C. Spices (eds.). 1957. *Microbial Ecology*. Cambridge University Press, Cambridge, England.	– Proceedings of the first symposium on microbial ecology held as the Seventh Symposium of the Society of General Microbiology in April 1957 in London, England. Now mainly of historical interest.
Wood, E. J. F. 1965. *Marine Microbial Ecology*. Chapman and Hall, London.	– Introductory-level text that assumes the reader has some background in microbiology but none in oceanography.
ZoBell, C. E. 1946. *Marine Microbiology*. Chronica Botanica, Waltham, Mass.	– Classic work that describes microorganisms in the oceans.

Table 1.3

General science and specialty journals containing reports of interest to the microbial ecologist

Subject area	Journal
General science	*BioScience*
	Nature (London)
	Science (AAAS)
General microbiology	*Archives for Microbiology (Archiv für Mikrobiologie)*
	CRC Critical Reviews in Microbiology
	Current Biology
	FEMS Microbiology Letters
	Journal of Bacteriology
	Journal of General and Applied Microbiology
	Journal of Industrial Microbiology
	Microbiological Reviews
Aquatic environment	*Experimental Marine Biology and Ecology*
	Hydrobiologia
	Journal of Freshwater Biology
	Journal of Plankton Research
	Limnology and Oceanography
Soil environment	*Soil Biology and Biochemistry*
	Soil Science
	Soil Science Society of America Journal
Molecular biology	*Biotechniques*
	Microbial Releases
	Molecular Ecology
	PCR: Methods and Applications
Environment and pollution	*Bulletin of Environmental Contamination and Toxicology*
	Environmental Pollution
	Environmental Science and Technology
	Marine Pollution Bulletin

This background is, of course, an idealized one. Because of the short history of the field, many of the currently active investigators were not formally trained as microbial ecologists. Many have acquired the necessary extension of their basic background by reading and by interacting with scientists from other disciplines. Subspecialization determines the most essential areas of background extension. The microbial ecologist who works on symbiotic nitrogen fixation by crop plants will need to extend his or her knowledge in microbial and plant biochemistry, genetics, and soil science but will have no pressing need for a background in zoology or oceanography.

Bear in mind, however, that research trends and opportunities shift rapidly. A broad and flexible background is the best insurance against premature scientific obsolescence.

OPPORTUNITIES FOR THE MICROBIAL ECOLOGIST

Professional opportunities for the microbial ecologist currently exist in academic, governmental, industrial, and consulting organizations. During recent academic retrenchments, microbial ecology fared reasonably well, and both faculty and grant-supported research positions have expanded in this area. Government agencies, such as the US Environmental Protection Agency (USEPA) and the US Department of Agriculture (USDA), provide employment opportunities in their research institutions and also on the administrative-regulatory level. Industrial research laboratories, voluntarily or under regulatory pressure, are conducting a rapidly increasing volume of research on the environmental effects of their products, wastes, and operational procedures. Private consulting firms contract with governmental or industrial organizations to perform similar environmental impact studies. Genetic engineering and the consequent accidental or intentional release of novel microorganisms confront the microbial ecologist with a new set of challenges related to biotechnology (Tiedje et al. 1989). The survival of engineered microorganisms after their release, their ecological interaction and/or genetic recombination with natural populations, and the resulting effects on

the environment are unsolved questions that are not only scientifically intriguing but of potentially vital significance (Alexander 1985).

SUMMARY

The essential role of microorganisms in the global cycling of materials (biogeochemical cycles) has emerged slowly in the shadow of more glamorous medical microbiology investigations. Since the early 1960s, environmental problems have focused attention on these essential cycling processes. By 1970, microbial ecology emerged as a field of specialization, and it has experienced dramatic growth in terms of research and publication volume. Judging from the range and magnitude of unsolved environmental problems, microbial ecology should continue to be a challenging field of specialization for microbiologists with appropriately broad training and interests.

REFERENCES & SUGGESTED READINGS

Alexander, M. 1981. Biodegradation of chemicals of environmental concern. *Science* 211:132–138.

Alexander, M. 1984. *Biological Nitrogen Fixation: Ecology, Technology and Physiology.* Plenum Press, New York.

Alexander, M. 1985. Ecological consequences: Reducing the uncertainties. *Issues in Science and Technology* 1:57–68.

Atlas, R., and R. Bartha. 1992. Hydrocarbon biodegradation and oil spill bioremediation. *Advances in Microbial Ecology.* in press.

Beijerinck, M. W. 1888. The root-nodule bacteria. *Botanische Zeitung* 46:725–804 (in German). English translation in T. D. Brock. *Milestones in Microbiology*, pp. 220–224.

Beijerinck, M. W. 1901. On oligonitrophilic microorganisms. *Zentralblatt für Bakteriologie,* Part II 7:561–582 (in German). English translation in T. D. Brock. *Milestones in Microbiology*, pp. 237–239.

Brefeld, O. 1881. *Botanische Untersuchungen über Schimmelpilze: Culturmethoden.* Leipzig.

Brock, T. D. 1961. *Milestones in Microbiology.* Prentice-Hall, Englewood Cliffs, N.J.

Brock, T. D. 1978. *Thermophilic Microorganisms and Life at High Temperatures.* Springer-Verlag, New York.

Bulloch, W. 1938. *The History of Bacteriology.* Oxford University Press, London.

Cagniard-Latour, C. 1838. Report on the fermentation of wine. *Annales de Chimie et de Physique* 68:206–207, 220–222 (in French). English translation in H. W. Doelle. *Microbial Metabolism,* pp. 26–32.

Cameron, R. E. 1971. Antarctic soil microbial and ecological investigations. In L. O. Quam (ed.). *Research in the Antarctic.* American Association for the Advancement of Science, Washington, D.C., pp. 137–189.

Cholodny, N. 1930. Über eine neue Methode zur Untersuchung der Bodenmikroflora. *Archiv für Mikrobiologie* 1:620–652.

Cohn, F. 1876. Studies on the biology of bacilli. *Beiträge zur Biologie der Pflanzen* 2:249–276 (in German). English translation in T. D. Brock. *Milestones in Microbiology*, pp. 49–56.

Deherain, P. P. 1897. The reduction of nitrate in arable soil. *Comptes Rendus Academie des Sciences* 124:269–273 (in French). English translation in H. W. Doelle. *Microbial Metabolism*, pp. 233–236.

Dobell, C. 1932. *Antony van Leeuwenhoek and His Little Animals*. Russell and Russell, New York.

Doelle, H. W. 1974. *Microbial Metabolism*. Benchmark Papers in Microbiology. Dowden, Hutchinson and Ross, Stroudsburg, Pa.

Doetsch, R. N. (ed.). 1960. *Microbiology—Historical Contributions from 1766 to 1908*. Rutgers University Press, New Brunswick, N.J.

Fleming, A. 1929. On the antibacterial action of cultures of a *Penicillium,* with special reference to their use in the isolation of *B. influenzae. British Journal of Experimental Pathology* 10:226–236.

Gause, G. F. 1934. *The Struggle for Existence*. Williams and Wilkins, Baltimore.

Grainger, T. H., Jr. 1958. *A Guide to the History of Bacteriology*. The Ronald Press Co., New York.

Häeckel, E. 1866. *Generelle Morphologie der Organismen*. Reimer, Berlin.

Hooke, R. 1665. *Micrographia*. Royal Society, London. (1961. Dover Publications, New York.)

Hungate, R. E. 1979. Evolution of a microbial ecologist. *Annual Reviews of Microbiology* 33:1–20.

Jannasch, H. W. and M. J. Mottl. 1985. Geomicrobiology of deep-sea hydrothermal vents. *Science* 229:717–725.

Kamp, A. F., J. W. M. La Riviere, and W. Verhoeven. 1959. *A. K. Kluyver, His Life and Work*. North Holland Publishers, Amsterdam, The Netherlands.

Kaserer, H. 1906. Die Oxydation des Wasserstoffs durch Mikroorganismen. *Zentralblatt für Bakteriologie,* Part II 16:681–696, 769–815.

Koch, R. 1883. The new methods for studying the microcosm of soil, air and water. *Deutsches Arztblatt* 137:244–250 (in German). English translation in R. N. Doetsch, (ed.). *Microbiology—Historical Contributions from 1776 to 1908*, pp. 122–131.

Kruif, Paul de. 1926. *Microbe Hunters*. Harcourt Brace Co., New York.

Küntzing, F. 1837. Mikroskopische Untersuchungen über die Hefe and Essigmutter nebst mehreren anderen dazu gehörigen vegetabilischen Gebilden. *Journal für Praktische Chemie* 11:385–409.

Lechevalier, H. A., and M. Solotorovsky. 1965. *Three*

Centuries of Microbiology. McGraw-Hill, New York.

Leeuwenhoek, A. van. 1683. Letters 39. *Philosophical Transactions of the Royal Society of London*, September 17.

Leisinger, T. , A. M. Cook, R. Hütter, and J. Nuesch. 1982. *Microbial Degradation of Xenobiotics and Recalcitrant Compounds*. Academic Press, New York.

Liebig, J. 1839. About the occurrence of fermentation, putrefaction and decay, and their causes. *Annales de Chimie et de Physique* 48:120–122, 130–136, 142–144 (in German). English translation in H. W. Doelle. *Microbial Metabolism*, pp. 434–439.

Pasteur, L. 1861. On the organized bodies which exist in the atmosphere; examination of the doctrine of spontaneous generation. *Annales des Sciences Naturelles*, 4th ser. 16:5–98 (in French). English translation in T. D. Brock. *Milestones in Microbiology*, pp. 43–48.

Pfennig, N. 1987. van Niel remembered. *ASM News* 53:75–77.

Rossi, G., S. Riccardo, G. Gesue, M. Stanganelli, and T. K. Wang. 1936. Direct microscopic and bacteriological investigations of the soil. *Soil Science* 41:53–66.

Saussure, N. de. 1839. *Memoires Societe de Physique et Histoire Naturelle de Geneve* 8:136.

Schloesing, J. J. T., and A. Müntz. 1879. On nitrification by organized ferments. *Comptes Rendus Academie des Sciences* 84:301–303 (in French). English translation in R. N. Doetsch (ed.). *Microbiology—Historical Contributions from 1776 to 1908*, pp. 103–107.

Schwann, T. 1837. A preliminary report concerning experiments on the fermentation of wine and putrefaction. *Annales de Chimie et de Physique* 11:184–193. English translation in H. W. Doelle. *Microbial Metabolism*, pp. 19–23.

Söhngen, N. L. 1906. *Het onstaan en verdwijnen von waterstof en methaan onder den invloed von het organische leven*. Dissertation, University of Delft, The Netherlands.

Tiedje, J. M., R. K. Colwell, Y. L. Grossman, R. E. Hodson, R. E. Lenski, R. N. Mack, and P. J. Regal. 1989. The planned introduction of genetically engineered organisms: Ecological considerations and recommendations. *Ecology* 70:298–315.

Tyndall, J. 1877. Further researches on the deportment and vital persistence of putrefactive and infective organisms from a physical point of view. *Philosophical Transactions of the Royal Society of London* 167:149–206. Reprinted in T. D. Brock. *Milestones in Microbiology*, pp. 56–58.

Van Iterson, G., Jr., L. E. Den Dooren de Jong, and A. J. Kluyver. 1940. *Martinus Willem Beijerinck, His Life and Work*. M. Nijhoff, The Hague, The Netherlands.

Waksman, S. A. 1953. *Sergei N. Winogradsky—His Life and Work*. Rutgers University Press, New Brunswick, N.J.

Winogradsky, S. N. 1887. Concerning sulfur bacteria. *Botanische Zeitung* 45:489–507 (in German). English translation in R. N. Doetsch (ed.). *Microbiology—Historical Contributions from 1776 to 1908*, pp. 134–145.

Winogradsky, S. N. 1890. Investigations on nitrifying organisms. *Annales de Institute Pasteur* 4:213–231 (in French). English translation in R. N. Doetsch (ed.). *Microbiology—Historical Contributions from 1776 to 1908*, pp. 146–154.

ZoBell, C. E. 1946. *Marine Microbiology*. Chronica Botanica, Waltham, Mass.

2

Microbial Evolution

THE ORIGINS OF LIFE

Until the first half of this century, the prevailing scientific view held that our planet was lifeless during the greater part of its long history. The oldest macrofossils of plants and animals are only 0.6–0.7 billion years old. If these were the remains of the original living organisms, the planet would have been lifeless for almost 4 billion years. However, we now have credible evidence that microbial life existed more than 3.5 billion years ago, that is, just 1 billion years after the formation of our planet and almost 3 billion years before the appearance of plants and animals (Nisbet 1980). Microscopic fossils of prokaryotic cells have been identified in 3.5-billion-year-old rocks (Figure 2.1). Additionally, rocks that are 3.4 billion years old have been found to contain organic matter enriched for ^{12}C; organisms selectively incorporate ^{12}C relative to ^{13}C into their biomass, thereby increasing the proportion of ^{12}C above that found in the atmosphere. The geochemical evidence of this observation indicates that living organisms were assimilating carbon into organic molecules from atmospheric methane and carbon dioxide to an extent that was changing the chemical composition of Earth within 1 billion years of Earth's formation. It thus appears that life has been present on Earth throughout most of its history, and that microorganisms had a profound influence in shaping the currently prevailing physicochemical conditions.

It is an oversimplification to hold that life developed on our planet because conditions were relatively mild and permissive as compared to other planets. It is true that Earth was favored by a sufficient mass and gravitational pull to retain most atmospheric gases and by a distance from the sun that allowed most of its water to remain in the liquid state. However, many other conditions that we consider favorable did not predate life, but rather were shaped by life during its long evolution on our planet. The role of life in forming the physicochemical environment of our planet and its profound importance in maintaining the environment in its current state was emphasized by the *Gaia hypothesis* (named for the Greek Earth goddess *Gaia*) formulated by James Lovelock (1979). The *Gaia hypothesis* is that Earth acts like a superorganism and

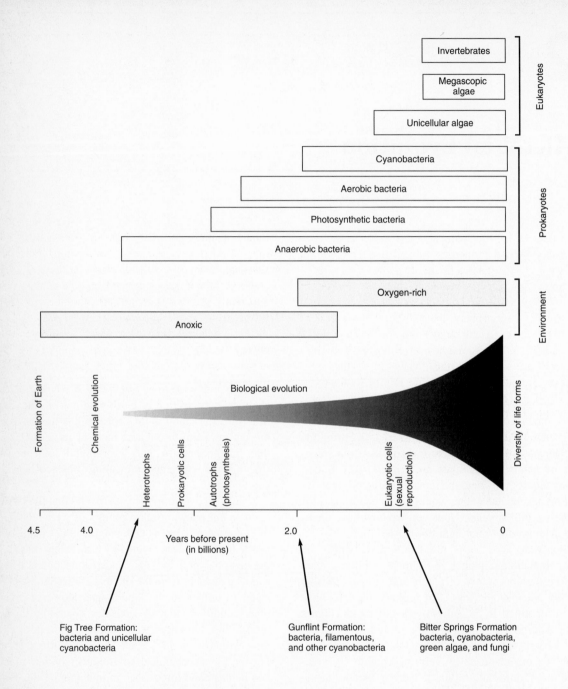

Figure 2.1
Time scale of evolutionary events discussed in the text. The most important fossil evidence is shown in relation to the presently accepted scientific interpretation of the major evolutionary events and appearance of organisms. (Source: Schopf 1978.)

Table 2.1

Comparison of atmospheric and temperature conditions

Atmosphere	Mars	Venus	Earth without life	Earth with life
Carbon dioxide (CO_2)	95%	98%	98%	0.03%
Nitrogen (N_2)	2.7%	1.9%	1.9%	79%
Oxygen (O_2)	0.13%	trace	trace	21%
Surface temperature (°C)	−53	477	290±50	13

Source: Lovelock 1979.

that through the biochemical activities of its biota its physicochemical properties are self-regulated so that they are maintained in a favorable range for life. As an example, the sun has heated up by 30% during the past 4–5 billion years. Given Earth's original carbon dioxide-rich atmosphere, the average surface temperature of a lifeless Earth today would be 290°C ± 50°C. By changing the original heat-trapping, carbon dioxide-rich atmosphere to the present oxidizing, low-carbon dioxide atmosphere, life has maintained the average surface temperature of our planet at a favorable 13°C (Table 2.1). If Lovelock's hypothesis is correct, *Gaia* exercises her powers principally through microbial processes.

Chemical Evolution

Our current understanding of conditions on prebiotic Earth and the idea of a gradual chemical evolution toward life were first proposed between 1925 and 1930 independently by the Russian scientist Alexander Ivanovich Oparin and the British scientist John B. S. Haldane (Haldane 1932; Oparin 1938, 1968; Dickerson 1978). According to their views, primitive prebiotic Earth had an anaerobic atmosphere consisting largely of nitrogen, hydrogen, carbon dioxide, and water vapor, with smaller amounts of ammonia, carbon monoxide, and hydrogen sulfide. In such a chemical mixture, organic compounds would have formed with relatively small inputs of energy. Oxygen was absent or present only in trace amounts. Consequently, the lack of an ozone shield allowed high

fluxes of ultraviolet (UV) light to reach Earth's surface. Temperature extremes, both geographical and seasonal, were probably greater than today. Large amounts of abiotically formed organic matter were apparently present, mainly in dissolved or suspended forms, from which life could evolve. Radiant, geothermal, electric-discharge, and radioactive-decay energy fueled the slow chemical evolution of this organic matter toward ever more complex and polymeric forms.

The resulting macromolecules were endowed with an inherent tendency to aggregate and form membranelike interfaces toward the surrounding liquid, foreshadowing a cellular organization. In an environment free of oxygen and microbial decomposers, this chemical evolution could proceed uninterrupted for millions of years.

This scenario, based largely on theoretical considerations, received experimental support in the 1950s from the work of Stanley L. Miller and Harold C. Urey (Miller 1957; Miller and Urey 1959; Miller and Orgel 1974), who, using a relatively simple apparatus containing water and reducing gas mixtures and receiving energy input as heat, electric discharges, or UV-radiation (Figure 2.2), were able to produce a surprising array of organic molecules, including most of the essential amino acids and nucleic acid bases. In the Miller-Urey experiments, formaldehyde and hydrogen cyanide formed first from the breakdown of methane. These compounds subsequently combined to form urea and formic acid. Later, amino acids were produced, including glycine, alanine, glutamic acid, valine, proline, and aspartic acid.

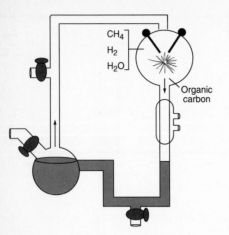

CH₄
H₂
H₂O
Organic
carbon

Figure 2.2

Apparatus used by Stanley L. Miller and Harold C. Urey to demonstrate the abiotic synthesis of organic compounds, including amino acids and nucleotides, under conditions prevalent on the prebiotic planet Earth. The apparatus is filled with reducing gas mixtures. Water vapor is created in a boiling flask and is condensed at another part of the apparatus. In the setup shown, electric discharges between electrodes serve as the principal energy input for organic synthesis.

In an aqueous environment, equilibrium conditions do not favor extensive polymerization, which requires both input of energy and removal of water. Suspended clay particles, however, consist of closely packed silicate and alumina sheets with both negative and positive charges that could provide an absorbing and ordering matrix with primitive catalytic reaction centers. Energy-rich organic molecules, such as amino acid adenylates, were demonstrated to polymerize into proteinlike polypeptide chains on such clay matrices (Katchalsky 1973). Proteins and nucleic acids could have formed in this manner in the reducing atmosphere of early Earth.

In support of an alternative theory, Sidney W. Fox demonstrated the spontaneous formation of "thermal proteinoids" by the moderate heating of amino acid mixtures in the dry state. Thermal proteinoids exhibit a self-ordering tendency, spontaneously aggregating into microspheres (Fox 1965; Fox and Dose 1977). Some abiotically formed proteinoids and other polymers show rudimentary catalytic activity, reminiscent of enzymes. Some of the abiotically formed proteinoids also may have served as templates that allowed their duplication, either by outside chemical forces or by the primitive enzymelike action of other proteinoids.

Cellular Evolution

In addition to his theoretical work on chemical evolution, Oparin and his coworkers performed some fascinating studies on the properties of microspheres that form spontaneously in the colloidal solution of two different polymeric substances, such as gum arabic and histone (Dickerson 1978). These microspheres, which Oparin called coacervates, develop spontaneously when phospholipids are placed into water. Coacervates form an outer boundary that resembles the membranes that surround all living cells, having two layers. Coacervates grow by accumulating additional lipids from the surrounding medium; they pinch off projections to form independent new coacervates, much like the division of some living cells; chemical reactions can occur within the cavities of coacervates. Coacervates behave as semipermeable membranes and vacuoles; they are capable of selective uptake of substances from their environment; and they mimic the growth and division of cells. The incorporation of enzymes, electron carriers, or chlorophyll into these coacervate droplets allows the modeling of some anabolic and catabolic processes, electron transport, and light energy utilization that we normally associate only with living cells. Admittedly, these model systems demonstrate only the surprising self-organizing capacity of polymeric molecules; it is not suggested that coacervates were the actual intermediates in the chemical evolution process.

More directly relevant are the microspheres formed by thermal proteinoids and their limited catalytic and self-replicating capabilities demonstrated by Fox (Fox 1965; Fox and Dose 1977). It is conceivable that similar self-replicating proteinaceous microspheres, as yet without nucleic acids, represented the first step toward cellular organization. These postulated primitive cell-like structures are referred to as progenotes or protobionts.

A further advance toward cellular organization was probably the acquisition and use of nucleic acids, first probably RNA and later both DNA and RNA, as templates for protein synthesis (Darnell and Doolittle 1986). Along with further development of enzymatic capabilities and membrane organization, this led to the development of eugenotes, primitive versions of the prokaryotic cell. Fossil evidence for the existence of such cells was discovered in sedimentary rock dated as approximately 3.5 billion years old (Knoll and Awromile 1983).

The scarcity of morphological detail in prokaryotes and the poor geological preservation of subcellular features make the fossil record of microbial evolution pitifully incomplete. The molecular record of evolution can to some extent compensate for this unfortunate circumstance. If all living forms had a common ancestor that already possessed the basic macromolecules of life, such as RNA, DNA, and protein, one may assume that those organisms with very similar macromolecular sequences were closely related, while those with very different ones diverged early and evolved independently. Comparisons were attempted using protein and DNA sequences, but comparisons of RNA sequences seem to be the most relevant and technically feasible. A detailed evolutionary tree for prokaryotes has been developed based on analyses of 16S ribosomal RNA (Fox et al. 1980; Woese et al. 1990). Analysis of ribosomal RNAs has also confirmed that the eukaryotic cell is a phylogenetic chimera (composed of two or more genetically distinct cell lines—named after a mythological monster constructed from diverse animal parts) in which several prokaryotic evolutionary lines have fused.

The evolutionary tree (Figure 2.3) based on 16S ribosomal RNA homologies assumes a common ancestral state on the level of the progenote or primitive eugenote. From this ancestor, three evolutionary lines diverged to form the archaebacteria (archaea), eubacteria (bacteria), and urkaryotes (predecessors of eukaryotes—eucarya) (Woese et al. 1990).

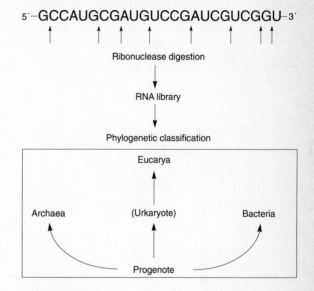

Figure 2.3
The descent of prokaryotic and eukaryotic organisms based on homologies of ribosomal RNA (rRNA) sequences. The rRNA is cut with a ribonuclease to form small oligonucleotides that are sequenced. The base sequences are used to form a library which is analyzed for homology with rRNA libraries from other organisms. Based upon similarities, a phylogenetic classification is made. (Source: Fox et al. 1980.)

The ancestral archaea gave rise to contemporary methanogens, halobacteria, and thermoacidophiles, such as *Sulfolobus* (Figure 2.4). The archaea are apparently primitive prokaryotes; the physiological adaptations permit most to grow at high temperatures and under strict anaerobic condition. Many transform sulfur, and there is a common theme of sulfur metabolism; some also tolerate the highly acidic and saline conditions postulated to have characterized early Earth. Crenarchaeota comprises a group of extreme thermophiles that represent the oldest recorded evolutionary branch; these bacteria have a phenotype that most resembles the ancestral phenotype of the archaea.

The archaea differ from other living organisms in that they have membranes that are chemically dis-

Figure 2.4

The evolution of Archaea (archaebacteria), which gave rise to two major groups, the crenarchaeota and the euryarchaeota. (Source: Woese et al. 1990.)

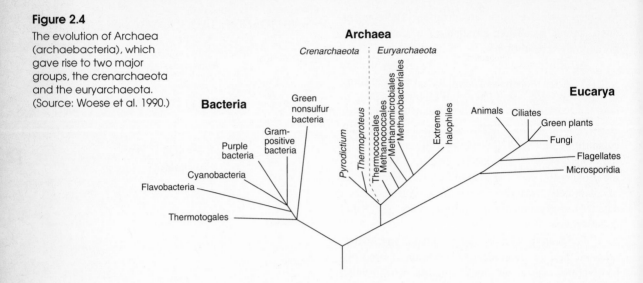

tinct; the membranes of archaea cells have branched hydrocarbons and ether linkages compared to the straight chain fatty acids and ester linkages found in the membranes of all other organisms; some form tetraethers and have monolayer membranes instead of the typical bilipids (Heathcock et al. 1985). The metabolic cofactors of the archaea also differ from those of bacteria and eukaryotes; cofactors of archaea include coenzyme M (involved in C_1 metabolism), factor F_{420} (involved in electron transport), 7-mercaptoheptanoylthreonine phosphate (involved in methanogenesis), tetrahydromethanopterin (instead of folate), methanofuran, and retinal.

The ancestral eubacteria gave rise to the rest of the contemporary prokaryotes (Figure 2.5). Notable is the relative antiquity of the photosynthetic forms, including the cyanobacteria. Ancestral cyanobacteria probably had only photosystem I and did not evolve oxygen. The spirochetes and Gram-positive bacteria also represent evolutionary lines that diverged early. Interestingly, the Gram-negative heterotrophic bacteria appear to be descendants of photosynthetic purple bacteria.

The fossil record indicates that eukaryotes appeared approximately 1.3–1.4 billion years ago. In the 1-billion-year-old Bitter Springs Formation in Australia, eukaryotic forms resembling green algae

and perhaps fungi are abundant. The appearance of eukaryotes coincides with a decline in stromatolites deposited by prokaryotic microbial mats. The unique characteristic of the eukaryotic line of descent is the 18S, rather than 16S, ribosomal RNA. This trait was presumably contributed by the cytoplasm of the hypothetical urkaryote, which appears to have been derived from the Archaea. Chloroplasts and mitochondria of eukaryotic organisms can be traced to the eubacterial line of descent, and at least one ribosomal protein is traceable to the archaebacteria (Zillig et al. 1982). The origin of the cell nucleus in eukaryotes is uncertain and may itself be chimeric. From the eukaryotic line descend the microbial groups protozoa, algae, and fungi, as well as the plant and animal kingdoms.

The advent of the eukaryotic cell, and presumably sexual reproduction, greatly hastened the pace of evolution. About 0.6 billion years ago, the first multicellular macrofossils of plants and animals appeared and the conventional Paleozoic geological age began. Until that time, evolution of life was synonymous with microbial evolution. By the time multicellular plants and invertebrates appeared, microorganisms had changed the atmosphere of Earth and adjusted most of its physicochemical parameters to their present state.

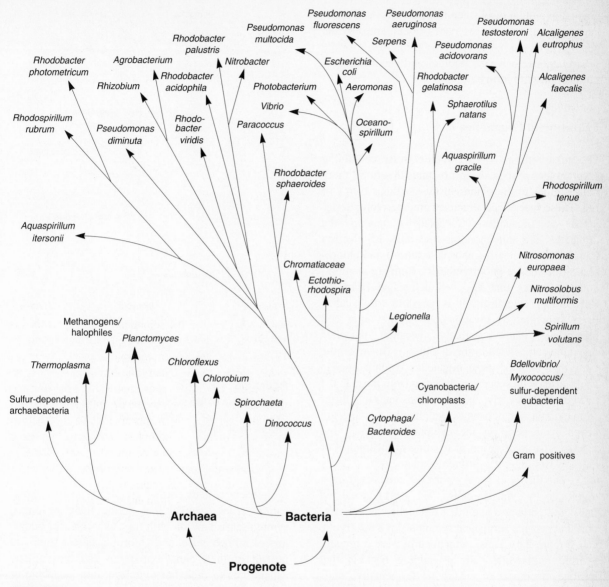

Figure 2.5

The evolution of the bacteria (eubacteria), which gave rise to diverse photosynthetic, chemoautotrophic, and heterotrophic bacteria. (Source: Woese et al. 1990.)

The 3 billion years of microbial evolution involved very limited changes in size and morphology. Compared to the evolutionary time scale of mul-ticellular organisms, microbial evolution appears to be excruciatingly slow. One is tempted to speculate that this long and seemingly uneventful period conceals a gradual evolution of biochemical pathways and regulatory mechanisms that are poorly documented by the geological record but that laid the groundwork for the subsequent explosive morphological diversification of multicellular life forms.

Evolution of Organelles

The eukaryotic cell is a complex entity containing several types of specialized structures that fulfill functions similar to the organs in multicellular forms of life. These specialized structures are, therefore, designated as organelles. Conspicuous among the organelles are mitochondria (the sites of electron transport and oxidative phosphorylation) and chloroplasts (the sites of photosynthesis). Both are membranous structures that are well delineated from the surrounding cytoplasm. It is possible to visualize these structures as invaginations of the outer cell membrane that eventually became detached organelles embedded in the cytoplasm. However, the facts that both of these organelles contain nucleic acids, often with sequences differing from those in the nucleus, that they have their own ribosomes that differ from cytoplasmic ribosomes, and, finally, that they are never synthesized in the cytoplasm *de novo* but always arise by division of existing mitochondria or chloroplasts suggest a different explanation.

The theory of serial symbiosis, advocated most vigorously by Lynn Margulis (1971, 1981), visualized these and perhaps some other organelles as prokaryotes that became permanent symbionts of a eukaryotic cell (Figure 2.6). Thus, mitochondria were originally aerobic prokaryotic cells resembling bacteria that became symbionts of eukaryotic fermentative cells and endowed them with a more efficient aerobic utilization system for organic substrates. Similarly, cyanobacteria or perhaps the phycobilin-free photosynthetic prokaryotes, the chloroxybacteria, were the ancestors of the chloroplast; this symbiotic association endowed heterotrophic eukaryotic cells with the capacity to live photosynthetically. The fact that numerous unicellular and multicellular organisms form intracellular associations with bacteria or

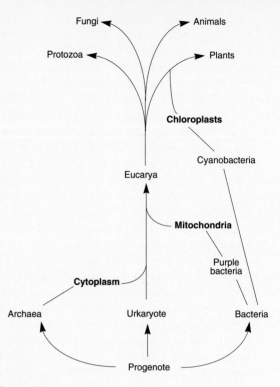

Figure 2.6

The evolution of the eukaryotes, involving acquisition of cytoplasm and organelles from prokaryotes (Source: Woese et al. 1990.)

cyanobacteria adds further credibility to this explanation. It is assumed that during their prolonged symbiotic association with the eukaryote, the ancestors of mitochondria and chloroplasts lost some but not all of their genetic material and biosynthetic capabilities to their hosts and became incapable of independent existence. Ultrastructural evidence indicates that some eukaryotes picked up cyanobacterial chloroplasts and, in turn, became endosymbionts of other eukaryotes themselves, with the consequent loss of their cellular independence and functions. The complex outer membrane structure and pigment composition of some chloroplasts preserve the record of such a multistage symbiotic evolution (Wilcox and Wedemayer 1985).

The case for the symbiotic derivation of eukaryotic flagella and cilia is weaker. These are clearly different from prokaryotic flagella; they are much thicker, have a complex "9 + 2" structure (nine peripheral fibers wound around two central ones), and are powered by ATP rather than by protonmotive force. They do not, however, have nucleic acid or ribosomes of their own and they do not multiply by division. The possibility of their prokaryotic and symbiotic origin is hinted at by a curious association of some protozoa with spirochetes (bacterial cells). The best-studied example is *Mixotricha paradoxa,* a flagellated protozoan living in the hindgut of some termites and involved in cellulose digestion. The flagella of *Mixotricha* are inactive, but it is mobile with what at first appeared to be cilia; on closer examination, these cilia turned out to be rows of spirochetes attached to the surface of *Mixotricha* by bracketlike basal bodies (Cleveland and Grimstone 1964). Their coordinated motion propels *Mixotricha*. Nonmotile, short, rod-shaped bacteria of unknown function are also attached to the surface of *Mixotricha* at regular intervals (Figure 2.7). Proponents of the serial symbiosis theory believe that the spirochetes attached in this manner may have lost their independent metabolic and genetic identity and become flagella or cilia (Margulis and McMenamin 1990). The thought is intriguing, but because of the lack of supportive evidence, it is not as widely accepted as the symbiotic origin of mitochondria and chloroplasts.

Even more speculative is the origin of the cell nucleus. As prokaryotes diversified and increased in biochemical and morphological complexity, a genome of increased size was needed to code for these traits and functions. The replication of this larger genome during cell division became more complex. Its spatial separation from the cytoplasm and its vegetative functions make good sense. It is conceivable that an endosymbiotic prokaryote consolidated its own genome with the genome of its host, lost its cytoplasm and vegetative functions, and became the nucleus of the new primitive eukaryotic cell. However, a nuclear membrane that compartmentalized the genome of the early eukaryote also could have evolved by invagination of the outer cell membrane or from membranes around vacuoles.

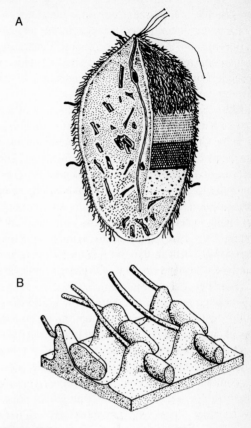

Figure 2.7

(A) *Mixotricha paradoxa,* apparently a ciliated flagellate. (B) The fine structure, revealing brackets and attached spirochetes. Rod-shaped bacteria of unknown function are also present (Source: Cleveland and Grimstone 1964.)

GENETIC BASIS FOR EVOLUTION

As proposed by Charles Darwin (1860), those organisms best adapted to survive in a given environment have a selective advantage, a principle known as natural selection or survival of the fittest. The introduction of diversity into the gene pools of microbial populations established the basis for selection and evolution (Clarke 1984). Mutations introduce variability into the genomes of microorganisms, resulting in changes in the enzymes the organism synthesizes. Variations in the genome are passed from one generation to another and are disseminated throughout the population. Because of their relatively short generation times, as compared to higher organisms, changes in the genetic information of microorganisms can be widely and rapidly disseminated.

Although in some cases the modification of the genome is harmful to the organism, some mutations being lethal or conditionally lethal, a mutation may change genetic information in a favorable way. The occurrence of favorable mutations introduces information into the gene pool that can make an organism more fit for surviving in its environment and for competing with other microorganisms for available resources. Over many generations, natural selective pressures may result in the elimination of unfit variants and the continued survival of organisms possessing favorable genetic information.

Although mutation is the basis for introducing variability into the genetic information of a cell, genetic exchange and recombination play the critical roles in redistributing genetic information (Hall 1984; Slater 1984). Recombination creates new allelic combinations that may be adaptive. Altering the organization of genetic information within populations provides a basis for directional evolutionary change. The exchange of genetic information can produce individuals with multiple attributes that favor the survival of that microbial population. The long-term stability of a population depends on its ability to incorporate adaptive genetic information into its chromosomes.

Mutation and general recombination provide the raw material for the natural selection of adaptive features. Reciprocal recombination produces evolutionary links between closely related organisms; nonreciprocal recombination provides a mechanism for rapid qualitative evolutionary changes. The fact that unrelated genomes can recombine suggests that different lines of evolution can suddenly converge.

Adaptive features within populations contribute to both change and stability within biological communities. The Darwinian principle of natural selection applies to microbial populations and determines which populations can successfully establish themselves within communities. According to the principle set forth by Martinus Beijerinck, "everything is everywhere, the environment selects" (Van Iterson et al. 1983). Some microorganisms possess features that make them better adapted for survival in a particular ecosystem. Microorganisms with features that make them poorly fitted for survival under changing conditions are soon eliminated by natural selection. Overspecialization in the gene pool can lead to temporary success of a population under a single set of conditions, but it will ultimately render the population nonadaptive. Genetic variability within populations and communities is necessary because the habitat of a microbial population is not static.

The evolution of biochemical pathways is driven by the basic Darwinian concepts of random mutation and natural selection, but the complexity and multistep nature of biosynthetic pathways makes it difficult to conceptualize how, for example, the ability to synthesize an essential amino acid from a sugar arose by this mechanism. The concept of the backward evolution of biochemical pathways formulated by Norman Horowitz (1965) makes this easier to understand. Generally speaking, a contemporary microorganism that needs compound A as a cell constituent has the ability to synthesize this compound from the very different compound E. Prior to the evolution of the pathways, the primitive microorganism took up compound A directly from its environment. As compound A became scarce, a mutant that possessed the capacity for the one-step conversion of the closely related and abundant compound B to compound A gained selective advantage and became dominant. As in turn compound B became scarce, the mutant that could convert compound C to compound B gained advantage because it already had the mechanism to convert compound B to the required compound A. This backward evolutionary sequence can be continued until the very different compound E can be converted to the essential compound A. Transfer and recombination of genetic information, especially if the pathway is coded on highly mobile plasmids or located on transposons, can greatly speed the evolution and diversification of biochemical capabilities.

EVOLUTION OF PHYSIOLOGICAL DIVERSITY

When life evolved on Earth, organic compounds that had formed abiotically served as initial substrates for growth. Cells developed the abilities to degrade these compounds and to derive energy from them for cell growth and maintenance. Methanogenic archaebacteria may have also used hydrogen and carbon dioxide in the reducing atmosphere to generate cellular energy. The central focus of biological energy transformations became ATP, still employed today by cells as their principal carrier of energy.

Gradually, organized sequences of enzymatically catalyzed degradation reactions (catabolic pathways) evolved that permitted cells to utilize the chemical energy of organic substrates for generating ATP more efficiently. The conversion of carbohydrates to pyruvate via glycolysis became such an ATP-generating pathway central to the energy transfers of many cells. Glycolysis appears to have evolved early and not to have changed in more than 3 billion years. Heterotrophic metabolism using glycolysis led to the formation of various organic end products, including alcohols and acids. These pathways were fermentative; they did not require external electron acceptors and occurred under strict anaerobic conditions. Relatively low amounts of ATP were generated per molecule of substrate utilized so that large amounts of organic substrates had to be transformed.

Early cells also developed the ability to utilize sulfur compounds. Sulfur metabolism is a central feature of the archaea. Cells appear to have evolved an early form of anaerobic respiration using sulfate in which hydrogen sulfide is formed. Rocks that are 2.7 billion years old show evidence of enrichment for the light isotope of sulfur (^{32}S). Living organisms use ^{32}S in preference to ^{34}S, causing the observed isotopic enrichment.

Because the nutritional resources of abiotically formed organic compounds were limited, selective pressure developed early for a more direct utilization of the radiant sun energy, a renewable energy source for life processes. Cells evolved with the ability to utilize solar energy for generating ATP. They used hydrogen sulfide, which was present in the oceans, as a source of electrons for the reduction of carbon dioxide. Early photosynthesis was most likely of the anoxygenic type (non-oxygen-producing) found today in the Rhodospirillaceae, Chromatiaceae, and Chlorobiaceae, anaerobic photosynthetic bacteria. These microorganisms lack photosystem II and are unable to use the hydrogen in water for the reduction of carbon dioxide; instead, reducing power is obtained from organic compounds, molecular hydrogen, or reduced sulfur compounds such as hydrogen sulfide. Early cyanobacteria (formerly known as blue-green algae) probably did not possess photosystem II either and utilized the same electron donors that can be split with a lower energy input than that needed for water. The use of chemiosmosis for ATP generation in the development of photosynthetic metabolism marked an important evolutionary step because it improved the efficiency of generating ATP.

The oldest assemblage of microfossils in the sedimentary rock of the 3.5-billion year-old Figtree Formation in Africa consists of prokaryotic forms, that is, bacteria. Carbon isotope ratios suggest that the microfossils resembling cyanobacteria were photosynthetic but that the more advanced oxygenic photosynthesis of cyanobacteria evolved only about 1 billion years later (Shopf 1978). The interim period was dominated by prokaryotic microbial mats, as evidenced by stromatolites, which are layered limestone pillars containing organic matter deposits. The recent discovery of living stromatolites in some warm and hypersaline environments, such as the shallow waters of Shark Bay, Australia, greatly aided the interpretation of these stromatolite fossils. It is now clear that they were deposited by filamentous and other prokaryotic microorganisms, predominantly cyanobacteria.

The evolution of oxygenic (oxygen-producing) photosynthesis in cyanobacteria is evidenced by the appearance of heterocyst-like structures and banded iron formations approximately 2.0–2.5 billion years ago. Heterocysts produced by cyanobacteria have the function of separating the oxygen-sensitive nitrogen-fixation system from oxygen-evolving photosynthesis and would have been superfluous in cyanobacteria with only anoxygenic photosynthesis. The dating of the development of oxygen-producing metabolism is based upon the observation that about 2.5 billion years ago virtually all iron was removed from the oceans. Over a very short period—a few million years—iron was deposited in sediments. Prior to this period, all iron deposits were reduced. Subsequent iron deposits are consistently oxidized unless formed in reducing sediment environments. This record indicates that about 2 billion years ago the originally reducing atmosphere of our planet changed to an oxidizing one. Oxygen produced by living organisms accumulated in the atmosphere.

The evolution of photosynthetic organisms provided a mechanism for carbon recycling and a renewable source of energy. Eventually, another abundant element in biomass—nitrogen—became a limiting factor for growth. Molecular nitrogen was abundant in the atmosphere, but cells could not directly utilize that form of nitrogen. Cells were dependent on organic nitrogen compounds or reduced inorganic forms of nitrogen. Under the reducing atmosphere, some cells developed the capacity of nitrogen fixation, enabling them to convert molecular nitrogen from the atmosphere to forms that could be used in cellular metabolism.

The geologic evidence suggests that just over 2 billion years ago photosynthetic microorganisms developed the ability to produce oxygen. Oxygen accumulation in the atmosphere halted abiotic generation of organic compounds, which only occurs under strictly anaerobic conditions. Ozone formed from molecular oxygen reduced the influx of ultraviolet radiation. Ultraviolet light, particularly of wavelengths less than 200 nm, was probably the major energy source driving the abiotic formation of organic matter prior to the biological introduction of oxygen into the atmosphere.

The evolution of photosystem II in cyanobacteria made available an almost inexhaustible source of reducing power in the form of water, although at the cost of a lower quantum efficiency in the utilization of sunlight. More solar energy is required to split the strong H–O–H bonds of water than to split the weaker H–S–H bonds of hydrogen sulfide. The oxygen evolved in this type of photosynthesis was undoubtedly toxic to most existing forms of anaerobic life; these either became extinct or were restricted to the specific environments that still harbor obligately anaerobic microorganisms. Nitrogen-fixing cells had to develop adaptations to protect nitrogenase, the enzyme involved in nitrogen fixation, because oxygen poisons this enzyme.

For the heterotrophic forms that were capable of adapting to the oxidizing atmospheric conditions, new and much more efficient modes of substrate utilization opened up, allowing great diversification in form and size. Some cells developed chemoautotrophic metabolic capabilities in which inorganic molecules are used to generate ATP. Others developed respiratory metabolism that greatly increased the efficiency of ATP generation from organic substrates. Respiratory metabolism is based upon the chemiosmotic generation of ATP. More efficient metabolic pathways and the need to store genetic information for an ever-increasing array of metabolic, regulatory, morphological, and behavioral traits probably encouraged the evolution of the eukaryotic cell.

SUMMARY

According to most theories on the origin of life on Earth, microorganisms (prokaryotes) were the first living inhabitants of the planet. Prokaryotes have evolved over a much longer time period than eukaryotes; that is, the evolution of bacteria has occurred over a much more extensive time period than that for higher organisms. The prokaryotes include the archaea (archaebacteria) and the eubacteria, which represent distinct evolutionary lines. The archaebacteria and eubacteria are as distantly related from each other as each from the eukaryotes. There are several theories that explain the evolution of modern eukaryotic microbial cells. One of the more interesting theories is that organelles and higher life forms evolved by serial symbiosis with genetically distinct cell lines uniting to form entirely new organisms. Various adaptive features have evolved in microbial populations that permit their survival in extreme environments.

REFERENCES & SUGGESTED READINGS

Carlisle, M. J., J. F. Collins, and B. E. B. Moseley (eds.). 1981. *Molecular and Cellular Aspects of Microbial Evolution.* Cambridge University Press, London.

Cavalier-Smith, T. 1987. The origin of eukaryotic and archaebacterial cells. *Annals of the New York Academy of Science* 503:17–54.

Clarke, P. H. 1984. Evolution of new phenotypes. In M. J. Klug and C. A. Reddy (eds.). *Current Perspectives in Microbial Ecology.* American Society for Microbiology, Washington, D.C., pp. 71–78.

Cleveland, L. R., and A. V. Grimstone. 1964. The fine structure of the flagellate *Mixotricha paradoxa* and its associated micro-organisms. *Proceedings of the Royal Society* (London), Series B 159:668–686.

Crawford, I. P. 1989. Evolution of a biosynthetic pathway: The tryptophan paradign. *Annual Reviews of Microbiology* 43:567–600.

Darnell, J. E., and W. F. Doolittle. 1986. Speculations on the early course of evolution. *Proceedings of the National Academy of Science USA* 83:1271–1275.

Darwin, C. 1860. *On the Origin of Species by Means of Natural Selection, or the Preservation of Favoured Races in the Struggle for Life.* D. Appleton and Co., New York.

Davis, B. D. 1989. Evolutionary principles and the regulation of engineered bacteria. *Genome* 31:864–869.

Dickerson, R. 1978. Chemical evolution and the origin of life. *Scientific American* 239(3):70–86.

Doolittle, W. F. 1987. The evolutionary significance of the archaebacteria. *Annals of the New York Academy of Science* 503:72–77.

Doolittle, W. F. 1988. Bacterial evolution. *Canadian Journal of Microbiology* 34:547–551.

Eberhard, W. G. 1990. Evolution in bacterial plasmids and levels of selection. *Quarterly Review of Biology* 65:3–22.

Fox, G. E., E. Stackebrandt, R. B. Hespell, J. Gibson, J. Maniloff, T. A. Dyer, R. S. Wolfe, W. E. Balch, R. S. Tanner, L. Magrum, L. Zablen, R. Blakemore, R. Gupta, L. Bonen, B. J. Lewis, D. A. Stahl, K. R. Luehrsen, K. N. Chen, and C. R. Woese. 1980. The phylogeny of prokaryotes. *Science* 209:457–463.

Fox, S. W. 1965. *The Origins of Prebiological Systems and Their Molecular Matrices.* Academic Press, New York.

Fox, S. W., and K. Dose. 1977. *Molecular Evolution and the Origin of Life.* Marcel Dekker, New York.

Freter, R. 1984. Factors affecting conjugal transfer in natural bacterial communities. In M. J. Klug and C. A. Reddy (eds.). *Current Perspectives in Microbial Ecology.* American Society for Microbiology, Washington, D.C., pp. 105–114.

Gray, M., and F. Doolittle. 1982. Has the endosymbiont hypothesis been proven? *Microbiological Reviews* 46:1–42.

Haldane, J. B. S. 1932. *The Causes of Evolution.* Harper & Row, New York.

Hall, B. G. 1984. Adaptation by acquisition of novel enzyme activities in the laboratory. In M. J. Klug and C. A. Reddy (eds.). *Current Perspectives in Microbial Ecology.* American Society for Microbiology, Washington, D.C., pp. 79–86.

Heathcock, C. H., B. L. Finkelstein, T. Aoki, and C. D. Poulter. 1985. Stereostructure of the archaebacterial C_{40} diol. *Science* 229:862–864.

Heinrich, M. R. (ed.). 1976. *Extreme Environments: Mechanisms of Microbial Adaptation.* Academic Press, New York.

Horowitz, N. H. 1965. The evolution of biochemical synthesis—Retrospect and prospect. In V. Bryson and H. J. Vogel (eds.). *Evolving Genes and Proteins.* Academic Press, New York, pp. 15–23.

Jensen, R. A. 1985. Biochemical pathways in prokaryotes can be traced backward through evolutionary time. *Molecular Biology and Evolution* 2:13–34.

Katchalsky, A. 1973. Probiotic synthesis of biopolymers on inorganic templates. *Naturwissenschaften* 60:215–220.

Knoll, A. H. 1985. The distribution and evolution of microbial life in the late Proterozoic era. *Annual Reviews of Microbiology* 39:391–417.

Knoll, A. H., and S. M. Awromile. 1983. Ancient microbial ecosystems. In W. E. Krumbein (ed.). *Microbial Geochemistry.* Blackwell Scientific Publications, Oxford, England, pp. 287–315.

Lake, J. A. 1985. Evolving ribosome structure: Domains in archaebacteria, eubacteria, eocytes and eukaryotes. *Annual Reviews in Biochemistry* 54:237–271.

Lovelock, J. E. 1979. *Gaia: A New Look at Life on Earth.* Oxford University Press, New York.

Lovelock, J. E. 1988. *The Ages of Gaia: A Biography of Our Living Earth*. W. W. Norton, New York.

Margulis, L. 1970. *Origin of Eukaryotic Cells*. Yale University Press, New Haven, Conn.

Margulis, L. 1971. The origin of plant and animal cells. *American Scientist* 59:230–235.

Margulis, L. 1981. *Symbiosis in Cell Evolution*. W. H. Freeman, San Francisco.

Margulis, L., and M. McMenamin. 1990. Marriage of convenience: The motility of the modern cell may reflect an ancient symbiotic union. *The Sciences* 30:30–37.

Miller, S. L. 1957. The formation of organic compounds on the primitive Earth. *Annals of the New York Academy of Science* 69:260–275.

Miller, S. L., and H. C. Urey. 1959. Organic compound synthesis on the primitive Earth. *Science* 130:245–251.

Miller, S. L., and L. E. Orgel. 1974. *The Origins of Life on Earth*. Prentice-Hall, Englewood Cliffs, N.J.

Nisbet, E. G. 1980. Archean stromatolites and the search for the earliest life. *Nature* 284:395–396.

Olsen, G. J., D. J. Lane, S. J. Giovannoni, N. R. Pace, and D. A. Stahl. 1986. Microbial ecology and evolution: A ribosomal RNA approach. *Annual Reviews of Microbiology* 40:337–365.

Oparin, A. I. 1938. *The Origin of Life*. Dover Publications, New York.

Oparin, A. I. 1968. *Genesis and Evolutionary Development of Life*. Academic Press, New York.

Pace, N. R., D. A. Stahl, D. J. Lane, and G. J. Olsen. 1985. Analyzing natural microbial populations by rRNA sequences. *ASM News* 51:4–12.

Schopf, J. W. 1978. The evolution of the earliest cells. *Scientific American* 239(3):110–138.

Slater, J. H. 1984. Genetic interactions in microbial communities. In M. J. Klug and C. A. Reddy (eds.). *Current Perspectives in Microbial Ecology*. American Society for Microbiology, Washington, D.C., pp. 87–93.

Sukhodolets, V. V. 1988. Organization and evolution of the bacterial genome. *Microbiological Sciences* 5:202–206.

Thayer, D. W. (ed.). 1975. *Microbial Interaction with the Physical Environment*. Dowden Hutchinson and Ross, Stroudsburg, Pa.

Van Iterson, G., Jr., L. E. Den Dooren de Jong and J. J. Kluyer. 1983. *Martinus Wilem Beijerinck: His Life and His Work*. Science Tech, Inc., Madison, Wis., pp. 106–132.

Vidal, G. 1984. The oldest eukaryotic cells. *Scientific American* 250(2):48–57.

Wilcox, C. E., and G. E. Wedemeyer. 1985. Dinoflagellates with blue-green chloroplasts derived from an endosymbiotic eukaryote. *Science* 227:192–194.

Woese, C. R. 1987. Bacterial evolution. *Microbiological Reviews* 51:181–192.

Woese, C. R., O. Kandler and M. L. Wheelis. 1990. Towards a natural system of organisms: Proposal for the domains archaea, bacteria, and eucarya. *Proceedings of the National Academy of Science* 87:4576–4579.

Zillig, W. 1987. Eukaryotic traits in Archaebacteria: Could the eukaryotic cytoplasm have arisen from archaebacterial origin? *Annals of the New York Academy of Science* 503:78-82.

Zillig, W., R. Schnabel, and J. Tu. 1982. The phylogeny of archaebacteria, including novel anaerobic thermoacidophiles, in the light of RNA polymerase structure. *Naturwissenschaften* 69:197–204.

Zillig, W., R. Schnabel, and K. O. Slatter. 1985. Archaebacteria and the origin of the eukaryotic cytoplasm. *Current Topics in Microbiology and Immunology* 114:1–18.

PART TWO

Population
Interactions

3

Interactions among Microbial Populations

Under natural conditions, a microorganism rarely exists in isolation. Even when a single microbial cell is successfully isolated in the laboratory, the individual normally will multiply and form a group (a clone) of similar individuals called a population. Typically, in a natural environment, numerous populations of different character coexist. The microbial populations that live together at a particular location, called a habitat, interact with each other to form a microbial community. The microbial community is structured so that each population contributes to its maintenance.

Both positive and negative interactions occur between individuals within a single microbial population and between the diverse microbial populations of a community. These interactions permit some populations to reach sizes optimal for the resources available within the habitat. The totality of the interactions between populations maintains the ecological balance of the community (Bull and Slater 1982). Ecological hierarchy and structure of microbial communities will be discussed in greater detail in chapter 6. Here we will focus on the interactions between populations that form the basis for community structure.

INTERACTIONS WITHIN A SINGLE MICROBIAL POPULATION

According to Allee's principle, both positive and negative interactions may occur even within a single population (Allee et al. 1949). Such interactions, originally observed on animals and plants, are interrelated and are population-density dependent. Subsequent examples will show that Allee's principle is highly suitable for explaining the density-dependent interactions that occur within microbial populations. Generally speaking, positive interactions increase the growth rate of a population, whereas negative interactions have the opposite effect (Figure 3.1). With increasing population density, positive interactions theoretically increase the growth rate to some asymptotic limit. In contrast, negative interactions decrease growth rate as the population density increases.

Commonly, positive interactions (cooperation) predominate at low population densities and negative ones (competition) at high population densities. As a result, there is an optimal population density for

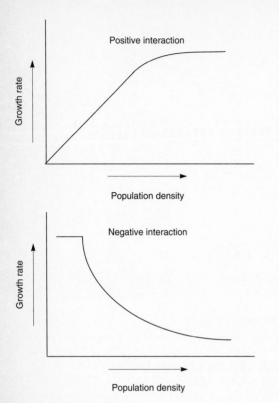

Figure 3.1

Effects of independent positive and negative interactions on growth rate with increasing population density.

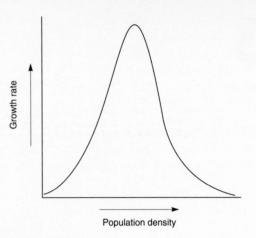

Figure 3.2

Combined effect of positive and negative interactions within a population. Growth rate exhibits an optimal population density. At low population densities, positive interactions predominate; at high population densities, negative interactions are dominant.

maximal growth rate (Figure 3.2). Growth rates below the optimal density are strongly affected by positive interactions; growth rates above the optimal population density are strongly influenced by negative interactions.

Positive Interactions

Positive interactions within a population are called cooperation. Cooperation within a single microbial population is evidenced by an extended lag period or a complete failure of growth when a very small inoculum is used in routine culture transfer procedures.

This is especially true for fastidious microorganisms and is a major obstacle to all isolation procedures that require the growth of single microbial cells into colonies. There is little doubt that populations of intermediate density are generally more successful than individual organisms in the colonization of natural habitats. A well-studied example of this is the "minimum infectious dose" of pathogenic microbial populations. Usually thousands of pathogens are required to cause disease, as a single pathogen is rarely able to overcome host defenses. Cooperation occurs within the population because the semipermeable cell membranes of microorganisms are imperfect and tend to "leak" low-molecular-weight metabolic intermediates that are essential for biosynthesis and growth. In a population, a significant extracellular concentration of these metabolites is established that counteracts further loss and facilitates reabsorption. For a single cell or a very low density population, however, losses may exceed replacement rates and prevent growth. A substantial inoculum, that is, a sufficiently large population, is able to adjust to an initially unfavorable culture medium redox potential, whereas a single cell or very small population may be unable to do so. Such isolation difficulties can be remedied by preparing a sterile filtrate of spent enrichment culture medium and including this filtrate as a major constituent of the new medium on which single cells of fastidious microorganisms are to be isolated.

Colony formation by microbial populations is probably an adaptation based on cooperative interactions within a population. Even motile bacteria that could move away from each other often remain in colonies. Microcolonies, rather than individual microorganisms, are normally observed adhering to particles. The association of a population within a colony allows for more efficient utilization of available resources. Some bacterial colonies even exhibit mass movement, occasionally rotating in place or migrating in a spiral path; such coordinated migration probably obtains beneficial results for the microorganisms within the colony as they seek new sources of nutrients.

Elaborate cooperative interactions within a population are evident in the case of the slime mold *Dictyostelium* (Clark and Steck 1979) (Figure 3.3). When food sources become limiting, the amoeboid cells of *Dictyostelium* swarm together to a central organism. This swarming action, in response to the chemical stimulus of cyclic AMP that is released, occurs as a pulsating wave motion as the stimulus to synthesize AMP is transmitted from proximal to distal cells. The cells unite to form a fruiting body and spores that subsequently disperse. Frequently, some spores reach habitats with a more abundant food supply via this mechanism, germinate, and resume an amoeboid life stage. Communication among members of this microbial population allows for cooperative searching and utilization of resources in the habitat. Similar cooperative interactions have been described for populations of myxobacteria (Shimkets 1990).

Cooperative interactions within populations can be particularly important when the population is utilizing insoluble substrates such as lignin and cellulose. The production of extracellular enzymes by individual members of the population makes such substrates available for all members of the population. At low population densities, the soluble products liberated by such enzymes would be rapidly lost from

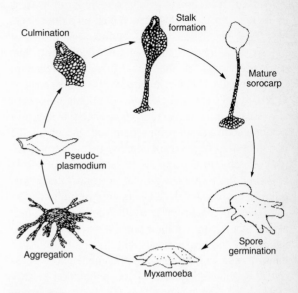

Figure 3.3

Life cycle of the slime mold *Dictyostelium* during which individual cells migrate together and form a single structure that aids in the survival of the population.

the population by dilution, whereas at higher population densities, the resulting soluble products can be utilized with higher efficiency. As an example, myxobacteria feed predominantly on insoluble substrates, and they use their exoenzymes in a communal manner to solubilize these substrates. Density-dependent growth of *Myxococcus xanthus* on insoluble casein was demonstrated in laboratory experiments (Rosenberg et al. 1977). No growth occurred on casein at cell densities lower than 10^3 per ml, and growth rate increased with increasing cell densities. This was not observed on prehydrolyzed casein. In addition to cooperative solubilization of substrates, organic acids produced by individuals within a population also can solubilize inorganic elements in certain habitats, such as soils or rock surfaces, making essential elements available for all members of a population.

Cooperation in a population can also function as a protective mechanism against hostile environmental factors. It is commonly observed in the laboratory that a given concentration of a metabolic inhibitor has less effect on a dense cell suspension than on a dilute one. Microbial populations within biofilms are orders of magnitude more resistant to antimicrobial agents than suspended cells of the same organisms (Shapiro 1991). In natural environments, with their exposures to ultraviolet light, high densities of populations probably shield some members from direct exposure, protecting them and allowing continued growth of the population. At high population densities, microorganisms are able to depress the freezing point of water, allowing for continued growth at temperatures that otherwise would preclude availability of liquid water.

Genetic exchange is another cooperative interaction that occurs within a population. Resistance to antibiotics and heavy metals and the ability to utilize unusual organic substrates are often genetically transmitted to other members of the population (Hardy 1981). This phenomenon allows the adaptive information that has arisen in one microorganism to be disseminated through a population. Genetic exchange is important in preventing overspecialization within a microbial population. Many exchange methods have evolved to permit this cooperative effort. These mechanisms include transformation, transduction, conjugation, and sexual spore formation. Even though such genetic exchange events occur only between two members of the population, they usually require high population densities. Bacterial populations, for example, generally require densities of greater than 10^5/mL for a genetic exchange by conjugation; at lower population densities, the probability of a successful genetic exchange is very low, and the process generally is not significant. Sometimes, however, aggregate formation at low population densities may facilitate genetic exchange. Recipient cells of *Enterococcus faecalis* produce pheromones that induce plasmid-bearing donor cells to synthesize agglutinins and thus to form mating aggregates with recipient cells (Shapiro 1991).

Negative Interactions

Negative interactions within a microbial population are called competition. Because members of a microbial population all utilize the same substrates and occupy the same ecological niche, if an individual within the population metabolizes a substrate molecule, then that molecule is not available for other members of the population. In natural habitats with low concentrations of available substrates, increased population densities cause increased competition for available resources. Competition within populations of predatory microorganisms occurs for available prey; competition within populations of parasitic microorganisms occurs for available host cells. Infection of a host cell by a member of a population effectively precludes further infection of that cell by other members of the population.

In addition to the direct vying for good sources of food, competition in its broad sense includes actual competition for available substrates, as well as other negative interactions, such as those that result from the accumulation of toxic substances produced by members of the population. Within a high-density population, metabolic products may accumulate to an inhibitory level. As discussed previously, the presence of "leaked" metabolic intermediates may represent a form of cooperation. However, the accumulation of some metabolic products, such as low–molecular-weight fatty acids and hydrogen sulfide (H_2S), as occurs under some conditions, constitutes a negative feedback mechanism; such accumulation may effectively limit further growth of some microbial populations even when substrate is still available. As examples of this type of competitive interaction, excessive H_2S accumulation can limit further sulfate reduction; accumulation of lactic acid and other fatty acids can halt the activity of *Lactobacillus;* and the accumulation of ethanol stops further fermentation by *Saccharomyces.* Fatty acid accumulation during hydrocarbon biodegradation similarly can block further microbial metabolism of the hydrocarbon substrates (Atlas and Bartha 1973), and the accumulation of dichloroaniline from 3,4-dichloropropionanilide (propanil) catabolism can stop further growth of *Penicillium piscarium* (Bordeleau and Bartha 1971).

An interesting genetic basis for negative interactions is the occurrence of genes that code for peptides or proteins that have lethal functions (Gerdes et al. 1986a, 1986b). Strains of *Escherichia coli,* for example, have the *hok* (host-killing) gene, which codes for a peptide (Hok) that damages transmembrane potential and leads to death of cells when *hok* is expressed. To prevent *hok* expression, these *E. coli* contain a gene called *sok* (suppression of killing) that codes for an antisense mRNA that blocks the expression of *hok*. As long as the cell is making the mRNA encoded by *sok,* the cell is protected. However, if the cell ceases to make the more labile *sok*-encoded mRNA, the more stable *hok* mRNA is expressed and the cell is killed. Both *hok* and *sok* are encoded on the same plasmids, so as long as the plasmids are retained, the cells of that population survive. If the plasmids are lost, however, the cells of that population self-destruct. This mechanism ensures that the surviving cells retain the plasmids, that is, it is a mechanism for plasmid survival. This may be important for the overall long-term benefit of the population since these plasmids also contain genes for antibiotic resistance that otherwise might be lost, but in the short term the population density is limited because some cells (those that lose the plasmids with *hok* and *sok*) kill themselves.

INTERACTIONS BETWEEN DIVERSE MICROBIAL POPULATIONS

When two different populations interact, one or both of them may benefit from the interactions, or one or both may be negatively affected by the interaction (Table 3.1). The categories used to describe these interactions represent a conceptual classification system; many specific cases are difficult to classify without ambiguity. Possible interactions between microbial populations can be recognized as negative interactions (competition and amensalism); positive interactions (commensalism, synergism, and mutualism); or interactions that are positive for one but negative for the other population (parasitism and predation). In simple communities, one or more of the above interactions can be observed and studied. Within a complex natural biological community, all of these possible interactions will probably occur

between different populations. In established communities, positive interactions among autochthonous (indigenous) populations are likely to be more developed than in newer communities. Invaders of established communities are likely to encounter severe negative interactions with autochthonous populations. The negative interactions between populations act as feedback mechanisms that limit population densities. Positive interactions enhance the abilities of some populations to survive as part of the community within a particular habitat.

The development of positive interactions permits microorganisms to use available resources more efficiently and to occupy habitats that otherwise could not be inhabited. Mutualistic relationships between microbial populations create essentially new organisms capable of occupying niches that could not be occupied by either organism alone. Positive interactions between microbial populations are based on combined physical and metabolic capabilities that enhance growth and/or survival rates. Interactions between microbial populations also tend to dampen environmental stress. The negative interactions that limit population densities represent self-regulation mechanisms that, in the long run, benefit the species by preventing overpopulation, destruction of the habitat, and extinction. Interactions between populations are a driving force in the evolution of community structure.

In the following sections, we shall consider in more detail the types of interactions that can occur between microbial populations. The examples given to illustrate the different types of interactions represent simplified cases. It is often difficult to classify unequivocally the full range of interactions that take place between two populations interacting in their natural environment.

Neutralism

The concept of neutralism implies a lack of interaction between two microbial populations. Neutralism cannot occur between populations having the same or overlapping functional roles within a community

Table 3.1

Types of interactions
between microbial
populations

	Effect of interaction	
Name of interaction	Population A	Population B
Neutralism	0	0
Commensalism	0	+
Synergism (protocooperation)	+	+
Mutualism (symbiosis)	+	+
Competition	-	-
Amensalism	0 or +	-
Predation	+	-
Parasitism	+	-

0 = no effect
+ = positive effect
− = negative effect

and is more likely between microbial populations with extremely different metabolic capabilities than between populations with similar capabilities. Because it is a negative proposition, it is difficult to demonstrate experimentally a total lack of interaction between two populations. Thus, examples for neutralism describe cases where interactions, if any, are of minimal importance.

Neutralism occurs between microbial populations that are spatially distant from each other. It is likely at low population densities where one microbial population does not sense the presence of another. Neutralism between microbial populations occurs in marine habitats and oligotrophic (low-nutrient) lake habitats where population densities are extremely low. In sediment or soil, it occurs when microbial populations occupy separate locations (microhabitats) on soil or sediment particles. Physical separation alone, however, does not ensure a relationship of neutralism. For example, a pathogenic microorganism may invade a plant root, resulting in the death of that plant and the destruction of the habitat of other microbial

populations on the leaves of that plant. Even though there has been no direct contact between the two populations, the destruction of the leaf habitat by the root pathogen precludes a relationship of neutralism between the two populations.

Neutralism occurs between two microbial populations when both populations are outside their natural habitats. For example, in the atmosphere, where population densities are low and all microorganisms can be considered allochthonous (nonindigenous), neutralism is probably the most common relationship between microbial populations.

Environmental conditions that do not permit active microbial growth favor neutralism. For example, microorganisms frozen in an ice matrix, such as in frozen food products, polar sea ice, or frozen freshwater lakes, are typically in a relationship of neutralism. Environmental conditions that favor the growth of microbial populations decrease or preclude the likelihood of neutralism.

Resting stages of microorganisms are more likely to exhibit a relationship of neutralism with other microbial populations than are actively growing vegetative cells. The low metabolic rates required for the maintenance of such resting structures do not force these organisms into competitive relationships for available resources with other microorganisms in the community. Production of spores, cysts, and similar resting bodies during periods of environmental stress, such as excessive heat or drought, allows microbial populations to enter relationships of neutralism. In such situations, the only significant interaction is that with the environmental stress. The strategy for survival is to avoid interpopulation interactions during these times. When environmental conditions become favorable, the resting stages of these organisms can germinate, forming vegetative cells that can then engage in positive or negative relationships.

Resting stages of some microbial populations allow for temporal and spatial niches within a habitat, where temporary coexistence of otherwise competitive populations may occur. Some of these populations would be eliminated by competitive exclusion if such mechanisms did not exist. The formation of resting bodies does not, however, ensure neutralism with other microbial populations. Some microbial populations produce enzymes that can degrade the resting stages of other microbial populations. Many resting microbial stages, though, are resistant to negative interactions with other microorganisms. The complex outer layers of endospores are resistant to enzymatic attack by most microorganisms, as are fungal spore walls with high melanin contents. Such resistant structures allow resting stages to persist within habitats for long periods of time without being affected by negative interactions with other microorganisms.

Commensalism

In a commensal relationship, one population benefits while the other remains unaffected. The term commensalism is derived from the Latin word *mensa* (table) and describes a relationship in which one organism lives off the "table scraps" of another one.

Although commensal relationships are common between microbial populations, they are usually not obligatory. Commensalism is a unidirectional relationship between two populations; the unaffected population, by definition, does not benefit from, nor is it negatively affected by, the actions of the second population. The recipient population may need the benefit provided by the unaffected population but it may also be able to receive the necessary assistance from other populations with comparable metabolic capabilities.

There are a number of physical and chemical bases for relationships of commensalism. Commensalism often results when the unaffected population, in the course of its normal growth and metabolism, modifies the habitat in such a way that another population benefits because the modified habitat is more suitable to its needs. For example, when a population of facultative anaerobes utilizes oxygen and lowers the oxygen tension, it creates a habitat suitable for the growth of obligate anaerobes. In such a habitat, the obligate anaerobes benefit from the metabolic activities of the facultative organisms. The facultative organisms remain unaffected by the relationship as long as the two populations do not compete for the same substrates. The occurrence of obligate anaerobes within microenvironments of predominantly aerobic habitats is dependent on such commensal relationships.

Production of growth factors forms the basis for many commensal relationships between microbial populations (Bell et al. 1974). Some microbial populations produce and excrete growth factors, such as vitamins and amino acids, that can be utilized by other microbial populations. For example, *Flavobacterium brevis* excretes cysteine, which *Legionella pneumophila* can use in aquatic habitats. In soil habitats, many bacteria depend upon the production of vitamins by other microbial populations. As long as the growth factors are produced in excess and excreted from the neutral organism in the relationship, the two populations can have a commensal relationship. The production of growth factors and their excretion into the environment allow fastidious microbial populations to develop in natural habitats.

The transformation of insoluble compounds to soluble compounds and the conversion of soluble compounds to gaseous compounds can also form the basis for a commensal relationship. The changes of state from solid to liquid and liquid to gas mobilize compounds so that they are moved to other habitats where they can benefit other microbial populations. For example, methane produced by bacterial populations in sediment can benefit methane-oxidizing populations in the overlying water column. Methanogenesis may result from a commensal relationship for some populations (Cappenberg 1975). Under certain conditions, *Desulfovibrio* can supply *Methanobacterium* with acetate and hydrogen from anaerobic respiration and fermentation, using sulfate and lactate to generate these products; *Methanobacterium* can then use the products from *Desulfovibrio* to reduce carbon dioxide to methane. Other examples include the production of hydrogen sulfide in buried sediment layers that can be used by photoautotrophic sulfur bacteria on the sediment surface or within a water column, and the transformation of soil-bound ammonia to nitrate by one bacterial population that allows the nitrogen to leach into the underlying soil column, where other soil microbial populations can then get their required nitrogen.

The activities of one microbial population can also make a compound available to another population without actually transforming the particular compound. For example, acids produced by one microbial population may release compounds that are bound or inaccessible to the second population. Such desorption processes are probably common in soil where many compounds can be bonded to mineral particles or humic materials.

Another basis for commensalism between populations is the conversion of organic molecules by one population into substrates for other populations. Some fungi, for example, produce extracellular enzymes that convert complex polymeric compounds, such as cellulose, into simpler compounds, such as glucose. The simpler compounds can be used by populations of other microorganisms that do not possess the enzymes needed to utilize the complex organic molecules. In some cases, a competitive relationship develops for the available simpler substrates. In other cases, the relationship may be truly commensal. Since the solubilized excess substrates would be lost for the first population by dilution in the environment, its use by a second population is not detrimental to the first one.

Cometabolism, whereby an organism growing on a particular substrate gratuitously oxidizes a second substrate that it is unable to utilize as nutrient and energy source, is the basis for various commensal relationships. According to a strict definition of cometabolism, the second substrate is not assimilated by the primary organism, but the oxidation products are available for use by other microbial populations. For example, *Mycobacterium vaccae* is able to cometabolize cyclohexane while growing on propane; the cyclohexane is oxidized to cyclohexanol, which other bacterial populations can then utilize (Beam and Perry 1974) (Figure 3.4) These bacterial populations benefit because they themselves are unable to metabolize intact cyclohexane; the *Mycobacterium* is unaffected since it does not assimilate the cyclohexane.

Yet another basis for commensal relationships is the removal or neutralization of a toxic material. The ability to destroy toxic factors is widespread in micro-

Figure 3.4

An example of commensalism based on cometabolism. Cyclohexane is cometabolized in the presence of propane by *Mycobacterium*, allowing for commensal growth of *Pseudomonas* on cyclohexane. (Source: Beam and Perry 1974.)

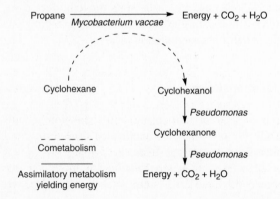

bial communities. The oxidation of hydrogen sulfide by *Beggiatoa* is an example of detoxification that benefits H_2S-sensitive aerobic microbial populations. *Beggiatoa* is not known to benefit from its relationship with the second population. Precipitation of heavy metals, such as mercury, by sulfate-reducers provides an additional example of detoxification. Production of volatile mercuric compounds by bacterial populations in aquatic habitats removes this toxic metal from the habitat (Jeffries 1982).

Some microbial populations are able to detoxify compounds by immobilization. As an example of such a commensal relationship, *Leptothrix* reduces manganese concentrations in some habitats, thereby permitting the growth of other microbial populations. Without the activities of the *Leptothrix* species, the manganese concentrations in these habitats would be toxic to other microbial populations.

In some cases, a microorganism may itself provide the suitable habitat that benefits a commensal partner. Bacteria are often observed on algal surfaces (Sieburth 1975) (Figure 3.5). Not only may the bacteria benefit from the metabolic activities of the algae, but their association with a surface enhances intrapopulation cooperation. In the particular example shown in Figure 3.5, the relationship may also be considered synergistic since this alga fails to grow normally in bacteria-free situations.

Synergism (Protocooperation)

A relationship of synergism between two microbial populations indicates that both populations benefit from the relationship, but unlike mutualism, to be described in the next section, the association is not an obligatory one. Both populations are capable of surviving in their natural environment on their own, although, when initially formed, the association offers some mutual advantages. Synergistic relationships are also loose in the sense that one member population is often easily replaced by another. In some cases, it is difficult to determine whether one of the populations is indeed benefiting, and thus whether the relationship should be considered commensal or synergistic. In

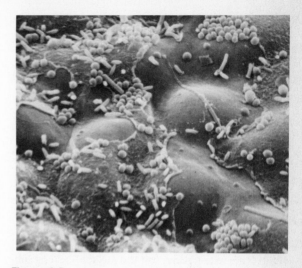

Figure 3.5

Epiphytic bacteria on the surface of *Ulva lactuca*. Such a relationship may be commensal with either population benefiting or synergistic with both populations benefiting. (Courtesy of P. W. Johnson and J. M. Sieburth. (Source: Sieburth 1975, reprinted by permission, copyright University Park Press.)

other cases, it is difficult to determine whether the relationship is obligatory and thus should be considered one of mutualism.

Synergism allows microbial populations to perform activities, such as the synthesis of a product, that neither population could perform alone. The synergistic activities of two microbial populations may allow completion of a metabolic pathway that otherwise could not be completed.

The term syntrophism is applied to the interaction of two or more populations that supply each other's nutritional needs. Figure 3.6 shows a theoretical example of cross-feeding, an example of syntrophism. Population 1 is able to metabolize compound A, forming compound B, but cannot go beyond this point without the cooperation of another population because it lacks the enzymes needed to bring about the next transformation in the pathway. Population 2 is unable to utilize compound A, but it can utilize compound B, forming compound C. Both populations

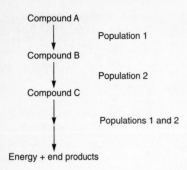

Figure 3.6
Synergistic relationship shown in cross-feeding.

1 and 2 are able to carry out the metabolic steps subsequent to the formation of compound C, producing needed energy and end products that neither population could produce alone.

A classic example of such syntrophism is exhibited by *Streptococcus faecalis* and *Escherichia coli* (Gale 1940) (Figure 3.7). Neither organism alone is able to convert arginine to putrescine. *Streptococcus faecalis* is able to convert arginine to ornithine, which can then be utilized by the *E. coli* population to produce putrescine; *E. coli* alone is able to utilize arginine, producing agmatine, but cannot produce putrescine without assistance. Once putrescine is produced, both *E. coli* and *S. faecalis* can use it.

Relationships of syntrophism are frequently based on the ability of one population to supply growth factors for another population. In a minimal medium, *Lactobacillus arabinosus* and *S. faecalis* are able to grow together but not alone (Nurmikko

1956) (Figure 3.8). The synergistic relationship is based on the fact that *S. faecalis* requires folic acid, which is produced by *Lactobacillus*, and *Lactobacillus* requires phenylalanine, which is produced by *Streptococcus*. Together the organisms grow quite well.

Similarly, cyclohexane can be degraded by a mixed population of a *Nocardia* species and a *Pseudomonas* species, but not by either population alone (Slater 1978) (Figure 3.9). The relationship is based on the ability of the *Nocardia* to metabolize cyclohexane, forming products that feed the *Pseudomonas* species. The *Pseudomonas* species produces biotin and growth factors required for the growth of the *Nocardia* species.

Other interesting examples involving *Chlorobium* are shown in Figure 3.10 (Wolfe and Pfennig 1977). If there is an available source of hydrogen sulfide and carbon dioxide, *Chlorobium* is able to utilize light energy, producing organic matter. Given a sup-

Figure 3.8
Synergistic growth of *Lactobacillus arabinosus* and *Streptococcus faecalis* in a medium lacking phenylalanine and folic acid. (a) Combined culture of *L. arabinosus* and *S. faecalis*. (b) *S. faecalis*, which requires folic acid, alone in culture. (c) *L. arabinosus*, which requires phenylalanine, alone in culture. The two organisms mutually supply each other with required growth factors. (Source: Nurmikko 1956. Reprinted by permission of Birkhauser Verlag.)

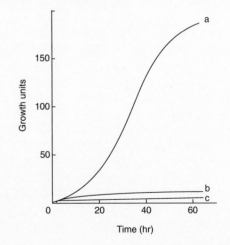

Figure 3.7
Synergistic relationship of *Streptococcus faecalis* and *Escherichia coli* that allows for production of putrescine from arginine. (Source: Gale 1940.)

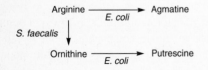

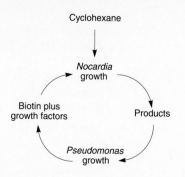

Figure 3.9

Synergistic degradation of cyclohexane by a *Nocardia* and a *Pseudomonas*. *Nocardia* supplies cyclohexane degradation products to *Pseudomonas* which supplies *Nocardia* with biotin. (Source: Slater 1978. Reprinted by permission of Institute of Petroleum, London.)

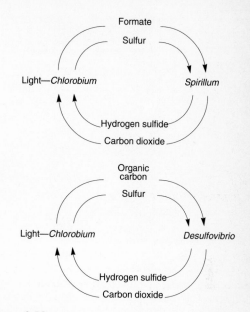

Figure 3.10

Synergistic relationships between *Chlorobium* and *Spirillum* (top) and *Chlorobium* and *Desulfovibrio* (bottom). The relationships are based on the cycling of C and S. *Chlorobium* reduces CO_2 and oxidizes H_2S; *Desulfovibrio* and *Spirillum* oxidize organic C and reduce sulfur to H_2S. (Source: Wolfe and Pfennig 1977. Reprinted by permission of American Society for Microbiology.)

ply of elemental sulfur and formate, *Spirillum* is able to produce hydrogen sulfide and carbon dioxide. Together, *Chlorobium* and *Spirillum* are able to supply each other's nutritional needs. The conversion of hydrogen sulfide to elemental sulfur by the *Chlorobium* species is a detoxifying step, as the hydrogen sulfide concentrations would otherwise kill the *Spirillum* species. *Chlorobium* and *Desulfovibrio* populations also exhibit such a relationship (Wolfe and Pfennig 1977). The *Desulfovibrio* population is able to supply *Chlorobium* with hydrogen sulfide and carbon dioxide; the *Chlorobium* population is able to supply *Desulfovibrio* with sulfate and organic matter required for metabolism by the *Desulfovibrio* and to remove toxic concentrations of hydrogen sulfide. A similar relationship should occur between sulfate-reducing and green sulfur bacteria. The cycling of carbon and sulfur from oxidized to reduced states by these organisms forms a closed loop and allows both organisms to metabolize vigorously in habitats where, on their own, they would be subject to substrate limitation and product inhibition.

There are similar synergistic relationships between bacterial populations involved in the cycling of nitrogen. For example, heterotrophic pseudomonads are chemotactically attracted to organic excretions formed by the heterocysts of *Anabaena oscillatorioides* and related species (Pearl 1987; Pearl and Gallucci 1985) (Figure 3.11). They form dense aggregates around the heterocyst, while few bacteria are associated with the photosynthetic cells of the filament. The pseudomonads oxidize the excreted organics and at the same time stimulate nitrogenase activity. The most likely mechanism for this stimulation is the lowering of oxygen concentration around the oxygen-sensitive nitrogenase. Such relationships are important in global biogeochemical cycling.

Epiphytic bacteria are often seen on the surfaces of algae. The relationship between epiphytic bacteria and algae may be based on the ability of the alga to utilize light and produce organic compounds and oxygen for the aerobic heterotrophic

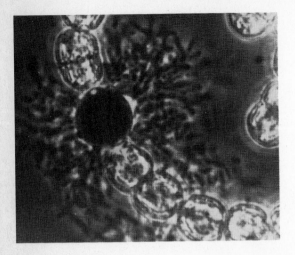

Figure 3.11

Anabaena spiroides heterocyst (dark cell) surrounded by attached heterotrophic bacteria. (Source: Pearl 1987, reprinted by permission of Springer-Verlag.)

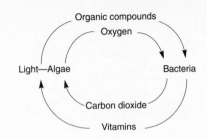

Figure 3.12

Synergistic relationship between algae and epiphytic bacteria based on carbon and oxygen cycling.

metabolism of the bacteria (Figure 3.12). The bacteria, in turn, mineralize excreted organic matter, supplying the algae with carbon dioxide and, sometimes, growth factors needed for phototrophic assimilation.

Chemotaxis can play an important role in the association of algae and bacteria in aquatic habitats (Chet and Mitchell 1976; Jones 1982). Like the heterocysts of cyanobacteria, algae excrete organic compounds, some of which attract bacterial populations. The bacteria produce vitamins that are used by the algae. Extracellular products of algae have been shown to stimulate, selectively, certain populations of marine bacteria, enabling algae to attract and form associations with specific bacterial populations. The bacterial populations that move to the algae can benefit from the relatively high concentrations of organic matter.

Some relationships of synergism are based on the ability of a second population to accelerate the growth rate of the first one. For example, some *Pseudomonas* species can grow on orcinol, but show a higher affinity for this substrate and grow more rap-

idly in the presence of other bacterial populations (Slater 1978) (Figure 3.13). The secondary populations, which cannot utilize orcinol alone, benefit because they are able to utilize the excreted organic compounds produced by the *Pseudomonas*. The *Pseudomonas* species benefits because its growth rate is accelerated, presumably by the removal of the excreted organic compounds that otherwise would act through a negative feedback mechanism to repress catabolic activity.

Some synergistic relationships allow microorganisms to produce enzymes that are not produced by either population alone. For example, populations of closely related *Pseudomonas* have been found to produce lecithinase when grown together, whereas neither strain produces lecithinase alone (Bates and Liu 1963). The production of lecithinase allows both populations to benefit from the utilization of lecithin. Similarly, some mixed microbial populations produce cellulase, whereas the individual populations are unable to attack cellulose.

Several important degradation pathways of agricultural pesticides involve synergistic relationships. *Arthrobacter* and *Streptomyces* strains isolated from soil are capable of completely degrading the organophosphate insecticide diazinon and can grow on this compound as the only source of carbon and energy (Gunner and Zuckerman 1968). This is accomplished by synergistic attack; alone neither culture can mineralize the pyrimidinyl ring of diazinon or grow on this compound. In a chemostat

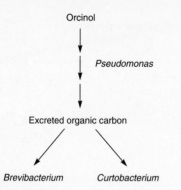

Figure 3.13
Synergistic relationship between an orcinol-utilizing *Pseudomonas* population and two secondary populations, based on acceleration of the growth rate of the primary population by the secondary ones. (Source: Slater 1978. Reprinted by permission of Institute of Petroleum, London.)

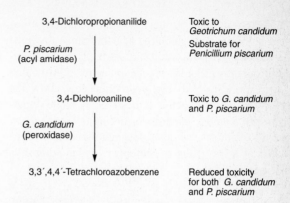

Figure 3.14
Synergistic degradation and detoxification of the herbicide 3,4-dichloropropionanilide (propanil) by two soil fungi. (Source: Bordeleau and Bartha 1971.)

enrichment culture, *Pseudomonas stutzeri* is capable of cleaving the organophosphate insecticide parathion to *p*-nitrophenol and diethylthiophosphate, but is not capable of utilizing either of the resulting moieties (Munnecke and Hsieh 1976). *Pseudomonas aeruginosa* can mineralize the *p*-nitrophenol but is incapable of attacking intact parathion. Synergistically, the two-component enrichment degrades parathion with high efficiency, *P. stutzeri* apparently utilizing products excreted by *P. aeruginosa*.

More complex examples of synergism are based on the simultaneous removal of toxic factors and production of usable substrates. Two soil fungi, *Penicillium piscarium* and *Geotrichum candidum,* are capable of synergistically degrading and detoxifying the agricultural herbicide propanil (Bordeleau and Bartha 1971) (Figure 3.14). *Penicillium piscarium* cleaves propanil to propionic acid and 3,4-dichloroaniline. It uses propionic acid as a carbon and energy source, but is unable to process the toxic 3,4-dichloroaniline product any further. *Geotrichum candidum,* while unable to attack propanil, detoxifies 3,4-dichloroaniline by peroxidatic con-

densation to 3,3´,4,4´-tetrachloroazobenzene and other azo products. These end products are less toxic to these soil fungi than either 3,4-dichloroaniline or the parent herbicide, permitting, in the presence of additional carbon sources, increased growth yields for both fungal populations. Such mutual synergistic growth stimulation occurs only in the presence of the herbicide. In the absence of the herbicide, the two fungi compete for available nutrient resources and consequently decrease each other's growth yield.

Mutualism (Symbiosis)

Mutualistic relationships between populations can be considered as extended synergism. Mutualism, often also called symbiosis, is an obligatory relationship between two populations that benefits both populations. A mutualistic relationship requires close physical proximity; it is highly specific, and one member of the association ordinarily cannot be replaced by another related species. Relationships of mutualism allow organisms to exist in habitats that could not be occupied by either population alone. This does not exclude the possibility that the populations may exist separately in other habitats. The metabolic activities and physiological tolerances of populations involved in mutualistic relationships are normally quite different from those of either population by itself. Mutualistic relationships between microorganisms allow the microorganisms to act as if they were a single organism with a unique identity. According to the theory of serial symbiosis, some endosymbiotic relationships had key roles in the evolution of higher organisms (Margulis 1971).

Lichens The relationships between certain algae or cyanobacteria and fungi that result in the formation of lichens are probably the most outstanding examples of mutualistic, intermicrobial population relationships (Lamb 1959; Ahmadjian 1963, 1967; Hale 1974) (Figure 3.15). Lichens are composed of a primary producer (the phycobiont) and a consumer (the mycobiont). The phycobiont utilizes light energy and produces organic compounds that are utilized by the mycobiont. The mycobiont provides a form of protection and mineral nutrient transport for the phycobiont. In some cases, the mycobiont also makes growth factors available to the phycobiont. The algal and fungal members of the lichen form distinct layers that function as primitive tissues. The phycobiont may be a member of the cyanobacteria, the Chlorophycophyta, or Xanthophycophyta; the green alga *Trebouxia* and the cyanobacterium *Nostoc* are very common photobionts in lichens. Specificity is evident but not absolute; a given algal species may be able to form a lichen association with any of several compatible fungal species and vice versa, and in some lichens multiple phycobiont and/or mycobiont populations may be present.

Morphologically, there are three major types of lichens. Crustose lichens adhere closely to their substrate; foliose lichens are leafy in form and are attached to their substrate more loosely; fruticose lichens often consist of hollow upright stalks and are the loosest attached to their substrates.

Although the lichen represents a tight association between phycobiont and mycobiont, it is possible to dissociate and, in some cases, to reassociate the mutualistic populations. During reassociation of the

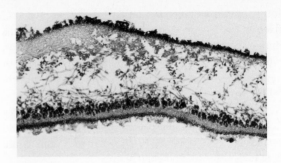

Figure 3.15

Photomicrograph of a lichen thallus showing outer fungal layers and inner algal layers. (Courtesy of Carolina Biological Supply Co.)

lichen, which must be carried out under carefully controlled conditions, the flattened portions of the fungal hyphae (appressoria) make contact with the algal cells; if the reassociation is successful, all the structures typical of the lichen thallus are reestablished, including fruiting bodies and soredia, which are algal cells surrounded by mycelial filaments and which act as "spores" of the lichen association (Ahmadjian et al. 1978). Studies on lichen association suggest that the lichen symbiosis is actually a case of controlled parasitism in which the alga has developed a degree of resistance to the parasitic fungus, such that the percentage of algal cells killed is balanced by the production of new cells (Ahmadjian 1982).

Lichens grow very slowly but are able to colonize habitats that do not permit the growth of other microorganisms. Most lichens are resistant to extremes of temperature and drying, enabling them to grow in hostile habitats such as on rock surfaces. Lichens also produce organic acids that solubilize rock minerals, aiding their growth on rocks.

Some lichens are able to fix atmospheric nitrogen, and in some habitats they are an important source of combined nitrogen. For example, in tundra soils, the lichen *Peltigera,* which contains the cyanobacterium *Nostoc,* provides a major source of fixed nitrogen for the biological community. Nitrogen-fixing lichens growing in the forest canopy provide fixed forms of nitrogen for the underlying vegetation; the fixed forms of nitrogen are leached from the lichens during rainstorms and are washed to the forest floor, where the nitrogen becomes available to the forest vegetation and is taken up by plant roots (Denison, 1973).

The mutualistic association in a lichen is very delicately balanced and can be disrupted by environmental changes (Gilbert 1969). Although lichens are able to occupy some hostile habitats, they are particularly sensitive to industrial air pollutants and have been disappearing from industrialized areas (Figure 3.16). Lichens have disappeared from cities with polluted air. It appears that the sulfur dioxide in the atmosphere is inhibitory to the lichens, probably because it inhibits the phycobiont. The reduced efficiency of photosynthetic activity of the phycobiont allows the mycobiont to overgrow it and leads to the elimination of the mutualistic relationship. Following destruction of the phycobiont, the fungus is unable to survive alone and is also eliminated from the habitat.

Endosymbionts of Protozoa   As with lichens, interesting mutualistic relationships exist between popula-

tions of algae and protozoa. *Paramecium* can host numerous cells of the alga *Chlorella* within its cytoplasm (Ball 1969). The alga provides the protozoan with organic carbon and oxygen, and the protozoan presumably provides the alga with protection, motility, carbon dioxide, and perhaps other growth factors. The presence of the *Chlorella* within the ciliate allows the protozoan to move into anaerobic habitats as long as there is sufficient light; without the alga the protozoan could not enter and survive in such habitats. Similarly, some foraminiferans enter into mutualistic relationships with Pyrrophycophyta or Chrysophycophyta; the presence of the algae confers a red color on the protozoa (Taylor 1982). Presumably, the mutualistic relationship is based on the ability of the alga to supply oxygen and photosynthate and on the ability of the foraminiferan to protect the algal partner from grazers. Under normal environmental conditions, such mutualistic relationships between algae and protozoa persist; but if stressed, for instance by a prolonged absence of light, the protozoan may digest the algal population.

Many other protozoa are hosts to endosymbiotic algae and cyanobacteria. Each protozoan can contain 50–100 algal cells. The algae found within freshwater protozoa normally belong to the Chlorophycophyta and are called zoochlorellae. The endosymbiotic algae of marine protozoa are most frequently dinoflagellates (Pyrrophycophyta) but can be Chrysophycophyta; these endozoic marine algae are called zooxanthellae. Endozoic cyanobacteria, which occur both in freshwater and marine protozoa, are called cyanellae.

The bacterial nature of many of the endosymbionts of protozoans has been established by electron microscopic examination, but many of the observable bacteria have yet to be isolated and cultured. Some multiply within the nucleus of the protozoan, others within the cytoplasm. The bacterial endosymbiont appears to contribute to the nutritional requirements of the protozoan. In the case of the flagellate protozoans *Blastocrithidia* and *Crithidia,* the bacterial endosymbiont provides hemin and other growth factors to the protozoa; populations of these protozoa lacking endosymbiotic bacteria require external sources of these growth factors (Chang and Trager 1974).

Figure 3.16

Disappearance of lichens from cities due to air pollution, showing changes in the lichen cover of ash trees moving away from Newcastle upon Tyne, England, to the west. (a) combined cover of all sensitive lichens. (b) cover of *E. prunastri.* (Source: Gilbert 1969. Reprinted by permission of Centre for Agricultural Publications and Documentation, Wageningen, The Netherlands.)

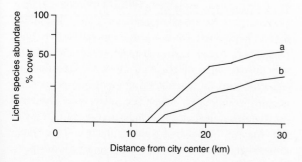

One of the most interesting mutualistic relationships between a protozoan and a bacterium is between *Paramecium aurelia* and a *Caedibacter* endosymbiont, formerly known as a kappa particle (Sonneborn 1959; Preer 1981). *Paramecium aurelia* populations occur in two forms: killers and sensitives. Killer *P. aurelia* populations contain endosymbionts; the sensitive strains lack the endosymbionts. The nature of the toxic substance associated with killer strains and the mechanisms by which the presence of the endosymbiont confers immunity are not fully known. All *Caedibacter* endosymbionts that confer the killer trait exhibit a strongly refractile inclusion body, known as an R body. The genetic information for the R body is coded on a *Caedibacter* plasmid. Although the R body itself does not appear to be the toxin, evidence indicates that the R body and the toxin are specified by the same coding region of the plasmid; it also appears that the R body is essential for the expression of the killer trait (Dilts and Quackenbush 1986). The presence of *Caedibacter* endosymbionts gives an advantage to killer strains in competition with sensitive strains. The obligately endosymbiotic *Caedibacter* strains are nutritionally dependent on their protozoan host.

Other Mutualistic Relationships   Bacterial populations involved in methane production have interesting mutualistic relationships. Some species of methanogenic bacteria, once believed to be pure cultures, have been shown to be mutualistic associations of bacteria. The mutualistic relationships among bacterial populations involved in the generation of methane appear to be based on electron transfer (Figure 3.17). For example, *Methanobacterium omelianskii* is associated with an "S" organism that utilizes acetate and supplies electrons to the *Methanobacterium* (Zeikus 1977; Wolin and Miller 1985). The *Methanobacterium* uses the electrons to reduce carbon dioxide to form methane. In the anaerobic digestion of a whey effluent, lactate and ethanol were converted to methane by a three-membered syntrophic association (Thiele and Zeikus 1988). As shown in Figure 3.18, in the floc (aggregate), *Desulfovibrio vulgaris* generates acetate and formate from ethanol and bicarbonate. The formate serves as an interspecies electron transfer agent to *M. formicicum*

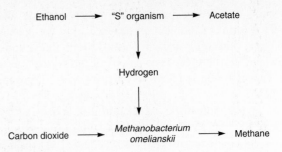

Figure 3.17
Mutualism based on hydrogen transfer in methanogens. (Source: Slater 1978. Reprinted by permission of Institute of Petroleum, London.)

that evolves methane, shuttling back bicarbonate to *D. vulgaris* for reduction to formate. The acetate generated in the ethanol oxidation process is cleaved by acetoclastic methanogens to methane and CO_2. In this system the cleavage of formate to CO_2 and H_2 plays only a negligible role and is performed by bacteria not associated with the floc.

Another relationship that can be viewed as mutualism is the interaction of lysogenic phage and bacterial populations. The genetic information of the phage is incorporated into the genome of the bacterial population. This provides a mechanism of survival for the phage in a dormant state over a long period of time. Lysogenic or defective phage do not result in the lysis of the host bacterial populations under normal conditions. The presence of DNA from a lysogenic phage can be difficult to detect in a host population. The addition of the phage DNA to the bacterial genome adds genetic information and capability to the bacterial population. Bacteria harboring lysogenic phage sometimes exhibit greater virulence (infectivity) or produce enzymes that are absent in uninfected bacterial populations. The relationship between temperate or lysogenic phage and bacteria also provides a mechanism for genetic exchange of bacterial DNA via transduction. Bacterial populations harboring lysogenic phage may acquire competitive advantage over other bacterial populations if the harbored phage acts toward these in a lytic manner. Survival of both the phage and the lysogenized bacterium are enhanced by the relationship.

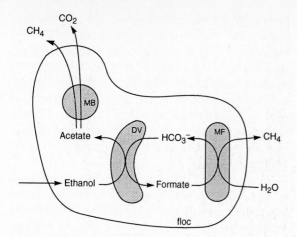

Figure 3.18

Model of a three-membered syntrophic association involved in the anaerobic digestion of a whey effluent to methane and carbon dioxide. Within the floc, *Desulfovibrio vulgaris* (DV) converts ethanol to acetate, coupled to the reduction of bicarbonate to formate. The formate is transferred to *Methanobacterium formicium* (MF) which releases methane. The acetate generated during ethanol oxidation is metabolized by an acetoclastic methanogen, such as *Methanosarcina barkeri* (MB), to methane and carbon dioxide. (Source: Thiele and Zeikus 1988.)

Competition

In contrast to positive interactions, competition represents a negative relationship between two populations in which both populations are adversely affected with respect to their survival and growth (Veldkamp et al. 1984). The populations may achieve lower maximal densities or lower growth rates than they would have in the absence of competition. Competition occurs when two populations utilize the same resource, whether space or a limiting nutrient. In this section, only competition for a resource will be discussed. Interference or inhibition via chemical substances will be discussed separately under amensalism. Competition may occur for any growth-limiting resource. Available sources of carbon, nitrogen, phosphate, oxygen, growth factors, water, and so on are all resources for which microbial populations may compete.

Competition tends to bring about ecological separation of closely related populations. This is known as the competitive exclusion principle (Fredrickson and Stephanopoulos 1981). Competitive exclusion precludes two populations from occupying exactly the same niche, because one will win the competition and the other will be eliminated; populations may coexist, however, if the populations can avoid absolute direct competition by using different resources at different times.

George F. Gause (1934) showed a classic example of competitive exclusion experimentally by using populations of two closely related ciliated protozoans, *Paramecium caudatum* and *P. aurelia* (Figure 3.19). Individually, with an adequate supply of bacterial prey, both protozoan populations were able to grow and maintain a constant population level. When the protozoans were placed together, however, *P. aurelia* alone survived after sixteen days. Neither organism attacked the other or secreted toxic substances; rather *P. aurelia* exhibited the more rapid growth rate and thus outcompeted *P. caudatum* for available food.

In contrast, a mixture of *P. caudatum* and *P. bursaria* were able to survive and reach a stable equilibrium even when grown together. The *P. caudatum* and *P. bursaria* competed for the same food source, but the two populations occupied different regions of the culture flask. Thus, the habitats of *P. caudatum* and *P. bursaria* were sufficiently different to allow the two species to coexist simultaneously although they had overlapping niches based on their common food source. The ability to occupy different habitats provided a mechanism that minimized competition and prevented extinction of one of the species.

Experiments using flow-through growth chambers called chemostats also have demonstrated the principle of competitive exclusion (Veldkamp et al. 1984). Under limiting conditions, a single bacterial population will survive in a chemostat, and other populations competing for the primary resource will be excluded from the system. The population with the highest intrinsic growth rate under the experimental conditions is the one that survives, and populations

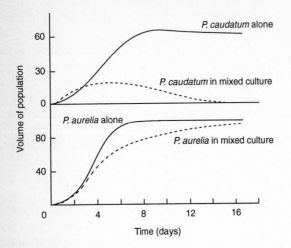

Figure 3.19

Competition between two protozoa that inhabit similar niches. In mixed culture *Paramecium aurelia* displaces *P. caudatum*. (Source: Gause 1934. Reprinted by permission, copyright 1934 Williams and Wilkins.)

with lower intrinsic growth rates become extinct. This does not occur, however, if there is a wall effect allowing a population to adhere to the experimental chamber, preventing it from being washed out of the system, or if synergistic or mutualistic interactions supersede competition.

Intrinsic growth rates of competing populations vary under different environmental conditions, explaining the coexistence of populations in the same habitat competing for the same resources. For example, in marine habitats, populations of psychrophilic and psychrotrophic bacteria are found together even though both populations may be competing for the same low concentrations of organic nutrients (Harder and Veldkamp 1971). At low temperatures, the psychrophilic populations exhibit higher intrinsic growth rates, and given sufficient time they would exclude the psychrotrophic populations. At higher temperatures, however, the psychrotrophic populations exhibit the higher intrinsic growth rate and the psychrophilic populations would be excluded. In habitats of varying temperature regimes, the advantage shifts back and forth, leading to seasonal shifts

in proportions of psychrotrophic and psychrophilic populations within natural aquatic habitats. In some instances, bacterial populations are excluded from the habitat during periods of time when temperature or other environmental factors do not allow them to compete successfully with other populations because of their lower intrinsic growth rates. Under different environmental conditions, these same bacterial populations may compete successfully and even become dominant.

Studies indicate that under varying environmental conditions, sulfur bacteria competing for the same substrates coexist (van Gemerden 1974) (Table 3.2). *Chromatium vinosum* was shown to outcompete, and lead to the exclusion of, *C. weissei* when growing in continuous light with sulfide as the growth-rate–limiting substrate. The specific growth rate of *C. vinosum* exceeded that of *C. weissei* regardless of the sulfide concentrations. However, with intermittent illumination, the organisms showed balanced coexistence when grown in continuous culture. The relative proportion of *C. vinosum* was found to be positively correlated to the length of light period, and the relative proportion of the *C. weissei* was positively correlated to the length of the dark period. The coexistence is explained by the fact that during the light period both strains grow with most of the sulfide being oxidized by the *C. vinosum;* during the dark period, sulfide accumulates, and upon illumination the *C. weissei* oxidizes the greater portion of the accumulated sulfide. The alternation of light and dark periods thus oscillates the balance between the two populations, allowing them to coexist. Similarly, the daily and seasonal variations in environmental conditions of a habitat lead to temporal oscillations in the success or displacement of competing populations. As long as competitive displacement does not go to completion, these fluctuations can continue.

Studies with algae have supported the resource-based competition theory (Tilman 1976). Possible outcomes of ecological competition, including stable coexistence, were observed in laboratory studies of two species of freshwater diatoms that were potentially limited by phosphate and silicate.

Table 3.2

Relative abundance of *Chromatium vinosum* and *Chromatium weissei* during steady state coexistence in relation to the light regime

Illumination regime	Organism	Relative abundance (%)	
		Inoculum	During steady state
Continuous light	C. vinosum	10	100
Continuous light	C. weissei	90	0
18 hr light	C. vinosum	30	100
6 hr dark	C. weissei	70	0
6 hr light	C. vinosum	20	63
6 hr dark	C. weissei	80	37
4 hr light	C. vinosum	60	30
8 hr dark	C. weissei	40	70

Source: van Gemerden 1974.

The relative abundances of these nutrients determined the outcome of competition, coexistence occurring only when the growth rates of both species were limited by different resources. If only one nutrient was limiting, competitive displacement occurred, with the population best able to acquire and utilize the limiting resource displacing the competing species. The algae used in these studies were the diatom *Asterionella formosa* and *Cyclotella meneghiniana*. If both species were limited by phosphate, *A. formosa* displaced *C. meneghiniana*. Under conditions of silicate limitation, the *C. meneghiniana* population displaced *A. formosa*. Competition of two suspension-feeding protozoan populations for growing bacterial populations in continuous culture can result in coexistence if one population grows faster than the second in one range of resource density and the opposite occurs in another range of resource density (Baltzis and Fredrickson 1984).

The development of dominant populations represents a case of competitive displacement. Abiotic parameters, such as temperature, pH, and oxygen, greatly influence the intrinsic growth rates of micro-bial populations and the outcome of a competitive struggle. For example, at high substrate concentrations, competition between a marine *Spirillum* and *Escherichia coli* results in the competitive exclusion of the *Spirillum*, whereas at low concentrations the reverse occurs and the *E. coli* is excluded (Jannasch 1968). Dominant microbial populations in sewage, which has a high organic substrate content, are rapidly displaced in competition with the autochthonous microbial populations of receiving streams and rivers, as the concentration of organic matter diminishes in the course of mineralization and dilution.

Competitive advantage need not be solely based on ability to utilize a substrate more rapidly. Tolerance to environmental stress also may be an important factor in determining the outcome of competition between microbial populations. For example, under conditions of drought, populations that can best tolerate and survive desiccation can displace less tolerant populations because they are better adapted or because they are more effective competitors under these conditions. Similarly, under other stress conditions, such as high temperatures or high salt concentrations, the population with the greatest tolerance to that factor may have the edge in the competition. These cases represent competition for survival during nongrowth conditions. During active growth conditions, the competitive advantage returns to the population with the highest growth rate under those conditions.

Amensalism (Antagonism)

Microorganisms that produce substances toxic to competing populations will naturally have a competitive advantage (Fredrickson and Stephanopoulos 1981). When one microbial population produces a substance that is inhibitory to other populations, the interpopulation relationship is called amensalism. The first population may be unaffected by the inhibitory substance or may gain a competitive edge that is beneficial. The terms antibiosis and allelopathy have been used to describe such cases of chemical inhibition. There are cases of complex amensalism between populations in natural habitats, for example, virucidal (virus-killing) factors in seawater and fungistasis (fungi-inhibiting) in soil (Lockwood 1964). The basis of these relationships is believed to be amensalism, since sterilization eliminates the inhibitory factors, but the chemical background of these complex antimicrobial activities remains to be explored.

Amensalism may lead to the preemptory colonization of a habitat. Once an organism establishes itself within a habitat it may prevent other populations from surviving in that habitat. The production of lactic acid or similar low-molecular-weight fatty acids is inhibitory to many bacterial populations (Pohunek 1961; Wolin 1969). Populations able to produce and tolerate high concentrations of lactic acid, for example, are able to modify the habitat so as to preclude the growth of other bacterial populations. *Escherichia coli* is unable to grow in the rumen, probably because of the presence of volatile fatty acids produced there by anaerobic heterotrophic microbial populations (Wolin 1969). Fatty acids produced by microorganisms on skin surfaces are believed to prevent the colonization of these habitats by other microorganisms. Populations of yeasts on skin surfaces are maintained in low

numbers by microbial populations producing fatty acids. Acids produced by microbial populations in the vaginal tract are probably responsible for preventing infection by pathogens such as *Candida albicans* (Pohunek 1961).

The oxidation of sulfur by *Thiobacillus thiooxidans* produces sulfuric acid. In aquatic habitats, this results in greatly lowered pH values. Growth of *T. thiooxidans* often occurs on reduced sulfur compounds associated with coal deposits; the resulting leaching of sulfuric acid produces acid mine drainage (Higgins and Burns 1975). The pH values of approximately 1 to 2 in streams receiving acid mine drainage preclude the growth of most microorganisms in the affected habitats. *Thiobacillus thiooxidans* is not known to benefit from this amensal relationship with the populations of microorganisms inhibited by the acid mine drainage.

Consumption or production of oxygen may alter the habitat in ways detrimental to microbial populations. The production of oxygen by algae precludes the growth of obligate anaerobes. Few obligate anaerobes are found in habitats where there is a significant production of oxygen by algal populations. Production of ammonium by some microbial populations is deleterious to other populations (Stojanovic and Alexander 1958). Ammonium is produced during the decomposition of proteins and amino acids at concentrations inhibitory to nitrite-oxidizing populations of *Nitrobacter*.

Some microbial populations produce alcohols. Low–molecular-weight alcohols, such as ethanol, are inhibitory to many microbial populations. The production of ethanol by yeasts prevents the growth of most bacterial populations in habitats where the ethanol can accumulate. Thus, few bacterial populations can grow in a habitat of fermenting grapes (wine). Ethanol production does not, however, eliminate all bacterial populations. *Acetobacter* is able to convert ethanol to acetic acid (vinegar) when oxygen is available. The acetic acid produced is also inhibitory to many bacterial populations. The presence in food products of such inhibitory compounds, such as lactic and propionic acids in cheese and acetic acid in vinegar, inhibits growth of spoilage-causing microbial populations in such products.

Inhibitory substances produced by microorganisms may also act as preservatives for organic compounds in natural habitats. For example, during the decomposition of cellulose in soil, organic acids are produced that prevent further breakdown of cellulose metabolites in subsurface soil.

Some microorganisms produce antibiotics. An antibiotic is a substance produced by one microorganism that, in low concentrations, kills or inhibits the growth of other microorganisms. The use of antibiotics in medicine has had a great influence on our ability to control disease and the distribution of some microbial populations. Although under favorable laboratory conditions antibiotics are produced by some microorganisms in high concentrations and can be demonstrated to be potent inhibitors of other microbial populations, their role in natural habitats is subject to debate.

Conditions that favor the production of antibiotics are not normally found in natural habitats. Antibiotics are secondary metabolites and are produced when excess substrate concentrations are available. The normal condition of most soil and aquatic habitats is one of limiting organic substrates; antibiotics are not found to accumulate in natural habitats. Many microbial populations are tolerant to antibiotics and/or are capable of degrading them. In aquatic habitats, antibiotics probably are rapidly diluted to ineffective concentration levels. Antibiotics produced in soil may be bound to clay minerals or other particulates and thus be inactivated. Antibiotic-producing microorganisms usually do not dominate in natural soil and aquatic habitats. Likewise, proportions of antibiotic-resistant strains are not exceptionally high in habitats where antibiotic-producing populations are found. On the other hand, if antibiotics were of no use in nature, they probably would not have been selected during evolutionary processes and would not be as widespread as, indeed, they are.

Production of antibiotics probably has a significant function in establishing amensal relationships within microenvironments and under certain environmental conditions (Park 1967). Opportunistic microorganisms, such as the zymogenous populations in soil habitats, grow under conditions of localized concentrations of high organic matter. These conditions may permit production of antibiotics. Microbial populations growing in such microhabitats have a distinct advantage if they can discourage competition for the available substrates by other populations through an amensal relationship involving antibiotic production. Antibiotic production may aid the initial colonization of a substrate, after which antibiotic production may no longer be required.

Although difficult to prove, antibiotic production by microbial populations associated with plant residues—where there is adequate energy to support the production of secondary metabolites—seems to play a role in interpopulation relationships (Bruehl et al. 1969). *Cephalosporium gramineum* is a pathogen of wheat that survives in dead wheat tissues between crops and is capable of secreting antifungal substances, which appear to exclude other fungal populations endeavoring to colonize the dead tissues. Populations of *C. gramineum* that do not produce antibiotics are less able to prevent colonization of dead wheat tissues by other fungal populations and thus are less able to survive in this habitat (Bruehl et al. 1969).

Another interesting relationship occurs on the skin of the New Zealand hedgehog (Smith and Marples 1964). *Trichophyton mentagrophytes* produces the antibiotic penicillin in its natural habitat on the skin of hedgehogs. Populations of *Staphylococcus* found in the same habitat are penicillin resistant. The production of penicillin by the *Trichophyton* population allows it to enter into amensal relationships with non-penicillin-resistant populations that attempt to colonize the skin of the hedgehog. The penicillin-resistant *Staphylococcus* has developed an adaptation that allows it to neutralize the amensal influence of the penicillin-producing *Trichophyton*. Adaptation here could reflect coevolution of the coexisting microbial populations.

Parasitism

In a relationship of parasitism, the population that benefits, the parasite, normally derives its nutritional requirements from the population that is harmed, the host. The host-parasite relationship is characterized by a relatively long period of contact, which may be direct physical or metabolic. Usually, but not always, the parasite is smaller in size than the host. Some parasites remain outside the cells of the host population and are called ectoparasites; other parasites penetrate the host cell and are called endoparasites.

Normally, the parasite-host relationship is quite specific. The available habitats for obligately parasitic populations are limited to available hosts. In some cases, host specificity depends on surface properties of the organisms that allow physical attachment of the parasite to the host cells.

Viruses are obligate intracellular parasites that exhibit great host cell specificity. There are viral parasites of bacterial, fungal, algal, and protozoan populations (Anderson 1957; Lemke and Nash 1974; van Etten et al. 1983). In some habitats, viruses are suspected of being responsible for the decline and disappearance of bacterial populations. Parasitism may contribute to the disappearance of other fecal organisms in sewage that enters aquatic habitats.

The interactions between bacteriophage and their host bacteria are subject to environmental modification (Roper and Marshall 1974). Host bacteria and phage may be adsorbed onto clay particles in sediment. In saline sediments, fecal bacteria, such as *Escherichia coli*, are protected from phage attack by sorption to clay particles. At lower salinities, *E. coli* and phage are desorbed from the particle, and the parasitic phage can attack and eliminate the host bacterial population. Protection of host populations by sorption onto particles provides an important mechanism for escape from parasitism.

Like bacteriophage, the bacterium *Bdellovibrio* is parasitic on Gram-negative bacterial populations (Stolp and Starr 1963, 1965; Starr and Seidler 1971; Rittenberg 1983) (Figure 3.20). *Bdellovibrio* is highly motile, attaining speeds up to one hundred cell lengths per second compared to only ten lengths per second for *E. coli,* a potential host species. The encounter between predator and prey appears to be random; no clear evidence has been found for chemotactic attraction. Only a small percentage of cell contacts result in permanent attachment of *Bdellovibrio* to the outer cell membrane of its Gram-negative host. Subsequently, *Bdellovibrio* penetrates the outer cell membrane of its

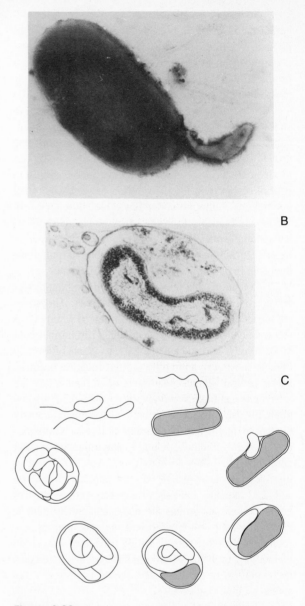

Figure 3.20

(A) Electron micrograph showing attachment of the parasitic bacterium *Bdellovibrio* to host cell. (Source: Marbach et al. 1976. Reprinted by permission of Springer-Verlag.) (B) Electron micrograph of thin cross section showing *Bdellovibrio* within host cell. (Source: Marbach et al. 1976. Reprinted by permission of Springer-Verlag.) (C) Life cycle of *Bdellovibrio bacteriovorus.*

host and enters the periplasmic space but not the cell proper; *Bdellovibrio* is an ectoparasite. During the parasitic interaction that spans approximately one hour, *Bdellovibrio* modifies the cell envelope of its host in both degradative and synthetic manners. The host cell loses its original shape and becomes spherical (bdelloplast), yet retains the cell contents for use by the parasite. The cell contents of the host are partially degraded and utilized by *Bdellovibrio* with high efficiency. When it enters the periplasmic space, *Bdellovibrio* loses its flagellum and grows into a filament without cell division. When the cell contents of the host are exhausted, the filament divides into individual cells, which develop flagella. The bdelloplast lyses and releases the progeny. The burst size (number of progeny per host cell) depends on the size of the host cell and may range from about four in *E. coli* to about twenty in the much larger *Spirillum serpens*. Wild strains of *Bdellovibrio* are obligately parasitic, but in contrast to bacteriophage, they have complete sets of catabolic, anabolic and energy-generating enzymes. Mutants have been obtained that grow on heat-killed host cells and even on rich synthetic media. While *Bdellovibrio* is capable of eradicating its host in a laboratory setting, the impact of this parasitic interaction in natural environments is strongly attenuated and of limited significance. The interaction of *Bdellovibrio* and host *E. coli* has been observed to be partially inhibited by the presence of montmorillonite clays (Roper and Marshall 1978) (Figure 3.21). The clay particles appear to form an envelope around the *E. coli* and inhibit the ability of the parasitic *Bdellovibrio* to reach the host cells.

Other ectoparasitic microorganisms can cause lysis without direct contact. Myxobacteria, for example, can cause lysis of susceptible microbial strains at some distance, apparently with the aid of exoenzymes. The myxobacteria derive their nutrition from material released by the lysed microorganisms. Soil myxobacteria are able to lyse Gram-negative and Gram-positive bacterial populations. Some myxobacteria, such as *Cytophaga* populations produce enzymes that cause lysis of susceptible algae (Stewart and Brown 1969). Some bacterial populations produce chitinase, an enzyme that causes lysis of

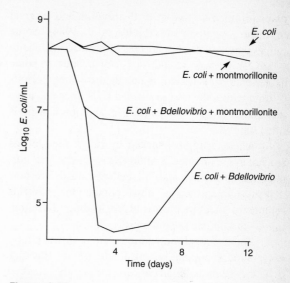

Figure 3.21

Effect of clay particles on the interaction of *Bdellovibrio* and *Escherichia coli*. The presence of clay particles (montmorillonite) attenuates the parasitism of *Bdellovibrio*. (Source: Roper and Marshall 1978. Reprinted by permission, copyright Springer-Verlag.).

chitin-containing fungal cell walls. Other bacterial populations produce laminarinase or cellulase enzymes that similarly attack the cell wall structures of some algal and fungal populations. In all of these cases, ectoparasitic populations cause lysis of host cells releasing nutrients that then can be utilized by the ectoparasites, even though many of these ectoparasites are not obligate parasites and can derive nutrition by other means.

Some microbial populations are more resistant to lysis than others. Microbial populations may produce resting stages, such as cysts and spores, that are more resistant to lytic activities of ectoparasites than the vegetative cells. This provides a mechanism for escaping the pressures of ectoparasitism that allow the host populations to persist.

Protozoan populations are subject to parasitism by a large number of fungi, bacteria, and other protozoa. The ability of the opportunistic human pathogen *Legionella pneumophila* to parasitize protozoa may play an important role in its distribution and survival in aquatic environments (Fields et al. 1984). Some fungal populations are parasites of protozoa (Madelin 1968). These may be endo- or ectoparasites. In some cases, endoparasites are maintained for long periods of time within the host protozoan cells. In such cases, the line between endoparasitism and mutualism is often blurred.

Algae, too, are attacked by fungi; the chitrids are notable examples. Uniflagellate chitrids are normally ectoparasites, and biflagellate chitrids are normally endoparasites of algal populations. Chitrids frequently infect freshwater algae, leading to decimation of susceptible populations.

Some fungal populations are parasitized by other fungal populations (Barnett 1963). Basidiomycetes, such as *Agaricus,* are frequently attacked by other fungi, such as *Trichoderma.* Such infections create difficulties in the commercial cultivation of mushrooms.

Microorganisms that are themselves parasites may also serve as host cells for other parasitic populations. This phenomenon is known as hyperparasitism. For example, *Bdellovibrio,* a bacterial parasite, may serve as a host cell for appropriate phage populations. Fungal populations that are parasitic on algae are themselves subject to bacterial and viral parasitism. Some rust fungi are parasitized by lytic bacteria.

Parasitism has benefits beyond the immediate ones derived by the parasites themselves. Parasitic interactions provide a mechanism for population control. Because the intensity of parasitism is population density dependent, parasites can only thrive as long as there are abundant host populations. Parasitism results in a reduction of host population density that then allows for the accumulation and renewal of environmental resources being utilized by the hosts. When host populations decline, the resources available for the parasites also diminish, leading to a decrease of the parasite population. As long as it has escaped complete extinction by the parasites, the host population can then recover. Without negative feedback control by mechanisms such as parasitism, host populations might continue to grow unchecked until all the resources needed for their growth were exhausted, which, in turn, would lead to a crash in the population of the parasite organisms and their possible extinction.

Predation

In the microbial world, the distinction between parasitism and predation is not sharp. For example, the interaction between *Bdellovibrio* and susceptible Gram-negative bacteria is regarded by some as parasitism, by others as predation. Predation typically occurs when one organism, the predator, engulfs and digests another organism, the prey. Normally, predator-prey interactions are of short duration and the predator is larger than the prey (Slobodkin 1968).

Early theoretical considerations of predator-prey relationships (Lotka 1925; Volterra 1926) led to the prediction that interactions of predator and prey species lead to regular cyclic fluctuations of the two populations (Figure 3.22). In each cycle, as the prey population increases, the predator population follows, overtakes, and overcomes it, producing a decline in the prey. The predator population then follows the declining prey supply downward until the predators reach such a low level that the prey population begins to rise again, whereupon the predator again follows it upward. Assuming that the environ-

Figure 3.22
Theoretical predator-prey fluctuations. (Source: Krebs 1972. Reprinted by permission of Harper and Row; copyright 1972 Charles Krebs.)

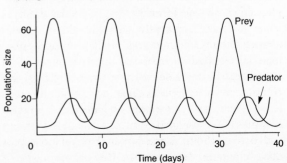

ment is constant, the amplitude of the cycle is determined by the initial population density and remains the same. Theoretically, such predator-prey populations could cycle forever.

According to the Lotka-Volterra model, the prey population adds individuals according to its intrinsic rate of increase (r_1) times its density (N_1) and loses individuals at a rate proportional to encounters of predator and prey individuals, hence to the product of prey density and predator density (N_2). Thus, the rate of change for the prey population is expressed mathematically as Equation 1.

(1) $dN_1/dt = r_1N_1 - PN_1N_2$

The predator adds individuals at a rate proportional to this same product of densities and loses individuals according to a death rate (d_2) times its own density, which mathematically is expressed as Equation 2.

(2) $dN_2/dt = aPN_1N_2 - d_2N_2$

P is a coefficient of predation that relates predator births to prey consumed. The equations describe a cyclic fluctuation that is stable unless disturbed.

Experimental models, however, have rarely supported the Lotka-Volterra model. In a classic series of experiments, Gause (1934) showed that when *Didinium nasutum* is introduced with *Paramecium caudatum,* it preys on the *Paramecium* until the *Paramecium* population becomes extinct. Lacking a food source, the *Didinium* population also becomes extinct (Figure 3.23). If a few members of the *Paramecium* population are able to hide and escape predation by the *Didinium,* then the *Paramecium* population can recover following extinction of the *Didinium.* Only by periodically introducing *Paramecium* and *Didinium* populations was Gause able to produce a sustained interaction with marked fluctuations in population numbers. Gause (1934) also was able to establish periodic oscillations using *P. bursaria* as a predatory species and the yeast *Schizosaccharomyces pombe* as a prey species but was unable to establish sustained oscillations. In this heterotrophic, nutrient-limited system, the amplitude of the oscillations decreased, leading to extinction of

both species. The above experiments indicate that intense predator-prey relations may lead to the extinction of both the predator and the prey populations.

The failure of experimental models, generally, to reveal persistent population oscillations has forced a reevaluation of the underlying assumptions of the Lotka-Volterra model. The error appears to be the invalidity of the inherent assumption that predatory encounter is random in time and space and, like a bimolecular collision, is proportional to the product of predator and prey populations (Williams 1980). Experi-

Figure 3.23

Predator-prey interactions between *Paramecium* and *Didinium.* (A) *Didinium* introduced at high *Paramecium* levels with no refuge site. Intense predation eliminates *Paramecium,* and *Didinium* subsequently starves. (B) *Didinium* introduced at low *Paramecium* levels with sediment refuge site. *Didinium* starves, and *Paramecium* that escape predation multiply after elimination of *Didinium.* (C) *Paramecium* and *Didinium* introduced every three days, preventing permanent extinction and producing predator-prey cycles. Arrows indicate times of protozoan introduction. (Source: Gause 1934. Reprinted by permission, copyright 1934 Williams and Wilkins.)

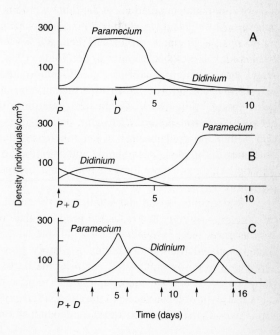

mental evidence (Curds and Bazin 1977; Bazin and Saunders 1978; Boraas 1980) indicates that predator-prey populations establish nonoscillatory steady-state conditions or cycles that become limiting, such that the magnitude of the oscillation progressively decreases. Theoretical models based on saturation kinetics and prey refuge are, for the most part, able to explain these observations (Whittaker 1975; Williams 1980).

Although predation destroys individuals, the prey population as a whole may benefit from accelerated nutrient recycling. The increase in phytoplankton growth rates due to nitrogen regeneration by predaceous zooplankton has been found to fully compensate for the mortality of the individual phytoplankton caused by the zooplankton (Sterner 1986). Moderate predatory pressure keeps prey populations from exhausting the carrying capacity of their environment and maintains the prey population in a dynamic state of growth.

It is important to recognize that the interactive nature of predator-prey relations is adaptive. There is an organizational control through negative feedback that regulates the population sizes of the predator and prey. If either the predator or the prey were completely eliminated, populations of the other would be deleteriously affected. In studies with *Tetrahymena pyriformis,* a predator protozoan, and *Klebsiella pneumoniae,* a prey bacterium, van den Ende (1973) found that a stable predator-prey relationship could be established at a level that ensured the survival of both species (Figure 3.24). In the relationship between *Tetrahymena* and *Klebsiella,* natural selection within the predator-prey system increases the efficiency of the predator in finding and engulfing its prey but also favors those individual prey that escape predation.

Under conditions of starvation, only the smallest predators survive; that is, as the prey population is consumed, there is selection for small predators. Bacterial populations under strong predatory pressure cease to produce capsules, enabling the bacterial populations to grow more rapidly and to adhere more readily to solid surfaces. In the case of *Tetrahymena* and *Klebsiella,* coexistence is dependent on the ability of the prey to find refuge. This demonstrates that coexistence of a predator and its prey can result from environmental heterogeneity, with the coexisting species being partially separated by their abilities to occupy separate niches, that is, by niche diversification. The diversity of real-world habitats may permit persistent population oscillations and coexistence.

Interactions with clay particles have been shown to provide a protection mechanism for prey bacteria from predatory populations. Crude clay provides a physical separation of predator and prey that slowed the rate of engulfment of *Escherichia coli* by *Vexillifera* (Roper and Marshall 1978) (Figure 3.25). This again shows that physical structures within natural habitats provide mechanisms for lessening the pressure of predation on prey populations, allowing coexistence.

In microbial predation, it is common to find excessive size differences between a predator and its prey. Populations of ciliate, flagellate, and amoeboid protozoa are predatory on bacterial populations. The

Figure 3.24

Predator-prey interactions between *Tetrahymena* and *Klebsiella* in sucrose-limited continuous culture showing out-of-phase cyclic oscillation of predator and prey populations. (Source: van den Ende 1973. Reprinted from *Science* by permission, copyright 1973 American Association for the Advancement of Science.)

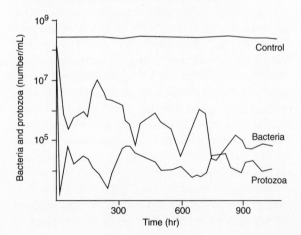

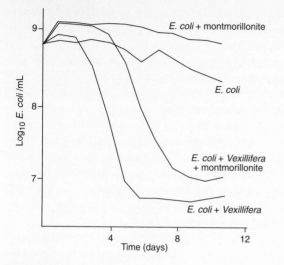

Figure 3.25

Effect of crude clay particles on the predation of *Vexillifera* on *Escherichia coli*. Montmorillonite reduces intensity of interaction. (Source: Roper and Marshall 1978. Reprinted by permission, copyright Springer-Verlag.)

Figure 3.26

Normal shape (A) of *Euplotes octocarinatus* and its response (B) to predation by *Lembladion lucens*. The change in prey morphology prevents its engulfment by the predator. (Source: Kuhlman and Heckmann 1985.)

nondiscriminatory consumption of bacterial populations by such protozoan predators is called grazing. The protozoan populations engulf the bacteria by phagocytosis, and the prey species are digested and degraded by lytic enzymes following ingestion (Hirsch 1965).

Predation by flagellate protozoa may be responsible for maintaining relatively stable populations of bacteria in aquatic systems. This grazing activity retains carbon within the food web (Fenchel and Jorgensen 1977; Fenchel 1980; Sherr et al. 1983; Wright and Coffin 1984). Ciliate protozoa, such as *Paramecium, Vorticella,* and *Stentor,* have cell mass 10^3 to 10^4 times larger than the average prey bacterium, such as *Enterobacter aerogenes*. Such relatively large predatory protozoa often employ a strategy of filter feeding so that they do not consume all their energy chasing small, low-calorie prey. Filter feeding works at an energy deficit when the prey density becomes too low. The strategy of many filter-feeding protozoa appears to be to stop filter feeding in such situations and thus conserve energy.

This seems to occur when the homogeneously suspended bacterial prey density falls between 10^5 to 10^6 per ml. Such a shutdown of filter feeding gives the prey population a good opportunity to recover.

Some microbial structures are resistant to predators. For example, soil amoebae engulf and consume vegetative cells of *Bacillus* species, but endospores of *Bacillus* are less subject to predation. The protozoan *Entodinium caudatum* is subject to predation by the larger protozoan *E. vorax*. *Entodinium caudatum* produces spineless and spined cells; *E. vorax* preferentially consumes the spineless populations, leading to the extinction of the spineless forms while the spiny populations of *E. caudatum* escape and reproduce. The ciliate protozoan *Euplotes octocarinatus* reacts to predation by *Lembladion lucens* and other ciliate protozoans that engulf their prey by changing their cell shape in a manner that interferes with engulfment (Kuhlman and Heckmann 1985) (Figure 3.26). This defense is so efficient that in axenic culture it leads to the starvation of *L. lucens*. In nature, this defense would induce the predator to engulf alternate prey species.

SUMMARY

Within a single population, cooperation is usually the predominant interaction at low population densities. Cooperation favors aggregation of individuals, resulting in spatial incongruities as evidenced by the distribution of microcolonies within natural habitats. At higher population densities, competition predominates, leading to dispersal.

Two different populations can exhibit a variety of interactions (Table 3.1). Neutralism can exist when the populations do not have the opportunity to interact. This can be accomplished by physical separation or by temporal separation of activities and is favored by low population densities and low levels of metabolic activities.

Commensal relationships, which are neutral to one population and favor another population, are frequently based on physical or chemical modifications of the habitat. Production of growth factors, production of substrates, mobilization of growth factors or substrates, and removal of inhibitory substances form the basis for many commensal relationships. Cometabolism is an important basis for commensalism. Other bases include provision of surfaces for growth and provision of transport mechanisms that enhance dispersal.

Synergism benefits both interacting populations; it allows for new or accelerated activities by microbial populations acting together, permitting microorganisms to combine their metabolic activities to perform transformations of substrates that could not be carried out by the individual populations. Synergistic relationships can result in the establishment of elemental cycling reactions and are an important basis for the development of community structure. Mutualism is an extension of synergism that allows populations to join in an obligate relationship, forming a single unit population that can occupy habitats that would be restrictive for the individual populations. Over evolutionary time it may become impossible to recognize the individual identities of populations that unite in mutualistic relationships.

Competition is a negative interaction between microbial populations. Competition is greatest between microbial populations attempting to occupy the same or overlapping niches. During growth conditions, competitive success is based on the highest intrinsic growth rate under given environmental parameters. Under nongrowth conditions, competitive success is based on tolerance and survival capabilities. Under constant environmental conditions, competition often will result in the establishment of dominant populations and the exclusion of populations of unsuccessful competitors. Exclusion may result in a spatial or temporal displacement that lessens the competitive interaction for available resources. Varying environmental parameters create conditions that allow for the coexistence of competing populations.

Microbial populations can enter into amensal relationships with other populations by modifying the habitat and by producing toxic chemicals. The basis of the amensal relationship in some cases results from the alteration of concentrations of inorganic compounds such as oxygen, ammonia, mineral acids, and hydrogen sulfide. In other cases, the amensal relationship is based on production of low–molecular-weight organic compounds such as fatty acids or alcohols. In most natural habitats, the role of antibiotics in amensal relationships is difficult to demonstrate, but this does not preclude the possibility that, under conditions that exist in some natural habitats, antibiotics may play a significant role in the establishment of amensal relationships.

Parasitism exerts a negative influence on susceptible host populations and benefits the parasite. The parasite population may be entirely dependent on the host for its nutritional requirements (an obligate parasite) or may have alternative mechanisms for meeting its nutritional needs. In an overall sense, parasitism is a mechanism for controlling population densities that provides long-range benefits to host populations and ecosystems alike.

As with parasitism, predation is a mechanism of population control that dampens population explosions and prevents exhaustion of nutritional resources of the habitat that would lead to severe population crashes. Coexistence of predator and prey species can occur with periodic population oscillations if the prey species can temporarily escape predation pressure and recover. The grazing pressure of predation places a stress on prey populations, which acts through natural selection to encourage evolution of features in prey species that provide mechanisms for escape. Such

adaptations include the abilities to reproduce rapidly, to develop resistant resting stages such as endospores, and to acquire surface structures, such as spines, that discourage predators.

REFERENCES & SUGGESTED READINGS

Ahmadjian, V. 1963. The fungi of lichens. *Scientific American* 208(2):122–132.

Ahmadjian, V. 1967. *The Lichen Symbiosis*. Blaisdell Publishing Co., Waltham, Mass.

Ahmadjian, V. 1982. The nature of lichens. *Natural History* 91:31–37.

Ahmadjian, V., J. B. Jacobs, and L. A. Russel. 1978. Scanning electron microscope study of early lichen synthesis. *Science* 200:1062–1064.

Alexander, M. 1971. *Microbial Ecology*. John Wiley and Sons, New York.

Allee, W. C., A. E. Emerson, O. Park, T. Park, and K. P. Schmidt. 1949. *Principles of Animal Ecology*. W. B. Saunders Co., Philadelphia.

Anderson, E. S. 1957. The relations of bacteriophages to bacterial ecology. In R. E. O. Williams and C. C. Spicer (eds.). *Microbial Ecology*. Cambridge University Press, Cambridge, England, pp. 189–217.

Atlas, R. M., and R. Bartha. 1973. Fatty acid inhibition of petroleum biodegradation. *Antonie van Leeuwenhoek* 39:257–271.

Ball, G. H. 1969. Organisms living on and in protozoa. In T. T. Chen (ed.). *Research in Protozoology*. Pergamon Press, Oxford, England, pp. 565–718.

Baltzis, B. C., and A. G. Fredrickson. 1984. Competition of two suspension-feeding protozoan populations for a growing bacterial population. *Microbial Ecology* 10:61–68.

Barnett, H. L. 1963. The nature of mycoparasitism by fungi. *Annual Review of Microbiology* 17:1–14.

Bates, J. L., and P. V. Liu. 1963. Complementation of lecithinase activities in closely related pseudomonads: Its taxonomic implication. *Journal of Bacteriology* 86:585–592.

Bazin, M. J., and P. T. Saunders. 1978. Determination of critical variables in a microbial predator-prey system by catastrophe theory. *Nature* 275:52–54.

Beam. H. W., and J. J. Perry. 1974. Microbial degradation of

cycloparaffinic hydrocarbons via cometabolism and commensalism. *Journal of General Microbiology* 82:163–169.

Bell, W. H., J. M. Lang, and R. Mitchell. 1974. Selective stimulation of marine bacteria by algal extracellular products. *Limnology and Oceanography* 19:833–839.

Boraas, M. E. 1980. A chemostat system for the study of rotifer-algal-nitrate interactions. In W. E. Kerfoot (ed.). A.S.L.O. Special Symposium III: *The Evolution and Ecology of Zooplankton Communities*. University Press of New England, Hanover, N.H.

Bordeleau, L. M., and R. Bartha. 1971. Ecology of a pesticide transformation: Synergism of two soil fungi. *Soil Biology and Biochemistry* 3:281–284.

Brock, T. D. 1966. *Principles of Microbial Ecology*. Prentice-Hall, Englewood Cliffs, N.J.

Bruehl, G. W., R. L. Miller, and B. Cunfer. 1969. Significance of antibiotic production by *Cephalosporium gramineum* to its saprophytic survival. *Canadian Journal of Plant Science* 49:235–246.

Bull, A. T., and J. H. Slater. 1982. Microbial interactions and community structure. In: A. T. Bull and J. H. Slater (eds.). *Microbial Interactions and Communities*. Academic Press, London, pp. 13–44.

Bungay, H. R., and M. L. Bungay. 1968. Microbial interactions in continuous culture. *Advances in Applied Microbiology* 1:269–290.

Cappenberg, T. E. 1975. A study of mixed continuous cultures of sulfate reducing and methane producing bacteria. *Microbial Ecology* 2:60–72.

Cappenberg, T. E., E. Jonejan, and J. Kaper. 1978. Anaerobic breakdown processes of organic matter in freshwater sediments. In M. W. Loutit and J. A. R. Miles (eds.). *Microbial Ecology*. Springer-Verlag, Berlin, pp. 91–99.

Chang, K. P., and W. Trager. 1974. Nutritional significance of symbiotic bacteria in two species of hemoflagellates. *Science* 183:351–352.

Chet, I., and R. Mitchell. 1976. Ecological aspects of microbial chemotactic behavior. *Annual Reviews of Microbiology* 30:221–239.

Clark, R. J., and T. L. Steck. 1979. Morphogenesis in *Dictyostelium*: An orbital hypothesis. *Science* 204:1163–1168.

Curds, C. R., and M. J. Bazin. 1977. Protozoan predation in batch and continuous culture. *Advances in Aquatic Microbiology* 1:115–176.

DeFreitas, M. J., and G. Frederickson. 1978. Inhibition as a factor in the maintenance of the diversity of microbial ecosystems. *Journal of General Microbiology* 106:307–320.

Denison, W. D. 1973. Life in tall trees. *Scientific American* 228(6):75–80.

Devay, J. E. 1956. Mutual relationships in fungi. *Annual Reviews of Microbiololgy* 10:115–140.

Dilts, J. A., and R. L. Quackenbush. 1986. A mutation in the R body sequence destroys expression of the killer trait in *P. tetraaurelia*. *Science* 232:641–643.

Fenchel, T. 1980. Suspension-feeding in ciliated protozoa: Functional response and particle eye selection. *Microbial Ecology* 6:1–11.

Fenchel, T. 1982. Ecology of heterotrophic microflagellates: Adaptations to heterogenous environments. *Marine Ecology Progress Series* 9:25–33.

Fenchel, T. M., and B. B. Jorgensen. 1977. Detritus food chains of aquatic ecosystems: The role of bacteria. *Advances in Microbial Ecology* 1:1–58.

Ferry, B. W. 1982. Lichens. In R. G. Burns and J. H. Slater (eds.). *Experimental Microbial Ecology*. Blackwell Scientific Publications, London, pp. 291–319.

Fields, B. S., E. B. Shotts, Jr., J. C. Feeley, G. W. Gorman, and W. T. Martin. 1984. Proliferation of *Legionella pneumophila* as an intracellular parasite of the ciliated protozoan *Tetrahymena pyriformis*. *Applied and Environmental Microbiology* 47:467–471.

Fredrickson, A. G. 1977. Behavior of mixed cultures of microorganisms. *Annual Review of Microbiology* 31:63–89.

Fredrickson, A. G., and G. Stephanopoulos. 1981. Microbial competition. *Science* 213:972–979.

Gale, E. F. 1940. The production of amines by bacteria. III. The production of putrescine from arginine by *Bacterium coli* in symbiosis with *Streptococcus faecalis*. *Journal of Biochemistry* 34:853–857.

Gause, G. F. 1934. *The Struggle for Existence*. Williams and Wilkins, Baltimore.

Gerdes, K., F. W. Bech, S. T. Jorgensen, A. Lobner-Olesen, P. B. Rasmussen, T. Atlung, L. Boe, O. Karlstrom, S. Molin, and K. von Meyenburg. 1986a. Mechanism of postsegregational killing by the *hok* gene product of the *parB* system of plasmid R1 and its homology with the *relF* gene product of the *E. coli relB* operon. EMBO Journal 5:2023–2029.

Gerdes, K., P. B. Rasmussen, and S. Molin. 1986b. Unique type of plasmid maintenance function: Postsegregational killing of plasmid-free cells. *Proceedings of the National Academy of Sciences USA* 83:3116–3120.

Gilbert, O. L. 1969. The effect of SO_2 on lichens and bryophytes around Newcastle upon Tyne. In *European Symposium on the Influences of Air Pollution on Plants and Animals*. Centre for Agricultural Publishing and Documentation, Wageningen, The Netherlands, pp. 223–235.

Gooday, G. W., and S. A. Doonan. 1980. The ecology of algal–inverterbrate symbioses. In D. C. Ellwood, J. N. Hedger, M. J. Latham, J. M. Lynch, and J. H. Slater (eds.). *Contemporary Microbial Ecology*. Academic Press, London, pp. 377–390.

Gunner, H. B., and B. M. Zuckerman. 1968. Degradation of diazinon by synergistic microbial action. *Nature* (London) 217:1183–1184.

Hale, M. E. 1974. *The Biology of Lichens*. Edward Arnold, London.

Harder, W., and H. Veldkamp. 1971. Competition of marine psychrophilic bacteria at low temperatures. *Antonie van Leeuwenhoek* 37:51–63.

Hardy, K. 1981. *Bacterial Plasmids*. Aspects of Microbiology, Series No. 4. American Society for Microbiology, Washington, D.C.

Higgins, I. J., and R. G. Burns. 1975. *The Chemistry and Microbiology of Pollution*. Academic Press, London, pp. 218–223.

Hirsch, J. G. 1965. Phagocytosis. *Annual Reviews of Microbiology* 19:339–350.

Jannasch, H. W. 1968. Competitive elimination of Enterobacteriaceae from seawater. *Applied Microbiology* 16:1616–1618.

Jannasch, H. W., and R. I. Mateles. 1974. Experimental bacterial ecology studied in continuous culture. *Advances in Microbial Physiology* 11:165–212.

Jeffries, T. W. 1982. The microbiology of mercury. *Progress in Industrial Microbiology* 16:23–75.

Jones, A. K. 1982. The interactions of algae and bacteria. In A. T. Bull and J. H. Slater (eds.). *Microbial Interactions and Communities*. Academic Press, London, pp. 189-248.

Kelly, D. P. 1978. Microbial ecology. In K. W. A. Chater and H. J. Somerville (eds.). *The Oil Industry and Microbial Ecosystems*. Heyden and Son, London, pp. 12–27.

Krebs, C. J. 1972. *Ecology: The Experimental Analysis of Distribution and Abundance*. Harper and Row, New York.

Kuenen, J. G., and W. Harder. 1982. Microbial competition in continuous culture. In R. G. Burns and J. H. Slater (eds.). *Experimental Microbial Ecology*. Blackwell Scientific Publications, London, pp. 342–367.

Kuenen, J. G., L. A. Robertson, and H. van Germerden. 1985. Microbial interactions among aerobic and anaerobic sulfur-oxidizing bacteria. *Advances in Microbial Ecology* 8:1–59.

Kuhlman, H. W., and K. Heckmann. 1985. Interspecific morphogens regulating prey-predator relationship in protozoa. *Science* 227:1347–1349.

Lamb, I. M. 1959. Lichens. *Scientific American* 201(4):144–156.

Lemke, P. A., and C. H. Nash. 1974. Fungal virus. *Bacteriological Reviews* 38:29–56.

Lockhead, A. G. 1958. Soil bacteria and growth-promoting substances. *Bacteriological Review* 22:145–153.

Lockwood, J. C. 1964. Soil fungistasis. *Annual Reviews of Phytopathology* 2:341–362.

Lotka, A. J. 1925. *Elements of Physical Biology*. Williams and Wilkins, Baltimore.

Madelin, M. F. 1968. Fungi parasitic on other fungi and lichens. In G. C. Ainsworth and A. S. Sussman (eds.). *The Fungi*. Vol. 3. Academic Press, New York, pp. 253–269.

Marbach, A., M. Varon, and M. Shilo. 1976. Properties of marine Bdellovibrios. *Microbial Ecology* 2:284–295.

Margulis, L. 1971. The origin of plant and animal cells. *American Scientist* 59:230–235.

Mitchell, K., and M. Alexander. 1962. Microbiological changes in flooded soil. *Soil Science* 93:413–419.

Mitchell, R. 1968. Factors affecting the decline of nonmarine microorganisms in seawater. *Water Research* 2:535–543.

Mitchell, R. 1971. Role of predators in the reversal of imbalances in microbial ecosystems. *Nature* 230:257–258.

Munnecke, D. M., and D. P. H. Hsieh. 1976. Pathway of microbial metabolism of parathion. *Applied and Environmental Microbiology* 31:63–69.

Nurmikko, V. 1956. Biochemical factors affecting symbiosis among bacteria. *Experientia* 12:245–249.

Odum, E. P. 1971. *Fundamentals of Ecology*. W. B. Saunders, Philadelphia.

Orenski, S. W. 1966. Intermicrobial symbiosis. In S. M. Henry (ed.). *Symbiosis*. Academic Press, New York, pp. 1–33.

Park, D. 1967. The importance of antibiotics and inhibiting substances. In A. Burges and F. Row (eds.). *Soil Biology*. Academic Press, New York, pp. 435–447.

Pearl, H. W. 1987. Role of heterotrophic bacteria in promoting N_2-fixation by *Anabaena* in aquatic habitats. *Microbial Ecology*. 4:215–231.

Pearl, H. W., and K. K. Gallucci. 1985. Role of chemotaxis in establishing a specific nitrogen-fixing cyanobacterial association. *Science* 227:647–649

Pohunek, M. 1961. Streptococci antagonizing the vaginal *Lactobacillus*. *Journal of Hygiene, Epidemiology, Microbiology, and Immunology* (Prague) 5:267–270.

Preer, L. B. 1981. Prokaryotic symbionts of *Paramecium*. In M. P. Starr, H. Stolp, H. G. Truper, A. Ballows, and H. G. Schlegel (eds.). *The Prokaryotes*. Springer-Verlag, Berlin, pp. 2127–2136.

Rittenberg, S. C. 1983. *Bdellovibrio*: Attack, penetration and growth on its prey. *ASM News* 49:435–439.

Roper, M. M., and K. C. Marshall. 1974. Modification of interaction between *Escherichia coli* and bacteriophage in saline sediment. *Microbial Ecology* 1:1–14.

Roper, M. M., and K. C. Marshall. 1978. Effects of a clay mineral on microbial predation and parasitism on *Escherichia*. *Microbial Ecology* 4:279–290.

Rosenberg, E., K. H. Keller, and M. Dworkin. 1977. Cell-density dependent growth of *Myxococcus xanthus* on casein. *Journal of Bacteriology* 129:770–777.

Seaward, M. R. D. 1977. *Lichen Ecology*. Academic Press, London.

Shapiro, J. A. 1991. Multicellular behavior of bacteria. *ASM News* 57:247–253.

Sherr, B. F., E. B. Sherr, and T. Berman. 1983. Grazing, growth, and ammonium excretion rates of a heterotrophic microflagellate fed with four species of bacteria. *Applied and Environmental Microbiology* 45:1196–1201.

Shimkets, L. J. 1990. Social and developmental biology of the myxobacteria. *Microbiological Reviews* 54: 473-501.

Sieburth, J. M. 1975. *Microbial Seascapes*. University Park Press, Baltimore.

Slater, J. H. 1978. Microbial communities in the natural environment. In K. W. A. Chater and H. S. Somerville (eds.). *The Oil Industry and Microbial Ecosystems*. Heyden and Sons, London, pp. 137–154.

Slater, J. H., and A. T. Bull. 1978. Interactions between microbial populations. In A. T. Bull and P. M. Meadow (eds.). *Companion to Microbiology*. Longman, London, pp. 181–206.

Slobodkin, L. B. 1968. How to be a predator. *American Zoologist* 8:43–51.

Smith, J. M. B., and M. J. Marples. 1964. A natural reservoir of penicillin resistant strains of *Staphylococcus aureus*. *Nature* 201:844.

Society for Experimental Biology. 1975. *Symbiosis*. Cambridge University Press, Cambridge, England.

Sonneborn, T. M. 1959. Kappa and related particles in *Paramecium*. *Advances in Virus Research* 6:229–356.

Starr, M. P., and R. J. Seidler. 1971. The Bdellovibrios. *Annual Reviews of Microbiology* 25:649–675.

Sterner, R. W. 1986. Herbivores' direct and indirect effect on algal populations. *Science* 231:605–607.

Stewart, J. R., and R. M. Brown, Jr. 1969. Cytophaga that kills or lyses algae. *Science* 164:1523–1524.

Stojanovic, B. J., and M. Alexander. 1958. Effect of inorganic nitrogen on nitrification. *Soil Science* 86:208–215.

Stolp, H., and M. P. Starr. 1963. *Bdellovibrio bacteriovorous* gen. et sp. n., a predatory ectoparasitic and bacteriolytic microorganism. *Antonie van Leeuwenhoek* 29:217–248.

Stolp, H., and M. P. Starr. 1965. Bacteriolysis. *Annual Reviews of Microbiology* 19:79–104.

Strelkoff, A., and G. Poliansky. 1937. On natural selection in some infusoria entodiniomorpha. *Zoological Journal* (Moscow) 16:77–84.

Taylor, F. J. R. 1982. Symbioses in marine microplankton. *Annals of the Institute of Oceanography* (Paris) 58:61–90.

Thiele, J. H., and G. Zeikus. 1988. Control of interspecies electron flow during anaerobic digestion: Significance of formate transfer versus hydrogen transfer during syntrophic methanogenesis in flocs. *Applied and Environmental Microbiology* 54:20–29.

Tilman, D. 1976. Ecological competition between algae: Experimental confirmation of resource-based competition theory. *Science* 192:463–465.

van den Ende, P. 1973. Predator-prey interactions in continuous culture. *Science* 181:562–564.

van Etten, J. L., D. E. Burbank, D. Kutzmarski, and R. H. Meints. 1983. Virus infection of culturable *Chlorella*-like algae and development of a plaque assay. *Science* 219:994–996.

van Gemerden, H. 1974. Coexistence of organisms competing for the same substrate: An example among the purple sulfur bacteria. *Microbial Ecology* 1:104–119.

Veldkamp, H., H. van Gemerden, W. Harder, and H. J. Laanbroek. 1984. Competition among bacteria: An overview. In A. J. Klug and C. A. Reddy (eds.). *Current Perspectives in Microbial Ecology*. American Society for Microbiology, Washington, D.C., pp. 279–280.

Volterra, V. 1926. Variazioni e fluttuazioni del numero d'individui in specie animali conviventi. *Memorie/Academia Nazionale dei Rome Italy Lincei* (Rome Italy) 2:31–113.

Waksman, S. A. 1961. The role of antibiotics in nature. *Perspectives of Biological Medicine* 4:271–287.

Weis, D. S. 1982. Protozoal symbionts. In R. G. Burns and J. H. Slater (eds.). *Experimental Microbial Ecology*. Blackwell Scientific Publications, London, pp. 320–341.

Wessenberg, H., and G. Antipa. 1970. Capture and ingestion of *Paramecium* by *Didinium nasutum*. *Journal of Protozoology* 17:250–270.

Whittaker, R. H. 1975. *Communities and Ecosystems*. Macmillan, New York.

Williams, F. M. 1980. On understanding predator–prey interactions. In D. C. Ellwood, J. N. Hedger, M. J. Latham, J. M. Lynch, and J. H. Slater (eds.). *Contemporary Microbial Ecology*. Academic Press, London, pp. 349–375.

Wireman, J. W., and M. Dworkin. 1975. Morphogenesis and developmental interactions in myxobacteria. *Science* 189:516–522.

Wolfe, R. S., and N. Pfennig. 1977. Reduction of sulfur by *Spirillum* 5175 and syntrophism with *Chlorobium*. *Applied and Environmental Microbiology* 33:427–433.

Wolin, M. J. 1969. Volatile fatty acids and the inhibition of *Escherichia coli* growth by rumen fluid. *Applied Microbiology* 17:83–87.

Wolin, M. J., and T. L. Miller. 1985. Interspecies hydrogen transfer: 15 years later. *ASM News* 48:561-565.

Wright, R. T., and R. B. Coffin. 1984. Measuring microzooplankton grazing on planktonic marine bacteria by its impact on bacterial production. *Microbial Ecology* 10:137–149.

Zeikus, J. G. 1977. The biology of methanogenic bacteria. *Bacteriological Reviews* 41:514–541.

4

Interactions between Microorganisms and Plants

Not only do microorganisms exhibit relationships of neutralism, commensalism, synergism, mutualism, amensalism, competition, and parasitism among microbial populations, they also exhibit these types of interactions with plants (Wheeler 1975; Agrios 1978; Dickinson and Lucas 1982; Campbell 1985; Fitter 1985). Some of these interactions are beneficial to both the plant and the microbial populations, whereas others are detrimental to the plant or the microbial populations. Plant surfaces provide important habitats for microorganisms, with some microorganisms growing only on plant surfaces, such as roots.

INTERACTIONS WITH PLANT ROOTS

The Rhizosphere

Plant roots provide suitable habitats for the growth of microorganisms, and particularly high numbers of many different microbial populations are found on and surrounding plant roots. Interactions between soil microorganisms and plant roots satisfy important nutritional requirements for both the plant and the associated microorganisms (Brown 1974, 1975; Bowen and Rovira 1976; Lynch 1976, 1982a, 1982b; Balandreau and Knowles 1978; Dommergues and Krupa 1978; Newman 1978; Harley and Russell 1979; Bowen 1980). This is apparent by the large numbers of microorganisms found in the rhizoplane, which is defined as the actual surface of the plant roots (Campbell and Rovira 1973; Rovira and Campbell 1974; Bowen and Rovira 1976) (Figure 4.1).

The density of microorganisms is also high in the rhizosphere soil, defined as the thin layer of soil adhering to a root system after shaking has removed the loose soil (Figure 4.2). The size of the rhizosphere depends on the particular plant root structure, but generally the contact area with soil is very large. Per total plant biomass, the fibrous root structure of grassy plants provides a larger surface area than root systems characterized by a taproot. For example, the root system of a single wheat plant can be more than 200

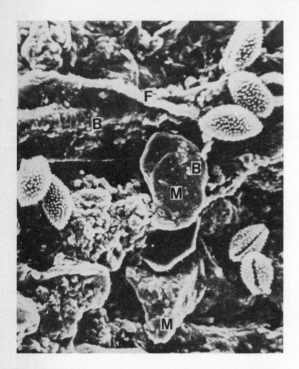

Figure 4.1

Scanning electron micrograph of rye grass root showing complexity of root surface and the rhizosphere. B = bacteria; F = fungal mycelia; M = mineral grains. Oval structures with granular surface appear to be fungal spores. (Source: Campbell and Rovira 1973. Reprinted by permission of Pergamon Press.)

Figure 4.2

Photograph of plant root system showing root hairs and soil particles attached to the rhizoplane.

meters in length. Assuming an average root diameter of 0.1 mm, we may calculate a root surface area in excess of 6 square meters. Only 4% to 10% of the actual rhizoplane, however, is in direct physical contact with microorganisms; most microorganisms associated with roots occur in the surrounding rhizosphere (Bowen 1980).

An interesting and as yet insufficiently explored modification of the rhizosphere is the rhizosheath, characterized as a relatively thick soil cylinder that adheres to the plant roots (Figure 4.3). The formation of rhizosheaths is typical of some desert grasses but also occurs in some grass species that grow under less extreme conditions (Wullstein et al. 1979; Duell and

Peacock 1985). The sand grains in the rhizosheath are cemented together by an extracellular mucigel, apparently excreted by the root cells. The rhizosheath appears to be an adaptation for moisture conservation, but it undoubtedly also provides an environment for extensive root-microbe interactions. Increased nitrogen-fixation activity has been measured in rhizosheath soil.

Plant Root Effects on Microbial Population The structure of the plant root system contributes to the establishment of the rhizosphere microbial population (Nye and Tinker 1977; Russell 1977; Bowen 1980; Lynch 1982a). The interactions of plant roots and rhizosphere microorganisms are based largely on interactive modification of the soil environment by processes such as water uptake by the plant system, release of organic chemicals to the soil by the plant roots, microbial production of plant growth factors, and microbially mediated availability of mineral nutrients. Within the rhizosphere, plant roots have a

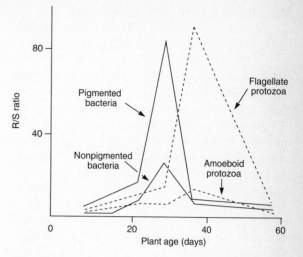

Figure 4.4

R/S ratios indicating rise and fall of bacterial and protozoan populations within the rhizosphere during development of *Sinapsis alba*. (Source: Campbell, 1977; based on Darbyshire and Greaves, 1967. Reprinted by permission of National Research Council of Canada.)

Figure 4.3

Rhizosheath on cereal rye (*Secale cereale*). The diameter of the soil cylinders is about 8 mm. Rhizosheaths represent an adaptation for moisture conservation and also provide an extended environment for plant-microbe interactions. (Source: R. W. Duell, Rutgers University, New Brunswick, New Jersey.)

direct influence on the composition and density of the soil microbial community, known as the rhizosphere effect. The rhizosphere effect can be seen by looking at the ratio of the number of microorganisms in the rhizosphere soil to the number of corresponding microorganisms in soil remote from roots (the R/S ratio). Generally R/S ratios range from 5 to 20, but it is common to find an R/S ratio of 100, that is, microbial populations 100 times higher in the rhizosphere than in the surrounding root-free soil (Gray and Parkinson 1968; Woldendorp 1978). The actual extent of the rhizosphere effect is dependent on the particular plant and its physiological maturity (Darbyshire and Greaves 1967) (Figure 4.4).

There is a higher proportion of Gram-negative, rod-shaped bacteria and a lower proportion of Gram-positive rods, cocci, and pleomorphic forms in the rhizosphere than in root-free soil (Rovira and Campbell 1974; Woldendorp 1978; Campbell 1985). There is also a relatively higher proportion of motile, rapidly growing bacteria, such as *Pseudomonas,* in the rhizosphere than elsewhere in soil. In many cases this increase represents a direct influence of plant root exudates on soil microorganisms, which favor microorganisms with high intrinsic growth rates. Organic materials released from roots include amino acids, keto acids, vitamins, sugars, tannins, alkaloids, phosphatides, and other unidentified substances (Rovira 1969). Roots surrounded by microorganisms excrete many times as much organic material as sterile roots. Although a few of these materials inhibit microorganisms, most stimulate microbial growth. The influence of materials released by the plant into the soil is evidenced by the observation that bacterial populations within the rhizosphere have markedly different nutritional properties than bacteria in root-free soil.

Many rhizosphere bacteria require amino acids for maximal growth, and it is likely that root exudates supply these acids.

Microorganisms in the rhizosphere may undergo successional changes as the plant grows from seed germination to maturity. During plant development, a distinct rhizosphere succession results in rapidly growing, growth-factor–requiring, opportunistic microbial populations. These successional changes correspond to changes in the materials released by the plant roots to the rhizosphere during plant maturation. Initially, carbohydrate exudates and mucilaginous materials support the growth of large populations of microorganisms within the grooves of the epidermal plant cells, on the root surface, and within mucilaginous layers surrounding the roots. As the plant matures, autolysis of some of the root material takes place as part of normal root development and simple sugars and amino acids are released into the soil, stimulating the growth of *Pseudomonas* and other bacteria with high intrinsic growth rates. R/S ratios usually decline when plants stop growing and become senescent.

The microaerophilic *Azospirillum* (formerly *Spirillum lipoferum*) and the aerobic *Azotobacter paspali* are nitrogen–fixing soil bacteria regularly associated with the rhizosphere of certain tropical grasses of the *Digitaria, Panicum,* and *Paspalum* genera. These bacteria use root exudates as the energy source to support significant nitrogen fixation. In field trials, this rhizospheric nitrogen fixation replaced up to 40 kg of nitrogen per ha per year (Smith et al. 1976). *Azospirillum* was found also in the rhizosphere of some temperate zone grasses and corn (*Zea mays*) (Lamm and Neyra 1981), but the rates of nitrogen fixation in these cases appear to have little practical significance. On the other hand, rhizospheric nitrogen fixation helps in meeting the nitrogen demand of rice crops (Swaminathan 1982). A side effect of biochemical nitrogen fixation is the evolution of hydrogen. Only some *Rhizobium* and *Bradyrhizobium* strains have the hydrogenase enzyme to utilize this side product; others wastefully evolve hydrogen gas. Nodulated root systems evolving hydrogen were found to be colonized by hydrogen-oxidizing *Acinetobacter* strains that utilized this resource not commonly available in aerobic environments (Wong et al. 1986).

Zostera marina (eel grass) and *Thalassia testudinum* (turtle grass) cover shallow coastal seafloor areas of temperate and tropical regions, respectively. The rhizosphere of these plants in the anaerobic marine sediments is the site of high nitrogen-fixation activity by *Desulfovibrio, Clostridium,* and other anaerobes (Patriquin and Knowles 1972; Capone and Taylor 1980; Smith 1980). Because of the projection of short-term measurements, the reported rates are controversial, but they approach or exceed those reported for *Rhizobium*-legume associations (100–500 kg of N per ha per year). This constitutes a significant input of combined nitrogen for coastal waters. Similarly high nitrogen-fixation rates were reported for the rhizosphere of the dominant semisubmerged salt-marsh grass *Spartina alterniflora* (Hanson 1977; Patriquin and McClung 1978), but little is known about the microorganisms involved. There is little doubt, however, that the reported rhizospheric nitrogen-fixation activities constitute significant combined nitrogen inputs for the coastal marine environment.

Effects of Rhizosphere Microbial Populations on Plants Just as plant roots have a direct effect on the surrounding microbial populations, microorganisms in the rhizosphere have a marked influence on the growth of plants. In the absence of appropriate microbial populations in the rhizosphere, plant growth may be impaired (Lynch 1976; Dommergues and Krupa 1978; Campbell 1985). Microbial populations in the rhizosphere may benefit the plant in a variety of ways, including increased recycling and solubilization of mineral nutrients; synthesis of vitamins, amino acids, auxins, and gibberellins, which stimulate plant growth; and antagonism with potential plant pathogens through competition and development of amensal relationships based on production of antibiotics.

Plants that grow in flooded sediments and soils have evolved adaptations for conducting oxygen from the shoot to the roots (Raskin and Kende 1985), but in such anaerobic environments the roots also have to cope with the toxic hydrogen sulfide generated by sulfate reduction (Drew and Lynch 1980). Rice, and perhaps other partially submerged plants, are protected against the toxic effect of hydrogen sulfide by a mutualistic association with *Beggiatoa* (Joshi and Hollis 1977). This microaerophilic, catalase-negative, sulfide-oxidizing, filamentous bacterium benefits from the oxygen and catalase enzyme provided by the rice roots, and in turn, *Beggiatoa* aids the rice plant by oxidizing the toxic hydrogen sulfide to harmless sulfur or sulfate, thus protecting the cytochrome system of the rice roots.

Organisms in the rhizosphere produce organic compounds that affect the proliferation of the plant root system (Lynch 1976). Microorganisms synthesize auxins and gibberellin-like compounds, and these compounds increase the rate of seed germination and the development of root hairs that aid plant growth (Brown 1974). *Arthrobacter, Pseudomonas,* and *Agrobacterium* populations found in the rhizosphere have been reported to be capable of producing organic chemicals that stimulate growth of plants. The rhizosphere of wheat seedlings, for example, contains a significant proportion of bacteria that produce indoleacetic acid (IAA), a plant growth hormone that can increase the growth of plant roots. In older wheat plants, a lower proportion of microorganisms in the rhizosphere is capable of producing IAA. This may be in response to a decline in root exudate production, but it also beneficially corresponds to a decreased need for the growth hormone by the plant.

Allelopathic (antagonistic) substances released by microorganisms in the rhizosphere may allow plants to enter amensal relationships with other plants. Such allelopathic substances surrounding some plants can prevent invasion of that habitat by other plants, and this may represent a synergistic relationship between a plant and its rhizosphere microbial community. Bacterial populations in the rhizosphere of young wheat plants have been shown to inhibit the growth of pea and lettuce plants. As a wheat plant matures, the proportion of these bacteria decreases and they are replaced by a higher proportion of microorganisms capable of producing growth-promoting substances similar to gibberellic acid.

Microorganisms in the rhizosphere influence the availability of mineral nutrients to the plants,

sometimes using limiting concentrations of inorganic nutrients before they can reach plant roots, and in other cases increasing the availability of inorganic nutrients to the plant (Barber 1978). Rhizosphere microorganisms increase the availability of phosphate through solubilization of materials that would otherwise be unavailable to plants. Plants have been shown to exhibit higher rates of phosphate uptake when associated with rhizosphere microorganisms than in sterile soils (Campbell 1985). The principal mechanism of increasing phosphate availability is the microbial production of acids that dissolve apatite, releasing soluble forms of phosphorus. Iron and manganese may be more available to plants because of rhizosphere microorganisms that produce organic chelating agents, thus increasing the solubility of iron and manganese compounds. It has also been shown that microorganisms on roots significantly increase the uptake rates of calcium by the roots. This increase may be due to high concentrations of carbon dioxide in the rhizosphere produced by microorganisms, which increase the solubility and thus the availability of calcium. Translocation of various radiolabelled organic compounds and heavy metals along mycelial filaments has also been demonstrated (Grossbard 1971; Campbell 1985).

Although increased uptake of minerals due to rhizosphere microorganisms is beneficial, the abundant microbial populations in the rhizosphere can sometimes create a deficiency of required minerals for the plants (Agrios 1978). For example, bacterial immobilization of zinc and oxidation of manganese cause the plant diseases "little leaf" of fruit trees and "grey speck" of oats, respectively. Microorganisms in the rhizosphere may immobilize limiting nitrogen, making it unavailable for the plant (Campbell 1985). The immobilization of nitrogen by microorganisms in the rhizosphere accounts for an appreciable loss of added nitrogen fertilizer intended for plant use. Part of the nitrogen is immobilized in the form of microbial protein, but some may also be lost to the atmosphere by denitrification.

Although diverse and complex, the majority of interactions in the rhizosphere are mutually beneficial to both plants and microorganisms and are synergistic in character. A further exploration and optimization of these interactions may lead to significant improvements in crop production.

Mycorrhizae

Some fungi enter into a mutualistic relationship with plant roots called mycorrhizae (literally, fungus root) in which the fungi actually become integrated into the physical structure of the roots (Hartley 1965; Cooke 1977; Dommergues and Krupa 1978; Powell 1982; Campbell 1985). The fungus derives nutritional benefits from the plant roots, contributes to plant nutrition, and does not cause plant disease. Mycorrhizal associations differ from other rhizosphere associations between plants and microorganisms by the greater specificity and organization of the plant-fungus relationship. The mycorrhizal association involves the integration of plant roots and fungal mycelia, forming integrated morphological units. The widespread existence of mycorrhizal associations between fungi and plant roots attests to the importance of this interaction.

Mycorrhizal associations exist for prolonged periods with the maintenance of a healthy physiological interaction between the plant and the fungus. The mycorrhizal associations of fungi and plant roots represent a diverse relationship—both in terms of structure and physiological function—that leads to a nutrient exchange favorable to both partners. Enhanced uptake of water and mineral nutrients, particularly phosphorus and nitrogen, has been noted in many mycorrhizal associations; plants with mycorrhizal fungi are therefore able to occupy habitats they otherwise could not (Smith and Daft 1978).

There are two basic types of mycorrhizal associations: ectomycorrhizae (Marks and Kozlowski 1973; Marx and Krupa 1978) and endomycorrhizae (Sanders et al. 1975; Hayman 1978). In ectomycorrhizae, the fungus (an ascomycete or basidiomycete) forms an external pseudoparenchymatous sheath more than 40 μm thick and constituting up to 40% of the dry weight of the combined root-fungus structure (Hartley 1965) (Figure 4.5). The fungal hyphae penetrate the intercellular spaces of the epidermis and of

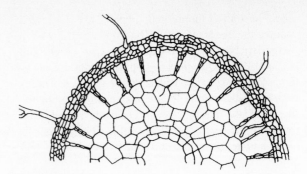

Figure 4.5

Cross section of ectotrophic mycorrhizal rootlet showing fungal sheath and intercellular penetration. (Source: Hartley 1965. Reprinted by permission, copyright University of California Press, Berkeley, California.)

the cortical region of the root but do not invade the living cells. The morphology of the root is altered, forming a shorter, dichotomously branching cluster with a reduced meristematic region.

In contrast to the predominantly exogenous ectomycorrhizae, endomycorrhizae invade the living cells of the root, which become filled with mycelial clusters (Hartley 1965) (Figure 4.6). In a widespread form of endomycorrhizae, the microscopic appearance

Figure 4.6

Cross-section of endotrophic mycorrhizal rootlet showing intracellular penetration of hyphae. (Source: Hartley 1965. Reprinted by permission; copyright University of California Press, Berkeley, California.)

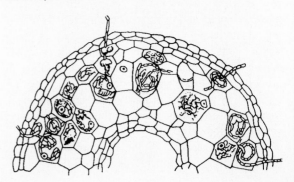

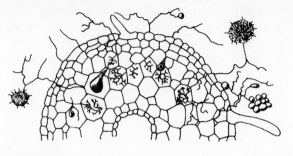

Figure 4.7

Cross-section of rootlet with vesicular-arbuscular mycorrhiza showing penetrating hyphae and "treelike" and "vesicle-like" hyphal structures. (Source: Hartley 1965. Reprinted by permission, copyright University of California Press, Berkeley, California.)

of intracellular hyphal clusters causes these to be called vesicular-arbuscular (VA) mycorrhizae (Hartley 1965) (Figure 4.7). In some cases, endo- and ectomycorrhizae may be combined, and are referred to as an ectendomycorrhizae.

Ectomycorrhizae Ectomycorrhizae are common in gymnosperms and angiosperms, including most oak, beech, birch, and coniferous trees (Marks and Kozlowski 1973). Most trees in temperate forest regions have ectomycorrhizal associations. Many fungi can enter into ectomycorrhizal associations, including ascomycetes, such as truffles, and basidiomycetes, such as *Boletus* and *Amanita*. Ectomycorrhizal fungi generally have optimal growth temperatures between 15°C and 30°C and are acidophilic with optimal growth at pH 4.0 to 6.0 or even as low as pH 3.0 (Hartley 1965). Most ectomycorrhizal fungi grow well on simple carbohydrates such as disaccharides and sugar alcohols. They generally utilize complex organic sources of nitrogen, amino acids, and ammonium salts; many require vitamins such as thiamine and biotin and are able to produce a variety of metabolites that they release to the plant, including auxins, gibberellins, cytokinins, vitamins, antibiotics, and fatty acids (Frankenberger and Poth 1987). Some ectomycorrhizal fungi are capable of producing enzymes such as cellulase, but such activity is normally suppressed within the host plant and, therefore, the fungi do not digest the plant roots.

The plant probably derives several benefits from its association with ectomycorrhizal fungi, including longevity of feeder roots; increased rates of nutrient absorption from soil; selective absorption of certain ions from soils; resistance to plant pathogens; increased tolerance to toxins; and increased tolerance ranges to environmental parameters, such as temperature, drought, and pH (Marks and Kozlowski 1973; Harley and Smith, 1983). The ectomycorrhizal fungi receive photosynthesis products from the host plant and thus escape intense competition for organic substrates with other soil microorganisms. The nutritional benefit to the fungus is demonstrated by the fact that many mycorrhizal fungi fail to form fruiting bodies outside of a mycorrhizal association with a plant, even though vegetative saprophytic growth is usually possible under these circumstances.

An ectomycorrhizal infection has a morphogenetic effect that leads to characteristic dichotomous branching and prolonged growth and survival of plant rootlets, probably due to production of growth hormones by the ectomycorrhizal fungi (Frankenberger and Poth 1987). Formation of root hairs is suppressed, and fungal hyphae overtake their function, thus greatly increasing the radius of nutrient availability for the plant. Ectomycorrhizal roots take up ions, such as phosphate and potassium, in excess of the rates displayed by uninfected roots. The mechanisms of uptake are dependent on fungal metabolic activity. There may be a primary accumulation of phosphate within the fungal sheath followed by transfer to the plant root. Nitrogen-containing compounds and calcium have also been found to be absorbed into the fungal mycelial sheath, followed by transfer to the plant root. Interdependence of the plant root and the ectomycorrhizal fungus is thus based in large part on their ability to supply each other with major and minor nutrients.

Plants with ectomycorrhizae also are able to resist pathogens that otherwise would attack the plant roots. The sheaths produced by ectomycorrhizal fungi present an effective physical barrier to penetration by plant root pathogens, and many basidiomycetes that are ectomycorrhizal have been shown to produce anti-biotics. Plants with ectomycorrhizal fungi survive in soils inoculated with pathogens that enter through plant roots; for example, inoculation of nursery soils with fungi that can enter into ectomycorrhizal associations has produced marked decrease in mortality of host trees such as Douglas firs (Neal and Bollen 1964). Ectomycorrhizal roots also produce a variety of volatile organic acids that have fungistatic effects. The increased production of such compounds by host cells infected with ectomycorrhizal fungi maintains a balance with the mutualistic fungus and deters infection by pathogenic fungi. Most plants appear to respond to mycorrhizal infection by producing inhibitors that also contribute to the resistance of the ecto-mycorrhizal roots to pathogenic infection.

Endomycorrhizae Endomycorrhizal associations, which are not of the vesicular-arbuscular (VA) type, occur in a few orders of plants, such as the Ericales, which include heath, arbutus, azalea, rhododendron, and American laurel (Sanders et al. 1975). The endomycorrhizae of the plant genera belonging to the Ericales are characterized by nonpathogenic penetration of the root cortex by septate fungal hyphae that often form intracellular coils. Although the fungi do not fix atmospheric nitrogen, the endomycorrhizal association may increase plant access to combined nitrogen in soil as demonstrated by better nitrogen nutrition in mycorrhizal as compared to nonmycorrhizal plants. There is greater phosphatase activity in mycorrhizal roots than in nonmycorrhizal roots, and the mycorrhizal fungi can transfer phosphate from external sources to the host plant. The association of endomycorrhizae in Ericales appears to improve the growth of the host plant in nutrient-deficient soils, and the widespread occurrence of endomycorrhizal infections in Ericales indicates that these plant root tissues provide a good ecological niche for these fungi.

Virtually all orchid roots are internally infected and attacked by fungal hyphae, which pass through surface cells into cortical cells to form mycorrhizae. The fungi form coils within the cells of the outer cortex, and later the hyphae lose their integrity and much of their contents passes into the host cell. Orchids are obligately mycorrhizal under natural conditions, often forming associations with the fungi *Armillaria mellea* and *Rhizoctonia solani*. These endomycorrhizal associations enhance the ability of orchid seeds to germinate, but the fungi can also be parasitic to the host plant. This fact and, conversely, the digestion of some fungal mycelium by the plant give this association the character of a precariously balanced mutual parasitism.

The VA type of endomycorrhizal association, which frequently goes unnoticed because it does not have a macroscopic effect on root morphology, occurs in more plant species than all other types of endo- and ectomycorrhizae combined (Mosse 1973; Sanders et al. 1975; Bowen 1984). Vesicular-arbuscular mycorrhizae occur in wheat, maize, potatoes, beans, soybeans, tomatoes, strawberries, apples, oranges, grapes, cotton, tobacco, tea, coffee, cacao, sugarcane, sugar maple, rubber trees, ash trees, hazel shrubs, honeysuckle, and various herbaceous plants. They also occur in angiosperms, gymnosperms, pteridophytes, and bryophytes and in most major agricultural crop plants. Vesicular-arbuscular endomycorrhizal fungi have not as yet been grown in pure culture. Lacking regular septa, VA fungi have traditionally been assigned to the single genus *Endogone,* but this genus has now been subdivided into several genera.

The chief diagnostic feature of VA mycorrhizae is the presence of vesicles and arbuscules in the root cortex (Hartley 1965) (Figure 4.7) . Inter- and intracellular hyphae are present in the cortex, and the infection inside the root is directly linked to an external mycelium that spreads into the soil. In general, the mycelium forms a loose network in the soil around the VA mycorrhizal root. The VA mycorrhizal association results in increased phosphate uptake by the plant and improved uptake of other ions, such as zinc, sulfur, and ammonium, from soil (Chiariello et al. 1982).

The VA mycorrhizal associations are thus similar physiologically to the ectomycorrhizae. These mycorrhizal associations enhance the ability of plants to recolonize soils that are barren of vegetation (Tinker 1975; Wills and Cole 1978). Plants with mycorrhizal fungi have an increased ability to take up nutrients from deficient soils, which is often essential for the survival of plants. The frequent and dismal failure of conventional agriculture on cleared tropical rainforest sites is directly attributable to the failure of farmers to account for this fact.

The lushness of tropical rainforests is deceiving. They often grow on highly leached and nutrient-deficient soils (Jordan 1982). In temperate forests, litter and soil humus constitute major nutrient reservoirs. Due to high temperatures and humidity, which enhance rapid biodegradation, there is little humus and litter associated with tropical rainforests, and the chief reservoir of nutrients is the living plant biomass. Because of the leaching by heavy daily rainfall, there are virtually no soluble mineral nutrient salts in the rainforest soil. In this situation, mycorrhizal associations are pivotal to nutrient conservation. Very often mycorrhizal fungus continues to act as a saprophyte and decomposes freshly fallen plant litter at high rates. The liberated inorganic nutrient salts do not enter the soil environment, where they would be rapidly lost by leaching, but are directly transmitted through mycorrhizal hyphae to the plant roots. This "closed circuit nutrient cycling" represents a highly efficient conservation mechanism.

In attempts to use rainforest sites for agricultural production, the natural vegetation is cut down and burned to release the mineral nutrients tied up in the biomass. The nutrients allow the raising of one crop, or at best of a few crops, but the soluble nutrients are rapidly leached from the soil. Only substantial input of synthetic fertilizer, usually unavailable or expensive to use, can achieve continued production. The depleted soil becomes barren and often erodes.

At low population densities, this "slash-and-burn" type of shifting cultivation can be maintained without undue damage because it involves only small patches surrounded by the forest. After two or three crops, the plot is abandoned and a new one is cleared. The abandoned plot gets reseeded and is gradually reclaimed by the forest. Traditionally, this land has been allowed decades for recovery before being used again for raising crops. Increasing population density, however, forces increasingly longer use and shorter recovery cycles, eventually inflicting irreversible damage on the land. Use of cleared tropical rainforest sites for sustainable agriculture with little or no syn-thetic fertilizer input continues to present a great scientific and agronomic challenge. Most likely any solution to the problem will include the raising of perennial woody crop plants with well-developed mycorrhizal associations. These plants are clearly preferable to annual herbaceous crops, which are unsuitable for continuous closed-circuit mineral nutrient cycling.

A better knowledge and practical use of mycorrhizal associations is also called for in temperate climates, especially in connection with land reclamation. Restoration of clearcut forests and revegetation of industrial wastelands, such as coal tips (tailings) and strip-mined areas, are often dependent on using plants with suitable mycorrhizal associations. Preinoculation of seeds of pine trees with a suitable ectomycorrhizal fungus (*Pisolithus tinctorius*) has been shown to enhance the ability of this plant to revegetate high-altitude and other unfavorable habitats (Marx et al. 1977; Ruehle and Marx 1979). If high fertilizer application rates are used, the VA-type endomycorrhizae of crop plants provide little or no benefit in terms of yield (Tinker 1975). When little or no fertilizer is added, however, the benefit of the VA mycorrhizae can be clearly demonstrated (Daft and Hacskaylo 1977).

SYMBIOTIC NITROGEN FIXATION IN NODULES

One of the most important mutualistic relationships between microorganisms and plants involves the invasion of the roots of suitable host plants by nitrogen-fixing bacteria, resulting in formation of a nodule within which the bacteria are able to fix atmospheric nitrogen (Dalton and Mortenson 1972; Brill 1975, 1979, 1980; Bergersen 1978; Nutman et al. 1978; Schmidt 1978; Dazzo 1982; Lynch 1982b; Postgate 1982; Smith 1982; Campbell 1985) (Figure 4.8). The symbiotic fixation of nitrogen is of extreme importance for the maintenance of soil fertility. In agricultural practices, it is utilized to increase crop yields.

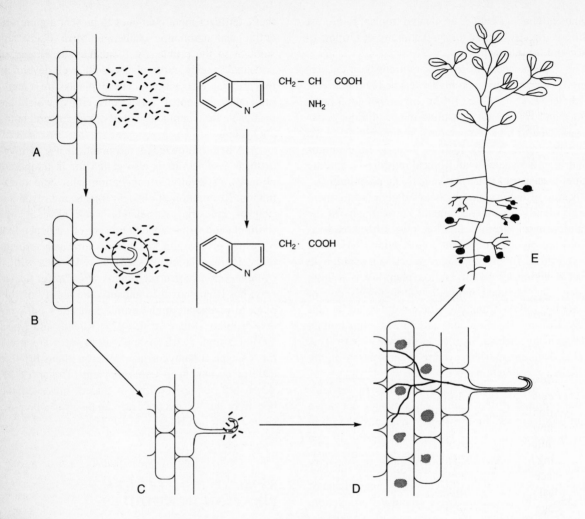

Figure 4.8

Interactions between rhizobia and leguminous plant roots leading to infection and nodule formation, (A) Rhizobia are chemotactically attracted to root hair. Mediated by lectins, some attach to the root hair cell wall. Tryptophan is a component of the root hair exudate. (B) Tryptophan transformed by the rhizobia to indolacetic acid (IAA). This plant growth hormone causes the root hair to curl or branch around the attached rhizobia. Polygalacturonase, secreted by the rhizobia or possibly by the plant, depolymerizes and softens the root hair cell wall. (C) Rhizobia gaining entry into the root hair cell. The root hair cell nucleus directs the development of the infection thread. (D) The infection thread, a tube consisting of cell membrane and surrounding cellulosic wall, growing into the root cortex and infecting some tetraploid cells that proliferate and form nodule tissue. The rhizobia are released from the infection thread, lose their rod shape, become irregularly formed bacteroids, and commence nitrogen fixation. (E) Nodulated leguminous plant.

Table 4.1

Characteristics of the genera *Rhizobium, Bradyrhizobium,* and *Azorhizobium*

Feature	*Rhizobium*	*Bradyrhizobium*	*Azorhizobium*
Flagella on			
liquid medium	None	None	One lateral
solid medium	Peritrichious	One polar or subpolar	Peritrichious
Growth on N_2 fixed *ex planta*	None	None	All strains
Growth rate in culture	Usually fast	Usually slow	Fast
Location of *nod* and *nif* genes	Mainly plasmid	Mainly chromosomal	Probably chromosomal
Host specificity range	Usually narrow	Often broad	Only one species so far identified
Agricultural significance	Most leguminous grain and forage crops	Soybeans	None

Source: Sprent and Sprent 1990.

Nitrogen–Fixing Associations between Rhizobia and Legumes The nitrogen-fixing (diazotrophic) associations of rhizobia with leguminous plants are of great importance both in global nitrogen cycling and in agriculture. Until recently, all nodulating and nitrogen-fixing bacteria associated with leguminous plants were placed into a single genus, *Rhizobium.* Now two additional genera, *Azorhizobium* and *Bradyrhizobium,* are recognized, and additional genera may be added as the numerous species of the legume family are investigated for their nodulating bacteria. Table 4.1 lists some of the characteristics of each genus (Sprent and Sprent 1990). *Azorhizobium* is a unique member of the group that forms stem nodules on tropical leguminous trees (*Sesbania rostrata*). In contrast to members of the two other genera, *Azorhizobium* is capable of growing with atmospheric nitrogen in its free-living state (*azo* refers to nitrogen). *Rhizobium* and *Bradyrhizobium* are not capable of doing so, although some nitrogen fixation by free-living bacteria can be demonstrated at reduced oxygen tension. *Bradyrhizobium* differs from *Rhizobium* by its slow growth in culture (*brady* refers to slow) and some additional

characteristics listed in Table 4.1. However, the infection and nodulation processes are similar for both genera. Unless noted otherwise, information provided for rhizobia pertains to both *Rhizobium* and *Bradyrhizobium.*

While rhizobia occur as free-living heterotrophs in soil, they are not dominant members of soil microbial communities and do not fix atmospheric nitrogen in this state. Under appropriate conditions, however, rhizobia can invade root hairs, initiate the formation of a nodule, and develop nitrogen-fixing activity. The association between rhizobia and plant roots is specific, with mutual recognition between the two compatible partners based upon chemotactic response and specific binding to the root hair prior to invasion and establishment of the root nodule. The legume plant root recognizes the right population of rhizobia, which in turn recognizes the right kind of leguminous root. Within the rhizosphere, plant roots supply the rhizobia with compounds that are transformed by them to substances involved in the initiation of the infection process and subsequent nodule development (Fahrareus and Ljunggren 1968). Establishment of an adequate rhizosphere population of rhizobia is an absolute prerequisite for infection.

Soil conditions have a marked effect on rhizobia in terms of their survival and ability to infect root hairs (Dixon 1969; Alexander 1985). Rhizobia are mesophilic, but some exhibit tolerance to low temperatures, down to 5°C, and others tolerate temperatures up to 40°C. Some rhizobia are sensitive to low pH and cannot establish root hair infections in acidic soils; nitrate and nitrite ions also inhibit nodule formations at relatively low concentrations.

The process of nodule formation is the result of a complex sequence of interactions between rhizobia and plant roots (Solheim 1984) (Figure 4.9). Initially, the plant root secretes unspecified compounds that chemotactically attract the bacteria to the rhizosphere and allow the bacteria to multiply. Lectins, plant proteins with high affinity to carbohydrate moieties on the surface of appropriate rhizobial cells, have been identified as specific mediators of the attachment of rhizobia to susceptible root hairs (Dazzo and Hubbell 1975; Dazzo and Brill 1979; Hubbell 1981). During the nodulation process, tryptophan, secreted by the plant roots, is oxidized to IAA by the rhizobia, and the IAA, together with unknown cofactors probably arising from the host plant roots, initiate hair curling or branching. The root hairs grow around the bacterial cells. Polygalacturonase, secreted by the rhizobia or possibly by the plant roots, depolymerizes the cell wall and allows bacteria to invade the softened plant root tissues (Hubbell 1981; Ridge and Rolfe 1985).

After penetration of the primary root hair wall, the infection proceeds by the development of an infection tube ("thread") that is surrounded by the cell membrane and a cellulosic wall. It contains *Rhizobium* cells lying end-to-end in a polysaccharide matrix. The infection thread penetrates through and between root cortex cells. As the infection thread grows, the cell's enlarged nucleus moves and directs the development of the infection thread. The first cells of the developing nodule contain twice the normal number of chromosomes. These tetraploid cells give rise to the central nodule cells in which the rhizobia develop to produce nitrogen-fixing tissue. Associated cells of normal ploidy give rise to unin-fected supporting tissues that connect the nodule to the root vascular system.

Within the infected tissue, rhizobia multiply, forming unusually shaped and sometimes grossly enlarged cells called bacteroids. Interspersed with the bacteroid-filled cells of the nodule are uninfected vacuolated cells that may be involved in the transfer of metabolites between the plant and microbe tissues. During transformation of normal rhizobial cells into bacteroids, the bacterial chromosomes degenerate, eliminating the bacteroids' capacity for independent multiplication. The bacteroid cells produce and contain active nitrogenase, but host-plant tissues appear to play a role in the initiation and control of nitrogenase synthesis. The bacteroid within the nodule carries out the fixation of atmospheric nitrogen. Under normal conditions, neither free-living rhizobia nor uninfected leguminous plants are able to bring about the fixation of atmospheric nitrogen.

For active nitrogen fixation, the plant-rhizobia association requires various organic and inorganic compounds. The trace element molybdenum is required and forms an important part of the nitrogenase enzyme. Nitrogenase also contains high amounts of sulfur and iron, which thus are requirements for active nitrogen fixation by nodules; cobalt and copper are also required in lower amounts.

Nodules have a characteristic red-brown color owing to the presence of leghemoglobin, which is a constant and prominent feature in the central tissue of all nitrogen-fixing leguminous nodules. The leghemoglobin serves as an electron carrier, supplying oxygen to the bacteroids for the production of ATP and at the same time protecting the oxygen-sensitive nitrogenase system. The heme portion of the leghemoglobin is coded for by the rhizobia and the globin portion by the plant. Leghemoglobins are unique for legume root nodules and occur nowhere else in the plant kingdom.

Specific expresssion of plant and bacterial genes accompanies the development of the rhizobial-plant symbiosis (Nap and Bisseling 1990). The genes invovled in root nodule formation are collectively called nodulin genes. In the fast-growing *Rhizobium* species, most of these genes are located on large *Sym* plasmids, whereas the slow-growing *Bradyrhizobium*

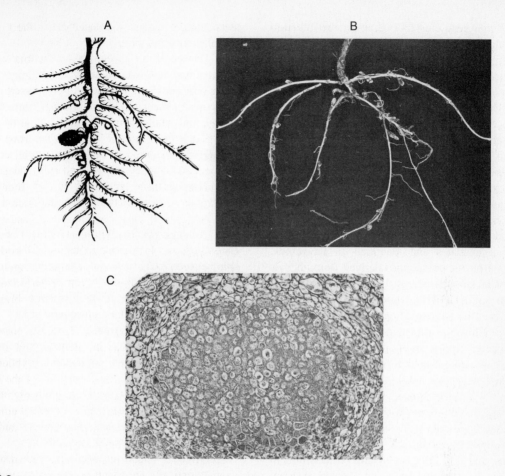

Figure 4.9

Root nodule: (A) Drawing showing root nodules, by Marcello Malipighi, 1679.
(B) Photograph of roots of clover showing nodulation. (C) Photomicrograph of root
nodule cross section showing inner core plant cells densely packed with *Rhizobium*.
(Source:Carolina Biological Supply Co.)

species carry the late nodulin genes on the bacterial
chromosome. Genes essential for the process of
nitrogen fixation include *nif* and *fix*, among which
are the structural genes for nitrogenase. The rhizobial genes required for nodule formation and nodulin
gene expression include the nodulation *(nod)* genes,
several groups of genes concerned with the structure
of the outer surface of the bacterium (the *exo, lps,*
and *ndv* genes), and a number of less well-defined
genes. The host-specific *nod* genes determine the
specificity of nodulation on a particular host. The
*nod*D gene is the only *nod* gene that is constitutively
expressed in both the free-living and symbiotic
states of *Rhizobium*. In combination with flavonoids
excreted by plant roots, the NodD protein probably
acts as a transcriptional activator for all other *nod*
genes, and the gene is essential for nodulation as the
common *nod*ABC genes.

The *nod* gene clusters coding for the infection and nodulaton process are generally located on the Sym plasmids, which also carry specificity genes. It is possible to transfer the Sym plasmids between rhizobial strains and thus to alter the range of host legumes they can infect. Sym plasmids may even be transferred to *Agrobacterium* and other related bacteria, which then acquire the ability to nodulate the specified plants. However, such associations usually do not fix nitrogen. The *nif* genes that code for the biochemical nitrogen-fixing mechanism may or may not be plasmid associated in various rhizobial strains (Postgate 1982; Sprent and Sprent 1990).

Nodule–inducing molecules have been purified from plant exudates and identified as flavonoids, three-ring aromatic compounds derived from phenylpropanoid metabolism (Long 1989). The NodD protein of a particular *Rhizobium* species proves most responsive to the flavonoids excreted by its homologous host. Rhizobia grown in the presence of root exudate or *nod* gene-inducing flavonoids contain factors that cause root hair deformation. In alfalfa and clover, the most active inducers are flavones, such as luteolin (3´,4´,5,7-tetrahydroxyflavone) (Long 1989). The alfalfa symbiont *R. meliloti* excretes a sulfated β–1,4–tetrasaccharide of D-glucosamine, in which three amino groups are acetylated and one is acylated with an unsaturated C_{16} fatty acid chain. In addition to the *nod* genes, rhizobia have other genes that are required for normal nodule development, including the genes involved in the production of bacterial outer surface components such as exopolysaccharides (*exo* genes), lipopolysaccharides (*lps* genes) and (cyclic) glucans (*ndv* genes), in addition to genes related to drug resistance, auxotrophy, and carbohydrate metabolism.

A side reaction of the N_2-reduction process is the evolution of hydrogen gas. The evolution of H_2 wastes photosynthetic energy and leads to lowered yields. Some Sym plasmids also carry *hup* genes coding for hydrogenase activity. The hydrogenase oxidizes H_2 to water and recovers energy by chemiosmotic coupling with ATP synthesis. This beneficial process saves photosynthetic energy that would otherwise go to waste. (Albrecht et al. 1979). The genetic manipulation of the association of rhizobia with legumes is likely to increase both the range and the efficiency of the system.

The economic incentives for such research are strong. Important leguminous grain crops include soybeans (*Glycine max*), peanuts (*Arachis hypogaea*), and many bean (*Phaseolus, Vigna*), pea (*Pisum, Cajanus, Cicer*), and lentil (*Lens*) varieties. Important leguminous forage crops include alfalfa (*Medicago*) and clover (*Trifolium*) varieties. Leguminous trees are critical components of many tropical and subtropical ecosystems ranging from semidesert acacias to important hardwoods of the rainforest (Sprent and Sprent 1990).

Nonleguminous Nitrogen-Fixing Mutualistic Relationships   In addition to the mutualistic relationship between *Rhizobium* and leguminous plants, other symbiotic relationships between bacteria and nonleguminous plants result in the formation of root nodules and the ability to fix atmospheric nitrogen (Evans and Barber 1977; Akkermans 1978). The formation of root-nodule symbiosis in nonleguminous plants occurs with *Rhizobium* populations, cyanobacteria, and actinomycetes.

Actinomycete (*Frankia*) symbioses are more frequent in temperate and circumpolar regions, whereas cyanobacteria and rhizobial symbioses are more common in tropical and subtropical regions. Actinomycete symbioses occur in angiosperms and characteristically are found in species of *Alnus* (such as alder), *Myrica* (such as bayberry), *Hippophae, Comptonia, Casuarina,* and *Dryas,* among others. Much of the soil nitrogen in high latitudes, as in Scandinavia, probably originates from the root-nodule symbiosis of actinomycetes in nonlegumes, particularly *Alnus* plants, and to a lesser extent in *Dryas, Myrica,* and *Hippophae* plants. *Casuarina* inhabits subtropical coastal sand dunes and semidesert environments (Morris et al. 1974; Callahan et al. 1978; Sprent and Sprent 1990).

When the hyphae of suitable actinomycetes penetrate the root, cortical cells are stimulated to divide (Berry 1984). The hyphae penetrate the dividing cells, forming clusters within the host, and vesicles form at the periphery of the infected hyphal tips. In the neighborhood of the primary infected tissue, a root primordium is initiated that grows into the cortex. The endophytic actinomycete invades the meristem cells of the primordium, and factors produced by the actinomycetes stimulate the development of the root primordium. The dichotomous division of the top meristem results in the formation of a cluster of lobes called a rhizothamnion, which is typical of all actinomycete nodules. Within the nodule, actinomycetes produce nitrogenase and fix atmospheric nitrogen.

Some liverworts, mosses, pteridophytes, gymnosperms, and angiosperms are able to establish mutualistic relationships with nitrogen-fixing cyanobacteria of the genera *Nostoc* or *Anabaena*. The gymnosperm *Cycas* has specialized root structures, called coralloid roots, that exhibit nodulelike structures even before invasion by the cyanobacteria. The angiosperm *Gunnera* has stem nodules. In these symbiotic relationships, the cyanobacteria are restricted to heterotrophic metabolism; the plant supplies suitable organic compounds through its photosynthetic activity, and the cyanobacteria within the underground roots produce fixed forms of nitrogen. There is a relatively high rate of exudation of nitrogenous compounds by the cyanobacteria within the plant roots as compared to free-living cyanobacteria. (Postgate 1982; Sprent and Sprent 1990).

INTERACTIONS WITH AERIAL PLANT STRUCTURES

Stems, leaves, and fruits of plants provide suitable habitats for microbial populations called epiphytic microorganisms; heterotrophic and photosynthetic bacteria, fungi (particularly yeasts), lichens, and some algae regularly occur on these aerial plant surfaces (Preece and Dickinson 1971; Collins 1976; Dennis 1976; Dickinson 1976, 1982; Dickinson and Preece 1976; Blakeman 1981). The habitat adjacent to the plant leaf surface is known as the phyllosphere, and the habitat directly on the surface of the leaf is the phylloplane. Various bacterial and fungal populations occupy the phylloplane and phyllo-

sphere habitats. The numbers of microorganisms on leaf surfaces depend on the season and age of the leaf.

Epiphytic microorganisms on plant surfaces are directly exposed to climatic changes. These populations must withstand direct sunlight, periods of desiccation, and periods of high and low temperatures. Most successful epiphytes are pigmented and have specialized protective cell walls, adaptive features for withstanding these adverse environmental conditions.

Epiphytic microorganisms also exhibit various spore discharge mechanisms that allow them to move from one plant surface to another. Insects play an important role in the dissemination of microorganisms on fruit surfaces. Fruit flies carry yeasts between fruit surfaces. Sometimes there is a close synergistic relationship among microorganisms, insects, and plants, such as among figs, yeasts, and fig wasps.

Numerical taxonomic studies have been used to identify the bacteria in the phyllosphere (Goodfellow et al. 1976a, 1976b, 1978). The principal populations in the phyllosphere of the green needles of some pine trees are *Pseudomonas* species, including *Pseudomonas fluorescens*. Populations from the phylloplane of pine trees are relatively proficient in utilizing sugars and alcohols as carbon sources compared to bacterial populations in the litter layer underlying these trees, which exhibit more lipolytic and proteolytic activity. Bacterial populations in the phylloplane of rye exhibit seasonal changes: xanthomonads and pink chromogens have high populations in May, xanthomonads and pseudomonads in July, xanthomonads in September, and listeriae and staphylococci in October.

Yeasts frequently inhabit the leaves of plants (Davenport 1976; Dickinson 1976, 1982). Populations of *Sporobolomyces roseus*, *Rhodotorula glutinis*, *Rhodotorula mucilaginosa*, *Cryptococcus laurentii*, *Torulopsis ingeniosa*, and *Aureobasidium pullulans* are normal inhabitants of the phyllosphere. There is an abundance of pigmented populations of yeasts and bacteria on leaf surfaces. The pigments of these microbial populations probably afford protection against exposure to direct sunlight on the leaf surface. *Sporobolomyces* is perhaps the most successful fungus that develops in the phyllosphere; it produces ballistospores that it can shoot from one leaf to another, facilitating its dispersal. Many other fungi, including Ascomycota, Basidiomycota, and Deuteromycota, have been isolated in the phyllosphere. Some of these are undoubtedly allochthonous populations; some are associated with plant disease. Populations of *Alternaria, Epicoccum,* and *Stemphylium* have been found to be phylloplane invaders that grow extensively only under favorable conditions. Populations of *Ascochytula, Leptosphaeria, Pleospora,* and *Phoma* found in the phyllosphere are primarily saprophytes that are unable to grow extensively until the onset of senescence. Allochthonous fungal populations frequently found in the phyllosphere include *Cryptococcus, Myrothecium, Pilobolus,* and various other fungal populations whose normal habitat is the soil.

Flowers form a group of short-lived habitats for epiphytic microorganisms (Dickinson 1976, 1982). *Candida reukaufii* and *C. pulcherrima* are found in flower habitats. The high sugar content of the flower nectar makes the flower a suitable habitat for these yeast populations. Many other yeast populations, including other *Candida* species, *Torulopsis, Kloeckera,* and *Rhodotorula,* have been found within flowers. More yeasts have also been found to inhabit stamens and stigmas than the corollas and calyxes of flowers. Following fertilization of flowers and during the ripening into mature fruits, the environmental conditions of this habitat change and populations of microorganisms undergo a succession, with populations of *Saccharomyces* sometimes becoming dominant.

There are positive and negative interactions between the microbial populations found on plant surfaces. The growth of osmophilic yeasts lowers the sugar concentration, making the habitat suitable for invasion by other microbial populations. Unsaturated fatty acids produced by yeasts may inhibit development of Gram-positive bacteria on fruit surfaces. Bacterial populations that develop on fruit surfaces are dependent on growth factors, such as thiamine and nicotinic acid, that are produced by yeasts. Yeast populations are also dependent on growth factors produced by bacteria on fruit surfaces.

Microorganisms grow on the bark of trees. Lichens and various fungi, such as bracket and shelf fungi, are conspicuous colonizers of tree bark

Figure 4.10

Photograph of a lichen growing on a tree trunk. Lichens commonly colonize tree surfaces. (Source: O. K. Miller.) Virginia Polytechnic Institute, Blacksburg, VA.

(Figure 4.10). Myxomycetes, including species of *Licea, Trichia,* and *Fuligo,* form fruiting bodies on tree bark. The fungi *Colletotrichum* and *Taphrina* grow on woody stems and tree bark. Lichens grow in the canopy of the forest within the phyllosphere. Some lichens fix atmospheric nitrogen that can be used by the surrounding vegetation.

There are several mutualistic relationships between microorganisms and plant structures other than roots that result in nitrogen fixation. The aquatic fern *Azolla* and the cyanobacterium *Anabaena* enter into such a relationship (Peters and Mayne 1974) (Figure 4.11). *Azolla* grows on the surface of quiet water in tropical and subtropical zones. On the lower surface of the *Azolla* leaves are mucilage-containing cavities that always contain *Anabaena.* During development of the *Azolla* leaf, the cavities form invaginations around the *Anabaena.* Within these cavities the *Anabaena* fix large quantities of atmospheric nitrogen. The relationship is mutualistic, since the *Azolla* supplies the *Anabaena* with nutrients and growth factors and the *Anabaena* supplies the *Azolla* with fixed forms of nitrogen.

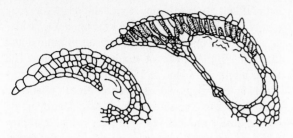

Figure 4.11
Drawing illustrating the symbiotic relationship between the cyanobacterium *Anabaena* and the aquatic fern *Azolla*. (Source: Smith 1938. Reprinted by permission; copyright McGraw-Hill.)

Azolla species thrive in subtropical rice paddies, and under favorable conditions, their cyanobacterial symbionts may fix several kg of nitrogen per ha per day. This may yield 50 to 150 kg of nitrogen per ha per year and is a major factor in the continued fertility of Asian rice fields that receive very little additional nitrogen fertilizer. The *Anabaena-Azolla* association continues to fix nitrogen even in the presence of added nitrogen fertilizer and is relatively insensitive to low pH and to salinity. For these reasons, it has great biotechnological potential for replacing scarce synthetic nitrogen fertilizer in developing countries (Swaminathan 1982).

Nitrogen-fixing bacteria are also found in the phyllosphere of other terrestrial plants, including the phylloplane of conifer trees (Fletcher 1976; Jones 1976). Some of the nitrogen fixed within the phylloplane is retained in the canopy and recycled by the microbial populations, some is leached to the soil, some is taken up directly by leaves, and some is grazed by herbivores. Some bacterial species can infect leaves of Myrsinaceae (such as marlberry) and Rubiaceae (such as partridgeberry), forming leaf nodules. Some leaf nodules are capable of nitrogen fixation. In *Ardisia*, a member of Myrsinaceae, a symbiotic relationship with bacteria is associated with the buds. Without the bacterial symbiont, *Ardisia* develops a "cripple" condition. Similar leaf-nodule associations have been reported in the tropical plants *Paretta*

and *Psychotria* with the nitrogen-fixing *Chromobacterium* and *Klebsiella*, respectively (Silver et al. 1963; Centifanto and Silver 1964).

In several interesting mutualistic associations, fungal endophytes grow intercellularly in wild and cultivated grasses of the Poaceae family and endow them with protection against herbivores (Johnson et al. 1985; Christensen et al. 1991; Shardl et al. 1992). The best described are associations of the tall fescue (*Festuca arundinacea*) and the perennial rye grass (*Lolium perenne*) with the fungi *Acremonium coenophialum* and *A. lolii*, respectively. The endophytes cause no disease symptoms in the grasses, but while receiving photosynthate from the host plant, they synthesize a range of alkaloids such as ergopeptides, lolines, lolitrems, and peramines. These act as poisons and/or feeding deterrents against nematodes, aphids, insects, and mammalian herbivores. The lolitrems are especially potent mammalian neurotoxins and can cause livestock losses (rye grass staggers) on heavily infected pastures.

The described *Acremonium* species have no free-living stage and are not infectious. They perpetuate themselves through the plant seeds that contain the endophyte. Experimentally, the seeds can be freed from the endophyte by high temperature storage. *Acremonium* has no sexual stage and is classified with Deuteromycota but shows strong morphological and biochemical similarities to the ascomycete *Epichloe typhina* that causes "choke disease" in some related grasses. *Epichloe typhina* is also an asymptomatic endophyte until the flowering stage of the infected grass. At that time profuse ectophytic (external) fungal growth with ascospore formation occurs and halts flower development. It is tempting to speculate that the clearly mutualistic *Acremonium* developed from the moderately parasitic *E. typhina*.

One of the most interesting ecological relationships between bacteria and plants involves the role of certain phyllosphere bacteria in initiating ice crystal formation that results in frost damage to the plant (Lindow et al. 1978, 1982a, 1982b; Lindow and Connell 1984). Some strains of *Pseudomonas syringae* and *Erwinia herbicola* produce a surface protein that can initiate ice crystal formation. Large epiphytic populations of ice-nucleation–active bacteria occur on many plants. Leaf surface populations of these ice-nucleation–active bacteria prevent supercooling in the plant parts on which they reside by initiating damaging ice formation when ambient temperatures reach –2°C to –4°C. The epiphytic bacteria are conditional plant pathogens, causing death due to frost damage only at temperatures that can initiate the freezing process. Laboratory experiments have demonstrated that ice crystals do not form until –7°C to –9°C, when ice-nucleation–active populations are replaced with mutant strains that do not produce the ice crystal initiating proteins, thereby limiting frost damage.

The development of genetically engineered strains of ice-minus *P. syringae,* that is, strains that do not form the ice-nucleating surface protein, and the proposal to apply such strains to field crops for frost protection have caused great public and scientific concern. Little is known about the importance of the normal role of ice-nucleation bacteria and the possible environmental consequences of modifying the normal ecological relationship between ice-nucleation–active bacteria and the plants upon which they reside. Various government agencies, scientific societies, and institutions of the legal system are currently involved in this debate over the safety and efficacy of applying ice-minus bacteria to protect agricultural crops against frost damage.

MICROBIAL DISEASES OF PLANTS—PLANT PATHOGENS

Microbial diseases of plants, whether caused by viruses, bacteria, fungi, or protozoa, are of ecologic and economic importance (Stevens 1974; Wheeler 1975; Robinson 1976; Agrios 1978; Strange 1982; Campbell 1985). Plant pathology is an extensive field; only an abbreviated discussion of the plant pathogens (called plant pests by the US Department of Agriculture) and the resulting plant diseases is possible in this chapter. In a broad sense, microbial diseases of plants cause malfunctions that result in the reduced capability of the plant to survive and maintain its ecological niche. Plant diseases may result in death or may greatly impair the growth yield of the plant. Occasionally, microbial diseases of plants result in famine and hence mass migration of human populations. The potato blight in Ireland in 1845 resulted in mass starvation and widespread emigration from Ireland to North America. Chestnut blight disease destroyed the native North American chestnut trees, which had provided an important cash crop, especially in the Appalachian area. In 1970 a leaf blight of maize caused the destruction of more than 10 million acres of corn crops during that one year.

The development of plant diseases due to microbial pathogens normally follows a pattern of initial contact of the microorganisms with the plant, entry of the pathogen into the plant, growth of the infecting microorganisms, and development of plant disease symptoms. Pathogenic microorganisms may contact the plant in the rhizoplane or the phylloplane. Because most fungal plant pathogens are dispersed through the air as spores, they often contact the leaves or stems of plants. Most viral plant diseases are transmitted by insect vectors; most of these pathogens contact the plant at the phylloplane. Some bacterial and fungal pathogens are also carried by insect vectors (Harris and Maramorosch 1981). Soilborne animals, such as nematodes, transmit some plant pathogens, contacting the plant at the rhizoplane. Motile pathogens in the soil, including plant pathogenic bacteria such as *Pseudomonas* species, and fungi, such as *Oidium,* are attracted to plant roots via chemotaxis and contact the plant in the rhizoplane.

Spores of plant pathogenic fungi distributed by air currents need to attach themselves to leaves or stems of susceptible plants. The conidiospores of the ascomycete *Magnaporthe grisea,* causative agent of the rice blast disease, efficiently attaches itself to hydrophobic surfaces such as the leaf cuticle or a cuticle-simulating teflon membrane by means of a mucilage stored in the spore tip for this purpose. Moist air or dew causes this mucilage to swell, rupture the spore tip, and glue the spore to the leaf cuticle (Hamer et al. 1988).

Plant pathogens may enter plants through wounds or natural openings, such as the stomata (Figure 4.12). When entrance is to be gained through the

Figure 4.12

Scanning electron micrograph of bacteria on leaf surface of *Zea mays* showing penetration through a stoma. (Source: A. Karpoff, University of Louisville.)

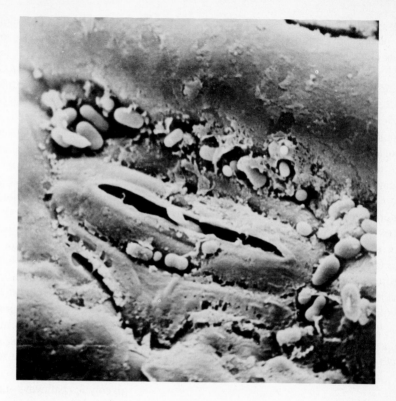

stomata, the hyphae of a germinating spore need a mechanism to locate and recognize such openings. For the germinating uredospores of the rust fungus *Uromyces appendiculatus,* the topographical signal was identified as a ridge approximately 0.5 μm in height, resembling the ridge formed by the stomatal guard cells of the bean (*Phaseolus vulgaris*) plant. When the hyphae encountered such a ridge, even when fabricated from polystyrene for this experiment, they underwent morphological differentiation forming appressoria (flattened hyphae) necessary for the penetration of the stomata. Ridges less than 0.25 μm or more than 1.0 μm in height did not elicit such response (Hoch et al. 1987). Many viruses enter plants through wounds caused by the vector carrying the virus, although others enter through the roots with the water being taken up by the plant. Some plant pathogens are able to penetrate the cuticle of the plant directly. Such penetration involves attachment of the pathogen to the plant surface, followed by for-

mation of a penetration peg, which passes through the cuticle and the cell wall. When a specific antibody blocked the cutinase of *Fusarium solani,* no infection of the susceptible bean plant took place, although the spore remained viable (Maiti and Kolattukudy 1979). Plant cuticles and tissues are often subjected to enzymatic attack by the pathogen that softens them at the site of penetration. Powdery mildews (*Erysiphe* species), the soft rot pathogen *Botrytis cinerea,* and *F. solani* are examples.

Microbial pathogens that have successfully entered the plant disrupt normal plant function by producing degradative enzymes, toxins, and growth regulators (Friend and Threlfall 1976). Soil plant pathogens produce pectinases, celluloses, and hemicelluloses that result in degeneration of the plant structure, producing soft rots and other lesions. Destruction of plant growth regulators by pathogens results in dwarfism, whereas microbial production of IAA, gibberellins, and cytokinins by some plant pathogens result in gall formation and excessive elongation of plant stems. Toxins produced or induced by pathogenic organisms in plants interfere with the normal metabolic activities of the plant. The toxin produced by *Pseudomonas tabaci,* which causes tobacco wildfire disease, has been characterized as β-hydroxy-diaminopimelic acid; it interferes with the metabolism of methionine.

Some plant pathogenic fungi produce highly selective toxins including: low–molecular-weight cyclic peptides and linear polyketols (Sheffer and Livingston 1984). These appear to interfere with mitochondria and cell membranes. Damage to cell membranes may facilitate the spread of the infection. Tolerance to the toxins is usually associated with resistance to the fungal disease. Resistance to the toxins seems to be based on modified receptor sites.

Plants can develop a variety of morphological or metabolic abnormalities as a result of microbial infections (Table 4.2). Invasion of plant cells by pathogenic microorganisms sometimes results in rapid death of the plant. In other cases, the plant may undergo slower changes. Pathogens that penetrate

Table 4.2

Some symptoms of microbial diseases of plants

Disease	Symptom
Necrosis (rot)	Death of plant cells; may appear as spots in localized areas
Canker	Localized necrosis resulting in lesion, usually on stem
Wilt	Droopiness due to loss of turgor
Blight	Loss of foliage
Chlorosis	Loss of photosynthetic capability due to bleaching of chlorophyll
Hypoplasia	Stunted growth
Hyperplasia	Excessive growth
Gall	Tumorous growth

plants directly often elicit a morphological response by the plant with the formation of structures called papillae. This response may be an attempt to block the spread of the pathogens. Cell walls of infected plant tissues are often modified, resulting in swelling or other distortions of the cell. Invasion by plant pathogens may disrupt cell permeability, causing imbalances in water relations that lead to leakage and death of plant cells. Changes in permeability may be caused by pectic enzymes or by toxins produced by the plant pathogens. Bacterial cankers, in diseases such as fire blight of pears and apples caused by *Erwinia amylovora,* involve damage to the water-conducting tissues of plants. Blockage of water transport can also lead to desiccation and symptomatic wilt, as in *Fusarium* wilt of tomatoes caused by a reduction of water flow through the xylem. Various bacterial pathogens cause wilts by blocking the stomata, which modifies transpiration and water transport in the plant.

Plant pathogens may alter the metabolic activities of the plant. Diseased plants sometimes show changes in respiratory activity, which may be caused by electron transport uncoupling or changes in the glycolytic pathways of carbohydrate metabolism. Plant pathogens may interfere with carbon dioxide fixation. Foliar pathogens sometimes produce chlorosis, which prevents plants from carrying out oxidative photophosphorylation and producing the ATP needed for carbon dioxide fixation. Plant pathogens also may cause changes in protein synthesis, resulting in impaired metabolism. Overgrowth and gall formation involve alterations in the nucleic acid function controlling protein synthesis. Changes in protein synthesis may cause alterations of metabolic pathways and activities of the enzymes produced.

Once plants exhibit disease symptoms due to invasion by a primary plant pathogen, they are subject to additional invasion by opportunistic secondary pathogens. Loss of integrity of surface structures and cell walls allows invasion by many opportunistic microorganisms. Zymogenous (opportunistic) soil microorganisms rapidly colonize dead plant material that falls to the soil surface and becomes part of the litter.

Plants under attack by microbial pathogens may react by synthesizing antimicrobial substances called phytoalexins (Deacon 1983). This defense mechanism may slow or even stop the infection process and may leave the plant with an increased systemic resistance to subsequent challenge by the pathogen. This subject will be discussed in greater detail later in connection with biological control.

Table 4.3

Some plant pathogenic viruses

DNA VIRUSES

Caulimovirus group: cauliflower mosaic virus

RNA VIRUSES

Rod-shaped viruses

    Tobavirus group: tobacco rattle virus;

    pea early browning virus

    Tobamovirus group: tobacco mosaic virus

Viruses with flexuous or filamentous particles

    Potexvirus group: potato virus X

    Carlavirus group: carnation latent virus

    Potyvirus group: potato virus

    Beet yellows virus

    Festuca necrosis virus

    Citrus tristeza virus

Isometric viruses

    Cucumovirus group: cucumber mosaic virus

    Tymovirus group: turnip yellow mosaic virus

    Nepovirus group: tobacco ringspot virus

    Bromovirus group: brome mosaic virus

    Tombushvirus group: tomato bushy stunt

    Alfalfa mosaic virus

    Pea enation mosaic virus

    Tobacco necrosis virus

    Wound tumor virus group: wound tumor virus; rice dwarf virus

    Tomato spotted wilt virus

Rhabdoviruses

    Lettuce necrotic yellows virus

Viral Diseases of Plants

Many plant pathogenic viruses are classified according to their ability to cause a particular disease (Table 4.3). Vectors are important in the transport of plant pathogenic viruses that occur in the soil or in diseased plant tissues of susceptible host plants. Various insects, including aphids, leafhoppers, mealybugs, ants, and nematodes, can act as vectors for viral diseases of plants.

A necessary attribute of plant pathogens in general, and viral pathogens of plants in particular, is the ability to survive outside infected host cells until susceptible living plants can be found (Baker and Snyder 1965; Esau 1968; Plumb and Thresh 1983). As obligate intracellular parasites, plant pathogenic viruses are dependent on finding suitable plant cells for their replication. Outside susceptible plant cells—including within vectors—viruses must maintain their integrity in order to retain their infective capability. Persistence of viruses within vectors, where they are not subject to inactivation by soil microbial enzymes as they would be if free in soil, is one means by which viruses may survive in soil. Environmental factors that affect the survival and movement of vector organisms, such as soil texture and moisture, determine in large part the patterns of dissemination of pathogenic viruses. The distribution of viral plant diseases often follows the spatial distribution pattern of the vectors.

Plant pathogenic viruses may also be transmitted within or on infected plant structures. Tobacco rattle virus, for example, is detectable on the pollen of infected petunia plants. Together with pollen, it is disseminated through the air to susceptible plants. Viruses are also transmitted on plant seeds. The spread of viruses on plant structures involved in the reproductive activity of the plant, such as pollen and seeds, ensures that viruses are maintained with susceptible host plant populations and that viral diseases are endemic to plant populations.

Viroids have been implicated as the cause of potato spindle disease, chrysanthemum stunt, and citrus exocortis disease (Diener 1979). It is not clear yet how viroids are successfully transmitted nor precisely how they code for their own replication.

Table 4.4

Some diseases of plants cause by bacteria

Organism	Disease	Organism	Disease
Pseudomonas		*Erwinia*	
P. tabaci	Wildfire of tobacco	E. amylovora	Fire blight of pears and apples
P. angulata	Leaf spot of tobacco	E. tracheiphila	Wilt of cucurbits
P. phaseolicola	Halo blight of beans	E. stewartii	Wilt of corn
P. pisi	Blight of peas	E. carotovora	Blight of chrysanthemum
P. glycinea	Blight of soybeans		
P. syringae	Blight of lilac	*Corynebacterium*	
P. solanacearum	Moko of banana	C. insidiosum	Wilt of alfalfa
P. caryophylli	Wilt of carnation	C. michiganese	Wilt of tomato
P. cepacia	Sour skin of onion	C. facians	Leafy gall of ornamentals
P. marginalis	Slippery skin of onion	*Agrobacterium*	
P. savastanoi	Olive knot disease	A. tumefaciens	Crown gall of various plants
P. marginata	Scab of gladiolus	A. rubi	Cane gall of raspberries
Xanthomonas		A. rhizogenes	Hairy root of apple
X. phaseoli	Blight of beans	*Mycoplasma*	
X. oryzae	Blight of rice	M. sp.	Aster Yellows
X. pruni	Leaf spot of fruits	M. sp.	Peach X disease
X. juglandis	Blight of walnut	M. sp.	Peach Yellows
X. citri	Canker of citrus	M. sp.	Elm phloem necrosis
X. campestris	Black rot of crucifers	*Spiroplasma*	
X. vascularum	Gumming of sugarcane	S. sp.	Citrus stubborn disease
Streptomyces		S. sp.	Bermuda grass witches'-broom
S. scabies	Scab of potato	S. sp.	Corn stunt
S. ipomoeae	Pox of sweet potato		

Bacterial Diseases of Plants

Plant pathogenic bacterial species occur in the genera *Mycoplasma, Spiroplasma, Corynebacterium, Agrobacterium, Pseudomonas, Xanthomonas, Streptomyces,* and *Erwinia*. These bacteria are widely distributed and cause a number of plant diseases, including hypertrophy, wilts, rots, blights, and galls (Table 4.4).

Because most plant pathogenic bacteria do not form resting stages, they must remain in intimate contact with plant tissues during all stages of their life history. Many plant pathogenic bacteria can remain viable on plant seeds and in other plant tissues during times of plant dormancy because they are resistant to desiccation. Many are obligate parasites and are unable to compete with saprophytic soil bacteria. Other non-

obligate bacterial pathogens of plants are able to reproduce in soil, often causing plant diseases only after sufficient populations have developed in the soil.

Some bacterial pathogens exhibit no significant soil phase. For example, during the winter, *Erwinia amylovora*, which causes fire blight in fruit trees, does not grow but remains dormant within infected tissues of the stems and branches of trees (Figure 4.13). In the spring, rain and insects distribute this organism to new plants.

Some bacterial diseases are caused by pathogens that have a permanent soil phase; for example, some fluorescent *Pseudomonas* species cause soft rots in plants. These organisms, which are abundant and occur as saprophytes in the rhizosphere, infect plants through the roots.

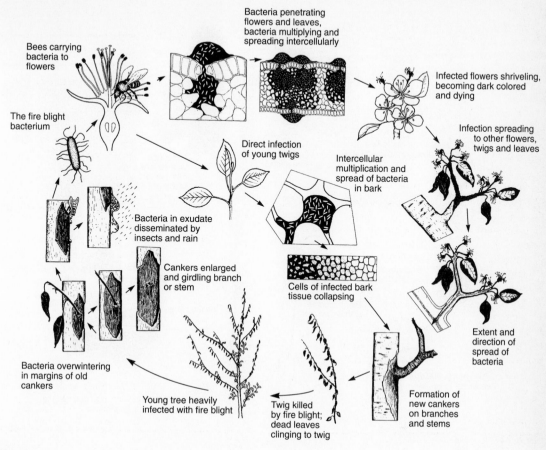

Bacteria penetrating flowers and leaves, bacteria multiplying and spreading intercellularly

Bees carrying bacteria to flowers

Infected flowers shriveling, becoming dark colored and dying

The fire blight bacterium

Infection spreading to other flowers, twigs and leaves

Direct infection of young twigs

Intercellular multiplication and spread of bacteria in bark

Bacteria in exudate disseminated by insects and rain

Cankers enlarged and girdling branch or stem

Cells of infected bark tissue collapsing

Extent and direction of spread of bacteria

Bacteria overwintering in margins of old cankers

Young tree heavily infected with fire blight

Twig killed by fire blight; dead leaves clinging to twig

Formation of new cankers on branches and stems

Figure 4.13

Life cycle of *Erwinia amylovora,* causing the fire blight of pears and apples. Note continuous association of the bacterium with plant tissue throughout its life cycle and lack of a free soil phase. (Source: Agrios 1978. Reprinted by permission, copyright Academic Press.)

Many bacterially caused plant diseases are seedborne. These pathogenic bacteria survive on seeds for a transient period in the soil. Bacteria may be carried on seeds as a surface contaminant or in the micropyle. *Pseudomonas phaseolicola,* for example, is carried in the micropyle and causes halo blight of beans, whereas *Xanthomonas malvacearum,* which causes blight of cotton, occurs externally on the cotyledon margins during germination of the seed. As this pathogen is maintained exter-

nally, it is subject to considerable influence by soil properties such as texture, moisture, and temperature.

Crown gall is an extremely interesting plant disease that occurs after viable *Agrobacterium tumefaciens* cells enter either through the root or through wounded surfaces of susceptible dicotyledonous plants, usually at the soil-plant stem interface (Lippincott and Lippincott 1975; Nester and Montoya 1979; Kosuge and Nester 1984) (Figure 4.14). Occurring in fruit trees, sugar beets, or other

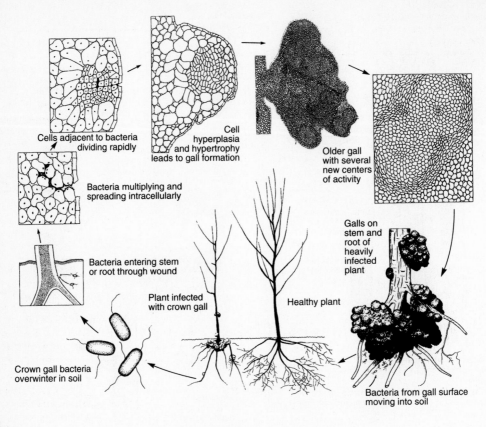

Cells adjacent to bacteria dividing rapidly

Cell hyperplasia and hypertrophy leads to gall formation

Older gall with several new centers of activity

Bacteria multiplying and spreading intracellularly

Bacteria entering stem or root through wound

Galls on stem and root of heavily infected plant

Plant infected with crown gall

Healthy plant

Crown gall bacteria overwinter in soil

Bacteria from gall surface moving into soil

Figure 4.14

Life cycle of crown gall caused by *Agrobacterium tumefaciens*. Note tumorlike growth which is symptomatic of this disease. (Source: Agrios 1978. Reprinted by permission, copyright Academic Press.)

broad-leaved plants, the disease is manifested by formation of a tumor growth, the crown gall. *Agrobacterium tumefaciens* is able to transform host plant cells into tumorous cells. In contrast to normal plant tissues, tumor tissues grow in the absence of added auxin and cytokinin and synthesize either octopine or nopaline, which are amino acid derivatives of arginine. Once established by the initial bacterial infection, the tumor continues to grow and manifest these characteristics even if viable *Agrobacterium* are eliminated. The tumor maintenance principle has been identified as a fragment of a

large, tumor-inducing plasmid called the Ti-plasmid. A fragment of this bacterial plasmid is transferred to the plant, where it is maintained in the tumor tissue. Strains of *Agrobacterium* lacking this plasmid fail to induce plant galls. A gene sequence on the Ti-plasmid codes for transformation of tryptophan, in two steps, to indole-3-acetamide and indole-3-acetic acid, thus making the transformed plant cells auxin independent (Thomashov et al. 1986).

The same plasmid fragment, which is responsible for tumor induction, also codes for the synthesis of the unusual amino acid derivatives octopine and

Table 4.5

Some diseases of plants caused by fungi

Organism	Disease	Organism	Disease
Myxomycota		Basidiomycota	
Plasmodiophora	Clubfoot of crucifers	*Ustilago*	Smut of corn and wheat
Polymyxa	Root disease of cereals	*Tilletia*	Stinking smut of wheat
Spongospora	Powdery scab of potatoes	*Sphacelotheca*	Loose smut of sorghum
		Urocystis	Smut of onion
Phycomycota		*Puccinia*	Rust of cereals
Olpidium	Root disease of various plants	*Cronartium*	Pine blister rust
Synchytrium	Black wart of potato	*Uromyces*	Rust of beans
Urophlyctis	Crown wart of alfalfa	*Exobasidium*	Stem galls of ornamentals
Physoderma	Brown spot of corn	*Fomes*	Heart rot of trees
Pythium	Seed decay, root rots	*Polyporus*	Stem rot of trees
Phytophthora	Blight of potato	*Typhula*	Blight of turf grasses
Plasmopara	Downy mildew of grapes	*Armillaria*	Root rots of trees
Rhizopus	Soft rot of fruits	*Marasmius*	Fairy ring of turf grasses
Ascomycota		Deuteromycota	
Taphrina	Peach leaf curl	*Phoma*	Black leg of crucifers
Erysiphe	Powdery mildew of grasses	*Colletotrichum*	Anthracnose of crops
Microsphaera	Powdery mildew of lilac	*Cylindrosporium*	Leaf spot of various plants
Podosphaera	Powdery mildew of apple	*Alternaria*	Leaf spot and blight of plants
Ceratocystis	Dutch elm disease	*Aspergillus*	Rot of seeds
*Diaporthe*t	Bean pod blight	*Botrytis*	Blight of various plants
Endothia	Chestnut blight	*Cladosporium*	Leaf mold of tomato
Claviceps	Ergot of rye	*Fusarium*	Root rot of many plants
Dibotryon	Black knot of cherries	*Helminthosporium*	Blight of cereals
Mycosphaerela	Leaf spots of trees	*Penicillium*	Blue mold rot of fruits
Ophiobolus	Take all of wheat	*Thielaviopsis*	Black root rot of tobacco
Venturia	Apple scab	*Verticillium*	Wilt of various plants
Diplocarpon	Black spot of roses	*Rhizoctonia*	Root rot of various plants
Lophodermium	Pine needle blight		
Sclerotinia	Soft rot of vegetables		

nopaline, also called "opines." The incorporation of this genetic information into the plant cell diverts some of the plant photosynthate into production of these opines.

The inducing *A. tumefaciens* strains have the unusual ability to use the respective opine as their sole source of carbon, energy, and nitrogen. Thus, through genetic recombination, many details of which remain to be elucidated, *A. tumefaciens* creates for itself a highly specific niche in which a constant supply of nutrients is assured. The unusual nature of the opine substrates largely excludes any competitors, ensuring successful competition of *A. tumefaciens* for a niche. As *A. tumefaciens* is closely related to *Rhizobium,* this natural genetic engineering system is currently under intense investigation for its potential ability to insert useful genetic information, such as nitrogen-fixation capability, into plant genomes.

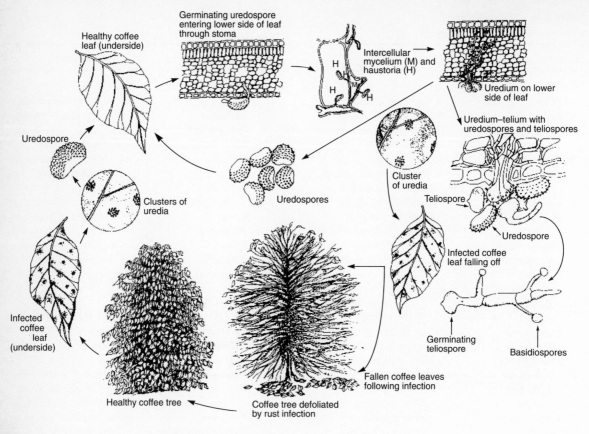

Figure 4.15

Life cycle of coffee rust caused by *Hemeleia vastatrix*. (Source: Agrios 1978. Reprinted by permission, copyright Academic Press.)

Fungal Diseases of Plants

Most plant diseases are caused by pathogenic fungi (Table 4.5). Perhaps the most important economic fungal diseases of plants are caused by the rusts and smuts. These fungi are Basidiomycota and have a complex life cycle. There are more than twenty thousand species of rust fungi and more than one thousand species of smut fungi. Rust fungi require two unrelated hosts for the completion of their normal life cycle (Figure 4.15). Important diseases caused by smut fungi include loose smut of oats, corn smut, bunt or stinking smut of wheat, and onion smut. Smut and rust fungi cause millions of dollars in crop damage every year.

Many fungi are well adapted to act as effective plant pathogens. The variety of spores produced by fungi permit aerial transmission between plants, and the production of resting spores allows plant pathogenic fungi to remain viable outside of host plants. Many plant pathogenic fungi exhibit a complicated life cycle, part of which is accomplished during plant infection and part of which is accomplished outside of host plants. Most plant pathogenic fungi spend part of their lives on host plants and part in the soil or on plant debris in the

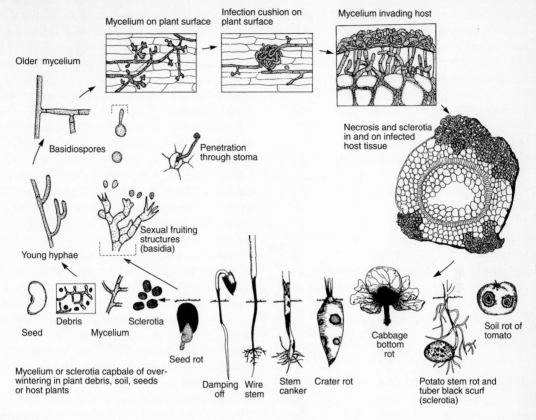

Figure 4.16

Life cycle of *Rhizoctonia solani* causing various diseases in plants. Invasion of plant occurs through roots. (Source: Agrios, 1978; reprinted by permission; copyright Academic Press.)

soil (Garrett 1970) (Figure 4.16). Some fungi, such as *Venturia,* pass part of their lives on the host as parasites and part on dead tissues on the ground as saprophytes. These fungi remain in intimate contact with host tissues, whether living or dead.

To invade susceptible plants and cause disease, all plant pathogens require favorable environmental conditions. Survival and infectivity of most plant pathogenic fungi depend on prevailing conditions of temperature and moisture. Germination of mycelia generally occurs only between −5°C and +45°C, with an adequate supply of moisture. Spores, though, can retain viability for long periods of time under environmental conditions that do not allow for germination.

Changes in environmental conditions may affect the pathogen, the host, or both. Most plant pathogens develop best and cause the most severe disease of plants during the warmer months of the year; during winter many are inactive. Some fungi, however, such as *Typhula* and *Fusarium,* which cause snow mold of cereals and turf grasses, respectively, thrive only in cool seasons or cool regions. In some cases, the optimum temperature for disease development is different than the optimal growth temperature of either the pathogen or the host. For example, the fungus *Thielaviopsis basicola,* which causes black root rot of tobacco, has an optimal growth temperature that is higher than the optimum temperature for the disease; in this case, the host is less able to resist the pathogen at temperatures of 17°C–23°C. In other cases, such as root rots of wheat and corn caused by the fungus *Gibberella zeae,* the optimum temperature for development of disease is higher than the optimal growth temperatures of both the pathogen and the wheat. Moisture, like temperature, influences the initiation and development of infectious plant diseases. Some plant pathogens are dispersed by rain droplets, initiating contact between the pathogen and the susceptible plant. The distribution of rainfall in some regions is closely correlated with the occurrence of plant diseases. Downy mildew of grapes and fire blight of pears, for example, are more severe when there is high rainfall or high relative humidity during the growing season. Soil pH also has a marked effect on the infectivity of soilborne plant pathogens. For example, *Plasmodiophora brassicae* causes club root of crucifers at a pH value of approximately 5.7, but the disease is completely checked at pH 7.8.

SUMMARY

Microorganisms interact with plant roots on the actual surface of the root (rhizoplane) and within the region directly influenced by the root (rhizosphere). Interactions of microorganisms and plant roots are especially important in providing nutritional requirements for the plant and its associated microorganisms. The rhizosphere is a complex zone of interactions among microbial populations and between microorganisms and plants. Microbial populations within the rhizosphere are higher in numbers and physiologically different from those in root-free soil. The differences are due to substances released from plant roots that chemically modify the adjacent soil region. The presence of rhizosphere microorganisms also is important to the plants, enabling some plants to grow better than otherwise would be possible.

In addition to interacting with numerous microorganisms in the rhizosphere, the roots of many plants establish a mutualistic relationship with specific fungi, in which the fungus actually becomes an integral part of the root. Such mutualistic relationships are called mycorrhizae. There are three types of mycorrhizae: ecto-, endo-, and vesicular-arbuscular, each of which has a distinct morphology. Mycorrhizae enhance the uptake of mineral nutrients and enable plants to grow in habitats that otherwise might not provide sufficient nutrients.

The establishment of mutualistic relationships of certain plants with nitrogen-fixing bacteria, most notably the relationship between legumes and *Rhizobium,* is extremely important to the ecology of other organisms in the biosphere. Without these nitrogen-fixing activities or human intervention through the introduction of chemical fertilizers, the abilities of all organisms to produce proteins would be highly constrained. Plants capable of forming mutualistic associations for the fixation of atmospheric nitrogen grow well even in nitrogen-deficient soils and leave the soil enriched in nitrogen for the subsequent crop. The relationship of *Rhizobium* and leguminous plants results in the formation of nodules within which the *Rhizobium* fix atmospheric nitrogen; the physiology of both the plant root and the *Rhizobium* are drastically altered by this association.

In addition to having root-microbe interactions, some microorganisms colonize the aerial structures of plants. Characteristic microbial populations develop on the phylloplane (leaf surface) of particular plants. Some such epiphytic microorganisms benefit the plant, in some cases protecting the plant against pathogens.

Plants are subject to many diseases caused by microorganisms, and plant pathogens have great ecologic and economic impact. The relationship between some plant pathogens and the host plant is greatly affected by the stationary nature of the plant, the periodicity of plant growth, the protective surfaces of the plant, and the nutritional state of the host plant. Plant pathogens must have an autonomous means of dispersal so they can reach plants, they must have some mechanism for survival between finding suitable host plants, and they must have some mechanism for entering the plant. Because obligately plant pathogenic microorganisms have a limited period of viability outside of host plant tissues, it is possible to control plant pathogens of agricultural crops by appropriate management procedures. These procedures include planting resistant species of crops and using crop rotation procedures. Plants have natural resistance mechanisms to defend against microbial pathogens, and such resistance can be genetically selected. Genetic breeding of crop plants allows for selection of resistant varieties. Planting resistant varieties removes available hosts for plant pathogenic microorganisms. Also, plant pathogens are normally specific for particular host plants. When crops are rotated, suitable plant hosts are periodically removed. Populations of plant pathogens that are high following infection of susceptible crops are greatly reduced due to the absence of suitable plant hosts, allowing for successful reestablishment of susceptible plant crops in successive years.

References & Suggested Readings

Agrios, G. N. 1978. *Plant Pathology.* Academic Press, New York.

Akkermans, A. D. L. 1978. Root nodule symbioses in non-leguminous N_2-fixing plants. In Y. R. Dommergues and S. V. Krupa (eds.). *Interactions Between Non-Pathogenic Soil Microorganisms and Plants.* Elsevier Publishing Co., Amsterdam, The Netherlands, pp. 335–372.

Albrecht, S .L., R. J. Maier, F. J. Hanns, S. A. Russell, D. W. Emericyh, and H. J. Evans. 1979. Hydrogenase in *Rhizobium japonicum* increases nitrogen fixation by nodulated soybeans. *Science* 203:1255–1257.

Alexander, M. 1985. Ecological constraints on nitrogen fixation in agricultural ecosystems. *Advances in Microbial Ecology* 8:163–183.

Amarger, N. 1984. Evaluation of competition in *Rhizobium* spp. In M. J. Klug and C. A. Reddy (eds.). *Current Perspectives in Microbial Ecology.* American Society for Microbiology, Washington, D.C., pp. 300–305.

Baker, K. F., and W. C. Snyder (eds.). 1965. *Ecology of Soil-Borne Plant Pathogens: Prelude to the Biological Control.* University of California Press, Berkeley.

Balandreau, J., and R. Knowles. 1978. The rhizosphere. In Y. R. Dommergues and S. V. Krupa (eds.). *Interactions Between Non-Pathogenic Soil Microorganisms and Plants.* Elsevier Publishing Co., Amsterdam, The Netherlands, pp. 243–268.

Barber, D. A. 1978. Nutrient uptake. In Y. R. Dommergues and S. V. Krupa (eds.). *Interactions Between Non-Pathogenic Soil Microorganisms and Plants.* Elsevier Publishing Co., Amsterdam, The Netherlands, pp. 131–162.

Barber, D. A., and K. B. Gunn. 1974. The effect of mechanical forces on the exudation of organic substances by the roots of cereal plants grown under sterile conditions. *New Phytologist* 73:39–45.

Barber, D. A., and J. M. Lynch. 1977. Microbial growth in the rhizosphere. *Soil Biology and Biochemistry* 9:305–308.

Bateman, D. F., and R. L. Miller. 1966. Pectic enzymes in tissue degradation. *Annual Review of Phytopathology* 4:119–146.

Bergersen, F. J. 1978. Physiology of legume symbiosis. In Y. R. Dommergues and S. V. Krupa (eds.). *Interactions Between Non-Pathogenic Soil Microorganisms and Plants.* Elsevier Publishing Co., Amsterdam, The Netherlands, pp. 304–334.

Berry, A. M. 1984. The actinorhizal infection process: Review of recent research. In M. J. Klug and C. A. Reddy (eds.). *Current Perspectives in Microbial Ecology.* American Society for Microbiology, Washington, D.C., pp. 222–229.

Blakeman, J. P. 1981. *Microbial Ecology of the Phylloplane.* Academic Press, London.

Bowen, G. D. 1980. Misconceptions, concepts and approaches in rhizosphere biology. In D. C. Ellwood, J. N. Hedger, M. J. Lathan, J. M. Lynch, and J. H. Slater (eds.). *Contemporary Microbial Ecology.* Academic Press, London, pp. 283–304.

Bowen, G. D. 1984. Development of vesicular-arbuscular mycorrhizae. In M. J. Klug and C. A. Reddy (eds.). *Current Perspectives in Microbial Ecology.* American Society for Microbiology, Washington, D.C., pp. 201–207.

Bowen, G. D., and A. D. Rovira. 1976. Microbial colonization of plant roots. *Annual Review of Phytopathology* 14:121–144.

Brill, W. J. 1975. Regulation and genetics of bacterial nitrogen fixation. *Annual Review of Microbiology* 29:109–129.

Brill, W. J. 1979. Nitrogen fixation: Basic to applied. *American Scientist* 67:458–466.

Brill, W. J. 1980. Biochemical genetics of nitrogen fixation. *Microbiological Reviews* 44:449–467.

Brill, W. J. 1981. Agricultural microbiology. *Scientific American* 245(3):198–215.

Brown, M. E. 1974. Seed and root bacterization. *Annual Review of Phytopathology* 12:181–197.

Brown, M. E. 1975. Rhizosphere micro-organisms—Opportunists, bandits or benefactors. In N. Walker (ed.). *Soil Microbiology*. John Wiley and Sons, New York, pp. 21–38.

Callaham, D., P. D. Tredici, and J. G. Torrey. 1978. Isolation and cultivation in vitro of the actinomycete causing root nodulation in *Comptonia*. *Science* 199:899–902.

Campbell, R. E. 1977. *Microbial Ecology*. Blackwell Scientific, Oxford, England.

Campbell, R. 1985. *Plant Microbiology*. Edward Arnold Publishers, London.

Campbell, R., and A. D. Rovira. 1973. The study of the rhizosphere by scanning electron microscopy. *Soil Biology and Biochemistry* 5:747–752.

Capone, D. G., and B. F. Taylor. 1980. N_2 fixation in the rhizosphere of *Thalassia testudinum*. *Canadian Journal of Microbiology* 26:998–1005.

Centifanto, Y. M., and W. S. Silver. 1964. Leaf nodule symbiosis. I. Endophyte of *Psychotria bacteriophila*. *Journal of Bacteriology* 88:776–781.

Chen, T. A., and C. H. Liao. 1975. Corn stunt *Spiroplasma*: Isolation, cultivation and proof of pathogenicity. *Science* 188:1015–1017.

Chiariello, N., J. C. Hickman, and H. A. Mooney. 1982. Endomycorrhizal role for interspecific transfer of phosphorus in a community of annual plants. *Science* 217:941–943.

Christensen, M. J., G. C. M. Latch, and B. A. Tappen. 1991. Variation within isolates of *Acremonium* endophytes from perennial rye grasses. *Mycological Research* 95:918–923.

Collins, M. A. 1976. Colonization of leaves by phylloplane saprophytes and their interactions in this environment. In C. H. Dickinson and T. F. Preece (eds.). *Microbiology of*

Aerial Plant Surfaces. Academic Press, London, pp. 401–418.

Cooke, R. 1977. *Biology of Symbiotic Fungi*. John Wiley and Sons, New York.

Crosse, J. E. 1968. Plant pathogenic bacteria in soil. In T. R. G. Gray and D. Parkinson (eds.). *The Ecology of Soil Bacteria*. University of Toronto Press, Toronto, Canada, pp. 552–572.

Daft, M. J., and E. Hacskaylo. 1977. Growth of endomycorrhizal and nonmycorrhizal red maple seedings in sand and anthracite spoil. *Forest Science* 23:207–216.

Dalton, H., and L. E. Mortenson. 1972. Dinitrogen (N_2) fixation (with a biochemical emphasis). *Bacteriological Reviews* 36:231–260.

Darbyshire, J. F., and M. R. Greaves. 1967. Protozoa and bacteria in the rhizosphere of *Sinapsis alba L.*, *Trifolium repens L.*, and *Lolium perenne L. Canadian Journal of Microbiology* 13:1057–1068.

Darbyshire, J. F., and M. R. Greaves. 1970. An improved method for the study of the interrelationships of soil microorganisms and plant roots. *Journal of Soil Biology Biochemistry* 2:166–171.

Davenport, R. R. 1976. Distribution of yeasts and yeast-like organisms on aerial surfaces of developing apples and grapes. In C. H. Dickinson and T. F. Preece (eds.). *Microbiology of Aerial Plant Surfaces*. Academic Press, London, pp. 325–360.

Dazzo, F. B. 1982. Leguminous root nodules. In R. G. Burns and J. H. Slater (eds.). *Experimental Microbial Ecology*. Blackwell Scientific Publishers, Oxford, England, pp. 431–436.

Dazzo, F. B., and D. H. Hubbell. 1975. Cross-reacting antigens and lectin as determinants of host specificity in the *Rhizobium*-clover association. *Applied Microbiology* 30:1017–1033.

Dazzo, F. B. ,and W. J. Brill. 1979. Bacterial polysaccharide which binds *Rhizobium trifolii* to clover root hairs. *Journal of Bacteriology* 137:1362–1373.

Deacon, J. W. 1983. *Microbial Control of Plant Pests and Disease*. Aspects of Microbiology No. 7., American Society for Microbiology, Washington, D.C.

Dennis, C. 1976. The microflora on the surface of soft fruit. In C. H. Dickinson and T. F. Preece (eds.). *Microbiology of Aerial Plant Surfaces*. Academic Press, London, pp. 419–432.

Deverall, B. J. 1977. *Defense Mechanisms of Plants*. Cambridge University Press, Cambridge, England.

Dickinson, C. H. 1976. Fungi on the aerial surfaces of higher plants. In C. H. Dickinson and T. F. Preece (eds.). *Microbiology of Aerial Plant Surfaces*. Academic Press, London, pp. 293–325.

Dickinson, C. H. 1982. The phylloplane and other aerial plant surfaces. In R. G. Burns and J. H. Slater (eds.). *Experimental Microbial Ecology*. Blackwell Scientific Publishers, Oxford, England, pp. 412–430.

Dickinson, C. H., and T. F. Preece (eds.). 1976. *Microbiology of Aerial Plant Surfaces*. Academic Press, London.

Dickinson, C. H., and J. A. Lucas. 1982. *Plant Pathology and Plant Pathogens*. Blackwell Scientific Publications, Oxford, England.

Diener, T. O. 1979. *Viroids and Viroid Diseases*. Wiley-Interscience, New York.

Dixon, R. O. D. 1969. Rhizobia (with particular reference to relationships with host plants). *Annual Reviews of Microbiology*. 23:137–158.

Dommergues, Y. R., and S. V. Krupa. 1978. *Interactions Between Non-Pathogenic Soil Microorganisms and Plants*. Elsevier Publishing Co., Amsterdam, The Netherlands.

Drew, M. C., and J. M. Lynch. 1980. Soil anaerobiosis, microorganisms and root function. *Annual Review of Phytopathology* 18:37–67.

Dreyfus, B. L., D. Alazard, and Y. R. Dommergues. 1984. Stem nodulating rhizobia. In M. J. Klug and C. A. Reddy (eds.). *Current Perspectives in Microbial Ecology*. American Society for Microbiology, Washington, DC., pp. 161–169.

Duell, R. W., and G. R. Peacock. 1985. Rhizosheaths on mesophytic grasses. *Crop Science* 25:880–883.

Esau, K. 1968. *Viruses in Plant Hosts: Form, Distribution, and Pathogenic Effects*. University of Wisconsin Press, Madison.

Evans, H. J., and L. Barber. 1977. Biological nitrogen fixation for food and fiber production. *Science* 197:332–339.

Fahrareus, G., and H. Ljunggren. 1968. Pre-infection phases of the legume symbiosis. In T. R. G. Gray and D. Parkinson (eds.). *The Ecology of Soil Bacteria*. University of Toronto Press, Toronto, Canada, pp. 396–421.

Fitter, A. H. 1985. *Ecological Interactions in the Soil Environment: Plants, Microbes and Animals*. Blackwell Scientific Publications, Oxford, England.

Fletcher, N. J. 1976. Bacterial symbiosis in the leaf nodules of Myrsinaceae and Rubiaceae. In C. H. Dickinson and T. F. Preece (eds.). *Microbiology of Aerial Plant Surfaces*. Academic Press, London, pp. 465–486.

Frankenberger, W. T., Jr., and M. Poth. 1987. Biosynthesis of indole-3-acetic acid by the pine ectomycorrhizal fungus *Pisolithus tinctorius. Applied* and *Environmental Microbiology* 53:2908–2913.

Friend, J., and D. R. Threlfall. 1976, *Biochemical Aspects of Plant-Parasite Relationships*. Academic Press, London.

Garrett, S. D. 1970. *Pathogenic Root-Infecting Fungi*. Cambridge University Press, Cambridge, England.

Goodfellow, M., B. Austin, and D. Dawson. 1976a. Classification and identification of phylloplane bacteria using numerical taxonomy. In C. H. Dickinson and T. F. Preece (eds.). *Microbiology of Aerial Plant Surfaces*. Academic Press, London, pp. 275–292.

Goodfellow, M., B. Austin, and C. H. Dickinson. 1976b. Numerical taxonomy of some yellow-pigmented bacteria isolated from plants. *Journal of General Microbiology* 97:219–233.

Goodfellow, M., B. Austin, and C. H. Dickinson. 1978. Numerical taxonomy of phylloplane bacteria isolated from *Lolium perenne. Journal of General Microbiology* 104:139–155.

Gray, T. R. G., and D. Parkinson (eds.). 1968. *The Ecology of Soil Bacteria*. University of Toronto Press, Toronto, Canada.

Grossbard, E. 1971. The utilization and translocation by micro-organisms of carbon-14 derived from decomposition of plant residues in soil. *Journal of General Microbiology* 60:339–348.

Hamer, J. E., R. J. Howard, F. G. Chumley, and B. Valent. 1988. A mechanism for surface attachment in spores of a plant pathogenic fungus. *Science* 239:288–290.

Hanson, R. B. 1977. Nitrogen fixation (acetylene reduction) in a salt marsh amended with sewage sludge and organic carbon and nitrogen compounds. *Applied and Environmental Microbiology* 33:846–852.

Harley, J. L., and R. S. Russell. 1979. *The Soil-Root Interface*. Academic Press, London.

Harley, J. L., and S. E. Smith. 1983. *Mycorrhizal Symbiosis*. Academic Press, London.

Harris, E., and K. Maramorosch. 1981. *Pathogens, Vectors, and Plant Diseases*. Academic Press, New York.

Hartley, J. L. 1965. Mycorrhiza. In K. F. Baker and W. C. Snyder (eds.). *Ecology of Soil-Borne Plant Pathogens*. University of California Press, Berkeley, pp. 218–229.

Hayman, D. S. 1978. Endomycorrhizae. In Y. R. Dommergues and S. V. Krupa (eds.). *Interactions Between*

Non-Pathogenic Soil Microorganisms and Plants. Elsevier Publishing Co., Amsterdam, The Netherlands, pp. 401–442.

Henry, S. M. (ed.). 1966. *Symbiosis. Vol. 1: Association of Microorganisms, Plants and Marine Organisms.* Academic Press, New York.

Hoch, H. C., R. C. Staples, B. Whitehead, J. Comeau, and E. D. Wolf. 1987. Signaling for growth orientation and cell differentiation by surface topography in *Uromyces. Science* 235:1639–1662.

Hubbell, D. H. 1981. Legume infection by *Rhizobium*: A conceptual approach. *BioScience* 31:832–837.

Jensen, H. L., and A. L. Hansen. 1968. Observations on host plant relations in root nodule bacteria on the Lotus-Anthyllis and the Lupinus-Ornithopus groups. *Acta Agriculturae Scandinavia* (Copenhagen) 18:135–142.

Johnson, M. C., D. L. Dahlman, M. R. Siegel, L. P. Busch, C. M. Latch, D. A. Potter, and D. R. Varney. 1985. Insect-feeding deterrents in endophyte-infected tall fescue. *Applied and Environmental Microbiology* 49:568–571.

Jones, K. 1976. Nitrogen fixing bacteria in the canopy of conifers in a temperature forest. In C. H. Dickinson and T. F. Preece (eds.). *Microbiology of Aerial Plant Surfaces.* Academic Press, London, pp. 451–464.

Jordan, C. F. 1982. Amazon rain forests. *American Scientist* 70:394–401.

Joshi, M. M., and J. P. Hollis. 1977. Interaction of *Beggiatoa* and rice plant: Detoxification of hydrogen sulfide in the rice rhizosphere. *Science* 195:179–180.

Kosuge, T., and E. W. Nester (eds.). 1984. *Plant-Microbe Interactions.* Macmillan, New York.

Kucey, R. M. N. 1987. Increased phosphorus uptake by wheat and field beans inoculated with phosphorus-solubilizing *Penicillium bilaji* strain and with vesicular-arbuscular mycorrhizal fungi. *Applied and Environmental Microbiology* 53:2699–2703.

Lamm, R. B., and C. A. Neyra. 1981. Characterization and cyst production of azospirilla isolated from selected grasses growing in New Jersey and New York. *Canadian Journal of Microbiology* 27:1320–1325.

Last, F. T., and D. Price. 1969. Yeasts associated with living plants and their environs. In A. H. Rose and J. S. Harrison (eds.). *The Yeasts.* Academic Press, New York, pp. 183–218.

Lindow, S. E., D. C. Arny, and C. D. Upper. 1978. Distribution of ice nucleation-active bacteria on plants in nature. *Applied and Environmental Microbiology* 36:831–838.

Lindow, S. E., D. C. Arny, and C. D. Upper. 1982a. Bacterial ice nucleation: A factor in frost injury to plants. *Plant Physiology* 70:1084–1089.

Lindow, S. E., S. S. Hiteno, W. R. Barcket, D. C. Arny, and C. D. Upper. 1982b. Relationship between ice nucleation, frequency of bacteria, and frost injury. *Plant Physiology* 70:1090–1093.

Lindow, S. E., and J. H. Connell. 1984. Reduction of frost injury to almond by control of ice nucleation-active bacteria. *Journal of the American Horticultural Society* 109:48–53.

Lippincott, J. A., and B. Lippincott. 1975. The genus *Agrobacterium* and plant tumorgenesis. *Annual Reviews of Microbiology* 29:377–406.

Lippincott, J. A., B. B. Lippincott, and J. J. Scott. 1984. Adherence and host recognition in *Agrobacterium* infection. In M. J. Klug and C. A. Reddy (eds.). *Current Perspectives in Microbial Ecology.* American Society for Microbiology, Washington, D.C., pp. 230–236.

Long, S. R. 1989. Rhizobium-legume nodulation: Life together in the underground. *Cell* 56:203-214.

Lynch, J. M. 1976. Products of soil microorganisms in relation to plant growth. *CRC Critical Reviews in Microbiology* 5:67–107.

Lynch, J. M. 1982a. The rhizosphere. In R. G. Burns and J. H. Slater (eds.). *Experimental Microbial Ecology.* Blackwell Scientific Publishers, Oxford, England, pp. 395–411.

Lynch, J. M. 1982b. *Soil Biotechnology: Microbiological Factors in Crop Productivity.* Blackwell Scientific Publications, Oxford, England.

Maas Geesteranus, H. P. (ed.). 1972. *Proceedings of the Third International Conference on Plant Pathogenic Bacteria.* University of Toronto Press, Toronto, Canada.

Maiti, I. B., and P. E. Kolattukudy. 1979. Prevention of fungal infection of plants by specific inhibition of cutinase. *Science* 205:507–508.

Marks, G. C., and T. T. Kozlowski. 1973. *Ectomycorrhizae—Their Ecology and Physiology.* Academic Press, New York.

Marx, D. H., and S. V. Krupa. 1978. Ectomycorrhizae. In Y. R. Dommergues and S. V. Krupa (eds.). *Interactions Between Non-Pathogenic Soil Microorganisms and Plants.* Elsevier Publishing Co., Amsterdam, The Netherlands, pp. 373–400.

Marx, H. M., W. C. Bryan, and C. E. Cordell. 1977. Survival and growth of pine seedlings with *Pisolithus ecto-*

mycorrhizae after two years on reforestation sites in North Carolina and Florida. *Forest Science* 23:363–373.

Morris, M., D. E. Eveleigh, S. C. Riggs, and W. N. Tiffney, Jr. 1974. Nitrogen fixation in the bayberry *(Myrica pennsylvania)* and its role in coastal succession. *American Journal of Botany* 61:867–870.

Mosse, B. 1973. Advances in the study of vesicular-arbuscular mycorrhiza. *Annual Review of Phytopathology* 11:171–196.

Nap, J. P., and T. Bisseling. 1990. Developmental biology of a plant-prokaryote symbiosis: the legume root nodule. *Science* 250:948-954.

Neal, J. L., Jr., and W. B. Bollen. 1964. Rhizosphere microflora associated with mycorrhizae of Douglas fir. *Canadian Journal of Microbiology* 10:259–265.

Nester, E. W., and A. Montoya. 1979. Crown gall: a natural case of genetic engineering. *ASM News* 45:283–286.

Newman, E. I. 1978. Root micro-organisms: their significance in the ecosystem. *Biological Reviews* 53:511–554.

Newman E. I., and A. Watson. 1977. Microbial abundance in the rhizosphere: A computer model. *Plant and Soil* 48:17–56.

Nutman, P. S., M. Dye, and P. E. Davis. 1978. The ecology of *Rhizobium*. In M. W. Loutit and J. A. R. Miles (eds.). *Microbial Ecology*. Springer-Verlag, Berlin, pp. 404–410.

Nye, P. H., and P. B. Tinker. 1977. *Solute Movement in the Soil-Root System*. Blackwell Scientific Publications, Oxford, England.

Patriquin, D., and R. Knowles. 1972. Nitrogen fixation in the rhizosphere of marine angiosperms. *Marine Biology* 16:49–58.

Patriquin, D. G., and C. R. McClung. 1978. Nitrogen accretion, and the nature and possible significance of the N_2-fixation (acetylene reduction) in Nova Scotian *Spartina alterniflora* stands. *Marine Biology* 47:227–242.

Peters, G. A., and B. C. Mayne. 1974. The *Azolla, Anabaena azollae* relationship. I. Initial characterization of the association. *Plant Physiology* 53:813–819.

Plumb, R. T., and J. M. Thresh. 1983. *Plant Virus Epidemiology*. Blackwell Scientific Publications, Oxford, England.

Postgate, J. R. 1982. *The Fundamentals of Nitrogen Fixation*. Cambridge University Press, Cambridge, England.

Powell, C. L. 1982. Mycorrhizae. In R. G. Burns and J. H. Slater (eds.). *Experimental Microbial Ecology*. Blackwell Scientific Publishers, Oxford, England, pp. 447–471.

Preece, T. F., and C. H. Dickinson (eds.). 1971. *Ecology of Leaf Surface Micro-Organisms*. Academic Press, London.

Raskin, I., and H. Kende. 1985. Mechanisms of aeration in rice. *Science* 228:327–329.

Ride, J. P. 1978. The role of cell wall alterations in resistance to fungi. *Annals of Applied Biology* 89:302–306.

Ridge, R. W., and B. G. Rolfe. 1985. *Rhizobium* sp. degradation of legume root hair cell wall at the site of infection thread origin. *Applied and Environmental Microbiology* 50:717–720.

Robinson, R. A. 1976. *Plant Pathosystems*. Springer-Verlag, Berlin.

Rovira, A. D. 1969. Plant root exudates. *Botanical Review* 35:35–57.

Rovira, A. D., and R. Campbell. 1974. Scanning electron microscopy of microorganisms on roots of wheat. *Microbial Ecology* 1:15–23.

Ruehle, J. L., and D. H. Marx. 1979. Fiber, food, fuel and fungal symbionts. *Science* 206:419–422.

Russell, R. S. 1977. *Plant Root Systems: Their Function and Interaction with the Soil*. McGraw-Hill, London.

Sanders, F. E., B. Mosse, and P. B. Tinker. 1975. *Endomycorrhizas*. Academic Press, London.

Schmidt, E. L. 1978. Ecology of the legume root nodule bacteria. In Y. R. Dommergues and S. V. Krupa (eds.). *Interactions Between Non-Pathogenic Soil Microorganisms and Plants*. Elsevier Publishing Co., Amsterdam, The Netherlands, pp. 269–304.

Schmidt, E. L. 1979. Initiation of plant root-microbe interactions. *Annual Reviews of Microbiology* 33:355–378.

Shardl, D. J., and D. O. Hall. 1988. The *Azolla-Anabaena* association: Historical perspective, symbiosis and energy metabolism. *Botanical Review* 54:353–386.

Shardl, D. J., J. -S. Liu, J.F. White, Jr., R. A. Finkel, Z. Au, and M. R. Siegel. 1992. Molecular phylogenetic relationships of non-pathogenic grass mycosymbionts and clavicipitaceous plant pathogens. *Plant Systematics and Evolution*. 179:In press.

Sheffer, R. P., and R. S. Livingston. 1984. Host-selective toxins and their role in plant disease. *Science* 223:17–21.

Siver, W. S., Y. M. Centifanto, and D. J. D. Nicholas. 1963. Nitrogen fixation by the leaf nodule endophyte of *Psychotria bacteriophila*. *Nature* (London) 199:396–397.

Smith, D. W. 1980. An evaluation of marsh nitrogen fixation. In V. S. Kennedy (ed.). *Estuarine Perspectives*. Academic Press, New York, pp. 135–142.

Smith, D. W. 1982. Nitrogen fixation. In R. G. Burns and J. H. Slater (eds.). *Experimental Microbial Ecology*. Blackwell Scientific Publishers, Oxford, England, pp. 212–220.

Smith, G. M. 1938. *Cryptogamic Botany*. Vol. II, Bryophytes and Pteriodophytes. McGraw-Hill, New York.

Smith, M. S., and J. M. Tiedje. 1979. The effect of roots on soil denitrification. *Journal of Soil Science Society America* 43:951–955.

Smith, R. L., J. H. Bouton, S. C. Schank, K. H. Quesenberry, M. E. Tyler, J. R. Milam, M. H. Gaskins, and R. C. Little. 1976. Nitrogen fixation in grasses inoculated with *Spirillum lipoferum*. *Science* 193:1003–1005.

Smith, S. E., and M. J. Daft. 1978. The effect of mycorrhizas on the phosphate content, nitrogen fixation and growth of *Medicago sativa*. In M. W. Loutit and J. A. R. Miles (eds.). *Microbial Ecology*. Springer-Verlag, Berlin, pp. 314–319.

Society for Experimental Biology. 1975. *Symbiosis*. Cambridge University Press, Cambridge, England.

Solheim, B. 1984. Infection process in the *Rhizobium-legume symbiosis*. In M. J. Klug and C. A. Reddy (eds.). *Current Perspectives in Microbial Ecology*. American Society for Microbiology, Washington, D.C., pp. 217–221.

Sprent, J. I., and P. Sprent. 1990. *Nitrogen Fixing Organisms—Pure and Applied Aspects*. Chapman and Hall, London.

Stevens, R. B. 1974. *Plant Disease*. The Ronald Press Co., New York.

Strange, R. N. 1972. Plants under attack. *Science Progress* (Oxford) 60:365–385.

Strange, R. N. 1982. Pathogenic interactions of microbes with plants. In R. G. Burns and J. H. Slater (eds.). *Experimental Microbial Ecology*. Blackwell Scientific Publishers, Oxford, England, pp. 472–489.

Strange, R. N. 1984. Molecular basis for the specificity of plant pathogenic microorganisms for their hosts. In M. J. Klug and C. A. Reddy (eds.). *Current Perspectives in Microbial Ecology*. American Society for Microbiology, Washington, D.C., pp. 208–216.

Swaminathan, M. S. 1982. Biotechnology research and third world agriculture. *Science* 218:967–972.

Tatum, L. A. 1971. The southern corn leaf blight epidemic. *Science* 171:1113–1116.

Ten Houten, J. G. 1974. Plant pathology: Changing agricultural methods and human society. *Annual Review of Phytopathology* 12:1–11.

Thomashov, M. F., S. Hugly, W. G. Buchholz, and L. S. Thomashov. 1986. Molecular basis for the auxin-independent phenotype of crown gall tumor tissues. *Science* 231:616–618.

Tinker, P. B. H. 1975. Effects of vesicular-arbuscular mycorrhizas on higher plants. In *Symbiosis* (Symposia of the Society for Experimental Biology). Cambridge University Press, Cambridge, England, pp. 325–350.

Vancura, V., and A. Hovaldik. 1965. Root exudates of some vegetables. *Plant and Soil* 22:21–32.

Walker, N. 1975. *Soil Microbiology*. John Wiley and Sons, New York.

Wallstein, L. H., M. L. Bruening, and W. B. Bollen. 1979. Nitrogen fixation associated with sand grain root sheaths (rhizosheaths) of certain xeric grasses. *Physiologia Plantarum* 46:1–4.

Wheeler, H. 1975. *Plant Pathogenesis*. Springer-Verlag, New York.

Williamson, D. L., and R. F. Whitcomb. 1975. Plant mycoplasmas: A cultivable *Spiroplasma* causes corn stunt disease. *Science* 188:1018–1020.

Wills, B. J., and A. L. J. Cole. 1978. The use of mycorrhizal fungi for improving establishment and growth of *Pinus* species used for high-altitude revegetation. In M. W. Loutit and J. A. R. Miles (eds.). *Microbial Ecology*. Springer-Verlag, Berlin, pp. 320–323.

Woldendorp, J. W. 1978. The rhizosphere as part of the plant-soil system. In *Structure and Functioning of Plant Population*. Verhandeligen der Koninklijke, Nederlandse Akademie van Wetsenschappen, Afdeling Natuurkunde, Twede Reeks, deel 70.

Wong, T.-Y., L. Graham, E. O'Hara, and R. J. Maier. 1986. Enrichment for hydrogen-oxidizing *Acinetobacter* species in the rhizosphere of hydrogen-evolving soybean root nodules. *Applied and Environmental Microbiology* 52:1008–1013.

Wullstein, L. H., M. L. Bruening, and W. B. Bollen. 1979. Nitrogen fixation associated with sand grain root sheaths (rhizosheaths) of certain xeric grasses. *Physiologia Plantarum* 46:1–4.

Chpt 1
Chpt 8
Chpt 2
Chpt 7
Chpt 6
Chpt 9   All
Chpt 11   323-332
_____
Chpt 3
Chpt 4
Chpt 10   289-292
Chpt 11   All
Chpt 12
Chpt 13
Chpt 14
Chpt 5

...ns with Animals

...s between ...ease and it ...introduc- ...he popular ...llpox and ...societies, ...We should ...he interac- ...result in animal and human disease, microorganisms enter into various relationships with animals that are beneficial to animal populations. Various animals feed directly on microbial populations, and many animals gain nutritional benefit from microorganisms. Indeed, without microbial contributions to their nutrition, the majority of animals could not survive. Humans depend upon microorganisms to provide vitamins and other metabolites. There are numerous beneficial relationships that benefit both the animal and microbial partners and without which neither the microbial nor the animal populations could survive. In this chapter we will examine both the positive and negative interactions between microbes and animals.

MICROBIAL CONTRIBUTIONS TO ANIMAL NUTRITION

Predation on Microorganisms by Animals

Predatory animals are usually unable to survive on disproportionately small prey because they would have to expend more energy capturing such prey than they would derive from consumption. Nevertheless, many invertebrate animals can satisfy part or all of their food requirements by preying on microorganisms 10^5 to 10^7 times smaller in biomass than themselves. This feat is accomplished by two feeding strategies: grazing on microbial aggregations or filter feeding.

Grazing   A common feeding strategy of aquatic Gastropoda (snails), Echinodermata (sea urchins), and Patellidae (limpets), for example, is to scrape and ingest the microbial crust from submerged surfaces where the microbial populations are able to reach high densities because of the physical absorption of dissolved nutrients on these surfaces (Marshall 1980). Scraping mouth organs, such as the radula of snails and the five-toothed "lantern of Aristotle" of sea urchins, are adaptations for this process. The size difference between predator and prey becomes relatively unimportant in this feeding process, called grazing, because the predator pursues coherent masses of millions of microbes rather than individual prey.

The microorganisms associated with fecal pellets are an important source of food for many aquatic and some terrestrial animals (Turner and Ferrante 1979). Digestion of food during passage through the alimentary canal is usually incomplete. Sugars, lipids, and proteins are preferentially digested and absorbed, while cellulosic and other fibrous material remains largely intact. The excreted fecal pellets are further decomposed by the remnants of intestinal microorganisms and additional microorganisms from the environment. In this process, recalcitrant plant polymers are solubilized and converted in part to microbial biomass. Reingestion of the fecal pellet by the same animal or other animal populations allows a more complete utilization of the food resource. In addition to converting indigestible plant fiber to more digestible microbial biomass, the microorganisms often supply critical vitamins that otherwise would be absent from the diet.

In the terrestrial environment, various soil microarthropods and certain rodents and lagomorphs, including rabbits, are coprophagous and regularly reingest some of their own fecal material. In aquatic environments, some invertebrates, such as snails, graze on the microbial populations developing on fecal pellets deposited by other animals (Frankenberg and Smith 1967). Various members of the marine meiofauna secrete slime trails, which bacterial, fungal, and algal populations colonize. These slime trails provide nutrients for the growth of the microbial populations. The microorganisms adhere to the slimes, which act as mucous traps. The animals then retrace their tracks and graze on the microbial populations that have grown on and are now entrapped in the mucous slime trails.

Many marine and freshwater invertebrates consume microbial populations growing on detrital particles (Fenchel and Jørgensen 1977; Turner and Ferrante 1979). These animals rely upon microorganisms to concentrate nutrients and to convert plant polymers in detrital particles that have a low N:C ratio into proteinaceous microbial biomass with a high N:C ratio, thereby enhancing the nutritional value of the detritus. Many detritus feeders digest mainly the microbial biomass associated with the detritus particle; microorganisms recolonize the undigested plant polymer, and then detritus feeders reingest the detrital particle with its new microbial biomass. In soil, the earthworm similarly digests mainly the microorganisms associated with ingested soil particles. In one rather unusual example, the desert snail *Euchondrus desertorum* grazes on lichens that grow under the surfaces of the rocks (endolithic lichens). In order to get to the lichens, the snails grind up approximately 1 metric ton of limestone per hectare per year, contributing substantially to the rock weathering process (Shachak et al. 1987).

Some invertebrates graze selectively on particular microorganisms. The ingestion of *Sphaerocystis* by *Daphnia,* for example, actually enhances the growth of the alga (Porter 1976). Most of the ingested *Sphaerocystis* cells survive passage through the gut of *Daphnia,* whereas other algal species, such as *Chlamydomonas,* are assimilated by *Daphnia* with greater efficiency. The *Sphaerocystis* contributes little to the nutrition of *Daphnia,* but within the gut of *Daphnia, Sphaerocystis* obtains nutrients such as phosphorus from the remains of other algal species. In this case, the digestive activities of the animal enhance the growth of *Sphaerocystis;* in the absence of zooplankton grazers, cyanobacterial populations replace the *Sphaerocystis* populations.

Filter Feeding Many sessile benthic and planktonic invertebrates exhibit a different feeding strategy, called filter feeding, to exploit suspended planktonic microbial prey (Jørgensen 1966; Tait and DeSanto 1972; Fenchel and Jørgensen 1977). These animals remain more or less stationary and filter the prey out of suspension; this strategy is energetically more advantageous because the prey are minute and in a relatively homogeneous suspension. The animals maintain a flow of water using cilia and/or various modified organs, such as legs, antennae, tentacles, gills, and tails. The microorganisms are filtered from the water through gills, tentacles, and mucous nets. Gills may serve the dual purpose of securing both a food source and an oxygen supply. The filter-feeding activity of the zooplankton is sometimes referred to as grazing, to indicate that most of the prey organisms are primary producers, but secondary producers (bacteria), predators (protozoa), and nonliving detrital particles are also ingested in the process, and therefore it is more accurately called filter feeding.

Benthic filter-feeding invertebrates include the Porifera (sponges), Bryozoa (moss animals), sessile crustacea such as barnacles, tube-building polychaete worms, Lamellibranchiata (bivalves), Brachiopoda (lamp shells or goose barnacles), and Tunicata (sea squirts). These filter feeders ingest planktonic algae, free-swimming microorganisms, and detrital particles with attached microbial biomass. The filter-feeding mechanisms of bivalves and sea squirts are particularly efficient, enabling these animals to capture particles as small as viruses from water at high rates. Important planktonic filter-feeding microcrustacea include Cladocera, which inhabit mainly freshwater environments; Copepoda, which live in both marine and freshwater environments; and Euphausia ("krill"), found mainly in marine habitats. Rotifers, pelagic snails (Pteropoda), Larvacia (planktonic tunicates), and various invertebrate larval forms (trochophore, nauplius, and zoea larvae) are also planktonic filter feeders, ingesting microalgae as well as protozoa, bacteria, and smaller members of the zooplankton.

Cultivation of Microorganisms by Animals for Food and Nutrition

Most herbivorous animals are unable to digest the cellulosic parts of the plant materials they consume. They rely upon the enzymatic capabilities of microorganisms to degrade this material and to produce substances that they can assimilate (McBee 1971). These substances include monomeric biodegradation products of cellulose and microbial biomass. Thus,

coprophagy and ingestion of microbe-covered detritus, discussed earlier in this section, can either be viewed as animal predation on microbes or as a loosely organized synergistic effort in food digestion. Other close-knit mutualistic relationships in food digestion can take the form of intestinal symbionts or a directed external cultivation of microbial biomass for subsequent consumption. Several plant-eating insects actually cultivate pure cultures of microorganisms on plant tissues in a mutualistic relationship (Buchner 1960; Brooks 1963; Batra and Batra 1967). The protein-rich microbial biomass is used as the principal food source by the insect population. In turn, the microorganisms are dispersed by the insects and are provided with a habitat in which they can proliferate.

Various leaf-cutting ant populations maintain mutualistic relationships with fungi (Weber 1966, 1972; Batra and Batra 1967) (Figure 5.1). The ants supply leaf tissue for the microorganisms, disperse the microorganisms by inoculating segments of the leaves, and shield the cultivated fungus from competitors, maintaining a virtual monoculture that breaks down rapidly if the cultivating ants are removed. The ability of myrmicine ants to cultivate and maintain a single species of fungi in fungal gardens is essential for the maintenance of the ant population. The cultivation and pruning of fungi within these gardens greatly alters the morphology of the fungi (Figure 5.2). The ant *Acromyrmex disciger,* and possibly *Atta,* cultivate the fungi *Leucocoprinus* or *Agaricus;* the ant *Cyphomyrmex rimoseus* cultivates *Tyridomyces formicarum; Cyphomyrmex costatus* and *Myricocrypta buenzlii* cultivate a *Lepiota* species; and the ant *Apterostigma mayri* maintains a relationship with an *Auricularia* species. The attine ants are important because they are responsible for introducing large amounts of organic matter into the soil in tropical rainforests. The organic matter thus produced forms the basis for food web interactions in which several other animal populations participate.

It is intriguing to consider how the ants maintain these pure fungal cultures. Selective inhibitors produced by ants or by the microbial culture have been suggested but not demonstrated. In the case of the basidiomycete cultivated by *Atta* species, the fun-

Figure 5.1

Cultivation of fungal gardens by ants.
(A) Surface view of an ant colony of the fungus-growing ant *Atta* showing anthill. (B) Tunnel of fungus-growing ant showing queen and progeny cultivating fungi on leaves. (C) Leaf-cutting behavior of *Atta*.
(D) An ant gathering fungi with mandibles. (Source: Weber 1972. Reprinted by permission of American Society of Zoologists.)

gus is deficient in proteases and competes poorly with other fungi if not under cultivation by the ants (Martin 1970). *Atta* chews off sections of green leaves and carries them to a subterranean nest where they are macerated, mixed with saliva and fecal discharge (both of which contain proteases), and inoculated with fungal mycelia. In this case, complementary enzymatic activity and preemptive colonization maintain the pure culture rather than antibiosis. In populations of ants that are not leaf cutters, woody particles or other plant debris are brought to the ant nest and are inoculated with fungi. Regardless of the means by

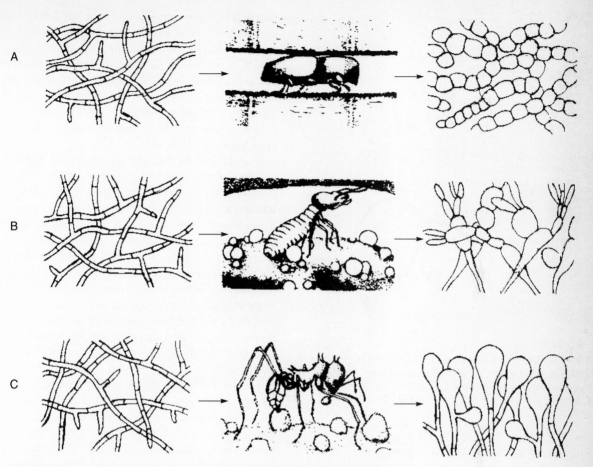

Figure 5.2
Appearance of uncultivated (left) and cultivated (right) fungal mycelia and the respective gardening insect (middle): (A) ambrosia beetle, (B) termite, and (C) attine ant. (Source: Batra and Batra 1967, The Fungus Gardens of Insects. Copyright 1967 Scientific American Inc. All rights reserved.)

which the fungi are cultivated, the ants later harvest a part of the fungal biomass and by-products and then ingest them. In addition to gaining the nutritional value of the fungal biomass itself, by feeding on the mycelia and the decomposing cellulosic substrate, the ants also acquire cellulase enzymes they are unable to produce themselves. These cellulases continue their depolymerizing activity in the gut of the ant, enabling it to digest cellulose with acquired enzymes (Martin 1979).

Various wood-inhabiting insects, such as ambrosia beetles, also maintain mutualistic relationships with fungal populations (Baker 1963; Batra and Batra 1967). Each species of ambrosia beetle is normally associated with only one species of ambrosia fungus. Various populations of ambrosia beetles maintain mutualistic relationships with populations of *Monilia*, *Ceratocystis*, *Cladosporium*, *Penicillium*, *Endomyces*, *Cephalosporium*, *Endomycopsis*, and

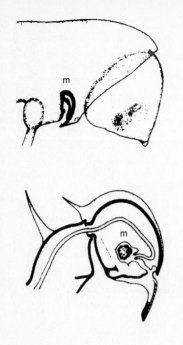

Figure 5.3

Mycetangia (m) of (A) the ambrosia beetle and (B) a fungus-growing ant. (Source: Batra and Batra 1967, The Fungus Gardens of Insects. Copyright 1967 Scientific American Inc. All rights reserved.)

several other fungal genera (Cooke 1977). Many ambrosia fungal species exhibit dimorphism between mycelial and yeastlike forms. The fungi are maintained and protected from desiccation within specialized organs, pocketlike invaginations known as mycangia or mycetangia (Figure 5.3). The mycetangia are present in only one sex of the ambrosia beetle. When an ambrosia beetle tunnels into wood, fungal spores are dislodged from the mycetangia and inoculated onto the wood surface along with secretions that provide nutrients for spore germination.

The growth of the cultivated fungi is highly dependent on temperature and moisture, requiring a wood moisture content in excess of 35%. The ambrosia beetles clean the passages of debris and feces and open and close the entrance to the hole in response to weather conditions, thereby maintaining favorable environmental growth conditions for the ambrosia

fungi. The development of what are essentially monocultures of ambrosia fungi within the tunnels appears to be due to antagonism of the ambrosia fungi toward opportunistic invading microbial populations and to secretions from the ambrosia beetles that may have selective antimicrobial properties. When ambrosia beetles abandon a series of tunnels, many other fungal species rapidly invade the habitat and overgrow the ambrosia fungi.

Ambrosia beetles are not capable of digesting cellulose themselves and depend upon the fungi to convert cellulose into protein-rich biomass. Some species, especially in their larval forms, are entirely dependent on ambrosia fungi as their food source. The mutualistic relationship between ambrosia beetles and their associated fungi is required for the survival of the beetle population, as the fungus provides the food required by the insect and its larvae. The fungal population also produces growth factors, such as vitamins, utilized by the beetles. Pupation in ambrosia beetles depends in part on production of ergosterol by the associated fungal population. Thus, the beetles supply a suitable habitat for the fungi consisting of wood fragments and fecal material within the humid atmosphere of the excavated tunnels, and the activities of the beetles maintain environmental conditions favorable for the growth of the ambrosia fungi.

Similar associations of insect and fungal populations have been found for bark-feeding beetles, ship timber worms, wood wasps, and gall midges (Batra and Batra 1967). In the first three cases, the insects physically excavate passages into wood substances, which they inoculate with fungal spores carried in exterior, pouchlike structures. The fungi grow in the wood structures, degrading plant materials, which then become available for ingestion by the insects. The fungi associated with bark beetles, blue stain fungi, have been shown to degrade sugars, starch, proteins, pectins, and, to a limited extent, cellulose. The larval galleries of the bark beetles are made in the phloem of the tree and usually result in rapid death of the tree. Fungi associated with ship timber worms ferment mainly xylem sap. In each case, the fungal biomass can serve as a food source for the insect populations, especially the larval forms. Some gall midges deposit their eggs, together with fungal spores, in leaves or buds. The fungi grow parasitically on the tissue of the developing gall, while the midge larva feeds, at least in part, on the fungal mycelium.

Various populations of termites maintain mutualistic relationships with external and internal microbial populations (Sands 1969). Some termites cultivate external fungal populations that contribute to their ability to live on wood. Without the associated fungal populations, these termite species are unable to survive on wood alone, and some termites can only use wood that has already been subjected to extensive fungal degradation. Similarly to the case of leaf-cutting ants, the fungal populations appear to contribute enzymes to the guts of the termites. In these cases, some cellulase enzymes are produced within the termite gut, but others are acquired from the ingestion of fungi growing within the termite nests; that is, the termites acquire digestive enzymes from externally grown fungal populations (Martin 1979).

Some higher termite populations cultivate species of the basidiomycete *Termitomyces,* much in the manner of the fungus-gardening ants. The termites actively gather and disseminate fungal spores to establish new nests. Lower termites maintain a mutualistic relationship with internal populations of protozoa, which are responsible for the degradation of cellulose and the production of metabolites that the termite can assimilate (Breznak 1975; Breznak and Pankratz 1977; Martin and Martin 1978; Yamin 1981) (Figure 5.4). These lower termites are not associated with external fungal gardens; higher termites that cultivate fungal populations lack internal cellulase-producing protozoan populations.

Protozoan and bacterial populations found within the guts of some termites and wood-eating cockroaches ferment cellulose anaerobically, producing carbon dioxide, hydrogen, and acetate. Some of the hydrogen and carbon dioxide is converted to methane, but this pathway, which is largely wasteful for the insect, involves only negligible amounts. Acetogenic bacteria convert most of the H_2 and CO_2 to acetate (Breznak and Switzer 1986). The acetate is absorbed through the wall of the ter-

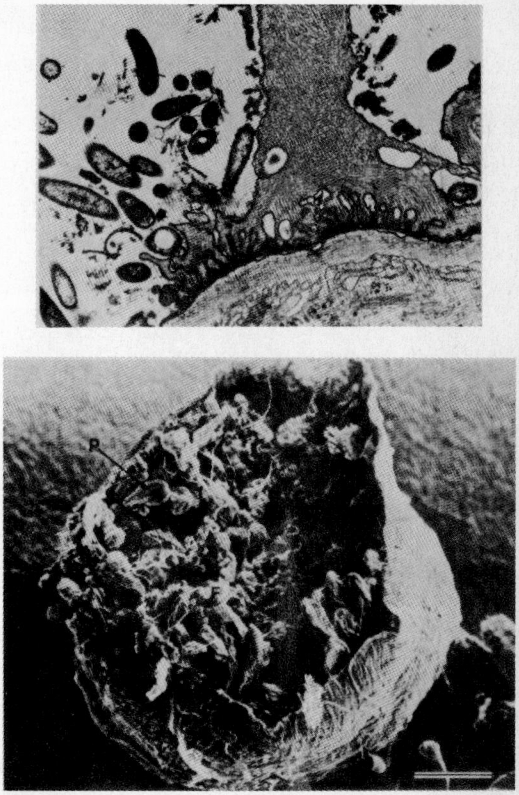

Figure 5.4

Electron micrographs (EM) of microorganisms in a termite gut. (A) Transmission EM of thin section showing presence of protozoa. (B) Scanning EM showing numerous protozoa (P) in the lumen of the gut. (Source: Breznak and Pankratz 1977. Reprinted by permission of American Society for Microbiology.)

mite hindgut and is oxidized aerobically, forming carbon dioxide and water.

Within their guts, termites also contain bacterial populations, some of which fix atmospheric nitrogen (Benemann 1973). *Enterobacter agglomerans* has been isolated from the guts of some termites and shown to be capable of nitrogen fixation under conditions of reduced oxygen tension (Potrikus and Breznak 1977). The nitrogen-fixing activities of *E. agglomerans* may be important in the nitrogen economy of some termite populations, especially

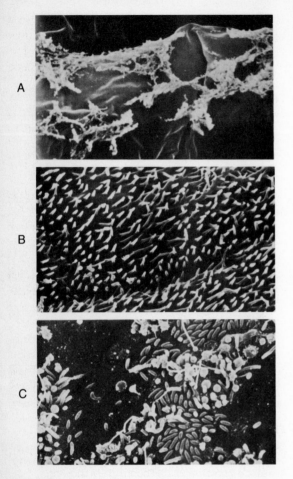

Figure 5.5
Scanning electron micrographs showing relationships between marine microorganisms and the wood-boring isopod *Limnoria tripuncata.* (A) Gut tissue showing lack of associated microorganisms. (B) Gut tissue showing villi and lack of microbial colonization. (C) External surface of the isopod showing presence of extensive microbial community. (Source: Sleeter et al. 1978. Reprinted by permission, copyright Springer-Verlag.)

during their developmental stages, because the cellulose-based diet of this termite is deficient in nitrogen.

During their early stages of life, bloodsucking species of insects nearly always feed on microorgan-

isms. Insects that live on such restricted diets also normally develop mutualistic relationships with microbial populations that are essential for their survival. The microbial populations, which are often maintained within mycetomes, supplement dietary deficiencies of the animal by producing growth factors. Removal of the microbial populations of mycetomes of lice, for example, results in the animal's failure to reproduce. Reproductive activity and growth can be restored if the animals are supplied with B vitamins and yeast extract.

Other animal populations exhibit less specific synergistic relationships with external microbial populations that contribute to food digestion. The marine wood-boring invertebrate *Limnoria,* for example, obtains some of its nutrients from marine fungi that grow in the tunnels it creates within wood pilings (Sleeter et al. 1978) (Figure 5.5). The fungi may provide growth factors and other metabolic products to *Limnoria. Limnoria* is capable of synthesizing its own cellulases, however, and is not dependent on cellulose digestion by fungi. As mentioned earlier, wood (cellulose) is a diet deficient in nitrogen. It is not surprising that some of wood-boring marine shipworms (*Teredo*) harbor nitrogen-fixing bacteria in their gut, and apparently this source of nitrogen enables them to grow at rapid rates (Carpenter and Culliney 1975).

Some plant-eating birds maintain intestinal populations of bacteria and fungi that produce cellulolytic enzymes (Henry 1967). These microbial populations are able to degrade cellulose within the intestinal tract of the bird, providing it with nutrients. Honey guides, which consume beeswax, contain populations of the bacterium *Micrococcus cerolyticus* and the yeast *Candida albicans* that are able to utilize the beeswax when supplied with cofactors from the bird. Birds can assimilate nutrients produced from the degradation of the beeswax.

Various fish and aquatic invertebrates contain microbial populations within their digestive tracts that contribute to food digestion (Trust et al. 1978; Sochard et al. 1979; Atlas et al. 1982). Amphipods, for example, contain high proportions of chitinase-producing *Vibrio* species; these bacterial populations partially degrade the chitin ingested by these animals, producing monomers that the animals can absorb and utilize. Some fish, such as catfish and carp, contain microbial populations that produce cellulase enzymes. The degradation of cellulose by bacterial populations within the fish gut produces metabolic products that the fish absorb.

Most warm-blooded animals, including humans, contain extremely complex microbial communities within their gastrointestinal tracts (Figure 5.6). In the lower intestine, each gram of feces contains approximately 10^{11} microorganisms, belonging to up to four hundred different species (Lee 1985). In the human intestine, the strict anaerobes belonging to the genera *Bacteroides, Fusobacterium, Bifidobacterium,* and *Eubacterium* are the most numerous, but no single

Figure 5.6

Scanning electron micrograph showing the yeast *Torulopsis pintolopesii* on the surface of the secreting epithelium in a monoassociated gnotobiotic mouse. *Torulopsis* is an indigenous yeast that colonizes the lining of the stomach of mice and rats following weaning. (Source: Savage 1978. Rreprinted by permission, copyright Springer-Verlag.)

species has a dominant role in this highly diverse microbial community. In some animals, such as pigs, the microbial populations of the gastrointestinal tract contribute to the nutrition of the animals by fermenting carbohydrates. There is some evidence that in older pigs microorganisms digest cellulose within the intestinal tract and that the animal can utilize the products of the cellulose degradation (Kenworthy 1973). Other microbial activities within the gastrointestinal tract, such as the degradation of amino acids, may be detrimental to the animal because of competition between the animal and the microbial populations for these nutrients.

In monogastric animals, the main contribution to digestion by intestinal microbial populations appears to be in the production of growth factors rather than in the production of partially degraded substrates. Although animals clearly absorb products derived from the microbial metabolism of ingested foods, it is not always clear whether any of these products are actually required. In some cases, microorganisms supply required vitamins; specific microorganisms synthesize vitamin K, for example, and germfree animals lacking appropriate microbial populations exhibit symptoms of vitamin deficiency (Luckey 1965). Besides their contributions to digestion and nutrition, normal gastrointestinal microbial populations, by their presence and preemptive colonization, constitute an important barrier to attack by intestinal pathogens. The high incidence of severe intestinal infections after prolonged antibiotic therapy and when germfree animals are exposed to normal nonsterilized food demonstrates this (Lee 1985).

Associations of Chemolithotrophs and Deep-Sea Hydrothermal Vent Animals

Investigations of deep-sea hydrothermal vent regions have revealed unusual communities containing high biomasses of various animals that appear to be fed by chemolithotrophic bacteria. Large tube worms and bivalves are prominently observed in these regions. The predominant part of the biomass observed at the warm deep-sea vents is generated by the symbiotic association of prokaryotic cells with the large (up to 2.5 meters in length) vestimentiferan tube worm *Riftia pachyptila* and a large clam, *Calpytogena magnifica* (Cavanaugh et al. 1981; Grassle 1985; Jannasch and Mottl 1985). *Riftia pachyptila*, which lacks both mouth and gut, depends upon its association with chemolithotrophic sulfur-oxidizing bacteria (Felbeck 1981; Jones 1981; Cavanaugh et al. 1981; Jannasch and Nelson 1984; Grassle 1985; Jannasch and Mottl 1985). The prokaryotic cell structure, DNA base ratio, genome size, and enzymatic activities clearly identify the symbionts as bacteria. They are found within a separate trophosome tissue (from Greek *trophein*, to nourish) within the body cavity of *R. pachyptila*. The trophosome may amount to 60% of the worm's wet weight. The animal's dependence on the microbial symbiont has developed to the point where all ingestive and digestive morphological features have been lost.

The worm has a striking red color from a special hemoglobin that is capable of transporting, through the bloodstream of the animal, both O_2 and H_2S to the bacteria in the trophosome for chemoautotrophic metabolism. Hydrogen sulfide binds to most other hemoglobins irreversibly and interrupts further oxygen transport. Key enzymes of the Calvin cycle of carbon dioxide reduction, including ribulose diphosphate carboxylase, were identified in the trophosome, along with elemental sulfur of light isotopic ratio, pointing to its geothermal rather than seawater origin. The described evidence leaves little doubt about the nature of this interesting nutritional relationship between a sulfide oxidizer and a previously unknown invertebrate although, to date, it has not been possible to grow the bacterial symbiont outside of its deep-sea host, making direct experimentation on this system difficult. *Calyptogena magnifica* and *C. ponderosa* are large clams (30 to 40 cm in length), also unique to the deep-sea geothermal vent environments. The body of the clams is again striking red in color from O_2/H_2S-transporting hemoglobin. Although these clams still have a digestive system, their gills contain large numbers of H_2S-oxidizing chemoautotrophs that fulfill similar symbiotic functions as described for *R. pachyptila*.

The reports on sulfide-oxidizing bacterial symbionts in animals of the deep-sea geothermal vent communities inspired investigations on invertebrates that live in shallow but sulfide-exposed waters. Estuarine sediments evolve high amounts of H_2S caused by sulfate reduction during organic matter degradation in these anoxic sediments. These investigations led to the discovery of sulfide-oxidizing symbiotic bacteria in the fills of several estuarine bivalves, such as the gutless clam *Solemya reidi* and other species of the *Solemya, Lucinoma, Myrtea,* and *Thyasira* genera. Some of the symbionts are intracellular; others are located in separate bacteriocysts between the cuticle and the animal tissue. Some of the bivalves have retained filter-feeding capacity; others rely on their symbionts exclusively (Cavanaugh 1983; Felbeck 1983; Southward 1986). Interestingly, in *S. reidi* the sulfide oxidation is mediated not only by the symbiotic bacteria in the gills, but also by the mitochondria of the host clam (Powell and Somero 1986).

Some deep-sea marine environments (for example, Louisiana Slope, Gulf of Mexico, and Origon Subduction Zone) are exposed to hydrocarbon seeps, usually in combination with sulfide. Recently discovered mussels of the family Mytilidae living in such environments were found to harbor within their gill tissue symbiotic methanotrophic (methane-oxidizing) bacteria with typical stacked internal membrane structures. The "light" carbon isotope ratio in both the mussels and their bacterial symbionts indicated that most of the carbon of this symbiotic association originated in fossil methane (Childress et al. 1986; Brooks et al. 1987). Subsequently, these mussels were also grown in the laboratory with methane serving as their only source of carbon and energy (Cary et al. 1988).

Digestion within the Rumen

The contribution of microbial populations to food digestion within ruminant animals has been intensively studied (Hungate 1966, 1975). Ruminant animals include deer, moose, antelope, giraffe, caribou, cow, sheep, and goat. These animals consume grasses, leaves, and twigs rich in cellulose. Mammals, including ruminants, do not produce cellulase enzymes themselves but depend on associated microbial populations for degrading cellulosic materials. Ruminants have a specialized chamber known as the rumen that contains large populations of protozoa and bacteria that contribute to food digestion. The rumen provides a relatively uniform and stable environment that is anaerobic, is 30°C to 40°C, and has a pH of 5.5 to 7.0. These conditions, optimal for the associated microorganisms, and the continuous supply of ingested plant materials permit the development of very dense communities of microorganisms.

The overall fermentation that occurs within the rumen can be described by Equation 1 (Wolin 1979).

$$(1) \quad 57.5 \ (C_6H_{12}O_6) \longrightarrow 65 \text{ acetate} + 20 \text{ propionate}$$
$$+ 15 \text{ butyrate} + 60 \ CO_2 + 35 \ CH_4 + 25 \ H_2O$$

Carbon flow and energy balances have been determined for the biochemical activities of the microbial populations in the rumen. Microorganisms within the rumen convert cellulose, starch, and other ingested nutrients to carbon dioxide, hydrogen gas, methane, and low–molecular-weight organic acids, such as acetic, propionic, and butyric acids. The organic acids are absorbed into the bloodstream of the animal where they are oxidized aerobically to produce energy. Ruminants are also able to utilize protein produced by the associated microbial populations. Methane produced by methanogenic bacteria within the rumen is expelled and does not appear to contribute to the nutrition of the animal.

The anaerobic environment in the rumen ensures that a relatively small percentage of the caloric value of food (about 10%) is lost to the animal during the microbial digestion process. Even some of this "lost" energy benefits the animal by helping to maintain its body temperature. Ruminants are excellent utilizers of low-grade, high-cellulose food but relatively inefficient utilizers of the high-grade proteinaceous feed used in feed lots. Crosslinking the proteins in high-grade feed by treatment with formaldehyde, dimethyloldurea, and other agents blocks their degradation by rumen microorganisms, ensuring that valuable protein is digested and absorbed in the lower portions of the gastrointestinal tract rather than being fermented to methane (Friedman et al. 1982)

The rumen harbors a great diversity of microorganisms (Hungate 1966, 1975) (Table 5.1). Bacterial populations within the rumen include cellulose digesters, starch digesters, hemicellulose digesters, sugar fermenters, fatty acid utilizers, methanogenic bacteria, proteolytic bacteria, and lipolytic bacteria. These populations include species of *Bacteroides, Ruminococcus, Succinimonas, Methanobacterium, Butyrivibrio, Selenomonas, Succinivibrio, Streptococcus, Eubacterium,* and *Lactobacillus.* Many of these bacterial populations produce acetate, the predominant acid within the rumen. The bacteria also produce propionate, the only fermentation acid that can be converted into carbohydrates by the ruminant. The diverse bacterial community of the rumen possesses the broad enzymatic capabilities required for digesting the various plant components ingested by ruminants. Some nitrogen-fixation activity was also noted in the rumen, but the amounts (approximately 10 mg of N per head of cattle per day) are too low to make a significant contribution to the nutrition of the animal (Hardy et al. 1968). One reason is that ammonia present in the rumen tends to repress nitrogen fixation. On the other hand, the ammonia in the rumen can be converted by the microbial community to microbial protein, subsequently digested by the ruminant. Theoretically, one could maintain ruminants like cattle on cellulose and ammonium salts alone, but ammonium toxicity precludes this from being a practical approach to raising cattle.

The rumen contains, in addition to bacteria, large populations of protozoa; most are ciliates, but some are flagellates. Rumen ciliates are a highly specialized group that grow anaerobically, ferment plant materials for energy, and tolerate the presence of dense bacterial populations. Some protozoan populations within the rumen are capable of digesting cellulose and starch; others ferment dissolved carbohydrates. Some are predators on bacterial populations. The proteins of the rumen protozoa are, in turn, digested by the ruminant's enzymes. Rumen protozoa store large amounts of carbohydrates, which the ruminant digests along with the proteins of the protozoan biomass. The digestion of protozoa occurs in the omasum and abomasum, compartments of the ruminant stomach located adja-

Table 5.1

Fermentation products and energy sources of some rumen bacteria

Organism	Energy sources*	Major fermentation products[†]
Bacteroides succinogenes	C, S, G	A, S
Bacteroides amylophilus	S	A, S, F
Bacteroides ruminicola	S, X, G	A, S, F
Ruminococcus flavefaciens	C, X, G	A, S, F, H
Succinivibrio dextrinosolvens	G	A, S
Succinimonas amylolytica	S, G	S
Ruminococcus albus	C, X, G	A, F, H, E
Butyrivibrio fibrisolvens	C, S, X, G	F, H, B
Eubacterium ruminantium	X, G	F, B, L
Selenomonas ruminantium	S, G, L, Y	A, P, L
Veillonella alcalescens	L	A, P, H
Streptococcus bovis	S, G	L
Lactobacillus vitulinus	G	L
Methanobacterium ruminantium	$H_2 + CO_2$, F	M

*Energy Sources: C = cellulose; S = starch; X = xylan; G = glucose; L = lactate; Y = glycerol; F = formate.
[†]Fermentation Products: A = acetate; S = succinate; F = formate; H = hydrogen; E = ethanol; B = butyrate; P = propionate; L = lactate; M = methane. Many also produce CO_2.
Source: Wolin 1979.

cent to the rumen. The transfer of carbon from bacteria to protozoa to the ruminant animal constitutes a short and efficient food chain. Protozoa are probably more efficiently digested than bacteria because the latter have resistant cell walls and high nucleic acid contents.

The relationship between the ruminant and the microbial populations of the rumen is a borderline case between synergism and mutualism. Both partners clearly derive benefits from the relationship. Some microbial populations are only found within the specialized environment of the rumen; others also occur in other environments. The microorganisms in the rumen digest plant materials, making low–molecular-weight fatty acids and microbial proteins available for utilization by the ruminant. Some of the bacterial populations in the rumen require growth factors, but others are able to produce vitamins to supply their own nutritional requirements and those of the ruminant.

The rumen provides a suitable environment and a constant supply of substrates for the fermentative activities of these microorganisms. The rumination process (rechewing previously ingested food) grinds the plant material and provides an increased surface area for microbial attack. The animal's saliva also contributes to rendering the ingested plant material susceptible to microbial attack. The movement of the ruminant stomach supplies sufficient mixing for optimal microbial growth and metabolic activities. The removal of low–molecular-weight fatty acids from the rumen by absorption into the animal's bloodstream allows continued prolific growth of the microbial population. An accumulation of these acids would be toxic to the microbes.

The high diversity of microbial populations in the rumen allows the microbial community to respond to changes in the ruminant's diet. Some ingested plant materials contain large amounts of cellulose, others large amounts of hemicellulose materials, and still others, large amounts of starches. The relative proportions of microbial populations within the rumen shift according to the nature of the plant materials ingested. Abrupt changes in diet, such as the change from dry hay in winter to pasture grass in spring, can upset the rumen fermentation system, resulting in excessive production of methane that can distend the rumen, sometimes to the extent that it compresses the lungs, suffocating the animal. This condition is known as bloat of sheep or cattle, and once it has developed, only puncturing the rumen to release excess methane can save the animal.

Other animals that exhibit ruminant-like digestion include colobid monkeys, sloths, hippopotamuses, camels, and macropod marsupials. In each of these animals, microbial populations associated with the foregut are capable of degrading cellulosic and other plant materials. The enzymatic activities of these microorganisms produce volatile fatty acids that the animals can utilize. These microbial populations contribute to partial food digestion. In nonruminant mammals that subsist primarily on plant materials, such as horses, pigs, and rabbits, microbial cellulose digestion occurs in an enlarged cecum with the production of volatile fatty acids. These fatty acids are absorbed through the intestinal lining, enter the bloodstream, and are ultimately oxidized within the animal's cells, producing carbon dioxide and water.

Symbiotic Associations with Photosynthetic Microorganisms

Some invertebrate animals enter into mutualistic relationships with photosynthetic microorganisms, including unicellular algae and cyanobacteria (Taylor 1973a, 1973b, 1975; Muscatine et al. 1975). Classically, these microorganisms are called endozoic algae. They are described as zooxanthellae if the algal cells are yellow to reddish brown, including the dinoflagellates, which are Pyrrophycophyta; zoochlorellae if the algal cells are pale to bright green; and cyanellae if the dominant pigment is blue-green.

Symbiotic associations with photosynthetic microorganisms have been reported for various species of polychaete worms, platyhelminth worms, molluscs (including bivalves such as the giant clam), tunicates, echinoderms, hydroids, jellyfish, anemones, corals, sea fans, and sponges. The most common occurrence of endozoic algae appears to be in the coelenterates, such as hydra, anemones, and corals. The marine sponges are the most common hosts of cyanobacterial sym-

bionts. Chlorophycophyta are found primarily in freshwater invertebrates. Dinoflagellates are the most frequent algal symbionts of marine invertebrates. Few of the endosymbiotic algae can be cultivated separately from their animal hosts, and consequently their classification is vague. Molecular genetic tools may remedy this situation. In a recent survey of twenty-two host taxa, closely related zooxanthellae were found in taxonomically distant hosts, indicating that associations were established in many independent events rather than in long phylogenetic coevolution between the zooxanthellae and their animal hosts (Rowan and Powers 1991).

The mutualistic relationship between primary-producer microorganisms and consumer animal populations is based on the microorganism's ability to supply the animal with organic nutrients and the animal's ability to provide a physiologically and nutritionally suitable environment for the microorganisms. In some cases, morphological adaptations facilitate close contact between the microbial and animal cells, allowing for efficient materials transfer between microorganisms and animals. The symbiotic relationship between *Convoluta roscoffensis,* a ciliated platyhelminth, and the green alga *Platymonas convolutae* has been extensively examined (Holligan and Gooday 1975). The alga supplies the animal with amino acids, amides, fatty acids, sterols, and oxygen; the animal supplies the alga with carbon dioxide and uric acid (Figure 5.7). The algal-animal association provides for an efficient closed system of mutual nutrient exchange, based on cycling of carbon, nitrogen, phosphorus, and oxygen in chemical forms that one partner can synthesize and another can utilize.

Corals, which are coelenterate animals, establish mutualistic relationships with endozoic dinoflagellates (Yonge 1963). Coral reefs also provide suitable habitats for the growth of external synergistic populations of algae, such as calcareous red algae. The growth of the coral depends upon the metabolic activities of the endozoic dinoflagellates living within the tissues of the coral polyp. Reef corals precipitate calcium from seawater mainly during periods of maximal algal photosynthesis. The assimilation of CO_2 shifts the equilibrium from the more soluble

Figure 5.7

Metabolic relationship
between the alga
Platymonas convolutae
and *Convoluta roscoffensis*.
The figure shows the shift in
this mutualistic relationship
between light and dark
growth, indicating changes
in carbon and utilization.
The *Platymonas* supplies
carbon dioxide and
ammonia from uric acid. In
the light, the alga produces
O_2 and organic carbon,
which is used by the
Platymonas. (Source:
Holligan and Gooday 1975.
Reprinted by permission,
copyright Cambridge
University Press.)

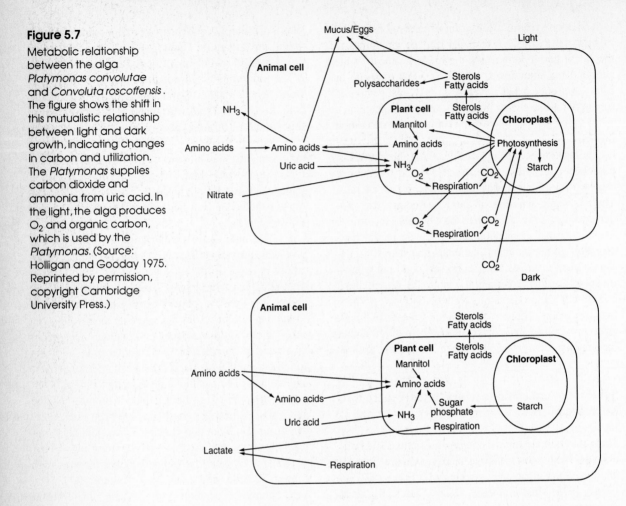

bicarbonate to the less soluble carbonate. The growing coral polyp benefits from algal production of organic matter and removal of ammonia that builds up within the animal. The mutualistic relationship of endozoic algae and corals appears to be based on the exchange of carbon, nitrogen, phosphorus, and oxygen-containing compounds.

Animals harboring endozoic algae exhibit behavioral characteristics that benefit the algal partner (McLaughlin and Zahl 1966). Animals with endozoic algae normally exhibit phototactic responses. For example, coral polyps extend toward the light, providing the algae with accessible illumination for photosynthetic activities. Anemones with zooxanthellae also show phototactic responses. Freeswimming animals with endozoic algae move to depths with optimal light penetration for the photosynthetic activities of the endozoic algae.

The mutualistic relationship between photosynthetic microorganisms and animals may be especially beneficial during some periods of the year. *Convoluta*, for example, probably requires an efficient mechanism for retaining and recycling nutrients during winter, when there is less terrestrial runoff to supply nutrients to this coastal organism (Holligan and Gooday 1975).

An interesting seasonal endosymbiotic association has been found between *Euglena* and nymphs of some species of damselfly (Willey et al. 1970). This symbiotic relationship occurs only during the winter, when members of *Euglena* occupy the lower digestive tract of damselfly nymphs. This association allows for the survival of both *Euglena* and damselfly populations during winter, when the lake habitats of these populations are frozen; during summer, the association is interrupted and both populations are independent.

FUNGAL PREDATION ON ANIMALS

Nematode- and Rotifer-Trapping Fungi

Some fungi prey on nematodes (Figure 5.8) as a source of nutrients (Pramer 1964; Nordbring-Hertz and Jansson 1984). The most common genera of nematode-trapping fungi are *Arthrobotrys, Dactylaria, Dactylella,* and *Trichothecium.* There are several

Figure 5.8

Photomicrographs showing examples of nematode-trapping fungi.
(A) *Arthrobotrys conoides* traps, clusters of adhesive hyphal loops or rings.
(B) *A. conoides* with trapped nematode.
(C) *Dactylella drechslerii* traps are knobs coated with an adhesive.
(D) *D. drechslerii* with trapped nematode.
(E) *A. dactyloides* traps, rings comprised of three cells each, which capture prey by occlusion. Open traps are shown on left, closed traps on right.
(F) *A. dactyloides* with trapped nematode.
(Source: D. Pramer, Rutgers, the State University.)

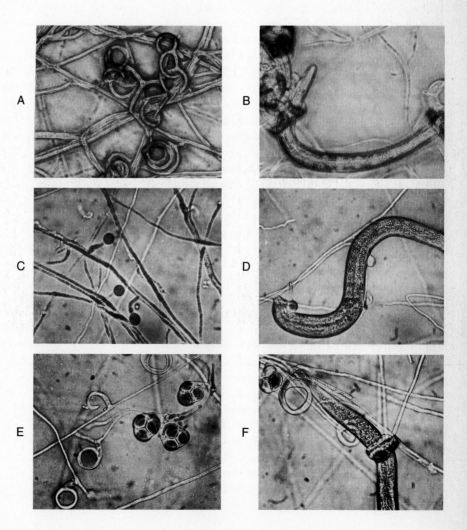

mechanisms by which these fungi capture nematode prey, including production of networks of adhesive branches, stalked adhesive knobs, adhesive rings, and constrictive rings. When a nematode attempts to move past an adhesive hyphal structure, it sticks to it and is trapped. When it tries to pass through a constricting ring, the fungal ring contracts by sudden osmotic swellings and traps the nematode (Figure 5.8). Violent movements and attempts by the nematode to escape generally fail. The fungal hyphae penetrate into the nematode, which is then enzymatically degraded. When growing in the absence of nematodes, some of these fungi fail to produce trap structures. The presence of the prey nematode appears to induce the formation of the morphological structures that trap the nematodes. This is a unique relationship in which the presence of the prey induces the formation of fungal structures that result in its capture and consumption.

Most nematode-trapping fungi, including the ones described above, are Deuteromycota, but a few Basidiomycota also have the ability to attack and digest nematodes. *Hohenbuehelia* and *Resupinatus* species were found to capture nematodes by means of adhesive knobs. The edible oyster mushroom *Pleurotus ostreatus* and related *Pleurotus* species form no trap structures. Instead, by means of some toxin, they rapidly paralyze the nematodes. Subsequently, the hyphae invade and digest the immobilized nematode. The described basidiomycota often grow on decaying wood, a nitrogen-poor substrate. It is suggested that the captured nematodes are a source of supplemental nitrogen for the fungi (Thorn and Barron 1984).

The parasitic oomycete *Haptoglossa mirabilis* attacks rotifer worms. Zoospores of this fungus, after a short swarming period, produce a special cyst that almost immediately germinates in the form of a "gun" cell. The gun cell attached to the cyst resembles a bowling pin in shape. When a rotifer contacts the tip of this structure (the muzzle of the "gun"), within a fraction of a second the gun cell discharges a missile that breaches the cuticle of the rotifer and injects an infectious sporidium. Inside the rotifer, the sporidium grows into a thallus, killing the host.

Eventually, new zoospores are released to perpetuate the cycle (Robb and Barron 1982).

Fungal–Scale Insect Associations

An interesting relationship exists between scale insects and fungi of the genus *Septobasidium* (Buchner 1960; Brooks 1963; Batra and Batra 1967). Scale insects are plant parasites that extract plant juices with their long sucking tubes (haustorium) (Figure 5.9). Scale insects become infected with fungi when they emerge from the parent scale on a plant; the fungi develop hyphae that surround the maturing scale insects, trapping but not immediately killing all of them. The insects live and produce young within the mass of hyphae covering the parent scale. The fungal hyphae retain the adults while the juvenile scale insects feed on the plant between the hyphae. In this relationship, the fungus affords the scale insects protection from other parasites and predators, and the scale insect provides nutrients utilized by the fungus. The movement of young scale insects from one plant to another ensures dissemi-

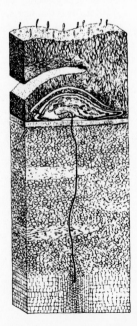

Figure 5.9

Scale insect embedded in the mycelium of *Septobasidium* fungus. (Source: Batra and Batra 1967, The Fungus Gardens of Insects. Copyright 1967 Scientific American Inc. All rights reserved.)

Figure 5.10
Symbiotic relationships of luminous bacteria with animals. (A) The cephalopod *Eurymma* with open mantle, showing enlarged section through the organs that contain luminous bacteria (arrow). (B) The flashlight fish *Anomalops*, with luminous organ containing bacteria in its folds (arrow) located under the eye. (C) The tunicate *Pyrosoma*, with luminous organs, showing an enlarged mycetocyte with bacterial cells (arrow). (Source: Buchner 1960. Reprinted by permission of Springer-Verlag.)

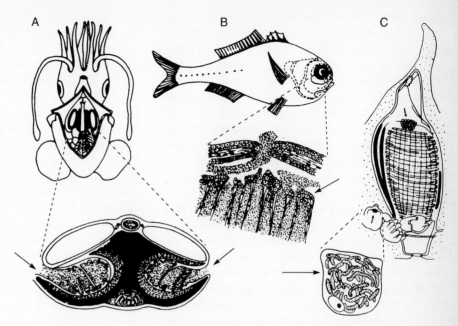

nation of the fungus. Some mature scale insects live through the winter and produce young before the fungus kills them; fungi invade others before they reach maturity so that they fail to reproduce. The fungus derives nutrients by hyperparasitizing the insect; the insect derives protection from the fungus. The relationship between the scale insect and the fungus allows both populations to survive and reproduce.

Interestingly, *Septobasidium* may also play a role in determining the sex of scale insect offspring. Microorganisms do not occur in association with all scale insect eggs; those eggs that possess symbiotic microorganisms develop into females, whereas eggs lacking microorganisms become males. The mutualistic association appears to prevent loss of the sex chromosome in the female, but the mechanism for this retention is not known.

SYMBIOTIC LIGHT PRODUCTION

Some marine invertebrates and fish establish mutualistic relationships with luminescent bacteria (Buchner 1960;

Morin et al. 1975; Hastings and Nealson 1977; McCosker 1977; Nealson and Hastings 1979; Leisman et al. 1980; Nealson et al. 1984). Depending upon the particular species, the organs containing the luminescent bacteria may be localized near the eye, abdomen, rectum, or jaw (Figure 5.10). In squid, for example, luminescent bacteria occur in a pair of glands in the mantle cavity near the ink sac. The luminescent bacteria, *Beneckea* and *Photobacterium* species, are contained within special saclike organs that generally have external pores that allow the bacteria to enter and provide for exchange with the surrounding seawater. The fish supplies the bacteria with nutrients and protection from competing microorganisms.

Luminescent bacteria generally emit light continuously, but some fish are able to manipulate the organs containing these bacteria so as to emit flashes of light. The flashlight fish *Photoblepharon* is capable of shutting off its light by drawing a dark curtain, like an eyelid, over the light organ. Another small, tropical, light-emitting marine fish, *Anomalops,* has a slightly different mechanism for shutting off its light. Here the light organ, lined on the inside with reflective guanine-containing cells, is rotated, like an eyeball, almost 180°. This way the light, now reflected inward, is obscured. Both *Photoblepharon* and *Anomalops* have their light organs situated under their eyes and may be used to some extent as a flashlight or headlamp by these nocturnal fishes. Both *Photoblepharon* and *Anaomalops* are gregarious and their light emission is also thought to aid their schooling behavior and perhaps repel predators. The light organs of both fishes are tightly packed with light-emitting bacteria (Figure 5.10B). To date, efforts to culture these have failed, although these fishes are readily accessible in the live state and have also been kept in the laboratory.

Although *Photoblepharon* and *Anomalops* live in shallow waters, most mutualistic associations with luminescent bacteria occur in deep-sea fishes that live below the level of light penetration. The light emitted by the associated luminescent bacteria permits species recognition among these fish. The pattern and location of the luminescent organs on the fish and the fact that they often occur in only one sex indicates that the luminescent bacteria may be critical for mate recognition. The fact that some light organs located near the eyes include a reflective concave mirror of guanine-containing cells and lenslike focusing structures indicates that some fish use the light as searchlights. Movement of the luminescent organs may also allow these fish to lure prey and to communicate with other fish. Deep-water species cannot be obtained in live state, making the study of these associations particularly difficult.

ECOLOGICAL ASPECTS OF ANIMAL DISEASES

Some microorganisms, including various viruses, bacteria, fungi, protozoa, and even algae, cause animal diseases (Dubos and Hirsch 1965; Merchant and Packer 1967). We can distinguish two types of disease-causing processes, one in which the microorganism grows on or within the animal, causing an infection that results in a disease condition, and a second in which the microorganism grows outside of the animal, producing toxic substances that cause animal diseases or altering the habitat of the animal so that the animal can no longer survive in a healthy condition.

Microorganisms growing in their natural habitats, such as water or soil, can alter environmental conditions, adversely affecting the animal populations occupying those habitats. Such conditions often arise when there are population imbalances within the microbial community. For example, when eutrophic conditions allow for large algal blooms in lakes, the growth of photoautotrophic populations adds large quantities of organic matter to the lake waters, and the subsequent degradation of this organic matter by heterotrophic microorganisms depletes the dissolved oxygen, creating anoxic conditions. This depletion produces disease symptoms and death in the oxygen-requiring animal populations, sometimes causing major fish kills.

Microbial populations may also produce inorganic chemicals or organic toxins that cause diseases in animals. Hydrogen sulfide produced by microbial populations in sediment can accumulate in concentrations toxic to animals burrowing in the sediment. Acid mine drainage, resulting from microbial oxidation of sulfur, can produce highly acidic conditions in drainage streams, causing disease conditions and death of aquatic animals. Toxins produced by microorganisms outside of susceptible host animals often enter the animal during ingestion of food, causing food poisoning (Shilo 1967; Ciegler and Lillehoj 1968). Many mushrooms contain toxins that are extremely poisonous to animals that ingest them, resulting in severe disease symptoms and sometimes death. Aflatoxins, produced by some *Aspergillus* species growing on grain products, cause disease in poultry and other populations that consume the toxin-containing food products.

Microbial toxins can remain active even under conditions that do not permit the survival and growth of the population that produced them. Red tides result in disease and death in susceptible animal populations exposed to toxins produced by blooms of dinoflagellates. Some dinoflagellates produce toxins that kill fish, whereas others kill primarily invertebrates. Toxins concentrated by filter-feeding animals also can cause disease in higher members of the food web. Humans acquire paralytic shellfish poisoning by consuming shellfish containing concentrated algal toxins.

When considering the ecological aspects of animal diseases caused by toxins, one must examine the conditions under which the toxin-forming microbial populations can grow and produce these substances, the ability of the toxin to remain in an active form in the environment, the concentrations of toxin required to produce disease; and the factors that dilute or concentrate toxins. Toxins are often not produced in sufficient concentrations or are degraded or diluted so that they do not produce diseases in animal populations. Only when conditions permit extensive toxin production by microbial populations do toxins accumulate in concentrations that cause animal diseases. In some cases, toxins produced by microorganisms are concentrated within the food web, producing diseases in higher consumer populations with no apparent effect on lower members of the food web. Environmental conditions that favor the growth of toxin-producing microbial populations increase the possible incidence of toxin-caused animal diseases.

In contrast to animal diseases caused by toxin-producing microorganisms that grow outside of the animal, infectious pathogens or parasitic microorganisms must be capable of growing in or on the animal's tissues. Some pathogenic microorganisms are obligate intracellular parasites and are entirely dependent for their existence and survival on their successful invasion and reproduction within susceptible animal cells. Although stable carrier states can occur, in which a balance between the pathogen and the host is established, disease-causing infectious microorganisms more commonly grow for only a limited period of time within susceptible animals, after which the host animal either dies, resulting in destruction of the habitat for the pathogenic microorganisms, or develops an immune response that precludes further growth of the pathogenic microorganisms. Populations of pathogenic microorganisms, therefore, must be transmitted to new susceptible members of the animal population for continued growth. It is important to consider the routes of transfer, how long the patho-

gens can survive outside of host animals, the environmental factors that contribute to survival outside of the host, and the reservoirs or alternate host populations for the particular pathogenic population.

Infectious microorganisms have a limited number of routes available for entering host animals. Pathogenic microorganisms usually enter animals through normal body openings, such as the respiratory and gastrointestinal tracts. Most pathogens are unable to penetrate the outer skin layers and establish infection, but there are exceptions, including a ciliate that is able to penetrate fish skin surfaces and causes ich (Figure 5.11). Breaks in the protective skin surface, caused by wounds or insect bites, however, permit various pathogens to infect animals. Most pathogenic microorganisms are restricted to one portal of entry for establishing disease; entering another part of the body leads to a different relationship between the microbe and the animal. In some cases, a microbial population is nonpathogenic under normal conditions but can become an opportunistic pathogen under particular circumstances. For example, *Escherichia coli* is a normal inhabitant of the human intestine, where it does not produce disease, but if it enters the urinary tract, it causes a urinary tract infection. The limitation of pathogenic microorganisms to a particular portal of entry is caused partly by environmental constraints, such as low pH, which the potential pathogen must tolerate; partly by the immune response system of the host animal, which can recognize and destroy many invading microorganisms before they can establish an infective population level; and partly by the antagonistic activities of the non-disease–causing microorganisms that comprise the normal microbiota inhabiting the surfaces of animal tissues.

While the tissues of healthy animals and humans are sterile, numerous microorganisms colonize their body surfaces. Best studied in this respect is the human skin (Marples 1969). The sloughing off of epidermal cells and the secretion of sebaceous and sweat glands provide keratin, lipids, and fatty acids as potential growth substrates. On the other hand, generally dry conditions, salinity, and inhibitory action of some fatty acids create a somewhat hostile environ-

Figure 5.11
Scanning electron micrographs of the ciliate protozoan *Ichthyophthirinus multifiliis.*
(A) On a fish fin surface prior to penetration. (B) Penetration into gill tissue. (Source: T. Kozel.)

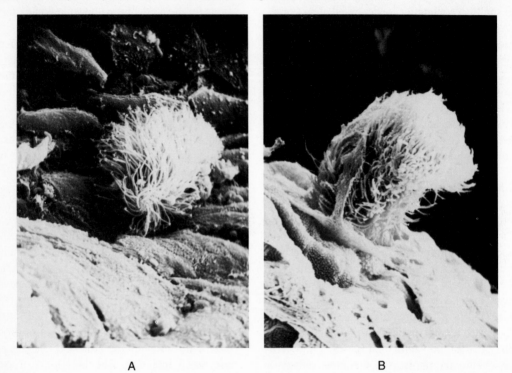

A   B

ment that is best tolerated by some Gram-positive bacteria and coryneforms and by some yeasts. Most of these microorganisms are harmless commensals, but some of them like *Staphylococcus aureus* and *Candida albicans,* have the potential of becoming opportunistic pathogens in debilitated or immunocompromised individuals or if the skin's integrity is disrupted by wounds or burns. Interestingly, there are large differences in the density and species composition of the microbial populations of various skin areas. This, in turn, is caused by variations in the dryness or moistness of different skin areas and in the density of secretory glands. The skin of the forearms and the back are the most sparsely populated "desert" regions of the skin, with only a few hundred to a few thousand microorganisms per cm. On the other hand, the axilla, groin, scalp, and areas between fingers and toes are heavily populated, with 10^5 to 10^6 microorganisms per cm^2. *Staphylococcus* and *Corynebacterium* are the most abundant bacterial genera; Gram negatives are restricted to the more moist areas of the skin. The most common yeasts are *Pityrosporum ovale* and *C. albicans.* The presence of these microorganisms is not harmful, and by their presence and competitive action, they tend to make an invasion by real pathogens more difficult.

The communicability of disease-causing microorganisms depends on the ability of the pathogen to escape from its host animal, to contact a new susceptible animal, and to enter successfully the tissues of that animal. If there is a delay between the time a pathogenic microorganism is released from one host and

Table 5.2

Examples of disease-causing microorganisms dispersed by animal vectors

Etiologic agent	Reservoir	Vector	Disease
Rickettsia rickettsii	Rodent	Tick	Rocky Mountain spotted fever
Rickettsia typhi	Rodent	Flea	Endemic typhus
Rickettsia prowazekii	Human	Louse	Epidemic typhus
Rhabdovirus	Rodent, dog	Rodent, dog	Rabies
Togavirus	Human, monkey	Mosquito (*Aedes aegypti*)	Yellow fever
Arbovirus	Horse, human, bird	Mosquito	Encephalitis
Trypanosoma cruzi	Human, various animals	Conebug	Chagas' disease
Yersinia pestis	Rodent	Flea	Bubonic plague
Trypanosoma gambiense and other *Trypanosoma* spp.	Human, various animals	Tsetse fly	African sleeping sickness
Leishmania donovani	Cat, dog, rodent	Sandfly	Dum-dum fever (kala-azar disease)
Borrelia burgdorferi	Mouse, deer	Deer tick (*Ixodes dammini*)	Lyme disease (borreliosis)

the time it contacts a new susceptible host, the pathogen must be able to survive this period. Pathogenic microorganisms are normally transmitted by direct contact, through airborne and waterborne dispersal, through ingestion with food, and via vectors. Pathogenic microorganisms transmitted via prolonged air travel must be resistant to desiccation, and normally high numbers of microorganisms must be released to ensure successful dissemination to the next susceptible host. Vectors are important in the distribution of pathogenic microorganisms that could not survive for extended periods outside host cells. There is often great specificity between the pathogenic microbes, the vectors, and the susceptible hosts (Table 5.2). The geographic distribution of some infectious diseases reflects the distribution of suitable vector populations. Environmental conditions often determine the distribution of pathogens, affecting host susceptibility, pathogen reproduction, pathogen survival, and possible routes of transmission. Temperature, pH, E_h, concentrations of organic nutrients, and concentrations of inorganic nutrients all influence the survival time of

pathogens in the environment; these same environmental factors are important in determining the size and distribution of susceptible animal populations and suitable vector populations.

An outbreak of kyasanur forest disease, a rare disease caused by a virus, provides an interesting example of how populations of pathogenic microorganisms, reservoirs for the pathogens, vectors, and animal hosts are interrelated (Smith 1978). In 1956, there was an outbreak of kyasanur forest disease associated with the following factors: The infection was naturally maintained within the forest in reservoir populations of birds and mammals with ticks acting as vectors; there was an increase in the human population in the region, which led to extensive cattle grazing in the forest; the presence of the cattle led to an increase in the tick population that was heightened because the adult stage of the tick, which controls the overall tick population, is dependent on large mammals to meet its nutritional requirements; the tick population became infected with the virus, transmitting it to monkeys, thus amplifying it; and, finally, the virus was transmitted to humans. According to one theory, the human immunodeficiency virus (HIV) that causes acquired immunodeficiency syndrome (AIDS) in humans, recognized in the 1960s and now becoming a worldwide devastating epidemic, may have originated from a related virus of the African green monkey. A mutant of this virus may have spread to the human population through a bite wound or via an unknown vector.

Another pathogen of current concern that has an interesting ecology is *Borrelia burgdorferi,* a spirochete that causes borreliosis, or Lyme disease, named from the Connecticut community Old Lyme, where the first outbreak in the early 1970s was observed (Miller 1987). Lyme disease is transmitted by the bite of the deer tick (*Ixodes dammini*). Initially Lyme disease has flulike symptoms with or without a characteristic circular rash around the bite. If recognized and treated early by antibiotics, the infection is relatively easy to cure. When left untreated, the disease may cause severe joint, heart, and central nervous system damage. The young tick feeds on small rodents like the white-footed mouse; the adult feeds mainly on deer. Ticks pick up *Borrelia* from these wild animals that are asymptomatically infected. When the infected tick feeds on humans or some domestic animals, *Borrelia* is transmitted and causes the described disease. The resurgence of the white-tailed deer population near suburban areas appears to be the main cause of increasingly frequent human infections.

Similar population balances are involved among causative, reservoir, vector, and host populations for other diseases, and similar considerations must be made when soil or water is involved as the reservoir or as the route of transmission. For example, outbreaks of Legionnaires disease, caused by *Legionella,* are associated with evaporating water bodies, such as air-conditioning systems, that provide a reservoir for the organism and a mechanism through which the organism can become airborne via aerosol formation.

When considering the ecology of infectious diseases, it is also necessary to recognize the importance of the immune response that renders some animals insusceptible to particular populations of pathogenic microorganisms. The immune system renders these animals unsuitable, for the most part, as habitats for invading microorganisms, and relatively few pathogens succeed in overcoming the immune defense system. Even pathogenic microorganisms that successfully invade and establish an infection within a susceptible animal are normally eliminated from the host animal when the immune response becomes fully activated. If the immune system is impaired, as occurs in AIDS, when human T-cells become infected with HIV, the person becomes unprotected against numerous microbial infections and is unable to survive in a nonsterile environment. The physiological state of the host animal often determines in part the effectiveness of the immune system in protecting against microbial invasion. Animals debilitated by poor nutrition or other stresses are extremely susceptible to pathogenic invasion.

SUMMARY

Microorganisms are important contributors to the nutrition of some animals. Detritus-feeding and coprophagous animals derive a substantial portion of their food from microbial biomass. Grazing and filter feeding on microorganisms by invertebrate animals is a crucial link in aquatic food webs, making the biomass of microbial primary and secondary producers available to higher trophic levels. Many animals lack the enzymes necessary to digest some or all of the food resources available to them and thus require the assistance of microbial populations. Of particular importance is the ability of various microbial populations to produce the extracellular enzymes that degrade complex plant polymers, such as cellulose, which otherwise would be unavailable to the animal. A number of animal populations depend on the microbial conversion of plant polymers to fatty acids that can be utilized for energy production. Some microbial populations also produce vitamins, which are required but not synthesized by the animal host. The activities of microbial populations in the gastrointestinal tract thus contribute to food digestion, supply vitamins, and provide considerable protection against pathogenic invaders.

In most cases, these activities seem essential to the well-being of the animal.

Some microbial and animal populations establish mutualistic relationships, many of which are largely based upon nutrient exchange and maintenance of a suitable habitat. Ants, for example, grow pure cultures of fungi on leaves and use the fungal biomass and metabolic products as their nutrient supply. Termites similarly maintain microbial populations to upgrade the food sources available to them. Various invertebrates contain photosynthetic partners, and there is a mutual exchange of nutrients. In some cases, microbes derive their nutrition by preying on animals, as in the case of nematode-trapping fungi. Other animal-microbe interactions have unusual foundations, such as the ability of luminescent bacteria to provide light in the deep ocean.

The spread of disease among animal populations is an ecological process that is dependent on the biological properties of the causative organism, the biological properties of the host organism, and abiotic and biotic factors that affect the transmission of the pathogen between hosts. The relationship between pathogenic microbial populations and host animal populations is important in determining the size and distribution of each population. Animal populations provide suitable habitats for the growth and continued existence of successful pathogens, and microbially caused diseases are important in determining the ability of animal populations to compete for and successfully occupy particular niches. Resistance to microbial diseases is a measure of fitness that acts through natural selection to determine the success of particular animal populations. Microbial diseases of animal populations are a selective force that acts to control the animal population, both qualitatively and quantitatively.

REFERENCES & SUGGESTED READINGS

Atlas, R. M., M. Busdosh, E. J. Krichevsky, and T. Kaneko. 1982. Bacterial populations associated with the Arctic amphipod *Boeckosimus affinis. Canadian Journal of Microbiology* 28:92–99.

Baker, J. M. 1963. Ambrosia beetles and their fungi, with particular reference to *Platypus culindrus* Fab. In P. S. Nutman and B. Mosse (eds.). *Symbiotic Associations.* Cambridge University Press, Cambridge, pp. 232–265.

Batra, S. W. T., and L. R. Batra. 1967. The fungus gardens of insects. *Scientific American* 217(5):112–120.

Beck, J. W., and J. E. Davies. 1976. *Medical Parasitology.* C. V. Mosby Co., St. Louis, Mo.

Benemann, J. R. 1973. Nitrogen fixation in termites. *Science* 181:164–165.

Bowden, G. H. W., D. C. Ellwood, and I. R. Hamilton. 1979. Microbial ecology of the oral cavity. *Advances in Microbial Ecology* 3:135–218.

Breznak, J. A. 1975. Symbiotic relationships between termites and their intestinal microbiota. In *Symbiosis*, Symposia of the Society for Experimental Biology, No. 29. Cambridge University Press, Cambridge, England, pp. 559–580.

Breznak, J. A., and H. S. Pankratz. 1977. In situ morphology of the gut microbiota of wood-eating termites. *Applied and Environmental Microbiology* 33:406–426.

Breznak, J. A., and J. M. Switzer. 1986. Acetate synthesis from H_2 plus CO_2 by termite gut microbe. *Applied and Environmental Microbiology* 52:623–630.

Brooks, M. A. 1963. Symbiosis and aposymbiosis in arthropods. In *Symbiotic Associations,* Proceedings of the Thirteenth Symposium of the Society for General Microbiology. Cambridge University Press, Cambridge, England, pp. 200–231.

Brooks, J. M., M. C. Kennicutt II, C. R. Fisher, S. A. Macko, K. Cole, J. J. Childress, R. R. Bidigare, and R. D. Vetter. 1987. Deep sea hydrocarbon seep communities: Evidence for energy and nutritional carbon sources. *Science* 187:1138–1142.

Buchner, P. 1960. Tiere als Mikrobenzuchter. Springer-Verlag, Heidelberg, Germany.

Carpenter, E. J., and J. L. Culliney. 1975. Nitrogen fixation in marine ship worms. *Science* 187:551–552.

Carr, D. L., and W. R. Kloos. 1977. Temporal study of the Staphylococci and Micrococci of normal infant skin. *Applied and Environmental Microbiology* 34:673–680.

Cary, S. C., C. R. Fisher, and H. Felbeck. 1988. Mussel growth supported by methane as sole carbon and energy source. *Science* 240:78–80.

Cavanaugh, C. M. 1983. Symbiotic chemoautotrophic bacteria in marine invertebrates from sulfide-rich habitats. *Nature* (London) 302:58–61.

Cavanaugh, C. M., S. L. Gardiner, M. L. Jones, H. W. Jannasch, and J. B. Waterbury. 1981. Prokaryotic cells in the hydrothermal vent tube worm *Riftia pachyptila*, Jones: Possible chemoauthotrophic symbionts. *Science* 213:340–342.

Childress, J. J., R. C. Fisher, J. M. Brooks, M. C. Kennicutt II, R. Bidigare, and A. E. Anderson. 1986. A methanotrophic marine molluscan (Bivalvia, Mytilidae) symbiosis: Mussels fueled by gas. *Science* 233:1306–1308.

Ciegler, A., and E. B. Lillehoj. 1968. Mycotoxins. *Advances in Applied Microbiology* 10:155–219.

Clarke, R. T. J., and T. Bauchop (eds.). 1977. *Microbial Ecology of the Gut*. Academic Press, London.

Coleman, G. S. 1975. The role of bacteria in the metabolism of rumen entodiniomorphid protozoa. In *Symbiosis*, Symposia of the Society for Experimental Biology, No. 29. Cambridge University Press, Cambridge, England, pp. 533–558.

Cooke, R. 1977. *Biology of Symbiotic Fungi*. John Wiley and Sons, New York.

Costerton, J. W. 1984. Direct ultrastructural examination of adherent bacterial populations in natural and pathogenic ecosystems. In M. J. Klug and C. A. Reddy (eds.). *Current Perspectives in Microbial Ecology*. American Society for Microbiology, Washington, D.C., pp. 115–124.

Dubos, R. J., and J. G. Hirsch (eds.). 1965. *Bacterial and Mycotic Infections of Man*. J. B. Lippincott Co., Philadelphia.

Felbeck, H. 1981. Chemoautotrophic potential of the hydrothermal vent tube worm, *Riftia pachyptila*, Jones (Vestimentifera). *Science* 213:336–338.

Felbeck, H. 1983. Sulfide oxidation and carbon fixation by the gutless clam *Solemya reidi:* An animal-bacteria symbiosis. *Journal of Comparative Physiology* 152:3–11.

Fenchel, T. M., and B. B. Jørgensen. 1977. Detritus food chains of aquatic systems. *Advances in Microbial Ecology* 1:1–58.

Frankenberg, D., and K. L. Smith, Jr. 1967. Coprotrophy in marine animals. *Limnology and Oceanography* 12:443–450.

Friedman, M., M. J. Diamond, and G. A. Bruderick. 1982. Dimethylolurea as a tyrosine reagent and protein protectant against ruminal degradation. *Journal of Agricultural and Food Chemistry* 30:72–77.

Geddes, D. A. M., and G. N. Jenkins. 1974. Intrinsic and extrinsic factors influencing the flora of the mouth. In F. A. Skinner and J. G. Carr (eds.). *The Normal Microbial Flora of Man*. Academic Press, London, pp. 85–100.

Gibbons, R. J., and J. van Houte. 1975. Bacterial adherence in oral microbiology. *Annual Reviews of Microbiology* 29:19–44.

Gordon, H. A., and L. Pesti. 1971. The gnotobiotic animal as a tool in the study of host microbial relationships. *Bacteriological Reviews* 35:390–429.

Grassle, J. F. 1985. Hydrothermal vent animals: Distribution and biology. *Science* 229:713–717.

Hardie, J. M., and G. H. Bowden. 1974. The normal microbial flora of the mouth. In F. A. Skinner and J. G. Carr (eds.). *The Normal Microbial Flora of Man*. Academic Press, London, pp. 47–84.

Hardy, R. W. F., R. D. Holsten, E. K. Jackson, and R. C. Burns. 1968. The acetylene-ethylene assay for nitrogen fixation: Laboratory and field evaluation. *Plant Physiology* 43:1185–1207.

Hastings, J. W., and K. H. Nealson. 1977. Bacterial bioluminescence. *Annual Reviews of Microbiology* 31:549–595.

Henry, S. M. (ed.). 1967. *Symbiosis*. Academic Press, New York.

Hoffman, H. 1966. Oral microbiology. *Advances in Applied Microbiology* 8:195–251.

Höfte, H. and H. Whiteley. 1989. Insecticidal crystal proteins of *Bacillus thuringiensis*. *Microbiological Reviews* 53:242–255.

Holligan, P. M., and G. W. Gooday. 1975. Symbiosis in *Convoluta roscoffensis*. In *Symbiosis*, Symposia of the Society for Experimental Biology, No. 29. Cambridge University Press, Cambridge, England, pp. 205–228.

Hungate, R. E. 1966. *The Rumen and Its Microbes*. Academic Press, New York.

Hungate, R. E. 1975. The rumen microbial ecosystem. *Annual Reviews of Microbiology* 29:39–66.

Hungate, R. E. 1978. Gut microbiology. In M. W. Loutit and J. A. R. Miles (eds.). *Microbial Ecology*. Springer-Verlag, Berlin, pp. 258–264.

Jannasch, H. W., and D. C. Nelson. 1984. Recent progress in the microbiology of hydrothermal vents. In M. J. Klug and C. A. Reddy (eds.). *Current Perspectives in Microbial Ecology*. American Society for Microbiology, Washington, D.C., pp. 170–176.

Jannasch, H. W., and M. J. Mottl. 1985. Geomicrobiology of deep-sea hydrothermal vents. *Science* 229:717–725.

Jones, G. W. 1984. Mechanisms of the attachment of bacteria to animal cells. In M. J. Klug and C. A. Reddy (eds.). *Current Perspectives in Microbial Ecology.* American Society for Microbiology, Washington, D.C., pp. 136–143.

Jones, M. L. 1981. *Riftia pachyptila* Jones: Observations on the vestimentiferan worm from the Galapagos Rift. *Science* 213:333–336.

Jørgensen, C. B. 1966. *Biology of Suspension Feeding.* Pergamon Press, Oxford, England.

Kenworthy, R. 1973. Intestinal microbial flora of the pig. *Advances in Applied Microbiology* 16:31–54.

Lee, A. 1985. Neglected niches: The microbial ecology of the gastrointestinal tract. *Advances in Microbial Ecology* 8:115–162.

Leisman, G., D. H. Cohn, and K. H. Nealson. 1980. Bacterial origin of luminescence in marine animals. *Science* 208:1271–1273.

Luckey, T. D. 1965. Effects of microbes on germfree animals. *Advances in Applied Microbiology* 7:169–223.

McBee, R. H. 1971. Significance of intestinal microflora in herbivory. *Annual Reviews of Ecological Systematics* 2:165–176.

McCosker, J. E. 1977. Flashlight fishes. *Scientific American* 236(3):106–114.

McFall-Ngai, M. J., and E. G. Ruby. 1991. Symbiont recognition and subsequent morphogenesis as early events in an animal-bacterial mutualism. *Science* 254:1491–1494.

McLaughlin, J. J. A., and P. A. Zahl. 1966. Endozoic algae. In S. M. Henry (ed.). *Symbiosis,* Vol. 1. Academic Press, New York, pp. 258–297.

Marples, M. J. 1969. Life on the human skin. *Scientific American* 220(1):108–119.

Marples, M. J. 1974. The normal microbial flora of the skin. In F. A. Skinner and J. G. Carr (eds.). *The Normal Microbial Flora of Man.* Academic Press, London, pp. 7–12.

Marshall, K. C. 1980. Reactions of microorganisms, ions and macromolecules at interfaces. In D. C. Ellwood, J. N. Hedger, M. J. Latham, J. M. Lynch, and J. H. Slater (eds.). *Contemporary Microbial Ecology.* Academic Press, London, pp. 93–106.

Martin, J. K. 1979. Biochemical implications of insect mycophagy. *Biological Reviews* 54:1–21.

Martin, M. M. 1970. The biochemical basis of the fungus-attine ant symbiosis. *Science* 169:16–20.

Martin, M. M., and J. S. Martin. 1978. Cellulose digestion in the midgut of the fungus-growing termite *Macrotermes natalensis*: The role of acquired digestive enzymes. *Science* 199:1453–1455.

Merchant, I. A., and R. A. Packer. 1967. *Veterinary Bacteriology and Virology.* Iowa State University Press, Ames, Iowa.

Miller, J. A. 1987. Ecology of a new disease. *BioScience* 37:11–15.

Morin, J. G., A. Harrington, K. Nealson, H. Krieger, T. O. Baldwin, and J. W. Hastings. 1975. Light for all reasons: Versatility in the behavioral repertoire of the flashlight fish. *Science* 190:74–76.

Mortensen, A. 1984. Importance of microbial nitrogen metabolism in the ceca of birds. In M. J. Klug and C. A. Reddy (eds.). *Current Perspectives in Microbial Ecology.* American Society for Microbiology, Washington, D.C., pp. 273–278.

Muscatine, L., C. B. Cook, R. L. Pardy, and R. R. Pool. 1975. Uptake, recognition and maintenance of symbiotic *Chlorella* by *Hydra viridis*. In *Symbiosis,* Symposia of the Society for Experimental Biology, No. 29. Cambridge University Press, Cambridge, England. pp. 175–204.

Nealson, K. H., and J. W. Hastings. 1979. Bacterial bioluminescence: Its control and ecological significance. *Microbiological Reviews* 43:496–518.

Nealson, K. H., M. G. Haygood, B. M. Tebo, M. Roman, E. Miller, and J. E. McCosker. 1984. Contribution by symbiotically luminous fishes to the occurrence and bioluminescence of luminons bacteria in seawater. *Microbial Ecology* 10:69–77.

Noble, W. C., and D. G. Pitcher. 1978. Microbial ecology of the human skin. *Advances in Microbial Ecology* 2:245–289.

Nordbring-Hertz, B., and H. Jansson. 1984. Fungal development, predacity, and recognition of prey in nematode-destroying fungi. In M. J. Klug and C. A. Reddy (eds.). *Current Perspectives in Microbial Ecology.* American Society for Microbiology, Washington, D.C., pp. 327–333.

Porter, K. G. 1976. Enhancement of algal growth and productivity by grazing zooplankton. *Science* 192:1332–1334.

Porter, K. G. 1984. Natural bacteria as food resources for zooplankton. In M. J. Klug and C. A. Reddy (eds.). *Current Perspectives in Microbial Ecology.* American Society for Microbiology, Washington, D.C., pp. 340–345.

Potrikus, C. J., and J. A. Breznak. 1977. Nitrogen-fixing *Enterobacter agglomerans* isolated from guts of wood-eating termites. *Applied and Environmental Microbiology* 33:392–399.

Powell, M. A., and G. N. Somero. 1986. Hydrogen sulfide oxidation is coupled to oxidative phosphorylation in mitochondria of *Solemya reidi*. *Science* 233:563–566.

Pramer, D. 1964. Nematode-trapping fungi. *Science* 144:382–388.

Rau, G. H. 1981. Hydrothermal vent clam and tube worm $^{13}C/^{12}C$: Further evidence on nonphotosynthetic food sources. *Science* 213:338–340.

Robb, E. J., and G. L. Barron. 1982. Nature's ballistic missile. *Science* 218:1221–1222.

Rosebury, T. 1962. *Microorganisms Indigenous to Man*. McGraw-Hill Book Company, New York.

Rowan, R., and D. A. Powers. 1991. A molecular genetic classification of zooxanthellae and the evolution of animal algal symbioses. *Science* 251:1348–1351.

Sands, W. A. 1969. The association of termites and fungi. In K. Krishna and F. M. Wessner (eds.). *Biology of Termites*, Vol. 1. Academic Press, New York, pp. 495–524.

Savage, D. C. 1977. Microbial ecology of the gastrointestinal tract. *Annual Reviews in Microbiology* 31:107–133.

Savage, D. C. 1978. Gastrointestinal microecology: One opinion. In M. W. Loutit and J. A. R. Miles (eds.). *Microbial Ecology*. Springer-Verlag, Berlin, pp. 234–239.

Shachak, M., C. G. Jones, and Y. Cranot. 1987. Herbivory in rocks and the weathering of a desert. *Science* 236:1098–1099.

Shilo, M. 1967. Formation and mode of action of algal toxins. *Bacteriological Reviews* 31:18–193.

Sleeter, T. D., P. J. Boyle, A. M. Cundell, and R. Mitchell. 1978. Relationships between marine microorganisms and the wood-boring isopod *Limnoria tripuncata*. *Marine Biology* 45:329–336.

Smith, C. E. G. 1978. "New" viral zoonoses: past, present and future. In M. W. Loutit and J. A. R. Miles (eds.). *Microbial Ecology*. Springer-Verlag, Berlin, pp. 170–174.

Smith, H. 1968. Biochemical challenge of microbial pathogenicity. *Bacteriological Reviews* 32:164–184.

Smith, K. M. 1976. *Virus-Insect Relationships*. Longman Group, Cambridge, England.

Sochard, M. R., D. F. Wilson, B. Austin, and R. R. Colwell. 1979. Bacteria associated with the surface and gut of marine copepods. *Applied and Environmental Microbiology* 37:750–759.

Southward, E. C. 1986. Gill symbionts in thiasirids and other bibalve molluscs. *Journal of Marine Biology* 66:889–914.

Tait, R. V., and R. S. DeSanto. 1972. *Elements of Marine Ecology*. Springer-Verlag, New York, pp. 18–33 and 155–169.

Tannock, G. W. 1984. Control of gastrointestinal pathogens by normal flora. In M. J. Klug and C. A. Reddy (eds.). *Current Perspectives in Microbial Ecology*. American Society for Microbiology, Washington, D.C., pp. 374–382.

Taylor, D. L. 1973a. Algal symbionts of invertebrates. *Annual Reviews of Microbiology* 27:171–187.

Taylor, D. L. 1973b. The cellular interactions of algal-invertebrate symbiosis. *Advances in Marine Biology* 11:1–56.

Taylor, D. L. 1975. Symbiotic dinoflagellates. In *Symbiosis*, Symposium of the Society for Experimental Biology, No. 29. Cambridge University Press, Cambridge, England, pp. 267–278.

Thorn, R. G., and G. L. Barron. 1984. Carnivorous mushrooms. *Science* 224:76–78.

Trust, T. J., J. I. MacInnes, and K. H. Bartlett. 1978. Variations in the intestinal microflora of salmonid fishes. In M. W. Loutit and J. A. R. Miles (eds.). *Microbial Ecology*. Springer-Verlag, Berlin, pp. 250–254.

Turner, J. T., and J. G. Ferrante. 1979. Zooplankton fecal pellets in aquatic ecosystems. *BioScience* 29:670–677.

Van Rie, J. W. H. McGaughey, D. E. Johnson, B. D. Barnett, and H. van Mellaert. 1990. Mechanism of insect resistance to the microbial insecticide *Bacillus thuringiensis*. *Science*. 247:72–74.

Weber, N. A. 1966. Fungus-growing ants. *Science* 153:587–604.

Weber, N. A. 1972. The fungus-culturing behavior of ants. *American Zoologist* 12:577–587.

Willey, R. L., W. R. Bowen, and E. Durban. 1970. Symbiosis between *Euglena* and damselfly nymphs is seasonal. *Science* 170:80–81.

Wolin, M. J. 1979. The rumen fermentation: A model for interactions in anaerobic ecosystems. *Advances in Microbial Ecology* 3:49–78.

Woodroffe, R. C. S., and D. A. Shaw. 1974. Natural control and ecology of microbial populations on skin and hair. In F. A. Skinner and J. G. Carr (eds.). *The Normal Microbial Flora of Man*. Academic Press, London, pp. 13–34.

Wu, M. M. H., C. S. Wu, M. H. Chiang, and S. F. Chou. 1972. Microbial investigations on the suffocation disease of rice in Taiwan. *Plant and Soil* 37:329–344.

Yamin, M. A. 1981. Cellulose metabolism by the flagellate *Trichonympha* from a termite is independent of endosymbiotic bacteria. *Science* 211:58–59.

Yonge, C. M. 1963. The biology of coral reefs. *Advances in Marine Biology* 1:209–260.

ZoBell, C. E., and C. B. Feltham. 1937. Bacteria as food for certain marine invertebrates. *Journal of Marine Research* 1:312–327.

6

Microbial Communities and Ecosystems

DEVELOPMENT OF MICROBIAL COMMUNITIES

The assemblage of microorganisms living together constitute a community. Studies that examine communities fall in the realm of synecology, which deals with the study of interactions between the various populations. Such studies are in contrast to autecological studies which examine an individual organism or population in relation to its environment; autecological studies emphasize life history and behavior of individual populations as a means of adaptation to their environment.

The community is the highest biological unit in an ecological hierarchy made up of individuals and populations (Figure 6.1). It is a unified assemblage of populations occurring and interacting at a given location called a habitat. The species assemblage that inhabits a volume of resource, such as a leaf, is called a unit community. The unit community is defined in a clear manner, such that populations within the community tend to interact with each other and not with populations in other communities (Swift 1984). Within the community, the populations using the

same resources comprise the guild structure. Each population within a community has a specialized functional role called a niche. There are a limited number of niches within a community, and these are filled by the indigenous populations of that community.

Figure 6.1

Levels of ecological organization.

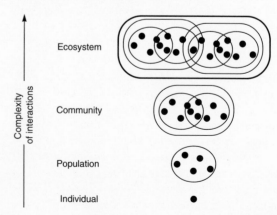

Population Selection within Communities: *r* and *K* Strategies

Microorganisms, like higher plants and animals, have evolved strategies that enable them to successfully survive and maintain themselves within communities. One artificial scheme for viewing these strategies classifies organisms along an *r–K* gradient (Andrews 1984). The *r* strategists rely upon high reproductive rates for continued survival within the community, whereas the *K* strategists depend upon physiological adaptations to the environmental resources or carrying capacity of the environment.

An *r* strategist microorganism would be one that, through rapid growth rates, takes over and dominates situations in which resources are temporarily abundant. While *r* strategists have high reproductive rates, they have few other competitive adaptations; they tend to prevail in situations that are not resource limited, that is, where nutrients are not severely limiting, and where high reproduction rates outweigh the advantages of other competitive adaptations. Fungi and bacteria, such as pseudomonads, which rapidly colonize and grow on sugar-rich plant materials that fall into soil, are *r* strategists. Cyanobacteria and dinoflagellates that respond to phosphate or other mineral nutrient enrichment with an explosive bloom also are *r* strategists. Similarly, *Aspergillus, Penicillium, Pseudomonas, Bacillus,* and similar heterotrophs are *r* strategists because they rapidly colonize and degrade easily available high concentrations of organic matter (Andrews and Hall 1986).

Populations of *r* strategists are subject to extreme fluctuations. They tend to prevail in uncrowded communities and devote a large portion of their resources to reproduction. When resources become scarce or conditions turn unfavorable, their populations crash, that is, they experience rapid reduction. Formation of abundant and resistant spores for dispersal and survival during long inactive periods is of great advantage to such microbial *r* strategists.

Although in comparison to macroorganisms all microorganisms might appear to be *r* strategists, and microorganisms have generally been viewed in this manner, compared to each other some of them can be considered *K* strategists. For example, Sergei Winogradsky's zymogenous (opportunistic) soil populations closely correspond to the concept of *r* strategy, and his autochthonous (humus-degrading) populations correspond to the concept of *K* strategy (Waksman 1953). *K* strategists, which reproduce more slowly than *r* strategists, tend to be successful in resource-limited situations. Populations of *K* strategists are usually more stable and permanent members of the community. They prevail under conditions of crowding and devote a smaller portion of their resources to reproduction. Soil streptomycetes that grow slowly on complex soil organic matter exemplify K strategists.

Other examples of *K* strategist microorganisms are desmids in oligotrophic lakes and ponds; marine spirilla and vibrios and freshwater prosthecate bacteria, which are able to use extremely dilute concentrations of organic matter; *Agrobacterium, Corynebacterium,* and similar humus-degrading soil bacteria; and Basidiomycota, which degrade the cellulosic and lignin components of forest litter.

Succession within Microbial Communities

The individual populations of a community fill the niches in that ecosystem. With time, some populations are replaced by other populations that are better adapted to fill a functional role (ecological niche) within the ecosystem; that is, the structure of the community evolves with time. The types of interrelationships among populations in a community, as well as adaptations within populations, contribute to the ecological stability of the community. Some interrelationships involving microbial populations are loose associations in which one microbial population can replace another; others are tight associations in which one microbial population cannot replace another.

Development of a more-or-less stable community usually involves a succession of populations, that is, an orderly sequential change in the populations of the community. Community succession begins with colonization or invasion of a habitat by microbial populations (Golley 1977). If the habitat has not been previously colonized—for example, the gastrointestinal tract of newborn animals—the process is known as primary succession. When succession occurs in a habitat with a previous colonization and succession history, it is called secondary succession. Secondary succession is the consequence of some catastrophic event that has disrupted and altered the course of primary succession.

The first colonizers of a virgin environment are called pioneer organisms. All pioneer microorganisms must be able to reach the virgin environment, so a common feature of pioneer microorganisms is effective dispersal mechanisms. Beyond this, pioneer characteristics vary with the environment to be colonized.

Preemptive colonization occurs when pioneer organisms alter conditions in the habitat in ways that discourage further succession. Preemptive colonization may extend the reign of pioneer organisms, but populations better adapted to the now-colonized and thus altered habitat usually replace the pioneers. As this habitat undergoes additional changes, secondary invaders are also replaced.

Succession ends when a relatively stable assemblage of populations, called a climax community, is achieved. According to classical ecological thinking, such climax communities represent a state of equilibrium; current ecological thinking is that equilibrium and climax communities rarely occur, but rather disturbances randomly disrupt the successional process, preventing the community from ever reaching equilibrium (Lewin 1983; Wiens 1983). It has always been difficult to apply the concept of climax to microbial communities. In some situations, though, regular successional population changes of microorganisms occur, leading to a relatively stable microbial community.

In some successional processes, microorganisms modify the habitat in a way that permits new populations to develop; for example, the creation of anaerobic conditions by facultative anaerobes allows the development of obligate anaerobic populations. This is known as autogenic succession. In contrast, allogenic succession occurs when a habitat is altered by environmental factors, such as seasonal changes. Even when the successional process follows a predictable sequence of population changes, the causative factors responsible for this orderly sequence are often poorly understood.

The rapid generation times of many microorganisms can lead to large population fluctuations, environmental changes may preclude the orderly succession of microbial communities, or initial random events may determine which microorganisms fill the niches of the ecosystem and direct the sequence of successional events that follow; hence, succession to a climax microbial community does not occur in many habitats. Even if an equilibrium in species diversity is attained, this can rarely persist indefinitely because disturbances intervene. Disturbance alters the terms of whatever interactive equilibrium has been attained among the populations in the community, promoting the accelerated extinction of some species and facilitating the immigration of new species (Swift 1984).

Autotrophic–Heterotrophic Succession   When gross photosynthesis (P) exceeds the rate of community respiration (R), organic matter accumulates. Autotrophic succession occurs in cases where P/R is initially greater than 1. As long as P is greater than R, biomass will accumulate during the autotrophic succession. As the P/R ratio approaches 1, succession toward a stable community is occurring. An autotrophic succession of microorganisms occurs in environments largely devoid of organic matter when there is a nonlimiting supply of solar energy. Autotrophic succession occurs in young pioneer communities, such as on newly exposed volcanic rock. In autotrophic succession within a mineral environment, such as on bare rock, the photosynthetic pioneer organisms have minimal nutritional requirements and high tolerance to adverse environmental conditions. The ability to use atmospheric nitrogen is an advantage; terrestrial cyanobacteria and lichens are good examples of pioneers in this type of environment.

As opposed to autotrophic succession, organic matter will disappear when P/R is less than 1 because consumption is greater than production. Succession in such a situation is called heterotrophic succession. In heterotrophic succession, the energy flow through the system decreases with time; there is insufficient organic matter input, and the community gradually dissipates its stored chemical energy. Heterotrophic succession is usually temporary, because it culminates in the extinction of the community when the stored energy supply is exhausted. Many microbial communities involved in decompositional processes exhibit such temporary heterotrophic succession. For example, the microbial communities on a fallen log disappear after the log is completely decomposed.

It is possible for heterotrophic succession to lead to a stable community if there is a continuous source of allochthonous organic matter, that is, organic matter from an external source. As an example, heterotrophic succession in the microbial community of the gut leads to a stable climax community as long as there is regular input of food. If the animal stops feeding, however, the microbial community is disrupted and rapidly eliminated. Pioneers in a heterotrophic succession need to have, above all, high metabolic and growth rates in order to stay ahead of secondary invaders.

Examples of Successional Processes   An interesting heterotrophic successional process occurs on detrital particles that enter aquatic habitats (Fenchel and Jørgensen 1977) (Figure 6.2). Fresh particulate detritus consists mainly of mechanically shredded tissue of dead leaves, roots, stems, or thallus of macrophytes mixed with smaller amounts of debris from other sources. Microbial communities associated with detritus are complex, but predictable population changes occur during succession. If sterilized natural detrital particles are placed in sea- or freshwater inoculated with a small amount of natural detritus, a characteristic succession of organisms occurs. This succession leads to a microbial community closely resembling that of natural detritus. Bacteria occur in small numbers on the particles after 6 to 8 hours and reach their maximal numbers after a period of 15 to 150 hours. The bacterial populations then decrease and become relatively stable after about 200 hours. Small zooflagellates appear about 20 hours after inoculation and reach maximal population sizes after 100 to 200 hours. Ciliates appear after about 100 hours and reach maximal numbers between 200 to 300 hours. Other groups of microorganisms, including rhizopods and diatoms, usually appear late in the succession.

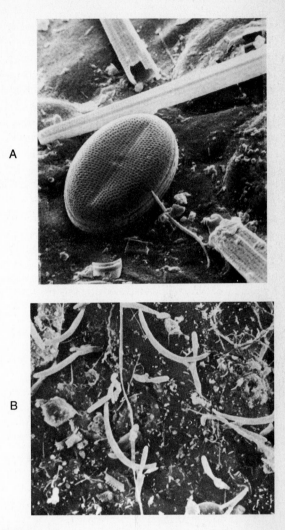

Figure 6.2

Scanning electron micrographs of detrital microorganisms on submerged surfaces: (A) Diatoms on leaf surface (Courtesy of C. Versfeld) (B) Large curved bacteria on leaf litter in a woodland stream. (Source: Suberkropp and Klug 1974. Reprinted by permission, copyright Springer-Verlag.)

Environmental factors, such as temperature, influence the time sequence of succession. The individual populations of the microbial community associated with detrital particles interact as a predator-prey system: The bacteria are the prey and the protozoa are the primary predators. The overall result of the microbial succession on detrital particles is the decomposition of some of the nitrogen-poor plant polymers and their partial replacement with nitrogen-rich microbial biomass. For most detritus-feeding invertebrates and vertebrates, this microbial upgrading of the detritus is essential. Few comparable examples for the succession of bacterial populations on plant material that enters the soil are known. The succession of fungal populations on leaf litter, however, has been studied (Kendrick and Burges 1962) (Figure 6.3). A parallel succession of bacterial populations probably occurs simultaneously and may affect the observed fungal succession.

An interesting observation about community succession in soil is that nitrification (the conversion of ammonium to nitrate) is inhibited in many climax ecosystems (Rice and Pancholy 1972). Populations of nitrifying bacteria decrease or disappear in forest and grassland ecosystems during succession to a climax community, resulting in the accumulation of ammonium nitrogen. This successional process appears to have adaptive value, since ammonium ions are less readily leached from soil than nitrate ions.

Interesting successions of microbial communities are associated with animal tissues (Marples 1965; Skinner and Carr 1974; Noble and Pitcher 1978). The sterile intestinal and skin tissues of newborn animals permits observation of community succession from the time of initial colonization. The population levels and types of microbes in climax communities in the gastrointestinal ecosystems are regulated by multifactorial processes. Some of the regulatory forces in these processes are exerted by the animal hosts, some by the microbes, some by diet, and some by the environment. Within the gastrointestinal tract are many niches filled by various microbial populations. The succession of bacterial populations in humans and other nonruminant mammals normally begins with colonization of the gastrointestinal tract by *Bifidobacterium* and *Lactobacillus* species. This is followed by a succession of facultative anaerobes, such as *Escherichia coli* and *Streptococcus faecalis*. Populations of strictly anaerobic bacteria, such as *Bacteroides*, appear late in the succession, after the beginning of solid food ingestion. These populations of obligate anaerobes become dominant. Climax microbial communities in gastrointestinal tracts are generally reached by the time of weaning.

As a specific example, the intestinal microbial community of mice initially is composed of lactic acid bacteria, *Flavobacterium*, and enterococci (Schaedler et al. 1965) (Figure 6.4). Populations of

Figure 6.3

Succession of fungal populations on pine-needle litter from the initial populations on the live needles (lower left) to the small decomposed pieces in the A_0 layer (upper right). The vertical width of the bar indicates the frequency of occurrence for observed portions of pine needles of the same size; the wider the bar, the greater the frequency. (Source: Kendrick and Burges 1962. Reprinted by permission of Braunschweig Verlag von J. Cramer.)

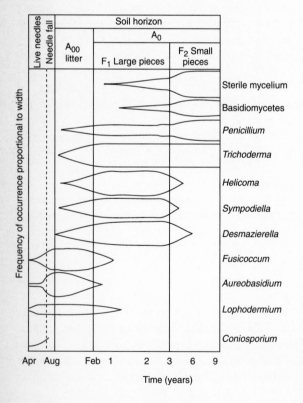

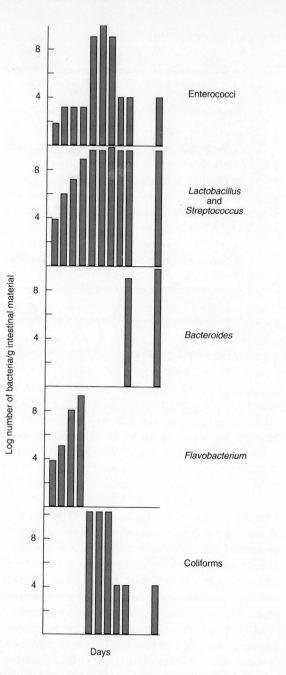

Figure 6.4

Succession of microbial populations in the large intestine of mice following birth. The data are expressed as log number per gram of large intestine homogenate. (Source: Schaedler et al. 1965.)

Flavobacterium increase for about eight days, after which they disappear from the intestinal microbial community. Populations of enterococci and coliforms increase dramatically with the disappearance of *Flavobacterium,* but decrease to lower population levels several days later. Lactic acid bacteria increase regularly for ten days, after which their numbers remain relatively constant. Populations of obligately anaerobic *Bacteroides* are absent or present in very low numbers in the initial colonizing microbial community. After eighteen days, however, there is a dramatic increase in the population of *Bacteroides,* which becomes the dominant population in the climax community.

In ruminants, succession leads to the development of a complex obligately anaerobic microbial community (Hungate 1975). Included in the climax community of the rumen are populations of cellulosedegrading bacteria, such as *Bacteroides* and *Ruminococcus*; starch-degrading bacteria, such as *Selenomonas;* protein-degrading bacteria, such as *Veillonella;* methanogenic bacteria, such as *Methanobacterium;* cellulose- and pectin-degrading protozoa, such as *Polyplastron;* and other populations. There are important predator-prey relations between the bacterial and protozoan populations. The protozoan populations, predominantly ciliates, appear late in the succession, following the development of the complex bacterial community. The pioneer bacterial community modifies the environment with the production of various volatile acids and the removal of oxygen, allowing succession to proceed to the climax community.

Homeostasis and Secondary Succession   Many established communities have a high degree of stability, that is, they are resistant to change. Part of this apparent stability is based upon homeostasis, a compensating mechanism that acts to maintain steady-state conditions and, by a variety of control mechanisms, to counteract perturbations that would upset this steady state. The concept of a stable community does not imply static conditions. Individual populations are subject to regular and irregular fluctuations. These fluctuations occur in response to internal or external conditions and contribute to the maintenance of overall ecosystem stability. As an example, an accumulation of nitrite or hydrogen sulfide in the ecosystem temporarily increases the populations that use these metabolic intermediates; in turn, these increased populations lower the concentrations of materials that otherwise would accumulate to toxic levels. Population shifts may also occur in response to overall environmental conditions, such as seasonal light and temperature changes. A metabolic niche filled by a mesophilic population during the summer season may be occupied by a psychrophilic population during the winter, but the metabolic function that may be vital for the ecosystem is performed in either case. Such seasonal shifts in populations are known to occur on a regular basis.

Some microbial populations exhibit annual rhythms. For example, *Vibrio parahaemolyticus* exhibits an annual cycle in estuaries (Kaneko and Colwell 1975). In temperate zones, this organism occurs in the water column during the spring and summer months but disappears during the winter. The cycle is based in part on low winter water temperatures. Some members of this bacterial population survive in sediment during the winter, allowing for continuance of the cycle.

The occurrence of regularly timed population fluctuations raises the concept of a temporal niche; microorganisms may occupy a niche in a habitat at one particular time but not at another. Various algal populations in many aquatic habitats, for example, exhibit seasonal succession (Stockner 1968) (Figure 6.5). Under the various temperature and light regimes that occur during each season, different populations fill the niches of the ecosystem. In some cases, the existence of temporal niches may act to diminish direct competition between populations, allowing the coexistence of some populations that appear to compete for identical resources within spatially overlapping habitats.

Some seasonal population fluctuations can be viewed as repetitive successions toward a stable community that is repeatedly upset by the abrupt environmental changes associated with the change of seasons. An example of such a succession occurs annually in the nearshore regions of the Arctic Ocean

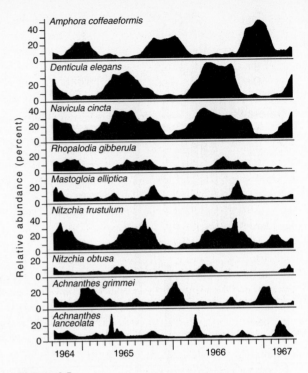

Figure 6.5

Fluctuations of diatom populations in a thermal spring showing regularity of seasonal diatom populations change, that is, annual succession. (Source: Stockner 1968. Reprinted by permission of British Phycology Society.)

(Kaneko et al. 1977). Each spring, there are large blooms of algae on the underside of the sea ice; the blooms occur at the same time each year and represent a regular seasonal successional event (Horner and Alexander 1972). Bacterial populations, predominantly *Flavobacterium* and *Microcyclus* species, flourish after the algal bloom under the ice (Kaneko et al. 1979). As the ice melts, the habitat is removed, ending community succession. The algal populations released into the water column diminish and are consumed by predators or are subjected to bacterial decomposition. During the winter, the pigmented bacterial populations also disappear from the surface waters, but some algal and pigmented bacterial populations survive in the sediment. When winter ice forms again, it establishes a suitable habitat for the recurrence of this seasonal succession. With the return of sufficient sunlight in the spring, algae colonize under the ice and the process repeats itself.

Occasionally, severe or catastrophic environmental changes may overwhelm the ecosystem's homeostatic controls, destroying the existing community and initiating of a secondary successional process. Examples of such catastrophic events include the introduction of pollutants into aquatic or terrestrial ecosystems, volcanic eruptions, fungicide applications to soils and plants, and many others.

After a disturbance, homeostasis acts to restore the disrupted community, and once the disrupting factor is removed, usually there is a secondary succession back to the original community. For example, washing the skin disrupts the microbial community of that habitat. Washing normally does not result in a new climax community but rather initiates a succession back to the original microbial community (Marples 1974) (Figure 6.6). Similarly, the microbial community in an agricultural soil returns to its original composition following tillage.

Homeostasis also acts to restore the initial community when alien or allochthonous microorganisms enter the habitat. For example, many of the microorganisms that enter the gastrointestinal tract on food particles are quickly eliminated, and even after the ingestion of microorganisms that cause disease, the climax microbial community of the gastrointestinal tract eventually returns to normal. Similarly, allochthonous microorganisms that enter soil and aquatic habitats are transient members of the microbial community, and antagonistic interactions (negative feedback) act to remove the alien microbial populations, thereby restoring the original composition of the community.

Population interactions that lead to the establishment of a defined community are most likely based on physiological interactions between the various populations. The functional roles of specific populations within the communities of certain ecosystems have been defined; the interspecies population relationships within the rumen ecosystem, for example, have been defined, leading to a relatively complete

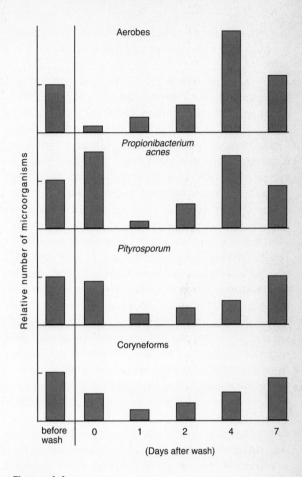

Figure 6.6

Succession of microorganisms on the human scalp following shampooing. The figure demonstrates the rapid recovery that occurs following this disturbance. (Source: Marples 1974. Reprinted by permission of Academic Press, London.)

understanding of community structure and ecosystem function (Hungate 1975). Work using chemostats has elucidated some of the interactions between microbial populations that lead to the establishment of stable community structures in aquatic ecosystems (Slater 1978, 1980). In chemostat studies, stability often occurs when several interacting populations cooperate to best exploit the available resources. In some cases two species can constitute a stable community structure, whereas in other experiments additional member populations are needed before community stability is achieved.

Assuming stable environmental conditions, organisms that do not successfully fill a niche or functional role tend to be eliminated from a community (MacArthur and Wilson 1967; Whittaker 1975; May 1976). Competition between two populations tends to eliminate one of the populations from the common habitat, especially when competition is focused on a single resource and when the populations do not otherwise interact; however, a number of factors mitigate the severity of competition, and thus competitors often coexist (Fredrickson and Stephanopoulos 1981; Lewin 1983). Changing conditions, such as seasonal and diurnal fluctuations (time-varying inputs), multiplicity of resources, spatial heterogeneity of habitats, regulation by predators, dormancy, and various other factors can preclude competitive exclusion and permit coexistence of populations apparently occupying the identical niche. W. Michael Kemp and William Mitsch (1979) developed a model to explain the paradox of the plankton and explained why diverse phytoplankton species apparently occupy the same niche. They found that turbulence prevented exclusion of competing populations; the underlying principle is that discontinuities in the environment permit the development of diverse stable phytoplankton communities with overlapping niches.

Genetic Exchange in Microbial Communities

When adaptive features are introduced into the gene pool, the rapid reproductive rates of microorganisms, which may be as short as fifteen to twenty minutes, allow for the quick and widespread expression of the features. The rapid spread of antibiotic resistance in bacteria is a good example of the action of natural selection. Antibiotic resistance arose by spontaneous mutation or recombination before the medical use of these substances, but during this period there was no great selective advantage for pathogenic microorganisms to possess the characteristic of antibiotic resis-

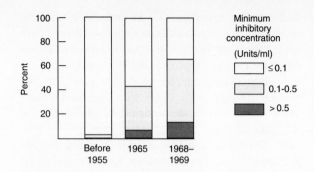

Figure 6.7
Development of antibiotic (penicillin) resistance in gonococcal populations during the period 1950–1969. The increasing occurrence of resistant strains coincides with the increased use of antibiotics and reflects selective pressure. (Source: Martin et al. 1970.)

tance. Subsequent to the 1950s, when antibiotic use became prevalent, microorganisms capable of continued growth, that is, infection, in an individual receiving antibiotics acquired a particular selective advantage. Within habitats receiving frequent dosages of antibiotics, such as hospitals, resistance to antibiotics has become widespread and an increasingly common characteristic of disease-causing microorganisms (Martin et al. 1970) (Figure 6.7).

A critical factor in determining the persistence of any population within a community is its genetic fitness, that is, the contribution of one or more gene alleles of the population to succeeding generations (Lenski 1992). Stability of a community depends upon the totality of the genes within the individual populations. Genes can be transferred to new populations within the community to form new allelic combinations with differing degrees of fitness (Levy and Miller 1989; Drahos and Barry 1992). Differences in fitness between alleles or genotypes reflect systematic differences in either mortality or reproduction, which in turn reflect systematic differences in ecological properties such as the ability to compete for limiting resources, susceptibility to predation, and so on. Processes that bring about a systematic change in the frequency of alleles include mutation, recombination, and genetic drift (a random change in the frequencies of alleles within a population) (Lenski 1992). Genetic drift differs from selection in that changes in the frequencies of alleles are due to chance events rather than to systematic differences in ecological properties such as competitive ability. Genetic drift may cause some change in the relative frequency of two selectively neutral strains, including even the extinction of one or the other.

Three principal mechanisms of genetic transfer and recombination lead to new combinations of alleles among bacteria: conjugation, a process that involves direct contact between donor and recipient cells; transduction, a process that involves bacteriophage-mediated transfer of DNA from donor to recipient bacterial cell; and transformation, a process that involves absorption of free DNA by a competent recipient cell. There is a relatively high potential for gene transfer by these mechanisms in the environment, particularly when population densities are high, but there are also significant restrictions to potential recombination (Freter 1984; Stotzky 1989; Fry and Day 1990; Miller 1992; Saye and O'Morchoe 1992; Stewart 1992) (Figure 6.8). Restriction enzymes degrade foreign DNA that is not protected by specific methylation of DNA nucleotides. Nevertheless, it now appears that genes are transferred in the environment at relatively higher rates than initially thought and that a major force for the change and evolution of natural populations is the introduction of new genetic elements into the gene pool through gene transfer and genetic recombination (Miller 1992).

Plasmids are especially important in the rapid transfer among populations in a microbial community (Beringer and Hirsch 1984; DeFlaun and Levy 1989). Hospital wastes, raw sewage, sewage effluents, fresh and marine water, animal feedlots, plants, and soils have all been shown to contain bacteria that transfer plasmids by conjugation. Antibiotic resistance genes are rapidly disseminated among bacterial populations, particularly under selective pressure. Plasmids with antibiotic resistance genes, as well as those with degradative pathway genes, can be transferred to a wide variety of bacterial species in many genera (Miller 1992). Plasmids can behave dynamically in the environment. While certain specialized plasmidborne genes

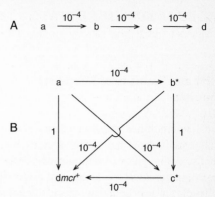

Figure 6.8

The effect of methylation and restriction on gene transfer among several bacterial strains. (A) Normal transfer frequencies. (B) Altered frequencies of gene exchange caused by methylation. The asterisk (*) denotes cytosine methylating host strain; mcr^+ denotes strain that cleaves C-methylated DNA. Average restricted frequencies of transfer are given as a proportion of the nonrestricted frequency. (Source: Saunders et al. 1990).

are conserved through many generations, others are rapidly lost (Drahos and Barry 1992).

Genes that contribute to the fitness of populations are usually maintained within the community, and those that do not are lost from that community. Nonessential genes should not be conserved, particularly if there is significant competition for an ecological niche, because the expression of a nonessential gene can have a marked inhibition on the relative growth rate of the host cell. Gene transfer, however, can maintain an allele or extrachromosomal element in a population in the face of opposing selection (Stewart and Levin 1977). One strain that is less fit than another can be maintained by recurring mutation or by migration from another source population. One allele that is less fit than another may also be maintained in a population by virtue of its association with a favorable allele elsewhere in the genome (Lenski 1992). Such linkage disequilibrium is prevalent in bacteria because of their asexual reproduction and recombination due to conjugation, transformation, and transduction.

STRUCTURE OF MICROBIAL COMMUNITIES

Diversity and Stability of Microbial Communities

Biological communities usually contain a few species with many individuals and many species with few individuals. Although a few dominant species normally account for most of the energy flow within a trophic level, the less abundant species determine in large part the species diversity of that trophic level and the whole community. Diversity generally decreases when one or a few populations attain high densities; high numbers signify successful competition and dominance of a single population.

The maturity of a community is reflected in both its diversity and its productivity. Mature ecosystems are complex; the presence of a high number of species (high diversity) allows for many interspecies relationships. A community that has a complex structure, rich in information as reflected by high species richness (number of different species), needs a lower amount of energy for maintaining such structure. This lowered energy requirement is reflected in a lower primary production rate per unit biomass, while a stable diversity level is maintained. The strong inverse relationship between diversity and productivity is especially pronounced when environmental changes favor rapid microbial growth (Margalef 1979; Revelante and Gilmartin 1980).

Species diversity of a community is somewhat like the genetic diversity of a population in that it allows for varied responses within a dynamic ecosystem. If an environment is dominated by a strong unidirectional factor, less flexibility is needed to maintain stability. In such cases, it is adaptive for populations to become stenotolerant (highly specialized) and for communities to be dominated by a few populations (low species diversity). Microbial populations in a salt lake, for example, normally are more stenohaline than populations in an estuary; microbial communities in a hot spring are less diverse than in an unpolluted river.

Species diversity tends to be low in physically controlled ecosystems because adaptations to the prevailing physicochemical stress are of the highest priority and leave little room for the evolution of closely balanced and integrated species interactions. Acid bogs, hot springs, and Antarctic desert habitats are examples of physically controlled habitats where species diversity is relatively low. Species diversity tends to be higher in biologically controlled ecosystems, that is, where the importance of interpopulation interactions outweighs that of abiotic stress. In such biologically accommodated communities the physicochemical environment allows greater interspecies adaptation, resulting in species-rich associations. Microbial diversity in many habitats, such as soil, is normally high; in contrast, under conditions of stress or disturbance, such as in infected plant or animal tissues, diversity is markedly low.

Although stability is associated with high diversity in biologically acclimated microbial communities, there is no established cause and effect relationship between diversity and stability. It is clear that no one population is all important in a community with high diversity, so that even if a population is eliminated the whole community structure will not be disrupted, but it is not clear what level of diversity is necessary to maintain community stability.

The fact that communities with high diversities are able to cope with environmental fluctuations within broad tolerance ranges does not imply that they are able to cope with severe and continued environmental disturbances. The diverse, stable community of activated sludge, for example, can tolerate the influx of low concentrations of many toxic chemicals, but high inputs of some toxic chemicals can cause the community to collapse. As a case in point, an input of hexachloropentadiene and octachloropentene to the municipal sewerage system in Louisville, Kentucky, devastated the microbial community of the treatment plant and caused the cessation of sewage treatment of this municipality for months.

Various investigations on the colonization of substrates by diatoms have examined the premise that the nature of the processes of substrate invasion and succession determines the composition of the community and the diversity of the stable community (Patrick 1963, 1967). The size of the area, the number of species in the pool capable of invading the substrate, and the invasion rate all influence species richness and overall diversity of the diatom community. A reduced invasion rate, with the size, area, and number of species in the species pool remaining the same, lowers the total number of species in the community. Those species that normally are low in numbers within diatom communities are particularly affected, and this decrease in species diversity is seen as lowered species richness. One of the main results of a high invasion rate is that species with relatively low numbers are maintained within the community; thus the presence of rare species is ensured. Such rare species increase diversity and may act to stabilize a community during variable environmental conditions. If environmental conditions change, they may be better adapted to survival than the common species, and species composition of the community, especially relative abundances, might shift. The size of the area available for invasion affects the composition of the community. As species invade the area, the rate of increase of species richness declines, but increasing numbers of species within the community need not alter the biomass of that community. Eventually the number of species reaches a level of stability.

During the early stages of community succession, the number of species tends to increase. Species diversity probably peaks during the early or middle stages of succession and may decline inordinately in the stable climax community. A major question in ecological theory concerns the optimal relationship between diversity and community stability.

Several investigators have used protozoa as experimental organisms to examine the influence of predator-prey interactions on community diversity and stability (Hairston et al. 1968; Luckinbill 1979). The stability of predatory protozoa (*Paramecium*) populations was increased by increasing diversity at the bacterium level, but three species of *Paramecium* were less stable than two-species communities; one pair of *Paramecium* species consistently had greater stability without the third species than with it, indicating that there were significant second-order effects

with two species having an interaction that was detrimental to the third species (Hairston et al. 1968). Stability of a predatory protozoa population depends on the characteristics of the particular species serving as prey and not simply on the diversity of species present (Luckinbill 1979).

Species Diversity Indices

Several mathematical indices that describe the species richness and apportionment of species within the community, called species diversity indices, are used to describe the assemblage of populations within a community (Pielou 1975) (Table 6.1). Species diversity indices have rarely been applied to microbial communities because of the technical difficulties in speciating the large numbers of microorganisms their use requires (Atlas 1984a). Microbial ecologists often use numerical taxonomy to determine the microbial species (taxonomic units) present in a sample (Kaneko et al. 1977; Holder-Franklin et al. 1981; Sørheim et al. 1989; Bascomb and Colwell 1992). In numerical taxonomy, a large number of characteristics—often phenotypic characteristics—are determined for organisms isolated from a sample, and cluster analyses are then performed to establish the similarities of the organisms. Similar organisms are considered to belong to the same species.

Species diversity indices relate the number of species and the relative importance of individual species. The two major components of species diversity are species richness, or variety, and evenness, or equitability. Species richness can be expressed by simple ratios between total species and total numbers. It measures the number of species in the community but not how many individuals species represent. Equitability measures the proportion of individuals among the species present; this indicates whether there are dominant populations.

A widely used measure of diversity is the Shannon-Weaver index (Shannon and Weaver 1963). This general diversity index is sensitive to both species richness and relative species abundance. Caution must be used in interpreting the Shannon-Weaver

Table 6.1

Examples of diversity indices

Species Richness (d)

$$d = \frac{S-1}{\log N}$$ where S = number of species

N = number of individuals

Shannon-Weaver Index of Diversity ($\overline{H}$)

$$\overline{H} = \frac{C}{N}(N\log N - \sum n_i \log n_i)$$

where $C = 2.3$

N = number of individuals

n_i = number of individuals in the ith species

Evenness (e)

$$e = \frac{\overline{H}}{\log S}$$ where $\overline{H}$ = Shannon-Weaver diversity index

S = number of species

Equitability (J)

$$J = \frac{\overline{H}}{H_{max}}$$ where $\overline{H}$ = Shannon-Weaver diversity index

H_{max} = theoretical maximal Shannon-Weaver diversity index for the population examined—assumes each species has only one member.

---

index, because it is sensitive to sample size, especially with small samples. Equitability, which is independent of sample size, can be calculated from the Shannon-Weaver index. Another approach, known as rarefaction, is to compare the observed number of species with those predicted by a computer model (Simberloff 1978). This approach has been applied to microbial communities (Mills and Wassel 1980).

Theoretically, diversity should increase during succession. This increase in diversity has been observed for the periphyton community of Lake Washington, in the state of Washington, using submerged grids and electron microscopic observations (Jordan and Staley 1976). Diversity, measured with the Shannon-Weaver index, increased during a ten-day period (Table 6.2). During this time, some pioneer populations disappeared, and the relative proportions of biomass shifted from heterotrophic bacteria to algae and cyanobacteria.

Diversity generally is lower in communities under stress (Atlas 1984b; Atlas et al. 1991). For example, diversity calculated with the Shannon-Weaver index is lower in surface water bacterial communities in the Arctic Ocean than in temperate oceans where low temperatures and ice do not cause the same degree of physical stress (Kaneko et al. 1977; Atlas 1984a, 1984b). Disturbances, such as the introduction of pollutants into aquatic ecosystems, have been found to reduce diversity in diatom communities (Patrick 1963) (Figure 6.9) and in bacterial communities (Sayler et al.

Table 6.2

Successional changes in microbial community on submerged grids

Time (days)	Diversity (H')	Biomass (bacteria/algae)
1	3.1	6.07
3	4.2	0.23
6	4.4	0.31
10	4.8	—

Source: Jordan and Staley 1976.

Table 6.3

Effect of exposure to oil on diversity of bacterial populations in arctic seawater

Time[*]	Diversity[†] ($\bar{H}$)	Equitability (J)
0	3.5	0.76
2	3.6	0.78
4	2.4	0.52
6	2.1	0.46

*Weeks of *in situ* exposure to Prudhoe Bay crude oil.
† Controls not exposed to oil had $\bar{H}$ values (Shannon–Weaver indices) of 3.4–3.7 and J values of 0.74–0.80 throughout this period.

1982, 1983; Atlas 1984a, 1984b; Atlas et al. 1991; Mills and Mallory 1987; Bej et al. 1991) (Table 6.3). A rapid decrease in the diversity of protozoan communities was observed after a temporary thermal shock, followed by a return to diversity levels of control communities (Cairns 1969) (Figure 6.10) Diversity of microbial communities is a sensitive index of pollution. Theoretically, stressed communities with low diversities are less well adapted to deal with further environmental fluctuations and stress than biologically accommodated communities with higher diversities.

Figure 6.9

Effects of pollution on diversity of aquatic diatom populations. Magnitude of pollution: Ridley Creek < Egg Harbor ≤ Back River. Decreases in diversity are due to pollution stress. (Source: Patrick 1963. Reprinted by permission of New York Academy of Sciences.)

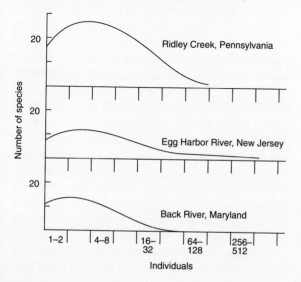

Figure 6.10

Effect of thermal shock on species diversity of a protozoan community. This experiment shows the rapid decrease in diversity that occurs following thermal shock and subsequent recovery. (Source: Cairns 1969. Reprinted by permission of J. Cairns, Jr.)

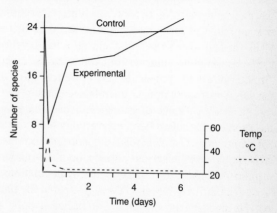

Genetic Diversity Indices

The genetic diversity of a microbial community can be assessed by measuring the heterogeneity of the DNA from the entire microbial community. To do this, DNA first must be extracted from the total microbial community in a sample collected from a given habitat. The recovered DNA represents the total gene pool of the community. DNA can be recovered from environmental samples by collecting the microbial cells—usually by centrifugation—and then lysing the cells, or by directly lysing the cells in the sample and then extracting the DNA (Torsvik 1980; Ogram et al. 1988; Fuhrman et al. 1988; Steffan et al. 1988; Somerville et al. 1989). It is important that impurities be removed from the DNA. Using polyvinylpyrrolidone during the extraction procedure helps remove humic materials that interfere with measurement of the heterogeneity of the DNA. The DNA is further purified using hydroxyapatite column chromatography and/or cesium chloride buoyant density centrifugation. The DNA is then sheared and heated to separate the double-stranded DNA into relatively short single strands. The temperature is lowered, and the rate of reannealing of the DNA to reform double-stranded DNA is determined by measuring the absorbance at 260 nm.

The rate of DNA reannealing is a measure of DNA heterogeneity (Wetmur and Davidson 1968; Britten et al. 1974) (Figure 6.11). The greater the similarity of DNA fragments (low genetic diversity), the more rapid the rate of DNA reannealing. The greater the heterogeneity of the DNA (high genetic diversity), the slower the rate of DNA reannealing. The initial concentration of DNA multiplied by the time that it takes for half of the DNA to reanneal, a value called $C_0t_{1/2}$, is a useful parameter for expressing the diversity of DNA in the population. Dividing the $C_0t_{1/2}$ of the DNA extracted from a microbial community by the mean $C_0t_{1/2}$ value of genomes of bacteria gives the number of totally different bacterial populations in the community. C_0t plots for DNA isolated from bacterial populations take into account both the amount of information and its distribution in the populations.

Vigdis Torsvik et al. (1990a) compared the phenotypic and genetic diversities of bacterial popula-

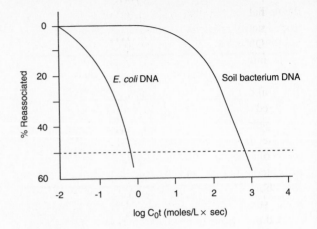

Figure 6.11

Reassociation of soil bacterial DNA (C_0t plot). The reassociation of *Escherichia coli* DNA is included for comparison. The abscissa gives the log initial concentration of single-stranded DNA (in mole-nucleotides per liter) multiplied by time in seconds. The ordinate gives the percent of reassociated DNA. (Source: Torsvik et al. 1990).

tions isolated from a soil community. The phenotypic diversity as determined by the Shannon–Weaver index, equitability, rarefaction, and cumulative differences was high but indicated some dominant biotypes. Genetic diversity was measured by reassociation of mixtures of denatured DNA isolated from the bacterial strains (C_0t plots). The observed genetic diversity was high. Reassociation of DNA from all bacterial strains together revealed that the population contained heterologous DNA equivalent to twenty totally different bacterial genomes, that is, genomes without homology. This study showed that reassociation of DNA isolated from a collection of bacteria gave a good estimate of the diversity of the collection and that there was good agreement among different phenotypic diversity measures and in particular the Shannon–Weaver diversity index. C_0t plots, like the Shannon–Weaver index, measure both the number of different populations (different genomes) and the relative proportions of those populations.

Torsvik et al. (1990b) also examined the heterogeneity of DNA extracted directly from soil. They found that a major part of DNA isolated from the bacterial fraction of soil is very heterogeneous. The $C_0t_{1/2}$ of the DNA recovered from the total soil microbial community was approximately 4,600, which is equivalent to 4,000 completely different genomes of standard soil bacteria. Thus, the soil microbial community appeared to have 4,000 different populations based upon genetic diversity measurements. These results indicate that the soil microbial community is composed of a large number of genetically separate clones of bacterial populations. The genetic diversity found was about 200 times higher than measured for isolated strains, indicating that the phenotypically defined populations isolated by standard plating techniques make up only a small fraction of the soil bacterial community. Most of the diversity is located in that part of the microbial community that cannot be isolated and cultured by standard techniques. Nonculturable K-selected populations appear to comprise a large number of different, highly specialized bacterial populations. These results are consistent with a theory that the majority of bacterial populations isolated and grown on agar plates are the r-selected bacterial populations that have high growth rates and grow on high nutrient concentrations.

ECOSYSTEMS

The community and its abiotic surroundings comprise the ecosystem, which is a self-sustaining ecological unit; energy flows through and materials are cycled within the ecosystem by various populations of the community (Odum 1983). The complexity of ecosystems requires that some degree of simplification and some application of theoretical considerations be made in order to understand the underlying forces and principles that govern the functioning of ecosystems (Pielou 1969, 1977; Walters 1971; Levin 1974; Smith 1974; May 1976; Hall and Day 1977a, 1977b). The natural environmental fluctuations that occur within ecosystems make this difficult without examining subsystem components and the interrelationships of the subsystems.

Experimental Ecosystem Models

Experimental models greatly simplify the interactions between microbial populations and between the microbial community and the environment, but each experimental model must be questioned as to how well it mimics the real ecosystem; that is, can the data gathered from the experimental model be extrapolated to the real world? Within experimental models, it is possible to define environmental conditions and biological populations. Models used in laboratory experiments are called microecosystems or microcosms and these multicomponent models are needed to understand the interactive relationship of microbial communities and the roles played by microbial populations within ecosystems (Figure 6.12). Microcosms are useful in examining microbial ecological processes (Armstrong 1992; Cripe and Pritchard 1992; Fredrickson et al. 1992; Hagedorn 1992; Hood 1992). When all biological populations are defined or known, the system is said to be axenic or gnotobiotic. The simplest gnotobiotic system is a single pure microbial strain aseptically inoculated into a sterile medium. More complex gnotobiotic systems involve multiple defined microbial populations and sometimes defined plant and/or animal populations.

Batch Systems In batch system models, the biological components and a supporting nutritive medium are added to a closed system. Batch systems are self-sustaining when there is a suitable input of radiant energy and when photoautotrophic organisms are present within the enclosed microcosm (Byers 1963, 1964; Pritchard and Bourquin 1984). In such self-sustaining systems, nutrients are recycled within the microecosystem; there is an input of light energy and a loss of heat. In experimental batch systems where there is insufficient input of light energy and/or a lack of photoautotrophs within the microcosm, the system will run downhill and the biological community eventually will be eliminated. Extrapolation of data from heterotrophic batch systems to real ecosystems is often difficult because excess nutrient concentrations are normally added to such systems to permit the continued growth of the enclosed microorganisms

Figure 6.12

Examples of flow-through and batch systems: (A) turbidostat, (B) chemostat, and (C and D) batch microcosms. (Sources: Carpenter 1968; Nixon 1969; and Byers 1963. Modified and reprinted with permission of American Society of Limnology and Oceanography.)

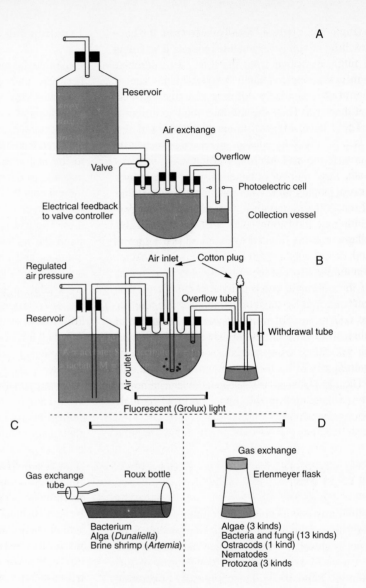

Flow-through Systems Flow-through systems have been developed and used in ecological studies as model systems for aquatic habitats where nutrient concentrations are often growth-rate limiting. The most commonly used flow-through system for ecological studies is the chemostat (Monod 1950; Jannasch 1965; Tempest 1970; Veldkamp and Kuenen 1973; Jannasch and Mateles 1974; Veldkamp 1977;

for a suitable period of time before the energy is dissipated and the biological community disappears. Batch systems, however, are suitable for modeling ecosystems that receive high inputs of allochthonous organic matter; for example, batch model systems can be used to study the fate of leaf litter in forest soils, since the addition of leaves to soils within natural forest habitats occurs within a limited time frame during the fall.

Veldkamp et al. 1984). A chemostat permits the continuous flow of growth medium through a chamber containing microbial populations. The growth medium contains some required nutrient in limiting concentrations that determines the specific growth rates of the microbial populations.

The chemostat is a self-regulating system that reaches a steady-state condition. At the steady-state condition, the specific growth rate is equal to the dilution rate, which is the rate of input of the growth medium divided by the volume of the culture chamber. If the growth rate is initially greater or less than the dilution rate, the concentration of microorganisms will change until steady-state conditions are achieved (Monod 1950; Jannasch 1969; Veldkamp 1977; Veldkamp et al. 1984) (Figure 6.13). At steady-state conditions the concentration of biomass and the concentration of growth-limiting substrate within the culture vessel are determined by the concentration of the substrate in the incoming growth medium and by a growth yield coefficient, biomass formed divided by substrate used.

The chemostat is well suited for use as a model system for studies on the interactions of aquatic microbial populations. Using chemostats, it has been demonstrated that the success of microbial populations competing for the same substrate depends on substrate concentration and maximal specific growth rate; at a low substrate concentration one population may outcompete another, while at a higher substrate concentration the results of competition may be just the opposite (Veldkamp 1970, 1977; Harder et al. 1977; Kuenen and Harder 1982; Dykhuizen and Hartl 1983; Veldkamp et al. 1984) (Figure 6.14). Phrased another way, some microorganisms are able to sustain higher specific growth rates at lower substrate concentrations than others. Such studies explain why particular microbial populations are more successful than others in ecosystems at particular times. These competition studies have revealed that the diversity within the microbial community can be explained, in part, in terms of selection by substrate concentration (Veldkamp 1977; Veldkamp et al. 1984). Within a constant environment and complete niche overlap, only one population generally will survive within a chemostat;

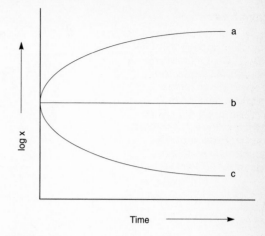

Figure 6.13

Relationship between growth rate (μ) and dilution rate (D) in a chemostat showing change of bacterial concentration following inoculation. (a) Initially μ>D; x (biomass) increases and s (concentration of growth-limiting substrate) decreases until μ=D. (b) Steady state is obtained immediately. (c) Initially μ<D; x decreases and s increases until μ=D. (Source: Veldkamp 1977, based on Jannasch 1969. Reprinted by permission, copyright Plenum Press.)

other unsuccessful competitor populations will be washed out. Multiple populations, however, can exist within a chemostat if the environmental conditions are varied in a cyclical manner so that alternate microbial populations periodically have the higher specific growth rates. Multiple populations can also exist within chemostats when the populations are not limited by the same substrate or if they interact positively in utilization of the substrate.

Microcosms More complex interactions can be examined using microcosms that include a variety of microbial, plant, and animal populations and also permit inclusion of multiple habitats and interfaces within one model system—for example, sediment, water, sediment-water interface, and plant and animal surfaces (Byers 1963, 1964; Nixon 1969; Bourquin et al. 1977; Nixon and Kremer 1977; Pritchard and Bourquin 1984; Armstrong 1992; Cripe and Pritchard

Figure 6.14

Growth rate (μ) at different substrate concentrations (S) for two different organisms (a and b). As shown, the μ-S relationships for the organisms are in (A) K_s^a < K_s^b and μ_{max}^a > μ_{max}^b and (B) K_s^a < K_s^b and μ_{max}^a < μmax^b. (Source: Veldkamp 1970. Reprinted by permission, copyright Academic Press.)

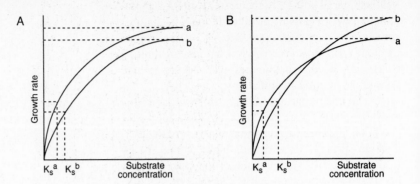

1992; Fredrickson et al. 1992; Hagedorn 1992; Hood 1992) (Figure 6.15). Flow-through microcosms permit studies on the interactions of microbial populations within complex communities and on the flow of energy and materials through the system.

Germfree animals and plants have provided suitable experimental models for investigating their interactions with microorganisms (Phillips and Smith 1959; Hsu and Bartha 1979). Using gnotobiotic systems, that is, systems in which the higher organism as well as all microbial populations are defined, it has been possible to examine the role of microorganisms in nutrition and general health of the animal population. Germfree animals generally have elongated and thinner-walled intestinal tracts as compared to normal controls and may show requirements for additional vitamins normally supplied by the intestinal bacterial community.

Germfree animals provide a direct means of investigating metabolic relationships between known mixtures of microorganisms in their natural habitat, such as the mouth, skin, or intestine. With the use of germfree animals, it has been possible to demonstrate the involvement of particular microbial populations in disease conditions, for example, the involvement of *Lactobacillus* species in the production of dental caries. It has also been possible to demonstrate the necessity of naturally beneficial relationships between microbial populations for the development of certain diseases in animals.

Mathematical Models

Examination of experimental model microecosystems and natural ecosystems makes possible the development of conceptual or hypothetical understanding of the functioning of the system that can be expressed by mathematical relationships (Patten 1971; Walters 1971; Williams 1973; May 1976; Pielou 1977; Shoemaker 1977; Christian et al. 1986; Robinson 1986; Wanner 1986). It is possible to express the interrelationships between microbial populations and between microorganisms and their surroundings using various mathematical expressions normally employing linear and/or differential equations. The use of differential equations of the form $dN/dt = f(x)$ permits the expression of a change in parameter N, such as microbial biomass, with time, according to some specified relationship $f(x)$, where $f(x)$ may represent various functions of one or more variables. Where ecological relationships follow general laws, such as the thermodynamic laws concerning entropy, and the preservation of matter, such relationships can be defined.

Mathematical models may be based on ecological theory or may be developed to simulate particular ecosystems or subsystems. Some models are developed to describe a particular system, others to predict dynamic change. Theoretical models do not attempt to mimic the details of real ecosystems, but rather to model fundamental or general ecological principles. Simulation models, on the other hand, attempt to

Figure 6.15

Microcosm for determining effects and fate of toxicants in a salt marsh. The apparatus contains core chambers implanted into the salt-marsh sediment; CO_2 is collected from the exit tube by trapping in KOH; volatile organics are trapped on a resin in the exit tube. (Source: Bourquin, et al. 1977. Reprinted by permission; copyright Society of Industrial Microbiology.)

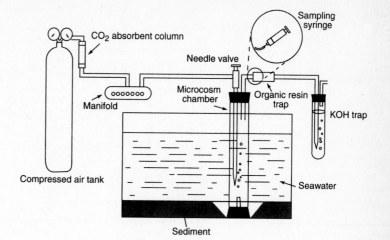

reflect accurately and predict detailed changes within the systems being modeled.

Selecting the Components of the Model   Whenever one constructs a mathematical model, some information about the system must be eliminated to allow increased understanding of the overall system (Hall and Day 1977a). There is an attempt to eliminate "noise," that is, random fluctuations in background data. Just as we do not process all perceptual information available to our brains at all times, models must filter the available data. What can be eliminated depends on the purpose of the model. In some ways, models are analogous to semipermeable membranes; some items are excluded, others retained. Sometimes the wrong information is screened out and eliminated; this is a risk in model building. Like students studying for an exam, sometimes important information is overlooked, while at other times the right material is studied and retained. Arbitrariness in filtering out information can lead to the development of alternative and conflicting models. The true test is how well the model predicts the actual system, that is, the verification of the model.

A flow diagram illustrating the activities involved in developing a mathematical model is shown in Figure 6.16. This diagram emphasizes the role of feedback in the model-building process. Feed-

back tends to stabilize both natural and model systems; the outputs from one process influence inputs to that process at a later time.

Within the model there will be state or condition indices, which are measures of particular conditions within the ecosystem, such as bacterial biomass, temperature, and nutrient concentrations. These condition or state indices are interrelated, and hence it is necessary to develop a conceptual model of the component parts and interrelationships between these parts before developing the mathematical model. Because of complex interrelationships between biological populations and abiotic components of an ecosystem, the development of such a conceptual model is quite difficult. Conceptual models are often constructed in a systems diagram as a first step, after which specific mathematical formulae can be used to express quantitative interrelationships between its components. The components of the model and their relationships are often illustrated by compartmentalized boxes and flow arrows, which represent the flow of energy or materials from one compartment to another. In developing the conceptual framework for a model, it is sometimes useful to express interrelationships between the components as causal loop diagrams.

A causal loop diagram indicates whether a compartment exerts positive or negative effects on other compartments. A simple causal loop diagram for a

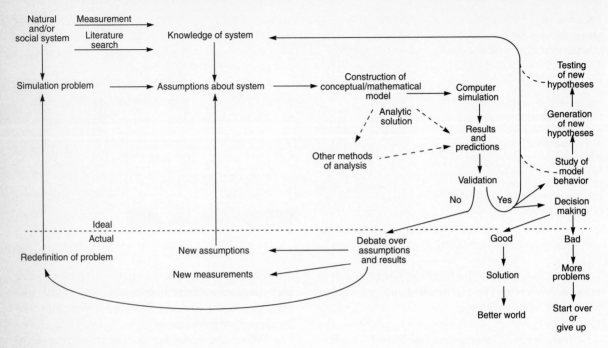

Figure 6.16

A diagram illustrating the process of building a mathematical model. (Source: Hall and Day 1977a. Reprinted by permission, copyright John Wiley and Sons.)

predator-prey relationship is shown in Figure 6.17. The prey population has a positive effect on the predator population; as the prey population increases, the predator population also increases. The predator population has a negative effect on the prey population. In positive feedback, if there is a deviation from a particular state, the deviation tends to increase in whatever direction it began. Presuming no other feedback loops, the numbers of microorganisms will increase geometrically; the more organisms there are, the more progeny they will produce. In negative feedback loops, deviations tend to be counteracted, and there is a return to stability.

Oscillatory behavior is a characteristic of negative feedback loops; the magnitude of the oscillations depends on the sensitivity and reaction time of the feedback loop. Indeed, ecosystems must contain negative feedback loops in order to maintain stability and

homeostasis. Without such negative feedback loops, there would be no control, and the system would follow a unidirectional path to some unstable end. An

Figure 6.17

A causal loop diagram for a predator-prey relationship.

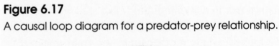
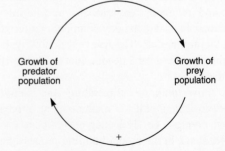

exception to this rule may occur within an ecosystem when there is unidirectional flow and the system is limited by the rate of input of an essential nutrient. In this case, negative feedback may be restricted by movement downstream out of the system; the limiting input may act to control the system and to maintain stability.

Mathematical relationships are particularly useful in describing the growth of populations and the interactions between populations (Whittaker 1975). Table 6.4 shows a series of theoretical mathematical relationships describing growth and interactions of microbial populations. Other equations can be used to describe changes in concentrations of nutrients within an ecosystem, rates of enzymatic activities, succession of populations within a community, and so on. These equations can be used to produce graphs of theoretical or expected results, often called simulations (Shoemaker 1977) (Figure 6.18). Sometimes, these equations accurately reflect observed growth interactions within experimental or real ecosystems. In other cases, additional sources of variance must be considered to mimic the real ecosystem. Testing the validity of a model involves determining how well the model explains experimental data. Often statistical tests are used to determine the accuracy of a mathematical model. The least-squares criterion is most often employed (Robinson 1986). A good model should yield the minimal sum of the squared errors for the experimental data about the curve generated by the mathematical model.

Mathematical equations that describe the interrelationships between system components can be used to develop larger models with complex interrelationships between populations (Nixon and Kremer 1977). We shall not consider the mathematics of such complex ecosystem models here, as these require an extensive mathematical background. The suggested readings (Patten 1971; Shoemaker 1977) should be consulted for a detailed examination of the mathematical formulae and the supportive computer programs involved in the development of holistic mathematical ecosystems models. The microbial ecologist, though, should be aware that such mathematical approaches to modeling exist and that the

development of models that accurately simulate ecosystems. The development of mathematical models for whole ecosystems has highlighted the fact that microorganisms play important roles within ecosystems and that their functions in mediating the flow of energy and materials through and within ecosystems cannot be ignored.

MICROBIAL COMMUNITIES IN NATURE

Microbes within Macro-communities

Naturalists, ecologists, and plant geographers characterize biological communities by their dominant macrophyte primary producers (Smith 1990). Thus, for example, they designate "tall grass prairie," "oak-hickory forest," and "*Spartina* salt marsh" communities. Obviously, these and similar macrophyte-dominated communities have their inconspicuous but essential microbial components that assure the biodegradation of organics and the biogeochemical cycling of minerals (see chapters 11 and 12). In addition, the microbial component of the community interacts with plant and animal life in other important ways (see chapters 4 and 5). However, the role of microorganisms as primary producers in such environments ranges from negligible to minor (for example, cyanobacterial and algal production in the *Spartina* salt marsh). In the offshore regions of deeper freshwater lakes and of the oceans, which are free of macrophytes, all primary production is due to cyanobacteria and microalgae. It would be logically justified to refer to these as "microbial primary production" communities, but instead, they are designated by tradition as limnetic (freshwater) and pelagic (marine) communities. Of course, in addition to their role as primary producers, microorganisms also serve as biodegraders and mineral cyclers in these communities. As zooflagellates and other protozoa, they are also amply represented among the first level of consumers. More than 70% of our globe is covered by pelagic and limnetic communities, and we

Table 6.4
Mathematical expressions of population growth and interactions

---

If there are no restricting factors, the growth of a microbial population is described as

$$N_t = N_0 e^{rt}$$

where N_0 = initial population density, r = intrinsic rate of growth, and N_t = density after any time period t.
If the rate of growth is limited,

$$N_t = \frac{K}{1 + [\,(K - N_0)\,/N_0\,]\,e^{-rt}}$$

where K = the carrying capacity or limiting density. e = base of natural logarithms, 2.71828.

The growth rates of competing populations are

$$\frac{dN_1}{dt} = r_1 N_1 \frac{(K_1 - N_1 - \alpha_1 N_2)}{K_1} \qquad \frac{dN_2}{dt} = r_2 N_2 \frac{(K_1 - N_1 - \alpha_2 N_1)}{K_2}$$

where the two populations have rates of increase r_1 and r_2 and competition coefficients $\alpha 1$ and $\alpha 2$.
If $\alpha_1 < K_1/K_2$ and $\alpha_2 > K_2/K_1$, only population 1 survives. If $\alpha_1 > K_1/K_2$ and $\alpha_2 < K_2/K_1$, only population 2 survives.
If $\alpha_1 > K_1/K_2$ and $\alpha_2 > K_2/K_1$, one or the other population survives. If $\alpha_1 < K_1/K_2$ and $\alpha_2 < K_2/K_1$, both populations survive.

The growth of a prey population is

$$\frac{dN_1}{dt} = r_1 N_1 - P N_1 N_2$$

where the prey adds individuals according to its intrinsic rate of growth (r_1) times its density (N_1) and loses individuals at a rate proportional to encounters of predator and prey—prey density times predator density (N_2). The predator adds individuals at a rate proportional to this same product of densities and loses individuals according to death rate (d_2) times its own density:

$$\frac{dN_2}{dt} = a P N_1 N_2 - d_2 N_2$$

where P = a coefficient of predation and a = a coefficient relating predator births to prey consumed. According to this description (Lotka-Volterra equations) there will be cyclic fluctuations of predator and prey populations.

If two populations have a mutually beneficial relationship

$$\frac{dN_1}{dt} = r_1 N_1 \left(\frac{K_1 - N_1 + b N_2}{K_1 + b N_2} \right) \qquad \frac{dN_2}{dt} = r_2 N_2 \left(\frac{K_2 - N_2 + a N_1}{K_2} \right)$$

where b = a coefficient for support of individuals of population 1 by individuals of population 2. a = a coefficient for the effect of population 1 on population 2. If population 1 becomes totally dependent on population 2, K_1 approaches zero and the carrying capacity for population 1 is defined by population 2 and the coefficient of support, b. Assuming a limit on the benefit per host individual from the interaction with the symbiont, in the form $C = D/(D + N_2)$, a stable mutualism is expressed by

$$\frac{dN_1}{dt} = r_1 N_1 \left(\frac{b N_2 - N}{b N_2} \right) \qquad \frac{dN_2}{dt} = r_2 N_2 \left(\frac{K_2 - N_2 + a N_1 C}{K_2} \right)$$

Source: Whittaker 1975.

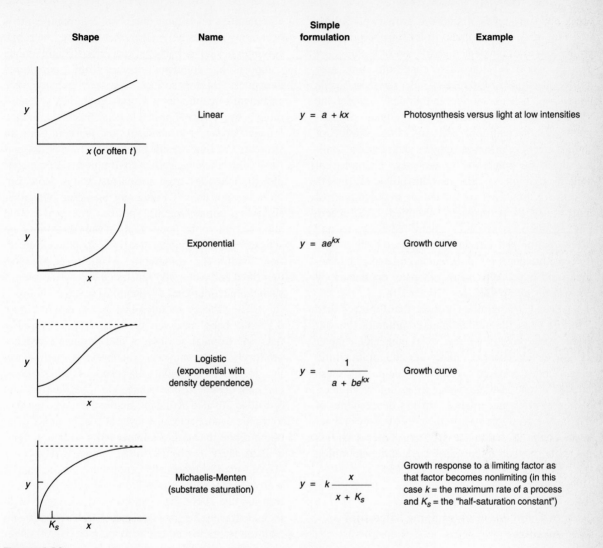

Shape	Name	Simple formulation	Example
	Linear	$y = a + kx$	Photosynthesis versus light at low intensities
	Exponential	$y = ae^{kx}$	Growth curve
	Logistic (exponential with density dependence)	$y = \dfrac{1}{a + be^{kx}}$	Growth curve
	Michaelis-Menten (substrate saturation)	$y = k\,\dfrac{x}{x + K_s}$	Growth response to a limiting factor as that factor becomes nonlimiting (in this case k = the maximum rate of a process and K_s = the "half-saturation constant")

Figure 6.18

Mathematical relationships commonly encountered in modeling. a, b, k, and K_s are constants. (Source: Hall and Day 1977a. Reprinted by permission, copyright John Wiley and Sons.)

should be aware that the dominant community members in these ecosystems are microorganisms rather than the more visible fishes and invertebrates.

In terrestrial and shallow-water environments, macrophytes tend to outcompete microorganisms in capturing the radiant energy of the sun. Only when unusually harsh physical conditions restrict the growth of macrophytes can microorganisms dominate or even monopolize primary production. Lichens dominate bare rock face, some hot and cold desert areas, and some types of tundra as primary producers. Lichens may form thick foliose-fruticose mats amounting to as much as a metric ton of biomass per hectare. Under less favorable conditions, they may adhere as a thin crust to rocks. Under the most hostile conditions, lichens survive endolithically, under the surface of the rocks. In some shallow aquatic environments, anoxic conditions, extremes of temperature, or salinity may exclude macrophytes and assure a dominance of the microbial community. Examples of aquatic environments that are unfavorable for growth of primary producers are the anoxic hypolimnions of temporarily or permanently stratified lakes where photosynthesis is primarily controlled by green and purple sulfur and nonsulfur bacteria, alkaline soda lakes with blooms of *Spirulina* and related cyanobacteria, and hypersaline solar salt pans dominated by *Halobacterium* species (see chapter 10).

Fossil stromatolites indicate that for more than 2 billion years the predominant communities on our planet were prokaryotic microbial mats (see chapter 2). The development of eukaryotes and multicellular invertebrates brought about the decline of these microbial mats, which appear to be highly vulnerable to invertebrate grazing, but the descendents of the early prokaryotes continue to exist as cyanobacterial crusts in certain desert environments and as microbial mats in hypersaline bays and evaporation flats (Campbell 1979; Cohen et al. 1984).

Structure and Function of Some Microbial Communities

Some microbial communities are fairly self-contained with microbe-microbe interactions dominating over the interactions of microbes with plants and animals. While many macro-communities have been studied in great detail and depth, methodological difficulties have, until recently, prevented similar studies on natural microbial communities. This fact shifted much emphasis to experimental and mathematical models of microbial communities. While it is still difficult to perform the detailed qualitative and quantitative species inventories used in the characterization of macro-

communities on natural microbial communities, it is becoming increasingly clear that diverse microbial communities have their own characteristic and unique composition, dominant species, and population dynamics. On the surface, one may perceive this statement to contradict Martinus Beijerinck's "everything is everywhere" principle (see chapter 1). In fact, efficient dispersal mechanisms and resistant dormant structures at times result in isolation of microorganisms from uncharacteristic environments (for example, thermophiles from permafrost soils). However, these are not active community members but rather surviving, allochthonous species. The active and abundant microbial populations of the community are selected by their habitat and by population interactions within the community. Margulis et al. have described and pictorially illustrated some examples of such unique microbial communities (1986a, 1986b).

The prokaryotic microbial mat shown in Figure 6.19 has been observed on hypersaline "evaporite flats" of tropical seashores. The dominant primary producers of the mat are cyanobacteria, particularly the insulated cablelike filaments of *Microcoleus chthonoplastes*. *Nostoc, Anabaena, Oscillatoria*, and *Spirulina* are also abundant. Below the oxygenic phototrophic cyanobacteria, a layer of purple anoxygenic phototrophic bacteria was observed, and below these a black layer of diverse sulfate reducers. The latter convert photosynthetically produced organic material with seawater sulfate to H_2S. Purple sulfur bacteria in turn use the H_2S. The cyanobacterial filaments with their extracellular mucopolysaccharide sheaths, in addition to photosynthetic activity, give overall cohesion to the mat and protect it both from disintegration and from drying out during low-tide cycles.

"Desert crusts" (Campbell 1979) are cyanobacterial mats, dominated again by *M. chthonoplastes* and by *M. vaginatus*. Both of these secrete common mucopolysaccharide sheath for several trichomes, as illustrated in Figure 6.19. In dry weather, the sheath material forms a brittle film over the desert sand. It protects the cyanobacterial mat and, at the same time, stabilizes the sand. In moist weather, the trichomes slide out of the hydrated sheath, spread, and secrete new sheath material. Associated with the mats are

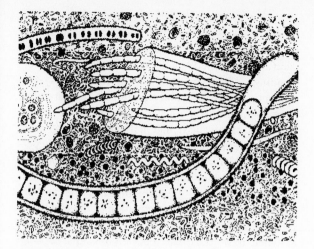

Figure 6.19

Drawing of a prokaryotic microbial mat from a marine evaporite flat. The mat is dominated by *Microcoleus chthonoplastes* trichromes in a common mucopolysaccharide sheath. Some of these are shown in cross section. A large *Oscillatoria* trichrome has gliding motility, as do the *Microcoleus* trichromes. Other coccoid and filamentous cyanobacteria, heterotrophs and sulfate reducers add to the complexity of this microbial mat community. (Source: Margulis et al. 1986a.; Reprinted by permission Elsevier.)

Figure 6.20

Drawing of the *Chromatium* layer of Lake Cisó's anoxic phototrophic community. The large flagellated *Chromatium* cells form a layer close to the surface and above the *Chlorobium* layer, giving the lake water a purple color. (Source: Margulis et al. 1986b; Reprinted by permission American Institute of Biological Sciences.)

additional cyanobacteria, including *Calothrix, Scytonema, Phormidium, Schizothrix, Gloeocapsa,* and *Nostoc.* Most of these are filamentous and display gliding motility; some have the ability to fix molecular nitrogen. Desert crusts have been described on undisturbed deserts of the southwestern United States. They fulfill an important role in stabilizing desert sand and pave the way for succession by lichens and mosses, but they are fragile and easily destroyed by grazing and trampling. The early Precambrian colonization of dry land was most likely due to similar cyanobacterial crusts.

Lake Cisó in northeastern Spain is fed by underground seepage that passes through gypsum ($CaSO_4 \cdot 2H_2O$) formations and is rich in sulfate. The lake is relatively small, of moderate depth, and protected from wind. These conditions promote strati-

fication. Organic matter from photosynthesis is decomposed at the anaerobic lake bottom predominantly by sulfate reducers such as *Desulfovibrio.* The resulting H_2S rises to the illuminated portion of the water column and is utilized by anaerobic phototrophic purple and green sulfur bacteria, *Chromatium* and *Chlorobium,* respectively. *Chromatium* is less sensitive to oxygen and more sensitive to high H_2S concentrations than *Chlorobium* and stratifies close to the lake surface (Figure 6.20). The green *Chlorobium* layer stratifies underneath, aided also by the fact that it can absorb light of a wavelength that is not efficiently utilized by *Chromatium.* Shifting light and temperature conditions in combination with predation on *Chromatium* by the *Vampirococcus* and *Daptobacter* induce bloom and die-off cycles of *Chromatium.* These cycles result in dramatic color changes of the lake from purple to green and back. Eukaryotes (protozoa and algae) in Lake Cisó are normally restricted to the upper few cm of oxygenated water. Occasionally, strong winds extend this oxygenated layer, causing algal blooms and additional color changes. Somewhat similar stratification phenomena may develop in eutrophic lakes and fjords during summer if light penetrates into the anoxic portion of the water column below the thermocline.

Some additional characteristic and interesting microbial communities are described in other chapters of this text, including the anaerobic cellulose-digesting microbial communities of the rumen and the termite hindgut (see chapter 5) and the chemoautotrophic microbial community of deep-sea hydrothermal vents (see chapters 9 and 12). Attempts to describe the composition and functioning of complex natural microbial communities are relatively recent with much research remaining to be performed.

SUMMARY

Microbial populations rarely occur alone but rather are normally found within communities of organisms. Communities are usually characterized by a high state of diversity. Authochthonous members of communities occupy the niches of the ecosystem. Within ecosystems, microorganisms compete for available resources. In some cases microorganisms are eliminated from a community by competitive exclusion, but a number of factors can permit coexistence. Current ecological theories emphasize disturbances and environmental heterogeneity as important factors in permitting coexistence of competitive populations within the same habitat. Successional processes occur, sometimes in characteristic patterns, but disturbances often disrupt the orderly development of microbial communities.

Batch, flow-through, laboratory, and *in situ* model systems permit the microbial ecologist to manipulate both abiotic and biotic parameters in a carefully controlled manner that allows for an understanding of how the components of the ecosystem interact. Substrate concentrations, temperatures, light intensities, and other parameters can be manipulated to mimic particular habitats. In many such model systems, macroorganisms are excluded, permitting examination of the interrelationships between microbial populations. Control of environmental fluctuations greatly simplifies the system, allowing unequivocal interpretation of the results. In some cases, results from these model systems have revealed new bases for interrelationships between microbial populations.

Perhaps the real value of both experimental and mathematical models is that they provide a tool for developing an understanding of ecosystem function and the factors that control the flow of energy and matter through an ecosystem and that they allow for the development of a predictive capability. Both experimental and mathematical models permit the microbial ecologist to test hypotheses concerning the function of microorganisms within the ecosystem and the interactions of microbial populations. The predictive capability developed in such models is especially useful in the proper management of ecosystems. Systems analysis and mathematical models have been applied in the management of waste disposal and control of pathogens. By understanding the function of the overall system, it is possible to predict the consequences of a perturbation to any part of the system. Within the near future, mathematics and modeling is likely to have an increased role in microbial ecology, just as it has now in general ecology.

REFERENCES & SUGGESTED READINGS

Andrews, J. H. 1984. Relevance of *r*- and *K*-theory to the ecology of plant pathogens. In M. J. Klug and C. A. Reddy (eds.). *Current Perspectives in Microbial Ecology.* American Society for Microbiology, Washington, D.C., pp. 1–7.

Andrews, J. H., and R. F. Hall. 1986. *r*- and *K* selection and microbial ecology. *Advances in Microbial Ecology* 9:99–147.

Armstrong, J. L. 1992. Persistence of recombinant bacteria in microcosms. In M. A. Levin, R. J. Seidler, and M. Rogul (eds.). *Microbial Ecology: Principles, Methods, and Applications.* McGraw-Hill, New York. pp. 495–509.

Atlas, R. M. 1984a. Diversity of microbial communities. *Advances in Microbial Ecology* 7:1–47.

Atlas, R. M. 1984b. Use of microbial diversity measurements to assess environmental stress. In M. J. Klug and C. A. Reddy (eds.). *Current Perspectives in Microbial Ecology.* American Society for Microbiology, Washington, D.C., pp. 540–545

Atlas, R. M., A. Horowitz, M. I. Krichevsky, and A. K. Bej. 1991. Response of microbial populations to environmental disturbance. *Microbial Ecology* 22:249–256.

Baross, J. A., J. W. Deming, and R. G. Becker. 1984. Evidence for microbial growth in high-pressure, high temperature environments. In M. J. Klug and C. A. Reddy (eds.). *Current Perspectives in Microbial Ecology.* American Society for Microbiology, Washington, D.C., pp. 186–195.

Bascomb, S., and R. R. Colwell. 1992. Application of numerical taxonomy in microbial ecology. In M. A. Levin, R. J. Seidler, and M. Rogul (eds.). *Microbial Ecology: Principles, Methods, and Applications.* McGraw-Hill, New York. pp. 113–137.

Bej, A. K., M. Perlin, and R. M. Atlas. 1991. Effect of introducing genetically engineered microorganisms on soil microbial community diversity. *FEMS Microbiology Ecology* 86:169–176.

Beringer, J. E. and P. R. Hirsch. 1984. The role of plasmids in microbial ecology. In M. J. Klug and C. A. Reddy (eds.). *Current Perspectives in Microbial Ecology.* American Society for Microbiology, Washington, D.C., pp. 63–70.

Bourquin, A. W., M. A. Hood, and R. L. Garnas. 1977. An artificial microbial ecosystem for determining fate and effects of toxicants in a salt marsh environment. *Developments in Industrial Microbiology* 18:185–191.

Britten, R. J., D. E. Graham, and S. R. Neufeld. 1974. Analysis of repeating DNA sequences by reassociation. *Methods in Enzymology* 29:363–418.

Bunnell, F. 1973. Decomposition: Models and the real world. *Bulletin of the Ecological Research Commission* (Stockholm) 17:407–415.

Byers, R. J. 1963. The metabolism of twelve aquatic laboratory microecosystems. *Ecological Monographs* 33:281–306.

Byers, R. J. 1964. The microcosm approach to ecosystem biology. *American Biology Teacher* 26:491–498.

Cairns, J. C. 1969. Factors affecting the number of species in fresh water protozoan communities. In J. C. Cairns (ed.). *The Structure and Function of Freshwater Microbial Communities.* Virginia Polytechnic Institute and State University, Blacksburg, Va., pp. 219–248.

Cameron, R. E., R. C. Honour, and F. A. Morelli. 1976. Antarctic microbiology: Preparation for Mars life detection, quarantine and back contamination. In M. R. Heinrich (ed.). *Extreme Environments: Mechanisms of Microbial Adaptation.* Academic Press, New York, pp. 57–82.

Campbell, S.E. 1979. Soil stabilization by a procaryotic desert crust: Implications for Precambrian land biota. *Origins of Life* 9:335–348.

Carpenter, E. J. 1968. A simple, inexpensive algal chemostat. *Limnology and Oceanography* 13:720–721.

Chapman, V. J., and D. J. Chapman. 1973. *The Algae.* Macmillan, London.

Christian, R. R., R. Wetzel, S. M. Harlan, and D. W. Stanley. 1986. Growth and decomposition in aquatic microbial systems: Approaches in simple models. In F. Megusar and M.Gantar (eds.). *Perspectives in Microbial Ecology.* Slovene Society for Microbiology, Ljubljana, Yugoslavia, pp. 38–45.

Clarke, P. H. 1984. Evolution of new phenotypes. In M. J. Klug and C. A. Reddy (eds.). *Current Perspectives in Microbial Ecology.* American Society for Microbiology, Washington, D.C., pp. 71–78.

Clesceic, L. S., R. A. Park, and J. A. Bloomfield. 1977. General model of microbial growth and decomposition in aquatic ecosystems. *Applied and Environmental Microbiology* 33:1047–1058.

Cleveland, L. R., and A. V. Grimstone. 1964. The fine structure of the flagellate *Mixotricha paradoxa* and its associated micro-organisms. *Proceedings of the Royal Society* (London) Series B 159:668–686.

Cohen, Y., R.W. Castenholz and H. Halvorson (eds.). 1984. *Microbial Mats: Stromatolites.* Alan R. Liss Pub., New York.

Cripe, C. R., and P. H. Pritchard. 1992. Site-specific aquatic microcosms as test systems for fate and effects of microorganisms. In M. A. Levin, R. J. Seidler, and M. Rogul (eds.). *Microbial Ecology: Principles, Methods, and Applications.* McGraw-Hill, New York. pp. 467–493.

Dale, M. B. 1970. Systems analysis and ecology. *Ecology* 51:2–16.

Darnell, J. E., and W. F. Doolittle. 1986. Speculations on the early course of evolution. *Proceedings of the National Academy of Science USA* 83:1271–1275.

DeFlaun, M. F., and S. B. Levy. 1989. Genes and their varied hosts. In S. B. Levy and R. V. Miller (eds.). *Gene Transfer in the Environment.* McGraw-Hill, New York, pp. 1–32.

Dickerson, R. 1978. Chemical evolution and the origin of life. *Scientific American* 239(3):70–86.

Drahos, D. J., and G. F. Barry. 1992. Assessment of genetic stability. In M. A. Levin, R. J. Seidler, and M. Rogul (eds.). *Microbial Ecology: Principles, Methods, and Applications.* McGraw-Hill, New York, pp. 161–181.

Dykhuizen, D. E., and D. L. Hartl. 1983. Selection in chemostats. *Microbiological Reviews* 47:150–168.

Fenchel, T. M., and B. B. Jørgensen. 1977. Detritus food chains of aquatic ecosystems: The role of bacteria. *Advances in Microbial Ecology* 1:1–58.

Fredrickson, A. G., and G. Stephanopoulos. 1981. Microbial competition. *Science* 213:972–979.

Fredrickson, J. K., H. Bolton, Jr., and G. Stotzky. 1992. Methods for evaluating the effects of microorganisms on biogeochemical cycling. In M. A. Levin, R. J. Seidler, and M. Rogul (eds.). *Microbial Ecology: Principles, Methods, and Applications.* McGraw-Hill, New York, pp. 579–606.

Freter, R. 1984. Factors affecting conjugal transfer in natural bacterial communities. In M. J. Klug and C. A. Reddy (eds.). *Current Perspectives in Microbial Ecology.* American Society for Microbiology, Washington, D.C., pp. 105–114.

Friedmann, E. I., and R. O. Friedmann. 1984. Endolithic microorganisms in extreme dry environments: Analysis of a lithobiotic microbial habitat. In M. J. Klug and C. A. Reddy (eds.). *Current Perspectives in Microbial Ecology.* American Society for Microbiology, Washington, D.C., pp. 177–185.

Fry, J. C., and M. J. Day. 1990. *Bacterial Genetics in Natural Environments.* Chapman and Hall, London.

Fuhrman, J. A., D. E. Comeau, A. Hagstrom, and A. M. Chan. 1988. Extraction from natural planktonic microorganisms of DNA suitable for molecular biological studies. *Applied and Environmental Microbiology* 54:1426–1429.

Golley, F. B. (ed.). 1977. *Ecological Succession.* Benchmark Papers in Ecology 15. Dowden Hutchinson and Ross, Stroudsburg, Pa.

Goodman, M. R. 1974. *Study Notes in System Dynamics.* Wright Allen Press, Cambridge, Mass.

Hagedorn, C. 1992. Experimental methods for terrestrial ecosystems. In M. A. Levin, R. J. Seidler, and M. Rogul (eds.). *Microbial Ecology: Principles, Methods, and Applications.* McGraw-Hill, New York, pp. 525–541.

Hairston, N. G., A. D. Allen, R. R. Colwell, D. J. Fustuyma, J. Howell, M. D. Lubin, J. Mahias, and J. H. Vandermeer. 1968. The relationship between species diversity and stability: An experimental approach with protozoa and bacteria. *Ecology* 49:1091–1101.

Hall, B. G. 1984. Adaptation by acquisition of novel enzyme activities in the laboratory. In M. J. Klug and C. A. Reddy (eds.). *Current Perspectives in Microbial Ecology.* American Society for Microbiology, Washington, D.C., pp. 79–86.

Hall, C. A. S., and J. W. Day, Jr. (eds.). 1977a. *Ecosystem Modeling in Theory and Practice: An Introduction with Case Histories.* John Wiley and Sons, New York.

Hall, C. A. S., and J. W. Day, Jr. 1977b. Systems and models: terms and basic principles. In C. A. S. Hall and J. W. Day (eds.). *Ecosystem Modeling in Theory and Practice.* John Wiley and Sons, New York, pp. 5–36.

Hall, C. A. S., J. W. Day, Jr., and H. T. Odum. 1977. A circuit language for energy and matter. In C. A. S. Hall and J. W. Day (eds.). *Ecosystem Modeling in Theory and Practice.* John Wiley and Sons, New York, pp. 37–48.

Harder, W., J. G. Kuenen, and A. Martin. 1977. Microbial selection in continuous cultures: A review. *Journal of Applied Bacteriology* 43:1–24.

Heinrich, M. R. (ed.). 1976. *Extreme Environments: Mechanisms of Microbial Adaptation.* Academic Press, New York.

Holder–Franklin, M. A., A. Thorpe, and C. J. Cormier. 1981. Comparison of numerical taxonomy and DNA–DNA hybridization in diurnal studies of river bacteria. *Canadian Journal of Microbiology* 27:1165–1184.

Hood, M. A. 1992. Experimental methods for the study of fate and transport of microorganisms in aquatic systems. In M. A. Levin, R. J. Seidler, and M. Rogul (eds.). *Microbial Ecology: Principles, Methods, and Applications*. McGraw-Hill, New York, pp. 511–523.

Horner, R., and V. Alexander. 1972. Algal populations in Arctic sea ice: An investigation of heterotrophy. *Limnology and Oceanography* 17:454–458.

Horowitz, N. H. 1965. The evolution of biochemical synthesis — Retrospect and prospect. In V. Bryson and H. J. Vogel (eds.). *Evolving Genes and Proteins*. Acedemic Press, New York, pp. 15–23.

Hsu, T. S., and R. Bartha. 1979. Accelerated mineralization of two-organophosphate insecticides in the rhizosphere. *Applied and Environmental Microbiology* 37:36–41.

Hulbert, E. M. 1963. The diversity of phytoplanktonic populations in oceanic, coastal, and estuarine regions. *Journal of Marine Research* 21:81–93.

Hungate, R. E. 1975. The rumen microbial ecosystem. *Annual Reviews in Microbiology* 29:39–66.

Hurlbert, S. H. 1971. The nonconcept of species diversity: A critique and alternative parameters. *Ecology* 52:577–586.

Jannasch, H. W. 1965. Use of the chemostat in ecology. *Laboratory Practice* 14:1162–1167.

Jannasch, H. W. 1969. Estimations of bacterial growth rates in natural waters. *Journal of Bacteriology* 99:156–160.

Jannasch, H. W., and R. I. Mateles. 1974. Experimental bacterial ecology studied in continuous culture. *Advanced Microbial Physiology* 11:165–212.

Jannasch, H. W., and D. C. Nelson. 1984. Recent progress in the microbiology of the hydrothermal vents. In M. J. Klug and C. A. Reddy (eds.). *Current Perspectives in Microbial Ecology*. American Society for Microbiology, Washington, D.C., pp. 170–176.

Johnson, C. F., and E. A. Curl. 1972. *Methods for Research on the Ecology of Soil-borne Plant Pathogens*. Burgess Publishing Co., Minneapolis, Minn.

Jordan, T. L., and J. T. Staley. 1976. Electron microscopic study of succession in the periphyton community of Lake Washington. *Microbial Ecology* 2:241–276.

Jost, J. L., J. F. Drake, A. G. Frederickson, and H. M. Tsuchiya. 1973. Interactions of *Tetrahymena pyriformis, Escherichia coli, Azotobacter vinelandii,* and glucose in a mineral medium. *Journal of Bacteriology* 113:834–840.

Kaneko, T., and R. R. Colwell. 1975. Incidence of *Vibrio parahaemolyticus* in Chesapeake Bay. *Applied Microbiology* 30:251–257.

Kaneko, T., R. M. Atlas, and M. Krichevsky. 1977. Diversity of bacterial populations in the Beaufort Sea. *Nature* 270:596–599.

Kaneko, T., M. I. Krichevsky, and R. M. Atlas. 1979. Numerical taxonomy of bacteria from the Beaufort Sea. *Journal of General Microbiology* 110:111–125.

Kemp, W. M., and W. J. Mitsch. 1979. Turbulence and phytoplankton diversity: A general model of the paradox of the plankton. *Ecological Modelling* 7:201–222.

Kendrick, W. B., and A. Burges. 1962. Biological aspects of the decay of *Pinus sylvestris* leaf litter. *Nova Hedwiga* 4:313–344.

Kuenen, J. G., and W. Harder. 1982. Microbial competition in continuous culture. In R. G. Burns and J. H. Slater (eds.). *Experimental Microbial Ecology*. Blackwell Scientific Publications, Oxford, England, pp. 342–367.

Lenski, R. E. 1992. Relative fitness: Its estimation and its significance for environmental applications of microorganisms. In M. A. Levin, R. J. Seidler, and M. Rogul (eds.). *Microbial Ecology: Principles, Methods, and Applications*. McGraw-Hill, New York, pp. 183–198.

Levin, S. A. (ed.). 1974. *Ecosystem Analysis and Prediction*. Proceedings of a Conference on Ecosystems, Alta, Utah, July, 1974. Sponsored by SIAMS Institute for Mathematics and Society, and supported by National Science Foundation.

Levy, S. B., and R. V. Miller (eds.). 1989. *Gene Transfer in the Environment*. McGraw-Hill, New York.

Lewin, R. 1983. Santa Rosalia was a goat. *Science* 221:636–639.

Luckinbill, L. S. 1979. Regulation, stability, and diversity in a model experimental microcosm. *Ecology* 60:1098–1102.

MacArthur, R. H., and E. O. Wilson. 1967. *The Theory of Island Biogeography*. Princeton University Press, Princeton, N.J.

Margalef, R. 1967. *Perspectives in Ecological Theory*. University of Chicago Press, Chicago.

Margalef, R. 1979. Diversity. In A. Sournia (ed.). *Monographs on Oceanographic Methodology*. UNESCO, Paris, pp. 251–260.

Margulis, L., L. L. Baluja, S. M. Awramik, and D. Sagan. 1986a. Community living long before man. *The Science of the Total Environment* 56:379–397.

Margulis, L., D. Chase, and R. Guerrero. 1986b. Microbial communities. *BioScience* 36:160–170.

Marples, M. J. 1965. *The Ecology of the Human Skin*. C. C. Thomas, Springfield, Ill.

Marples, R. P. 1974. Effect of germicides on skin flora. In F. A. Skinner and J. G. Carr (eds.). *The Normal Microbial Flora of Man*. Academic Press, London, pp. 35–46.

Martin, J. E., A. Lester, E. V. Price, and J. D. Schmale. 1970. Comparative study of gonococcal susceptibility to penicillin in the United States, 1955–1969. *Journal of Infectious Disease* 122:159–161.

May, R. M. 1976. *Theoretical Ecology: Principles and Applications*. W. B. Saunders, Philadelphia.

Miller, R. V. 1992. Overview: Methods for the evaluation of genetic transport and stability in the environment. In M. A. Levin, R. J. Seidler, and M. Rogul (eds.). *Microbial Ecology: Principles, Methods, and Applications*. McGraw-Hill, New York, pp. 141–159.

Mills, A. L. and R. A. Wassel. 1980. Aspects of diversity measurement for microbial communities. *Applied and Environmental Microbiology* 40:578–586.

Mills, A. L., and L. M. Mallory. 1987. The community structure of sessile heterotrophic bacteria stressed by acid mine drainage. *Microbial Ecology* 14:219–232.

Monod, J. 1950. La technique du culture continue: Theorie et applications. *Annals of the Institute Pasteur* (Paris) 79:390–410.

Munson, R. J. 1970. Turbidostats. In J. R. Norris and D. W. Ribbons (eds.). *Methods in Microbiology*, Vol. 2. Academic Press, New York, pp. 349–376.

Nixon, S. W. 1969. A synthetic microcosm. *Limnology and Oceanography* 14:142–145.

Nixon, S. W., and J. N. Kremer. 1977. Narragansett Bay: The development of a composite simulation model for a New England estuary. In C. A. S. Hall and J. W. Day (eds.). *Ecosystem Modeling in Theory and Practice*. John Wiley and Sons, New York, pp. 621–673.

Noble, W. C., and D. G. Pitcher. 1978. Microbial ecology of the human skin. *Advances in Microbial Ecology* 2:245–290.

Odum, E. P. 1983. *Basic Ecology*. Saunders College Publishing. Philadelphia, Pa., pp. 13–17.

Odum, H. T. 1960. Ecological potential and analogue circuits for the ecosystem. *American Scientist* 48:1–8.

Odum, H. T. 1971. *Environment, Power and Society*. John Wiley and Sons, New York.

Odum, H. T., S. W. Nixon, and L. H. Disalvo. 1970. Adaptation for photoregenerative cycling. In J. Cairns, Jr. (ed.). *The Structure and Function of Freshwater Microbial Communities*. Research Division Monograph #3. Virginia Polytechnic Institute and State University, Blacksburg, Va., pp. 1–29.

Odum, H. T., and E. Odum. 1978. *Energy Basis of Man and Nature*. McGraw-Hill Book Co., New York.

Ogram, A., G. S. Sayler, and T. Barkay. 1988. DNA extraction and purification from sediments. *Journal of Microbiological Methods* 7:57–66.

Overton, W. S. 1977. A strategy of model construction. In C. A. S. Hall and J. W. Day (eds.). *Ecosystem Modeling in Theory and Practice*. John Wiley and Sons, New York, pp. 49–74.

Pace, N. R., D. A. Stahl, D. J. Lane, and G. J. Olsen. 1985. Analyzing natural microbial populations by rRNA sequences. *ASM News* 51:4–12.

Patrick, R. 1963. The structure of diatom communities under varying ecological conditions. *Annals of the New York Academy of Science* 108:359–365.

Patrick, R. 1967. The effect of invasion rate, species pool, and size of area on the structure of the diatom community. *Proceedings of the National Academy of Sciences* USA. 58:1335–1342.

Patrick, R. 1976. The formation and maintenance of benthic diatom communities. *Proceedings of the National Academy of Sciences* USA. 120:475–484.

Patten, B. C. 1961. Competitive exclusion. *Science* 134:1599–1601.

Patten, B. C. (ed.). 1971. *Systems Analysis and Simulation in Ecology*. Vols. 1–3. Academic Press, New York.

Phillips, A. W., and J. E. Smith. 1959. Germ-free animal techniques and their applications. *Advances in Applied Microbiology* 1:141–174.

Pielou, E. C. 1969. *An Introduction to Mathematical Ecology*. John Wiley and Sons, New York.

Pielou, E. C. 1975. *Ecological Diversity*. John Wiley and Sons, New York.

Pielou, E. C. 1977. *Mathematical Ecology*. John Wiley and Sons, New York.

Pritchard, O. H., and A. W. Bourquin. 1984. The use of microcosms for evaluation of interactions between pollutants and microorganisms. *Advances in Microbial Ecology* 7:133–215.

Revelante, N., and M. Gilmartin. 1980. Microplankton diversity indices as indicators of eutrophication in the northern Adriatic Sea. *Hydrobiolgia* 70:277–286.

Rice, E. L., and S. K. Pancholy. 1972. Inhibition of nitrification by climax ecosystems. *American Journal of Botany* 59:1033–1040.

Robinson, J. A. 1986. Approaches and limits to modeling microbiological processes. In F. Megusar and M. Gantar (eds.). *Perspectives in Microbial Ecology*. Slovene Society for Microbiology, Ljubljana, Yugoslavia, pp. 20–29.

Saunders , J. R., J. A. Morgan, C. Winstanley, F. C. Raitt, J. P. Carter, R. W. Pickup, J. G. Jones, and V. A. Sanders. 1990. Genetic approaches to the study of gene transfer in microbial communities. In J. C. Fry and M. J. Day (eds.). *Bacterial Genetics in Natural Environments*. Chapman and Hall, New York, pp. 3–21.

Savage, D. C. 1977. Microbial ecology of the gastrointestinal tract. *Annual Reviews of Microbiology* 31:107–134.

Saye, D. J., and S. B. O'Morchoe. 1992. Evaluating the potential for genetic exchange in natural freshwater environments. In M. A. Levin, R. J. Seidler, and M. Rogul (eds.). *Microbial Ecology: Principles, Methods, and Applications*. McGraw-Hill, New York, pp. 283–309.

Sayler, G. S., T. W. Sherrill, R. E. Perkins, L. M. Mallory, M. O. Shiaris, and D. Petersen. 1982. Impact of coal-coking effluent on sediment microbial communities: A multivariate approach. *Applied and Environmental Microbiology* 44:1118–1129.

Sayler, G. S., R. E. Perkins, T. W. Sherrill, B. K. Perkins, M. C. Reid, M. S. Shields, H. L. Kong, and J. W. Davis. 1983. Microcosm and experimental pond evaluation of microbial community response to synthetic oil contamination in freshwater sediments. *Applied and Environmental Microbiology* 46:211–219.

Schaedler, R. W., R. Dubos, and R. Costello. 1965. The development of the bacterial flora in the gastrointestinal tract of mice. *Journal of Experimental Medicine* 122:59–66.

Shannon, C. E., and W. Weaver. 1963. *The Mathematical Theory of Communication*. University of Illinois Press, Urbana, Ill.

Shoemaker, C. 1977. Mathematical construction of ecological models. In C. A. S. Hall and J. W. Day (eds.). *Ecosystem Modeling in Theory and Practice*. John Wiley and Sons, New York, pp. 75–114.

Simberloff, D. 1978. Use of rarefaction and related methods in ecology. In K. L. Dickson, J. Cairns, Jr., and R. J.

Livingston (eds.). *Biological Data in Water Pollution Assessment: Quantitative and Statistical Analyses*. ASTM STP 652. American Society for Testing and Materials, Philadelphia, pp 150–165.

Skinner, F. A., and J. G. Carr (eds.). 1974. *The Normal Microbial Flora of Man*. Academic Press, London.

Slater, J. H. 1978. The role of microbial communities in the natural environment. In K. W. A. Chater and H. J. Somerville (eds.). *The Oil Industry and Microbial Ecosystems*. Heyden and Sons, London, pp. 137–154.

Slater, J. H. 1980. Physiological and genetic implications of mixed population and microbial community growth. In D. Schlessinger (ed.). *Microbiology—1980*. American Society for Microbiology, Washington, D.C., pp. 314–316.

Slater, J. H. 1984. Genetic interactions in microbial communities. In M. J. Klug and C. A. Reddy (eds.). *Current Perspectives in Microbial Ecology*. American Society for Microbiology, Washington, D.C., pp. 87–93.

Smith, J. M. 1974. *Models in Ecology*. Cambridge University Press, Cambridge, England.

Smith, L. 1990. *Ecology and Field Biology*. 4th ed. Harper and Row, New York.

Somerville, C. C., I. T. Knight, W. L. Straube, and R. R. Colwell. 1989. Simple, rapid method for direct isolation of nucleic acids from aquatic environments. *Applied and Environmental Microbiology* 55:548–554.

Sørheim, R., V. L. Torsvik, and J. Goksøyr. 1989. Phenotypic divergences between populations of soil bacteria isolated on different media. *Microbial Ecology* 17:181–192.

Steffan, R. J., J. Goksøyr, A. K. Bej, and R. M. Atlas. 1988. Recovery of DNA from soil and sediments. *Applied and Environmental Microbiology* 54:2908–2915.

Stewart, F. M., and B. R. Levin. 1977. The population biology of bacterial plasmids: A priori conditions for the existence of conjugationally transmitted factors. *Genetics* 87:209–228.

Stewart, G. J. 1992. Natural transformation and its potential for gene transfer in the environment. In M. A. Levin, R. J. Seidler, and M. Rogul (eds.). *Microbial Ecology: Principles, Methods, and Applications*. McGraw-Hill, New York, pp. 253–282.

Stockner, J. G. 1968. The ecology of a diatom community in a thermal stream. *British Phycological Bulletin* 3:501–514.

Stotzky, G. 1989. Gene transfer among bacteria in soil. In S. B. Levy and R. V. Miller (eds.). *Gene Transfer in the Environment*. McGraw-Hill, New York, pp. 165–222.

Suberkropp, M. J., and M. J. Klug. 1974. Decomposition of deciduous leaf litter in a woodland stream: A scanning electron microscopic study. *Microbial Ecology* 1:96–103.

Summers, A. O. 1984. Genetic adaptations involving heavy metals. In M. J. Klug and C. A. Reddy (eds.). *Current Perspectives in Microbial Ecology*. American Society for Microbiology, Washington, D.C., pp. 94–104.

Swift, M. J. 1984. Microbial diversity and decomposer niches. In M. J. Klug and C. A. Reddy (eds.). *Current Perspectives in Microbial Ecology*. American Society for Microbiology, Washington, D.C., pp. 8–16.

Tempest, D. W. 1970. The continuous cultivation of microorganisms. I. Theory of the chemostat. In J. R. Norris and D. W. Ribbons (eds.). *Methods in Microbiology*, Vol. 2. Academic Press, New York, pp. 259–276.

Thayer, D. W. (ed.). 1975. *Microbial Interaction with the Physical Environment*. Dowden Hutchinson and Ross, Stroudsburg, Pa.

Torsvik, V. L. 1980. Isolation of bacterial DNA from soil. *Soil Biology and Biochemistry* 12:15–21.

Torsvik, V., K. Salte, R. Sørheim, and J. Goksøyr. 1990a. Comparison of phenotypic diversity and DNA heterogeneity in a population of soil bacteria. *Applied and Environmental Microbiology* 56:776–781.

Torsvik, V., J. Goksøyr, and F. L. Daae. 1990b. High diversity in DNA of soil bacteria. *Applied and Environmental Microbiology* 56:782–787.

Veldkamp, H. 1970. Enrichment cultures of prokaryotic organisms. In J. R. Norris and D. W. Ribbons (eds.). *Methods in Microbiology*, Vol. 3A. Academic Press, New York, pp. 305–361.

Veldkamp, H. 1977. Ecological studies with a chemostat. *Advances in Microbial Ecology* 1:59–94.

Veldkamp, H., and J. G. Kuenen. 1973. The chemostat as a model system for ecological studies. *Bulletin of the Ecological Research Commission* (Stockholm) 17:347–355.

Veldkamp, H., H. van Gemerden, W. Harder, and H. J. Laanbroek. 1984. Competition among bacteria: An overview. In M. J. Klug and C. A. Reddy (eds.). *Current Perspectives in Microbial Ecology*. American Society for Microbiology, Washington, D.C., pp. 279–290.

Walters, C. J. 1971. Systems ecology: The systems approach and mathematical models in ecology. In E. P. Odum (ed.). *Fundamentals of Ecology*. W. B. Saunders, Philadelphia, Pa., pp. 276–292.

Wanner, O. 1986. Analysis of biofilm dynamics. In F. Megusar and M. Gantar (eds.). *Perspectives in Microbial Ecology*. Slovene Society for Microbiology, Ljubljana, Yugoslavia, pp. 30–37.

Wetmur, J. G., and N. Davidson. 1968. Kinetics of renaturation of DNA. *Journal of Molecular Biology* 31:349–370.

Whittaker, R. H. 1975. *Communities and Ecosystems*. Macmillan, New York.

Wiens, J. A. 1983. Competition or peaceful coexistence. *Natural History* 92(3):30–34.

Wilcox, C. E., and G. J. Wedemayer. 1985. Dinoflagellate with blue-green chloroplasts derived from an endosymbiotic eukaryote. *Science* 227:192–194.

Williams, F. M. 1973. Mathematical modelling of microbial populations. *Bulletin of the Ecological Research Commission* (Stockholm) 17:417–426.

PART THREE

Quantitative And Habitat Ecology

7

Measurement of Microbial Numbers, Biomass, and Activities

Understanding the structure and functioning of ecosystems depends upon more than simply recognizing the interrelationships among populations; it requires quantitative information about numbers of organisms, biomass of populations, rates of activity, rates of growth and death, and cycling and transfer rates of materials within ecosystems. This quantitative approach, more than anything else, distinguishes ecologists from naturalists. In this chapter we will examine how microbial ecologists quantitatively examine the microbial components of ecosystems. The methods used in these determinations are critical in defining microbial ecology as a scientific discipline.

Numbers, biomass, and activity represent distinct ecological parameters. Though normally correlated to each other, these parameters should not be used in an interchangeable manner. The nature of the ecological problem dictates which is the most appropriate one to measure. On occasion, technical problems force us to measure a less relevant parameter, such as numbers, and to calculate from this a more relevant parameter, such as biomass. Whenever this

is done, it is important to assess critically whether such extrapolation is justified under the prevailing circumstances.

It stands to reason that a change in microbial numbers usually correlates to similar changes in biomass and activity. Since such correlations appear to be so self-evident, one needs to be alert to situations when they are either not proportional or entirely absent. Due to life cycles or nutrient availability, the cell size of microorganisms may change. In such situations, biomass may increase or decrease without a corresponding change in microbial numbers. Similarly, in the case of a sporulating fungus or actinomycete, numbers may increase dramatically with little or no change in biomass. Microbial activity will correlate with numbers and biomass only as long as environmental conditions remain constant. Any change in temperature, nutrient availability, or other environmental determinants will alter microbial activity without necessarily changing either numbers or biomass. Since unwarranted extrapolations abound in literature, microbial ecologists need to evaluate those critically and on a case-by-case basis.

To enumerate microorganisms in natural systems—that is, to measure the numbers of total microorganisms or those of specific populations—representative subsamples are analyzed and used to project the whole community and/or ecosystem (Board and Lovelock 1973). The term representative means that the sample must reflect the diversity and density of organisms in the entirety of the sampled environment. In many environments, the distribution of microorganisms is not homogeneous, but patchy. Any single sample is, of necessity, minute as compared to the whole environment being sampled and therefore may easily lead to an over- or underestimate of the true abundance. This is particularly true because microorganisms live within microhabitats that are not recognized at the time of sample collection. Processing composite samples, prepared from individual samples collected by the use of a suitable sampling grid, can minimize this error. Confidence limits of extrapolations from samples to the entire environment should be statistically determined. It is important to understand that each determination consists of three phases: sample collection, sample processing, and actual measurement. The manner of sample collection and sample processing can profoundly influence the outcome of the measurements; and therefore, all three procedures must be considered when the results are interpreted.

SAMPLE COLLECTION

Different approaches must be used for sampling microorganisms from such diverse ecosystems as the guts of animals, the surface layer of soils, waters of a lake, and the sediment of the deep sea. The lack of direct access to some environments requires the use of remote control sampling devices. In some enumeration procedures, sample processing and even counting procedures are coupled with sampling. The method of obtaining samples is determined by the physical and chemical properties of the ecosystem being examined, by the expected abundance of microorganisms, and by the enumeration or measurement procedures to be performed. Different sampling approaches are used for different environments (Table 7.1).

Sampling procedures must ensure that numbers or activities of microorganisms are not altered, either positively or negatively, in a nonquantifiable manner during collection of the sample. Sampling procedures also must ensure that samples are representative and are not contaminated with foreign microorganisms—that is, sampling procedures must ensure that the microorganisms come only from the ecosystem being examined.

Soil Samples

When collecting soil samples, microbiologists often do not use aseptic technique but pragmatically rely on a shovel and pail because of the abundance of microorganisms in the soil relative to possible contaminants from air or nonsterile containers. When soil must be collected from a particular depth, a soil corer is used. The soil corer, a hand- or motor-driven hollow tube with a sharp cutting edge, is used with or without a liner to hold the core of collected soil. Soil corers are designed to minimize compaction during sample recovery.

The sampling of microorganisms present in soil in relatively low numbers may require the use of attractants or baits. In some cases it is sufficient to provide a surface that will, by absorption, concentrate the naturally occurring dissolved nutrients. The Cholodny-Rossi buried slide technique has been used extensively in the sampling and observation enumeration of microorganisms in soils and sediments (Cholodny 1930; Rossi et al. 1936). In this technique, a glass microscope slide is implanted into the soil or sediment (Figure 7.1). Assuming that the clean glass slide surface is nonselective and acts like the surface of mineral particles in soil, the types and proportions of the organisms that adhere to the slide can be considered representative of the soil community in general. A modern variation on the buried slide technique is the exposure of electron microscope grids into natural environments (Hirsch and Pankratz 1970). Observation of the retrieved grids with the electron microscope reveals more microbial morphological detail than light microscopy can.

Table 7.1

Comparison of microbial sampling approaches in major natural environments

Environment	Access	Numbers	Sampling devices	Sample processing
Air	Direct	Low	Filters, Andersen samplers	Concentration on filters
Water	Direct or remote	High or low	Nets, containers, filters	Dilution or concentration
Sediment	Remote	High	Grabs, corers	Serial dilution
Soil	Direct	High	Shovels, corers	Serial dilution

Another variation of the buried slide technique is the replacement of glass slides with flattened glass capillaries, called pedoscopes when used in soil and peloscopes when used in sediments (Aristovskaya and Parinkina 1961; Perfilev and Gabe 1969; Aristovskaya 1973) (Figure 7.2). The capillaries resemble sediment or soil pore spaces, and microorganisms may enter them freely. The flattened optical surface of the capillaries facilitates microscopic observation and counting after retrieval. The capillaries may also be filled with nutrient solutions to attract microorganisms, in which case the technique resembles an enrichment culture and, therefore, will not reflect the original composition of the natural microbial community of the environment. Major achievements attributed to the use of capillary techniques in soils include (1) the observation of microbial landscapes of silt deposits, such as the abundance of *Caulobacter,* algae, and flagellate protozoa in wet soils, the development of *Metallogenium* indicative of iron-manganese deposition, and the extensive growth of *Gallionella* when ferrous iron is present; (2) the enumeration of new genera that had remained unknown due to their unusual characteristics; and (3) the observation of life-cycle stages in natural habitats.

Water Samples

Sampling problems are typically greater in aquatic habitats than in soils because remote collection normally is required and because microorganisms from other habitats could potentially contaminate the relatively lower numbers of microorganisms in water samples. Various sampling devices are used for collecting water and sediment samples; each has its advantages and limitations. Many of these devices are designed to ensure that the sample really comes from

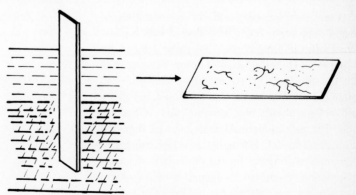

Figure 7.1

Schematic representation of buried slide technique for collection and enumeration of microorganisms.

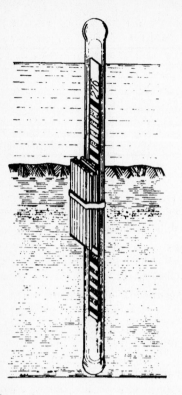

Figure 7.2

Peloscope for *in situ* observation of microorganisms in sediment. Pedoscope is a similar device for observation in soil. (Source: Aristovskaya 1973. Reprinted by permission of Swedish National Research Council.)

depth, allowing water to fill the chamber (Bordner and Winter 1978). The J-Z sampler, for example, consists of an evacuated glass bottle or a compressed rubber bulb with a sealed glass tube inlet (Figure 7.3). After the device is lowered to the desired depth, a weighted messenger is sent down the line to break the glass inlet tube, allowing the water to enter the flask or bulb. The Niskin sterile bag water sampler consists of an evacuated polyethylene bag with a sealed rubber tube inlet. The sampling bag is mounted on a spring-loaded holder, and lowered to the desired depth; a messenger is then lowered and releases a

Figure 7.3

J-Z sampler for collection of water samples at shallow depths. (a) Bracket. (b) Messenger. (c) Glass tube to be broken by messenger. (d) Rubber inlet tube. (e) Partially evacuated sample bottle.

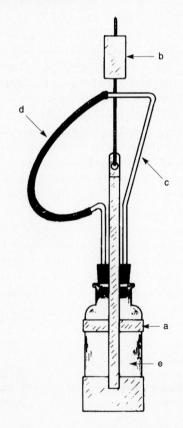

a particular location—from exactly 100 m of depth, for example.

Water samples collected for enumeration of algae and/or protozoa often are collected with Nansen and VanDorn sampling bottles, but these devices are not suitable for bacteriological sampling because they cannot be sterilized. Planktonic algae and protozoa are often collected in nets towed by a boat, but many microflagellates will not survive this collection method. The planktonic organisms may be funneled into collection bottles, where they are concentrated.

Samplers designed for the collection of bacterial populations from waters include ones consisting of an evacuated chamber that can be opened at a given

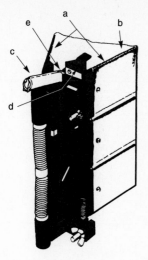

Figure 7.4

Sterile Niskin butterfly water collecting bag. (a) Spring-loaded holder. (b) Sterile plastic bag. (c) Rubber inlet tube. (d and e) Knife blade for opening inlet when triggered by messenger.

knife blade that cuts the rubber inlet hose, causing springs to spread the sampling bag, thus drawing in a water sample (Figure 7.4). Because the sampling devices are lowered through the water column to reach the desired depth, the collected sample may be contaminated with water from a different depth. Contamination may also originate from the sampling wire used to lower the device; more elaborate sampling devices are needed to preclude such contamination (Figure 7.5). Specialized sampling devices also are required for collecting water samples from great depths because of decompression problems during recovery; devices have been designed to open at a specific depth, allow a sample to enter, and close again in order to maintain the appropriate pressure during recovery (Jannasch and Wirsen 1977).

Sediment Samples

For collecting sediment samples from marine or freshwater ecosystems there are a variety of grab samplers or corers. Grab samplers have two spring-loaded jaws that are triggered when the sampler reaches bottom or when a messenger is lowered. When triggered, the jaws of the grab sampler snap shut, dig into the sediment, and collect a sample. Most grab samplers do not prevent contamination of the sample by overlying water. Box corers, frequently used by benthic biologists but only rarely by marine microbiologists, recover sediment in a relatively undisturbed state. Core samples are also used for collection of sediments for microbiological analyses. Some caution must be used to prevent compression of the sample so that the vertical stratification of the microbial community can be preserved. Coring devices can have a sterile lining that aids in handling the sample and permits dividing the core sample into vertical sections.

In addition to these remote sampling devices deployed from the surface of the water, divers and submersibles are sometimes used to collect samples for microbiological analyses. These means of collection have the advantage of enabling the collectors to see what they are sampling, to make rational decisions about what to collect, and to describe the samples. Deep-diving submersibles such as the Alvin, which has been used to investigate the deep ocean thermal vent regions, have been particularly valuable because, with their aid, some samples can be collected, processed, left *in situ* for incubation, and later recovered for analysis.

Air Samples

Sampling microorganisms from the air requires that processing be coupled with sample collection. Microbial numbers in air are generally quite low, and it is impractical to collect large volumes of air for later processing. Air samples are, therefore, generally collected by passing a measured volume of air through a membrane filter (Gregory 1973). The pore size of the membrane filter may vary for the collection of different types of microorganisms. In the Andersen air sampler, air is drawn through a graded series of grids of decreasing pore size (Andersen 1958).

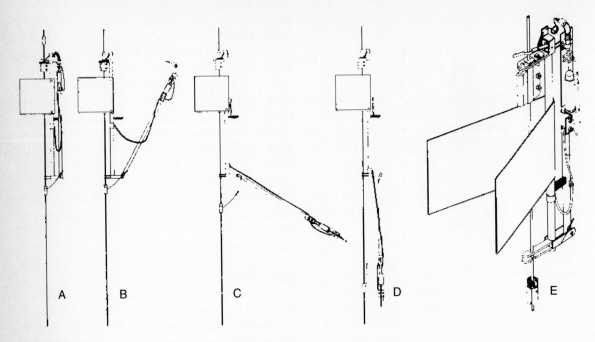

Figure 7.5

Microbial water sampling device designed to avoid contamination from the hydrographic wire. (A) Vane maintaining direction into current. (B) Sampling arm swinging out when triggered by a messenger; sterile cover of sample tube pulling away. (C) Syringe plunger retracting to draw in sample. (D) Sampling inlet closing. (E) Detail of sampling apparatus. (Source: Jannasch and Maddux 1967. Reprinted by permission of *Journal of Marine Research*.)

An agar plate is placed under each grid to collect microorganisms and other airborne particles that, by their inertia, impact on the agar surface. Air velocity increases with decreasing grid openings; therefore, increasingly smaller particles land on successive agar plates.

Biological Samples

Retrieving microorganisms from plants and animals involves collecting a vital fluid, such as sap or blood; dissecting to recover a particular tissue, such as liver; collecting excreted products, such as fecal matter; or sampling a surface tissue. In cases where the surface is exposed, microorganisms may be scraped or washed from the tissue. Microorganisms can be recovered from plant surfaces by washing with sterile solutions, from the oral cavity by scraping and washing tooth surfaces, and from human skin by swabbing. Such nondestructive sampling requires that the microorganisms not be tightly bonded to the plant or animal, and concern must be given to the efficiency of such recovery methods. In other cases, the tissue or cells may be preserved together with the associated microorganisms. This is especially useful when examination of the anatomical relationship, or enumeration of microbial populations, is required. Such approaches are useful when considering associations between microbial populations, such as between phage and bacteria, as well as when examining associations between plants or animals and microbiota.

SAMPLE PROCESSING

Only rarely are naturally occurring microbial populations found in concentrations that are convenient for measurements or counting; therefore, the microorganisms in the sample normally must be either concentrated or diluted. Samples with too many microorganisms may be brought to appropriate concentrations by serial dilutions (the successive diluting of the sample with an appropriate diluent), and samples with too few microorganisms can be concentrated by centrifugation or by passage through membrane filters. The conditions employed for processing affect the eventual numbers of microorganisms that are enumerated.

If counts are to reflect accurately the numbers of viable microorganisms present in the sample at the time of collection, then processing must be accomplished quickly because microorganisms reproduce rapidly in collection vessels, yielding artificially elevated counts. This problem, known as the bottle effect, is particularly observed with seawater, which naturally lacks adsorptive surfaces; when placed into a glass, contained nutrients are concentrated by adsorption on the wall, becoming more concentrated and available to microbes, which are then able to reproduce and greatly increase their numbers. Conditions used in processing samples can also kill some of the microorganisms present in the original sample, thus yielding artificially depressed counts.

If samples are to be diluted for viable count procedures, microorganisms must be evenly dispersed in the diluent. This is a difficult task, especially for soil and sediment samples where microorganisms are usually bound to particles. The efficiency of recovery of microorganisms is highly dependent on the chemical composition and osmotic strength of the diluent, the time of mixing, the temperature, the chemical composition of a dispersant if one is used, and the degree of agitation (Tables 7.2 and 7.3). The optimal recovery process varies from sample to sample, precluding an absolute standardization of methodology.

If samples must be concentrated by filtration, appropriate filters must be selected. Filters with pores that are too large fail to collect the microorganisms; filters with pores that are too small clog easily and

cannot filter sufficient volumes. In general, polycarbonate filters produced by Nuclepore are considered superior to nitrocellulose filters produced by Millipore because of the flatness of the surface and the uniformity of the pores. The particular chemical composition of the filter may also affect the viable counts of microorganisms.

Table 7.2

Effect of mixing time on viable counts of soil bacteria

Mixing time in blender	Viable count (#/g) on soil extract agar incubated 12 days at 25°C	
	Soil 1	Soil 2
15 sec	7.4×10^6	2.9×10^6
1 min	8.1×10^6	4.2×10^6
2 min	$*8.7 \times 10^6$	3.9×10^6
4 min	8.4×10^6	7.0×10^6
8 min	7.1×10^6	$*7.1 \times 10^6$

*Optimal mixing time.
Source: Jensen 1968.

Table 7.3

Effect of chemical composition of diluent on viable counts of soil bacteria

Diluent	Viable count (#/g) on soil extract agar—8-day incubation at 25°C with 1-min delay between preparation of dilutions and planting
Tap water	1.9×10^7
Pyrophosphate 0.1%	2.3×10^7
Ringer's solution	1.6×10^7
Winogradsky's solution	3.0×10^7

Source: Jensen 1968.

It is obvious that conditions not compatible with the physiological requirements of particular groups of microorganisms will lead to their underestimation because of loss of viability. For example, enumeration of obligate anaerobes requires that diluents be sparged free of oxygen and that all processing be carried out under an oxygen-free atmosphere. The enumeration of obligate psychrophiles requires that the diluent and all glassware, such as pipettes, be prechilled to avoid killing these microorganisms.

When the enumeration procedure is not designed to discriminate between living and dead microorganisms, the samples may be preserved by adding formaldehyde or glutaraldehyde. Such preparations are suitable for direct microscopic observations.

The recovery of viruses from environmental samples requires special processing to concentrate the viruses (Sobsey 1982; Gerba 1983). Viruses can be concentrated from water by repetitive adsorption and elution processes; in one such procedure, after acidification, viruses in large volumes of water can be adsorbed onto epoxy, fiberglass, or nitrocellulose filters. Adsorbed viruses can then be eluted with an alkaline solution. The sorption and elution process can be repeated, resulting in a many-thousandfold concentration of the viruses. Viruses can also be eluted from particles such as occur in sewage sludge, using flocculating materials and adsorptive filters that maintain the infectivity of viruses during these concentration procedures. Somewhat different procedures are required for recovery of viruses from animal or plant tissues (Figure 7.6).

Figure 7.6

An example of a viral recovery and concentration method that has been used for examination of shellfish. (Source: Sobsey et al. 1978. Reprinted by permission of American Society for Microbiology.)

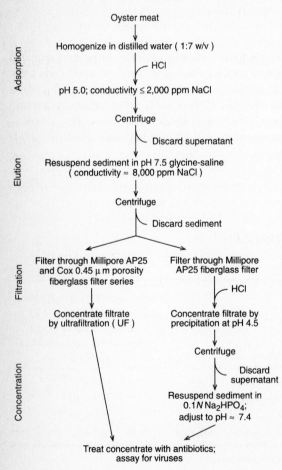

DETECTION OF MICROBIAL POPULATIONS

Phenotypic Detection

The detection of microorganisms based upon phenotypic characteristics requires that those microorganisms to be specifically detected must be recovered from environmental samples and that recognizable and distinctive phenotypes must be expressed during *in vitro* culture (Atlas 1991). In some cases, the observation of a single unique characteristic may be sufficient to recognize the specifically targeted microbe. A microorganism with a unique phenotype, such as the failure to produce a peptide that initiates ice crystal formation (ice-minus phenotype), could be selected based upon such a single distinguishing characteristic.

In other cases, it may be necessary to determine a pattern of multiple characteristics in order to distinguish the specific targeted microbe from all others. The unambiguous detection of a microbe based upon phenotype is a difficult task because even distantly related organisms can have similar phenotypes for some characteristics.

The classical approaches for detecting microorganisms are to place viable microbial cells onto a solid medium (plating procedures) or into a liquid broth (enrichment procedures), containing all of the nutrients essential for the growth of the targeted microorganisms, and to incubate the inoculated cultures under conditions that favor the growth of those microbes so that their phenotypes can be observed.

Plating procedures rely upon the ability to separate microorganisms so that individual microbial cells are deposited at discrete locations on a solid medium; when the deposited microorganisms reproduce, they form discrete colonies that comprise clones of the original microorganisms. Plating procedures accomplish two things. First, they separate individual microbes from the multitude of microorganisms in a natural microbial community so that the phenotypes of individual microbes can be determined, which is the basis of the pure culture methods that are the mainstay of microbiology. Second, they amplify the signal (phenotype) to be observed through cell reproduction. The phenotypic characteristics of the millions of presumably identical cells in a colony can be observed much more readily than those of individual microbial cells.

Plating procedures are effective for detecting a particular microbe when that target microbe constitutes a significant portion of the microbial community, but microorganisms constituting extremely low proportions of a community can be overlooked by plating procedures. In cases where a microbe makes up only a small fraction of a community, growth in liquid cultures under conditions that specifically favor the growth of the target microbe over the growth of nontarget microbes can be used to enrich for the target microbe.

With adjustments for the chemical composition of the growth medium and incubation conditions, plating and enrichment procedures are designed to be differential and/or selective to detect specific microorganisms. Because the ability of a particular microorganism to grow on or in a particular medium is based on its ability to utilize specific organic and inorganic nutrients, a culture medium can be made selective on the basis of its constituents (such as the particular carbon sources, the salt concentration, and so on). Also the incubation conditions can be made selective by adjusting the physical growth conditions (such as temperature, oxygen concentration, and so on). The addition of inhibitory compounds to a medium suppresses the development of the majority of microorganisms, enabling the development of the desired species to occur.

Because the phenotype of the microorganism depends upon the specific pathways for the metabolism of the nutrients in the medium, specific dyes, redox indicators, and so on, can be included in the medium to reflect the metabolism of specific nutrients by different populations and hence the differentiation of target microorganisms.

Methods for detecting microorganisms use cultivation media that are usually formulated to take advantage of specific traits of the organism, such as nutritional capabilities and/or resistances to specific antibiotics, so that target microorganism are favored for growth over other strains. Generally, a visible phenotypic characteristic (such as pigmentation), the ability to use a specific substrate, or resistance to a certain antibiotic or heavy metal is used to differentiate the target microorganism from nontarget microorganisms.

Immunological Detection

Direct immunofluorescence has been used to detect microorganisms in environmental samples (Wright 1992), including in water (Grimes and Colwell 1983), on buried slides, on root surfaces, in nodules, and in soil (Bohlool and Schmidt 1980; Bohlool 1987). This method has also been used to detect viable but nonculturable bacteria (Xu et al. 1982) and hence may be useful for detecting microorganisms in aquatic and marine ecosystems. Distribution of *Vibrio cholerae* in the natural environment of a cholera-endemic region has been studied using immunofluorescence and has been detected in environmental samples when cultural methods were ineffective (Brayton and Colwell 1987). Detection of rhizobia by enzyme-linked immunosorbant assays (ELISA) in peat and in soil has been reported (Nambairan and Anjaiah 1985). Fluorescent ELISA results compare with those from the antibiotic-resistance technique when soil populations exceed 10^4 cells per milliliter (Renwick and Jones 1985). Rhizobial strains in root nodule material can also be identified by ELISA (Wright et al. 1986).

Gene Probe Detection

Gene probes and nucleic acid hybridization techniques have been employed for the detection of target nucleic acid sequences that are diagnostic of specific microorganisms in environmental samples (Sayler and Layton 1990; Knight et al. 1992). With the use of gene probes, specific microbial populations can be detected in environmental samples. A gene probe is a relatively short nucleotide sequence that can bind (hybridize) to nucleotide sequences with homologous sequences in the target microorganism. Hybridization of the gene probe is diagnostic of the presence of a specific microbial population. Temperature, contacting time, salt concentration, the degree of mismatch between the base pairs, and the length and concentration of the target and probe sequences are factors

affecting the hybridization or reassociation of the complementary DNA strands (Berent et al. 1985; Britton and Davidson 1985; Sambrook et al. 1989).

Most hybridization protocols are based on the hybridization of an immobilized target or probe molecule on a solid phase such as a nitrocellulose or nylon filter surface (Figure 7.7). First, single-stranded target nucleic acids are attached to the filter surface. Next, the filters are prehybridized to block nonspecific nucleic acid binding sites. Labeled probe DNA is added to the filters, and the probe is allowed to hybridize. The labeled probe establishes a double-stranded molecule with the complementary target sequences. Typically, the gene probe is labeled with ^{32}P, and following hybridization, detection is by autoradiography. ^{32}P can be incorporated using ^{32}P-labeled free nucleotides and nick translation or end labeling using T4-polynucleotide kinase to incorporate the ^{32}P into the gene probe. After hybridization, excess unbound labeled probe is washed off and the hybrid (target-probe) sequences are detected.

Two approaches have been used to recover DNA from environmental samples: isolation of microbial cells followed by cell lysis and nucleic acid purification (cell extraction) and direct lysis of microbial cells in the environmental matrix followed by nucleic acid purification (direct extraction) (Steffan et al. 1988).

For water samples, cells can be collected by filtration and then lysed to isolate their nucleic acids (Bej et al. 1990, 1991a; Fuhrman et al. 1988; Giovan-

Figure 7.7

Illustration of nucleic acid hybridization, which is the basis for gene probe detection.

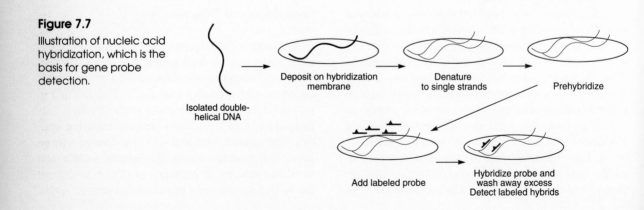

Isolated double-helical DNA

Deposit on hybridization membrane

Denature to single strands

Prehybridize

Add labeled probe

Hybridize probe and wash away excess Detect labeled hybrids

noni et al. 1990; Sommerville et al. 1989). Alternately, filtered cells may first be subjected to enzymatic lysis and/or phenol-chloroform extraction procedures to yield adequate DNA. C. C. Sommerville et al. (1989) have demonstrated a simple method for isolating nucleic acids from aquatic samples that allows filtering relatively large volumes of water. Cell collection and lysis are performed in a single filter cartridge, and chromosomal DNA, plasmid DNA, and RNA can be selectively recovered. J. H. Paul et al. (1990) reported that with certain algal DNA preparations and with all preparations of environmental DNA, purification by cesium chloride-ethidium bromide ultracentrifugation was required for isolation of purified DNA. In other studies, multiple phenol or phenol-chloroform extractions have produced sufficiently pure DNA from cyanobacteria (Zehr and McReynolds 1989) or planktonic microorganisms (Fuhrman et al. 1988; Lee and Fuhrman 1990).

Cell extraction methods for soils were developed by D. L. Balkwill et al. (1975) and J. Goksøyr and colleagues (Faegri et al. 1977; Torsvik and Goksøyr 1978). A number of groups have used this approach with some modification to isolate DNA from soil (Torsvik 1980; Bakken 1985; Holben et al. 1988; Steffan et al. 1988; Jansson et al. 1989; Torsvik et al. 1990) or sediment (Steffan et al. 1988; Steffan and Atlas 1990). One simple but significant improvement of this approach has been the inclusion of a polyvinylpolypyrrollidone (PVPP) treatment to decrease sample humic content prior to cell lysis (Holben et al. 1988; Steffan et al. 1988), thereby simplifying DNA purification.

Direct extraction of DNA from environmental samples was first described by A. Ogram et al. (1987). In this method, cells are lysed while still within the soil matrix by incubation with sodium dodecyl sulfate followed by physical disruption with a bead beater. The DNA is then extracted with alkaline phosphate buffer. This method has been successfully used to recover DNA from sediments and soils (Ogram et al. 1987; Steffan et al. 1988). Purification procedures can be any combination of cesium chloride-ethidium bromide ultracentrifugation (Holben et al. 1988; Steffan et al. 1988; Paul et al. 1990), hydroxylapatite or affin-

ity chromatography (Torsvik and Goksøyr 1978; Ogram et al. 1987; Steffan et al. 1988; Paul et al. 1990), phenol-chloroform extractions, ethanol precipitations (Fuhrman et al. 1988), dialysis, or repeated PVPP treatments (Holben et al. 1988; Steffan et al. 1988; Weller and Ward 1989; Paul et al. 1990). Significantly higher yields of DNA are recovered with the direct extraction method than with the cell recovery procedure, but the DNA may contain impurities that can inhibit enzymatic manipulation (Steffan et al. 1988).

Amplification of DNA—The Polymerase Chain Reaction   The polymerase chain reaction (PCR) (Mullis et al. 1986; Mullis and Faloona 1987; Mullis 1990) has begun to be applied to environmental detection of microorganisms (Bej et al. 1991c; Steffan and Atlas 1991). It permits the *in vitro* replication of defined sequences of DNA. One application of this technique is to enhance gene probe detection of specific gene sequences. By exponentially amplifying a target sequence, PCR significantly enhances the probability of detecting rare sequences in heterologous mixtures of DNA.

The polymerase chain reaction involves melting the DNA to convert double-stranded DNA to single-stranded DNA, annealing primers to the target DNA, and extending the DNA by nucleotide addition from the primers by the action of DNA polymerase (Figure 7.8). The oligonucleotide primers are designed to hybridize to regions of DNA flanking a desired target gene sequence. The primers are then extended across the target sequence by using Taq DNA polymerase (a heat-stable enzyme derived from the thermophilic bacterium *Thermus aquaticus*) in the presence of free deoxyribonucleotide triphosphates, resulting in a duplication of the starting target material. When the product DNA duplexes are melted and the process is repeated many times, an exponential increase in the amount of target DNA results. The essential components of the PCR reaction mixture are Taq DNA polymerase, oligomer primers, deoxyribonucleotide (dNTPs), template DNA, and magnesium ions. The temperature cycling for PCR can be achieved using an automated thermal cycler. Annealing temperatures in the range of 55°C

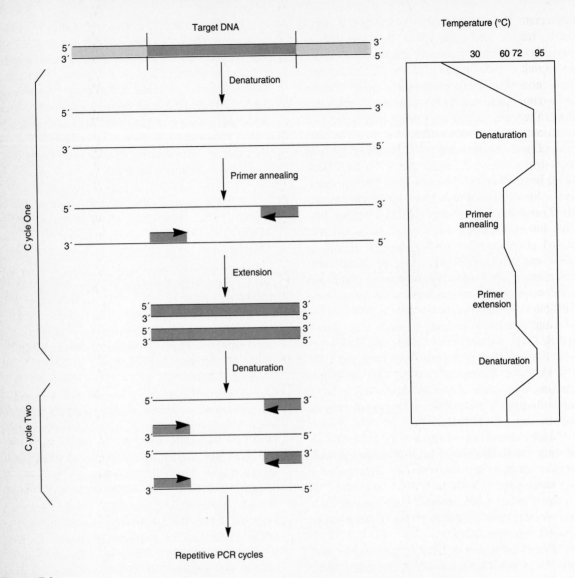

Figure 7.8
Illustration of the polymerase chain reaction (PCR) for the amplification of a target gene sequence.

to 72°C generally yield the best results. Typical denaturation conditions are 94°C to 95°C for 30 to 60 seconds. Lower temperatures may result in incomplete denaturation of the target template and/or the PCR product and failure of the PCR. Too many cycles can increase the amount and complex-ity of nonspecific background products. Too few cycles give low product yield.

The polymerase chain reaction is useful for environmental surveillance (Atlas and Bej 1990). A. K. Bej et al. (1990, 1991a) used PCR amplification and gene probe detection to detect coliform bacteria in environmental water samples. This PCR amplification and radiolabeled gene probes detected as little as 1 to 10 fg of genomic *E. coli* DNA and as few as one to five viable *E. coli* cells in 100 ml of water. Thus, they demonstrated the potential use of PCR amplification as a method to detect indicators of fecal contamination of water.

Both viable culturable and viable nonculturable cells of *Legionella pneumophila,* formed during exposure to hypochlorite, showed positive PCR amplification, whereas nonviable cells did not (Bej et al. 1991b). Viable cells of *L. pneumophila* were also specifically detected by using mRNA from the gene that codes for macrophage infectivity potentiator (*mip* mRNA) as the target, reverse transcription to form cDNA, and PCR to amplify the signal. When cells were killed by elevated temperature, only viable culturable cells were detected, and detection of viable culturable cells corresponded precisely with positive PCR amplification.

R. J. Steffan and R. M. Atlas (1988) used PCR to amplify a 1.0-kilobase (kb) probe-specific region of DNA from the herbicide-degrading bacterium *Pseudomonas cepacia* AC1100 in order to increase the sensitivity of detecting the organism by dot-blot analysis. The 1.0-kb region was an integral portion of a larger 1.3-kb repeat sequence present as fifteen to twenty copies on the *P. cepacia* AC1100 genome. Amplified target DNA was detectable in samples initially containing as little as 0.3 pg of target. The addition of 20 µg of nonspecific DNA isolated from sediment samples did not hinder amplification or detection of the target DNA. The detection of 0.3 pg of target DNA was at least a 10^3-fold increase in the sensitivity of detecting gene sequences compared with dot-blot analysis of nonamplified samples. After bacterial DNA was isolated from sediment samples, PCR permitted the detection of as few as 100 cells of *P. cepacia* AC1100 per 100 g of sediment sample against a background of 10^{11} diverse nontarget organisms; that is, *P. cepacia* AC1100 was positively detected at a concentration of 1 cell per g of sediment (Figure 7.9). This represented a 10^3-fold increase in sensitivity compared with nonamplified samples.

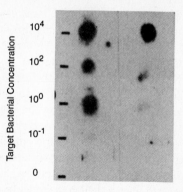

Figure 7.9

Dot-blot hybridization following PCR amplification of a gene segment diagnostic of *Pseudomonas cepacia* AC1100 showing increased sensitivity (left) over nonamplified case (right). (Source: Steffan and Atlas 1988. Reprinted by permission.)

Reporter Genes Genetic markers, or bioreporters, can be used to track specific microbial populations in the environment. A genetic marker is a genetic element that permits the detection of an unrelated biological function. The *lacZ* gene, which encodes β-galactosidase, is a useful biomarker. The β-galactosidase enzyme cleaves the disaccharide lactose into glucose and galactose. When the synthetic compound 5-bromo-4-chloro-3-indolyl-ß-D-galactoside (X-gal) is substituted, the β-galactosidase is able to cleave the compound, creating an insoluble blue pigment that is very noticeable on most agar media. The versatility of the *lacZ* system makes it the most widely used reporter gene. It has been used as a reporter of recombinant *Pseudomonas* in studies to determine whether genetically engineered microorganisms will survive and disperse if released at an environmental site (Drahos et al. 1986). *Pseudomonas* strains that incorporate the *lacZY* genes are capable of cleaving the X-gal substrate to produce the characteristic blue color. This trait makes them distinguishable from the nonrecombinant *Pseudomonas* species that are indigenous to the site.

The *xylE* gene of the TOL plasmid has been cloned for use as a transcriptional fusion reporter gene. This gene encodes catechol 2,3 oxygenase (C 2,3 O), and a simple aerosol spray technique has been used to find recombinant colonies. Colonies that turn yellow in the presence of the catechol spray are expressing C 2,3 O and produce 2-hydroxymuconic semialdehyde (Zukowski et al. 1983). C. Ingram et al. (1989) utilized the *xylE* reporter gene for a quantitative description of promoters from *Streptomyces*. R. J. King et al. (1991) recently described a similar assay system using the 2,4-dichlorophenoxyacetate monooxygenase (*tfdA*) gene. The *tfdA* gene product converts phenoxyacetate to phenol, which then can be reacted with a dye to form a colored compound that can be analyzed spectrophotometrically. P. B. Lindgren et al. (1989) isolated the gene for ice nucleation (*inaZ*) and incorporated it into a transposon. This transposon was then used to mutagenize plant pathogenic bacterial strains, and the mutants were tested using a droplet freezing technique. Expression of the *inaZ* gene results in rapid freezing of the droplets at –9°C. Dilutions of the cell suspension can be tested in the same way, allowing a sensitivity that is reported to be far greater than that of the *lacZ* system.

The production of visible light (bioluminescence) by various organisms is another unusual trait. It has been studied in great detail in recent years, and its use as a reporter system has increased with genetic and physiological analysis of the *lux* genes, and with the availability of cloning vectors and other genetic constructions (Boivin et al. 1988) that facilitate their use. Several bacterial species can produce light, including members of *Vibrio, Photobacterium, Alteromonas,* and *Xenorhabdus; Vibrio* has received the most attention. The first practical application of bacterial bioluminescence was in 1985 when J. Engebrecht et al. (1985) constructed fusions between *E. coli* promoters from the *lac* and *ara* operons and the *lux* genes of *Vibrio fischeri*. A mini-Mu-*lux* vector was used to facilitate these constructions. Induction of the *lac-lux* or *ara-lux* fusion resulted in a significant increase in bioluminescence. This report promoted light measurement as a convenient *in vivo* assay for gene expression.

The *lux* gene constructions have been used to investigate gene expression of plant pathogenic bacteria (Shaw and Kado 1986; Rogowsky et al. 1987). In this work, the induction of gene expression of bacterial virulence genes could be followed in plants by examining light production. R. S. Burlage et al. (1990) described a fusion between a promoter (*nah*) for the degradation of naphthalene in *Pseudomonas* and the *lux* genes. Using this plasmid construction, they were able to demonstrate gene expression on a continuous basis, revealing an unexpected pattern.

Environmental analysis using a bioluminescent reporter strain now appears to be a practical technique. J. M. King et al. (1990) used a *nah-lux* construction to demonstrate naphthalene degradation in soil slurries. Despite the quenching of bioluminescence that occurs in particulate and contaminated samples, this work was successful in detecting both the presence and bioavailability of specific contaminants in the complex slurry. E. A. S. Rattray et al. (1990) have also described the great sensitivity of this technique in their analysis of bioluminescent *E. coli* cells that were added to soil and liquid media. Their results indicate that bioluminescent bacteria can be detected using a nondestructive assay with a sensitivity matched only by hybridization analysis. These reports support the concept of using bioluminescence as an indicator of presence and activity of bacteria in environmental samples.

DETERMINATION OF MICROBIAL NUMBERS

In the case of a pure bacterial culture, the question "How many microorganisms are there?" is relatively simple to answer. When dealing with the mixed communities of environmental samples, however, the problem becomes exceedingly complex and, in most cases, defies a simple and absolute answer. Quantification is possible, but the numbers obtained need to be qualified by the technique used; enumeration techniques need to be chosen with care to assure that results are relevant to the question being asked.

Microorganisms are extremely diverse, so the methods used to enumerate one group of microorganisms may be inappropriate for the enumeration of another group. The methods used for enumerating viruses (Primrose et al. 1982), bacteria (Herbert 1982), fungi (Parkinson 1982), algae (Round 1982), and protozoa (Finlay 1982) are quite different. Special techniques must be used for enumerating specific physiological groups, such as psychrophilic bacteria or obligately anaerobic bacteria. Enumeration of such defined groups requires special sampling and processing procedures. Precise differential criteria must be used for enumerating an individual microbial species such as *Escherichia coli,* which is often used as an indicator of sewage contamination.

The enumeration process must begin with a definition of which microorganisms are to be enumerated; all microorganisms, or only certain groups? Are the numbers of microorganisms to be converted to estimates of biomass? What are the characteristics of the habitat being examined? No universal method can be applied to all microorganisms and all habitats; the diversity of microorganisms and their habitats requires the use of a variety of methodological approaches.

The enumeration process can be broken down into three distinct phases: sampling, sample processing, and the actual counting procedure. All phases of the process must be considered when interpreting the results.

Two principal approaches are used for enumerating microorganisms: direct observation (or direct count) procedures and indirect (or viable count) procedures. Occasionally, numbers are also calculated from procedures that measure specific biochemical constituents of microorganisms, though it should be recognized that such procedures actually measure biomass. Each approach has its advantages and limitations (Atlas 1982).

Direct Count Procedures

Microscopic Count Methods Microorganisms can be counted by direct microscopic observations (Jones and Mollison 1948; Frederick 1965; Gray 1967; Gray et al. 1968; Harris et al. 1972; Daley and Hobbie 1975; Byrd and Colwell 1992). Direct count

procedures yield the highest estimates of numbers of microorganisms and are occasionally used for the indirect calculation of biomass. There are, however, several major drawbacks to direct observational methods, including the inability to distinguish living from dead microorganisms, the underestimation of microorganisms in samples containing high amounts of background debris, and the inability to perform further studies on the observed microorganisms.

For relatively large microorganisms, such as protozoa, algae, and fungi, direct counts can be conveniently accomplished using a counting chamber (Finlay et al. 1979). A variety of counting chambers that hold specified volumes, such as a hemocytometer or Petroff-Hauser chamber, can be used for counting cells (Parkinson et al. 1971).

A modified agar film technique of P. C. T. Jones and J. E. Mollison (1948) often is used for direct observation and enumeration of fungi (Skinner et al. 1952; Thomas et al. 1965; Parkinson 1973). In this technique, the sample is mixed with agar and pipetted onto glass slides, where it forms a thin film. The dried agar film can be stained with phenolic aniline blue, and the slides then can be viewed. Length of mycelia can be estimated by counting intersections with a superimposed grid.

Alternately, the microscope image can be projected onto a screen and traced. The length can be determined by running a distance-measuring device (map measure) along the traced lines (Table 7.4). If the average width of the mycelial filaments is also measured, the fungal biomass can be calculated (Table 7.5). A problem with this technique is that unless adequate dispersion is achieved, hyphae remain hidden among the soil particles and can only be seen when soil particles are broken (Skinner et al. 1952). This problem can be overcome by establishing the optimum grinding time and using a modern high-speed homogenizer.

The agar film technique for enumeration of fungi can be combined with fluorescence microscopy using fluorescein diacetate as the stain to estimate the biomass of living fungi (Babuik and Paul 1970; Soderstrom 1977). Only metabolically active cells are stained in this procedure because fluorescence occurs

Table 7.4

Length of fungal mycelium in tundra soils

Location	Sites	Mean mycelium length (m/g oven dry soil)
Devon Island, Canada	Mesic meadow	1,005
	Raised beach	199
Moor House, United Kingdom	Blanket bog	4,968
	Juncus moor	1,667
	Limestone grass	826
Signy Island, Antarctica	Hut bank	6,328
	Mountain moss	2,783
	Grassland	288
	Old moraine	84
	New moraine	4
	Marble knolls	144
	Marble schist soils	44
Hardangervidda, Norway	Wet meadow	5,000
	Dry meadow	1,000
	Birch wood	7,000
Ireland	Blanket bog	995
Sweden	Mire	3,260
United Kingdom	Wood	341
Mt. Allen, Canada	1,900 m elevation	580
	2,500 m elevation	801
	2,800 m elevation	50

Source: Dowding and Widden 1974.

only after enzymatic cleavage of the stain molecule; thus, use of fluorescein diacetate stain in conjunction with a vital stain that does not distinguish living from dead fungal hyphae permits the estimation of both living and total fungal biomass. A problem with this procedure is the high background fluorescence that occurs because esterases are released and react with the dye.

Epifluorescence microscopy with stains such as acridine orange (AODC), 4′,6-diamidino-2-phenyl-indole (DAPI), and fluorescein isothiocyanate (FITC) are widely used for direct counting of bacteria (Strugger 1948; Babuik and Paul 1970; Zimmerman and Meyer-Reil 1974; Daley and Hobbie 1975; Russel et al. 1975; Daley 1979; Geesey and Costerton 1979; Coleman 1980; Porter and Feig 1980). K. G. Porter and Y. S. Feig (1980) adapted DAPI for counting aquatic microorganisms and found it to work better than acridine orange for eutrophic and seston-rich water samples. Clays, colloids, and detritus of such samples produce a red orange background, which impairs acridine orange counting. Acridine orange and

Table 7.5

Length and biomass of fungal mycelium in soil

Sample	Length of hyphae (m/g)	Fungal biomass (g/m^2)
Soil 1	174	0.31
Soil 2	619	2.69
Soil 3	457	1.39
Soil 4	1,602	7.66

Source: O. K. Miller, Virginia Polytechnic Institute, unpublished data.

Table 7.6

Comparison of Nuclepore (polycarbonate) and Sartorius (cellulose nitrate) filters for direct counts

Site	Count (#/mL)	
	Sartorius (0.5 μm)	Nuclepore (0.2 μm)
Estuary	8.6 x 10^5	1.7 x 10^6
Reservoir	4.0 x 10^5	7.2 x 10^5
Pond	1.7 x 10^6	4.0 x 10^6

Source: Hobbie et al. 1977.

DAPI were comparable in efficiency of enumeration of total organisms in aquatic samples. Porter and Feig (1980) reported that DAPI-stained slide counts remained constant for up to 24 weeks when the stained slides were stored at 4°C in the dark, compared with acridine-orange-stained slides, which yielded decreased counts after storage for a week under the same conditions.

Another useful DNA-specific fluorochrome is Hoechst 33258 (bisbenzimide). Bisbenzimide binds to adenine and thymine-rich regions of the DNA, increasing in fluorescence after binding (Latt and Statten 1976). Hoechst 33258 has been recommended for enumeration of bacteria on surfaces, especially if the surface tends to bind acridine orange (Paul 1982). An advantage of the bisbenzimide stain is that detergent solutions, laboratory salts, or biological materials associated with DNA do not affect it. Fluorescent staining to detect microorganisms in environmental samples by direct microscopy has several applications in microbial ecology.

Use of low-fluorescing immersion oil is important to minimize background autofluorescence. Polycarbonate Nuclepore filters are superior to cellulose filters for direct counting of bacteria because they have uniform pore size and a flat surface that retains all the bacteria on top of the filter (Hobbie et al. 1977) (Table 7.6). The Nuclepore filters can be stained with irgalan black or other suitable dye to create a black background against which fluorescing microorganisms can be counted (Jones and Simon 1975). When acridine orange is used as a stain, bacteria and other microorganisms fluoresce orange and green. The green or orange color correlates with the physiological state of the microorganism, but attempts to separate living from dead microorganisms by the color of fluorescence are often misleading and unsatisfactory. Use of DAPI, which stains the DNA of bacterial cells and produces an intense blue fluorescence, has been found to be superior to use of acridine orange for visualizing small bacterial cells (Coleman 1980).

Counts obtained by direct epifluorescence microscopy are typically two orders of magnitude higher than counts obtained by cultural techniques (Table 7.7). Many small and unusually shaped bacteria are observable and countable with epifluorescence microscopy. These forms are usually impossible to

Table 7.7

Comparison of direct epifluorescent counts and viable plate counts

Sample	Soil		Marine water	
	Direct count	Viable count	Direct count	Viable count
A	5.0 x 10^8	3.1 x 10^7	2.2 x 10^3	1.3 x 10^1
B	1.1 x 10^9	6.2 x 10^7	8.2 x 10^4	7.6 x 10^2
C	2.0 x 10^9	1.7 x 10^8	1.3 x 10^6	2.1 x 10^4

Source: Atlas, unpublished data.

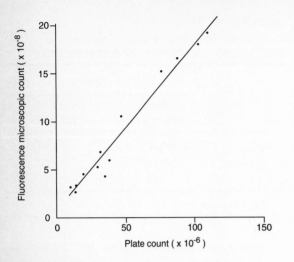

Figure 7.10

Correlation between direct microbial counts by fluorescence microscopy and viable plate counts for soil samples. There is a high correlation coefficient (r) of 0.97 for the linear regression line described by the equation y = 900.2 + 15.7(x − 49.5). (Source: Trolldenier 1973. Reprinted by permission of Swedish National Research Council.)

cultivate on laboratory media, however, and it is not clear whether these are "degenerate" nonviable representatives of known bacteria or forms with yet-unknown physiological requirements. In some cases, positive correlation between direct fluorescence microscopic counts and viable plate counts is high (Figure 7.10); in other cases, correlation is low (Table 7.8). The differences between direct and plate counts may simply reflect the selectivity of the media and incubation conditions used, the proportion of living and dead or injured bacteria, and the particular species in the sample.

The value of the direct count epifluorescence microscopy approach to enumeration is that it is applicable to a variety of habitats without the bias inherent in viable plate count procedures. It allows the estimation of numbers of microorganisms in marine, freshwater, and soil habitats despite the great differences in population sizes and physiological types that

occur in these various habitats. Direct counts are often directly proportional to biomass and thus can be used to estimate microbial biomass. Tedious size measurements must be made, however, to convert the direct microscopic counts to biomass estimates accurately.

Additionally, if the numbers of dividing cells are counted, it is possible to estimate *in situ* growth rates (Hagstrom et al. 1979). The frequency of dividing cells has been shown to correlate with other measures of microbial growth such as rates of RNA synthesis measured by incorporation of labeled adenine (Christian et al. 1982). This frequency of dividing cells (FDC) method assumes that the time between the initiation of cell constriction and cell separation is constant—an assumption that is not always valid (Staley and Konopka 1985). It is also difficult to recognize dividing cells using light microscopic methods.

The numbers of specific types of microorganisms can be estimated using fluorescent antibody techniques (Hill and Gray 1967; Schmidt 1973; Strayer and Tiedje 1978; Fliermans et al. 1979; Bohlool and Schmidt 1980). The fluorescent antibody technique is extremely specific for an individual microbial species and permits autecological studies, that is, studies of individual microorganisms in their natural environments (Bohlool and Schmidt 1973) (Table 7.9). In the FITC method, antibodies are conjugated with a fluorochrome, such as fluorescein isothiocyanate. Problems with the fluorescent antibody technique include nonspecific fluorescence of the background in some samples, the high degree of specificity of the antigen-antibody reaction that may preclude staining of even different strains of the same species, and possible reactivity between different organisms. Fluorescent antibody techniques have been applied to studies on selected microbial species in their natural habitats, including ecologically important organisms (Eren and Pramer 1966); monoclonal antibodies have been used to specifically detect methanogens in environmental samples (Conway de Macario et al. 1981, 1982), and such specific fluorescent dyes are applicable to enumerate many other specific populations.

Epifluorescence microscopy can be converted to automatic data analysis to relieve the sheer tedium involved in examining a large number of samples. As

Table 7.8

Counts of bacteria (numbers/mL)

Location	Season/Year	Direct count	Viable count	
			4°C	20°C
Beaufort Sea				
Ice	Winter 1976	9.9×10^4	6.6×10^1	4.5×10^1
Water	Summer 1975	8.2×10^5	9.6×10^3	7.3×10^3
	Winter 1976	1.8×10^5	6.1×10^2	1.1×10^1
	Summer 1976	5.2×10^5	5.0×10^4	2.7×10^4
NW Gulf of Alaska				
Water	Summer 1975	3.0×10^5	1.0×10^2	5.2×10^2
NE Gulf of Alaska				
Water	Winter 1976	1.4×10^5	1.3×10^2	2.4×10^2

Source: Kaneko et al. 1978.

early as 1952, image analysis was utilized to count microscopic coal particles (Walton 1952). G. L. Pettipher and U. M. Rodrigues (1982) applied image analysis to epifluorescence microscopy by counting acridine orange-stained bacteria in milk. M. E. Sieracki et al. (1985) used DAPI stain to detect and enumerate planktonic bacteria. They found that counts

done by standard visual analysis and by image analysis gave statistically equal results.

Image analysis can be a powerful technique, but some constraints apply. The results of image analysis will be only as good as the image provided. Staining of each organism must be strong, with little or no background or extraneous fluorescence. This method has limited application to soil or sediment samples without extreme care toward removal of particulates (Van Wambeke 1988). Another constraint is that a highly sensitive camera is necessary to detect low levels of light emitted from epifluorescent-stained samples. There is also some question as to the ability of the camera to differentiate between debris and cells when fluorescent counts are recorded. Overall, the expense of the system is compensated for by the welcome automation and reduction in time needed to analyze individual samples and by the increased productivity in analysis.

Fluorescent staining methods can also be used in combination with other procedures to determine numbers of living organisms. By combining the AODC method with INT (2-[p-iodophenyl]-3-[p-nitrophenyl]-5-phenyl tetrazolium chloride) staining, for example, it is possible to determine both total numbers of microorganisms and the numbers of

Table 7.9

Quantification of *Rhizobium japonicum*, USDA 110, in the rhizosphere of soybean plants grown in the field

Sample (weeks after plating)	Number *R. japonicum*/g soil*	
	Uninoculated	Inoculated
2	4.1×10^3	6.2×10^6
3	4.9×10^3	1.9×10^6
4	4.9×10^3	3.6×10^7
5	NT	6.4×10^5
6	NT	1.7×10^6

(NT = not tested)
* = Based on counts of 20 microscopic fields per sample.
Source: Bohlool and Schmidt 1973.

microorganisms carrying out respiration (Zimmerman et al. 1978). The method is based on the fact that electron transport systems in respiring microorganisms reduce INT to INT-formazan, so that respiring bacteria accumulate intracellular dark red spots. When applied to Baltic Sea water samples, 6%–12% of the bacteria were found to be active; in freshwater samples, 5%–36% of the bacteria were active (Zimmerman et al. 1978). This method can be difficult to interpret, depending on the size of the bacterial cells and INT grains that develop.

Numbers of actively growing bacteria can also be determined by incubation with nalidixic acid, an antibiotic that inhibits cell division (Kogure et al. 1979). This method is referred to as a direct viability count (DVC). Microscopic observation after fluorescent staining reveals elongated cells (actively growing microbes) and cells of normal size and shape (dormant or dead microbes). This method exploits the finding of Goss et al. (1965) that nalidixic acid is an inhibitor of DNA synthesis. The nalidixic acid prevents cell division in Gram-negative bacteria by suppressing DNA replication. Subsequently, cross-wall formation is obstructed because of lack of replication, so the cells elongate instead of dividing. K. Kogure and associates (1979) used yeast extract as a nutrient source, and incubation was for six hours. This allowed detection of viable bacteria in seawater samples. Incubation of seawater samples beyond twelve hours often allows either growth of Gram-negative, nalidixic acid-resistant bacteria (Quinn 1984) or a possible inactivation of nalidixic acid by high salt concentrations. Therefore, it is impossible to quantitate, by extended incubation, slow-metabolizing bacteria in samples, and these may go unnoticed in the six hours of incubation usually employed in the DVC procedure.

One can detect dormant cells—that is, cells physiologically responsive but not dividing—by DVC, technically a straightforward procedure. The disadvantage of the procedure is the subjective nature of what constitutes an elongated cell. D. B. Roszak and R. R. Colwell (1987a, 1987b) incorporated the DVC procedure and found that about 90% of those cells responsive by the DVC were metabolically active by microautoradiography.

Obviously, there are various methods to detect viability of bacteria in the environment. Which method is the most accurate has been the subject of debate for many years. Each method measures a different property of the cell, and this should be considered in the method's value. If possible, more than one method should be used to determine viability in environmental samples. However, what should be done in environmental experiments, and what actually can be done are often two different things.

Methodology for determination of viability is important, since many microorganisms that can no longer be cultured by viable plating techniques are now known to remain viable in nature. Nonculturable yet viable microorganisms may regenerate and still cause disease (Roszak and Colwell 1987a).

Another technique for determining bacterial numbers and the spectrum of actively metabolizing cells is autoradiography combined with direct microscopic observation (Brock and Brock 1968; Waid et al. 1973; Ramsay 1974; Fliermans and Schmidt 1975; Hoppe 1976; Faust and Correll 1977; Meyer-Reil 1978). In this method, bacteria incubated with a radiolabeled substrate, such as tritiated glucose, are subsequently collected on a bacteriological filter, which is placed on a glass slide and coated with a photographic emulsion. The actively metabolizing bacteria are radioactive and can be identified by the dark silver grains surrounding the cell. Applying this method to nearshore water samples indicates that 2.3%–56.2% of the total bacteria are metabolically active (Meyer-Reil 1978) (Table 7.10).

Electron microscopy instead of light microscopy can be used for direct counting of microorganisms (Gray 1967; Nikitin 1973; Todd et al. 1973; Bowden 1977; Larsson et al. 1978). Results of enumerations with scanning electron microscopy are comparable to those obtained by epifluorescent microscopy (Bowden 1977) (Table 7.11). Caution must be observed in the use of electron microscopic techniques because of the possibility of producing artifacts when specimens are metal coated to increase contrast for observation and when specimens are placed under high vacuum. As magnification increases, many viewing fields may have to be

Table 7.10

Microbiological variables measured in water samples taken from above sandy sediments at beaches of the Kiel Fjord and the Kiel Bight (4 to 13 July 1977)

Station*	Colony-forming units/mL x 10^3 (plate counts)	Total no. of bacteria/mL x 10^5 (direct counts)	Total biomass (mg/mL x 10^{-5})	Number of active bacteria/mL x 10^5 (autoradiography)	Active bacteria / mL x total no. of bacteria	Actual uptake rate of glucose (g/mL per h, x 10^{-3})
A	261.0	41.7	50.2	19.8	47.5	18.9
A´	188.0	57.8	65.4	32.5	56.2	23.5
B	81.4	26.5	32.1	11.0	41.5	11.9
B´	187.0	64.2	62.4	29.9	46.6	10.3
C	433.0	68.8	74.2	3.1	4.5	2.6
C´	98.2	67.2	70.9	23.8	35.4	6.6
D	3.4	52.2	58.0	1.2	2.3	1.2
D´	3.1	51.4	40.9	6.1	11.9	7.8
E	5.5	39.4	43.6	16.6	42.1	6.9
E´	2.9	65.8	60.1	19.2	29.2	4.8
F	7.4	56.1	60.8	28.3	50.5	3.8
F´	1.7	50.0	44.7	3.9	7.8	5.2

*A to F mark stations at the west side; A´ to F´ are corresponding stations at the east side; Stations A/A´ are located at the inner part, B/B´ at the center part, and C/C´ at the outer part of the Kiel Fjord; Stations D/D´, E/E´, and F/F´ are located at the Kiel Bight.
Source: Meyer-Reil 1978.

Table 7.11

Comparison of direct count enumerations of bacterial populations by epifluorescence and scanning electron microscopy for estuary water samples collected on various types of filters

Sample	Direct count		
		Epifluorescence	
	Scanning EM 0.2 mm Nuclepore	0.2 µm Nuclepore	0.2 µm Sartorius
Estuary 1	4.34 x 10^6	4.81 x 10^6	2.00 x 10^6
Estuary 2	3.30 x 10^6	2.92 x 10^6	0.78 x 10^6

Source: Bowden 1977.

scanned before a microorganism is observed; therefore, electron microscopic observation has been applied mainly to samples with naturally dense or artificially enriched microbial communities.

Particle Count Methods A particle counter, such as a Coulter counter, can also be used to estimate numbers of microorganisms directly (Kubitschek 1969). The Coulter counter electronically measures the number of particles within a fixed size range. Separate estimates of bacteria and protozoa can be obtained by setting appropriate size ranges. The problem with Coulter counter analysis is that small nonliving particles are counted together with microorganisms, which has limited the method's usefulness. However, recent advances in image analysis permit the specific recognition of microorganisms in complex samples, and increased usage of cell sorters in the enumeration of microorganisms is expected.

Viable Count Procedures

There are two basic approaches to viable count procedures: (1) the plate count technique; and (2) the most probable number (MPN) technique. All viable count procedures require separation of microorganisms into individual reproductive units. All viable count procedures are selective for certain microorganisms; the degree of selectivity varies with the particular viable count procedure. This selectivity is a disadvantage when trying to estimate the total viable microbial biomass within an ecosystem, but it permits the estimation of numbers of particular types of microorganisms.

Plate Count Methods: Selective and Differential Plating for Phenotypes The agar plate count method, which has been widely used for the enumeration of viable microorganisms, especially bacteria, has been severely criticized (Postgate 1969; Buck 1979). The problem lies in the misuse of the method and/or the misinterpretation of the results. All too often users fail to recognize that this technique can never be used to achieve "total counts."

Plate count procedures employ various media and incubation conditions. Agar is most often used as the solidifying agent because most bacteria lack the enzymes necessary for depolymerizing agar. Dilutions of samples can be spread on the top of the agar (surface spread method), or the sample suspension can be mixed with the agar just before the plates are poured (pour plate method). One must consider whether the microorganisms can survive the plating procedure. Some microbes are killed upon exposure to air in the surface spread plate method; others cannot tolerate the temperature needed to maintain melted agar in the pour plate method. Because the agar used for making media for bacteriological enumeration may contain organic contaminants, it is sometimes replaced with an alternative solidifying agent. When specific nutritional groups of microorganisms are to be enumerated, silica gel may be used for solidification. Because silica gel plates are more difficult to prepare than agar plates, they are used only when unavoidable.

The roll-tube method, used for enumeration of obligate anaerobes, is an extension of the pour plate method (Hungate 1969; Hungate and Macy 1973). The plates or tubes are incubated under specified conditions for a period of time to allow the bacteria to multiply and form macroscopic colonies, after which the colonies are counted. It is assumed that each colony originated from a single bacterial cell. For this assumption to be valid, the bacteria must be widely dispersed in or on the agar. Plates with too many colonies cannot be counted accurately because one colony may represent more than one original bacterium. Plates with too few colonies also must be discarded from the counting procedure for statistical reasons.

The main considerations for doing plate count procedures are the composition of the medium, the incubation conditions, and the period of incubation. Media for the enumeration of heterotrophs not capable of nitrogen fixation must contain usable sources of carbon, nitrogen, and phosphorus and required cations and anions, such as iron, magnesium, sodium, calcium, chloride, and sulfate. It is not usually clear why a particular medium yields the highest numbers of microorganisms from a particular ecosystem. Clearly, the media must meet nutritional requirements, including growth factors. General purpose media have nutrient concentrations much higher than those found in the natural ecosystem being studied. Concentrations of nutrients in many of these media used for total enumerations are high enough to be toxic to at least some microorganisms.

The real advantage of viable enumeration procedures is that conditions can be adjusted so that only members of a defined group are enumerated. Plate count media may be designed to be selective or differential. Selective plate count procedures are designed to favor the growth of the desired group of microorganisms. Growth of other groups is precluded by media composition and/or incubation conditions. Differential media do not preclude the growth of other microorganisms but permit detection of the desired group by some distinguishing characteristic.

Media can be designed for the selective enumeration of fungi. The viable plate count technique, though, is generally not the method of choice for enumerating fungi, because this technique favors enumeration of nonfilamentous fungi and spores (Menzies 1965). Plate counts, however, are suitable for enumeration of yeasts. To prevent overgrowth of fungal colonies by bacteria, which are likely to be more numerous than fungi in the sample, bacterial inhibitors are added to the media. The dye rose bengal and the antibiotics streptomycin and neomycin are often added as bacterial inhibitors. A simple technique for suppression of bacteria involves the lowering of the pH of the medium to 4.5–5.5. Most fungi are unaffected by this pH, while most bacteria are suppressed.

Media designed to inhibit the growth of one group of organisms while permitting growth of another group are known as selective media. For example, Sabouraud dextrose agar is used to enumerate fungi based on its low pH and carbohydrate source of carbon. Incorporation of penicillin or methylene blue, which inhibit Gram-positive organisms, into a medium is often used to select for Gram-negative bacteria. Antibiotic-resistant microorganisms can be enumerated on media with added antibiotics.

Differential media may be designed in a variety of ways. Reagents may be incorporated into the media that permit immediate visual differentiation of the desired bacteria, or reagents may be added after incubation to detect the desired bacteria. The key advan-

tage of differential media is that the procedure permits one to distinguish between the microorganisms being enumerated and others present in the sample.

Eosin methylene blue (EMB) agar and MacConkey's agar are media widely used to determine water quality. These media are both selective and differential. They select for the growth of Gram-negative bacteria by incorporating an inhibitor of Gram-positive bacteria. They differentiate bacteria capable of utilizing lactose by formation of characteristically colored colonies. On EMB agar, coliform bacteria, which are Gram-negative and utilize lactose, form a characteristic colony with a green metallic sheen. Estimates of coliform counts determined in this manner are often used as an indicator of water quality and for quality control in the food industry.

Colony Hybridization   Colony hybridization is an application of nucleic acid hybridization that is combined with conventional environmental microbiological sampling and viable plating procedures (Figure 7.11). Following initial growth, bacterial colonies or phage plaques are transferred from the surfaces of the cultivation media to hybridization filters. The colonies or phage containing plaques are then lysed by

Figure 7.11

Colony hybridization procedure.

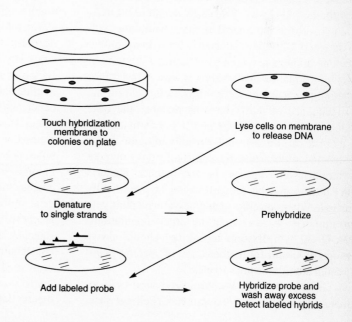

Touch hybridization membrane to colonies on plate

Lyse cells on membrane to release DNA

Denature to single strands

Prehybridize

Add labeled probe

Hybridize probe and wash away excess Detect labeled hybrids

alkaline or enzymatic treatment, after which hybridization is conducted. These methods depend on the ability of the target microorganisms to grow on the primary isolation medium and not to be totally overgrown by nontarget populations. Growth on the isolation medium increases the number of copies of the target gene to a level detectable by a gene probe.

The original protocol developed by M. Grunstein and D. S. Hogness (1975) was shown to be suitable for high-density plate screening for pure cultures (Hanahan and Meselson 1980). In cultures containing both nontarget *E. coli* and target *Pseudomonas putida*, one *P. putida* colony containing a toluene (TOL) catabolic plasmid in a background of approximately 1 million *E. coli* colonies was detected using a whole TOL plasmid probe (Sayler et al. 1985). In general for this type of environmental assay, the organism of interest must be relatively abundant in the population so that at least one positive colony on an agar plate of 100 to 1,000 colonies can be found. Additional sensitivity can be achieved by plating the isolated bacteria onto selective agar before colony hybridization.

In the colony hybridization procedure, bacterial cells growing on a viable culture medium are transferred to a suitable solid support, such as a nitrocellulose filter, and lysed, releasing denatured DNA, which adsorbs to the filter. The filter membrane with the attached DNA is incubated with a gene probe. The gene probe is prepared by isolating a segment of genetic information and labeling it with either ^{32}P or something such as biotin. Hybridization occurs if the base sequence of the lysed cells matches the base sequence of the gene probe and if the formation of the hybrid can be detected by autoradiography for ^{32}P-labeled probes and by enzymatic development for biotin-labeled probes. In this manner, bacteria with specific genetic properties can be specifically detected. When specific genotypes are in relatively low numbers as compared to other populations, DNA-DNA hybridization techniques should prove particularly useful for the detection of specific populations in environmental samples.

Gene probes can be hybridized with primary isolates from environmental samples (Echeverria et al. 1982; Fitts et al. 1983; Hill et al. 1983; Miliotis et al. 1989) or with secondary cultivation of already described strains (Datta et al. 1988; Falkenstein et al. 1988; Nortermans et al. 1989). The rationale for direct colony hybridization on primary cultivation include (1) avoiding a cultivation bias encountered by selective media, which may underestimate total abundance of a given genotype; (2) assuring that a given genotype is present in the population sampled, even if the genes are poorly expressed or poorly selected; (3) providing optimum permissive growth conditions for stressed organisms that may be unculturable on selective media; and (4) reducing analysis time for cultivation, presumptive quantification, and confirmation of a genotype/phenotype. Colony hybridization on secondary cultivation of pure cultures is usually used to confirm a specific genotype or to test a unique DNA sequence for gene probe development.

The major uses of colony hybridization in environmental studies have been for the detection, enumeration, and isolation of bacteria with specific genotypes and/or phenotypes, and for development of gene probes. Various microorganisms with specific metabolic activities of ecological importance have been detected using colony hybridization. These have included *Pseudomonas fluorescens* (Festl et al. 1986), 4-chlorobiphenyl degraders (Pettigrew and Sayler 1986), toluene degraders (Jain et al. 1987), naphthalene degraders (Blackburn et al. 1987), and mercury-resistant bacteria (Barkay 1985; Barkay and Olson 1986; Barkay et al. 1985). R. K. Jain et al. (1987) used colony hybridization and gene probes to study the maintenance of catabolic and antibiotic resistance plasmids in groundwater aquifer microcosms, and, using this method, were able to demonstrate that introduced catabolic plasmids or organisms can be maintained in groundwater aquifers without selective pressure. In another study, colony hybridization with a naphthalene gene probe analysis for the catabolic genotype was nearly two orders of magnitude more sensitive than the standard plate assay for naphthalene degradation (Blackburn et al. 1987). Colony hybridization also has been used with environmental samples to enumerate mercury-resistant bacteria in contaminated environments (Barkay et al. 1985; Barkay and Olson 1986).

Most Probable Number The most probable number (MPN) method, an alternative to plate count methods for determination of viable organisms, uses statistical analyses and successive dilution of the sample to reach a point of extinction (Cochran 1950; Alexander 1965; Melchiorri-Santolini 1972; Colwell 1979). Replicate dilutions, usually three to ten replicates per dilution, are scored as positive or negative, and the pattern of positive and negative scores is used in connection with appropriate statistical tables to obtain the most probable number of viable microorganisms. The MPN procedure gives a statistically based estimate of the number of microorganisms in the sample, and when relatively few replicate tubes are used, the confidence intervals are quite large. The MPN method for enumeration of bacteria has the advantage of permitting the use of liquid culture, avoiding the need to add a solidifying agent such as agar, with its possible contaminants, but it is more laborious and less precise than the plate count.

It is essential that criteria be established for differentiating positive from negative replicates. In many cases, the MPN test procedures are designed so that increases in turbidity (growth) can be seen and scored as positive. In other procedures, more elaborate tests, such as protein or chlorophyll determination, are required to score positive tubes. As with plate count procedures, MPN determinations can use selective and differential media.

A modified MPN method designed by B. N. Singh (1946) is often used for estimating numbers of protozoa. Multiple glass or polypropylene rings are embedded in sodium chloride agar in petri dishes, forming wells. The agar surface is inoculated with a heavy suspension of a bacterium, such as *Enterobacter aerogenes*. Dilutions of an environmental sample are inoculated into each of the wells, and protozoa present in the sample dilutions feed on the provided bacterial paste, producing a readily recognizable clearing. Clearing of the bacteria within a ring is scored positive, and the lack of clearing is scored negative.

An MPN method can be used for the enumeration of enteric viruses. In this method, serial dilutions of the samples containing viruses are added to tubes of suitable host cells in tissue cultures (Farrah et al. 1977). Following incubation, the tubes are examined for cytopathic effect (CPE), that is, the death of the infected cells. Quantitation can be achieved using MPN tables scoring tubes showing CPE as positive. The numbers of viruses may also be quantitated as the $TCID_{50}$ (tissue culture infectious dose—50%), the lowest dilution of the virus suspension that caused CPE in 50% of the tubes.

As in the plate count procedures, media and incubation conditions can be adjusted in the MPN procedure to select for particular groups of microorganisms or to differentiate microorganisms with desired characteristics. Obviously, the combinations of incubation conditions and media that can be used for enumerating specific groups of bacteria are infinite. Each procedure must be carefully selected and tested to permit the correct interpretation of results.

DETERMINATION OF MICROBIAL BIOMASS

Biomass is an important ecological parameter because it measures the quantity of energy being stored in a particular segment of the biological community. Measurement of biomass is used to determine the standing crop of a population and the transfer of energy between trophic levels within an ecosystem. Biomass literally means "mass of living material" and can be expressed in units of weight (grams) that can be converted to units of energy (calories). Unfortunately, the direct measurements of microbial biomass, such as by filtration and dry weight or by centrifugation and packed cell volume measurements as practiced on pure cultures, are rarely applicable to environmental samples. These techniques tend to measure mineral and detritus particles and nonmicrobial biomass along with microorganisms and fail to discriminate between trophic levels, that is, between producers and consumers. Consequently, the determination of microbial biomass is often imprecise.

Biochemical Assays

The most practical approach to the determination of microbial biomass is to assay for specific biochemicals that indicate the presence of microorganisms. Ideally, all the microbial biomass to be determined should have the same quantity of the biochemical being assayed, so that there is a direct correlation between the amount of the biochemical measured and the biomass of microorganisms. Also, the biochemical being assayed should be present only in the biomass to be determined. These two conditions are rarely, if ever, met, so the results of quantifying a particular biochemical must be extrapolated with caution in order to estimate the biomass of microorganisms that are present.

ATP and Total Adenylate   Present in all microorganisms, ATP can be measured with great sensitivity. Though dependent on physiological state, ATP concentrations are fairly uniform relative to cell carbon for many bacteria, algae, and protozoa (Figure 7.12). Because ATP is lost rapidly following the death of cells, a measurement of ATP concentrations can be used to estimate living biomass (Holm-Hansen 1969; Paerl and Williams 1976; Deming et al. 1979; Stevenson et al. 1979). The luciferin-luciferase assay can be used to detect ATP; in this assay, reduced luciferin reacts with oxygen to form oxidized luciferin in the presence of luciferase enzyme, magnesium ions, and ATP. Light is emitted in this reaction in an amount directly proportional to the ATP concentration. High-pressure liquid chromatography can also be used to measure ATP quantitatively. The method used to extract the ATP has a marked effect on the sensitivity and reliability of the assay (Deming et al. 1979; Stevenson et al. 1979). The ATP must be extracted rapidly and in a manner that maintains the integrity of the ATP, because ATP is easily converted to AMP. A variety of methods are used, including extraction with various organic solvents or hot buffers; the efficiency of a particular procedure depends upon the nature of the sample and the microbial community.

Measurements of ATP can be used to estimate microbial biomass. A factor of 250–286 is often used

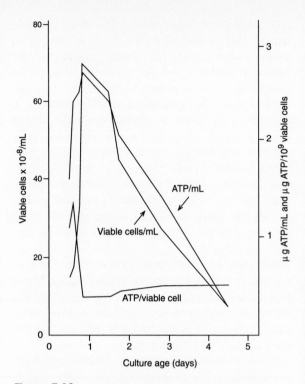

Figure 7.12

Correlation of ATP and viable count. In cultures older than 1 day, the ATP content per viable cell remains rather constant even during large changes in population size. (Source: Holm-Hansen 1973. Reprinted by permission, copyright 1973a, University of South Carolina Press.)

for conversion of ATP to cellular carbon for aquatic samples (Hamilton and Holm-Hansen 1965; Holm-Hansen and Booth 1966; Holm-Hansen 1973a, 1973b; Paerl and Williams 1976); for soil, a factor of 120 is used to convert ATP to biomass carbon (Ausmus 1973; Jenkinson and Oades 1979; Oades and Jenkinson 1979). There are some difficulties, however, in accurately estimating microbial biomass based upon ATP measurements. Some microorganisms alter their ATP concentrations radically when nutritional or physiological conditions change. Also, in some ecosystems, such as soil, sediment, and nearshore aquatic areas, ATP may be absorbed on particles (Bancroft et al. 1976; Jenkinson and Oades 1979). The sorption of

ATP interferes with its extraction and quantification. In addition, the presence of plant or animal cells, which also contain ATP, limits applicability of this method in some ecosystems.

Perhaps a better measure of microbial biomass, but one that is not universally accepted, is the total adenylate pool (A_T). The total adenylate pool (Equation 1) is insensitive to metabolic state:

$$(1) \quad A_T = ATP + ADP + AMP.$$

The total adenylate pool is determined in order to calculate energy charge, as described later in this chapter, and can conveniently be used simultaneously to estimate numbers and biomass of microorganisms. The advantage of using A_T is that it does not vary greatly during changes in metabolic activities of the organisms, whereas ATP is highly dependent on the physiological state of organism. When determinations of the total adenylate pool are used in conjunction with ATP measurements, the energy charge, which is a measure of the physiological state of the organisms within the sample, can be calculated (Wiebe and Bancroft 1975; Karl and Holm-Hansen 1978). The energy charge (EC) (Equation 2), which is independent of population size, is calculated as

$$(2) \quad EC = (ATP + 1/2 \ ADP)/(ATP + ADP + AMP).$$

Actively growing cells have an EC ratio of 0.8–0.95, cells in stationary growth phase have an EC ratio of about 0.6, and senescent or resting cells have an EC ratio of less than 0.5.

Cell Wall Components   Most bacteria contain muramic acid in their cell walls, and the specific relationship between murein and bacteria makes quantitation of this cell wall component useful for estimating bacterial biomass (Millar and Casida 1970; Moriarty 1975, 1977, 1978; King and White 1977; White et al. 1979). The assay for muramic acid is based on the release of lactate from muramic acid and either enzymatic analysis (Moriarty 1977) or chemical analysis (King and White 1977) to determine the concentration of lactate.

The conversion of muramic acid measurements to biomass assumes that all Gram-positive bacteria have a ratio of 44 µg MA/mg C and that all Gram-negative bacteria have 12 µg MA/mg C. In reality, there is a gradient of concentrations of muramic acid in Gram-positive and Gram-negative bacteria. To accurately use this method, it is necessary to estimate the proportions of Gram-negative and Gram-positive bacteria in the sample; erroneous estimates of these proportions will yield inaccurate estimates of biomass.

Gram-negative bacteria contain lipopolysaccharide as part of their cell wall complex. Lipopolysaccharide can be quantitated using the *Limulus* amoebocyte lysate (LPS) method (Watson et al. 1977; Watson and Hobbie 1979). This method uses an aqueous extract from the blood cells of horseshoe crabs (*Limulus polyphemus*) that reacts specifically with lipopolysaccharide to form a turbid solution; the degree of turbidity is directly proportional to the lipopolysaccharide concentration and can be quantitatively assayed. Because lipopolysaccharide occurs only in the cell walls of Gram-negative bacteria and occurs in a relatively constant proportion, it is a useful method for estimating numbers of bacteria in ecosystems dominated by Gram-negative forms. The LPS method is very sensitive, capable of detecting concentrations as low as ten cells/mL, and correlates well with counts obtained by other measures of biomass (Table 7.12, Figure 7.13).

To estimate fungal biomass, chitin can be measured (Swift 1973; Sharma et al. 1977; Willoughby 1978), because chitin occurs in the cell walls of many fungi but not in plants or other soil microorganisms.

Table 7.12

Relative amounts of cell carbon, ATP, and muramic acid in 10^9 cells of Gram-positive and Gram-negative bacteria

Bacteria	Cell C (mg)	ATP (µg)	Muramic acid (µg)
Gram-positive *Bacillus*	9×10^{-2}	1.1×10^1	3×10^0
Gram-negative *Pseudomonas*	1×10^{-1}	0.5×10^0	0.5×10^0

Source: Moriarty 1977.

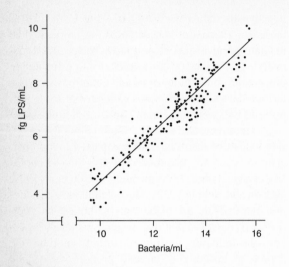

Figure 7.13
Correlation of LPS and direct counts of bacteria (note fg or femtogram = 10^{-15} g). (Source: Watson et al. 1977. Reprinted by permission of American Society for Microbiology.)

Measurement of chitin can therefore be used to estimate fungal biomass in the presence of other microbial populations and can be used to determine fungal biomass associated with plants, including plant roots in soil. The presence of microarthropods, though, does interfere with such determinations.

Chlorophyll and other Photosynthestic Pigments
Assuming the absence of plants, it is possible to estimate numbers of photosynthetic algae and bacteria by measuring chlorophyll or other photosynthetic pigments (Edmondson 1974; Cohen et al. 1977; Holm-Hansen and Riemann 1978). Chlorophyll a, the dominant photosynthetic pigment in algae and cyanobacteria, is a useful measure of the biomass of these photosynthetic microorganisms even though there may not be a constant relationship between biomass and chlorophyll content (Banse 1977). Estimates of the biomass of photosynthetic microorganisms based upon chlorophyll determinations have been found to correlate well with such estimates based upon ATP determinations (Paerl et al. 1976).

Chlorophyll a can be extracted with solvents, such as acetone or methanol, and quantified by measuring absorbance with a spectrophotometer using a wavelength of 665 nm. This wavelength can be adjusted to the corresponding absorption maxima to measure bacterial chlorophylls; for example, 850 nm is used for determining the biomass of purple photosynthetic bacteria. It is thus possible to estimate the biomass of different photosynthetic microbial populations in the same sample by estimating various chlorophylls that absorb at different wavelengths (Stanier and Smith 1960; Caldwell and Tiedje 1975).

The chlorophyll concentration in the extract can also be determined by spectrofluorometry (Lorenzen 1966; Loftus and Carpenter 1971; Sharabi and Pramer 1973; Caldwell 1977). Chlorophyll a, for example, has an excitation wavelength of approximately 430 nm and a wavelength of maximum fluorescence of approximately 685 nm. The spectrofluorometric determination is more selective than the spectrophotometric method. The double selectivity of excitation and fluorescence maxima tend to reduce interference of nonspecific absorption that may affect spectrophotometric determinations (Sharabi and Pramer 1973).

DNA Concentrations of DNA are maintained in relatively constant proportions within microorganisms, and determination of DNA can be used as a measure of microbial biomass (Holm-Hansen et al. 1968; Hobbie et al. 1972; Paul and Meyers 1982; McCoy and Olson 1985). For environmental samples, where sensitivity is critical for accurate DNA determinations, reaction with fluorescent dyes, such as ethidium bromide or Hoechst 33258, and spectrofluorometry are generally employed. Careful purification of the DNA is necessary in these assays to prevent interference, and control for the presence of eukaryotic DNA is also necessary. Estimates of biomass based upon DNA determinations and those by direct counts show significant correlations (McCoy and Olson 1985).

Protein Protein is easily quantified (Lowry et al. 1951), and bacterial heme proteins can be specifically detected by chemiluminescence (Oleniacz et al. 1968), but the use of protein measurements for estimating numbers of microorganisms is limited to situations where background protein levels are negligible. Also, different microorganisms contain different amounts of protein; thus, protein determinations are best used for estimation of microbial biomass when only a single population is present. Hence, total protein measurements have only limited applicability to environmental samples.

Physiological Approaches to Biomass Determination

Physiological approaches for estimating microbial biomass have been developed based upon respiratory activities. One such approach, which has been used to measure microbial biomass in soil, involves fumigating soil with chloroform and then measuring the CO_2 released from the mineralization of the microorganisms killed by the fumigation (Jenkinson 1976; Jenkinson and Powlson 1976; Anderson and Domsch 1978a). In this procedure, the soil microbes are killed by fumigation with $CHCl_3$ and subjected to mineralization by reinoculation of the soil with a small amount of the original soil; nonchloroform-treated soils incubated under the same conditions act as controls, and the amount of microbial carbon is calculated from the difference between the CO_2–C evolved from fumigated and nonfumigated samples. Conversion factors of 0.4–0.5 are recommended based on determinations made with various radiolabeled bacterial and fungal cultures. Average mineralization is 44% for fungi and 33% for bacteria; assuming a bacterial-to-fungal biomass ratio of 1:3 for soil, the combined average mineralization should be 41%. A high correlation has been found between soil ATP and biomass determined by this fumigation method (Oades and Jenkinson 1979) (Table 7.13).

Another physiological approach for estimating microbial biomass involves measurement of respira-

Table 7.13

Comparison of biomass estimated by soil fumigations and ATP in various soils

Soil sample	CO_2–C evolved by unfumigated soil in 10 days ($\mu g\ g^{-1}$)	CO_2–C evolved by fumigated soil in 10 days ($\mu g\ g^{-1}$)	Biomass C ($\mu g\ g^{-1}$)	ATP in soil ($\mu g\ g^{-1}$)
1	70	114	88	0.75
2	205	315	220	2.07
3	131	240	218	3.14
4	20	43	48	0.64
5	236	613	754	9.03
6	108	158	99	0.90
7	117	341	448	2.97
8	264	565	603	4.56
9	76	143	135	1.37
10	259	642	766	7.00
11	257	646	778	6.55
12	85	189	208	1.32
13	229	525	591	4.20

Source: Oades and Jenkinson 1979.

tion rates after substrate addition (Anderson and Domsch 1973, 1975, 1978b). The peak respiration rate measured during a short period is assumed to be proportional to the numbers of viable microorganisms in the samples. Microbial inhibitors can be used in the procedure to obtain separate estimates of bacterial and fungal biomass. Good correlation has been reported between this peak respiration rate method and the chloroform fumigation method.

MEASUREMENT OF MICROBIAL METABOLISM

Recent advances in analytical methodology provide the sensitivity needed to measure natural rates of microbial metabolic activities, but it remains difficult to make such rate measurements that reflect actual activities within natural ecosystems (Christian et al. 1982; Van Es and Meyer-Reil 1982; Staley and Konopka 1985). The difficulty lies in the fact that most methods require manipulation of the system in ways that can alter rates of microbial activity. Even minor manipulations can greatly alter the microenvironmental parameters that determine rates of microbial activity; as examples, simply enclosing a water sample produces a wall effect that can change the rates of microbial activity, and removal of samples for activity determinations can change various parameters, such as oxygen concentrations, thereby changing the rates of microbial activities being measured.

Heterotrophic Potential

One approach to measuring natural rates of heterotrophic microbial activity is to measure uptake rates for tracer levels of radiolabeled substrates to determine the heterotrophic potential for the utilization of that substrate (Wright and Hobbie 1965, 1966; Hobbie and Crawford 1969; Wright 1973, 1978). This approach assumes that microorganisms take up solutes from solution according to first order or saturation kinetics and that the rates of uptake increase with

increasing concentrations of the substrate to a maximal uptake rate (V_{max}). V_{max} can be calculated by plotting the rates of uptake of ^{14}C-radiolabeled substrates against their concentrations. The concentrations of substrates used must be low and must approximate their natural concentrations in the environment. In addition to measuring the maximal uptake potential, this method can be used to calculate turnover time for a given substrate and a transport constant that reflects the affinity for the given substrate. Natural turnover time is defined as the number of hours required for the existing heterotrophic population to take up and/or respire a quantity of substrate equal in concentration to the existing *in situ* concentration.

The procedure for determining heterotrophic potential consists of adding various concentrations of radiolabeled substrate to samples and incubating the samples under conditions that simulate the real environment. After incubation, the cells are collected on a filter and the incorporated radioactivity is counted by liquid scintillation. Originally, uptake was only measured by incorporation into cells; however, this method neglected the amount of substrate that had been taken up and metabolized. Therefore, the method was modified so that $^{14}CO_2$ produced during respiration is trapped, and this amount is added to the counts of ^{14}C incorporated into the cells (Hobbie and Crawford 1969).

If one assumes that uptake by a heterogeneous heterotrophic population acts like a single enzyme because uptake by different bacteria is functionally equivalent, the results should be comparable to Michaelis-Menten enzyme kinetic analyses. Using the linearized form of the Michaelis-Menten equation (Equation 3):

(3) $T/F = (K_t + S_n)/V_{max} + A/V_{max}$,

where

T/F in hours = time divided by the fraction of added substrate assimilated into the cells or converted into CO_2,

A = the concentration of substrate added,

$K_t + S_n$ = the transport constant plus the natural concentration of that substrate,

V_{max} = the maximal rate of substrate uptake.

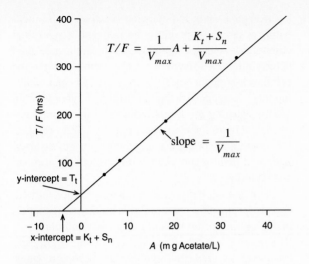

$$T/F = \frac{1}{V_{max}}A + \frac{K_t + S_n}{V_{max}}$$

slope $= \dfrac{1}{V_{max}}$

y-intercept $= T_t$

x-intercept $= K_t + S_n$

A (m g Acetate/L)

T / F (hrs)

Figure 7.14

Measurement of heterotrophic potential showing modified Lineweaver-Burke plot of uptake kinetics. T_t = turnover time, V_{max} = maximal uptake velocity; $K_t + S_n$ = transport constant + natural substrate concentration; T/F = velocity; A = substrate concentration. (Source: Wright and Hobbie 1966. Reprinted by permission, copyright 1966, Ecological Society of America.)

A plot of *T/F* versus *A* is linear (Figure 7.14). Using *T/F* as the Y axis and *A* as the X axis, the slope of the resulting line is equal to $1/V_{max}$; the Y intercept is equal to T_t (the turnover time), and the X intercept is equal to $K_t + S_n$. If the *in situ* concentration of the substrate is measured independently, then the transport constant K_t for that substrate can be determined. Conversely, if K_t can be determined independently, the equation may be used to calculate S_n, which is frequently too low to be detectable by direct chemical analysis.

The value of measuring V_{max} is that it gives a real estimate of *in situ* heterotrophic activities because the V_{max} value is a function of the existing population. It is sensitive to environmental changes, even when these changes do not affect population size. V_{max} is of great value when comparing seasonal and spatial differences of the metabolic activities of indigenous microorganisms.

The estimation of rates of microbial activity with this method is limited by the substrate(s) used.

Using replicate samples, it is possible to determine the relative V_{max} values and turnover times for a variety of substrates. Studies of heterotrophic potentials have utilized acetate, glucose and other carbohydrates, glutamate and other amino acids, or mixed ^{14}C-labeled photosynthate produced by algae incubated with $^{14}CO_2$ in light.

Many problems with the performance and interpretation of heterotrophic potential measurements should be mentioned. The substrate concentrations added must be within the right range to give a straight line slope that allows for calculation of V_{max}, T_t, and $K_t + S_n$. This generally requires adding many substrate concentrations, some of which are not utilized in the calculation. Adding too high a concentration of substrate saturates the system and should yield a rate equal to V_{max}. Excessive concentrations of added substrate, however, may be inhibitory. A more refined mathematical treatment is necessary to account for competitive and noncompetitive inhibitory effects. At present, the measurement of heterotrophic potential does not account for competitive inhibition and assumes a unique transport system for each substrate being measured, but this assumption is not valid, as common transport enzymes carry many substrates.

The measurement of heterotrophic potential also assumes that all members of the microbial population respond in the same way to variations in solute concentration. Again, this need not be true (Vaccaro and Jannasch 1967). In some cases, there is evidence for differential uptake rates by various bacterial populations, and by bacterial versus algal populations. Lower concentrations of added substrate generally give satisfactory slopes for calculation of V_{max} for bacterial populations and higher concentrations of substrates give satisfactory slopes for calculation of V_{max} for algal populations. Although heterotrophic potential measurements correct for $^{14}CO_2$ production, no correction is made for production of other volatile compounds, and this can cause a serious error when methane or volatile fatty acids are major metabolic products.

In addition to measurement of V_{max} and T_t, the percent respiration (^{14}C assimilated and $^{14}CO_2$ produced/$^{14}CO_2$) can be calculated. The percent respiration is indicative of the energy required for maintenance; the greater the percent respiration, the greater the metabolic energy used to maintain the microbial population; the lower the percent respiration, the greater the proportion of metabolic energy used for assimilation and growth.

The specificity of the heterotrophic potential measurement is both a problem and an advantage for the substrate being measured. The method gives an estimate for specific heterotrophic activity for a particular substrate; it does not measure overall heterotrophic activity. Measurement of heterotrophic potentials permits studies on the effects of factors such as pollutants on microbial activities.

Growth Rate Based upon Nucleotide Incorporation

Because DNA is synthesized in growing cells at a rate proportional to biomass, the rate of DNA synthesis reflects the growth rates of microorganisms (Van Es and Meyer-Reil 1982). Microbial growth rates have been determined in environmental samples by incubating samples with tritiated thymidine, using autoradiography of samples to determine rates of nucleotide incorporation (Brock and Brock 1968; Brock 1969). This technique is only applicable to organisms that can be recognized microscopically and for which the relation between growth rate and accumulation of radiolabel can be determined independently. Various other analyses have been used to determine the incorporation of radiolabeled nucleotides into RNA and DNA (Karl 1979, 1980, 1981; Fuhrman and Azam 1980; Karl et al. 1981a, 1981b; Staley and Konopka 1985). If the concentrations of RNA and DNA are known, the growth rates of the indigenous microbial community can be determined by this method.

Productivity and Decomposition

Photosynthesis In addition to heterotrophic potential measurements, which assess secondary productivity, rates of primary production (synthesis) can be made using radiotracer and other methods. Both heterotrophic and autotrophic assimilation of CO_2 can

be measured using radiolabeled bicarbonate by incubating a sample containing the indigenous microbial community with the radiolabeled substrate and then determining the amount of $^{14}CO_2$ assimilated into the cellular organic matter by filtering the cells and counting the ^{14}C trapped on the filters (Hubel 1966; Gorden 1972). Washing the filters eliminates any unincorporated ^{14}C-radiolabeled bicarbonate. The residual ^{14}C-containing organic compounds can be oxidized with acid dichromate and the released $^{14}CO_2$ trapped and quantitated.

In actual field measurements, photosynthetic and heterotrophic-chemolithotrophic incorporation of $^{14}CO_2$ are differentiated using light and dark bottle sets. In a typical procedure, both clear (light) and opaque or covered (dark) bottles are filled with water samples, radiolabeled bicarbonate is added, and the bottles are incubated *in situ* for several hours or an entire photoperiod. Incorporation results from the dark bottles are subtracted from incorporation results in the light bottles to obtain the net photosynthetic incorporation. Unlabeled CO_2 (bicarbonate) in the samples is measured to calculate the actual specific $^{14}CO_2$ activity in the water samples, and with sufficient measurements, rates of productivity for daily or annual periods can be determined.

The time frame of $^{14}CO_2$ incorporation experiments is relevant to the interpretation of the results. Short experimental periods of one or two hours give values close to the gross primary productivity, because there is little chance for the incorporated ^{14}C to be mineralized again by respiration. The values obtained for long exposure times will be closer to net primary productivity (photosynthesis minus respiration), because some of the incorporated ^{14}C will be mineralized under these conditions.

Respiration Radiolabeled carbon dioxide ($^{14}CO_2$) release from labeled substrates can also be used to determine decomposition rates for specific substrates. The complete degradation of a compound to its mineral components in which the organic carbon of the compound is converted to CO_2 by respiration is called mineralization. Rates of decomposition of both synthetic and natural organic compounds can

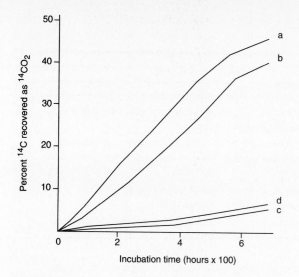

Figure 7.15

Decomposition of ^{14}C-labeled hemlock lignocellulose by soil microorganisms. The use of lignocellulose labeled specifically in different portions of the molecule permits identification of the origin of the CO_2. (a) ^{14}C cellulose-labeled lignocellulose in Soil #1. (b) ^{14}C cellulose-labeled lignocellulose in Soil #2. (c) ^{14}C lignin-labeled ligno-cellulose in Soil #1. (d) ^{14}C lignin-labeled lignocellulose in Soil #2. (Source: Crawford et al. 1977. Reprinted by permission of American Society for Microbiology.)

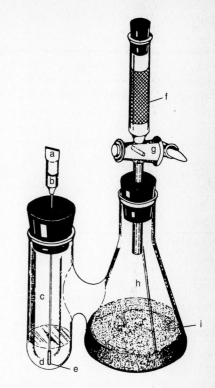

Figure 7.16

The biometer flask, a compact, commercially available enclosed unit used for measuring CO_2 production from soil. The biometer flask has the advantage that one can withdraw and replace the CO_2-absorbing alkali without exposing the system to contamination by atmospheric CO_2. (a) Rubber closure. (b) Syringe needle. (c) Sidearm. (d) Alkali. (e) Needle guard. (f) CO_2 absorbent (ascarite). (g) Stopcock. (h) Sample compartment. (i) soil. (Source: Bartha and Pramer 1965. Reprinted by permission, copyright Williams and Wilkins, Baltimore.)

be determined with this method (Figure 7.15). Measurements of $^{14}CO_2$ release from synthetic organic chemicals, such as pesticides, are essential for evaluating the environmental safety of such compounds, because unless mineralized by microorganisms, they will accumulate in the environment. In performing such analyses, it is essential to consider the position of the ^{14}C label, because one portion of a molecule can be degraded while another portion remains intact (Bartha and Pramer 1970).

Various other methods that do not employ radiolabeled substrates can also be used to measure rates of microbial respiration. These methods normally measure either rates of oxygen consumption or rates of carbon dioxide production. As long as aerobic conditions are maintained, CO_2 evolution

can be used as a measure of microbial respiratory activity with reasonable accuracy (Pramer and Bartha 1972). For long-term studies, rates of carbon dioxide production are often determined using specially designed enclosed flasks, such as the biometer flask (Bartha and Pramer 1965) (Figure 7.16) or using flow-through incubation systems, such as "gas trains," that pass a stream of CO_2-free air through the incubation flask and trap CO_2 from the effluent

Table 7.14

Some specific enzyme assays

Enzyme	Substrate	Description of assay
Dehydrogenase	Triphenyltetrazolium	Dehydrogenases convert triphenyltetrazolium chloride to triphenylformazan; the triphenylformazan is extracted with methanol and quantitated spectrophotometrically.
Phosphatase	p-Nitrophenol phosphate	Phosphatases convert the p-nitrophenol phosphate to p-nitrophenol, which is extracted in aqueous solution and quantitated spectrophotometrically.
Protease	Gelatin	Gelatin hydrolysis, as an example of proteolytic activity, can be measured by the determination of residual protein.
Amylase	Starch	The amount of residual starch is quantitated spectrophotometrically by the intensity of the blue color resulting from its reaction with iodine.
Chitinase	Chitin	Production of reducing sugars is measured using anthrone reagent.
Cellulase	Cellulose Carboxymethyl cellulose	Production of reducing sugars is measured using anthrone reagent. Cellulases alter the viscosity of carboxymethyl cellulose that can be measured.
Nitrogenase	Acetylene	Nitrogenase, besides reducing dinitrogen gas ($N{\equiv}N$) to ammonia (NH_3), is also capable of reducing acetylene ($HC{\equiv}CH$) to ethylene (C_2H_2); the rate of formation of ethylene can be monitored using a gas chromatograph with flame ionization detector, and the rate of nitrogen fixation can be calculated using an appropriate conversion factor.
Nitrate reductase	Nitrate	Dissimilatory nitrate reductase can be assayed by the disappearance of nitrate or by measuring the evolution of denitrification products, such as nitrogen gas and nitrous oxide, from samples using a gas chromatograph; denitrification can be blocked at the nitrous oxide level by the addition of acetylene, permitting a simpler assay procedure.

gas stream (Atlas and Bartha 1972). The CO_2 that is trapped can be quantitated by titration of the trapping base solution with acid of known concentration to reach a point of neutrality.

Rates of oxygen consumption can also be used as a measure of cellular respiratory activity. Various oxygen electrodes are available, including automated electrolytic respirometers. Newly developed microprobes can be used for *in situ* measurements (Revsbech 1983; Revsbech et al. 1983).

Specific Enzyme Assays

A variety of enzyme assays also can be used for measuring the metabolic activities of indigenous microorganisms. Some, such as measurement of dehydrogenase, esterase and phosphatase activities, assay general activities of a relatively large portion of the microbial community; others, such as measurement of cellulase, chitinase, and nitrogenase activities, assay the metabolic functions of small but important segments of the microbial community (Table 7.14). Enzymes involved in biogeochemical cycling are of interest to microbial ecologists; enzymes responsible for carbon and nitrogen cycling are of particular importance in the maintenance of communities and ecosystems. Various assays are used to measure specific enzymes, such as the acetylene reduction method for determining nitrogenase activity, have been developed and are widely employed in microbial ecological studies. When assaying enzymatic activities, it is important that the microbial community not be altered during the assay procedure if the measurement of enzymatic activity is to reflect *in situ* activities accurately. Caution must be given to maintenance of *in situ* conditions, particularly with reference to temperature, moisture content, and E_h, and incubation periods must be short enough to preclude changes in numbers of microorganisms, which could alter the levels of enzymes present in the sample.

SUMMARY

Numbers, biomass, and metabolic activity are the fundamental biotic parameters of microbial ecosystems. Much remains to be done to improve our accuracy and sensitivity in measuring these parameters. Further advances in microanalytical techniques are likely to improve these measurement procedures and reduce interferences. It is equally important to be aware of the true meaning and limitations of each measurement procedure. Knowing that a "total viable count" typically enumerates only about 1% of a microbial community places such data in the proper perspective. Although under some conditions numbers, biomass, and metabolic activity show proportional correlations, under many realistic circumstances they do not. This dictates caution and critical thinking when one parameter is calculated or projected from another directly measured parameter.

REFERENCES & SUGGESTED READINGS

Alexander, M. 1965. Most-probable-number method for microbial populations. In C. A. Black (ed.). *Methods of Soil Analysis.* Part 2, Chemical and Microbiological Properties. American Society of Agronomy, Madison, Wis., pp. 1467–1472.

American Public Health Association. 1975. *Standard Methods for the Examination of Water and Wastewater.* 14th ed. American Public Health Association, New York.

Andersen, A. A. 1958. A new sampler for the collection, sizing, and enumeration of the viable airborne bacteria. *Journal of Bacteriology* 76:471–484.

Anderson, J. P. E., and K. H. Domsch. 1973. Quantification of bacterial and fungal contributions to soil respiration. *Archiv für Mikrobiologie* (Berlin) 93:113–127.

Anderson, J. P. E., and K. H. Domsch. 1975. Measurement of bacterial and fungal contributions to respiration of selected agricultural and forest soils. *Canadian Journal of Microbiology* 21:314–322.

Anderson, J. P. E., and K. H. Domsch. 1978a. Mineralization of bacteria and fungi in chloroform-fumigated soils.

Soil Biology and Biochemistry 10:207–213.

Anderson, J. P. E., and K. H. Domsch. 1978b. A physiological method for the quantitative measurement of microbial biomass in soil. *Soil Biology and Biochemistry* 10:215–221.

Aristovskaya, T. V. 1973. The use of capillary techniques in ecological studies of microorganisms. *Bulletin from the Ecological Research Committee* (Stockholm) 17:47–52.

Aristovskaya, T. V., and O. M. Parinkina. 1961. New methods of studying soil microorganism associations. *Soviet Soil Science* 1:12–20.

Atlas, R. M. 1982. Enumeration and estimation of microbial biomass. In R. G. Burns and J. H. Slater (eds.). *Experimental Microbial Ecology.* Blackwell Scientific Publications, Oxford, England, pp. 84–104.

Atlas, R. M. 1992. Detection and enumeration of microorganisms based upon phenotype. In M. A. Levin, R. J. Seidler, and M. Rogul (eds.). *Microbial Ecology: Principles, Methods, and Applications.* McGraw-Hill, New York pp. 29–43.

Atlas, R. M., and R. Bartha. 1972. Degradation and mineralization of petroleum by two bacteria isolated from coastal water. *Biotechnology and Bioengineering* 14:297–308.

Atlas, R. M., and J. S. Hubbard. 1974. Applicability of radiotracer methods of measuring $^{14}CO_2$ assimilation for determining microbial activity in soil including a new *in situ* method. *Microbial Ecology* 1:145–163.

Atlas, R. M., D. Pramer, and R. Bartha. 1978. Assessment of pesticide effects on nontarget soil microorganisms. *Soil Biology and Biochemistry* 10:231–239.

Atlas, R. M., and A. K. Bej. 1990. Detecting bacterial pathogens in environmental water samples by using PCR and gene probes. In M. Innis, D. Gelfand, D. Sninsky, and T. White (eds.). *PCR Protocols: A Guide to Methods and Applications.* Academic Press, New York, pp. 399–407.

Ausmus, B. S. 1973. The use of the ATP assay in terrestrial decomposition studies. *Bulletin from the Ecological Research Committee* (Stockholm) 17:223–234.

Babuik, L. A., and E. A. Paul. 1970. The use of fluorescein isothiocyanate in the determination of the bacterial biomass in grassland soil. *Canadian Journal of Microbiology* 16:57–62.

Bakken, L. R. 1985. Separation and purification of bacteria from soil. *Applied and Environmental Microbiology* 49:1482–1487.

Balderston, W. L., B. Sherr, and W. J. Payne. 1976. Blockage by acetylene of nitrous oxide reduction in *Pseudomonas perfectomarinus. Applied and Environmental Microbiology* 31:504–508.

Balkwill, D. L., D. P. Labeda, and L. E. Casida, Jr. 1975. Simplified procedure for releasing and concentrating microorganisms from soil for transmission electron microscopy viewing as thin-section and frozen-etched preparations. *Canadian Journal of Microbiology* 21:252–262.

Bancroft, K., E. A. Paul, and W. J. Wiebe. 1976. The extraction and measurement of adenosine triphosphate from marine sediments. *Limnology and Oceanography* 21:473–480.

Banse, K. 1977. Determining the carbon to chlorophyll ratio of natural phytoplankton. *Marine Biology* 41:199–212.

Barkay, T. 1987. Adaptation of aquatic microbial communities to Hg^{2+} stress. *Applied and Environmental Microbiology* 53:2725–2732.

Barkay, T., D. L. Fouts, and B. H. Olson. 1985. The preparation of a DNA gene probe for the detection of mercury resistance genes in Gram-negative communities. *Applied*

and Environmental Microbiology 49:686–692.

Barkay, T., and B. H. Olson. 1986. Phenotypic and genotypic adaptation of aerobic heterotrophic sediment bacterial communities to mercury stress. *Applied and Environmental Microbiology* 52:403–406.

Baross, J. A., J. Liston, and R. Y. Morita. 1978. Incidence of *Vibrio parahaemolyticus* bacteriophages and other *Vibrio* bacteriophages in marine samples. *Applied and Environmental Microbiology* 36:492–499.

Bartha, R., and D. Pramer. 1965. Features of a flask for measuring the persistence and biological effects of pesticides. *Soil Science* 100:68–70.

Bartha, R., and D. Pramer 1970. Metabolism of acylanilide herbicides. *Advances in Applied Microbiology* 13:317–341.

Bej, A. K., R. J. Steffan, J. DiCesare, L. Haff, and R. M. Atlas. 1990. Detection of coliform bacteria in water by polymerase chain reaction and gene probes. *Applied and Environmental Microbiology* 56:307–314.

Bej, A. K., S. C. McCarty, and R. M. Atlas. 1991a. Detection of coliform bacteria and *Escherichia coli* by multiplex polymerase chain reaction: Comparison with defined substrate and plating methods for water quality monitoring. *Applied and Environmental Microbiology* 57:2429–2432.

Bej, A. K., M. H. Mahbubani, and R. M. Atlas. 1991b. Detection of viable *Legionella pneumophila* in water using PCR and gene probe methods. *Applied and Environmental Microbiology* 57:597–600.

Bej, A. K., M. H. Mahbubani, and R. M. Atlas. 1991c. Amplification of nucleic acids by polymerase chain reaction (PCR) and other methods and their applications. *Critical Reviews in Biochemistry and Molecular Biology* 26:301–334.

Berent, S. L., M. Mahmoudi, R. M. Torczynski, P. W. Bragg, and A. P. Bollon. 1985. Comparison of oligonucleotide and long DNA fragments as probes in DNA and RNA dot, Southern, northern, colony and plaque hybridizations. *BioTechniques* 3:208–220.

Black, C. A. 1965. *Methods of Soil Analyses.* Part 2, Chemical and Microbiological Properties. American Society of Agronomy, Madison, Wis.

Blackburn, J. W., R. L. Jain, and G. S. Sayler. 1987. Molecular microbial ecology of a naphthalene-degrading genotype in activated sludge. *Environmental Science and Technology* 21:884–890.

Board, R. G., and D. W. Lovelock. 1973. *Sampling—*

Microbiological Monitoring of Environments. Academic Press, London.

Bohlool, B. B. 1987. Fluorescence methods for study of *Rhizobium* in culture and *in situ.* In G. H. Elkan (ed.), *Symbiotic Nitrogen Fixation Technology.* Marcel Dekker, New York, pp. 127–155.

Bohlool, B. B., and E. L. Schmidt. 1973. A fluorescent antibody technique for determination of growth rates of bacteria in soil. *Bulletin from the Ecological Research Committee* (Stockholm) 17:336–338.

Bohlool, B. B., E. L. Schmidt, and C. Beasley. 1977. Nitrification in the intertidal zone: Influence of effluent type and effect of tannin on nitrifiers. *Applied and Environmental Microbiology* 34:523–528.

Bohlool, B. B., and E. L. Schmidt. 1980. The immunofluorescence approach in microbial ecology. *Advances in Microbial Ecology* 4:203–242.

Boivin, R., F. Chalifour, and P. Dion. 1988. Construction of a Tn5 derivative encoding bioluminescence and its introduction in *Pseudomonas, Agrobacterium,* and *Rhizobium. Molecular and General Genetics* 213:50–55.

Bordner, R., and J. Winter (eds.). 1978. *Microbiological Methods for Monitoring the Environment.* Part 1, Water and Wastes. Environmental Protection Agency, Cincinnati, Ohio.

Bowden, W. B. 1977. Comparison of two direct-count techniques for enumerating aquatic bacteria. *Applied and Environmental Microbiology* 33:1229–1232.

Brayton, P. R., and R. R. Colwell. 1987. Fluorescent antibody staining method for enumeration of viable environmental *Vibrio cholerae* O1. *Journal of Microbiological Methods* 6:309–314.

Britton, R. J., and E. H. Davidson. 1985. Hybridization strategy. In B. P. Hames and S. J. Higgins (eds.). *Nucleic Acid Hybridization: A Practical Approach.* IRL Press, New York, pp. 3–15.

Brock, M. L., and T. D. Brock. 1968. The application of micro-autoradiographic techniques to ecological studies. *Mitteilungen der Internationale Vereiningung für Theoretische und Angewandte Limnologie* 15:1–29.

Brock, T. D. 1969. Bacterial growth rate in the sea: Direct analysis by thymidine autoradiography. *Science* 155:81–83.

Brock, T. D. 1971. Microbial growth rates in nature. *Bacteriological Reviews* 35:39–53.

Brown, C. M. 1982. Nitrogen mineralisation in soils and sediments. In R. G. Burns and J. H. Slater (eds.). *Experimental Microbial Ecology.* Blackwell Scientific Publica-

tions, Oxford, England, pp. 154–163.

Buck, J. D. 1979. The plate count in aquatic microbiology. In J. W. Costerton and R. R. Colwell (eds.). *Native Aquatic Bacteria: Enumeration, Activity and Ecology.* ASTM Special Technical Publication No. 695. American Society for Testing Materials, Philadelphia, pp. 19–28.

Burlage, R. S., G. S. Sayler, and F. W. Larimer. 1990. Monitoring of naphthalene catabolism by bioluminescence with *nah-lux* transcriptional fusions. *Journal of Bacteriology* 172:4749–4757.

Byrd, J. J., and R. R. Colwell. 1992. Microscopy applications for analysis of environmental samples. In. M. A. Levin, R. J. Seidler, and M. Rogul (eds.). *Microbial Ecology: Principles, Methods, and Applications.* McGraw-Hill, New York, pp. 93–112.

Caldwell, D. E. 1977. Accessory pigment fluorescence for quantitation of photosynthetic microbial populations. *Canadian Journal of Microbiology* 23:1594–1597.

Caldwell, D. E., and J. M. Tiedje. 1975. The structure of anaerobic bacterial communities in the hypolimnia of several Michigan lakes. *Canadian Journal of Microbiology* 21:377–385.

Casida, L. E. 1977. Microbial metabolic activity in soil as measured by dehydrogenase determinations. *Applied and Environmental Microbiology* 34:630–636.

Cholodny, N. 1930. Über eine neue Methode zur Untersuchung der Bodenmikroflora. *Archiv für Mikrobiologie* (Berlin) 1:650–652.

Christian, R. R., R. B. Hanson, and S. Y. Newell. 1982. Comparison of methods for measurement of bacterial growth rates in mixed batch cultures. *Applied and Environmental Microbiology* 43:1160–1165.

Chrzanowski, T. H., L. H. Stevenson, and B. Kjerfve. 1979. Adenosine 5′-triphosphate flux through the North Inlet marsh system. *Applied and Environmental Microbiology* 37:841–848.

Cochran, W. G. 1950. Estimation of bacterial densities by means of the "most probable number." *Biometrics* 2:105–116.

Cohen, U., W. Krumbein, and M. Shilo. 1977. Solar Lake (Sinai): Bacterial distribution and production. *Limnology and Oceanography* 22:621–634.

Coleman, A. W. 1980. Enhanced detection of bacteria in natural environments by fluorochrome staining of DNA. *Limnology and Oceanography* 25:948–951.

Colwell, R. R. 1979. Enumeration of specific populations by the most-probable-number (MPN) method. In J. W.

Costerton and R. R. Colwell (eds.). *Native Aquatic Bacteria: Enumeration, Activity and Ecology*. ASTM Special Technical Publication No. 695. American Society for Testing Materials, Philadelphia, pp. 56–64.

Conway de Macario, E., M. J. Wolin, and A. J. L. Macario. 1981. Immunology of archaebacteria that produce methane gas. *Science* 214:74–75.

Conway de Macario, E., A. J. L. Macario, and M. J. Wolin. 1982. Specific antisera and immunological procedures for characterization of methanogenic bacteria. *Journal of Bacteriology* 149:320–328.

Crawford, C. C., J. E. Hobbie, and K. L. Webb. 1973. Utilization of dissolved organic compounds by microorganisms in an estuary. In L. H. Stevenson and R. R. Colwell (eds.). *Estuarine Microbial Ecology*. University of South Carolina Press, Columbia, pp. 169–180.

Crawford, D. L., R. L. Crawford, and A. L. Pometto III. 1977. Preparation of specifically labelled ^{14}C-(lignin)- and ^{14}C-(cellulose)-lignocelluloses and their decomposition by the microflora of soil. *Applied and Environmental Microbiology* 33:1247–1251.

Daley, R. J. 1979. Direct epifluorescence enumeration of native aquatic bacteria: Uses, limitations and comparative accuracy. In J. W. Costerton and R. R. Colwell (eds.). *Native Aquatic Bacteria: Enumeration, Activity and Ecology*. ASTM Special Technical Publication No. 695. American Society for Testing Materials, Philadelphia, pp. 29–45.

Daley, R. J., and J. E. Hobbie. 1975. Direct counts of aquatic bacteria by a modified epifluorescence technique. *Limnology and Oceanography* 20:875–882.

Datta, A. R., B. A.Wentz, D. Shook, and M. W. Trucksess 1988. Synthetic oligodeoxyribonucleotide probes for detection of *Listeria monocytogenes*. *Applied and Environmental Microbiology* 54:2933–2937.

Dazzo, F. B., and W. J. Brill. 1977. Receptor sites on clover and alfalfa roots for *Rhizobium*. *Applied and Environmental Microbiology* 33:132–136.

Deming, J. W., G. L. Picciolo, and E. W. Chappelle. 1979. Important factors in adenosine triphosphate determination using firefly luceiferace: Applicability of the assay to studies of native aquatic bacteria. In J. W. Costerton and R. R. Colwell (eds.). *Native Aquatic Bacteria: Enumeration, Activity and Ecology*. ASTM Special Technical Publication No. 695. American Society for Testing Materials, Philadelphia, pp. 89–98.

Dowding, P., and P. Widden. 1974. Some relationships between fungi and their environment in tundra regions. In *Soil Organisms and Decomposition in Tundra*. Tundra Biome Steering Committee, Swedish IBP, Stockholm, pp. 123–150.

Drahos, D. J., B. C. Hemming, and S. McPherson. 1986. Tracking recombinant organisms in the environment: ß-galactosidase as a selectable non-antibiotic marker for fluorescent Pseudomonads. *Bio/Techniques* 4:439–444.

Echeverria, P., J. Seriwatana, O. Chityothin, W. Chaicumpa, and C. Tirapat. 1982. Detection of enterotoxigenic *Escherichia coli* in water by filter hybridization with three enterotoxin gene probes. *Journal of Clinical Microbiology* 16:1086–1090.

Edmondson, W. T. 1974. A simplified method for counting phytoplankton. In R. A. Vollenweider (ed.). *A Manual on Methods for Measuring Primary Production in Aquatic Environments*. IBP Handbook No. 12, 2d ed. Blackwell Scientific Publications, Oxford, England, pp. 14–16.

Engebrecht, J., M. Simon, and M. Silverman. 1985. Measuring gene expression with light. *Science* 227:1345–1347.

Eren, J., and D. Pramer. 1966. Applications of the immunofluorescent staining to study of the ecology of soil microorganisms. *Soil Science* 101:39–45.

Eriksson, K. E., and S. C. Johnsrud. 1982. Mineralisation of carbon. In R. G. Burns and J. H. Slater (eds.). *Experimental Microbial Ecology*. Blackwell Scientific Publications, Oxford, England, pp. 134–153.

Erlich, H. A. (ed.). 1989. *PCR Technology: Principles and Applications for DNA Amplification*. Stockton, New York.

Faegri, A., V. L. Torsvik, and J. Goksøyr. 1977. Bacterial and fungal activities in soil: Separation of bacteria by a rapid centrifugation technique. *Soil Biology Biochemistry* 9:105–112.

Falkenstein, H., P. Bellemann, S. Walter, W. Zeller, and K. Geider. 1988. Identification of *Erwinia amylovora,* the fireblight pathogen, by colony hybridization with DNA from plasmid pEPA29. *Applied and Environmental Microbiology* 54:2798–2802.

Farrah, S. R., S. M. Goyal, C. P. Gerba, C. Wallis, and J. L. Melnick. 1977. Concentration of enteroviruses from estuarine water. *Applied and Environmental Microbiology* 33:1192–1196.

Faust, M. A., and D. L. Correll. 1977. Autoradiographic study to detect metabolically active phytoplankton and bacteria in Rhode River Estuary. *Marine Biology* 41:293–305.

Festl, H., W. Ludwig, and K. H. Schleifer. 1986. DNA

hybridization probe for the *Pseudomonas fluorescens* group. *Applied and Environmental Microbiology* 52:1190–1194.

Finlay, B. J. 1982. Procedures for the isolation, cultivation, and identification of bacteria. In R. G. Burns and J. H. Slater (eds.). *Experimental Microbial Ecology*. Blackwell Scientific Publications, Oxford, England, pp. 44–65.

Finlay, B. J., J. Laybourn, and I. Strachan. 1979. A technique for the enumeration of benthic ciliated protozoa. *Oecologia* (Berlin) 39:375–377.

Fitts, R., M. Diamond, C. Hamilton, and M. Neri. 1983. DNA–DNA hybridization assay for the detection of *Salmonella* spp. in foods. *Applied and Environmental Microbiology* 46:1146–1151.

Fliermans, C. B., and E. L. Schmidt. 1975. Autoradiography and immunofluorescence combined for autecological study of single cell activity with *Nitrobacter* as a model system. *Applied Microbiology* 30:674–677.

Fliermans, C. B., W. D. Cherry, L. H. Orrison, and L. Thacker. 1979. Isolation of *Legionella pneumophila* from nonepidemic-related aquatic habitats. *Applied and Environmental Microbiology* 37:1239–1242.

Focht, D. D. 1982. Denitrification. In R. G. Burns and J. H. Slater (eds.). *Experimental Microbial Ecology*. Blackwell Scientific Publications, Oxford, England, pp. 194–211.

Frederick, L. R. 1965. Microbial populations by direct microscopy. In C. A. Black (ed.). *Methods in Soil Analysis*. Part 2, Chemical and Microbiological Properties. American Society of Agronomy, Madison, Wis., pp. 1452–1459.

Fuhrman, J. A., and F. Azam. 1980. Bacterioplankton secondary production estimates for coastal waters of British Columbia, Antarctica and California. *Applied and Environmental Microbiology* 39:1085–1095.

Fuhrman, J. A., D. E. Comeau, A. Hagstrom, and A. M. Cham. 1988. Extraction from natural planktonic microorganisms of DNA suitable for molecular biological studies. *Applied and Environmental Microbiology* 54:1426–1429.

Geesey, G. G., and J. W. Costerton. 1979. Bacterial biomass determination in silt-laden river: Comparison of direct count epifluorescence microscopy and extractable adenosine triphosphate techniques. In J. W. Costerton and R. R. Colwell (eds.). *Native Aquatic Bacteria: Enumeration, Activity and Ecology*. ASTM Special Technical Publication No. 695. American Society for Testing Materials, Philadelphia, pp. 117–130.

Gerba, C. P. 1983. Methods for recovering viruses from the water environment. In G. Berg (ed.). *Viral Pollution of the Environment*. CRC Press, Boca Raton, Fla., pp. 19–35.

Giovannoni, S. J., T. B. Britschgi, C. L. Moyer, and K. G. Field. 1990. Genetic diversity in Sargasso Sea bacterioplankton. *Nature* (London) 345:60–62.

Gorden, R. W. 1972. *Field and Laboratory Microbial Ecology*. William C. Brown Company, Dubuque, Iowa, pp. 47–50.

Goss, W. A., W. H. Deitz, and T. M. Cook. 1965. Mechanism of action of nalidixic acid on *Escherichia coli*. Part II, Inhibitor of DNA synthesis. *Journal of Bacteriology* 89:1068–1074.

Graham, L. B., A. D. Colburn, and J. C. Burke. 1976. A new, simple method for gently collecting planktonic protozoa. *Limnology and Oceanography* 21:336–341.

Gray, T. R. G. 1967. Stereoscan electron microscopy of soil microorganisms. *Science* 155:1668–1670.

Gray, T. R. G. 1973. The use of the fluorescent-antibody technique to study the ecology of *Bacillus subtilis* in soil. *Bulletin from the Ecological Research Committee* (Stockholm) 17:119–122.

Gray, T. R. G., P. Baxby, I. R. Hall, and M. Goodfellow. 1968. Direct observation of bacteria in soil. In T. R. G. Gray and D. Parkinson (eds.). *The Ecology of Soil Bacteria*. University of Toronto Press, Toronto, Canada, pp. 171–197.

Gregory, P. H. 1973. *The Microbiology of the Atmosphere*. John Wiley and Sons, New York.

Griffiths, R. P., S. S. Hayasaka, T. M. McNamara, and R. Y. Morita. 1977. Comparison between two methods of assaying relative microbial activity in marine environments. *Applied and Environmental Microbiology* 34:801–805.

Grimes, D. J., and R. R. Colwell. 1983. Survival of pathogenic organisms in the Anacostia and Potomac rivers and the Chesapeake Bay Estuary. *Journal of the Washington Academy of Science* 73:45–50.

Grunstein, M., and D. S. Hogness. 1975. Colony hybridization: A method for the isolation of cloned DNAs that contain a specific gene. *Proceedings of the National Academy of Sciences USA* 72:3961–3965.

Hagstrom, A., V. Larsson, P. Horstedt, and S. Normark. 1979. Frequency of dividing cells, a new approach to the determination of bacterial growth rates in aquatic environments. *Applied and Environmental Microbiology* 37:805–812.

Hamilton, R. D., and O. Holm-Hansen. 1965. Adenosine triphosphate content of marine bacteria. *Limnology and Oceanography* 12:319–324.

Hanahan, D., and M. Meselson. 1980. Plasmid screening at high colony density. *Gene* 10:63–67.

Hanson, R. B. 1977. Nitrogen fixation (acetylene reduction) in a salt marsh amended with sewage sludge and organic carbon and nitrogen compounds. *Applied and Environmental Microbiology* 33:846–852.

Hardy, R. W. F., R. Holsten, E. K. Jackson, and R. C. Burns. 1968. The acetylene-ethylene assay for N_2 fixation: Laboratory and field evaluation. *Plant Physiology* 43:1185–1207.

Harris, J. E., T. R. McKee, R. C. Wilson, and U. G. Whitehouse. 1972. Preparation of membrane filter samples for direct examination with an electron microscope. *Limnology and Oceanography* 17:784–787.

Hayes, F. R., and E. H. Anthony. 1959. Lake water and sediment. Part VI, The standing crop of bacteria in lake sediments and its place in the classification of lakes. *Limnology and Oceanography* 4:299–315.

Heal, O. W. 1970. Methods of study in soil protozoa. In J. Phillipson (ed.). *Methods of Study in Soil Ecology.* UNESCO, Paris, pp. 119–126.

Herbert, R. A. 1982. Procedures for the isolation, cultivation and identification of bacteria. In R. G. Burns and J. H. Slater (eds.). *Experimental Microbial Ecology.* Blackwell Scientific Publications, Oxford, England, pp. 3–21.

Herrero, M., V. DeLorenzo, and K. N. Timmis. 1990. Transposon vectors containing non-antibiotic resistance selection markers for cloning and stable chromosomal insertion of foreign genes in Gram-negative bacteria. *Journal of Bacteriology* 172:6557–6567.

Hill, I. R., and T. R. G. Gray. 1967. Application of the fluorescent antibody technique to an ecological study of bacteria in soil. *Journal of Bacteriology* 93:1888–1896.

Hill, W. E., J. M. Madden, B. A. McCardell, D. B. Shah, J. A. Jagow, W. L. Payne, and B. K. Boutin. 1983. Foodborne enterotoxigenic *Escherichia coli* detection and enumeration by DNA colony hybridization. *Applied and Environmental Microbiology* 45:1324–1330.

Hill, W. E., W. L. Payne, and C. C. G. Aulisio. 1983. Detection and enumeration of virulent *Yersinia enteroclitica* in food by DNA colony hybridization. *Applied and Environmental Microbiology* 46:636–641.

Hirsch, P., and S. H. Pankratz. 1970. Study of bacterial populations in natural environments by use of submerged electron microscope grids. *Zeitschrift für Allgemeine Mikrobiologie* (Berlin)10:589–605.

Hobbie, J. E., and C. C. Crawford. 1969. Respiration corrections for bacterial uptake of dissolved organic compounds in natural waters. *Limnology and Oceanography* 14:528–532.

Hobbie, J. E., O. Holm-Hansen, T. T. Packard, L. R. Pomeroy, R. W. Sheldon, J. P. Thomas, and W. J. Wiebe. 1972. A study of the distribution and activity of micro-organisms in ocean water. *Limnology and Oceanography* 17:544–555.

Hobbie, J. E., R. J. Daley, and S. Jasper. 1977. Use of Nuclepore filters for counting bacteria by fluorescence microscopy. *Applied and Environmental Microbiology* 33:1225–1228.

Holben, W. E., J. K. Jansson, B. K. Chelm, and J.M. Tiedje. 1988. DNA probe methods for the detection of specific microorganisms in the soil bacterial community. *Applied and Environmental Microbiology* 54:703–711.

Holm-Hansen, O. 1969. Determination of microbial biomass in ocean profiles. *Limnology and Oceanography* 19:31–34.

Holm-Hansen, O. 1973a. Determination of total microbial biomass by measurement of adenosine triphosphate. In L. H. Stevenson and R. R. Colwell (eds.). *Estuarine Microbial Ecology.* University of South Carolina Press, Columbia, pp. 73–89.

Holm-Hansen, O. 1973b. The use of ATP determinations in ecological studies. *Bulletin from the Ecological Research Committee* (Stockholm) 17:215–222.

Holm-Hansen, O., and C. R. Booth. 1966. The measurement of adenosine triphosphate in the ocean and its ecological significance. *Limnology and Oceanography* 11:510–519.

Holm-Hansen, O., W. H. Sutcliffe, and J. Sharpe. 1968. Measurement of deoxyribonucleic acid in the ocean and its ecological significance. *Limnology and Oceanography* 13:507–514.

Holm-Hansen, O., and B. Riemann. 1978. Chlorophyll a determination: Improvement in methodology. *Oikos* 30:438–447.

Hoppe, H. G. 1976. Determination and properties of actively metabolising heterotrophic bacteria in the sea, measured by means of microautoradiography. *Marine Biology* 36:291–302.

Hsu, T. S., and R. Bartha. 1979. Accelerated mineralization of two organophosphate insecticides in the rhizosphere. *Applied and Environmental Microbiology* 37:36–41.

Hubel, H. 1966. Die ^{14}C Methode zur Bestimmung der Primärproduktion des Phytoplanktons. *Limnologica* (Berlin) 4:267–280.

Hungate, R. E. 1969. A roll tube method for cultivation of strict anaerobes. In J. R. Norris and D. W. Ribbons (eds.). *Methods in Microbiology,* Vol. 3B. Academic Press, London, pp. 117–132.

Hungate, R. E., and J. Macy. 1973. The roll-tube method for cultivation of strict anaerobes. *Bulletin from the Ecological Research Committee* (Stockholm) 17:123–125.

Hurst, C. J., S. R. Farrah, C. P. Gerba, and J. L. Melnick. 1978. Development of quantitative methods for the detection of enteroviruses in sewage sludges during activation and following land disposal. *Applied and Environmental Microbiology* 36:81–89.

Ingram, C., M. Brawner, P. Youngman, and J. Westpheling. 1989. *xylE* functions as an efficient reporter gene in *Streptomyces* spp.: Use for the study of *galP1*, a catabolite-controlled promoter. *Journal of Bacteriology* 171:6617–6624.

Innis, M., D. Gelfand, D. Sninsky, and T. White (eds.). 1990. *PCR Protocols: A Guide to Methods and Applications.* Academic Press, San Diego.

Jain, R. K., G. S. Sayler, J. T. Wilson, L. Houston, and D. Pacia. 1987. Maintenance and stability of introduced genotypes in groundwater aquifer material. *Applied and Environmental Microbiology* 53:996–1002.

Jannasch, H. W., and G. E. Jones. 1959. Bacterial populations in sea water as determined by different methods of enumeration. *Limnology and Oceanography* 4:128–139.

Jannasch, H. W., and W. S. Maddux. 1967. A note on bacteriological sampling of seawater. *Journal of Marine Research* 25:185–189.

Jannasch, H. W., and C. O. Wirsen. 1977. Retrieval of concentrated and undecompressed microbial populations from the deep sea. *Applied and Environmental Microbiology* 33:642–646.

Jansson, J. K., W. E. Holben, and J. M. Tiedje. 1989. Detection in soil of a deletion in an engineered DNA sequence by using gene probes. *Applied and Environmental Microbiology* 55:3022–3025.

Jenkinson, D. S. 1976. The effects of biocidal treatments on metabolism in soil. Part IV, The decomposition of fumigated organisms in soil. *Soil Biology and Biochemistry* 8:203–208.

Jenkinson, D. S., and D. S. Powlson. 1976. The effects of biocidal treatments on metabolism in soil. Part V, A method for measuring soil biomass. *Soil Biology and Biochemistry* 8:209–213.

Jenkinson, D. S., and J. M. Oades. 1979. A method for measuring adenosine triphosphate in soil. *Soil Biology and Biochemistry* 11:193–199.

Jensen, V. 1968. The plate count technique. In T. R. G. Gray and D. Parkinson (eds.). *The Ecology of Soil Bacteria.* University of Toronto Press, Toronto, Canada, pp. 158–170.

Jones, J. G., and B. M. Simon. 1975. An investigation of errors in direct counts of aquatic bacteria by epifluorescence microscopy, with reference to a new method for dying membrane filters. *Journal of Applied Bacteriology* 39:317–329.

Jones, P. C. T., and J. E. Mollison. 1948. The technique for the quantitative estimation of soil microorganisms. *Journal of General Microbiology* 2:54–69.

Kaneko, T., J. Hauxhurst, M. Krichevsky, and R. M. Atlas. 1978. Numerical taxonomic studies of bacteria isolated from Arctic and subarctic marine environments. In M. W. Loutit and J. A. R. Miles (eds.). *Microbial Ecology.* Springer-Verlag, Berlin, pp. 26–30.

Karl, D. M. 1979. Measurement of microbial activity and growth in the ocean by rates of stable ribonucleic acid synthesis. *Applied and Environmental Microbiology* 38:850–860.

Karl, D. M. 1980. Cellular nucleotide measurements and applications in microbial ecology. *Microbiological Reviews* 44:739–796.

Karl, D. M. 1981. Simultaneous rates of RNA and DNA synthesis for estimating growth and cell division of aquatic microbial communities. *Applied and Environmental Microbiology* 42:802–810.

Karl, D. M., and O. Holm-Hansen. 1978. Methodology and measurement of adenylate energy charge ratios in environmental samples. *Marine Biology* 48:185–197.

Karl, D. M., C. D. Winn, and D. C. L. Wong. 1981a. RNA synthesis as a measure of microbial growth in aquatic environments: Evaluation, verification, and optimization of methods. *Marine Biology* 64:1–12.

Karl, D. M., C. D. Winn, and D. C. L. Wong. 1981b. RNA synthesis as a measure of microbial growth in aquatic environments: Field measurements. *Marine Biology* 64:13–21.

King, J. D., and D. C. White. 1977. Muramic acid as a measure of microbial biomass in estuarine and marine samples. *Applied and Environmental Microbiology* 33:777–783.

King, J. M. H., P. M. DiGrazia, B. Applegate, R. Burlage, J. Sanseverino, P. Dunbar, F. Larimer, and G. S. Sayler. 1990. Bioluminescent reporter plasmid for naphthalene exposure and biodegradation. *Science* 249:778–791.

King, R. J., K. A. Short, and R. J. Seidler. 1991. Assay for detection and enumeration of genetically engineered microorganisms which is based on the activity of a deregulated 2,4-dichlorophenoxyacetate monooxygenase. *Applied and Environmental Microbiology* 57:1790–1792.

Knight, I. T., W. E. Holben, J. M. Tiedje, and R. R. Colwell. 1992. Nucleic acid hybridization techniques for detection, identification, and enumeration of microorganisms in the environment. In M. A. Levin, R. J. Seidler, and M. Rogul (eds.). *Microbial Ecology: Principles, Methods, and Applications.* McGraw-Hill, New York, pp. 65–91.

Kogure, K., U. Simidu, and N. Taga. 1979. A tentative direct microscopic method for counting living marine bacteria. *Canadian Journal of Microbiology* 25:415–420.

Kubitschek, H. E. 1969. Counting and sizing microorganisms with the Coulter counter. In J. R. Norris and D. W. Ribbons (eds.). *Methods in Microbiology,* Vol. 1. Academic Press, London, pp. 593–610.

Larsson, K., C. Wenbull, and G. Cronberg. 1978. Comparison of light and electron microscopic determinations of the number of bacteria and algae in lake water. *Applied and Environmental Microbiology* 35:397–404.

Latt, S. A., and G. Statten. 1976. Spectral studies on 33258 Hoechst and related bisbenzimadazole dyes useful for fluorescent detection of DNA synthesis. *Journal of Histochemistry and Cytochemistry* 24:24–32.

Lee, S., and J. A. Fuhrman. 1990. DNA hybridization to compare species compositions of natural bacterioplankton assemblages. *Applied and Environmental Microbiology* 56:739–746.

Lindgren, P. B., R. Frederick, A. G. Govindarajan, N. J. Panopoulos, B. J. Staskawicz, and S. E. Lindow. 1989. An ice nucleation reporter gene system: Identification of inducible pathogenicity genes in *Pseudomonas syringae* pv. *phaseolicola. EMBO Journal* 8:1291–1301.

Lloyd, A. B. 1973. Estimation of actinomycete spores in air. *Bulletin from the Ecological Research Committee* (Stockholm) 17:168–169.

Loftus, M. E., and J. H. Carpenter. 1971. A fluorometric method for determining chlorophylls a, b and c. *Journal of Marine Research* 29:319–338.

Lorenzen, C. J. 1966. A method for the continuous measurement of *in vivo* chlorophyll concentration. *Deep-Sea Research* 13:223–227.

Lowry, O. H., N. J. Rosebrough, A. L. Farr, and R. J. Randall. 1951. Protein measurement with the Folin phenol reagent. *Journal of Biological Chemistry* 193:265–275.

Maniatis, T., E. F. Fritsch, and J. Sambrook. 1982. *Molecular Cloning: A Laboratory Manual.* Cold Spring Harbor Laboratory, Cold Spring Harbor Press, N.Y.

McCoy, W. F., and B. H. Olson. 1985. Fluorometric determination of the DNA concentration in municipal drinking water. *Applied and Environmental Microbiology* 49:811–817.

Melchiorri-Santolini, U. 1972. Enumeration of microbial concentration of dilution series (MPN). In Y. I. Sorokin and H. Kadota (eds.). *Techniques for the Assessment of Microbial Production and Decomposition in Fresh Waters.* IBP Handbook No. 23. Blackwell Scientific Publications, Oxford, England, pp. 64–70.

Menzies, J. D. 1965. Fungi. In C. A. Black (ed.). *Methods of Soil Analysis.* Part 2, Chemical and Microbiological Properties. American Society of Agronomy, Madison, Wis., pp. 1502–1505.

Meyer-Reil, L. A. 1978. Autoradiography and epifluorescent microscopy combined for the determination of number and spectrum of actively metabolizing bacteria in natural waters. *Applied and Environmental Microbiology* 36:506–512.

Miliotis, M. D., J. E. Galen, J. B. Kaper, and J. G. Morris, Jr. 1989. Development and testing of a synthetic ologonucleotide probe for the detection of pathogenic *Yersinia* strains. *Journal of Clinical Microbiology* 27:1667–1670.

Millar, W. N., and L. E. Casida, Jr. 1970. Evidence for muramic acid in soil. *Canadian Journal of Microbiology* 16:299–304.

Moriarty, D. J. W. 1975. A method for estimating the biomass of bacteria in aquatic sediments and its application in trophic studies. *Oecologia* (Berlin) 20:219–229.

Moriarty, D. J. W. 1977. Improved method using muramic acid to estimate biomass of bacteria in sediments. *Oecologia* (Berlin) 26:317–323.

Moriarty, D. J. W. 1978. Estimation of bacterial biomass in water and sediments using muramic acid. In M. W.

Loutit and J. A. R. Miles (eds.). *Microbial Ecology*. Springer-Verlag, Berlin, pp. 31–33.

Morris, I. 1982. Primary production of the oceans. In R. G. Burns and J. H. Slater (eds.). *Experimental Microbial Ecology*. Blackwell Scientific Publications, Oxford, England, pp. 239–254.

Mullis, K. B. 1990. The unusual origin of the polymerase chain reaction. *Scientific American* 262(4):56–65.

Mullis, K., F. Faloona, S. Scharf, R. Saiki, G. Horn, and H. Ehrlich. 1986. Specific enzymatic amplification of DNA in vitro: The polymerase chain reaction. *Cold Spring Harbor Symposium on Quantitative Biology* 51:263–273.

Mullis, K. B., and F. A. Faloona. 1987. Specific synthesis of DNA in vitro via a polymerase-catalyzed chain reaction. *Methods in Enzymology* 155:335–351.

Nambair, P. T. C., and V. Anjaiah. 1985. Enumeration of rhizobia by enzyme-linked immunosorbent assay (ELISA). *Journal of Applied Bacteriology* 58:187–193.

Nannipieri, P., R. L. Johnson, and E. A. Paul. 1978. Criteria for measurement of microbial growth in soil. *Soil Biology and Biochemistry* 10:223–229.

Nikitin, D. I. 1973. Direct electron microscopic techniques for the observation of microorganisms in soil. *Bulletin from the Ecological Research Committee* (Stockholm) 17:85–92.

Nortermans, S.,T. Chakrobarty, M. Leimeister-Wachter, J. Dufrenne, and K. J. Heuvelman. 1989. Specific gene probe for detection of biotyped and serotype *Listeria* strains. *Applied and Environmental Microbiology* 55:902–906.

Oades, J. M., and D. S. Jenkinson. 1979. Adenosine triphosphate content of the soil microbial biomass. *Soil Biology and Biochemistry* 11:201–204.

Ogram, A., G. S. Sayler, and T. Barkay. 1987. The extraction and purification of microbial DNA from sediments. *Journal of Microbiological Methods* 7:57–66.

Oleniacz, W. S., M. A. Pisano, M. H. Rosenfeld, and R. L. Elgart. 1968. Chemiluminescent method for detecting micro-organisms in water. *Environmental Science and Technology* 2:1030–1033.

Olson, B. H. 1978. Enhanced accuracy of coliform testing in seawater by a modification of the most probable number method. *Applied and Environmental Microbiology* 36:438–444.

Paerl, H. W., M. M. Tilzer, and C. R. Goldman. 1976. Chloro-phyll a versus adenosine triphosphate as algal biomass indicators in lakes. *Journal of Phycology* 12:242–246.

Paerl, H. W., and N. J. Williams. 1976. The relation between adenosine triphosphate and microbial biomass in diverse aquatic ecosystems. *Internationale Revue der gesamten Hydrobiologie* 61:659–664.

Parkinson, D. 1970. Methods for the quantitative study of heterotrophic soil microorganisms. In J. Phillipson (ed.). *Methods of Study in Soil Ecology*. UNESCO, Paris, pp. 101–105.

Parkinson, D. 1973. Techniques for the study of soil fungi. *Bulletin from the Ecological Research Committee* (Stockholm) 17:29–36.

Parkinson, D. 1982. Procedures for the isolation, cultivation and identification of fungi. In R. G. Burns and J. H. Slater (eds.). *Experimental Microbial Ecology*. Blackwell Scientific Publications, Oxford, England, pp. 22–30.

Parkinson, D., T. R. G. Gray, and S. T. Williams. 1971. *Methods for Studying the Ecology of Soil Micro-Organisms*. IBP Handbook No. 19. Blackwell Scientific Publications, Oxford, England.

Paul, E. A., and R. L. Johnson. 1977. Microscopic counting and adenosine 5´-triphosphate measurement in determining microbial growth in soils. *Applied and Environmental Microbiology* 34:263–269.

Paul, J. H. 1982. Use of Hoechst dyes 33258 and 33342 for enumeration of attached and planktonic bacteria. *Applied and Environmental Microbiology* 43:939–944.

Paul, J. H., and B. Meyers. 1982. Fluorometric determination of DNA in aquatic microorganisms by use of Hoechst 33258. *Applied and Environmental Microbiology* 43:1393–1399.

Paul, J. H., L. Cazares, and J. Thurmond. 1990. Amplification of the *rbc*L gene from dissolved and particulate DNA from aquatic environments. *Applied and Environmental Microbiology* 56:1963–1966.

Peele, E. R., and R. R. Colwell. 1981. Application of a direct microscopic method enumeration of substrate-responsive marine bacteria. *Canadian Journal of Microbiology* 27:1071–1075.

Perfilev, B. V., and D. R. Gabe. 1969. *Capillary Methods of Investigating Microorganisms*. Oliver and Boyd, Edinburgh, Scotland.

Pettigrew, C. A., and G. S. Sayler. 1986. The use of DNA:DNA colony hybridization in the rapid isolation of 4-chlorobiphenyl degradative bacterial phenotypes. *Jour-

nal of Microbiological Methods 5:205–213.

Pettipher, G. L., and U. M. Rodrigues. 1982. Semi-automated counting of bacteria and somatic cells in milk using epifluorescence microscopy and television image analysis. *Journal of Applied Bacteriology* 53:323–329.

Porter, K. G., and Y. S. Feig. 1980. The use of DAPI for identifying and counting aquatic microflora. *Limnology and Oceanography* 25:943–948.

Postgate, J. R. 1969. Viable counts and viability. In J. R. Norris and D. W. Ribbons (eds.). *Methods in Microbiology*. Vol. 1. Academic Press, London, pp. 611–628.

Pramer, R., and R. Bartha. 1972. Preparation and processing of soil samples for biodegradation studies. *Environmental Letters* 2:217–224.

Pratt, D., and J. Reynolds. 1972. Selective media for characterizing marine bacterial populations. In R. R. Colwell and R. Y. Morita (eds.). *Effect of the Ocean Environment on Microbial Activities*. University Park Press, Baltimore, pp. 258–267.

Primrose, S. B., N. D. Seeley, and K. B. Logan. 1982. Methods for the study of virus ecology. In R. G. Burns and J. H. Slater (eds.). *Experimental Microbial Ecology*. Blackwell Scientific Publications, Oxford, England, pp. 66–83.

Prosser, J. I., and D. J. Cox. 1982. Nitrification. In R. G. Burns and J. H. Slater (eds.). *Experimental Microbial Ecology*. Blackwell Scientific Publications, Oxford, England, pp. 178–193.

Quinn, J. P. 1984. The modification and evaluation of some cytochemical techniques for the enumeration of metabolically active heterotrophic bacteria in the aquatic environment. *Journal of Applied Bacteriology* 57:51–57.

Ramsay, A. J. 1974. The use of autoradiography to determine the proportion of bacteria metabolising in an aquatic habitat. *Journal of General Microbiology* 80:363–373.

Rattray, E. A. S., J. I. Prosser, K. Killham, and L. A. Glover. 1990. Luminescence-based nonextractive technique for *in situ* detection of *E. coli* in soil. *Applied and Environmental Microbiology* 56:3368–3374.

Renwick, A., and D. G. Jones. 1985. A comparison of the fluorescent ELISA and antibiotic resistance identification techniques for use in ecological experiments with *Rhizobium trifolii. Journal of Applied Bacteriology* 58:199–206.

Revsbech, N. P. 1983. *In situ* measurement of oxygen profiles of sediment by use of oxygen microelectrodes. In E. G. Grainger and H. Forstner (eds.). *Polarographic Oxygen Sensors: Aquatic and Physiological Applications*.

Springer-Verlag, Heidelberg, Germany, pp. 265–273.

Revsbech, N. P., B. B. Jorgenson, T. H. Blackburn, and Y. Cohen. 1983. Microelectrode studies of the photosynthesis and O_2, H_2S, and pH profiles of a microbial mat. *Limnology and Oceanography* 28:1062–1074.

Rogowsky, P. M., T. J. Close, J. A. Chimera, J. J. Shaw, and C. I. Kado. 1987. Regulation of the *vir* genes of *Agrobacterium tumefaciens* plasmid pTi58. *Journal of Bacteriology* 169:5101–5112.

Rossi, G., S. Riccardo, G. Gesue, M. Stanganelli, and T. K. Wang. 1936. Direct microscopic and bacteriological investigations of the soil. *Soil Science* 41:53–66.

Roszak, D. B., and R. R. Colwell. 1987a. Metabolic activity of bacterial cells enumerated by direct viable count. *Applied and Environmental Microbiology* 53:2889–2983.

Roszak, D. B., and R. R. Colwell. 1987b. Survival strategies of bacteria in the natural environment. *Microbiological Reviews* 51:365–379.

Round, F. E. 1982. Procedures for the isolation, cultivation, and identification of algae. In R. G. Burns and J. H. Slater (eds.). *Experimental Microbial Ecology*. Blackwell Scientific Publications, Oxford, England, pp. 31–43.

Rowe, R., R. Todd, and J. Waide. 1977. Microtechnique for most-probable-number analysis. *Applied and Environmental Microbiology* 33:675–680.

Russell, W. C., C. Newman, and D. H. Williamson. 1975. A simple cytochemical technique for demonstration of DNA in cells infected with mycoplasmas and viruses. *Nature* (London) 253:461–462.

Sambrook, J., E. F. Fritsch, and T. Maniatis. 1989. In N. Ford, C. Nolan, and M. Feruson (eds.). *Molecular Cloning: A Laboratory Manual*, Vols. I–III. Cold Spring Harbor Laboratory Press, Cold Spring Harbor, New York

Sayler, G. S., M. S. Shields, E. T. Tedford, A. Breen, S. W. Hooper, K. M. Sirotkin, and J. W. Davis. 1985. Application of DNA-DNA colony hybridization to the detection of catabolic genotypes in environmental samples. *Applied and Environmental Microbiology* 49:1295–1303.

Sayler, G. S., and A. C. Layton. 1990. Environmental application of nucleic acid hybridization. *Annual Review of Microbiology* 44:625–648.

Schmidt, E. L. 1973. Fluorescent antibody techniques for the study of microbial ecology. *Bulletin from the Ecological Research Committee* (Stockholm) 17:67–76.

Schmidt, E. L., R. O. Bankole, and B. B. Bohlool. 1968. Fluorescent-antibody approach to study of rhizobia in soil. *Journal of Bacteriology* 95:1987–1992.

Shapton, D. A., and R. G. Board (eds.). 1971. *Isolation of Anaerobes*. The Society for Applied Bacteriology Technical Ser. No. 5. Academic Press, London.

Sharabi, E. D., and D. Pramer. 1973. A spectrophotofluorometric method for studying algae in soil. *Bulletin from the Ecological Research Committee* (Stockholm) 17:77–84.

Sharma, P. D., P. J. Fisher, and J. Webster. 1977. Critique of the chitin assay technique for estimation of fungal biomass. *Transactions of the British Mycological Society* 69:479–483.

Shaw, J. J., and C. I. Kado. 1986. Development of a *Vibrio* bioluminescence gene-set to monitor phytopathogenic bacteria during the ongoing disease process in a non-disruptive manner. *Bio/Techniques* 4:560–564.

Shuval, H. I., and E. Katzenelson. 1972. The detection of enteric viruses in the water environment. In R. Mitchell (ed.). *Water Pollution Microbiology*. John Wiley and Sons, New York, pp. 347–361.

Sieracki, M. E., P. W. Johnson, and J. M. Sieburth. 1985. Detection, enumeration, and sizing of planktonic bacteria by image-analyzed epifluorescence microscopy. *Applied and Environmental Microbiology* 49:799–810.

Simidu, U. 1972. Improvement of media for enumeration and isolation of heterotrophic bacteria in seawater. In R. R. Colwell and R. Y. Morita (eds.). *Effect of the Ocean Environment on Microbial Activities*. University Park Press, Baltimore, pp. 249–257.

Singh, B. N. 1946. A method of estimating the numbers of soil protozoa, especially amoeba, based on their differential feeding on bacteria. *Annals of Applied Biology* 33:112–119.

Skinner, F. A., P. C. T. Jones, and J. E. Mollison. 1952. A comparison of a direct and a plate counting technique for quantitative estimation of soil microorganisms. *Journal of General Microbiology* 6:261–271.

Skujins, J. 1967. Enzymes in soil. In A. D. McLaren and G. H. Peterson (eds.). *Soil Biochemistry*. Vol. 1. Marcel Dekker, New York, pp. 371–414.

Skujins, J. 1973. Dehydrogenase: An indicator of biological activities in arid soils. *Bulletin from the Ecological Research Committee* (Stockholm) 17:235–241.

Smith, D. W. 1982. Nitrogen fixation. In R. G. Burns and J. H. Slater (eds.). *Experimental Microbial Ecology*. Blackwell Scientific Publications, Oxford, England, pp. 212–220.

Smith, D. W., C. B. Fliermans, and T. D. Brock. 1972.

Technique for measuring $^{14}CO_2$ uptake by soil microorganisms *in situ*. *Applied Microbiology* 23:595–600.

Sobsey, M. D. 1982. Quality of currently available methodology for monitoring viruses in the environment. *Environment International* 7:39–51.

Sobsey, M. D., R. J. Carrick, and H. R. Jensen. 1978. Improved methods for detecting enteric viruses in oysters. *Applied and Environmental Microbiology* 36:121–128.

Soderstrom, D. E. 1977. Vital staining of fungi in pure cultures and in soil with fluorescein diacetate. *Soil Biology and Biochemistry* 9:59–63.

Sommerville, C. C., T. T. Knight, W. L. Straub, and R. R. Colwell. 1989. Simple, rapid method for direct isolation of nucleic acids from aquatic environments. *Applied and Environmental Microbiology* 55:548–554.

Sorenson, J. 1978. Denitrification rates in a marine sediment as measured by the acetylene inhibition technique. *Applied and Environmental Microbiology* 36:139–143.

Staley, J. T., and A. Konopka. 1985. Measurement of *in situ* activities of nonphotosynthetic microorganisms in aquatic and terrestrial habitats. *Annual Review of Microbiology* 39:321–346.

Stanier, R. Y., and J. H. C. Smith. 1960. The chlorophylls of green bacteria. *Biochimica et Biophysica Acta* 41:478–484.

Steffan, R. J., and R. M. Atlas. 1988. DNA amplification to enhance the detection of genetically engineered bacteria in environmental samples. *Applied and Environmental Microbiology* 54:2185–2191.

Steffan, R. J., J. Goksøyr, A. K. Bej, and R. M. Atlas. 1988. Recovery of DNA from soils and sediments. *Applied and Environmental Microbiology* 54:2908–2915.

Steffan, R. J., and R. M. Atlas. 1990. Solution hybridization assay for detecting genetically engineered microorganisms in environmental samples. *Bio/Techniques* 8:316–318.

Steffan, R. J., and R. M. Atlas. 1991. Polymerase chain reaction: Applications in environmental microbiology. *Annual Reviews of Microbiology* 45:137–161.

Stevenson, L. H., C. E. Millwood, and B. H. Hebeler. 1972. Aerobic, heterotrophic bacterial populations in estuarine water and sediments. In R. R. Colwell and R. Y. Morita (eds.). *Effect of the Ocean Environment on Microbial Activities*. University Park Press, Baltimore, pp. 268–285.

Stevenson, L. H., T. H. Chrazanowski, and C. W. Erkenbrecher. 1979. The adenosine triphosphate assay: Conceptions and misconceptions. In J. W. Costerton and R. R. Colwell (eds.). *Native Aquatic Bacteria: Enumeration,*

Activity and Ecology. ASTM Special Technical Publication No. 695, American Society for Testing and Materials, Philadelphia, pp. 99–116.

Strayer, R. F., and J. M. Tiedje. 1978. Application of fluorescent-antibody technique to the study of a methanogenic bacterium in lake sediments. *Applied and Environmental Microbiology* 35:192–198.

Strugger, S. 1948. Fluorescence microscope examination of bacteria. *Canadian Journal of Research, Ser. C* 26:188–193.

Sutcliffe, W. H., Jr., E. A. Orr, and O. Holm-Hansen. 1976. Difficulties with ATP measurements in inshore waters. *Limnology and Oceanography* 21:145–149.

Swift, M. J. 1973. Estimation of mycelial growth during decomposition of plant litter. *Bulletin from the Ecological Research Committee* (Stockholm) 17:323–328.

Thomas, A., D. P. Nicholoas, and D. Parkinson. 1965. Modification of the agar film technique for assaying lengths of mycelium in soil. *Nature* (London) 205:105.

Todd, R. L., K. Cromack, Jr., and R. M. Knutson. 1973. Scanning electron microscopy in the study of terrestrial microbial ecology. *Bulletin from the Ecological Research Committee* (Stockholm) 17:109–118.

Torsvik, V. L. 1980. Isolation of bacterial DNA from soil. *Soil Biology Biochemistry* 12:15–21.

Torsvik, V. L., and J. Goksøyr. 1978. Determination of bacterial DNA in soil. *Soil Biology Biochemistry* 10:7–12.

Torsvik, V. L., J. Goksøyr, and F. I. Daae. 1990. High diversity in DNA of soil bacteria. *Applied and Environmental Microbiology* 56:782–787.

Trolldenier, G. 1973. The use of fluorescence microscopy for counting soil microorganisms. *Bulletin from the Ecological Research Committee* (Stockholm) 17:53–59.

Tsernoglon, D., and E. H. Anthony. 1971. Particle size, water-stable aggregates, and bacterial populations in lake sediments. *Canadian Journal of Microbiology* 17:217–227.

Umbreit, W. W., R. H. Burris, and J. F. Stauffer. 1964. *Manometric Techniques.* Burgess Publishing Co., Minneapolis, Minn.

Vaccaro, R. F., and H. W. Jannasch. 1967. Variations in uptake kinetics for glucose by natural populations in seawater. *Limnology and Oceanography* 12:540–542.

Van Es, F. B., and L. A. Meyer-Reil. 1982. Biomass and metabolic activity of heterotrophic marine bacteria. *Advances in Microbial Ecology* 6:111–170.

Van Wambeke, F. 1988. Numeration et taille des bacteries planctoniques au moyen de l'analyse d'images couplee a l'epiflourescence. *Annales de l'Institut Pasteur—Microbiology* (Amsterdam) 139:261–272.

Waid, J. S., K. J. Preston, and P. J. Harris. 1973. Autoradiographic techniques to detect active microbial cells in natural habitats. *Bulletin from the Ecological Research Committee* (Stockholm) 17:317–322.

Wallis, C., J. L. Melnick, and C. P. Gerba. 1979. Concentration of viruses from water by membrane chromatography. *Annual Review of Microbiology* 33:413–438.

Walton, W. H. 1952. Automatic counting of microscopic particles. *Nature* (London) 169:518-520.

Watson, S. W., T. J. Novitsky, H. L. Quinby, and F. W. Valois. 1977. Determination of bacterial number and biomass in the marine environment. *Applied and Environmental Microbiology* 33:940–946.

Watson, S. W., and J. E. Hobbie. 1979. Measurement of bacterial biomass as lipopolysaccharide. In J. W. Costerton and R. R. Colwell (eds.). *Native Aquatic Bacteria: Enumeration, Activity and Ecology.* ASTM Special Technical Publication No. 695, American Society for Testing and Materials, Philadelphia, pp. 82-88.

Weller, R., and D. M. Ward. 1989. Selective recovery of 16S rRNA sequences from natural microbial communities in the form of cDNA. *Applied and Environmental Microbiology* 55:1818–1822.

White, D. C., R. J. Bobbie, J. S. Herron, J. D. King, and S. J. Morrison. 1979. Biochemical measurements of microbial mass and activity from environmental samples. In J. W. Costerton and R. R. Colwell (eds.). *Native Aquatic Bacteria: Enumeration, Activity and Ecology.* ASTM Special Technical Publication No. 695, American Society for Testing and Materials, Philadelphia, pp. 69–81.

Wiebe, W. J., and K. Bancroft. 1975. Use of the adenylate energy charge ratio to measure growth state of natural microbial communities. *Proceedings of the National Academy of Science USA* 72:2112–2115.

Willoughby, L. G. 1978. Methods for studying microorganisms on decaying leaves and wood in fresh water. In D. W. Lovelock and R. Davies (eds.). *Techniques for the Study of Mixed Populations.* Academic Press, London, pp. 31–50.

Wolf, H. W. 1972. The coliform count. In R. Mitchell (ed.). *Water Pollution Microbiology.* John Wiley and Sons, New York, pp. 333–345.

Wright, R. T. 1973. Some difficulties in using ^{14}C-organic solutes to measure heterotrophic bacterial activity. In L. H. Stevenson and R. R. Colwell (eds.). *Estuarine Microbial Ecology.* University of South Carolina Press, Columbia, pp. 199–217.

Wright, R. T. 1978. Measurement and significance of specific bacteria of natural waters. *Applied and Environmental Microbiology* 36:297–305.

Wright, R. T., and J. E. Hobbie. 1965. The uptake of organic solutes in lake water. *Limnology and Oceanography* 10:23–28.

Wright, R. T., and J. E. Hobbie. 1966. Use of glucose and acetate by bacteria and algae in aquatic ecosystems. *Ecology* 47:447–464.

Wright, S. F. 1992. Immunological techniques for detection, identification, and enumeration of microorganisms in the environment. In M. A. Levin, R. J. Seidler, and M. Rogul (eds.). *Microbial Ecology: Principles, Methods, and Applications.* McGraw-Hill, New York, pp. 45–63.

Wright, S. F., J. G. Foster, and O. L. Bennett. 1986. Production and use of monoclonal antibodies for identification of strains of *Rhizobium trifolii. Applied and Environmental Microbiology* 52:119–123.

Xu, H. S., N. Roberts, F. L. Singleton, R. W. Attwell, D. J. Grimes, and R. R. Colwell. 1982. Survival and viability of nonculturable *Escherichia coli* and *Vibrio cholerae* in the estuarine and marine environment. *Microbial Ecology* 8:313–323.

Yoshinari, T., and R. Knowles. 1976. Acetylene inhibition of nitrous oxide reduction by denitrifying bacteria. *Biochemical and Biophysical Research Communications* 69:705–710.

Zehr, J. P., and L. A. McReynolds. 1989. Use of degenerate oligonucleotides for the amplification of the *nifH* gene from the marine cyanobacterium *Trichodesmium thiebautii. Applied and Environmental Microbiology* 55:2522–2526.

Zimmerman, R., and L.A. Meyer-Reil. 1974. A new method for fluorescence staining of bacterial populations on membrane filters. *Kieler wissenschaftliche Meeresforschungen* (Kiel) 30:24–27.

Zimmerman, R., R. Iturriaga, and J. Becker-Birck. 1978. Simultaneous determination of the total number of aquatic bacteria and the number thereof involved in respiration. *Applied and Environmental Microbiology* 36:926–935.

Zukowski, M. M., D. F. Gaffney, D. Speck, M. Kauffmann, A. Findeli, A. Wisecup, and J. P. Lecocq. 1983. Chromogenic identification of genetic regulatory signals in *Bacillus subtilis* based on expression of a cloned *Pseudomonas* gene. *Proceedings of the National Academy of Sciences USA* 80:1101–1105.

8

Effects of Abiotic Factors and Environmental Extremes on Microorganisms

ABIOTIC LIMITATIONS TO MICROBIAL GROWTH

Liebig's Law of the Minimum

Justus Liebig, a German agricultural chemist active during the middle of the nineteenth century, recognized that, like atoms in a molecule, elements in a living organism are present in distinct proportions (Liebig 1840). According to Liebig's law, the total yield or biomass of any organism will be determined by the nutrient present in the lowest (minimum) concentration in relation to the requirements of that organism. In any given ecosystem, there will be some limiting nutritional factor. Hence, for example, crop yield cannot be increased by the addition of excess phosphorus if the soil is suffering from a shortage of nitrogen.

Liebig's Law of the Minimum applies to microorganisms just as it applies to plants and animals. Increase in the concentration of a particular limiting nutrient allows the affected population to grow or

reproduce until another factor becomes limiting. In a given ecosystem, the growth of one microbial population may be limited by concentrations of available phosphorus, and adding nitrogen will not permit additional growth of that population. Within the same ecosystem, however, growth of another microbial population may be limited by concentrations of available nitrogen, and the addition of nitrogen would permit further growth of those organisms.

Shelford's Law of Tolerance

The occurrence and abundance of organisms in an environment are determined not only by nutrients, but also by various physicochemical factors such as temperature, redox potential, pH, and many others. Shelford's Law of Tolerance describes how such abiotic parameters control the abundance of organisms in an ecosystem and states that for survival and growth each organism requires a complex set of conditions (Shelford 1913). For an organism to succeed in a given environment, each of these conditions must remain within the tolerance range of that organism; if any condition, such as temperature, exceeds the minimum or maximum tolerance of the organism, the organism will fail to thrive and will be eliminated. Consequently, psychrophilic microorganisms cannot grow in ecosystems with high temperatures; obligately anaerobic microorganisms cannot survive conditions of high oxygen tension; obligately halophilic microbes cannot grow in freshwater lakes; and so forth. The tolerance ranges of microorganisms and the fluctuations of chemical and physical factors in an ecosystem do not determine which microorganisms are present at any one moment, but rather which microorganisms can be present on a sustained basis in that ecosystem.

In reality, the presence and success of an organism or a group of organisms in an ecosystem depends both on nutrient requirements and on tolerance (Odum 1971). The population levels of most organisms are controlled in an ecosystem by the quantity and diversity of materials for which they have a minimum requirement, by critical physical factors, and by the tolerance limits of the organisms themselves to these and other components of the environment. Tolerance ranges for a given parameter are somewhat interactive with other parameters; thus, a microorganism that is not able to survive at a particular temperature in an ecosystem with a particular hydrogen ion concentration may be able to survive at that same temperature in another ecosystem with a different hydrogen ion concentration. The interactive nature of environmental determinants complicates the task of microbial ecologists in defining precisely the controlling or limiting factors in natural ecosystems. Nevertheless, from a bewildering variety of physicochemical conditions, often a single one can be identified that, by exceeding the tolerance limit, excludes an organism from a given environment. For microorganisms with complex life cycles, the tolerance of the most sensitive stage is the relevant one.

Autecology studies the interaction of isolated microbial populations with environmental determinants, while synecology considers the interaction of populations within ecosystems. Autecology includes the study of environmental determinants, their effect on microbial populations, and the mechanisms microorganisms have evolved. Our discussion of autecology will include adaptations to environmental extremes and environments where these extremes prevail.

The tolerance range to an environmental determinant that emerges from an autecological laboratory study is typically broader than the one that actually exists in nature. It stands to reason that a population existing close to the edge of its upper or lower tolerance limit grows rather slowly under these stressful conditions. In this state, it becomes vulnerable to negative synecological interactions such as competition, parasitism, or predation, while these factors are excluded from autecological studies. Synecological interactions may also cause a microbial population to do better under some suboptimal environmental conditions if these effectively exclude its main competitors or predators.

Most measures of environmental determinants record the average condition prevailing in a large sample and fail to consider the existence of microhabitats where microorganisms actually live within the samples. As an example, the oral cavity is often considered to be an aerobic environment because of the continuous passage of air through it, but obligate anaerobes survive and grow within anaerobic microhabitats within the oral cavity. One must also consider the fact that interactions among microbial populations can alter the environment in such a way as to allow for or prevent the growth of other microorganisms. For example, a microorganism may produce acidic metabolic products, preventing the survival in that ecosystem of microorganisms that cannot tolerate conditions of low pH, or conversely a microorganism may carry out a metabolic activity, such as nitrogen fixation, whose by-products are required for growth by another microorganism.

Another important consideration in examining limiting or tolerance factors is availability. A nutrient may be present in an ecosystem but in a form not available to microorganisms, that is, the microbial populations cannot use it. A microbial ecologist must consider not only the presence of materials but also their availability to the microbes. Toxic compounds, for example, may be bonded to particles such as clay and be unavailable for interaction with microorganisms; microorganisms are protected from the direct effects of the toxic material in such situations and can therefore tolerate higher concentrations of the toxic substance. Reduced availability of required substances can prevent microbial growth. Water or nutrients, for example, may become bound and unavailable for microbial uptake. Lack of availability may also occur because the chemical is in the wrong form. For example, nitrogen gas is present in virtually all ecosystems, but most microorganisms cannot use this form of nitrogen; thus, in many ecosystems, available nitrogen is a limiting factor. Similarly, phosphate is unavailable in environments that have a high pH and high bivalent cation concentrations. Iron in its oxidized form (Fe^{3+}) is insoluble and thus unavailable for microbial uptake.

A wide or a narrow tolerance range to an environmental determinant is described by the *eury-* or *steno-* prefixes, respectively. A euryhaline microorganism tolerates wide fluctuations of salt concentrations that typically occur in estuaries and tidal rivers. In contrast, a stenohaline microorganism is confined to a freshwater or a marine habitat. Preference for a particular determinant is expressed by the *-phile* suffix, as in halophile or thermophile. Microorganisms that have predominantly broad (eury-) tolerance ranges grow in a wide variety of habitats, while microorganisms with one or more narrow (steno-) tolerances grow only in a few special environments. Stenothermal psychrophiles or thermophiles will grow only in permanently cold or hot environments, respectively, and will not thrive in common environments where the temperatures fluctuate within the mesophilic range.

Some microorganisms are particularly well adapted for survival in extreme habitats where other organisms cannot survive. Their limits of tolerance to specific stress factors determine, in large part, the inhabitants of extreme environments, such as salt ponds, hot springs, alkaline soda lakes, and desert soils (Table 8.1). Many of these organisms have physiological properties that restrict their growth to extreme habitats. In less severe habitats, the sources of nutrients and the interactions among populations become more important in selecting for the populations that become established within the community.

ENVIRONMENTAL DETERMINANTS

Having considered the underlying reasons for examining the physical and chemical properties of an ecosystem, we shall now consider some specific environmental determinants that control microbial growth and activity in an ecosystem. The extremes of environmental conditions that microorganisms may have to tolerate include high temperatures approaching that of boiling water, low temperatures approaching freezing, low acidic pH values, high alkaline pH values, high salt concentrations, low water availability, high irradiation levels, low concentrations of usable nutrients, and high concentrations of toxic compounds. Many microorganisms that inhabit extreme environments, including hot springs, salt lakes, and Antarctic desert soils, possess specialized adaptive physiological features that permit them to survive and function within the physicochemical constraints of these ecosystems (Alexander 1976; Kushner 1978, 1980; Gould and Corry 1980; Brock 1985; Edwards 1990). The membranes and enzymes of microorganisms inhabiting extreme environments often have distinct modifications that permit them to function under conditions that would inhibit active transport and metabolic activities in organisms lacking these adaptive features. Because relatively few microbes possess these adaptations, diversity in extreme environments generally is low.

Survival and growth under extreme environmental conditions have always intrigued microbial ecologists. In addition to being "records," generally of fascination, microorganisms that survive and metabolize under extremely adverse conditions have practical importance as spoilage organisms that defy human efforts to kill or inhibit them. More recently, they have also become potential sources of enzymes that can be used as biocatalysts under conditions that would denature most proteins. The key enzyme for

Table 8.1

Some extreme physiological tolerance limits for microbial activity

Factor	Lower tolerance limit	Upper tolerance limit
Temperature	−12°C (psychrophilic bacteria)	>100°C (sulfate-reducing bacteria at 1000 atm; sulfur oxidizers in deep sea thermal vent regions)
E_h	−450 mV (methanogenic bacteria)	+850 mV (iron bacteria)
pH	0 (*Thiobacillus thiooxidans*)	13 (*Plectonema nostocorum*)
Hydrostatic pressure	0 (various microorganisms)	1,400 atm (barophilic bacteria)
Salinity	0 (*Hyphomicrobium*)	Saturated brines (*Dunaliella*, obligate halophilic bacteria)

the polymerase chain reaction (PCR) that allows the amplification of minute amounts of DNA is derived from a thermophilic bacterium isolated from thermal vents. Its enzyme withstands repeated heating cycles to 80°C.

Space exploration and our ability to soft-land instrument packages on planets like Mars have given additional impetus to research on microbes that can live under extreme environmental conditions. The development of life detection probes in turn sparked interest in the most hostile terrestrial environments, such as the dry valleys of Antarctica, where these devices could be tested.

Temperature

All microorganisms have a characteristic optimal growth temperature at which they exhibit their highest growth and reproduction rates. Microorganisms also have minimal growth temperatures below which they are metabolically inactive and do not demonstrate growth and upper temperature limits beyond which they fail to grow. In many ecosystems, temperature fluctuates on a daily and seasonal basis. Measurement of temperature at the time of sample collection does not reflect the range of temperatures

that the organism must tolerate, nor does it indicate whether that single temperature prevails during the entire time of microbial growth and activity. Considering the temperature fluctuations that occur in most habitats and the abilities of most microorganisms to tolerate some fluctuation around their optimal growth temperatures, the recording accuracy for temperature in most cases needs only be to the degree centigrade.

Both the optimal growth temperature and the range of temperatures that a particular microorganism can tolerate determine whether that microorganism will survive and what role it will play in a given ecosystem. Some microorganisms can grow below 0°C; others can grow at over 100°C. Microorganisms are classified as psychrophiles, mesophiles, and thermophiles, respectively, if their optimal growth temperatures are low (<0°C–<20°C), moderate (20°C–<45°C), or high (45°C–90°C). Extreme thermophiles (90°C–110°C) have been reported with growth temperature optima at 105°C and no growth below 85°C. The extreme thermophiles reported to date are all archaebacteria (Stetter 1982; Fischer et al. 1983), though most archaebacteria are mesophilic. A typical growth curve for a marine psychrophile (Figure 8.1) indicates that optimal growth occurs at 4°C, that the organism is capable of growth

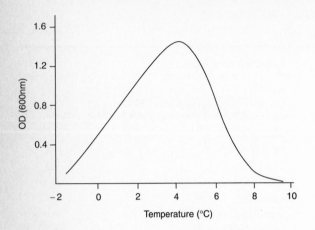

Figure 8.1

Growth attained by an Antarctic psychrophile during 80-hour incubation, measured by optical density, showing low optimal growth temperature and restricted range of temperature tolerance. (Source: Morita 1975. Reprinted by permission of American

at −1.8°C (the freezing point of seawater), and that the bacterium will not grow at or above 10°C (Morita 1975).

It is of interest to consider the adaptations that allow psychrophilic microorganisms to grow in temperature ranges that are prohibitive to most mesophilic organisms. In most mesophilic organisms, a control mechanism shuts off protein synthesis if temperatures fall to approximately 5°C or lower; psychrophilic organisms apparently have ways to override these control mechanisms (Inniss and Ingraham 1978). A higher proportion of unsaturated and/or short chain-length lipids keep the membranes of psychrophiles flexible at low temperatures. It is not known whether there is an absolute low-temperature limit to the metabolic activity of psychrophiles. The practical limit seems to be the freezing temperature of the cell contents and the surrounding water; a liquid phase is essential for metabolic processes. In an Antarctic pond kept from freezing by its high $CaCl_2$ content, microorganisms were observed and active metabolism was implied at temperatures as low as −24°C. It appears now that the observed microorganisms were allochthonous and did not metabolize actively. The salt concentrations required to keep water from freezing appear to prevent microbial growth at temperatures lower than −10°C (Heywood 1984).

Microbial Growth at High Temperatures High-temperature adaptations of thermophilic microorganisms include high proportions of saturated lipids in membranes, which prevent melting at high temperatures (Campbell and Pace 1968; Brock 1978; Castenholz 1979; Edwards 1990). Many thermophilic microorganisms produce enzymes that are not readily denatured at high temperatures. Sometimes, unusual amino acid sequences occur within the proteins of thermophiles, stabilizing these proteins at elevated temperatures (Zuber 1979). Some thermophiles, when growing at temperatures near the maximal growth temperature, exhibit amino acid and growth factor requirements that are not apparent at the optimal growth temperature. This temperature-dependent auxotrophy is evidence for selective heat inactivation of certain enzymatic pathways. The maximum heat tolerance of vegetative thermophilic cells coincides with the heat stability of their ribosomes. When the maximum growth temperature is exceeded, the ribosomes melt and protein synthesis ceases. Thermophiles have relatively high proportions of guanine and cytosine in their DNA that raise the melting point and add stability to the DNA molecules of these organisms.

A particular challenge for microorganisms is to maintain the integrity and fluidity of their cell membranes under varying temperature conditions. High temperature can disintegrate the cell membranes, while low temperature can freeze or gel them, interrupting their vital functions. Even in eurythermal microorganisms, the temperature range of their active metabolism rarely exceeds 30°C–40°C, and this limitation is most likely associated with membrane integrity and fluidity. A cell membrane suitably fluid at low temperature will disintegrate under high-temperature conditions; a membrane stable under high-temperature conditions will freeze up at low temperatures. Mesophilic bacteria have been reported to adjust their membrane composition to some extent to the prevailing growth temperature (Edwards 1990). As growth temperatures increase, the fatty acids in the cell membrane increase in chain length and branching, while their degree of unsaturation decreases. It stands to reason that such adjustments have their limits, and microorganisms cannot change their cell membrane composition completely. Generally, thermophilic eubacteria have a higher proportion of high-molecular-weight branched and saturated fatty acids than psychrophiles and mesophiles. Instead of fatty acid esters, archaebacteria have isoprenoid phytanyglycerol diether and biphytanyldiglycerol tetraether cell membrane constituents. These have high thermal stabilities, enabling some archaebacteria to have optimal growth temperatures above 90°C. Thermophilic eubacteria are restricted, probably because of their membrane composition, to somewhat lower temperatures.

Natural habitats for thermophiles are provided by volcanic activity. Steam vents with temperatures up to 500°C occur, but microbial life requires water in the liquid state. Microorganisms inhabit hot springs with temperatures up to the boiling point, but as the temperature rises, diversity generally decreases. Plants and animals are excluded above 50°C. Eukaryotes (protozoa, fungi, and algae) are excluded above 60°C. Photosynthetic prokaryotes (cyanobacteria and anoxyphotobacteria) are excluded above 90°C, while thermophilic archaebacteria actually increase in diversity between 90°C and 100°C (Brock 1978, 1985) (Figure 8.2). At normal atmospheric pressure, boiling restricts aquatic habitat temperatures to 100°C. However, hydrostatic pressure on the ocean floor raises the boiling point, and water temperatures up to 350°C have been measured in deep-sea hydrothermal vents. Bacterial growth at temperatures up to 250°C by a natural microbial community collected from a hot deep-sea vent has been reported (Baross and Deming 1983; Baross et al. 1984) but was not confirmed and is now considered to have been an experimental artifact (Trent et al. 1984). Growth at temperatures up to 110°C has been unquestionably measured in the case of archaebacterial strains isolated from deep and shallow marine hydrothermal vents (Stetter 1982; Fischer et al. 1983; Jannasch and Mottl 1985; Deming and Baross 1986). In spite of great technical difficulties, these extremely thermophilic archaebacteria have been intensively studied.

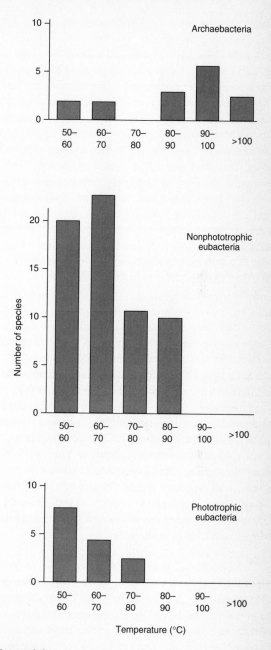

Figure 8.2

Effect of temperature on species diversity of bacteria growing in hot springs. The diversity of eubacteria decreases with increasing temperature, whereas the diversity of archaebacteria increases up to 100°C. (Source: Brock 1985.)

These archaebacteria tend to be anaerobic, and often chemolithotrophic. Some of them use dissolved hydrogen in the hydrothermal fluid to reduce elemental sulfur to H_2S (sulfur respiration) and CO_2 as a carbon source. *Pyrodictium, Thermoproteus,* and *Thermodiscus* are examples of archaebacteria that use this type of metabolism. Other heterotrophic anaerobic archaebacteria, such as *Desulfurococcus* and *Thermococcus,* use organic compounds as electron donors and elemental sulfur as an electron sink. *Methanomonas fervidus,* an extremely thermophilic methanogenic archaebacterium found in hydrothermal vents, reduces CO_2 with H_2 to methane. *Sulfolobus acidocaldarius,* an aerobic thermoacidophilic archaebacterium, oxidizes H_2S and elemental sulfur to sulfate. In addition to archaebacteria, a number of novel anaerobic thermophilic eubacteria—such as *Clostridium, Thermoanaerobacter, Thermodesulfobacterium,* and *Thermobacteroides* strains—were also isolated from hydrothermal vents. The optimal growth temperatures for these are lower (65°C–70°C) as compared to those for the extremely thermophilic archaebacteria. They have either a fermentative or a sulfate-reducing type of heterotrophic metabolism (Brock 1985).

In addition to the thermal environments of volcanic origin, domestic and industrial hot water storage tanks provide stable habitats for harmless oligotrophic thermophiles like *Thermus aquaticus* (Brock 1978). The thermophilic spore former *Bacillus stearothermophilus* is an abundant problem organism for the food industry. Composting (see chapter 12) provides a rich but usually short-lived environment for thermophilic heterotrophic bacteria and fungi.

Survival and Death of Microorganisms at High and Low Temperatures Inability to grow at a particular temperature does not necessarily lead to the death of a microorganism. The temperature tolerance of endospores, cysts, and other resting stages is well known, and even vegetative cells can survive for periods of time at temperatures outside their active growth range. Survival is often greatest at temperatures below the organism's growth range. Thermophiles are frequently isolated from Antarctic soils, for example (Horowitz et al. 1972).

Although microorganisms may survive low temperatures, their growth rates and metabolic activities decline below the optimal temperature. In the microbial ecology of foods, this fact is the basis of preservation and increasing shelf life of products by freezing or refrigeration (Derosier 1970). Many microorganisms in natural soil and aquatic habitats exhibit lower growth rates and metabolic activities during winter when temperatures are low. Temperatures above the growth range of the organism, however, frequently result in death of the organism. Some microorganisms, such as bacterial endospores and sclerotia of fungi, produce resistant resting stages that permit survival at elevated temperatures at which vegetative cells of that organism would not survive. Even such resistant resting stages as the bacterial endospore have upper temperature limits for survival.

The fact that elevated temperatures kill microorganisms is used in the preservation of many food products and the heat sterilization of various items. In the case of food products, the main concern is with the relatively resistant organisms that can cause spoilage and disease (Frazier and Westhof 1978). The resistance of organisms to elevated temperatures is expressed as the thermal death time (TDT), the time required at a particular temperature to kill a specified number of organisms. Examples of thermal death times for several bacteria, including endospore formers, are listed in Table 8.2. The TDT shows the equivalent times and temperatures required to cause an identical reduction in numbers of viable microorganisms. The food industry uses these to determine the appropriate processes for the pasteurization (reduction of numbers of viable microorganisms) or sterilization (complete killing of viable microorganisms) of a product.

The effects of temperature on microorganisms are interactive with other environmental parameters. For example, moisture has a marked influence on the thermal death of microorganisms. Moist heat is more effective in killing microorganisms than dry heat. Whereas 15 to 20 minutes at 120°C with steam is adequate to sterilize many materials, such as bacteriological media, dry heat at 160°C–180°C for three to four

Table 8.2

Approximate thermal death times for bacteria

Organisms	Time (min)	Temperature (°C)
Escherichia coli	20–30	57
Staphylococcus aureus	19	60
Bacillus subtilis (spores)	50–20	100
Clostridium botulinum (spores)	100–330	100

Source: Frazier and Westhoff 1978.

hours is often needed to achieve similar results. The pH also influences thermal killing; microorganisms tend to be more resistant to high temperatures at neutral pH than at acid or alkaline pH values.

Figure 8.3

Relationship of temperature and bacterial respiration in lake sediments. Increased O_2 consumption occurs at higher temperatures. (Source: Hargrave 1969. Reprinted by permission of Journal of the Fisheries Research Board of Canada.)

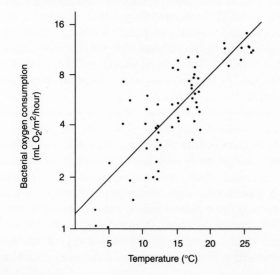

Effect of Temperature on Microbial Activities In addition to affecting survival and growth, temperature influences the metabolic activities of microorganisms. In general, higher temperatures that do not kill microorganisms result in higher metabolic activities. For example, increased O_2 consumption occurs as temperature increases (Figure 8.3). The increased respiratory activity reflects the underlying influence of temperature on enzymes up to the temperature where protein denaturation occurs (Hargrave 1969).

The change in enzyme activity caused by a 10°C rise is known as the 10° temperature quotient or Q_{10} value. In general, enzymes have Q_{10} values near 2—that is, an increase in temperature of 10°C within the tolerance range of the enzyme results in a doubling of activity. Table 8.3 shows examples of some Q_{10} values for microbial enzymes. In addition to Q_{10} values for individual enzymes, the ecologist must consider the effects of temperature on the rates of metabolism by certain microbial populations. Populations of sulfate reducers in salt marsh sediment, for example, have been reported to have a Q_{10} of 3.5 for the reduction of sulfate (Abdollahi and Nedwell 1979) (Figure 8.4). Thus sulfate reduction rates appear to be more sensitive to temperature change than most other metabolic processes.

Although biocatalytic reactions run faster at elevated than at mesophilic temperatures, the growth rate (generation time) of thermophilic microorganisms is not faster and often is considerably slower than that of

Table 8.3

Q_{10} values for some enzymes

Enzyme	Q_{10}
Catalase	2.2 (10°C–20°C)
Maltase	1.9 (10°C–20°C)
Maltase	1.4 (20°C–30°C)
Succinic oxidase	2.0 (30°C–40°C)
Urease	1.8 (20°C–30°C)

$$Q_{10} = \frac{\text{activity at temp } T + 10°C}{\text{activity at temp } T}$$

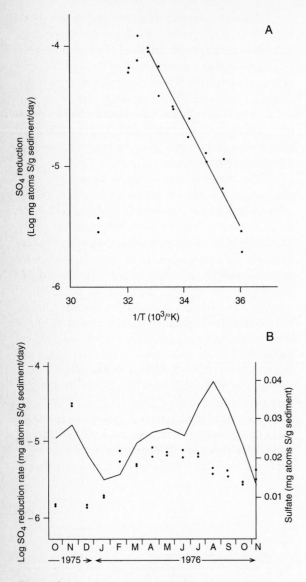

Figure 8.4

(A) Arrhenius plot of sulfate reduction in sediment showing linear decrease in activity with decreasing temperatur e. (B) Seasonal *in situ* rates of sulfate reduction calculated from the Arrhenius plot (A) and actual field temperatures. Closed circles indicate actual *in situ* concentrations of sulfate. High rates of sulfate reduction occurred during summer and fall and showed a negative correlation with actual sulfate concentrations. (Source: Abdollahi and Nedwell 1979. Reprinted by permission; copyright Springer-Verlag.)

the mesophiles (Brock 1978). The reason for this apparent paradox is the considerable amount of repair a thermophilic microorganism needs to perform in order to replace thermally denatured enzymes and other cell components. For biotechnological use, thermophiles are generally inferior to mesophiles for production of single-cell protein or antibiotics. They offer advantages, however, if the aim is degradation and bulk reduction, as in composting. It is also advantageous to run depolymerization, fermentation and methanogenesis reactions using thermophiles. In addition to speeding reaction rates, elevated reaction temperatures reduce cooling costs and contamination problems. Thermophiles are promising sources of heat-resistant enzymes such as proteases and lipases used in laundry detergent formulations, amylases and cellulases used in producing oligomeric sugar substrates for further fermentations, and other thermophilic enzymes used in industry, reagent kits, or scientific research (Brock 1985; Edwards 1990).

Radiation

The spectrum of electromagnetic radiation is continuous from extremely energetic, short-wavelength gamma rays to long-wavelength, low-energy radio waves. Portions of this continuous spectrum are defined as ultraviolet, visible, infrared radiation, and so forth. The radiant energy of the sun drives the entire global ecosystem; photosynthetic interaction with visible light radiation is of profound importance for terrestrial life. The other portions of the electromagnetic spectrum also are ecologically important and have marked effects on life.

Ionizing Radiation Radiations are designated as ionizing radiations if their interaction with matter produces unstable ions and free radicals that interact with living matter in a destructive manner. Gamma rays and x-rays range in wavelength from 10^{-11} cm to 10^{-6} cm, with gamma rays at the shorter, more energetic end of the range. Both gamma rays and x-rays are highly penetrating, and their energy levels are destructive to microorganisms. Low-level irradiation may cause mutations, and high-exposure doses destroy both nucleic acids and enzymes and kill microorganisms. Exposure levels from cosmic and natural radioactive sources are low; nevertheless, these low-level radiations may have contributed to the spontaneous mutations that form the basis for natural selection over geologic ages.

The resistance of microorganisms to gamma radiation varies. As is true with other environmental extremes, microorganisms tend to be more tolerant of ionizing radiation than macroorganisms. As a practical consequence, it is even necessary to add biocides to the cooling waters of nuclear reactors, because the high ionizing radiation levels fail to prevent microbial proliferation.

Bacterial endospores are highly resistant to gamma radiation; it takes 0.3–0.4 Mrads to effect a 90% kill, whereas one-tenth of this dose effects the same percentage kill in most vegetative bacteria. Vegetative cells of *Micrococcus radiodurans,* though, display high radiation resistance, tolerating as much as 1 Mrad without a reduction in viable count (Anderson et al. 1956). Extremely efficient DNA repair mechanisms, rather than any unusual protective substances, are responsible for the radiation resistance of this bacterium.

Ultraviolet Radiation Solar radiation includes ultraviolet (UV) light radiation, visible light radiation, and infrared radiation. Ultraviolet radiation (1–320 nm) is too energetic for photosynthetic use and is destructive to microorganisms (Norman 1954) (Figure 8.5). The most strongly germicidal wavelength (260 nm) coincides with the absorption maximum of DNA, suggesting that damage to DNA is the principal mechanism of the germicidal effect. Ultraviolet-induced dimerization of thymine bases in DNA strands causes faulty transcription, resulting in defective enzymes. Ultraviolet-damaged DNA can be repaired enzymatically by excision of the damaged portions, but this process requires time. The survival of the UV-damaged cells is enhanced by conditions that are adverse to rapid growth; optimal growth conditions allow for rapid expression of the UV damage with resulting cell death. Visible light also has a reactivation effect on UV-damaged cells (Rupert 1964;

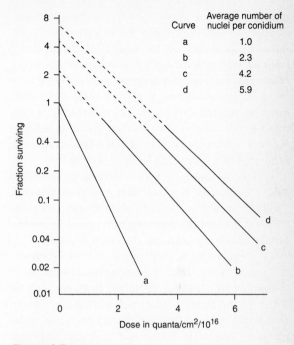

Figure 8.5

The inactivation of *Neurospora* conidia by UV radiation. Exposure was at 254 nm. Conidia with higher numbers of nuclei were less susceptible to UV inactivation, probably because UV acts by interference with DNA replication, and the presence of redundant DNA information prevents lethal damage. (Source: Norman 1954. Reprinted by permission, copyright Wistar Institute of Biology.)

Halldal and Taube 1972). Part of the effect may be photochemical; blue light tends to break the UV-induced thymine dimers. In addition, visible light triggers a DNA repair mechanism that is, paradoxically, not triggered by the more damaging UV radiation. In nature, microorganisms are exposed to UV radiation only in combination with high visible light intensities, which makes this adaptation understandable.

Visible Light Radiation The visible light spectrum is a relatively narrow band in the total electromagnetic spectrum. It ranges from 320 nm (violet) to the far red end of the spectrum at 800 nm. The quantity of light in a given habitat depends upon the obstructions the light beam has to penetrate. Light intensity decreases by the square of the distance from the source, primarily because of absorption and scattering. The sum of absorption and scattering processes, called extinction, is high in ecosystems with many suspended particles. The term turbidity, which refers to the cloudiness or number of suspended particles in a water body, often is used to describe the resistance of a water body to light penetration.

The intensity of light reaching a particular habitat influences rates of photosynthesis (Halldal 1980). In the absence of light, photosynthesis ceases. Up to some optimal light intensity, rates of photosynthetic activity increase (Steeman-Nielson et al. 1962) (Figure 8.6). As with many other environmental factors, the response of an organism to changes in light intensity will depend on the light intensity to which the organism is adapted. Figure 8.6 is clearly shows this by the different response curves of algae that were adapted to 3 Klux and those that had been growing at 30 Klux. In most cases, organisms can

tolerate a wider range of an environmental factor, such as light, if the change is gradual, allowing the organism more time to adapt, than if the change occurs abruptly. Interestingly, some of the cyanobacteria can alternate between oxygen-evolving photosynthesis at high light intensities and anoxygenic photosynthesis, using hydrogen sulfide as electron donor, at low light intensities (Cohen et al. 1975).

The quality (the color) of light, as well as the intensity of light, influences rates of photosynthesis. Not all wavelengths of light penetrate equally well (Figure 8.7). Energy of shorter wavelengths penetrates further than longer-energy wavelengths. Blue light, for example, can penetrate further into aquatic habitats than red light, a factor that leads to a spatial

Figure 8.7

Penetration of light of varying wavelengths into a water column, generally showing that shorter wavelengths of visible light (such as blue) penetrate to greater depths than lower wavelengths (such as red). Upper graph: arithmetic scale. Lower graph: semilog scale. (Source: Wetzel 1975. Reprinted by permission, copyright W. B. Saunders Co.)

Figure 8.6

The rate of photosynthesis per unit cell as a function of light intensity for *Chlorella vulgaris* adapted to high (30 Klux) and low (3 Klux) light intensities. The graph shows higher rates of photosynthesis at low light intensities by low-light-adapted cells. Maximal rates of photosynthesis (saturation) occur at 5 Klux for low–light-adapted cells and at 20–25 Klux for high–light-intensity-adapted cells. (Source: Steeman-Nielson, et al. 1962. Reprinted by permission of Scandinavian Society of Plant Physiology.)

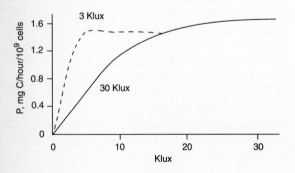

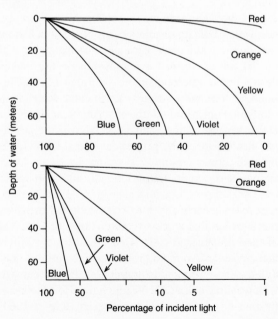

separation of algae and bacteria with different pigments that can absorb light energy of differing wavelengths.

Some microorganisms exhibit phototactic behavior; they move toward or away from a light source (Carlile 1980; Konopka 1984). This behavior leads to vertical migrations of motile microorganisms that adjust their depth in the water column to that of optimal light intensity. Buoyancy regulation is ecologically important to planktonic cyanobacteria that inhabit thermally stratified lakes (Konopka 1984). In these lakes, the depths at which light energy and inorganic nutrients are most abundant are very different. Cyanobacteria with gas vacuoles, which can migrate between these depths, may have a competitive advantage over phytoplankton, which are incapable of vertical migration.

Buoyancy in planktonic cyanobacteria is regulated in response to the levels of light energy and inorganic nutrients such as nitrogen and phosphorus. These factors may alter buoyancy as a result of their effects upon the rates of energy generation and utilization. When growth rate is nutrient limited, the light intensity at which neutral buoyancy occurs is hypothesized to be directly related to the concentration of the limiting nutrient. Species can mediate this response by either raising cell turgor pressure to collapse some existing gas vesicles or increasing cellular amounts of ballast molecules such as polysaccharide (Konopka 1984).

Light also appears to be a key stimulus in the establishment of circadian rhythms in some eukaryotic microorganisms (Sweeney and Haxo 1961) (Figure 8.8). Thus, light affects not only the growth and survival but also the behavior of microorganisms.

In addition to light energy as a limiting factor for primary production, solar radiation may be considered as a stress factor. Solar radiation contains radiation in the UV spectrum, which can cause cell death or mutation. Even excessive intensity of visible light is detrimental, because it causes photooxidative changes. Surface ecosystems—such as the tops of leaves; surfaces of oceans, rocks, soils, and animals; airborne particles; and droplets in clouds—receive high amounts of UV radiation. Microorganisms on snow, on ice, and in brines with crystalline salts receive unusually high UV

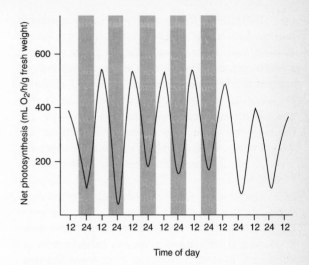

Figure 8.8
Circadian rhythms in *Acetabularia*. Shading shows light and dark periods. Periodicity of photosynthesis continues in constant light. (Source: Sweeney and Haxo 1961. Reprinted from *Science* by permission, copyright 1961 by the American Association for the Advancement of Science.)

and visible light exposures because of the reflectivity of their environments. Microorganisms in these environments are selected in part for their tolerance to UV radiation. Pigmentation for shielding the sensitive DNA and efficient repair mechanisms constitute adaptations that provide such tolerance.

Pressure

Atmospheric Pressure Atmospheric pressure is the pressure exerted by the weight of the air column, which at the surface of the Earth (sea level) is approximately 1 atmosphere (atm) (760 mm mercury). In general, changes in atmospheric pressure do not affect microorganisms. At extremely low atmospheric pressures, such as in the upper atmosphere or in an artificial vacuum, however, water may evaporate, and oxygen becomes limiting, rendering microorganisms metabolically inactive (Gregory 1973).

Hydrostatic Pressure Whereas atmospheric pressure reflects the weight of the air column, hydrostatic pressure is the pressure exerted by the weight of a water column. Hydrostatic pressure increases by approximately 1 atm for every 10 meters of depth. Because the atmospheric pressure at sea level is 1 atm, the pressure at 10 m is approximately 2 atm. The precise pressure at a given depth depends on the weight of the water, and hence pressure is greater at a given depth in saline water than at the same depth in fresh water.

Hydrostatic pressure can be calculated by measuring the density and the depth of the water. In routine hydrographic measurements, readings from shielded and unshielded reversing thermometers are compared; hydrostatic pressure compresses the mercury in the unshielded instrument, resulting in a lower reading at a given temperature. Using suitable conversion tables, the difference between the readings indicates the hydrostatic pressure at the point of thermometer reversion.

To study the growth and metabolism of microorganisms under high hydrostatic pressure, simple culture tubes in cylinders can be pressurized by a hydraulic press (ZoBell and Hittle 1967) (Figure 8.9). Precautions must be taken to avoid hyperbaric oxygenation and compression heat. Sudden decompression may rupture microbial cell membranes by releasing gas bubbles, but steady hydrostatic pressure in the range of 1 to 400 atm has little or no effect on the growth and metabolism of most microorganisms. Higher hydrostatic pressures tend to inhibit or stop the growth of terrestrial and shallow-water organisms, but not that of barotolerant deep-sea organisms, which may be exposed to pressures as high as 1,000 atm.

In response to pressures, organisms are either sensitive (they grow optimally at pressures of about 1 atm, and growth rates decline as pressure increases), barotolerant (they tolerate elevated pressures up to given limits without a decline in growth), or barophilic (they grow optimally at elevated pressures and may be completely inhibited or even killed at atmospheric pressure). The existence of obligate barophiles has been a question of prolonged controversy. Sampling and isolation of deep-sea bacteria without

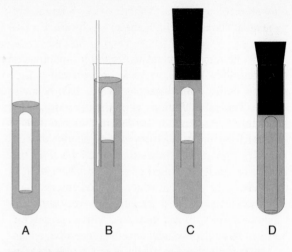

A   B   C   D

Figure 8.9

Culture tube for use in pressure cylinders for prevention of hyperbaric oxygenation. A measured amount of oxygen is introduced into an oxygen-free solution prior to pressurization. A rubber stopper, acting as a piston, is forced into the tube by external hydrostatic pressure created by a hydraulic press. (Source: ZoBell and Hittle 1967. Reprinted by permission of National Research Council Canada.)

loss of *in situ* pressure is required in order to determine the existence of decompression-sensitive strains (Yayanos and Dietz 1983). Early reports about barophilic deep-sea microorganisms (ZoBell 1964) could not be readily confirmed by other laboratories. Barotolerant organisms were found, but not obligate barophiles, probably because deep-sea samples underwent decompression and extensive temperature changes during retrieval.

In 1968, the deep-sea research submersible Alvin of the Woods Hole Oceanographic Institution on Cape Cod, Massachusetts, accidentally sank prior to a mission in 1500-m-deep water. The crew escaped, but their lunches, including broth and meat sandwiches, sank with the submersible. When the Alvin was raised ten months later, the surprisingly unspoiled condition of the food called new attention to microbial metabolic rates under high hydrostatic pressure and prompted Holger Jannasch and associates to investigate the *in situ* activities of deep-sea microorganisms (Jannasch and Wirsen 1977; Jannasch 1979; Jannasch and Nelson 1984). Various microbial media, some containing radiolabeled carbon sources, were incubated *in situ* on the deep-sea floor and in refrigerated incubators at various pressures and temperatures in the laboratory. These tests concluded that low temperatures of 2°C–4°C only moderately limit the activity of deep-sea microorganisms, but that the combination of low temperature and high hydrostatic pressure imposes a severe limitation on microbial activity. Typically, microbial metabolism at deep-sea pressures was an order of magnitude or more lower than the activity of the same deep-sea organisms at 1 atm and identical temperature.

When it became possible to retrieve and incubate deep-sea samples without decompression, the majority of isolates recovered using an apparatus that prevented decompression showed highly barotolerant characteristics, while the remainder (four out of fifteen) exhibited barophilic characteristics (Jannasch et al. 1982). This limited evidence seemed to indicate that the deep-sea microbial community is only moderately barotolerant rather than predominantly barophilic, and its activity is severely constrained, though not halted, by the prevailing high hydrostatic pressures.

More extensive investigations based on about one hundred isolates, some of them retrieved from the extreme depth of 10,500 m, have changed this initial impression somewhat (Yayanos 1986). It appears now that bacteria retrieved from a depth greater than 6,000 m are predominately barophilic. Their pressure optimum for growth showed definite correlations with the *in situ* pressure of their collection depth. The deep-sea isolates were also psychrophilic, and their temperature and pressure optima for growth influenced each other in a complex manner. One isolate from about 3,600 m of depth was clearly barophilic at its *in situ* temperature of 4°C. At its habitat pressure of approximately 360 atm, an increase in temperature up to 12°C increased its growth rate, while the same temperature increase at 1 atm decreased the growth rate drastically. Other isolates showed similar patterns with some quantitative differences. When incubated at *in situ* pressures, most isolates grew better at tem-

peratures somewhat higher than the *in situ* ones. At deep-sea pressures and *in situ* temperature of 2°C in nutrient-rich media, generation times of seven to thirty-five hours were measured for barophilic marine bacteria. Under more nutrient-limited conditions, generation times increased up to 200 hours. While these growth rates are very slow, deep-sea bacteria clearly make significant contributions to biodegradation and to marine food webs.

Considerable difficulty has been experienced in trying to maintain barophilic bacteria in the laboratory. Some of the enrichment cultures from the deep sea lose their viability after two or three transfers to new media, although some barophiles have survived in sediment samples for several months when stored at *in situ* pressures and temperatures. Even at refrigeration temperatures, deep-sea barophiles die off much more rapidly at 1 atm than when compressed to *in situ* pressures. The die-off was accompanied by ultrastructural changes in cell morphology (Chastain and Yayanos 1991). Despite the technical difficulties, many barophilic bacteria have now been cultured. A. Yayanos and coworkers have isolated a clearly barophilic deep-sea *Spirillum* that grows at least fifteen times faster at pressures between 300 and 600 atm than at 1 atm (Yayanos et al. 1979). At least some of the barophilic bacteria thus far studied have survived decompression during recovery from the deep sea, so temporary decompression may not always cause irreversible damage.

An explanation for the distribution of more-or-less temperature- and pressure-adapted bacteria at intermediate depth zones of the sea might be found in accumulating data on the considerable particle flux from surface waters. Shallow-water bacteria are attached to sinking organic detritus and fecal pellets. Thus, at intermediate depth, bacteria in the sediments and water column may represent in part relatively recent arrivals from areas of low or moderate pressure.

In nonbarophilic microorganisms, high hydrostatic pressure appears to inhibit microbial synthesis of RNA, DNA, and protein, membrane transport functions, and rates of activity of various enzymes (Baross et al. 1972; Hardon and Albright 1974). Enzyme proteins, in order to be active, must be in proper tertiary configuration. Some researchers have suggested that high hydrostatic pressures distort this configuration, leading to low enzyme activity. In the case of a barotolerant organism, the distortion would be minimal. Most enzymes appear actually to be capable of normal functioning over a wide range of hydrostatic pressures, permitting enzymatic activities even in the deep sea. Theoretically, an obligately barophilic microorganism might have its enzyme in proper tertiary configuration only at elevated hydrostatic pressure and would not function well at lower pressures (Kim and ZoBell 1972) (Figure 8.10).

Osmotic Pressure   Whereas high hydrostatic pressure is restricted to special habitats, all microorganisms must cope with osmotic pressure, which results from differences in solute concentrations on opposite sides of a semipermeable membrane. In order to equilibrate and equalize solute concentrations, water molecules attempt to move across the membrane in the direction of the higher solute concentration. In some hypotonic habitats, water molecules attempt to move into microbial cells, which could expand and rupture the cells; in hypertonic habitats, water molecules attempt to move out of the cells, which could dehydrate and shrivel the cells.

Microorganisms have evolved adaptive mechanisms that permit them to tolerate osmotic pressure within certain ranges. The rigid cell walls of bacteria and other microorganisms protect these organisms from osmotic shock by preventing the pressure exerted by incoming water molecules from expanding and bursting the cells. The pressure buildup within the cell at equilibrium prevents the entry of additional water. The contractile vacuoles of some protozoa prevent osmotic swelling in hypotonic solution by pumping out excess incoming water.

The majority of microorganisms can withstand the low osmotic pressures that prevail in distilled water, rainwater, and freshwater habitats either by building up intracellular pressure or by actively pumping out excess water from the cell. Fewer microorganisms can withstand the high osmotic pressure of concentrated solutions. In traditional terminology, microorganisms that tolerate or prefer high concentra-

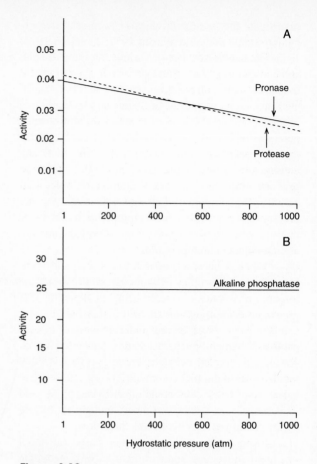

Figure 8.10

(A) Slight inhibition of pronase (from *Bacillus subtilis*) and protease (from *Streptomyces griseus*) activity by increased hydrostatic pressure. (B) Lack of inhibition of cell-free alkaline phosphatase activity by elevated hydrostatic pressure. (Source: Kim and ZoBell 1972. Reprinted by permission, copyright University Park Press.)

tions of organic solutes (usually sugars) are called osmotolerant or osmophilic, respectively (Jennings 1990). Honey, sap flows, flower nectar, molasses, and other sugar syrups provide habitats for osmotolerant and osmophilic microorganisms. These may cause spoilage of fruit preserves, syrups, and other food items normally stabilized by their high sugar concentration. *Debaromyces hansenii* and *Zygosaccharomy-*

ces rouxii are osmophilic yeasts; *Aspergillus* and *Penicillium* molds are osmotolerant.

The osmotic pressure of concentrated sugar solutions tends to dehydrate microbial cells. Osmotolerant and osmophilic microorganisms avoid this fate by building up balancing intracellular concentrations of "compatible solutes." These are low–molecular-weight organics such as glycerol, various sugars, glutamate, glycine betaine, and similar compounds. These balance the osmotic pressure and prevent water loss from the cell. The enzymes of osmotolerant and osmophilic microorganisms must be capable of functioning in the presence of these high-solute concentrations. The more adapted osmophiles pay a price for this capability. Their enzymes function suboptimally at low solute concentrations (Jennings 1990) (Figure 8.11). Osmotic concentrations tend to change. An osmotolerant or osmophilic microorganism with high intracellular solute concentration might burst when the osmotic pressure of the surrounding medium drops. These microorganisms, however, have the capability of sensing the changes in outside osmotic pressure and reacting by rapidly degrading or polymerizing their intracellular solutes to preserve their osmotic balance.

Figure 8.11

Effect of water activity and solute on the radial growth of the xerophilic fungus *Basipetospora halophila*. Optimum growth occurs in the range of a_w 0.85–0.90. High water activity inhibits growth. (Source Jennings 1990.)

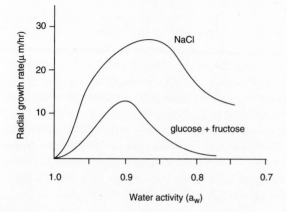

Salinity

Microorganisms that tolerate or require high salt concentrations are called halotolerant and halophilic, respectively (Gilmour 1990). At high salt concentrations, the hypertonic environment tends to dehydrate nonhalotolerant microorganisms.

In addition to affecting osmotic pressure, high salt concentrations tend to denature proteins; that is, they disrupt the tertiary protein structure, which is essential for enzymatic activity. High concentrations of salt dehydrate cells and denature enzymes. High-salinity habitats include salt lakes, which occur in arid regions where evaporation exceeds freshwater inflow or where a lake is fed by a salt spring. Partially landlocked marine lagoons and tidal evaporation flats can also develop high salt concentrations. Relatively few organisms can grow in highly saline waters, and often the biota of salt lakes are restricted to a few halophilic and halotolerant algal and bacterial species (Kushner 1968, 1980; Post 1977).

Halotolerant and halophilic microorganisms tend to exclude from their cell interiors the high and relatively toxic sodium ion concentrations that usually prevail in their environments. They achieve osmotic balance with their environment by mechanisms similar to those used by the osmophiles. The halophilic unicellular green alga *Dunaliella* that lacks a rigid cell wall builds up high intracellular glycerol concentrations for osmotic balance. The obligately halophilic *Halobacterium,* however, achieves osmotic balance with high intracellular concentrations of potassium chloride. The ribosomes of *Halobacterium* require high concentrations of potassium ions for stability. Its enzymes also require high salt concentrations for maintaining their active configuration and functions and are inactivated at low salinity (Larsen 1962, 1967; Gilmour 1990) (Figure 8.12). *Halobacterium,* a heterotrophic aerobic archaebacterium, lacks murein, and its cell wall appears to require sodium ions for stability. Some strains of *Halobacterium* have bacteriorhodopsin bilayer membrane components that serve as light-driven proton pumps. Light energy is used to pump protons out of the cell and thereby generate an elec-

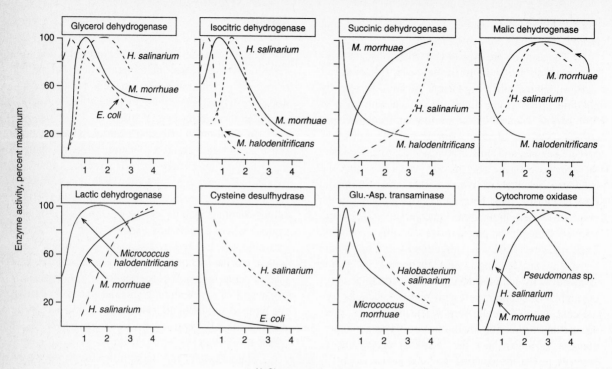

Figure 8.12

Effects of sodium chloride concentrations on various enzymes and nonhalophilic bac-teria. Some of the enzymes of halotolerant bacteria (such as succinic dehy-drogenase of *Halobacterium salinarium*) require high salt concentrations for maximal activity. Enzyme activities of nonhalotolerant organisms (such as glycerol dehydrogenase of *Escherichia coli*) decrease at high salt concentrations. (Source: Larsen 1962. Reprinted by permission, copyright Academic Press.)

trochemical potential. This, in turn, drives ATP syn-thesis. The described features make *Halobacterium* superbly adapted to live in highly saline, often satu-rated brine environments. *Halobacterium* is unable to tolerate low salinities and requires at least 3.0 M NaCl for growth.

Halotolerance is a characteristic of the enteric bacteria *Staphylococcus* and *Halomonas*. Some green algae (*Chlorella, Dunaliella*), some diatoms, and other unicellular algae are also halotolerant. Most algae and bacteria from the marine environments are moderately halophilic. Some marine microbiologists actually define marine bacteria as bacteria that require 3% NaCl for growth. True marine bacteria appear to require sodium for all membrane transport systems, whereas sodium-dependent transport in nonmarine bacteria has so far been shown to occur only in case of specific and nonessential metabolites (MacLeod 1985). Extreme halophiles are *Halobacterium, Halo-coccus, Natronobacter,* the anoxyphototrophic bacte-rium *Ectothiorhodospira,* and the green algae *Dunaliella* (Orem 1988; Gilmour 1990).

Water Activity

Liquid water is essential for all biochemical processes. For microorganisms, the critical factor is the available liquid water rather than the total amount of water present in the environment. The amount of water actually available for microbial use can be expressed as the water activity (a_w) (Brown 1976; Reid 1980; Griffin 1981). Water availability is also expressed sometimes as water potential (Ψ). Water potential is a thermodynamic term and is the chemical potential of the water in the system divided by the partial molar volume. Water potential is related to water activity according to Equation (1).

$$(1) \quad \Psi = RT \log_e a_w/V$$

where R = the gas constant, T = absolute temperature (°K) and V = partial molar volume. Water potential is measured in negative mega-pascals ($-MP_a$) or bars (bar = 0.987 atm), expressing the suction pressure necessary to withdraw water from the system. The water potential of pure water is 0. Water potential is frequently used in scientific literature in relation to soil moisture status. Crop plants are unable to withdraw water from soil when its water potential sinks to approximately -15 MP_a or -15 bar; therefore, this is their permanent wilting point. This corresponds to approximately 0.90 water activity (Jennings 1990). To avoid confusion, we will consistently use the simpler water activity (a_w) term in the rest of this text. Soil moisture status will be described in terms of water-holding capacity (WHC) percentage.

Water activity, the index of the water actually available for microbial use, depends on the number of moles of water, the number of moles of solute, and the activity coefficients for water and the particular solute. Water activity can be decreased not only by solutes (osmotic forces) but also by absorption to solid surfaces (matric forces). By definition, a_w of free distilled water is 1.0. Osmotic and matric forces usually lower a_w to some fraction of this value. Most microorganisms require a_w values above 0.96 for active metabolism, but some filamentous fungi and lichens are capable of growth at a_w values as low as 0.60 (Table 8.4). Such microorganisms are described as xerotolerant.

Table 8.4

Approximate water activities required for the growth of various microorganisms

Water activity (a_w)	Bacteria	Fungi
1.00	Caulobacter Spirillum	
0.90	Lactobacillus Bacillus	Fusarium Mucor
0.85	Staphylococcus	Debaromyces
0.80		Penicillium
0.75	Halobacterium	Aspergillus
		Saccharomyces rouxii
0.60		Xeromyces bisporus

Water availability in soils is traditionally measured in relation to WHC of the soil. A measured amount of dry soil is allowed to soak up water. After excess water is allowed to drain in a humid chamber, the weight increase of the soil sample is recorded. This value is 100% of the WHC. Optimal conditions for activity of aerobic soil microorganisms occur between 50% and 70% WHC, corresponding to 0.98–0.99 a_w. A higher water content, although not inhibitory by itself, starts to interfere with oxygen availability. The actual water content of a soil can be determined by measuring the wet and dry weights of the soil sample. A sample is weighed and then placed in an oven, usually for 12–14 hours at 105°C, after which the weight is again measured. The difference in weight between the initial (wet weight) and final (dry weight) readings represents the loss of water. The water content of a soil is most meaningfully expressed as a percentage of its total WHC.

In field situations, it is sometimes more convenient to determine water content on a volume basis. This is done by measuring the bulk density (Black 1965). A known volume of soil is collected, the soil dried, and the weight determined. Soil with a high water content will have a low weight per unit volume following drying, assuming that the soil pores were filled with water and not gases.

Microbial activities in the soil fluctuate with the water content. During periods of drought, microbial activities may be severely limited. The water content of soils affects microbial activity not only directly, but also indirectly by affecting the diffusion of gases; in waterlogged soils, diffusion of air is limited, generally resulting in anaerobic conditions that may result in a shift from aerobic to anaerobic metabolism.

The water content in the atmosphere is generally low and limiting for microbial growth (Gregory 1973). In some regions, such as in tropical areas, the humidity in the air is sufficiently high to allow microbial growth. In such humid areas, microorganisms may grow on exposed surfaces using water from the air. Lichens tolerate desiccation and grow receiving only intermittent precipitation. A number of fungi and lichens grow on exposed plant surfaces. Various fungi grow on canvas and leather in humid regions. In addition to influencing survival in the atmosphere, water as rainfall causes a flux of microorganisms into and out of the atmosphere. Rainfall washes some organisms out of the atmosphere; raindrops trigger the release to the atmosphere of others (Figure 8.13) (Hirst et al. 1955).

Desert soils, by definition, receive less than 25 cm of rainfall per year. Many deserts are extremely dry and are subject to extreme diurnal variations in temperature because of the lack of moisture in the atmosphere overlying them to moderate the loss of heat. Microorganisms living in dry desert soils must be able to tolerate long periods of desiccation (Skujins 1984).

In the dry valleys of Antarctica, microorganisms must also tolerate very low temperatures and, during part of the year, high irradiation levels (Horowitz et al. 1972). In such environments, many microorganisms have developed adaptations that allow them to survive in a dormant state during unfavorable conditions and to grow actively only during those brief periods of time when conditions are favorable, such as after a rainstorm. Many of the bacteria and fungi living in desert soils form spores that allow them to persist, if necessary, for decades between growth periods. When there is adequate moisture, the spores germinate, and for a brief period the organisms can actively grow and reproduce.

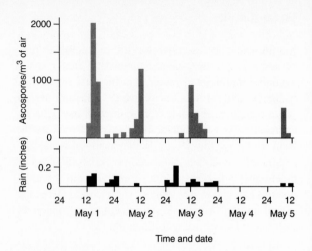

Figure 8.13

Concentration of ascospores of *Venturia inaequalis* in air related to rainfall. High numbers of ascospores were found shortly following the rainfall. (Source: Hirst et al. 1955. Reprinted by permission of Administration A—Fish, Food, and Plant Pathology—of England and the Controller of Her Britannic Majesty's Stationery Office.)

Rock-inhabiting (lithobiontic) microorganisms colonize rock surfaces, or the rock fabric adjacent to the surface, in a wide variety of climates and environments. Conspicuous among microorganisms on rock surfaces are lichens, which often form brightly colored crusts. These organisms are particularly resistant to desiccation. Other microorganisms colonize cracks and fissures within rocks that provide some protection against desiccation and irradiation. These are called chasmendolithic microorganisms.

Some microorganisms, called cryptoendolithic, also grow within the rock matrix (Friedmann and Ocampo 1976; Friedmann 1982; Friedmann and Ocampo-Friedmann 1984; Wynn-Williams 1990). The morphologies of endolithic colonization in cold and hot deserts are remarkably similar. The uppermost 1 to 3 mm of rock are free from microorganisms. Colonies are present in the next few millimeters, where the microorganisms adhere to or grow between rock crystals. Endolithic microorganisms in hot deserts and in Antarctica share a common survival strategy. They are capable of switching their metabolic activities on and off in response to rapid changes in environmental conditions.

There are, however, major differences between the endolithic microorganisms of hot and cold deserts (Friedmann and Ocampo-Friedmann 1984). In hot deserts, all endolithic samples examined to date contained only prokaryotic primary producers, which form a narrow green zone a few millimeters below and parallel to the rock crust. They are cyanobacteria, nearly always members of the genus *Chroococcidiopsis*. In hot deserts, a favorable combination of available water and moderate temperature range occurs in early morning after dewfall. Later in the day, the temperature rises and the relative humidity drops drastically. The combination of temperature increase in a hydrated state and the concomitant loss of water imposes a severe environmental stress that eukaryotic algae do not seem to tolerate. This is probably the reason that, in hot deserts, endolithic primary producers are exclusively cyanobacteria.

The endolithic microbial communities in polar deserts show a higher level of complexity than those in hot deserts. In most cases, polar endolithic communities contain a number of organisms, including eukaryotes, that are dominant. The fungi and algae together form a symbiotic lichen association in which the fungi and algae form segregated bands within the rock matrix that are very different from the usual integrated thalloid lichen organization. The coherent, protective pseudotissue, termed plectenchyma, is missing. Instead, loose filaments and cell clusters grow between rock crystals. In endolithic lichens, the rock matrix prevents the development of and eliminates the need for a coherent plectenchyma. The activity of the cryptoendolithic lichen community results in the mobilization of iron compounds in the rock as well as in a characteristic exfoliative weathering pattern.

Due to a partial dissolution of the rock in the microbially colonized layer, the surface layer peels off, exposing and causing the destruction of the microorganisms. The cryptoendolithic community reestablishes itself below the rock surface, and the process repeats itself with a gradual erosion of the rock over geologic time periods. The growth rate of the Antarctic cryptoendolithic microbial communities appears to be so slow that it strains the imagination. Recent measurements of photosynthetic radiocarbon incorporation at *in situ* temperatures divided by the existing microbial biomass indicated turnover times around 20,000 years (Johnston and Vestal 1991). Even allowing for methodological difficulties and uncertain assumptions, there is little doubt that microbial generation times in such cryptoendolithic microbial communities are in the order of magnitude of hundreds and perhaps even thousands of years.

Movement

The movement of air and water aids the passive dispersal of microorganisms. Of equal importance is the role of movement in importing and distributing nutrients for microbial growth and in removing metabolic by-products. Factors such as solubility, diffusion, viscosity, specific gravity, porosity, and the flow characteristics of the ecosystem control the movement of materials in and out of ecosystems.

Some ecosystems are characterized by extensive flow, such as in rivers, or by turbulence, such as in oceans. Other ecosystems are generally quiescent, such as ponds, or static, such as soil. Ecosystems with extensive flow or turbulence have greater mixing capacity. Materials are moved into and out of such ecosystems in part by the movement or mixing of the ecosystem. Even in quiescent and static systems, thermal convection, evapotranspiration, and leaching move materials.

Diffusion   In ecosystems that lack extensive mixing, materials may move by diffusion (Koch 1990). Diffusion results in a spreading out of a substance from its source, lowering concentration. Diffusion of materials through ecosystems depends in part on temperature. The molecular weight of the solute and the viscosity (flow characteristics) of the solvent determine in part the ability of a solution to diffuse through an ecosystem. Diffusion is often augmented by mixing due to differences in specific gravity and by thermal convection. Gaseous molecules, with low specific gravities, for example, tend to rise, generally moving upward through aquatic ecosystems. In the atmosphere, diffusion of gases is augmented by convection along temperature gradients, with warmer gases rising and cooler gases descending through the atmosphere. The movement of gases along thermal gradients results in turbulence and mixing.

In terrestrial ecosystems, diffusion of materials is a function of porosity. Porosity refers to the number and volume of pores in a soil or sediment particle matrix. The pores, sometimes referred to as interstitial spaces, may be filled with liquid or gases. Diffusion of materials occurs through the pores, and exchange rates between material in the interstitial spaces and external sources affect diffusion rates and the availability of materials essential for microbial growth and activity.

Adsorption Surface adsorption and tension may restrict movement of materials through an ecosystem. Surface tension may restrict movement of materials into an ecosystem—for example, from the atmosphere into an aquatic ecosystem—resulting in the accumulation of nutrients and pollutants. Adsorption may bind materials to particles, decreasing the availability of those materials within the ecosystem. If the substance is an essential nutrient, this can decrease productivity, but if the substance is toxic, adsorption can increase microbial growth. Because of adsorption, even excessive levels of microbial inhibitors fail to suppress completely microbial activity in soil.

In environments with very dilute nutrient concentrations, the local enrichment of nutrients on adsorptive surfaces increases their microbial utilization. This phenomenon appears to aid the survival of copiotrophic microorganisms in oligotrophic environments (Kjelleberg 1984). The concentration of organic and inorganic nutrients by adsorption onto particles accounts in part for the preferential growth of many microorganisms on surfaces. The adsorptive properties of an ecosystem depend on available particle surface areas, charged binding sites, hydrophilic binding sites, hydrophobic binding sites, and other factors such as pH. In some cases, adsorption is essen-

tial for microbial reproduction. For example, many microorganisms only grow or reproduce on surfaces such as leaves, skin, teeth, and so on. Some bacterial viruses must adsorb onto specific receptor sites located on pili before penetrating into bacterial cells, where they reproduce.

The mechanisms of adherence of microorganisms to a variety of biotic and abiotic surfaces have been intensively examined (Marshall 1976; Fletcher and Marshall 1982; Costerton 1984; Dazzo 1984; Fletcher and McEldowney 1984; Rosenberg and Kjelleberg 1986). Bacteria growing in a wide variety of environmental and pathogenic situations have been shown to be surrounded by extracellular glycocalyx structures that mediate the attachment process (Costerton 1984); in some cases, proteins known as lectins mediate the specific attachment processes (Dazzo 1984). Charge interactions also bring about adsorption. Adhesion holds the bacterial cell in a suitable habitat, positioning the cell within the surface zone, which is rich in nutrients, because organic molecules are attracted to these interfaces. Development of a fibrous, anionic, exopolysaccharide matrix (glycocalyx) acts as an ion-exchange resin, attracting and concentrating charged nutrients. With cell growth, this matrix burgeons into a thick, coherent biofilm. This sessile biofilm mode of growth may protect adherent cells from surfactants, bacteriophages, phagocytic amoebae, chemical biocides, antiseptics, antibiotics, disinfectants, antibodies, and phagocytic leukocytes.

Hydrogen Ion Concentration

Microorganisms generally cannot tolerate extreme pH values. Under highly alkaline or acidic conditions, some microbial cell components may be hydrolyzed or enzymes may be denatured. There are, however, some acidophilic and alkalophilic bacteria that tolerate or even require extreme pH conditions for growth. Table 8.5 shows the optimal pH values and the tolerance of ranges of pH for various bacteria.

Amino acids are zwitterions, having both basic and acidic portions. The pH affects the dissociation of

Table 8.5

Minimum, optimum, and maximum pH in multiplication of various bacteri

Organism	pH		
	Minimum	Optimum	Maximum
Escherichia coli	4.4	6.0–7.0	9.0
Proteus vulgaris	4.4	6.0–7.0	8.4
Enterobacter aerogenes	4.4	6.0–7.0	9.0
Pseudomonas aeruginosa	5.6	6.6–7.0	8.0
Erwinia carotovora	5.6	7.1	9.3
Clostridium sporogenes	5.5–5.8	6.0–7.6	8.5–9.0
Nitrosomonas spp.	7.0–7.6	8.0–8.8	9.4
Nitrobacter spp.	6.6	7.6–8.6	10.0
Thiobacillus thiooxidans	1.0	2.0–2.8	6.0
Lactobacillus acidophilus	4.0	4.6–5.8	6.8
Bacillus acidocaldarius	2.0	3.5	6.0
Thermoplasma acidophilus	1.0	1.5	4.0
Sulfolobus acidocaldarius	1.0	2.5	4.0
Bacillus alcalophilus	8.5	9.5	11.5

Source: Doetch and Cook 1973. Added data from Ingledew 1990 and Kroll 1990.

these functional groups on protein molecules; in order to perform enzymatic activity, enzymes must be in a particular state of dissociation. Certain pH values will be optimal for activities of specific enzymes. Optimal pH values may depend on other factors, such as salt concentrations (Lieberman and Lanyi 1972) (Figure 8.14). Extreme pH values irreversibly denature most proteins.

The pH of an environment affects microorganisms and microbial enzymes directly and also influences the dissociation and solubility of many molecules that indirectly influence microorganisms. The pH determines in part the solubility of CO_2, influencing the rates of photosynthesis; the availability of required nutrients, such as ammonium and phosphate, which limit microbial growth rates in many ecosystems; and the mobility of heavy metals, such as copper, which are toxic to microorganisms.

A common feature of microorganisms that tolerate or even require pH extremes for growth is that their cytoplasm is maintained close to neutrality.

Their cell walls and cell membranes need to be adapted to keep their integrity under the pH extremes, maintain the cell interior near neutrality, and perform chemiosmotic ATP synthesis under these unusual conditions. The precise structural and biochemical adaptations remain insufficiently known at this time. W. J. Ingledew (1990) and R. G. Kroll (1990) discuss the biophysical background to such adaptations for acidophiles and for alkalophiles, respectively.

Many fungi are acidotolerant, while most bacteria are not. Some acidotolerant bacteria, like *Lactobacillus,* and acidophiles, like *Thiobacillus* and *Sulfolobus* (Table 8.5), create their own low pH environment by producing acids. *Lactobacillus* is a mixed acid fermenter and *Sulfolobus* produces sulfuric acid. *Bacillus acidocaldarius* and *Thermoplasma acidophilus* are heterotrophic thermoacidophiles that live in acidic environments created by chemolithotrophic sulfur oxidizers but do not produce acids themselves. The *Mycoplasma*-like *Thermoplasma* lacks a cell wall

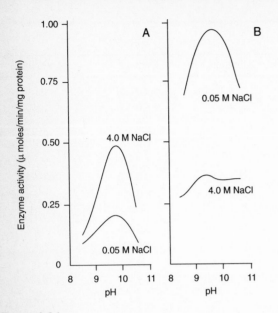

Figure 8.14

The pH dependence of threonine deaminase from *Halobacterium cutirubrum* assayed at 0.05 and 4.0 M NaCl concentrations in (A) the absence of ADP and (B) presence of ADP. In the absence of ADP, an increase of the NaCl concentration caused a narrowing of the pH optimum; in the presence of ADP, when the cooperative substrate kinetics were no longer observed, the pH response of the enzyme broadened with increased salt concentration. (Source: Lieberman and Lanyi 1972. Reprinted from *Biochemistry* by permission of American Chemical Society.)

and occurs in hot, acidic coal refuse piles. *B. acidocaldarius* occurs in acidic hot springs.

Lactobacillus has great practical importance in preparation of silage and fermented foods. *Thiobacillus* and *Sulfolobus* are used in bioleaching of low-grade copper and uranium ores (see chapter 14).

Many bacteria and fungi tolerate alkaline pH up to 9.0 (Table 8.5) but have pH optima near neutrality. True alkalophiles are some *Bacillus* strains such as *B. alcalophilus* and *B. pasteurii.* Alkalophilic are also some cyanobacteria such as *Microcystis aeruginosa, Plectonema nostocorum,* and several species of *Spirulina.* Microorganisms listed earlier as halophilic, such as *Halobacterium, Natronobacterium,* and

Natronococcus, are also alkalophiles and live in saline lakes with high pH. *Bacillus* strains produce proteases and lipases that are stable at high temperature, at alkaline pH, and in the presence of detergents. These are used in some laundry detergent products to enzymatically remove fat and protein-based stains (Kroll 1990).

Redox Potential

Many enzymatic reactions are oxidation-reduction reactions in which one compound is oxidized and another compound is reduced. The ability of an organism to carry out oxidation-reduction reactions depends on the oxidation-reduction state of the environment. In a solution, the proportion of oxidized to reduced components constitutes the oxidation-reduction potential or redox potential (E_h). Some microorganisms can be active only in oxidizing environments, while others can exist only in reducing environments (Baas-Becking and Wood 1955; Rheinheimer 1974) (Figure 8.15).

The redox potential is a relative value measured against the arbitrary 0 point of the normal hydrogen electrode. Any system or environment that accepts electrons from a normal hydrogen electrode is a half-cell that is defined as having a positive redox potential; any environment that donates electrons to this half-cell is defined as having a negative redox potential. The redox potential is measured in millivolts (mV). A high positive E_h value indicates an environment that favors oxidation reactions; a low negative E_h value indicates a strongly reducing environment. Strictly aerobic microorganisms can be metabolically active only at positive redox potentials, while strict anaerobes are active only at negative redox potentials. Facultative anaerobes can operate over a wide range of E_h values using oxygen as an electron sink at high redox potentials and fermenting or using alternate electron acceptors, such as ferric iron and nitrate ions, at low redox potentials. The solubility of certain nutrients is strongly influenced by the prevailing redox potential. At high redox potentials, iron and manganese exist in their trivalent (ferric) and tetravalent forms, respectively. These are insoluble and generally

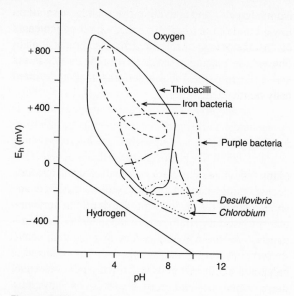

Figure 8.15

The pH and E_h tolerance contours of various bacteria involved in several cycling processes. The thiobacilli and iron bacteria are extremely tolerant to highly acidic and aerobic conditions, which are characteristic of their natural habitat. *Chlorobium* and *Desulfovibrio* are restricted to anaerobic and neutral to slightly alkaline pH zones. Note the interaction of pH and E_h as signified by the sloping hydrogen and oxygen potentials. (Source: Rheinheimer 1974. Based on Baas-Becking and Wood 1955. Reprinted by permission of John Wiley and Sons Ltd. and VEB Gustav Fischer Verlag.)

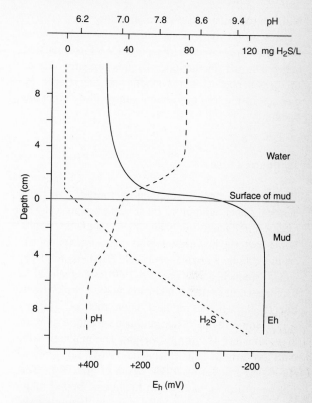

Figure 8.16

A sulfur-rich lakeshore habitat showing reduced pH and E_h below the surface of mud. H_2S concentrations were low in water with high E_h (aerobic zone) and high in sediment with low E_h (anaerobic zone). (Source: Rheinheimer 1974. Reprinted by permission John Wiley and Sons Ltd. and VEB Gustav Fischer Verlag.)

unavailable for microbial uptake. The bivalent forms of these metals that exist at low redox potentials are soluble and readily available.

The redox potential is greatly influenced by the presence or absence of molecular oxygen. Environments in equilibrium with atmospheric oxygen have E_h around +800 mV; environments at reduced oxygen tensions have lower redox potentials. Heterotrophic activity keeps the redox potential of aerobic natural waters at 400–500 mV, lower than might be expected for water fully equilibrated with the atmosphere. Below the surface of waterlogged soil and sediment, E_h values are usually negative (Figure 8.16). Low redox potentials may be caused by extensive growth

of heterotrophic microorganisms that scavenge all available oxygen. Such is often the case in highly polluted ecosystems where microorganisms utilize the available oxygen for decomposition processes. Sediments rich in organic matter can have E_h values as low as –450 mV. At these low E_h values, sulfate reduction yielding H_2S and CO_2 reduction yielding CH_4 can readily occur. It should be pointed out that it is difficult to measure E_h in natural habitats, and that heterogeneity is great; the development and use of microelectrodes, however, now makes it possible to measure redox potentials on a scale relevant to microbial microhabitats (Revsbech and Jørgensen 1986). Because of the dynamics at interfaces, one can have a low redox potential in the presence of oxygenated water. Diurnal fluctuations of redox potential due to the prevalence of photosynthetic oxygen release by day and respiratory oxygen consumption by night are common.

Magnetic Force

The effects of magnetism upon microorganisms have generally been ignored as most bacteria appear to be totally unaffected by changes in magnetic fields. Some bacteria, however, exhibit magnetotaxis, motility directed by a geomagnetic field (Blakemore 1975, 1982; Moench and Konetzka 1978). Magnetotactic bacteria contain dense inclusion bodies, which may impart a magnetic moment upon the cell (Figure 8.17). Some of these magnetotactic bacteria contain approximately 4% iron by weight. The magnetic iron is in the form of either magnetite (Fe_3O_4) or greigite (Fe_3S_4) (Farina et al. 1990; Mann et al. 1990). The magnetic iron is enclosed in membrane-bound magnetosomes. These allow cells to orient themselves in

magnetic fields and actively move by flagellar motion toward one of the magnetic poles. Initial enrichments of magnetotactic bacteria from sediments were obtained by placing magnets next to the culture flasks and collecting the microorganisms that aggregated near the magnetic poles.

The adaptive value of magnetotactic behavior is as yet unclear. Because the magnetotactic microorganisms that have been described are microaerophilic or anaerobic, it has been suggested that magnetotactic orientation may help these motile organisms to locate deeper, more reduced organic sediments. However, since some of the bacteria move toward the north and others toward the south pole of the magnets, this argument is not very convincing. The investigation of this and other aspects of magnetotactic behavior has been hampered by the difficulty in growing pure culture of these organisms, and most observations have been made only on mixed enrichments.

Experimentation on a pure culture of magnetotactic bacterium *Aquaspirillum magnetotacticum* (Ricci et al. 1991) revealed that even in a pure culture some bacteria are attracted to the north and others to the south magnetic pole. Pulse magnetization can alter and even reverse the orientation of *A. magnetotacticum* cells. However, the ecological value of magnetic orientation for microorganisms, if any, remains to be determined.

Organic Compounds

Organic matter in an ecosystem is frequently the limiting factor for growth of heterotrophic microorganisms. Specific organic substrates often favor the growth of particular populations with specific catabolic activities. Some microorganisms require relatively high levels of organic matter, whereas others grow only at relatively low concentrations.

In aquatic ecosystems, the chemical oxygen demand (COD) is often used as an index of the total organic carbon in the sample. The COD is determined by measuring the amount of oxidizing reagent consumed during oxidation of organic matter with dichromate or permanganate. The biological oxygen

Figure 8.17

Electron micrographs of magnetotactic bacteria showing accumulation and regular arrangement of iron-rich inclusions. (Courtesy of R. Blakemore.)

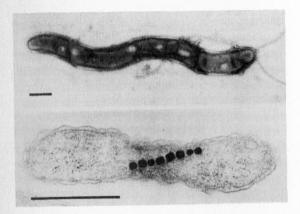

demand (BOD) yields an estimate of the readily usable concentrations of organic matter. The BOD is determined by allowing microorganisms to consume oxygen in solution during oxidation of the organic matter present in the sample. A high BOD indicates the abundance of organic substrates that can be used by microorganisms aerobically. When the BOD is high, however, microorganisms often use much of the available oxygen for the degradation of organic matter, creating conditions of oxygen depletion and resulting in the death of higher organisms, such as fish, that require O_2 for survival.

Most natural ecosystems are not characterized by high concentrations of usable organic matter; rather, microorganisms in many ecosystems, such as the oceans, live at very low concentrations of nutrients. Most free-living bacteria probably experience periods of starvation. Starvation survival is important in microbial ecology because it is a mechanism that perpetuates the species, permitting the genome to express itself once the environmental conditions become favorable (Kjelleberg 1984).

There is growing evidence of the existence of bacteria especially suited for the exploitation of low-nutrient flux habitats that can be distinguished from organisms especially suited for high-nutrient flux environments (Poindexter 1981; Button 1985; Fry 1990). As opposed to copiotrophs (bacteria growing on high-nutrient concentrations), whose natural distribution implies that nutrient abundance favors their survival and is an important factor in their competitiveness, oligotrophic (low-nutrient) bacteria possess physiological properties that permit them to efficiently use the limited nutrient resources available to them. Substrate uptake characteristics of oligotrophs must be suitable for acquisition of growth substrates against steep gradients between the cell and its surrounding, and acquired nutrients must be so managed that the cell can maintain its integrity and growth. To achieve a high uptake capacity relative to nutrient-consuming processes, oligotrophs ideally should be small spheres, slender rods, or envelope-appendaged bacteria that have high surface areas and high affinities (low K_m) for uptake for a variety of utilizable substrates. Indeed, many oligotrophic microorganisms,

such as *Caulobacter,* form appendages and are more efficient at transporting substrates into their cells than at excreting by-products.

Adaptations required for oligotrophs not only make them poor competitors at high substrate concentrations, but make high substrate concentrations outright inhibitory. Numerous obligate oligotrophs have been isolated that cease to grow when substrate concentrations reach 500–1,000 mg per liter (0.5%–1.0%). Optimal growth occurs around 50–100 mg/L substrate concentration. In contrast, copiotrophs grow optimally around substrate concentrations of 50,000 mg/L (Fry 1990).

Organic compounds can be potential nutrients, but some of them act as inhibitors or poisons. Various metabolic products, such as carboxylic acids, alcohols, and phenolics, are toxic to microbial populations, especially when they accumulate in some anaerobic environments where they are not readily metabolized to less toxic degradation products. Some microorganisms specifically excrete allelopathic substances (allelochemics), which are chemicals (other than food substances) produced by one species that affect growth, health, behavior, or population dynamics of other species (Whittaker and Feeny 1971); production of antibiotics and bacteriocins are prime examples of allelochemics.

Inorganic Compounds

Many inorganic compounds are essential nutrients for microorganisms (Stotzky and Norman 1964). There are also many inorganic substances that are toxic to microorganisms (Ehrlich 1978). It is generally necessary to consider the chemical form of inorganic compounds when considering such compounds as microbial nutrients or inhibitors. Inorganic compounds of importance in examining the ecology of microorganisms include the gases oxygen, carbon dioxide, carbon monoxide, hydrogen, nitrogen, nitrous oxide, and hydrogen sulfide; the elemental form of sulfur; the cations ammonium, ferrous iron, ferric iron, magnesium, calcium, silicon, sodium, potassium, boron, cobalt, copper, manganese, molybdenum, vanadium, nickel, zinc, mercury, cadmium, and lead; and the anions phosphate, carbonate, bicarbonate, sulfide, sulfate, nitrite, nitrate, chloride, chlorate, bromide, fluoride, silicate, selenite, and arsenate.

Of the inorganic ions, the heavy metals deserve special attention. While some heavy metals are required in trace amounts as nutrients, they become strongly inhibitory for microorganisms at relatively low concentrations (Gadd 1990). A general mechanism of heavy metal toxicity is their binding to the sulfhydryl (—SH) groups of essential microbial proteins and enzymes. Important in terms of heavy metal toxicity is the pH and the organic matter concentration of the environment. Low pH mobilizes heavy metals, while high pH precipitates them and reduces their toxicity. Organic matter tends to bind and chelate heavy metals, also reducing their toxicity.

Seawater is a large natural environment with substantial heavy metal concentrations and low levels of organic matter. Marine microorganisms are under considerable heavy metal stress. Industrial pollution can result in locally high levels of lead, tin, cadmium, mercury, chromium, copper, zinc, nickel, and so on. Microorganisms have evolved a number of defense mechanisms against heavy metal toxicity. These include extracellular precipitation or complexing, impermeability to or reduced transport of the metals across the cell membrane, and intracellular compartmentalization and detoxification. One or more of these defense mechanisms allows some microorganisms to function metabolically in environments highly polluted by heavy metals. A widespread response by microorganisms to heavy metal stress is the synthesis of special metal-binding compounds, the metallothioneins. These are low–molecular-weight cystine-rich proteins that bind and render harmless heavy metals like zinc, cadmium, and copper.

Oxygen and Its Reactive Forms   Oxygen exists in several forms with varying degrees of toxicity (Fridovich 1977). Molecular oxygen normally exists as O_2 in the atmosphere. Singlet oxygen (O^*) has a higher energy and is more toxic to microorganisms. It can be formed both biochemically and chemically. Peroxidase enzyme systems can form singlet oxygen

from hydrogen peroxide. The reactivity of singlet oxygen is a key factor in some biological interactions, such as the ability of blood phagocytes to kill invading microorganisms. Singlet oxygen also forms in the upper atmosphere photochemically and reacts with normal molecular oxygen to produce ozone. Ozone (O_3) is a strong oxidizing agent that kills microorganisms. Ozone is sometimes used as a sterilizing agent—for example, in some water treatment operations.

Another highly reactive oxygen species is the superoxide anion. Superoxide anions are generated by reactions of oxygen with many biochemicals such as flavins, quinones, and thiols. The superoxide anion can destroy lipids and other biochemical components of the cell. It is relatively long-lived and may explain the oxygen sensitivity of some obligate anaerobes. Various obligate anaerobes are not killed by O_2 directly, but are prevented from growing by the presence of oxygen, apparently because of the production and lack of effective removal of superoxide anions (Morris 1975). The superoxide anion can be converted to hydrogen peroxide plus O_2 by the action of the enzyme superoxide dismutase. Peroxide anions, as in hydrogen peroxide, are also quite toxic to microorganisms. Hydrogen peroxide can be decomposed by catalase enzymes to $H_2O + O_2$ or by peroxidase enzymes to H_2O with the coupled oxidation of a broad range of electron donors.

Carbon Dioxide and Monoxide   Carbon dioxide is required for both autotrophic and heterotrophic metabolism. It may be available in solution as carbonate or bicarbonate. Carbon dioxide consumption is often used as a measure of primary productivity. Production of CO_2 is a useful measure of the mineralization of organic matter.

Carbon monoxide is another form of inorganic carbon; it strongly inhibits microbial respiration pathways. Carbon monoxide binds with iron-containing proteins and molecules such as cytochromes, blocking electron transport. Some microorganisms can metabolize CO, but most are inhibited by it.

Nitrogen   For most microorganisms, nitrogen gas is an inert form of nitrogen. Only a restricted

Figure 8.18

Limiting effects of nitrogen and phosphorus on the biodegradation of petroleum in seawater. Solid portion of histogram indicates mineralization of oil measured by CO_2 evolution. Total height of histogram indicates the biodegradation of oil as measured by weight loss. Height of solid plus cross-hatched portion indicates biodegradation measured by gas chromatography. Natural levels of available N and P in seawater limit extent of biodegradation of oil spills. (Source: Atlas and Bartha 1972. Reprinted from *Biotechnology and Bioengineering* by permission of John Wiley and Sons.)

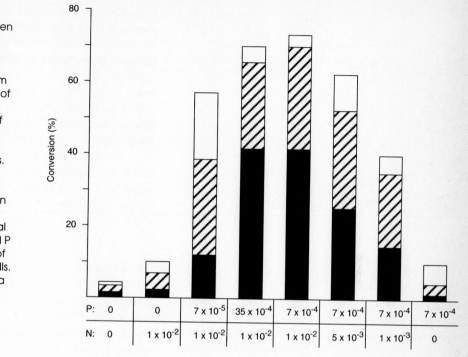

| P: | 0 | 0 | 7×10^{-5} | 35×10^{-4} | 7×10^{-4} | 7×10^{-4} | 7×10^{-4} | 7×10^{-4} |
| N: | 0 | 1×10^{-2} | 1×10^{-2} | 1×10^{-2} | 1×10^{-2} | 5×10^{-3} | 1×10^{-3} | 0 |

Concentration (M)

group of nitrogen-fixing bacteria are able to convert gaseous nitrogen into combined forms of nitrogen. The most important inorganic nitrogen-containing nutrients are ammonium, nitrate, and nitrite ions. Nitrogen is required for protein formation. In addition to being required for protein synthesis, ammonium or nitrite are oxidized by some chemolithotrophic microorganisms for generating energy. Some microorganisms use nitrate as the terminal electron acceptor in respiration pathways, resulting in denitrification, which returns nitrogen gas to the atmosphere. The fact that ammonium is a cation and nitrate and nitrite are anions results in differential leaching of these latter nitrogen-containing ions from environments such as soil.

Phosphorus Phosphate is required for microbial generation of ATP and for nucleic acid and membrane phospholipid synthesis. High concentrations of phosphate, however, may inhibit microbial growth. Concentrations of available nitrogen and phosphorus often limit both productivity and decomposition in aquatic habitats. Additions of available nitrogen and phosphorus are often used to increase productivity in aquaculture practices. The elimination of the phosphate limitation is sometimes undesirable, as it can lead to prolific algal or cyanobacterial growth. This growth has occurred following addition of phosphates, in the form of household detergents, to lakes where productivity was phosphate limited. On the other hand, controlled addition of nitrogen and phosphorus can enhance decomposition of polluting organic compounds. For example, the extent of petroleum biodegradation in seawater has been found to be limited by both nitrogen and phosphorus (Figure 8.18); this limitation can be overcome by application of oleophilic nitrogen and phosphorus fertilizer (Atlas and Bartha 1972; Atlas 1977).

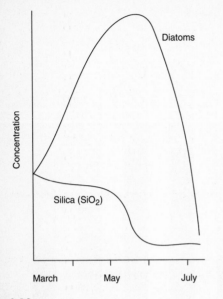

Figure 8.19

A graph illustrating that the amount of silica available in a lake can be the limiting factor in the growth of diatoms. Note silica depletion following diatom bloom and subsequent crash in the diatom population. Concentrations of other nutrients were measured, but none showed the type of correlation shown in this graph. (Source: Jensen and Salisbury 1972. Reprinted by permission, copyright Wadsworth Publishing Co., Belmont, Calif.)

Sulfur Microorganisms require inorganic sulfur for formation of proteins and other molecular components. Some oxidized forms of sulfur, such as sulfur dioxide, are toxic to microorganisms and may even be used as disinfecting agents. Some photoautotrophic bacteria use hydrogen sulfide as an electron donor. Hydrogen sulfide is also toxic to many enzyme systems that involve heavy metal atoms. Numerous microorganisms produce hydrogen sulfide as a metabolic product of protein metabolism.

Other Elements Iron is required in cytochromes and other iron proteins. The oxidation state of iron depends on the pH and the redox potential. Ferrous and ferric ions demonstrate different degrees of solubility, which affects their availability for biological uptake.

Chloride ions are essential for maintenance of membrane function. Chlorate and hypochlorite ions are strong oxidizing agents and are toxic to microorganisms. Hypochlorite is often used as a disinfectant for drinking water and as a final step in sewage treatment. Thus, it enters ecosystems, such as rivers, as part of the sewage effluent. Measurements of residual chlorine are made on such effluents to determine the concentrations of these toxic compounds.

Diatoms, silicoflagellates, and radiolaria require silicon for building cell wall structures. Figure 8.19 shows an example of the depletion of silicate during blooms of diatoms.

SUMMARY

A large number of abiotic physical and chemical factors influence the growth, survival, and metabolic activity of microorganisms. A variety of physical measurements and analytical procedures are needed in order to characterize the abiotic factors that govern the microbial inhabitants in a habitat. In some instances, one or a few factors can be identified that regulate the growth and activities of individual microbial populations in a given ecosystem. Often the effect of one abiotic factor on a microbial population is influenced by other abiotic factors. In some cases, microorganisms are limited in their activities by the availability of some essential nutrient; in other cases, some environmental factor may exceed the maximum or minimum tolerance of a particular microbial population.

Factors that have major influences on microorganisms include electromagnetic radiation, which in the form of visible light is the principal energy source for the global ecosystem, but which as high-energy ionizing radiation is destructive for life; temperature, which in many cases determines rates of growth and activity as well as the particular microorganisms that occur within an ecosystem and what functions they perform; pressure due to atmospheric, hydrostatic, and osmotic forces, which affects the survival of organisms; redox potential, which affects what forms of microbial metabolism may occur; pH, which influences survival and nutrient supply; availability of liquid water, which is essential for all active life processes; and the concentrations and chemical forms of organic and inorganic compounds, which may be essential nutrients or may act as inhibitors and poisons.

Microorganisms are enormously resilient and versatile in adapting to environmental extremes. New research results keep expanding our notions about the outer limits for active microbial life. Microbial metabolism and growth have been reported in the temperature range of $-10°C$ to $110°C$, in the pH range of 1.0 to 11.5, and in the osmotic range of distilled water to saturated brines and syrups. Some microorganisms tolerate and even require a hydrostatic pressure of 1,000 atm, while others survive at several megarads of ionizing radiation. Even in the most hostile hot and cold deserts, microorganisms form active communities inside porous rocks and leave almost no terrestrial environment uninhabited.

References & Suggested Readings

Abdollahi, H., and D. B. Nedwell. 1979. Seasonal temperature as a factor influencing bacterial sulfate reduction in a salt-marsh sediment. *Microbial Ecology* 5:73–79.

Alexander, M. 1976. Natural selection and the ecology of microbial adaptation in a biosphere. In M. R. Heinrich (ed.). *Extreme Environments: Mechanisms of Microbial Adaptation.* Academic Press, New York, pp. 3–25.

Anderson, A. W., H. D. Hordan, R. F. Cain, G. Parrish, and D. Duggan. 1956. Studies on a radio-resistant *Micrococcus.* Part I, Isolation, morphology, cultural characteristics and resistance to gamma irradiation. *Food Technology* 10:575–578.

Atlas, R. M. 1977. Stimulated petroleum biodegradation. *Critical Reviews in Microbiology* 5:371–386.

Atlas, R. M., and R. Bartha. 1972. Degradation and mineralization of petroleum in seawater: Limitation by nitrogen and phosphorus. *Biotechnology and Bioengineering* 14:309–318.

Baas-Becking, L. M. G., and E. J. F. Wood. 1955. Biological processes in the estuarine environment. *Koninkligke Nederlandse Akademie Van Wetenschappen Proceedings B* (Amsterdam) 58:160–181.

Baross, J. A., F. J. Hanus, and R. Y. Morita. 1972. Effects of hydrostatic pressure on uracil uptake, ribonucleic acid synthesis, and growth of three obligately psychrophilic marine vibrios, *Vibrio alginolyticus* and *Escherichia coli.* In R. R. Colwell and R. Y. Morita (eds.). *Effects of the Ocean Environment on Microbial Activities.* University Park Press, Baltimore, pp. 180–222.

Baross, J. A., and J. W. Deming. 1983. Growth of "black smoker" bacteria at temperatures of at least 250°C. *Nature* (London) 303:423–426.

Baross, J. A., J. W. Deming, and R. R. Becker. 1984. Evidence for microbial growth in high-pressure, high-temperature environments. In M. J. Klug and C. A. Reddy (eds.). *Current Perspectives in Microbial Ecology.* American Society for Microbiology, Washington, D.C., pp. 186–195.

Black, C. A. (ed.). 1965. *Methods of Soil Analysis: Part 2, Chemicals and Microbiological Properties.* American Society of Agronomy, Madison, Wis.

Blakemore, R. 1975. Magnetotactic bacteria. *Science* 190:377–379.

Blakemore, R. P. 1982. Magnetotactic bacteria. *Annual Review of Microbiology* 36:217–238.

Bott, T. L., and T. D. Brock. 1969. Bacterial growth rates above 90°C in Yellowstone hot springs. *Science* 164:1411–1412.

Brock, T. D. 1978. *Thermophilic Microorganisms and Life at High Temperatures.* Springer-Verlag, New York.

Brock, T. D. 1985. Life at high temperatures. *Science* 230:132–138.

Brown, A. D. 1976. Microbial water stress. *Bacteriological Reviews* 40:803–846.

Button, D. K. 1985. Kinetics of nutrient-limited transport and microbial growth. *Microbiological Reviews* 49:270–297.

Campbell, L. L., and B. Pace. 1968. Physiology of growth at high temperatures. *Journal of Applied Bacteriology* 31:24–35.

Carlile, M. J. 1980. Positioning mechanisms—the role of motility, taxis and tropism in the life of microorganisms. In D. C. Ellwood, M. J. Latham, J. N. Hedger, and J. M. Lynch (eds.). *Contemporary Microbial Ecology.* Academic Press, London, pp. 55–74.

Castenholz, R. W. 1979. Evolution and ecology of thermophilic microorganisms. In M. Shilo (ed.). *Strategies of Microbial Life in Extreme Environments*. Verlag Chemie, Weinheim, N.Y , pp. 373–392.

Chastain, R. A. and A. A. Yayanos. 1991. Ultra-structural changes in an obligately barophilic marine bacterium after decompression. *Applied and Environmental Microbiology* 57:1489–1497.

Cohen, Y., E. Padan, and M. Shilo. 1975. Facultative anoxygenic photosynthesis in the cyanobacteria, *Oscillatoria limnetica*. *Journal of Bacteriology* 123:855–861.

Colwell, R. R., and R. Y. Morita (eds.). 1972. *Effect of the Ocean Environment on Microbial Activity*. University Park Press, Baltimore.

Costerton, J. W. 1984. Direct ultrastructural examination of adherent bacterial populations in natural and pathogenic ecosystems. In M. J. Klug and C. A. Reddy (eds.). *Current Perspectives in Microbial Ecology*. American Society for Microbiology, Washington, D.C., pp. 115–123.

Dazzo, F. B. 1984. Attachment of nitrogen-fixing bacteria to plant roots. *Current Perspectives in Microbial Ecology*. American Society for Microbiology, Washington, D.C., pp. 130–135.

Deming, J. W., and J. A. Baross. 1986. Solid medium for culturing black smoker bacteria at temperatures to 120°C. *Applied and Environmental Microbiology* 51:238–243.

Derosier, N. W. 1970. *The Technology of Food Preservation*. The AVI Publishing Co., Westport, Conn.

Doetch, R. N., and T. M. Cook. 1973. *Introduction to Bacteria and their Ecobiology*. University Park Press, Baltimore.

Ducklow, H. W. 1984. Geographical ecology of marine bacteria: Physical and biological variability at the Mesoscale. In M. J. Klug and C. A. Reddy (eds.). *Current Perspectives in Microbial Ecology*. American Society for Microbiology, Washington, D.C., pp. 22–31.

Edwards, C. 1990. Thermophiles. In C. Edwards (ed.). *Microbiology of Extreme Environments*. McGraw-Hill, New York, pp. 1–32.

Ehrlich, H. L. 1978. How microbes cope with heavy metals, arsenic and antimony in their environments. In D. J. Kushner (ed.). *Microbial Life in Extreme Environments*. Academic Press, London, pp. 381–408.

Farina, M., S. Esquival and H. G. P. L. deBarros. 1990. Magnetic iron-sulfur crystals from a magnetotactic microorganism. *Nature* (London) 343:256–258.

Fischer, F., W. Zillig, K. D. Stetter, and G. Schreiber.

1983. Chemolithoautotrophic metabolism of anaerobic extremely thermophilic archaebacteria. *Nature* (London) 301:511–513.

Fletcher, M., and K. C. Marshall. 1982. Are solid surfaces of ecological significance to aquatic bacteria? *Advances in Microbial Ecology* 6:199–236.

Fletcher, M., and S. McEldowney. 1984. Microbial attachment to nonbiological surfaces. In M. J. Klug and C. A. Reddy (eds.). *Current Perspectives in Microbial Ecology*. American Society for Microbiology, Washington, D.C., pp. 124–129.

Frazier, W. C., and D. C. Westhoff. 1978. *Food Microbiology*. McGraw-Hill, New York.

Fridovich, I. 1977. Oxygen is toxic! *BioScience* 27:462–466.

Friedmann, E. I. 1982. Endolithic microorganisms in the Antarctic cold desert. *Science* 215:1045–1053.

Friedmann, E. I., and R. Ocampo. 1976. Endolithic blue-green algae in the dry valleys: Primary producers in the Antarctic ecosystem. *Science* 193:1247–1249.

Friedmann, E. I., and R. Ocampo-Friedmann. 1984. Endolithic microorganisms in extreme dry environments: Analysis of a lithobiontic microbial habitat. In M. J. Klug and C. A. Reddy (eds.). *Current Perspectives in Microbial Ecology*. American Society for Microbiology, Washington, D.C., pp. 177–185.

Fry, J. C. 1990. Oligotrophs. In C. Edwards (ed.). *Microbiology of Extreme Environments*. McGraw-Hill, New York, pp. 93–116.

Gadd, G.M. 1990. Metal tolerance. In C. Edwards (ed.) *Microbiology of Extreme Environments*. McGraw-Hill, New York, pp. 178-210.

Gilmour, D. 1990. Halotolerant and halophilic microorganisms. In C. Edwards (ed.). *Microbiology of Extreme Environments*. McGraw-Hill, New York, pp. 147–177.

Gould, G. W., and J. E. Corry. 1980. *Microbial Growth and Survival in Extremes of Environment*. Academic Press, New York.

Gray, T. R. G., and J. R. Postgate (eds.). 1976. *The Survival of Vegetative Microbes*. Cambridge University Press, New York.

Gregory, P. H. 1973. *The Microbiology of the Atmosphere*. John Wiley and Sons, New York.

Griffin, D. M. 1981. Water and microbial stress. *Advances in Microbial Ecology* 5:91–136.

Halldal, P. 1980. Light and microbial activities. In D. C.

Ellwood, M. J. Latham, J. N. Hedger, and J. M. Lynch (eds.). *Contemporary Microbial Ecology*. Academic Press, London, pp. 1–14.

Halldal, P., and P. O. Taube. 1972. Ultraviolet action and photoreactivation in algae. In A. C. Giese (ed.). *Photophysiology*. Vol. 7. Academic Press, New York, pp. 163–188.

Hardon, M. J., and L. J. Albright. 1974. Hydrostatic pressure effects on several stages of protein synthesis in *E. coli. Canadian Journal of Microbiology* 20:359–365.

Hargrave, B. T. 1969. Epibenthic algal production and community respiration in the sediments of Marion Lake. *Journal of the Fisheries Research Board of Canada* 26:2003–2026.

Heywood, R. B. 1984. Antarctic inland waters. In R. M. Laws (ed.). *Antarctic Ecology*. Vol. 1. Academic Press, New York, pp. 279–344.

Hirst, J. M., I. F. Storey, W. C. Wood, and J. J. Wilcox. 1955. The origin of apple scab epidemics in the Wisbeck area in 1953–1954. *Plant Pathology* 4:91–96.

Horowitz, N. H., R. E. Cameron, and J. S. Hubbard. 1972. Microbiology of the dry valleys of Antarctica. *Science* 176:242–245.

Hugo, W. B. (ed.). 1971. *Inhibition and Destruction of the Microbial Cell*. Academic Press, London.

Ingledew, W. J. 1990. Acidophiles. In C. Edwards (ed.). *Microbiology of Extreme Environments* McGraw-Hill, New York, pp. 33–54.

Inniss, W. E., and J. L. Ingraham. 1978. Microbial life at low temperatures: Mechanisms and molecular aspects. In D. J. Kushner (ed.). *Microbial Life in Extreme Environments*. Academic Press, London, pp. 73–103.

Jannasch, H. W. 1979. Microbial turnover of organic matter in the deep sea. *BioScience* 29:228–232.

Jannasch, H. W., and C. W. Wirsen. 1977. Microbial life in the deep sea. *Scientific American* 236:42–52.

Jannasch, H. W., C. O. Wirsen, and C. D. Taylor. 1982. Deep-sea bacteria: Isolation in the absence of decompression. *Science* 216:1315–1317.

Jannasch, H. W., and D. C. Nelson. 1984. Recent progress in the microbiology of hydrothermal vents. In M. J. Klug and C. A. Reddy (eds.). *Current Perspectives in Microbial Ecology*. American Society for Microbiology, Washington, D.C., pp. 170–176.

Jannasch, H. W., and M. J. Mottl. 1985. Geomicrobiology of deep-sea hydrothermal vents. *Science* 229:717–725.

Jennings, D. H. 1990. Osmophiles. In C. Edwards (ed.). *Microbiology of Extreme Environments*. McGraw-Hill, New York, pp. 117–146.

Jensen, W. A., and F. B. Salisbury. 1972. *Botany: An Ecological Approach*. Wadsworth Publishing Co., Belmont, Calif.

Johnston, C. G., and J. R. Vestal. 1991. Photosynthetic carbon incorporation and turnover in Antarctic cryptendolithic microbial communities: Are they the slowest growing communities on Earth? *Applied and Environmental Microbiology* 57:2308–2311.

Kim, J., and C. E. ZoBell. 1972. Occurrence and activities of the cell-free enzymes in oceanic environments. In R. R. Colwell and R. Y. Morita (eds.). *Effects of the Ocean Environment on Microbial Activities*. University Park Press, Baltimore, pp. 365–385.

Kjelleberg, S. 1984. Effects of interfaces on survival mechanisms of copiotrophic bacteria in low-nutrient habitats. In M. J. Klug and C. A. Reddy (eds.). *Current Perspectives in Microbial Ecology*. American Society for Microbiology, Washington, D.C., pp. 151–160.

Koch, A. L. 1990. Diffusion: The crucial process in many aspects of the biology of bacteria. *Advances in Microbial Ecology* 11:3770.

Konopka, A. 1984. Effect of light-nutrient interactions on buoyancy regulation by planktonic cyanobacteria. In M. J. Klug and C. A. Reddy (eds.). *Current Perspectives in Microbial Ecology*. American Society for Microbiology, Washington, D.C., pp. 41–48.

Kroll, R. G. 1990. Alkalophils. In C. Edwards (ed.). *Microbiology of Extreme Environments*. McGraw-Hill, New York, pp. 55–92.

Kushner, D. J. 1968. Halophilic bacteria. *Advances in Applied Microbiology* 10:73–97.

Kushner, D. J. (ed.). 1978. *Microbial Life in Extreme Environments*. Academic Press, London.

Kushner, D. J. 1980. Extreme environments. In D. C. Ellwood, M. J. Latham, J. N. Hedger, and J. M. Lynch (eds.). *Contemporary Microbial Ecology*. Academic Press, London, pp. 29–54.

Larsen, H. 1962. Halophilism. In I. C. Gunsalus and R. Y. Stanier (eds.). *The Bacteria: A Treatise on Structure and Function*. Vol. 4, *The Physiology of Growth*. Academic Press, New York, pp. 297-342.

Larsen, H. 1967. Biochemical aspects of extreme halophilism. *Advances in Microbial Physiology* 1:97–132.

Lieberman, M. M., and J. K. Lanyi. 1972. Threonine deaminase from extremely halophilic bacteria: Cooperative substrate kinetics and salt dependence. *Biochemistry* 11:211–216.

Liebig, J. 1840. *Chemistry in Its Applicaton to Agriculture and Physiology*. Taylor and Walton, London.

Lockwood, J. L., and A. B. Filonow. 1981. Responses of fungi to nutrient-limiting conditions and to inhibitory substances in natural habitats. *Advances in Microbial Ecology* 5:1–61.

MacLeod, R. A. 1985. Marine microbiology far from the sea. *Annual Review of Microbiology* 39:1–20.

Mann, S., N. H. C. Sparks, R. B. Frankel, D. A. Bazylinski, and H. Jannasch. 1990. Biomineralization of ferrimagnetic greigite (Fe$_3$S$_4$) and iron pyrite (FeS$_2$) in a magnetotactic bacterium. *Nature* (London) 343:258–261.

Marshall, K. C. 1976. *Interfaces in Microbial Ecology*. Harvard University Press, Cambridge, Mass.

Moench, T. T., and W. A. Konetzka. 1978. A novel method for the isolation and study of a magnetotactic bacterium. *Archive für Mikrobiology* (Berlin) 119:203–212.

Morita, R. Y. 1975. Psychrophilic bacteria. *Bacteriological Reviews* 39:144–167.

Morris, J. G. 1975. The physiology of obligate anaerobiosis. *Advances in Microbial Physiology* 12:169–246.

Norman, A. 1954. The nuclear role in the ultraviolet inactivation of *Neurospora* conidia. *Journal of Cellular Comparative Physiology* 44:1–10.

Odum, E. P. 1971. *Fundamentals of Ecology*. W. B. Saunders, Philadelphia.

Orem, A. 1988. The microbial ecology of the Dead Sea. *Advances in Microbial Ecology* 109:193–229.

Padan, E. 1984. Adaptation to bacteria to external pH. In M. J. Klug and C. A. Reddy (eds.). *Current Perspectives in Microbial Ecology*. American Society for Microbiology, Washington, D.C., pp. 49–55.

Poindexter, J. S. 1981. Oligotrophy: Fast and famine existence. *Advances in Microbial Ecology* 5:63–89.

Poindexter, J. S. 1984. Role of prostheca development in oligotrophic aquatic bacteria. In M. J. Klug and C. A. Reddy (eds.). *Current Perspectives in Microbial Ecology*. American Society for Microbiology, Washington, D.C., pp. 33–40.

Post, F. J. 1977. The microbial ecology of the Great Salt Lake. *Microbial Ecology* 3:143–165.

Reid, D. S. 1980. Water activity as the criterion of water availability. In D. C. Ellwood, M. J. Latham, J. N. Hedger, and J. M. Lynch (eds.). *Contemporary Microbial Ecology*.

Academic Press, London., pp. 15–28.

Revsbech, N. P., and B. B. Jørgensen. 1986. Microelectrodes: Their use in microbial ecology. *Advances in Microbial Ecology* 9:293–352.

Rheinheimer, G. 1974. *Aquatic Microbiology*. John Wiley and Sons, London.

Ricci, J. C. D., B. J. Woodford, J. L. Kirschvink, and M. R. Hoffmann. 1991. Alteration of the magnetic properties of *Aquaspirillum magnetotacticum* by a pulse magnetization technique. *Applied and Environmental Microbiology* 57:3248–3254.

Rosenberg, M., and S. Kjelleberg. 1986. Hydrophobic interactions: Role in bacterial adhesion. *Advances in Microbial Ecology* 9:353–393.

Rupert, C. S. 1964. Photoreactivation of ultraviolet damage. In A. C. Giese (ed.). *Photophysiology*. Vol. 2. Academic Press, New York, pp. 283–327.

Shelford, V. E. 1913. *Animal Communities in Temperate America*. University of Chicago Press, Chicago.

Skujins, J. 1984. Microbial ecology of desert soils. *Advances in Microbial Ecology* 7:49–91.

Steeman-Nielson, E., V. K. Hansen, and E. G. Jorgensen. 1962. The adaptation to different light intensities in *Chlorella vulgaris* and the time dependence on transfer to a new light intensity. *Physiologia Plantarum* (Copenhagen) 15:505–517.

Stetter, K. O. 1982. Ultrathin mycelia-forming organisms from submarine volcanic areas having an optimum growth temperature of 105°C. *Nature* (London) 300:258–260.

Stotzky, G., and A. G. Norman. 1964. Factors limiting microbial activities in soil. Part III. Supplementary substrate additions. *Canadian Journal of Microbiology* 10:143–149.

Sweeney, B. M., and F. T. Haxo. 1961. Persistence of a photosynthetic rhythm in enucleated *Acetabularia*. *Science* 134:1361–1363.

Trent, J. D., R. A. Chastain, and A. A. Yayanos. 1984. Possible artifacutal basis for apparent bacterial growth at 250°C. *Nature* (London) 307:737–740.

Wetzel, R. G. 1975. *Limnology*. W. B. Saunders, Philadelphia.

Whittaker, R. H., and P. P. Feeny. 1971. Allelochemics: Chemical interaction between species. *Science* 171:757–770.

Wynn-Williams, D. D. 1990. Ecological aspects of Antarctic microbiology. *Advances in Microbial Ecology* 11:71-146.

Yayanos, A. A. 1986. Evolutional and ecological implications of the properties of deep-sea barophilic bacteria. *Proceedings of the National Academy of Science USA* 83:9542–9546.

Yayanos, A., A. S. Dietz, and R. V. Boxtel. 1979. Isolation of a deep-sea barophilic bacterium and some of its growth characteristics. *Science* 205:808–810.

Yayanos, A., and A. S. Dietz. 1983. Death of a hadal deep-sea bacterium after decompression. *Science* 220:497–498.

ZoBell, C. E. 1964. Hydrostatic pressure as a factor affecting the activity of marine microbes. In M. Miyahe and T. Koyama (eds.). *Recent Researches in the Field of Hydrosphere, Atmosphere and Nuclear Geochemistry.* Maruzen Co., Tokyo, pp. 83–116.

ZoBell, C. E., and L. L. Hittle. 1967. Some effects of hyperbaric oxygenation on bacteria at increased hydrostatic pressures. *Canadian Journal of Microbiology* 13:1311–1319.

Zuber, H. 1979. Structure and function of enzymes from thermophilic microorganisms. In M. Shilo (ed.). *Strategies of Microbial Life in Extreme Environments.* Verlag-Chemie, Weinheim, New York, pp. 393–415.

9

Microorganisms in Their Natural Habitats: Air, Water, and Soil Microbiology

THE HABITAT AND ITS MICROBIAL INHABITANTS

The ecosphere, or biosphere, which constitutes the totality of living organisms on Earth and the abiotic surroundings they inhabit, can be divided into atmo-, hydro-, and litho-ecospheres; these divisions respectively describe the portions of the global expanse inhabited by living things in air, water, and soil environments.

Each of the major divisions of the ecosphere contains numerous habitats. A habitat is the physical location where an organism is found. The physical and chemical characteristics of a habitat influence the growth, activities, interactions, and survival of the microorganisms found in it. For microorganisms, one must consider not only the overall characteristics of the general habitat, but also the fine features of the microhabitats in which the microorganisms live. The habitat is one component of a broader concept known as the ecological niche, which includes not only where an organism lives but also what it does there; the niche is the functional role of an organism within an ecosystem.

Some microorganisms are autochthonous or indigenous within a given habitat. These autochthonous microbes, which are capable of survival, growth, and metabolic activity in that habitat, occupy the environmental niches available to the microbial populations in the given ecosystem. Each autochthonous microorganism must be viewed in terms of its ability to grow, to carry out active metabolism, and to compete successfully with the other autochthonous members of the microbial community. Autochthonous microorganisms generally exhibit adaptive features that make them physiologically compatible with their physical and chemical environment; they must be functional and competitive with the other living organisms present in that habitat.

In contrast to the indigenous members of a microbial community, some microorganisms may be foreign; these organisms are called allochthonous. Allochthonous microorganisms are transient members of their habitat and do not occupy the functional niches of that ecosystem. Typically, these microorganisms have grown elsewhere and have been transported into a foreign ecosystem.

Allochthonous microorganisms exhibit great variation in the lengths of time that they can survive in foreign ecosystems; some disappear quite rapidly, as exemplified by *Escherichia coli,* which generally survives less than twenty-four hours when it enters marine habitats (Mitchell 1968). In contrast, some microorganisms possess adaptations, such as the ability to form spores, which allow them to survive for long periods of time in foreign habitats. Pertinent points that make organisms autochthonous are the ability to adapt physiologically to the physical and chemical environment, thus enhancing survival, and the ability to escape predation and competition pressures.

Although the definitions of autochthonous and allochthonous microorganisms are mutually exclusive, it is often difficult to determine whether a microorganism found in a particular ecosystem is indeed autochthonous or allochthonous. Ecosystems are dynamic and exhibit continuous change; autochthonous microorganisms may be active during one period of time but dormant during another. Thus, even when a microorganism found in an ecosystem does not appear to be capable of growth and active metabolism at a given point in time, one must consider whether the environment exhibits cyclic changes that will render it suitable for the growth of that microorganism at some other time. It is also pertinent to know whether the microorganism has the capacity to survive until conditions become favorable. In a few cases, allochthonous microorganisms that enter an ecosystem are able to survive, grow, and carry out active metabolism, allowing them to fill unoccupied niches and perhaps to become autochthonous microorganisms; this is most likely to occur in ecosystems that are in a state of change due to stress or disturbance.

ATMO-ECOSPHERE

Characteristics and Stratification of the Atmosphere

The atmosphere consists of 79% nitrogen, nearly 21% oxygen, 0.032% carbon dioxide, and trace amounts of some other gases. It is saturated with water vapor to varying degrees, and it may contain droplets of liquid water, crystals of ice, and particles of dust. The atmosphere is divided into regions defined by temperature minima and maxima (Rumney 1968) (Figure 9.1). The troposphere, the region nearest Earth, interfaces with both the hydrosphere and the lithosphere. Above the troposphere is the stratosphere, and above this lies the ionosphere.

For the most part, the chemical and physical parameters of the atmosphere do not favor microbial growth and survival. Temperatures decrease with increasing height in the troposphere. At the top of the troposphere, temperatures are −43°C to −83°C, which are below the minimal growth temperatures for microorganisms. With increasing height in the atmosphere, the atmospheric pressure declines and concentrations of available oxygen decrease to a point that precludes aerobic respiration. The low concentrations of organic carbon are insufficient to support heterotrophic growth; available water is also scarce, lim-

Figure 9.1

Divisions of the atmosphere showing temperature and pressure gradients. The two lines indicate seasonal shifts in temperature. (Source: Rumney 1968.)

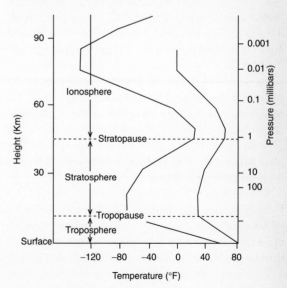

iting even the possibility of autotrophic growth of microorganisms in the atmosphere.

Microorganisms in the atmosphere are exposed to high intensities of light radiation. Exposure to ultraviolet (UV) light, which increases with height as the atmosphere thins and offers less filtering, or shielding, from UV radiation, causes lethal mutations and the death of microorganisms.

The stratosphere contains a layer of high ozone concentration, which acts to absorb UV light, protecting Earth's surface from excessive UV radiation (Thrush 1977). There is a justified concern today that certain human activities, such as flying of supersonic military and commercial jets, excessive use of fluorocarbons, and increased use of fertilizers (which results in increased release of N_2O from microbial denitrification), will decrease concentrations of ozone in the stratosphere, thus allowing increased amounts of UV light to reach Earth's surface. The seasonal development of an Antarctic ozone hole is a clear symptom of the lessening atmospheric concentration of ozone. An increased flux of UV radiation reaching the surface of Earth would be detrimental to the survival of organisms living above ground.

The stratosphere represents a barrier to the transport of living microorganisms to or from the troposphere and is characterized by slow mixing of gases. Organisms in the stratosphere are thus transported slowly and are exposed for prolonged periods of time to the prevailing concentrations of ozone and high UV light intensities. Only microorganisms shielded from these conditions in the stratosphere—as perhaps within a spacecraft—could survive passage out of Earth's atmosphere. For all practical purposes, the atmo-ecosphere does not extend above the troposphere.

The Atmosphere as Habitat and Medium for Microbial Dispersal

Even though the atmosphere is a hostile environment for microorganisms, there are substantial numbers of microorganisms in the lower troposphere, where, because of thermal gradients, there is rapid mixing of air (Gregory 1973). Movement through air represents a major pathway for the dispersal of microorganisms. Some microorganisms have evolved specialized adaptations that favor their survival in and dispersal by the atmosphere. Several viral, bacterial, and fungal diseases are spread through the atmosphere; outbreaks of disease from such microorganisms often follow prevailing winds.

Temporary locations in the troposphere may provide habitats for microorganisms. Clouds possess concentrations of water that permit growth of microorganisms. Light intensities and carbon dioxide concentrations in cloud layers are sufficient to support growth of photoautotrophic microorganisms, and condensation nuclei may supply some mineral nutrients. In industrial areas, there may even be sufficient concentrations of organic chemicals in the atmosphere to permit growth of some heterotrophs. Nevertheless, such "life in the sky" is only a fascinating possibility; conclusive proof is lacking, and practical importance of such life appears to be negligible.

Although many microorganisms that grow in the hydrosphere or lithosphere can become airborne, there are no known autochthonous atmospheric microorganisms. During dispersal, aquatic and soil microorganisms may enter and pass through the atmosphere before reaching other favorable aquatic or terrestrial ecosystems. Dispersal through the atmosphere ensures continued survival of many microorganisms.

Some microorganisms become airborne as growing vegetative cells, but more commonly, microorganisms in the atmosphere occur as spores. To facilitate discussion, the term spore is used here in an extended sense and includes cysts, soredia, and other nonvegetative resistant structures. Spores are metabolically less active than vegetative cells and are generally better adapted to survival in the atmosphere. Spores whose primary function is dispersal are known as xenospores. Fungi, algae, some protozoa, some bacteria (especially actinomycetes), and lichens produce spores that occur in the atmosphere. Viruses are metabolically inactive outside of host cells and thus may be considered spores while in the atmosphere.

Several properties of spores contribute to their ability to withstand transport through the atmosphere. First, their low metabolic rates mean that they do not require external nutrients and water to generate suffi-

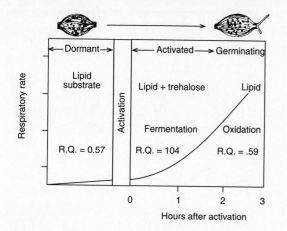

Figure 9.2

Metabolic events during germination of *Neurospora* ascospores. The dormant spores exhibit low respiratory rates as compared to germinating spores or vegetative cells. Respiratory quotients indicate lipid utilization in the dormant and late germination stages and carbohydrate utilization during activation period of germination. (Source: Sussman 1961. Reprinted by permission, copyright Burgess Publishing Co.)

cient energy for maintenance over long periods of time. This is essential for survival in the atmosphere with its paucity of water and nutrients. Figure 9.2 shows the change in metabolic activities when a resting spore germinates (Sussman 1961). The successful germination of a spore requires a favorable environment for growth. The metabolic activities of vegetative cells that are required for the maintenance of cell integrity cannot be supported for long in the atmosphere. Once the cells' internal reserve materials are used up, vegetative cells in the atmosphere cannot generate enough energy to maintain vital cellular functions, and the cells die.

Spores are produced in very high numbers; some fungi, for example, produce in excess of 10^{12} spores per single fruiting body per year (Ingold 1971). A large percentage of spores does not survive transport through the atmosphere to habitats that favor their germination; the success of only a few spores, however, ensures the survival and dispersion of the

microorganism. Various additional adaptations increase the ability of a spore to survive in the atmosphere. Some spores have extremely thick walls, which protect them against severe desiccation. Some spores are pigmented, which adds protection against exposure to damaging UV radiation. The relatively small size and low density of spores permit them to remain airborne for long periods of time before they sediment from the atmosphere. Usually, spores are relatively light; they may even contain gas vacuoles. They come in a variety of shapes; some are aerodynamically favorable for extended lateral travel through the atmosphere (Gregory 1973) (Table 9.1).

The passive liberation of spores into the atmosphere with air current movements is common among microorganisms that produce dry spores on aerial mycelia (Ingold 1971). Such microorganisms include the actinomycetes and many fungi. Some spores are transmitted upward from microbial fruiting bodies by convection currents; others move both vertically and laterally with wind currents (Figure 9.3). In general, the higher the wind speed and the lower the humidity, the greater the movement of spores. Wind-driven spore movement is especially important in the dispersal of microorganisms that occur on plant surfaces; many plant pathogenic fungi spread from one plant to another by this mechanism.

Although some dry spores are readily liberated into the air by wind currents, other spores remain

Table 9.1

Estimated distance traveled by several fungi, calculated from terminal velocities assuming nonturbulent wind moving at 1 m per second

Fungus	Assumed liberation height	Distance (m)
Helminthosporium	1 m	50
Puccinia	1 m	80
Agaricus	5 cm	40
Lycoperdon	5 cm	100

Source: Gregory 1973.

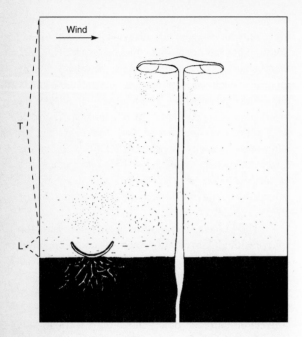

Figure 9.3

The discharge of ascospores into the air from a cup fungus and an agaric. Laminar air flow above the ground (horizontal dashes) and turbulent air flow above this (dashes in spirals) are illustrated. The spores of the cup fungus have just been discharged as a puff into the turbulent air zone. The agaric spores are steadily dropping into the turbulent air. (Source: Ingold 1971. Reprinted by permission of Oxford University Press.)

Figure 9.4

Splashing of peridiole from a basidiocarp of *Cyathus striatus*. (A and B) Raindrop landing in cup. (C) The peridiole splashing out with hapteron extended. (D) Hapteron sticking to a plant as the peridiole is carried forward by its momentum and the funiculus is extended by a pull. (E) Peridiole jerked backward when the funiculus is extended to its full length. (F) Peridiole swinging around the plant stem as another raindrop falls. (Source: Brodie 1951. Reprinted by permission of National Research Council Canada.)

attached even in air currents with high velocities. Some of these spores are liberated when water droplets in the air collide with the spore-bearing bodies (Figure 9.4). Raindrops may liberate spores to the atmosphere by vibrating the structures they are attached to, breaking the adhering forces. The "splash cups" of the bird's-nest fungi are adaptations designed to utilize the force of landing raindrops to release spores (Brodie 1951).

Spores and even vegetative microorganisms often enter the atmosphere as aerosols. Aerosols may form from a variety of sources, including splash from falling raindrops; spray from breaking waves; rapidly moving water currents striking solid objects, such as rocks in rivers; gas movements through water columns, such as gas bubbles rising from sediment through the water column; forced air streams in sewage treatment plants; and so on (Teltsch and Katznelson 1978). Some microorgan-

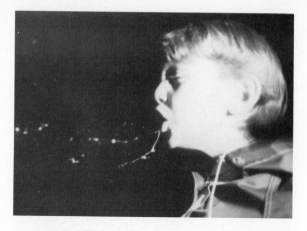

Figure 9.5

Moisture droplets released during a violent sneeze, showing aerosol dispersal.

Figure 9.6

Release of a spore cluster by *Pilobolus*. The whole sporangium is shot off as a unit. The spore mass, represented by black area, adheres to expelled fluid droplet (light area). (Source: Buller 1909. Reprinted by permission of Hafner Publishing Co.)

isms are released on droplets by mechanisms such as coughing and sneezing by humans or animals (Figure 9.5). This type of release is especially important in the dispersal through the atmosphere of some pathogenic bacteria and animal viruses.

In addition to the passive mechanisms that allow microorganisms to enter the atmosphere, there are a number of adaptive active mechanisms that discharge microbial spores into the atmosphere. In *Pilobolus,* the entire spore cluster is ejected when a vacuole in the sporangium base becomes turgid through an increase in osmotic pressure and then causes the structure to burst and the spores to be carried away in a jet of water (Buller 1909) (Figure 9.6). *Pilobolus* shoots its spore cluster a distance of 1–2 m and orients the direction toward the highest light intensity, that is, to an open area where air currents are most likely to cause further dispersal.

In most ascomycetes, the ascospores are actively discharged. Typically, the ascus swells at maturity and bursts at the tip, propelling the spores into the air to a distance ranging from several millimeters to several centimeters. The bursting of the ascus is caused by the change in osmotic pressure when glycogen is converted to soluble sugars. Some

ascomycetes exhibit the phenomenon of puffing, that is, the simultaneous breaking of a large number of asci with the release of a visible cloud of spores (Buller 1934) (Figure 9.7). Changes in environmental conditions, such as light, humidity, or temperature, can trigger the puff of spores. In *Sphaerobolus,* a basidiomycete that grows on old dung in succession to *Pilobolus* and ascomycetes, the basidiospores are released when an inner turgid layer of vegetative hyphae flips inside out, propelling the spherical spore mass upward several meters (Figure 9.8).

Having become airborne, both spores and vegetative microbial cells face the problem of survival (Stetzenbach et al. 1992). Most microorganisms can survive a short passage (a few millimeters) through the atmosphere, but relatively few survive long-distance transport since desiccation can cause microorganisms in the atmosphere to lose viability, particularly in the lower layers of the atmosphere during the day.

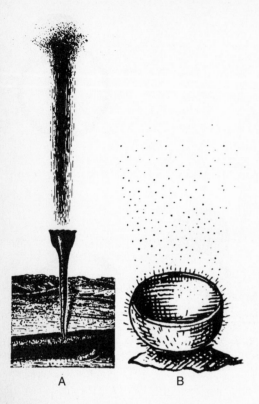

Figure 9.7

Spore release by puffing as illustrated (A) by Buller (1934) for *Sarcoscypha protracta*; and (B) by Micheli (1729) for a small cup fungus. (Source: Ingold 1971. Reprinted by permission of Oxford University Press.)

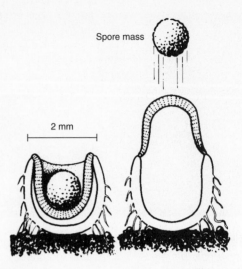

Figure 9.8

Spore discharge in *Sphaerobolus*. The spore mass is propelled upward several meters facilitating dispersal. (Source: Jensen and Salisbury 1972. Reprinted by permission of Wadsworth Publishing Co., Belmont, California.)

Figure 9.9

Exposure to direct sunlight of normal *Micrococcus luteus* (*Sarcina lutea*) (yellow) and pigmentless mutant (white) in air and a nitrogen atmosphere. The yellow wild type (a) is resistant to light both in air and under an atmosphere of nitrogen. The white mutant (b) is resistant to light under nitrogen but (c) is rapidly killed by exposure to light in the presence of oxygen, demonstrating the protective effect of pigmentation against photooxidation. (Source: Mathews and Sistrom 1959. Reprinted from *Nature* by permission, copyright Macmillan Journals Ltd.)

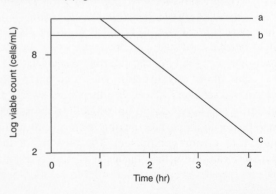

Exposure to short-wavelength radiation, such as UV light, is another major cause of loss of viability of microorganisms in the atmosphere. Microorganisms carried through the atmosphere on "rafts," such as dust or soil particles, may be protected from the harmful effects of UV radiation. Some microorganisms are protected from these lethal effects by pigments. Pigmented fungi and bacteria suffer less damage when exposed to UV light than colorless species. Exposure to sunlight in the air is lethal to nonpigmented strains of *Micrococcus luteus,* for example, but not to yellow-pigmented strains (Mathews and Sistrom 1959) (Figure 9.9). The presence of the yellow pigment

Table 9.2

Quantitative estimates of microorganisms (%) in the troposphere

Microorganism	N. Canada 17 m	N. Canada 3,000 m	Quebec 3,000 m
Bacteria			
Gram-positive pleomorphic rods	46	24	20
Gram-negative rods	20	5	4
Spore formers	18	38	33
Cocci	15	23	41
Fungi			
Cladosporium	—	73	82
Alternaria	—	7	3
Penicillium	—	3	2
Other	—	17	13

Source: Gregory 1973.

appears to protect these bacteria from exposure to sunlight. Death does not occur in the absence of air, indicating that light-induced killing is a photooxidation process that requires oxygen. Similarly, a colorless (carotenoid-free) mutant of *Halobacterium salinarium* was inhibited by high light intensities; the pigmented wild strain was free of such inhibition (Dundas and Larsen 1962).

Microorganisms in the Atmo-Ecosphere

There have been attempts to quantitate numbers of individual microbial genera and/or types in the atmosphere (Gregory 1973) (Table 9.2). Quantitative sampling of microorganisms in the atmosphere typically employs either a viable plate count procedure or a direct count procedure using a modified contact slide (a slide with a sticky substance passed through the atmosphere by an aircraft). These methodological approaches give very different results. Plate count procedures favor the enumeration of bacteria, yeasts, and some fungal species; contact slide procedures favor the enumeration of fungal spores. In contemporary air samplers, a pump draws a calibrated volume of air through a membrane filter with 0.5 μm or smaller pore size (Andersen 1958). For direct counts, the filter is dissolved or clarified; for viable counts, the filter is placed on an appropriate solid medium and incubated.

Cladosporium is frequently reported to be the major fungal constituent of the atmosphere (Gregory 1973). Typically, concentrations of microorganisms in the atmosphere, up to a height of 3,000 m, range from 10^1 to 10^4 per m^3. The numbers of microorganisms in the atmosphere vary by season. Fungi are generally more abundant in June, July, and August than during the rest of the year. On the other hand, bacteria have been reported to be most abundant during spring and autumn (Bovallius et al. 1978) (Figure 9.10).

Microorganisms are removed from the atmosphere in a variety of ways. They may settle due to gravitational forces or they may be removed by rain and other forms of precipitation (Gregory 1973). Following a rainstorm, concentrations of microorganisms in the atmosphere are typically reduced. Despite the factors that tend to remove microorganisms from the atmosphere and reduce their viability during transport, some microorganisms are accomplished air travelers. *Puccinia graminis*, for example, is known to be transported over long distances

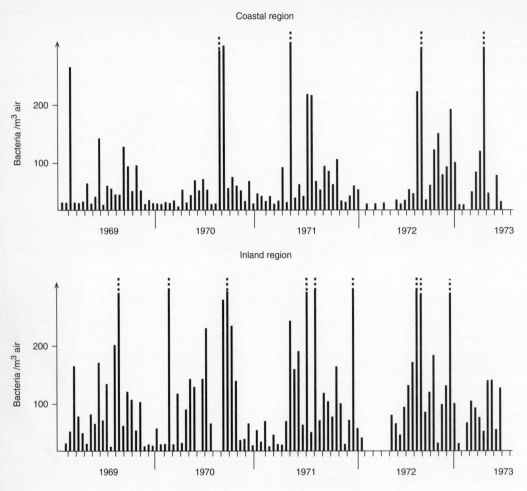

Figure 9.10
Bacterial concentration in air showing seasonal variation at two locations in Sweden. Bars with dashed ends in these histograms indicate off-scale readings. Generally, lower concentrations of bacteria occur in coastal regions due to paucity of bacteria in marine air masses. Seasonally low concentrations occur during winter and higher concentrations during summer and fall. (Source: Bovallius, et al. 1978. Reprinted by permission of American Society for Microbiology.)

through the atmosphere while maintaining viability, an important dispersal mechanism for this wheat rust fungus (Bowden et al. 1971) (Table 9.3). Other microorganisms, including some viruses, bacteria, and fungi, are probably capable of surviving atmospheric transport across oceans (Hirst et al. 1967).

The fact that many microorganisms appear to be ubiquitous in nature is largely due to the effectiveness of air transport. Geographical discontinuities in the distribution of some microbial populations are due primarily to the location of suitable habitats for growth.

Table 9.3

Dissemination of *Puccinia graminis*

Distance from source (km)	Concentration relative to source (%)
0	100
300	5
560	6
840	2
970	0.2

Source: Stakman and Hamilton 1939.

HYDRO-ECOSPHERE

The hydrosphere is a more suitable habitat for microbial growth than the atmosphere. By its very definition, the hydrosphere contains water which is necessary for microbial metabolism. Unlike the atmosphere, the hydrosphere contains autochthonous microbiota, to which one can ascribe certain limited general characteristics. These microbiota are able to grow at low nutrient concentrations. Most aquatic microorganisms are motile, by means of either flagella or other mechanisms. Some aquatic bacteria, such as the prosthecates, exhibit unusual shapes that increase the ratio of surface area to volume, allowing for more efficient uptake of the low levels of nutrients available in most freshwater bodies.

The hydrosphere is divided into freshwater and marine habitats. Freshwater habitats include lakes, ponds, swamps, springs, streams, and rivers. These habitats are collectively designated as limnetic. Their study is referred to as limnology. Marine habitats are the world's oceans and the estuarine habitats that occur at the interface between freshwater and marine ecosystems.

Freshwater Habitats

Freshwater habitats are classified based on their chemical and physical properties. Those with standing water, such as lakes and ponds, are called lentic habitats; those with running water are lotic habitats (Wetzel 1975).

The Neuston The uppermost layer of the hydrosphere, the surface microlayer, represents the interface between the hydrosphere and the atmosphere. It is characterized by high surface tension, a property arising from the interfacing of water with a gas (Marshall 1976). Under quiescent conditions, microorganisms form a surface film known as the neuston. The neuston layer is a favorable habitat for photoautotrophic microorganisms, since primary producers have unrestricted access to carbon dioxide from the atmosphere and to light radiation. Some mineral nutrients and metals become enriched in the surface microlayer. Secondary producers also proliferate here, using nonpolar organic compounds that accumulate in the surface tension layer and the high concentrations of oxygen available from the atmosphere. Microbial numbers in the surface microlayer are often 10- to 100-fold higher than in the underlaying water column. Bubbles rising through the neuston layer and bursting play an important role in the water-to-air transfer of bacteria and viruses (Blanchard and Syzdek 1970; Baylor et al. 1977). Various devices, including skimmers, touch-on screens, and rotating drums, have been developed for the qualitative and quantitative sampling of the neuston community (Parker and Barsom 1970; Norkrans 1980).

Characteristic autochthonous neuston microbiota include algae, bacteria, fungi, and protozoa (Valkanov 1968). Among the characteristic representative bacterial genera are *Pseudomonas, Caulobacter, Nevskia, Hyphomicrobium, Achromobacter, Flavobacterium, Alcaligenes, Brevibacterium, Micrococcus,* and *Leptothrix.* These bacterial genera include Gram-positive and Gram-negative, pigmented and nonpigmented, motile and nonmotile, rod and coccus, stalked and nonstalked forms. Cyanobacteria also occur in the neuston; typical genera include *Aphanizomenon, Anabaena*, and *Microcystis.* The filamentous fungus *Cladosporium* and various yeasts are frequently associated with the neuston. Algal genera found in the neuston include *Chromulina,*

Botrydiopsis, Codosiga, Navicula, Nautococcus, Proterospongia, Sphaeroeca, and *Platychrisis. Nautococcus* has been reported to have adaptive features particularly suited for its existence in the neuston. Protozoa found in the neuston include *Difflugia, Vorticella, Arcella, Acineta, Clathrulina, Stylonychia,* and *Codonosigna.* Some of these same organisms are also found in the underlying water column.

Swamps and Bogs Swamps and bogs are shallow aquatic environments dominated by emergent plants. Swamps develop under various climatic conditions in poorly drained shallow basins, often as a result of the gradual filling in of a lake by silt and vegetation. The water surface may be largely obscured by *Phragmites, Juncus, Typha,* and other semisubmerged plants. Primary production is generally high, while anoxic conditions in the submerged sediment hinder rapid organic matter decomposition. Consequently, large amounts of humified plant material (peat) accumulate.

Bogs develop only in cool, wet climates, usually in shallow rock pans. The dominant plant in bogs is the *Sphagnum* moss, which forms thick mats. The lower portions of the *Sphagnum* mat die off, but anoxic and acidic conditions generally restrict biodegradation and peat accumulates. The *Sphagnum* mat holds water by capillary forces so effectively that the bog can actually grow much higher than the rim of the rock pan. It can be visualized as a water-soaked sponge resting on a shallow plate or bowl. As the bog rises, it can no longer receive nutrients from water running over its surface and is effectively isolated from soil and bedrock through the thick accumulated peat layer. Such a bog depends upon nutrients only from the atmosphere and the rain; it becomes ombrotrophic (cloud nourished).

Both swamps and bogs are interesting environments because carbon recycling is severely restricted in them. Anoxic and acidic conditions partially explain this restricted recycling. Phenolic and polyphenolic substances from incomplete plant tissue degradation and probably additional, yet-unrecognized factors also contribute to low carbon recycling and fossil-fuel accumulation.

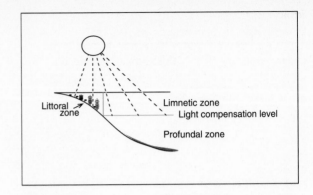

Figure 9.11

Zonation of a lake habitat based on light penetration. (Source: Odum 1971. Reprinted by permission, copyright W. B. Saunders Co., Philadelphia.)

Lakes Lakes are divided into three zones based on the penetration of light (Figure 9.11). The combined littoral and limnetic zones are known as the euphotic zone; here photosynthetic activity can occur. The profundal zone is the area of deeper water beyond the depth of effective light penetration; it does not exist in shallow ponds. In deep lakes the profundal zone extends from the light compensation level to the bottom.

The littoral zone is the region of the lake where light penetrates to the bottom; it is an area of shallow water usually dominated by submerged or partially submerged higher plants and attached filamentous and epiphytic algae. The limnetic zone, where the dominant primary producers are planktonic algae, is an area of open water away from the shore that descends to a level known as the compensation depth. The compensation depth is considered to be the lowest level having effective light penetration, where photosynthetic activity balances respiratory activity. Approximately 1% penetration of full sunlight intensity generally occurs at the compensation depth. Photosynthesis occurs below the compensation depth but at rates that usually are lower than the consumption rates of respiratory activities.

Microorganisms exhibit different absorption spectra determining which wavelengths of light can

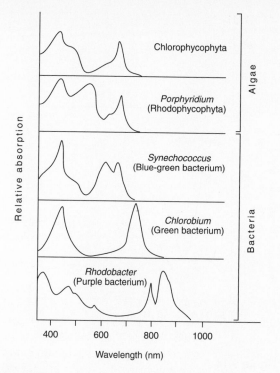

Figure 9.12

Absorption spectra of representative photosynthetic microorganisms. Note that *Chlorobium* and *Rhodopseudomonas* can absorb light in the far red part of the spectrum, which is not utilized by algae. Since these bacteria live below the algal layer, this ability offers ecological advantage. (Source: Stanier and Cohen-Bazire 1957. Reprinted by permission, copyright Cambridge University Press.)

be utilized for photosynthesis (Stanier and Cohen-Bazire 1957) (Figure 9.12). Green and purple sulfur bacteria normally grow at the sediment–water interface, as within the anoxic portion of the water column, below the layers of short-wavelength-absorbing algal and cyanobacterial growth. They do this by utilizing wavelengths of light not absorbed by the overlying phytoplankton. The purple and green sulfur bacteria obtain reducing electrons from H_2S at lower energy cost than H_2O-splitting photoautotrophs and thus require lower light intensities for carrying out photosynthesis.

The bottom of the lake, or benthos, represents the interface between the hydrosphere and the lithosphere. The lithosphere underlying a lake is referred to as sediment and is traditionally studied by aquatic microbiologists and geochemists. Conditions in the euphotic zone are favorable for the growth of photoautotrophs. Organisms in the profundal zone are largely secondary producers and are, for the most part, dependent on transport of organic compounds from the overlying zone. The benthic sediment habitat favors proliferation of microorganisms. Particulate nutrients sediment by gravitational forces and concentrate on the surface of the benthic sediment (Fletcher 1979; Marshall 1980; Paerl 1985). Surface sediments may be aerobic, allowing aerobic decomposition of accumulated organic nutrients. Anaerobic decomposition of organic compounds primarily occurs under the surface of lake sediments, where oxygen is depleted. This is because warm water is less dense than cold water. In spring, as the sun warms the water of the lake, a warm surface layer, the epilimnion, is formed. This layer is separated from the cold deep hypolimnion by the thermocline (Rigler 1964) (Figure 9.13).

Figure 9.13

Typical stratification of a lake in the temperate climate zone during the summer and winter periods, showing the profile of the temperature with depth. During spring and fall, stratification breaks down, resulting in complete mixing (turnover) of the lake water.

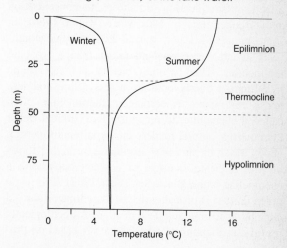

The epilimnion is the upper zone of water that lies above the thermocline. The thermocline is a zone characterized by rapid decrease in temperature, across which there is little mixing of water. The epilimnion is typically warm and oxygen rich during the summer. Vigorous photosynthetic production tends to deplete mineral nutrients in this layer to the extent that they limit primary production. The hypolimnion is the zone of water below the thermocline, which is generally characterized by low temperatures and low oxygen concentrations—poor light penetration restricts photosynthesis, and respiration depletes existing oxygen. Mineral nutrients tend to be relatively abundant. In the fall, as the epilimnion cools and reaches the temperature of the hypolimnion, the thermocline breaks down; this fall turnover results in a complete mixing of the lake. In this way, the hypolimnion is reoxygenated and mineral nutrients are replenished in the epilimnion.

Water has the thermal anomaly of being densest at 4°C. Once the whole lake cools to this temperature, water below 4°C does not sink but stays at the surface and freezes. This effectively insulates the lake from further cooling and prevents deeper bodies of water from freezing to their bottom. A weak winter stratification with 0°C–4°C epilimnion over a 4°C hypolimnion may develop. This stratification breaks down in the spring (spring turnover), followed by development of the summer stratification. Thermal stratification has a strong effect on the seasonal availability of and demand for mineral nutrients. Figure 9.14 shows the effects of stratification on phosphate concentration and turnover time in lake water (Rigler 1964). In some bodies of water, special hydrographic conditions maintain a permanent rather than a seasonal thermocline. This is often due to a dense saline hypolimnion overlayered by a less saline and less dense epilimnion. A permanent thermocline, often called a chemoline, to recognize its chemical rather than thermal nature, renders the hypolimnion permanently anaerobic and thus unsuitable for higher forms of life. A prime example of a water body with a chemocline is the Black Sea (Karl 1978). The name given this sea is probably due to its sediments stained dark by anaerobic sulfate reduction. An ecologically useful classification of lake habitats is based on pro-

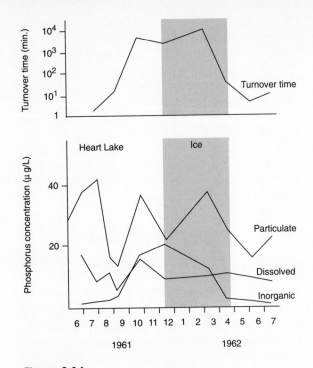

Figure 9.14

Seasonal changes in turnover time of inorganic phosphorus and amounts of particulate dissolved and inorganic phosphorus in a lake in Ontario. The turnover times are longer in winter than in summer; during summer, typical turnover times are of the order of minutes. Dissolved phosphorus is available in relatively high concentrations at the beginning of the spring phytoplankton blooms. (Source: Rigler 1964. Reprinted by permission of American Society of Limnology and Oceanography.)

ductivity and nutrient concentrations (Wetzel and Allen 1970). Oligotrophic lakes have low concentrations of nutrients. Typically, they are deep, have a larger hypolimnion than epilimnion, and have relatively low primary productivity. In contrast, eutrophic lakes have high nutrient concentrations, are usually shallower than oligotrophic lakes, and have higher rates of primary production. Oxygen concentrations are usually lower in eutrophic lakes than in oligotrophic lakes because of extensive aerobic decomposition of organic nutrients.

A number of important chemical parameters make lakes more-or-less suitable habitats for microorganisms (Hutchinson 1957). Concentrations of organic nutrients and oxygen, already mentioned, are important factors. Additionally, concentrations of inorganic nutrients, especially those that contain nitrogen and phosphorus, are important in determining the ability of the habitat to support microbial growth and metabolism. Concentrations of such nutrients are often influenced by exchange processes between the surrounding lithosphere and the lake waters. The availability of these essential nutrients is also highly dependent on biological activities that may sequester or mobilize inorganic substances.

The pH is another important factor influencing which microorganisms inhabit a particular lake. Some lakes are alkaline, some neutral, and still others acidic. Other conditions being equal, a higher pH favors primary production through higher availability of CO_2 in the forms of HCO_3^- and CO_3^{2-}. Salt concentrations also influence the characteristic autoch–thonous microorganisms of some lakes. Some land-locked inland lakes have high salt concentrations. The Great Salt Lake in Utah has a high salt concentration of almost 28%, eight times higher than that found in oceans.

Rivers Rivers differ from lakes in several major ways (Hynes 1970). Rivers are characterized by flowing waters; they have zones of rapid water movement and pools with reduced currents. Zones of rapid water movement tend to be shallower than pools. In pool zones, the decreased current velocity allows for deposition of silt. Silt is generally not deposited in zones of high current velocity; therefore, firm, rocky substrates underlie the rapid zones. Rivers have a high degree of interfacing with the lithosphere along the riverbanks, and there is a great deal of transfer of chemicals from the lithosphere into river waters through rainwater runoff and erosion of riverbanks.

The upper course of a river is usually characterized by swift flow, a high degree of oxygenation, and low temperature. Shading by forests generally keeps primary production in the upper course low, and organic material input is derived mostly from the sur-

rounding lithosphere. The middle course of a river is characterized by decreasing flow velocity, higher temperature, and less shading, resulting in significant intrinsic primary production. The lower course is subject to excessive silt deposition and tidal influences resulting in a die-off of stenohaline freshwater microorganisms and their replacement by salt-tolerant estuarine organisms.

Most microbial and many microscopic organisms in rivers are attached to surfaces such as on submerged rocks. Dissolved nutrients are rapidly absorbed by these attached organisms and are liberated upon death and decay only to be absorbed again a small distance downstream. As a consequence, nutrients do not move with the speed of the current but exhibit a much slower movement. The cycling of nutrients in a river does not occur in place, but rather nutrient cycling involves some downstream transport before a cycle is completed. The path of a nutrient can be viewed as a spiral rather than a cycle, a phenomenon known as nutrient spiraling (Webster and Patten 1979; Newbold et al. 1981).

Rivers often receive high amounts of effluents from industries and municipalities. Municipal sewage disposal into rivers introduces high concentrations of organic compounds (Figure 9.15). Oxygen concentrations are generally depleted below municipal sewage outfalls because of utilization of available oxygen during microbial decomposition of the organic compounds. Industrial effluents may also introduce toxic chemicals, such as heavy metals, which may adversely

Figure 9.15

Relationship of organic carbon and dissolved oxygen in a river above and below a sewage outfall.

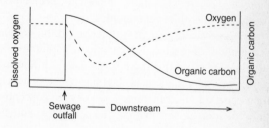

affect microbial activities and survival. Rivers receive high concentrations of agricultural chemicals through runoff. Some agricultural chemicals act as toxicants, adversely affecting microbial activities; others, such as fertilizers, may support uncontrolled proliferation of microorganisms.

Composition and Activity of Freshwater Microbial Communities

The microbial populations of lakes have been much more extensively studied than those of rivers (Hutchinson 1957; Rheinheimer 1980; Cole 1983). Many similarities exist, but rivers by their nature always contain higher proportions of allochthonous microorganisms. For these reasons, most of the following discussion concerns lakes.

Members of the genera *Achromobacter*, *Flavobacterium*, *Brevibacterium*, *Micrococcus*, *Bacillus*, *Pseudomonas*, *Nocardia*, *Streptomyces*, *Micromonospora*, *Cytophaga*, *Spirillum*, and *Vibrio* are reported as occurring widely in lake water (Taylor 1942). Stalked bacteria, such as *Caulobacter*, *Hyphomicrobium*, and other genera, are associated with submerged surfaces (Rheinheimer 1980).

Autotrophic bacteria are autochthonous members of the microbiota of lakes and play an important role in nutrient cycling (Caldwell 1977). Photoautotrophic bacteria normally found in lakes include the cyanobacteria and the purple and green anaerobic photosynthetic bacteria (Rheinheimer 1980). The cyanobacteria *Microcystis*, *Anabaena*, and *Aphanizomenon* can be dominant plankton in freshwater habitats. Chemolithotrophs play an important role in nitrogen, sulfur, and iron cycling within lakes; members of the genera *Nitrosomonas*, *Nitrobacter*, and *Thiobacillus* are especially important members of freshwater microbial communities.

There are important differences in the vertical distribution of bacterial populations within a lake (Rheinheimer 1980) (Figure 9.16). These differences reflect vertical variations in abiotic parameters such as light penetration, temperature, and oxygen concentration. Microorganisms that may be considered autoch-

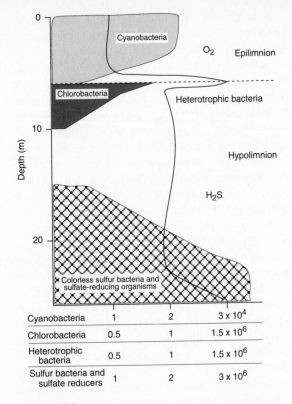

Cyanobacteria	1	2	3×10^4
Chlorobacteria	0.5	1	1.5×10^6
Heterotrophic bacteria	0.5	1	1.5×10^6
Sulfur bacteria and sulfate reducers	1	2	3×10^6

Figure 9.16

Vertical distribution of bacteria in a lake. Cyanobacteria (formerly blue-green algae) are abundant in the epilimnion. Sulfate reducers are abundant in the lower hypolimnion. Maximal concentrations of heterotrophs occur just below the zone of maximal photosynthetic production and at water-sediment interface. (Source: Rheinheimer 1980. Reprinted by permission of John Wiley and Sons Ltd., and VEB Gustav Fischer Verlag.)

thonous at the surface of a lake can be considered, in many cases, allochthonous in the benthic regions. For example, cyanobacteria are found typically in high numbers near the surface, where light penetration is adequate to support their photoautotrophic metabolism. Cyanobacteria that settle below the compensation depth are unable to compete; below the compensation depth they are allochthonous microorganisms. Photoautotrophic members of Chlorobiaceae and Chromatiaceae are autochthonous members of freshwater microbial communities at greater depths, where oxygen tensions are reduced and sufficient hydrogen sulfide is present but where there is still sufficient light penetration. Rhodospirillaceae occupy similar environments but rely on reduced organic electron donors instead of sulfur compounds. Heterotrophic bacterial populations are distributed throughout the vertical water column but usually reach maximum concentrations near the thermocline and near the bottom, where concentrations of available organic matter are high.

Microorganisms found in the sediment of freshwater lakes are usually different from those in the overlying waters. In shallow ponds and lakes, anaerobic photoautotrophic bacteria occur on the surface of the sediment, often conferring characteristic colors on these water bodies. Fungi are found on the debris that accumulates on the sediment surface. These fungi include cellulolytic forms. Bacteria capable of anaerobic respiration are important members of sediment microbiota and include *Pseudomonas* species capable of denitrification activities. Within the sediment, obligately anaerobic bacteria occupy important niches. These bacteria include endospore-forming *Clostridium* species, methanogenic bacteria that produce methane gas, and *Desulfovibrio* species that produce hydrogen sulfide.

Lakes vary in terms of the fungal genera that are present; the differences reflect variations in the organic substrates available for fungal utilization and in the biota that can be attacked by fungal parasites (Sparrow 1968). Many fungi in freshwater lakes, rivers, and streams are associated with foreign organic matter and thus should be considered as allochthonous members of such ecosystems. Many ascomycetes and fungi imperfecti are found on wood and dead plant materials in rivers and lakes; when the plant material is degraded, the associated fungi disappear. Replacement occurs when foreign material enters the hydrosphere. Yeasts are found in many freshwater bodies. Weakly fermentative members of *Torulopsis, Candida, Rhodotorula*, and *Cryptococcus* are the yeasts most commonly found in rivers, streams, and lakes.

Algae are clearly important autochthonous members of freshwater ecosystems. In large, deep lakes, phytoplankton contribute most of the organic carbon, which supports the growth of the heterotrophic organisms in freshwater ecosystems. Much information has been gathered on the distribution of freshwater algae; many treatises have described the algae of particular water bodies (Prescott 1962). The freshwater algae include members of Chlorophycophyta, Euglenophycophyta, Phaeophycophyta (few species), Chrysophycophyta, Cryptophycophyta, Pyrrhophycophyta, and Rhodophycophyta. Species of green algae, dinoflagellates, and diatoms dominate in most freshwater ecosystems.

Protozoa graze on phytoplankton and bacteria in aquatic habitats. The phagotrophic flagellates are especially important grazers of bacterial populations. Amoeboid, ciliated, and flagellated protozoa are found in streams, rivers, and lakes. Numerous genera of protozoa are found in freshwater habitats, including the common protozoans *Paramecium, Didinium, Vorticella, Stentor,* and *Amoeba*. The flagellate protozoan *Bodo* is common in polluted, low-oxygen waters (Westphal 1976).

In addition to autochthonous microbial populations, many allochthonous terrestrial microorganisms are carried by erosion and runoff from soils into freshwater aquatic ecosystems. Allochthonous microorganisms also enter when leaves from adjacent plants fall into these water bodies and when municipal sewage enters freshwater environments together with high concentrations of organic matter. Heterotrophic microbial populations in areas that receive high amounts of organic matter are generally elevated, but as the concentrations of imported organic matter decrease, populations of heterotrophic microorganisms also decline (Taylor 1942). The allochthonous microorganisms generally disappear in a relatively short time, being consumed by autochthonous members of the freshwater ecosystem. The numerous sources of input of allochthonous microorganisms, however, mean that microorganisms found in freshwater bodies often closely resemble terrestrial forms.

Microorganisms play a key role in lake productivity and the transformation of organic compounds within a lake (Kuznetsov 1959) (Figure 9.17). Rates of microbial metabolic activities, though, vary greatly

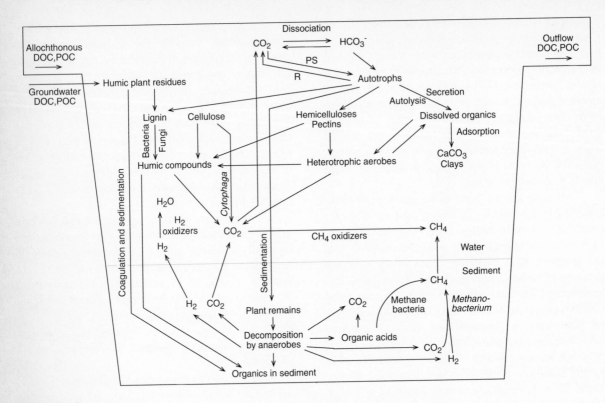

Figure 9.17

Simplified carbon cycle of a typical freshwater lake. DOC=dissolved organic carbon; POC=particulate organic carbon; PS=photosynthesis; R=respiration. The figure illustrates the key role of microorganisms in carbon cycling of a lake habitat. (Source: Wetzel 1975. Based on Kuznetsov 1959. Reprinted by permission, copyright W. B. Saunders Co., Philadelphia.)

seasonally and temporally both within individual freshwater bodies and between different freshwater habitats (Overbeck 1966; Wright and Hobbie 1966) (Table 9.4). There is a clear trend of increasing productivity rates from oligotrophic to eutrophic lakes. The annual ranges of rates of turnover show great variation within lakes and overlap between lakes. Comparison of summer and winter rates, however, shows that rapid turnover occurs during the summer when productivity is also high. Turnover times are slower during winter when productivity is also reduced.

The capacity of microorganisms, especially bacteria, to utilize dissolved organic compounds at very low concentrations is important in oligotrophic lakes that have low concentrations of available organic compounds (Kuznetsov et al. 1979; Ormerod 1983). A comparison of rates of algal and bacterial uptake of glucose (Figure 9.18) shows that at low substrate levels bacteria exhibit the greater uptake velocities, and at higher substrate levels algae show the higher uptake rates. Turnover times for carbon at very low substrate concentrations (less than 5 µg C/L) are much shorter for bacteria than for algae (Wright and Hobbie 1966). The ability of bacteria and some nanoplanktonic flagellate protozoa to utilize low concentrations of organic matter permits introduction of dissolved organic carbon into food webs that support the growth of higher organisms.

Table 9.4

Some examples of rates of primary and secondary productivity in freshwater lakes

Lake*	Primary productivity (mg C/m^3/day)
Schönsee, Germany	16
Schluensee, Germany	17
Grosser Plöner See, Germany	35
Plußsee, Germany	57

Lake†	Turnover T_1 (hours)	Secondary productivity P max (mg C/m^3/day)
Oligotrophic Lawrence, Mich.	40–300	1–80
Mesotrophic Crooked, Ind.	80–470	63–110
Eutrophic Little Crooked, Ind.	36–232	190–205
Lötsjön, Sweden		
Summer	0.4–5	<100
Winter	20–300	<20

*Source: Overbeck 1966.
†Source: Wright and Hobbie 1966.
Substrate = glucose.

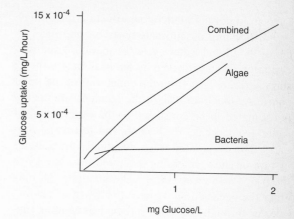

Figure 9.18

The velocity of glucose uptake by a bacterium and a green alga. At low substrate concentrations, uptake of glucose is primarily due to bacteria with little algal activity. At high substrate concentrations, bacteria exhibit saturation kinetics and uptake is greater for algae. (Source: Wright and Hobbie 1966. Reprinted by permission, copyright 1966 Ecological Society of America.)

Fungi and bacteria in freshwater ecosystems are also largely responsible for the decomposition of allochthonous organic matter (Barnes and Mann 1980; Cole 1983). Microorganisms are the initial colonizers of detrital particles. They initiate a food web that results in the recycling of the organic nutrients of detritus within the ecosystem. Microorganisms mediate the transfer of allochthonous organic carbon into the cell biomass carbon of the autochthonous members of freshwater ecosystems; they also play key roles within freshwater habitats in the transformations and cycling of numerous other elements.

The principal ecological functions of microorganisms in freshwater environments can be summarized as follows: (1) They decompose dead organic matter, liberating mineral nutrients for primary production. (2) They assimilate and reintroduce into the food web dissolved organic matter. (3) They perform mineral cycling activities. (4) They contribute to primary production. (5) They serve as a food source for grazers (Kuznetsov 1970). In large, deep lakes where most of the lake floor is disphotic and devoid of higher plants, planktonic microorganisms are the principal primary producers.

MARINE HABITATS

The oceans occupy 71% of Earth's surface, with a volume of 1.46 x 10^9 km^3, an average depth of 4,000 m, and a maximal depth of approximately 11,000 m (Rumney 1968). The huge water masses of the oceans have an important moderating effect on the climate of Earth, being the ultimate reservoir and receptacle of the global water cycle. Evaporation and precipitation of water driven by the heat energy of the sun distribute heat from the equatorial to the polar zones. As much as 50% of the incident sun energy is estimated to be consumed in the evaporation of water from ocean, freshwater, and terrestrial surfaces. The evaporated water eventually precipitates as rain or snow, releasing its stored heat energy in the process (Stewart 1969). The precipitation may enter the oceans directly or indirectly after passing over or through the terrestrial environment as runoff. In the latter process, minerals are leached into the oceans at much higher rates than they can be returned—for example, by ocean spray—to land. Consequently, the oceans are the ultimate sink for all water-soluble minerals and are saline, though not to the same degree as some landlocked lakes in hot, dry climates.

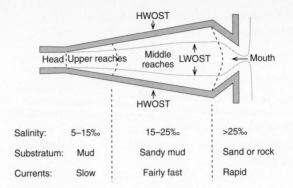

Figure 9.19

Principal environmental characteristics of an estuary. An estuary is characterized by a gradient from low salinity near the upper reaches to high salinity near the mouth. HWOST = high water line of spring tide; LWOST = low water line of spring tide. (Source: Perkins 1974. Based on Day 1950. Reprinted by permission, copyright Academic Press Inc. Ltd., London, and Royal Society of South Africa.)

Estuaries

Freshwater runoff in the form of rivers and groundwater seepage interfaces with marine waters in estuaries, which are characteristically more productive than either the ocean or the freshwater input (Day 1950). They are areas of highly variable environmental parameters in terms of temperature, salinity, pH, organic loading, and other factors. Estuaries are areas of mixing of fresh water and marine water and are typically highly productive regions because nutrients carried by rapidly flowing rivers are deposited at the river mouths or deltas. Estuaries are subject to tides and exhibit tidal flushing. Materials entering estuaries from rivers oscillate with the tides through the estuary with a net movement to the open sea.

Photosynthesis in estuaries almost always exceeds respiratory activities (Wood 1967). Large portions of estuaries are overgrown with semisubmerged higher plants; characteristic among these plants are the various grassy *Spartina* species in temperate-zone salt marshes and the mangrove forests in tropical estuaries. Tide flats are often overgrown with eelgrass (*Zostera*) in the temperate zone and with turtle grass (*Thalassia*) in the tropical zone.

Salt-marsh estuaries may receive nutrients through upwelling of deeper water masses along the continental shelf, but the larger portion of nutrient input usually comes from the adjacent land in the form of runoff. The physical construction of the estuary tends to trap nutrients that enter the estuary or that are produced within the estuary. Salt marsh-estuaries tend to recycle nutrients rapidly within the area with little relative loss to the deeper ocean.

In an ideal estuary, there is a salinity gradient from <5% at the upper end to >25% at the mouth (Perkins 1974) (Figure 9.19). The distinction between autochthonous and allochthonous organisms is particularly difficult in such transition zones (Stevenson et al. 1974). Organisms best adapted for continued survival in such habitats must be eurytolerant to many environmental factors. Both true freshwater and true marine organisms are only transitional members of estuaries.

Characteristics and Stratification of the Ocean

For the most part, environmental conditions in the marine ecosphere are remarkably uniform (Sverdrup et al. 1942; Stewart 1969; Tait and DeSanto 1972; Anikouchine and Sternberg 1973). This great uniformity is brought about by various mixing mechanisms, including tidal movements, currents, and thermohaline circulation.

Tides, which are produced by the gravitational pull of the moon and the sun, have a periodicity of about 12-1/2 hours; high and low tides occur twice daily in most locations. The differences in high and low tides are greatest every two weeks when spring tides occur and the smallest during the alternate weeks when neap tides occur (Figure 9.20).

Ocean currents arise from the frictional drag of the wind blowing across the surface of the water and the rotation of Earth. The rotational (coriolis) force of Earth and land obstacles results in largely circular current systems. Deep currents arise from variations in temperature and salinity, which create differences in water mass densities. These water mass densities cause thermohaline currents that mix the water masses vertically.

The oceans contain almost every naturally occurring chemical element, but most are in extremely low concentrations (Table 9.5) (Harvey 1957; Wenk 1969; Ross 1970; Broecker 1974). The major elements in oceans, aside from the hydrogen

Table 9.5

Chemical composition of seawater

Elements in seawater (ppm)					
Major		**Minor**		**Trace**	
H	1.1×10^5	Br	6.5×10^1	N	5.0×10^{-1}
O	8.6×10^5	C	2.8×10^1	Li	1.7×10^{-1}
Cl	1.9×10^4	Sr	8.0×10^0	Ru	1.2×10^{-1}
Na	1.1×10^4	B	4.6×10^0	P	7.0×10^{-2}
Mg	1.4×10^3	Si	3.0×10^0	I	6.0×10^{-2}
SO_4^{2-}	8.9×10^2	F	1.0×10^0	Fe	1.0×10^{-2}
Ca	4.0×10^2			Zn	1.0×10^{-2}
K	3.8×10^2			Mo	1.0×10^{-2}

Source: Ross 1970.

and oxygen that compose water, are sodium, chlorine, magnesium, sulfate, calcium, and potassium; minor components include carbon, bromine, strontium, boron, silica, and fluorine. Nitrogen, phosphorus, and iron, which are essential for microbial growth, occur in seawater only as trace elements at concentrations less than one ppm.

Salinities of marine habitats are normally in the range of 33%–37%, with an average of 35%. The pH of seawater is generally 8.3–8.5. Temperatures below 100 m of depth are usually between 0°C and 5°C; seasonal temperature fluctuations at any location are usually no more than 20°C, and variations of temperature over all of the oceans are within 35°C.

Vertical and Horizontal Zones of Marine Habitats
Although the environmental conditions of the ocean are highly uniform, definite zones can be recognized (Figure 9.21). The littoral zone, or intertidal zone, which represents the interface between the marine ecosphere and the lithosphere, occurs at the seashore. This zone is subjected to alternate periods of flooding and drying at high and low tides, respectively. The sublittoral zone extends from the low tide mark to the edge of the continental shelf. This region is also

Figure 9.20
General characteristics of a beach habitat. The shoreline profile shows the high and low tide lines.

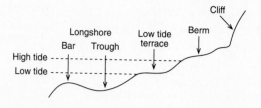

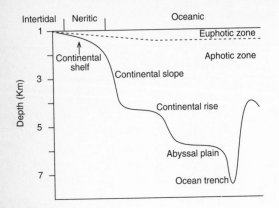

Figure 9.21

Major horizontal zonation in an ocean profile. (Source: Odum 1971. Reprinted by permission, copyright W. B. Saunders, Philadelphia.)

known as the neritic, or nearshore, zone. The average depth in the neritic zone is less than 200 m. The oceanic province extends seaward from the edge of the continental shelf. The term pelagic is used to designate open water or the high sea and includes portions of the neritic and the entirety of the oceanic provinces. The benthos or benthic region refers to the bottom, regardless of the overlying zone (Isaacs 1969).

The benthic region begins at the intertidal zone and extends downward. The continental shelf is a gently sloping benthic region that extends away from the land mass. At the continental shelf edge, the slope greatly increases. The continental slope, also known as the bathyl region, drops down to the sea floor. The deep-sea floor is known as the abyssal plain and usually lies at about 4,000 m. The ocean floor is not flat but has deep ocean trenches and submarine ridges. The deep ocean trenches are known as the hadal region (from the Greek *Hades*, underworld). Ocean trenches extend down to 11,000 m in depth.

As with the freshwater environment, the uppermost layer of the marine ecosphere is the surface tension layer, which interfaces the marine ecosphere with the atmosphere. The seawater-air interface is the habitat for the pleuston, the marine equivalent of the neuston, which includes bacterial and algal inhabitants

(Cheng 1975). *Pseudomonas* and unidentified pigmented genera are major bacterial populations (Wood 1967; Kaneko et al. 1979). A higher proportion of pleuston bacteria have been reported to utilize carbohydrates than bacteria in the underlying water. Populations of primary producers, including cyanobacteria (*Trichodesmium*), diatoms (*Rhizosolenia*), and drifting Phaeophycophyta (*Sargassum*), are sometimes found in the pleuston layer.

There does not appear to be a uniquely adapted algal genus in the pleuston that is equivalent to the *Nautococcus* of the neuston. Representatives of fungi and protozoa are occasionally found in the pleuston, along with various macroscopic invertebrates. Light intensities at the surface of tropical seas are often phototoxic for algae, and here the pleuston community is dominated by heterotrophs that live off organic matter, enriched severalfold in the surface tension layer (Sieburth et al. 1977). The pleuston layer also attracts invertebrates that can become conspicuous during periods of extended calm. The lines "Yea, slimy things did crawl. With legs upon the slimy sea" in Samuel Coleridge's "Rime of the Ancient Mariner" are poetic reference to such a condition (David 1965).

The marine ecosphere may also be divided into vertical zones (Figure 9.22). The euphotic zone is the area of effective light penetration to the compensation level. Below the euphotic zone lies the disphotic (aphotic) zone. The pelagic zone can be divided into an epipelagic zone of 0 to 200 m, which is typically

Figure 9.22

Major vertical zones of an ocean profile.

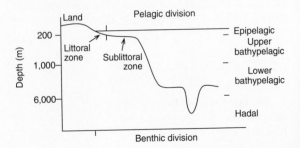

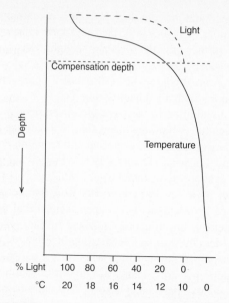

Figure 9.23

Vertical distribution of light and temperature in an ocean profile showing the compensation depth. As shown in this f igure, there is a thermocline just above the compensation depth. Note that temperature scale is not linear.

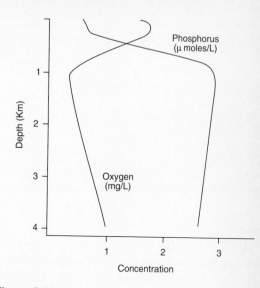

Figure 9.24

Vertical distribution of oxygen and phosphorus in an ocean profile.

euphotic and warm; a bathypelagic zone, extending from 200 to 6,000 m, which is normally disphotic and cold; and a hadal zone, below 6,000 m, which is cold and subject to extreme pressure.

The level of light intensity in the ocean depends largely on turbidity (Figure 9.23); in any case, little light penetrates below 25 m. Temperatures usually drop rapidly in the top 50 m of water, and below 50 m they are normally less than 10°C. Due to the concentrations of salt in seawater, normal freezing temperature is −1.8°C. Oxygen concentrations normally decrease below the surface, reaching a minimum at about 1,000 m with a gradual increase to near surface concentrations between 1,000 and 4,000 m (Figure 9.24). Phosphorus concentrations reach a maximum at 1,000 m and remain in fairly constant concentrations to 4,000 m.

Recycling of mineral nutrients is extremely slow in the pelagic environment. Dead organisms from the euphotic epipelagic zone sink into the great depths of the bathypelagic and ultimately the benthic zone. They carry with them essential nutrients, mainly nitrogen and phosphorus, that are liberated in the perpetual darkness of the deep ocean. From here they are returned to the surface water by upwelling currents at extremely slow rates. It takes several thousand years for the average mineral nutrient molecule to be returned to the warm euphotic surface waters. Consequently, primary production in the euphotic zone of the pelagic environment is severely limited by the lack of mineral nutrients, whereas the nutrient-rich deep waters lack light energy for photosynthetic primary production. As a result, more than 95% of the oceans is characterized by an extremely low rate of productivity, averaging 50 g carbon fixed per m^2 per year (Ryther 1969). In addition, most primary producers in the open oceans are extremely small planktonic algae (Round 1984). This condition results in pelagic food chains that are long and inefficient, supporting few sizable fish per water mass.

Virtually all commercial fishing is restricted to less than 5% of the oceans, primarily to coastal regions and specific upwelling zones (Ryther 1969). The nutrient-

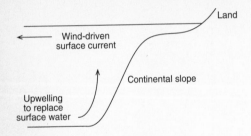

Figure 9.25

Upwelling of deep ocean water along continental slope to replace surface waters driven offshore by wind.

rich waters in these zones allow higher rates of primary production, the producers tend to be larger planktonic forms, and the food chains tend to be shorter and more efficient. Upwelling phenomena (Figure 9.25) often occur along the continental slope, caused by surface currents running rapidly away from the shore and being replaced by deeper, nutrient-rich water.

Although of limited significance to overall oceanic productivity, coral reefs are islands of strikingly high biomass and productivity, surrounded by the "biological desert" of unproductive tropical oceans. The contrast between the coral reef, teeming with fish and invertebrates, and the emptiness of the surrounding sea impresses even the most casual observer and continues to fascinate naturalists and ecologists. Ideal conditions for formation of coral reefs exist in warm tropical seas where the euphotic zone, at least initially, extends to the ocean floor. Central to the coral reef ecosystem is the association between the various species of coral polyps and their endozoic algae (see chapter 5). The algae supply photosynthate (organic carbon) and oxygen to the polyp and assist in the calcium carbonate precipitation that builds the reef. The algae receive protection, CO_2, and mineral nutrients from the plankton capturing and heterotrophic metabolism of the polyp (Johannes 1967). If the seafloor sinks or the water level rises slowly, the annual growth of the reef keeps it in the euphotic zone. If the growth of the reef cannot keep up and the reef sinks below the euphotic zone, the alga-polyp association is disrupted and the reef dies. If there is no sinking, the reef grows to the ocean surface, and wave erosion balances further growth. A change in light transmission of the water—for example, because of suspended sediment from dredging—can also kill a coral reef by interfering with algal photosynthesis.

In addition to the close-circuit nutrient cycling between the polyps and algae, the coral reef has additional mechanisms for capturing and retaining crucial mineral nutrients (DiSalvo 1971). Free-living algae and cyanobacteria associated with reef surfaces extract trace levels of nutrients from the seawater. Various invertebrates graze upon them. Fecal pellets and detritus are efficiently retained in the coral reef ecosystem by various filter-feeding invertebrates. The reef acts as a nutrient trap, while in the surrounding sea nutrients are rapidly lost to the disphotic and unproductive bathypelagic zone. This nutrient-capturing ability of the coral reef was tested using radiotracer experiments and was found to be highly efficient. The nutritional status of coral reefs is further bolstered by the nitrogen-fixing activity of cyanobacteria, principally *Calothrix crustacea* (Wiebe et al. 1975).

Marine sediments exhibit a profile of zonation (Fenchel 1969) (Figure 9.26). The surface sediment has a relatively high O_2 concentration, high NO_3^- concentration, high Fe^{3+} concentration, and low CO_2 concentration. These parameters change slightly

Figure 9.26

Profile of a marine sediment showing E_h, pH, O_2/H_2S, Fe^{3+}, CO_2/CH_4 and $NO_3^-/NO_2^-/NH_3$ concentrations. In the deeper anaerobic layers, the E_h decreases and reduced forms of minerals replace oxidized forms. (Source: Fenchel 1969. Reprinted by permission of Ophelia.)

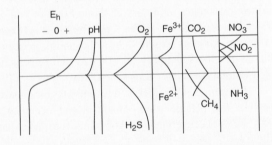

below the surface as the sediment changes from an oxidized zone, through a transitional zone, to a reduced zone. In the reduced zone, O_2 concentrations are virtually zero, and there is reduced sulfide as H_2S. Concentrations of Fe^{3+} disappear and are replaced by Fe^{2+}; CO_2 is, in part, replaced by CH_4; NO_3^- and NO_2^- are replaced by NH_4^+. These chemical changes with depth in the sediment are connected to a change in habitat conditions from aerobic to anaerobic.

Composition and Activity of Marine Microbial Communities

The pelagic marine habitat is a unique environment for both macro- and microorganisms (Isaacs 1969). It completely lacks higher plants; all primary production is carried out by microscopic algae and bacteria. In the euphotic zone, invertebrates and fishes have no opportunity to take refuge in a homogeneous and transparent environment. Glassy transparency and silvery ventral and dark dorsal surfaces offer minimal camouflage. For most fishes, speed is the most important attribute, for escape as well as for predation. Fishes of the epipelagic zone are highly streamlined and powerfully built. Schooling behavior offers some protection and is exhibited by many species. Deep-sea fishes are protected by darkness and are generally not adapted for high-speed swimming. They have to cope with a scarce and unpredictable food supply and often show adaptations for large, infrequent meals (Stockton and DeLaca 1982). Metabolic rates are slow under the prevailing low-temperature, high-pressure conditions (Jannasch and Wirsen 1977; Jannasch 1979).

Microbial numbers are relatively high in nearshore, upwelling, and estuarine waters, but sink as low as 1–100 per ml in pelagic waters (ZoBell 1946; Oppenheimer 1963; Kriss et al. 1967; Wood 1967; Karl and Holm-Hansen 1978). Here, heterotrophic bacteria tend to be associated with algal surfaces or detrital particles, which offer a nutritional advantage compared to the extremely low concentrations of dissolved organic nutrients in the pelagic seawater. Relatively high numbers of microorganisms occur in the first few centimeters of most marine sediments (10^7–10^8/g), but the

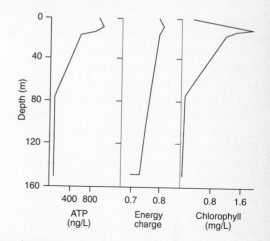

Figure 9.27

Distribution of ATP, energy charge, and chlorophyll in a marine profile approximately 5 km offshore in the California bight. Greatest microbial concentrations and potential metabolic activities occur at approximately 10 m of depth. There is a continuous decline in microbial biomass and energy charge below 20 m (Source: Karl and Holm-Hansen 1978. Reprinted by permission, copyright Springer-Verlag.)

numbers taper off in deeper sediment layers (Karl 1978). The reason seems to be more the depletion of available nutrients than the anaerobic conditions. The highest biomass of microorganisms in marine waters is normally near the surface and decreases with depth (Figure 9.27).

Aspects of Marine Microbial Populations One can ascribe certain features to the autochthonous microorganisms of the marine environment (Macleod 1965). Marine microorganisms should exhibit growth at salinities between 20 and 40 parts per thousand. A salt concentration of 33–35 parts per thousand represents the optimal salt concentration for genuine marine microorganisms. By an admittedly arbitrary definition, true marine bacteria will not grow in the absence of sodium chloride. Marine bacteria require the ions in marine waters to maintain proper membrane functions; for example, sodium and chloride are required for active transport. Some marine bacteria have multiple membranes surrounding their cells. Exposure to fresh water disrupts these membrane layers, causing a loss of viability in these bacteria.

Marine bacteria must be capable of growth at the low-nutrient concentrations found in the oceans. Many marine bacteria, though, will absorb onto detrital particles and grow well under localized conditions that are relatively nutrient rich. Because 90%–95% of the marine environment is below 5°C, marine bacteria must be capable of growth at low temperatures. Except in tropical surface waters, most marine bacteria should be psychrophilic or psychrotrophic. Bacteria found in deep ocean trenches are exposed to great hydrostatic pressures. In such areas, barotolerant bacteria are important members of the autochthonous community.

Most marine bacteria are Gram-negative and motile (ZoBell 1946). Marine bacteria are generally aerobes or facultative anaerobes; relatively few obligate anaerobes are found in marine waters. There is a relatively high proportion of proteolytic bacteria in marine habitats as compared to freshwater or soil habitats. Many bacterial genera contain marine species. *Pseudomonas* or *Vibrio* species are often found to be the dominant genera in marine environments; *Flavobacterium* species are also found in relatively high numbers. In Chesapeake Bay, for example, T. E. Lovelace et al. (1967) found 56% *Vibrio* species, 18% *Pseudomonas* species, and 6% *Flavobacterium* species. *Spirillum, Alcaligenes, Hyphomicrobium, Cytophaga, Microcyclus,* and actinomycetes are also frequently found in marine water samples.

Besides various Gram-negative bacteria, some Gram-positive bacteria, such as *Bacillus* species, are normally found in marine sediments. Below the surface of marine sediments, anaerobic bacteria comprise the autochthonous microorganisms. Marine sediments receive accumulations of organic matter from overlying waters that favor the growth of heterotrophic bacteria. Anaerobic *Desulfovibrio* species are found in marine sediments where these organisms reduce sulfate to hydrogen sulfide. Anaerobic methanogens are normally found in sediment below the layer of available sulfate.

Important populations of chemolithotrophic bacteria are involved in nitrogen cycling in marine waters (Belser 1979). These include members of genera *Nitrosococcus, Nitrosomonas, Nitrospina, Nitrococcus,* and *Nitrobacter.*

Fungal populations in marine ecosystems have been overlooked in the past, but several major treatises have now been compiled on the occurrence of fungi in the ocean (Johnson and Sparrow 1961; Hughes 1975). Some fungi found in marine ecosystems require sodium chloride for growth, others are salt tolerant. *Labyrinthula* species, which form a net plasmodium, are representative slime molds found in marine ecosystems; they are normally found in association with marine algae and plants. Yeasts are frequently found in marine waters (Lodder 1971). The most commonly found yeast genera are *Candida, Torulopsis, Cyrptococcus, Trichosporon, Saccharomyces,* and *Rhodotorula. Rhodosporidium,* a basidiomycete-related yeast, also has been found in marine habitats. Filamentous basidiomycetes, however, are rarely found in marine ecosystems. Occasionally, blooms of yeasts are encountered in discrete marine water bodies, such as within areas of the North Sea (Phaff et al. 1968).

Marine algae supply an essential input of carbon to the marine ecosystem (Taylor 1957; Boney 1966; Dawes 1974). The marine algae include members of Chlorophycophyta, Euglenophycophyta, Phaeophycophyta, Chrysophycophyta, Cryptophycophyta, Phyrrhophycophyta, and Rhodophycophyta. Most Phaeophycophyta, or brown algae, are marine. There are more than 1,500 species of marine brown algae; these algae are a conspicuous intertidal component extending from the upper littoral zone into the sublittoral zone to depths greater than 220 m in clear tropical waters. The marine brown algae include the kelps, such as *Fucus* and *Sargassum.* Members of Chlorophycophyta and Chrysophycophyta are prominent members of plankton. Marine plankton is found in maximal concentrations in the upper regions of the ocean, usually at 0–50 m in depth (ZoBell 1946; Holm-Hansen 1969) (Table 9.6). In very clear tropical waters, because of the high light intensity, the phytoplankton maximum is not found at the surface but at 10–15 m of depth. Green algae are found in greater numbers near the

Table 9.6

Concentration of phytoplankton off the coast of southern California

Depth (m)	Biomass µg C/L		Diatoms/L
	Diatoms	Total phytoplankton	
0	2.1	12.1	10
30	—	—	500
50	5.5	24.4	180
100	0.2	3.3	2
200	0.0	0.7	0

Sources: Holm-Hansen 1969; ZoBell 1946.

surface, usually disappearing below 30 m. Red algae and golden brown algae occur at somewhat greater depths.

Planktonic diatoms can be described as either holoplanktonic, which are either pelagic or littoral species but live an oceanic existence in the sense that they do not depend on the bottom to complete their life cycle; meroplanktonic, species that are pelagic only a portion of their life cycle and spend the remainder of their existence on the bottom; or tychopelagic, species that actually spend the major portion of their life cycle attached to a fixed substratum but enter the surface layers of the sea when forcefully torn from their usual habitat. Planktonic diatoms exhibit structural and physiological adaptations to the marine environment. Structurally, planktonic diatoms have elaborate projections that enlarge the surface area relative to the volume of the cell, slowing down their sinking. Some diatoms selectively adsorb monovalent ions over heavier divalent ions, a physiological adaptation that contributes to buoyancy. Other buoyancy mechanisms utilized by diatoms and other planktonic organisms are oil storage materials and gas bubbles, such as the gas vacuoles of *Trichodesmium* (Denton 1963).

There are occasional blooms of Pyrrophycophyta in marine waters. Blooms of dinoflagellates produce the so-called red tides, when concentrations of these algae become so great as to color the ocean red-brown; these blooms occasionally cover several square kilometers. The toxins produced by some of these dinoflagellates kill fish and other marine organisms. The causes of red tide dinoflagellate blooms are not well documented but may be associated with the surfacing of nutrients by upwelling of deep ocean currents (Round 1984).

Protozoa are an important component of marine zooplankton (Westphal 1976). Marine protozoa may exhibit adaptations to salt concentrations, sometimes exhibiting tolerances of up to 10% NaCl concentrations. Marine protozoa include flagellates, rhizopods, and ciliates. The flagellate Coccolithophoridae are the smallest planktonic protozoa in the sea and the chief constituent of marine plankton. Species of *Radiolaria* and *Acantharia* are also important components of marine plankton. Most *Radiolaria* species only live to a depth of 350 m, but other radiolarians occur at depths down to 4,000–5,000 m. *Tintinnidium* species are marine ciliates that thrive in the upper layers of the sea. Marine protozoa and microcrustacea graze on bacteria, phytoplankton, and smaller forms of zooplankton. The grazing by zooplankton provides a critical link in the marine food web between the very small primary producers and the higher members of the marine food web.

LITHO-ECOSPHERE

Lithosphere habitats occur as land masses, consisting of rocks and soil, and as sediments, already discussed in the section on the hydro-ecosphere. The soil, which arises from the weathering of parent rock materials, is by definition capable of acting as a habitat for biological organisms (Brady 1984).

Rocks

The rocks of Earth's crust are commonly classified as igneous, sedimentary, or metamorphic. Igneous rocks are formed directly by solidification of molten lava and include granite, basalt, and diorite. Quartz and feldspars are the primary minerals of igneous rocks. Sedimentary rocks arise from deposition and consolidation of weathered products of other rocks. For example, sandstone is a sedimentary rock that arises from quartz sands; shale arises from consolidation of clay. Sedimentary rocks include limestone, dolomite, sandstone, and shale.

Sedimentary rocks differ in their chemical composition based on the parent materials. Like other rocks, sedimentary rocks can undergo weathering, leading to the formation of soil. Metamorphic rocks are formed by the metamorphosis, or change, in form of other rocks. Igneous and sedimentary rocks that have been subjected to high pressure and temperatures change their physical form. Igneous rocks may be converted to coarse-grained crystalline gneisses and schists. Sedimentary rocks may be converted to quartzite, from sandstone; slate, from shale; or marble, from limestone.

Rock surfaces provide a suitable habitat for a limited number of microorganisms. Bacteria, algae, fungi, and lichens colonize many rock surfaces. A high proportion of bacteria and fungi found on terrestrial rock surfaces are able to solubilize silicates and other minerals through production of organic acids and chelating agents (Silverman and Munoz 1970). Bacteria and fungi on terrestrial rock surfaces are often found in crevices, which can retain water.

Along the shores of the oceans, large populations of cyanobacteria and algae inhabit the rocky coasts. These microorganisms occupy the supralittoral, or subaerial, zone. This interface zone between marine and terrestrial habitats is washed by the spray of the sea. Microorganisms that colonize the rock surfaces in this zone include cyanobacteria, such as *Calothrix*, Chlorophycophyta, such as *Enteromorpha*, and the Rhodophycophyta, such as *Porphyra* (Round 1984).

Soils

The weathering of rocks that results from physical, chemical, and biological forces reduces rock first to regolith (rock rubble) and then to soil. The five main factors involved in soil formation are parent material, climate, topography, biological activity, and time.

Soil Horizons When soil forms from regolith, a series of distinct horizons develops (Figure 9.28) as a result of the weathering process (Brady 1984). The O groups, or organic horizons, develop above the mineral soil and contain the soil organic matter known as humus; these horizons are formed from plant and animal materials deposited on the surface. The O horizon is divided into an O_1 horizon, where the original plant and animal forms are recognizable, and an O_2 horizon, where the plants and animals have decayed to a point beyond recognition. The A, or eluvial, horizon is a mineral horizon that lies near the soil surface. The A horizon is characterized as a zone of maximal leaching; it is divided into an A_1 layer, where mineral soil is mixed with humus, an A_2 horizon, with maximal leaching of silicate clays, iron oxides, and aluminum oxides, and an A_3 horizon, a transition to the underlying B horizon. The B, or illuvial, horizon is where deposition has taken place and there is the maximal accumulation of materials such

Figure 9.28

Vertical soil profile showing soil horizons. (Source: Buckman and Brady 1969. Reprinted by permission, copyright Macmillan Publishing Co.)

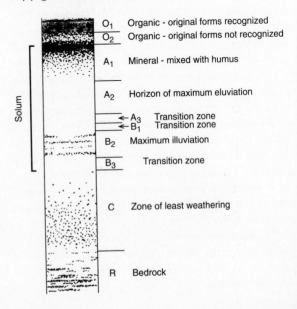

as iron oxides, aluminum oxides, and silicate clays. The combined A and B horizons are known as the solum. Biological activities do not greatly affect the C horizon, beneath the solum; the C horizon may contain accumulations of calcium and magnesium carbonates. Beneath the C horizon lie regolith and bedrock.

Soil Texture   Soils may also be classified by the relative proportions of clay, silt, and sand particles they contain (Brady 1984). With the aid of a soil triangle, relative proportions of these particles are used to determine the texture type of the soil (Figure 9.29). Soils that are dominated by one size class of particles are named according to that class, that is, as sand soils, silt soils, or clay soils; soils not dominated by any specific particle size are called loams. Intermediate classes of soil texture are recognized— for example, sandy loam soils. Soil texture is an important descriptive measure of the microbial habitat, as it describes in part the spatial interactions that can occur among microorganisms occupying that habitat. Soils are frequently described by their tex-

Figure 9.29
Soil texture triangle showing classes of soil based on proportions of sand, silt, and clay. To determine the class name, one line is drawn from the percent silt of the soil parallel to the percent clay side of the triangle, and another is drawn from the percent clay of the soil parallel to the percent sand side of the triangle. The class is given by the segment in which the two lines intersect. (Source: Alexander 1977. Reprinted by permisssion of John Wiley and Sons.)

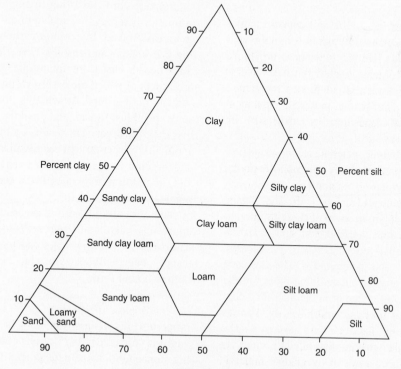

Table 9.7

Selected properties of major clay colloids

Property	Montmorillonite	Illite	Kaolinite
Size (μm)	0.01–1.0	0.1–2.0	0.1–5.0
Surface (m^2g)	700–800	100–200	5–20
Cation exchange capacity (meq/100 g)	80–100	15–40	3–15

ture and location, such as a New Jersey sandy loam or a Georgia clay.

When considering soil particles as habitats for microorganisms, it is also important to consider the nature of the clay particles (Marshall 1980). Clay colloids differ significantly in their physical and chemical properties (Table 9.7). These differences influence how many and what types of microorganisms can occupy the particular soil habitat (Hattori and Hattori 1976).

Soil Organic Matter The soil organic matter derives from the remains of plants, animals, and microbes (Bear 1964). Humic substances are those portions of the soil organic matter that have undergone sufficient transformation to render the parent material unrecognizable. Humic materials present in mineral soils typically constitute less than 10% by weight of the soil.

The genesis of humic material is a two-stage process involving the predominantly microbial degradation of organic polymers to monomeric constituents, such as phenols, quinones, amino acids, and sugars, and the subsequent polymerization of these by spontaneous chemical reactions, autoxidation, and oxidation catalyzed by microbial enzymes such as laccases, polyphenoloxidases, and peroxidases. The aromatic ring structures that serve as the building blocks of the humic acid core may originate from the microbial degradation of lignin or may be synthesized by various microorganisms from other carbon substrates. The humic material is in a dynamic state of equilibrium, its synthesis being compensated for by gradual mineralization of the existing material (Stevenson 1982).

According to their solubility characteristics, humic substances can be fractionated into fulvic acid, humic acid, and humin. None of these fractions can be assigned a definite chemical structure; all three are randomly assembled irregular polymers. The main differences between fulvic and humic acids are the lower molecular weight, higher oxygen to carbon ratio, and higher ratio of acidic functional groups per weight of the former as compared to the latter, but the spectrum is continuous and the dividing line is arbitrary. Molecular weights range from ~700 to ~300,000. Humin is regarded as a strongly bound complex of fulvic and humic acids to mineral material rather than a class of compounds by itself. The alcohol-soluble hymatomelanic acid, a minor fraction, consists of esterified or methylated humic acids.

Humic compounds are random polymers, and at best we can establish only their type structures. The theories on humic type structures are controversial and subject to constant revision and refinement. Perhaps the most accepted current theory visualizes an aromatic core consisting of single and condensed aromatic, heterocyclic, and quinoidal rings, linked and cross-linked by carbon-carbon, ether, amino, and azo bonds (Stevenson 1976) (Figure 9.30). The rings bear a variety of functional groups, the more prominent of which are carboxyl, phenolic hydroxyl, and carbonyl groups. Attached to this core are amino acids, peptides, sugars, and phenols, which form further cross linkages. The result is a three-dimensional spongelike structure that readily absorbs water, ions, and organic molecules in an exchangeable manner and, in addition, may chemically bind suitable compounds to its reactive functional groups. As a consequence, virtually all natural organic compounds and apparently also numerous human-made chemicals can occur in bound or absorbed form in humic substances; even active enzymes have been recovered in humus-bound form.

Chemical Properties of Soils The chemical properties of soils are important factors for microorganisms (Alexander 1977). Soils contain vastly differing concentrations and chemical forms of organic carbon, inorganic and organic nitrogen, and available

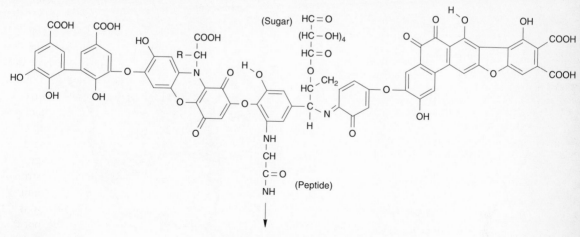

Figure 9.30

Proposed type structure for humic acid. (Source: Stevenson 1976, *Bound and Conjugated Pesticide Residues.* Reprinted by permission, copyright American Chemical Society.)

inorganic phosphorus (Brady 1984) (Table 9.8). The composition of the soil atmosphere (atmosphere-lithosphere interface) also varies greatly between soils. The soil atmosphere exists in the porous spaces between soil particles. Bulk density is a measure of packing of soil particles and indicates the extent of space that may be occupied by the soil atmosphere. At

Table 9.8

Ranges of nutrient concentration in temperate surface mineral soils

Nutrient	Concentration range (%)
Organic matter	0.40–10.0
Nitrogen	0.02–0.5
Phosphorus	0.01–0.2
Potassium	0.17–3.3
Calcium	0.07–3.6
Magnesium	0.12–1.5
Sulfur	0.01–0.2

Source: Brady 1984.

times the soil pores are filled with water, which displaces the soil atmosphere. Some soils or soil layers are aerobic, that is, the soil atmosphere contains oxygen, whereas others are anaerobic, that is, there is no free oxygen in the soil atmosphere. Even in aerobic soil layers, there are anoxic regions devoid of free oxygen (Sexstone et al. 1985) (Figure 9.31). The oxygen content of the soil atmosphere determines in large part the types of metabolism that can occur and the chemical transformations that the indigenous microorganisms can carry out.

Concentrations of CO_2 in the soil atmosphere commonly are one to two orders of magnitude higher than in the above-ground air. The concentrations of both CO_2 and O_2 in the soil atmosphere are affected by gas diffusion and microbial respiration. Concentrations of CO_2 generally increase, while O_2 concentrations decrease with depth in the soil column. In oxygen-deficient soils, other gases, including N_2O and N_2 from denitrification, CH_4 from methanogenesis, and H_2S from anaerobic sulfate reduction, occur in high concentrations in the soil atmosphere.

The plant cover of the soil is an important factor in determining the types and numbers of microorgan-

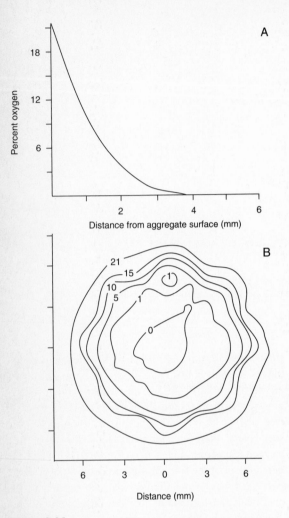

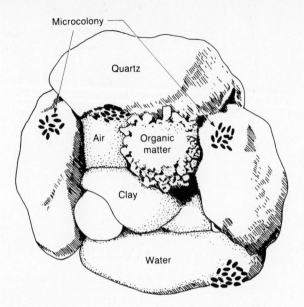

Figure 9.32

Section through a soil crumb showing microhabitats and patchy distribution of bacterial microcolonies. The section also shows the occurrence of water and air within pore spaces. (Source: Brock 1979. Reprinted by permission of Prentice-Hall, Englewood Cliffs, N.J.)

Figure 9.31

Oxygen concentrations surrounding soil aggregates. (A) Oxygen profile from a cultivated silt loam soil aggregate showing the rapid decrease in oxygen concentration beneath the surface and the occurrence of an anaerobic zone. (B) Map of oxygen concentration within an aggregate from a cultivated soil. The profile reveals an anaerobic zone near the center of the aggregate. The numbers show percent oxygen contours. (Source: Sexstone et al. 1985.)

interactions, as well as plant susceptibility to pathogens, exert selective influences on the soil microbial community.

Soil Microbial Communities Soil is generally a favorable habitat for the proliferation of microorganisms, with microcolonies developing on soil particles (Figure 9.32). Numbers of microorganisms in soil habitats (Table 9.9) normally are much higher than those found in freshwater or marine habitats (Mishustin 1975). Typically, 10^6–10^9 bacteria per gram are found in soil habitats. Microorganisms found in soil include viruses, bacteria, fungi, algae, and protozoa. Concentrations of organic matter are relatively high in soils, which favor the growth of heterotrophic microorganisms.

Sergei Winogradsky designated as "autochthonous" that part of the soil microbial community capable of utilizing refractory humic substances. Slow but

isms in that soil. Plant root exudates and senescent parts of plants are important sources of nutrients for soil microorganisms. Rhizosphere and mycorrhizal

Table 9.9

The average number of major groups of microorganisms in different soil types of the USSR

Zone	Soils	Total number of microorganisms ($\times 10^6$/g)	Non-spore-forming bacteria (%)	Bacilli (%)	Actinomycetes (%)	Fungi (%)
Tundra and taiga	Tundra-gley and gley-podzolic	2.1	94.9	0.7	1.5	2.9
Forest meadow	Podzols and soddy-podzolic	1.1	77.2	12.0	8.1	2.7
Meadow steppe	Chernozems	3.6	42.4	21.4	35.4	0.8
Dry steppe	Chestnut	3.5	45.4	19.4	34.6	0.6
Desert steppe and desert	Brown and sierozems	4.5	45.7	17.7	36.1	0.5

Source: Mishustin 1975.

constant activity is characteristic of these organisms, most of which are Gram-negative rods and actinomycetes. Winogradsky contrasted zymogenous or opportunistic soil organisms with autochthonous microorganisms. The former generally are not able to utilize humic compounds, but exhibit high levels of activity and rapid growth on easily utilizable substrates that become available in the form of plant litter, animal droppings, and carcasses. Intermittent activity with inactive resting stages is characteristic of such zymogenous organisms. *Pseudomonas, Bacillus, Penicillium, Aspergillus*, and *Mucor* are some of the bacterial and fungal genera of typical zymogenous organisms.

Zymogenous is not, however, synonymous with allochthonous. Although only intermittently active, zymogenous organisms are true indigenous soil forms. The term allochthonous should be reserved for those organisms, such as human and animal pathogens, that do not find suitable growth conditions in soil.

It is difficult to ascribe generalized adaptive features to the indigenous soil microbiota. Soils have many microhabitats, and at a particular location there may even be several microenvironmental situations that would favor different indigenous populations. In the rich litter horizons of surface soils, indigenous

microorganisms tolerate and grow on high concentrations of organic nutrients. Bacteria found in soils may be obligate aerobes, facultative anaerobes, microaerophiles, or obligate anaerobes. Individual soils may favor bacterial populations with a particular type of metabolism. For example, anoxic conditions in flooded soils favor proliferation of facultative and obligate anaerobes.

Although it may not be possible to describe the general features of indigenous soil bacteria, the abiotic parameters of some soils restrict the microbial populations that can develop there. For example, some soils have extremely alkaline pH values, while others are extremely acidic. In such soils, the indigenous microbial populations must possess adaptive features that allow them to grow there. Polar soils are frozen much of the year; the indigenous microorganisms in such soils must be psychrophilic or psychrotrophic. Desert soils are usually hot and arid. Indigenous microorganisms in desert soils must be capable of withstanding long periods of desiccation and high temperatures. Bacteria with adaptive features, such as the endospores of *Bacillus* species, are well suited for survival in such desert soils between the sporadic rains that provide sufficient water for growth.

Many bacterial genera occur in soil (Gray and Parkinson 1968). A higher proportion of Gram-positive bacteria is found in soil than in marine or freshwater habitats, but in absolute numbers Gram negatives predominate also in soil. A higher proportion of indigenous soil bacteria utilize substrates such as carbohydrates than bacterial populations found in the hydrosphere. Common bacterial genera found in soil include *Acinetobacter, Agrobacterium, Alcaligenes, Arthrobacter, Bacillus, Brevibacterium, Caulobacter, Cellulomonas, Clostridium, Corynebacterium, Flavobacterium, Micrococcus, Mycobacterium, Pseudomonas, Staphylococcus, Streptococcus,* and *Xanthomonas.* There are wide differences in the relative proportions of individual bacterial genera found in particular soils (Alexander 1977) (Table 9.10).

Actinomycetes can comprise a significant proportion of the bacterial population in the soil, about 10%–33% (Hattori and Hattori 1973; Alexander 1977). The genera *Streptomyces* and *Nocardia* are the most abundant actinomycetes found in the soil. *Micromonospora, Actinomyces,* and many other actinomycetes are indigenous to soils but generally are present in low numbers. Actinomycetes are relatively resistant to desiccation and can survive under conditions of drought in desert soils. They favor alkaline or neutral pH and are sensitive to acidity. Myxobacteria are found in soils and on forest litter. Representative genera of myxobacteria found in soils include *Myxococcus, Chondrococcus, Archangium,* and *Polyangium.*

Important photoautotrophic bacterial populations in soil include the cyanobacterial species *Anabaena, Calothrix, Chroococcus, Cylindrospermum, Lyngbya, Microcoleus, Nodularia, Nostoc, Oscillatoria, Phormidium, Plectonema, Schizothrix, Scytonema,* and *Tolypothrix.* Some of these species, such as *Nostoc,* provide both fixed forms of nitrogen and organic carbon in some soil habitats. Availability of fixed forms of nitrogen is an important limiting factor in soil for microbial activities and growth of higher plants. Cyanobacteria form surface crusts on soils bare of plant growth and contribute to soil stabilization. *Azotobacter* is an important heterotrophic free-living soil bacterium, capable of converting atmospheric nitrogen to fixed forms of nitrogen. Some anaerobic *Clostridium* species also fix nitrogen in soil; *Rhizobium* and the slower growing *Bradyrhizobium* fix atmospheric nitrogen within the nodules of certain plant roots. Several chemolithotrophic bacteria perform transformations of inorganic compounds that are essential for the maintenance of soil fertility.

A variety of sources for the allochthonous microorganisms can be found in soil habitats. Allochthonous organisms may enter from the air, from the hydrosphere, or in association with plant or animal material. For example, some plant pathogens enter with diseased plant tissue. Species of *Agrobacterium, Corynebacterium, Erwinia, Pseudomonas,* and *Xanthomonas* frequently enter the soil together with infected plant material. Some allochthonous microorganisms enter with animal droppings or sewage. Some of these bacteria are pathogens. For example, fecal streptococci and *Salmonella* species are found in soils contaminated with sewage. Such allochthonous bacteria normally are eliminated rapidly from soil ecosystems. In some cases, however, allochthonous

Table 9.10

Relative proportions of aerobic-facultatively anaerobic bacterial genera commonly found in soils

Genus	Percentage
Arthrobacter	5–60
Bacillus	7–67
Pseudomonas	3–15
Agrobacterium	1–20
Alcaligenes	1–20
Flavobacterium	1–20
Corynebacterium	2–12
Micrococcus	2–10
Staphylococcus	<5
Xanthomonas	<5
Mycobacterium	<5

Source: Alexander 1977.

microorganisms are able to survive for prolonged periods of time in soils. Endospores and other resistant forms of microorganisms are able to remain dormant, even in dry soils.

Fungi constitute a high proportion of the microbial biomass in soil (Gilman 1945; Domsch et al. 1980). Most types of fungi can be found in soil, either as indigenous or as allochthonous organisms. Soil fungi may occur as free-living organisms or in mycorrhizal association with plant roots. Fungi are found primarily in the top 10 cm of the soil and are rarely found below 30 cm. They are most abundant in well-aerated soils.

The most frequently isolated fungi from soils are members of fungi imperfecti, such as species of *Aspergillus, Geotrichum, Penicillium,* and *Trichoderma,* but numerous ascomycetes and basidiomycetes also occur in high numbers (Table 9.11). Some soil fungi, especially those found in association with plant roots, are difficult if not impossible to isolate and identify.

Table 9.11

Some representative genera of filamentous soil fungi

Lower fungi	Fungi imperfecti
Rhizopus	*Phoma*
Mucor	*Cephalosporium*
Allomyces	*Geotrichum*
Saprolegnia	*Aspergillus*
Pythium	*Penicillium*
	Aureobasidium
Basidiomycetes	*Cladosporium*
Agaricus	*Helminthosporium*
Amanita	*Alternaria*
Coprinus	*Fusarium*
Rhizoctonia	*Trichoderma*
Russula	*Arthrobotrys*
Boletus	
Ascomycetes	
Morchella	

Yeasts are common in most soils. As with the filamentous fungi, most indigenous soil yeasts are fungi imperfecti (Deuteromycota). Species of the genera *Candida, Rhodotorula,* and *Cryptococcus* are probably the most abundant indigenous yeasts in the soil. Species of *Lipomyces, Schwanniomyces, Kluyveromyces, Schizoblastosporion, Hansenula, Candida,* and *Cryptococcus* have been isolated exclusively from soils. The fact that these organisms have been found only in soils implies, but does not prove, that soil is their natural habitat. Many other yeasts are allochthonous organisms in soil, often entering with plant materials.

Most fungi in soil are opportunistic. They grow and carry out active metabolism when conditions are favorable, which implies adequate moisture, adequate aeration, and relatively high concentrations of utilizable substrates. Many soil fungi metabolize carbohydrates, including polysaccharides, and even allochthonous fungi that enter the soil often can grow and metabolize the major components of plant residues; relatively few fungal species, though, are able to degrade lignin (Garrett 1981).

Dormancy is a typical condition of soil fungi. Some fungi have been shown to remain dormant but viable for decades. In the absence of available substrates, they are inactive. Numerous fungi occur in soil as specialized dormant structures, which include sporangiospores, conidia, oospores, ascospores, basidiospores, chlamydospores, and sclerotia. Fungal mycelia may also be metabolically inactive in soil.

In soil, widespread inhibitory effects on germination of fungal spores, known as soil fungistasis, are evident (Lockwood 1977; Lockwood and Filonow 1981). Fungistasis occurs in most soils except some deep subsoils, highly acidic soils, and cold-dominated soils. The addition of readily decomposable organic materials reverses soil fungistasis. With the reversal of fungistasis, fungal spores and other propagules germinate, and fungal mycelia resume active growth. Soil fungistasis appears to be associated with microbial activity; sterilization eliminates fungistasis. The nature of the fungistatic agent(s), however, remains to be identified.

The great absorptive capacity of soil and its numerous microhabitats strongly influence interac-

Table 9.12

Annual decomposition by fungi and productivity of plants in Alaska

Site	Decomposition		Measured primary productivity g/m^2/yr	Indicated rate of organic accumulation g/m^2/yr
	g/m^2/yr	% wt loss cellulose		
Eagle	235	3.4	203	−32
Pt. Barrow	310	3.3	275	−35
Birch Forest	350	5.2	375	+25
Spruce Forest	198	3.4	180	−18

Source: Flanagan 1978.

tions between soil bacteria and fungi. For example, the presence or absence of montmorillonite-type clay, with its great absorptive and water-holding capacity, was shown to be the decisive factor in the outcome of some survival and competition experiments. G. Stotzky (1965) devised an ingenious adaptation of the replica-plating technique to study microbial interactions directly in soil. Sterilized soil was enclosed in a Petri dish, inoculated at marked points with selected microorganisms, and incubated. Periodically, the soil was sampled by an array of sterile stainless steel needles mounted on a disc to make a brush- or nailboard-type of device. This sampling device was subsequently touched to an agar medium, on which a colony pattern developed, replicating the distribution of the microorganisms in the soil. If colony morphology was insufficient to distinguish the interacting microorganisms, agar plates of diagnostic media were used. The spatial and temporal pattern of microbial growth obtained using this technique has revealed numerous positive and negative interactions to be influenced by the soil matrix.

A number of genera of algae inhabit the soil, living both on the soil surface and within the soil. Algae found in soil include members of Chlorophycophyta, Rhodophycophyta, Euglenophycophyta, and Chrysophycophyta. Most soil algae can be found growing on the soil surface or within the top millimeters of the soil, where up to 10^6 algal cells per gram sometimes are found. Algae that are indigenous

at the soil surface may move into the subsurface soil horizons, where they become allochthonous organisms and where other organisms consume them.

Free-living protozoa are found in most soil samples. Soil protozoa are small in size and low in diversity compared to those in aquatic environments. Before a protozoan is designated as a soil form, its vegetative stage must be found in the soil; the cysts of nonsoil protozoans often enter soil ecosystems. Flagellate protozoa are dominant in terrestrial habitats. Protozoan populations of soils are generally 10^4–10^5 organisms per gram. Protozoa are generally found in greatest abundance near the surface of the soil, within the top 15 cm. Most protozoa require relatively high concentrations of oxygen; this limits their distribution within the soil column. Protozoa are important predators on soil bacteria and algae.

As in aquatic environments, microorganisms in soil are agents of biodegradation and mineral recycling. Important plant polymers such as cellulose and lignin are generally recycled by microbial activity (Flanagan 1978) (Table 9.12). Because soil is a nutrient-rich environment, the numbers and diversity of heterotrophic microorganisms, especially of bacteria and fungi, are unusually high. Of great importance are nitrogen-fixing and mycorrhizal associations between microorganisms and higher plants. On the other hand, in the soil environment, where higher plants dominate, microorganisms play a subordinate role in primary production.

SUMMARY

Within habitats, some microorganisms are autochthonous or indigenous; these organisms fill the functional niches of the ecosystem. Other organisms are allochthonous, or foreign, to the ecosystem; they may survive for varying periods of time but are viewed as transients that do not fill the ecological niches of that ecosystem.

The atmosphere is not known to support autochthonous microbial populations, but it serves as a medium for the rapid and global dispersal of many types of microorganisms. There are important transfers of microorganisms and gaseous metabolites among the atmosphere, hydrosphere, and lithosphere. The hydrosphere and lithosphere, in contrast, contain large and diverse autochthonous microbial populations. These populations generally have physiological adaptations that allow them to survive and carry out metabolic activities that provide for energy flow through the ecosystem and for materials cycling within the system. Microorganisms are the principal producers as well as decomposers in aquatic ecosystems. In marine habitats, for example, planktonic microorganisms have a nearly exclusive role in primary production. Higher plants and benthic macroalgae contribute significantly to primary production only in estuaries and littoral areas. In soils, microorganisms play a subordinate role to plants as primary producers but have a critical role in organic matter decomposition and mineral cycling.

REFERENCES & SUGGESTED READINGS

Ainsworth, G. C., and A. S. Sussman. 1968. *The Fungi.* Vol. 3, *The Fungal Population.* Academic Press, New York.

Alexander, M. 1977. *Introduction to Soil Microbiology.* 2d ed. John Wiley and Sons, New York.

Andersen, A. A. 1958. New sampler for the collection, sizing, and enumeration of viable air-borne particles. *Journal of Bacteriology* 76:471–484.

Anikouchine, W. A., and R. W. Sternberg. 1973. *The World Ocean.* Prentice-Hall, Englewood Cliffs, N.J.

Barnes, R. S. K., and K. H. Mann. 1980. *Fundamentals of Aquatic Ecosystems.* Blackwell Scientific Publications, Oxford, England.

Baylor, E. R., V. Peters, and M. B. Baylor. 1977. Water-to-air transfer of virus. *Science* 197:763–764.

Bear, F. E. 1964. *Chemistry of the Soil.* Rheinhold Publishing Co., New York.

Belser, L. W. 1979. Population ecology of nitrifying bacteria. *Annual Review of Microbiology* 33:309–334.

Blanchard, D. C., and L. Syzdek. 1970. Mechanism for the water-to-air transfer and concentration of bacteria. *Science* 170:626–628.

Boney, A. D. 1966. *The Biology of Marine Algae.* Hutchinson Educational, London.

Bovallius, A., B. Bucht, R. Roffey, and P. Anas. 1978. Three-year investigation of the natural airborne bacterial flora at four locations in Sweden. *Applied and Environmental Microbiology* 35:847–852.

Bowden, J., P. H. Gregory, and C. G. Johnson. 1971. Possible wind transport of coffee leaf rust across the Atlantic Ocean. *Nature* (London) 229:500–501.

Brady, N. C. 1984. *The Nature and Properties of Soils.* Macmillan, New York.

Brock, T. D. 1978. *Thermophilic Microorganisms and Life at High Temperatures.* Springer-Verlag, New York.

Brock, T. D. 1979. *Biology of Microorganisms.* Prentice Hall, Englewood Cliffs, NJ.

Brodie, H. J. 1951. The splash-cup dispersal mechanism in plants. *Canadian Journal of Botany* 29:224–231.

Broecker, W. S. 1974. *Chemical Oceanography.* Harcourt Brace Jovanovich, New York.

Buller, A. H. R. 1909. *Researches on Fungi.* Vol. I. Longmans, Green, and Co., London.

Buller, A. H. R. 1934. *Researches on Fungi.* Vol. VI. Longmans, Green, and Co., London.

Burns, R. G. 1979. Interaction of microorganisms, their substrates and their products with soil surfaces. In D. C. Ellwood, J. Melling, and P. Rutter (eds.). *Adhesion of Microorganisms to Surfaces.* Academic Press, London, pp. 109–138.

Cairns, J. (ed.). 1970. *The Structure and Function of Freshwater Microbial Communities.* American Microscopical Society Symposium, Research Division Monograph 3. Virginia Polytech Institute and State University, Blacksburg.

Caldwell, D. E. 1977. The planktonic microflora of lakes. *Critical Reviews in Microbiology* 5:305–370.

Cheng, L. 1975. Marine pleuston: Animals at the sea-air interface. *Annual Reviews in Oceanography and Marine Biology* 13:181–212.

Cole, G. A. 1983. *Textbook of Limnology.* C. V. Mosby Co., St. Louis, Mo.

Colwell, R. R., and R. Y. Morita (eds.). 1972. *Effects of the Ocean Environment on Microbial Activities.* University Park Press, Baltimore.

David, P. M. 1965. The surface fauna of the ocean. *Endeavor* 24:95–100.

Dawes, C. J. 1974. *Marine Algae of the West Coast of Florida.* University of Miami Press, Coral Gables, Fla.

Day, J. H. 1950. The ecology of South African estuaries. Part 1 of a review of estuaries in general. *Transactions of the Royal Society of South Africa* 33:53–91.

Denton, E. J. 1963. Buoyancy mechanisms of sea creatures. *Endeavour* 22:3–8.

DiSalvo, L. H. 1971. Regenerative functions and microbial ecology of the coral reefs: Labeled bacteria in a coral reef microcosm. *Journal of Experimental Marine Biology and Ecology* 7:123–136.

Domsch, K. H., W. Gams, and T. H. Anderson. 1980. *Compendium of Soil Fungi.* Academic Press, New York.

Dundas, I. D., and H. Larsen. 1962. The physiological role of carotenoid pigments of *Halobacterium salinarium.* *Archiv für Mikrobiologie* (Berlin) 44:233–239.

Edmonds, R. L. (ed.). 1979. *Aerobiology:The Ecological Systems Approach.* Dowden, Hutchinson and Ross, Stroudsburg, Penn.

Fenchel, T. 1969. The ecology of marine microbenthos. Part IV. *Ophelia* 6:1–182.

Flanagan, P. W. 1978. Microbial decomposition in Arctic tundra and subarctic taiga ecosystems. In M. W. Loutit and J. A. R. Miles (eds.). *Microbial Ecology.* Springer-Verlag, New York, pp. 161–169.

Fletcher, M. 1979. The attachment of bacteria to surfaces in aquatic environments. In D. C. Ellwood, J. Melling, and P. Rutter (eds). *Adhesion of Microorganisms to Surfaces.* Academic Press, London, pp. 87–108.

Garrett, S. D. 1981. *Soil Fungi and Soil Fertility: An Introduction to Soil Mycology.* Pergamon Press, Elmsford, N.Y.

Gilman, J. C. 1945. *A Manual of Soil Fungi.* Collegiate Press, Ames, Iowa.

Gray, T. R. G., and D. Parkinson. 1968. *The Ecology of Soil Bacteria.* University of Toronto Press, Toronto, Canada.

Gregory, P. H. 1973. *The Microbiology of the Atmosphere.* John Wiley and Sons, New York.

Harvey, H. W. 1957. *The Chemistry and Fertility of Sea Waters.* Cambridge University Press, Cambridge, England.

Hattori, T., and R. Hattori. 1973. *Microbial Life in the Soil: An Introduction.* Marcel Dekker, New York.

Hattori, T., and R. Hattori. 1976. The physical environment in soil microbiology: An attempt to extend principles of microbiology to soil microorganisms. *Critical Reviews in Microbiology* 4:423–462.

Hirst, J. M., O. J. Stedman, and G. W. Hurst. 1967. Long-distance spore transport: Vertical sections of spore clouds over the sea. *Journal of General Microbiology* 48:357–377.

Holm-Hansen, O. 1969. Determination of microbial biomass in ocean profiles. *Limnology and Oceanography* 14:740–747.

Hughes, G. C. 1975. Studies on fungi in oceans and estuaries since 1961. Part I. *Lignicolous, Caulicolous,* and *Foliicolous* species. *Annual Review of Oceanography and Marine Biology* 13:69–180.

Hutchinson, G. E. 1957. *A Treatise on Limnology.* John Wiley and Sons, New York.

Hynes, H. B. N. 1970. *Ecology of Running Waters.* Liverpool University Press, Liverpool, England.

Ingold, C. T. 1971. *Fungal Spores: Their Liberation and Dispersal.* Clarendon Press, Oxford, England.

Isaacs, J. P. 1969. The nature of oceanic life. *Scientific American* 221:147–162.

Jannasch, H. W. 1979. Microbial turnover of organic matter in the deep sea. *BioScience* 29:228–232.

Jannasch, H. W., and C. O. Wirsen. 1977. Microbial life in the deep sea. *Scientific American* 236(6):42–52.

Jensen, W. A., and F. B. Salisbury. 1972. *Botany: An Ecological Approach.* Wadsworth Publishing Co., Belmont, Calif.

Johannes, R. E. 1967. Ecology of organic aggregates in the vicinity of coral reef. *Limnology and Oceanography* 12:189–195.

Johnson, T. W., and F. K. Sparrow, Jr. 1961. *Fungi in Oceans and Estuaries.* Hafner Publishing Co., New York.

Kaneko, T., M. I. Krichevsky, and R. M. Atlas. 1979. Numerical taxonomy of bacteria from the Beaufort Sea. *Journal of General Microbiology* 110:111–125.

Karl, D. M. 1978. Distribution, abundance, and metabolic state of microorganisms in the water column and sediments of the Black Sea. *Limnology and Oceanography* 23:936–949.

Karl, D. M., and O. Holm-Hansen. 1978. Methodology and measurement of adenylate energy charge ratios of environmental samples. *Marine Biology* 48:185–197.

Kriss, A. E., I. E. Mishustina, N. Mitskerich, and E. U. Zemetsora. 1967. *Microbial Population of Oceans and Seas.* Edward Arnold, London.

Kuznetsov, S. I. 1959. *Die Rolle der Mikroorganismen im Stroffkreislauf der Seen.* VEB Deutscher Verlag für die Wissenschaften, Berlin.

Kuznetsov, S. I. 1970. *Microflora of Lakes and Their Geochemical Activity* (in Russian). Izdatel'otro Nauka, Leningrad.

Kuznetsov, S. I., G. A. Dubinina, and N. A. Lapteva. 1979. Biology of oligotrophic bacteria. *Annual Review of Microbiology* 33:377–388.

Litchfield, C. D. 1976. M*arine Microbiology.* Benchmark Papers in Microbiology, Vol. 11. Academic Press, New York.

Lockwood, J. L. 1977. Fungistasis in soils. *Biological Reviews* 52:1–43.

Lockwood, J. L., and A. B. Filonow. 1981. Responses of fungi to nutrient-limiting conditions and to inhibitory substances in natural habitats. *Advances in Microbial Ecology* 5:1–62.

Lodder, J. 1971. *The Yeasts.* North Holland Publishing Co., Amsterdam.

Lovelace, T. E., H. Tubiash, and R. R. Colwell. 1967. Quantitative and qualitative commensal bacterial flora of *Crassostrea virginica* in Chesapeake Bay. *Proceedings of the National Shellfisheries Association* 58:82–87.

McKissich, G. E., L. G. Wolfe, R. L. Farrell, R. A. Greisemer, and A. Hellman. 1970. Aerosol transmission of oncogenic viruses. In I. H. Silver (ed.). *Aerobiology: Proceedings of the Third International Symposium.* Academic Press, New York, pp. 233–237.

McLean, R. C. 1918. Bacteria of ice and snow in Antarctica. *Nature* (London) 102:35–39.

Macleod, R. A. 1965. The question of the existence of specific marine bacteria. *Bacteriological Reviews* 29:9–23.

Marshall, K. C. 1976. *Interfaces in Microbial Ecology.* Harvard University Press, Cambridge, Mass.

Marshall, K. C. 1980. Adsorption of microorganisms to soils and sediments. In G. Bitton and K. C. Marshall (eds.). *Adsorption of Microorganisms to Surfaces.* John Wiley and Sons, New York, pp. 317–329.

Mathews, M. M., and W. R. Sistrom. 1959. Function of carotenoid pigments in non-photosynthetic bacteria. *Nature* (London) 184:1892–1893.

Meier, F. C. 1935. Microorganisms in the atmosphere of Arctic regions. *Phytopathology* 25:27.

Meredith, D. S. 1962. Spore discharge in *Cordana musae* (Zimm) Hohnel and *Zygosporium oscheoides* Mont. *Annals of Botany* (London), Part II 26:233–241.

Micheli, P. A. 1729. *Novum Plantarum Genera Florentine.* Florence.

Miller, O. K. 1972. *Mushrooms of North America.* E. P. Dutton Co., New York.

Mishustin, E. N. 1975. Microbial associations of soil types. *Microbial Ecology* 2:97–118.

Mitchell, R. 1968. Factors affecting the decline of nonmarine microorganisms in seawater. *Water Research* 2:535–543.

Moore-Landecker, E. 1972. *Fundamentals of the Fungi.* Prentice-Hall, Englewood Cliffs, N.J.

Newbold, J. D., J. W. Elwood, R. V. O'Neill, and W. van Winkle. 1981. Measuring nutrient spiralling in streams. *Canadian Journal of Aquatic Sciences* 38:860–863.

Norkrans, B. 1980. Surface microlayers in aquatic environments. *Advances in Microbial Ecology* 4:51–85.

Odum, E. P. 1971. *Fundamentals of Ecology.* W. B. Saunders, Philadelphia.

Oppenheimer, C. H. 1963. *Symposium on Marine Microbiology.* Charles C. Thomas, Springfield, Ill.

Ormerod, J. G. 1983. The carbon cycle in aquatic ecosystems. In J. H. Slater, R. Whittenburg, and J. W. T. Wimpenny (eds.). *Microbes in Their Natural Environments.* Thirty-fourth Symposium of the Society for General Microbiology. Cambridge University Press, Cambridge, England, pp. 463–482.

Overbeck, J. 1966. Primärproduktion und Gewässerbakterien. *Naturwissenschaften* (Berlin) 52:145.

Overbeck, J., and H. D. Babenzien. 1964. Bakterien und Phytoplankton eines Kleingewässers im Jahreszyklus.

Journal der Angewandter Mikrobiologie 4:49–76.

Padan, E. 1979. Impact of facultatively anaerobic metabolism on ecology of cyanobacteria (blue-green algae). *Advances in Microbial Ecology* 3:1–48.

Paerl, H. W. 1985. Influence of attachment on microbial metabolism and growth in aquatic ecosystems. In D. C. Savage and M. Fletcher (eds.). *Bacterial Adhesions: Mechanisms and Physiological Significance*. Plenum Press, New York, pp. 363–400.

Parker, B., and G. Barsom. 1970. Biological and chemical significance of surface microlayers in aquatic ecosystems. *BioScience* 20:87–93.

Perkins, E. J. 1974. *The Biology of Estuaries and Coastal Waters*. Academic Press, New York.

Pfister, R. M., and P. R. Burkholder. 1965. Numerical taxonomy of some bacteria isolated from Antarctic and tropical seawaters. *Journal of Bacteriology* 90:863–872.

Phaff, H. J., M. W. Miller, and E. M. Mrak. 1968. *The Life of Yeasts*. Harvard University Press, Cambridge, Mass.

Polunin, N. 1951. Arctic aerobiology: Pollen grains and other spores observed on sticky slides exposed in 1947. *Nature* (London) 168:718–721.

Prescott, G. W. 1962. *Algae of the Western Great Lakes Area*. William C. Brown Co., Dubuque, Iowa.

Reid, G. K. 1961. *Ecology of Inland Waters*. Van Nostrand Rheinhold Co., New York.

Rheinheimer, G. 1980. *Aquatic Microbiology*. John Wiley and Sons, New York.

Rigler, F. H. 1964. The phosphorus fractions and the turnover time of inorganic phosphorus in different types of lakes. *Limnology and Oceanography* 9:511–578.

Ross, D. A. 1970. *Introduction to Oceanography*. Appleton-Century-Crofts, New York.

Round, F. E. 1984. *The Ecology of the Algae*. Cambridge University Press, New York.

Rumney, G. R. 1968. *Climatology and the World's Climate*. Macmillan, New York.

Ryther, J. H. 1969. Photosynthesis and fish production in the sea. *Science* 166:72–76.

Sexstone, A. J., N. P. Revsbeck, T. B. Parkin, and J. M. Tiedje. 1985. Direct measurement of oxygen profiles and denitrification rates in soil aggregates. *Soil Science Society of America Journal* 49:645–651.

Sieburth, J. M. 1979. *Sea Microbes*. Oxford University Press, New York.

Sieburth, J. M., P. J. Willis, K. M. Johnson, C. M Burney, D. M. Lavoie, K. R. Hinga, D. A. Caron, F. W. French III, P. W. Johnson, and P. G. Davis. 1977. Dissolved organic matter and heterotrophic microneuston in the surface microlayers of the North Atlantic. *Science* 194:1415–1418.

Silverman, M. P., and E. F. Munoz. 1970. Fungal attack on rock: Solubilization and altered infrared spectra. *Science* 169:985–987.

Simidu, U., T. Kaneko, and N. Taga. 1977. Microbiological studies of Tokyo Bay. *Microbial Ecology* 3:173–191.

Slater, J. H., R. Whittenbury, and J. W. T. Wimpenny. 1983. *Microbes in Their Natural Environments*, Thirty-fourth Symposium of the Society for General Microbiology. Cambridge University Press, Cambridge, England.

Sparrow, F. K. 1968. Ecology of freshwater fungi. In G. C. Ainsworth and A. S. Sussman (eds.). *The Fungi*. Academic Press, New York, pp. 41–93.

Stakman, E. C., and L. M. Hamilton. 1939. Stem rust in 1939. *U.S. Department of Agriculture Plant Disease Reporter Supplement* 117:69–83.

Stanier, R. Y., and G. Cohen-Bazire. 1957. The role of light in the microbial world: Some facts and speculations. In *Microbial Ecology*. Seventh Symposium of the Society for General Microbiology. Cambridge University Press, Cambridge, England, pp. 56–89.

Stetzenbach, L. D., B. Lighthart, R. J. Seidler, and S. C. Hern. 1992. Factors influencing the dispersal and survival of aerosolized microorganisms. In M. A. Levin, R. J. Seidler, and M. Rogul (eds.). *Microbial Ecology: Principles, Methods and Applications*. McGraw-Hill, New York, pp. 455–465.

Stevenson, F. J. 1976. Organic matter reactions involving pesticides in soils. In D. D. Kaufman, G. G. Still, G. D. Paulson, and S. K. Bandal (eds.). *Bound and Conjugated Pesticide Residues*. ACS Symposium Ser. 29. American Chemical Society, Washington, D.C., pp. 180–207.

Stevenson, F. J. 1982. *Humus Chemistry*. Wiley-Interscience, New York.

Stevenson, L. H., C. E. Millwood, and B. H. Hebeler. 1974. Aerobic heterotrophic bacterial populations in estuarine water and sediments. In R. R. Colwell and R. Y. Morita (eds.). *Effects of the Ocean Environment on Microbial Activity*. University Park Press, Baltimore, pp. 268–285.

Stewart, R. W. 1969. The atmosphere and the ocean. *Scientific American* 221(3):76–86.

Stockton, W. L., and T. E. DeLaca. 1982. Food falls in the

deep sea: Occurrence, quality and significance. *Deep-Sea Research* 29:157–169.

Stotzky, G. 1965. Replica plating technique for studying microbial interactions in soil. *Canadian Journal of Microbiology* 11:629–636.

Sussman, A. S. 1961. The role of endogenous substrates in the dormancy of *Neurospora*. In H. O. Halvorson (ed.). *Spores II.* Burgess Publishing Co., Minneapolis, Minn., pp. 198–217.

Sverdrup, H. O., M. W. Johnson, and R. H. Fleming. 1942. *The Oceans.* Prentice-Hall, Englewood Cliffs, N.J.

Tait, R. V., and R. S. DeSanto. 1972. *Elements of Marine Ecology.* Springer-Verlag, New York.

Taylor, C. B. 1942. Bacteriology of freshwater. Part III The types of bacteria present in lakes and streams and their relationship to the bacterial flora of soil. *Hygiene* 42:284–296.

Taylor, W. R. 1957. *Marine Algae of the Northeastern Coast of North America.* University of Michigan Press, Ann Arbor.

Teltsch, B., and E. Katznelson. 1978. Airborne enteric bacteria and viruses from spray irrigation with waste water. *Applied and Environmental Microbiology* 35:290–296.

Thrush, B. A. 1977. The chemistry of the stratosphere and its pollution. *Endeavour* 1:3–6.

Trainor, F. R. 1978. *Introductory Phycology.* John Wiley and Sons, New York.

Valkanov, A. 1968. Das Neuston. *Limnologica* (Leipzig) 6:381–403.

Waksman, S. A. 1961. *The Actinomycetes.* Williams and Wilkins, Baltimore.

Webster, J. R., and B. C. Patten. 1979. Effects of watershed perturbation on stream potassium and calcium dynamics. *Ecological Monographs* 49:51–72.

Wenk, E., Jr. 1969. The physical resources of the ocean. *Scientific American* 221(3):167–176.

Westphal, A. 1976. *Protozoa.* Blackie, Glasgow, Scotland.

Wetzel, R. G. 1975. *Limnology.* W. B. Saunders, Philadelphia.

Wetzel, R. G., and H. L. Allen. 1970. Functions and interactions of dissolved organic matter and the littoral zone in lake metabolism and eutrophication. In Z. Kajak and A. Hillbricht-Ilkowska (eds.). *Productivity Problems of Freshwaters.* PWN Polish Scientific Publishers, Warsaw, pp. 333–347.

Wiebe, W. J., R. E. Johannes, and K. L. Webb. 1975. Nitrogen fixation in a coral reef community. *Science* 188:257–259.

Wolf, F. T. 1943. The microbiology of the upper air. *Bulletin of the Torrey Botanical Club* 70:1–14.

Wood, E. J. F. 1967. *Microbiology of Oceans and Estuaries.* Elsevier Publishing Co., New York.

Wright, R. T., and J. E. Hobbie. 1966. Use of glucose and acetate by bacteria and algae in aquatic ecosystems. *Ecology* 47:447–464.

ZoBell, C. E. 1946. *Marine Microbiology.* Chronica Botanica, Waltham, Mass.

PART FOUR

Biogeochemical Cycling

10

Biogeochemical Cycling: Carbon, Hydrogen, and Oxygen

BIOGEOCHEMICAL CYCLING

Biogeochemical cycling describes the movement and conversion of materials by biochemical activities within the ecosphere. This cycling occurs on a global scale, profoundly affecting the geology and present environment of our planet. Biogeochemical cycles include physical transformations, such as dissolution, precipitation, volatilization, and fixation; chemical transformations, such as biosynthesis, biodegradation, and oxidoreductive biotransformations; and various combinations of physical and chemical changes. These physical and chemical transformations can cause spatial translocations of materials—from the water column to the sediment, for example, from the soil to the atmosphere. All living organisms participate in the biogeochemical cycling of materials, but microorganisms, because of their ubiquity, diverse metabolic capabilities, and high enzymatic activity rates, play a major role in biogeochemical cycling (Pomeroy 1974; Jorgensen 1989).

Biogeochemical cycling is driven directly or indirectly by the radiant energy of the sun (Woodwell 1970). Energy is absorbed, converted, temporarily stored, and eventually dissipated, which is to say that energy flows through ecosystems. This flow of energy is fundamental to ecosystem function. Whereas energy flows through the ecosystem, materials undergo cyclic conversions that tend to retain materials within the ecosystem.

Through the geological ages, biogeochemical activities have altered conditions on Earth in a unidirectional manner. The most crucial of these unidirectional changes were the decomposition of abiotically formed organic matter on primitive Earth by early heterotrophic forms of life and the change of the originally reducing atmospheric conditions to oxidizing ones by the first oxygen-producing phototrophs. Contemporary biogeochemical processes, however, tend to be cyclic. The cyclic nature of material conversions leads to dynamic equilibria between various forms of cycled materials. Without the existence of these equilibria, the present physiological diversity of life could not exist. Not all biogeochemical activities, though, resemble closed cycles. Materials can be imported into or exported from ecosystems, thereby becoming available or inac-

Table 10.1

Distribution of biogenic elements within a periodic system

	Period Number				
	1	2	3	4	5
	1 H	3 Li	11 Na*	19 K*	37 Rb
		4 Be	12 Mg*	20 Ca*	38 Sr
				21 Sc	39 Y
				22 Ti	40 Zr
				23 V*	41 Nb
				24 Cr*	42 Mo*
				25 Mn*	43 Tc
				26 Fe*	44 Ru
				27 Co*	45 Rh
				28 Ni*	46 Pd
				30 Zn*	48 Cd
		5 B*	13 Al	31 Ga	49 In
		6 C	14 Si*	32 Ge	50 Sn*
		7 N	15 P	33 As	51 Sb
		8 O	16 S	34 Se*	52 Te
		9 F*	17 Cl*	35 Br	53 I*
	2 He	10 Ne	18 Ar	36 Kr	54 Xe

Atomic number and element

Major biogenic elements are shown with shading; minor and trace elements are marked with an asterisk. Most biogenic elements cluster in the first four periods, with only three trace elements falling into the fifth period. No known biogenic elements occur above atomic weight 53. Therefore, the sixth and seventh period, the lantanides and actinides, are omitted from this table.

cessible to microbial activities. Some materials, such as fossil fuel and limestone deposits, may be removed from active microbial cycling for many millions of years. Ecosystems vary greatly in the efficiency with which they retain specific materials, such as essential nutrient elements. Habitats capable of retaining nutrients, such as coral reefs and tropical rainforests, are capable of sustaining high rates of productivity even in generally nutrient-poor surroundings. Habitats that have a low capacity to retain essential nutrients, such as the epipelagic habitats, tend to have low, nutrient-limited primary production rates, even when light and temperature favor high productivity (Odum 1983).

Most elements are subject to some degree of biogeochemical cycling. As may be expected, elements that are essential components of living organisms, the so-called biogenic elements, are most regularly subject to biogeochemical cycling. Because the biogenic elements need to meet definite criteria of atomic weight and chemical reactivity, they are not randomly distributed in the periodic table of elements but form definite groups within the first five periods (Frieden 1972; Mertz 1981) (Table 10.1). This fact made the biogenic significance of certain trace elements, like nickel, predictable even before their requirement and function were experimentally established.

The intensity or rate of biogeochemical cycling for each element roughly correlates to the amount of the element in the chemical composition of biomass. The major elemental components of living organisms

(C, H, O, N, P, and S) are cycled most intensely. Minor elements (Mg, K, Na, and halogens) and trace elements (B, Co, Cr, Cu, Mo, Ni, Se, Sn, V, and Zn), which are required in small quantities and not by every form of life, are cycled less intensely. The minor and trace elements Fe, Mn, Ca, and Si are exceptions to this rule. Iron and manganese are cycled extensively in an oxidoreductive manner. Calcium and silicon, while minor components of protoplasm, form important exo- and endoskeletal structures in both micro- and macroorganisms and are consequently cycled on the global scale at the rate of many billions of tons per year. Nonessential and even toxic elements are also cycled to some extent, as evidenced by the bioaccumulation of radioactive strontium and cesium isotopes and by the microbial methylation of mercury, lead, and arsenic (Deevey 1970; Hutchinson 1970; Underwood 1977).

Microorganisms are sources of particular compounds in the ecosphere and sinks for others. Transfer rates between pools vary and generally are enzymatically mediated. The critical enzymatic activities involved in a particular elemental transformation within a habitat may be associated with one particular microbial population or with multiple microbial, plant, and/or animal populations. Some elemental transformations also occur chemically—that is, abiotically—and are not enzymatically mediated.

Reservoirs and Transfer Rates

In terms of biogeochemical cycling, the various chemical forms of a particular element constitute so-called pools, or reservoirs. When we refer to such reservoirs, we generally mean reservoirs on the global scale. Global reservoirs tend to be stable in the time frame of human history but may undergo shifts during geological ages. Within a particular habitat, elements also occur in reservoirs of a distinct size, and these reservoir sizes can vary greatly from habitat to habitat. Some chemical forms accumulate within a particular habitat; others are depleted (Odum 1983).

Reservoir size is an extremely important parameter to be considered in connection with possible per-

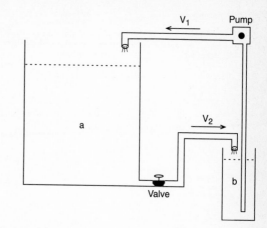

Figure 10.1

Mechanical model of reservoirs and biogeochemical cycling. a = large reservoir of material; b = small reservoir of material; V_1 = transfer rate from b to a; V_2 = transfer rate from a to b. When $V_1 = V_2$, the system is in equilibrium and the levels in the two reservoirs remain steady. When $V_1 \neq V_2$ the levels in the reservoirs will change; this change will occur more rapidly in the smaller reservoir than in the larger reservoir.

turbations of a cycling system. To make this point clear, let us consider a simple cycling system (Figure 10.1) in which water is pumped from the small reservoir b at rate V_1 into the large reservoir a, from which it returns by gravity flow at rate V_2. In equilibrium state, V_1 equals V_2, and the water levels in each reservoir remain steady. It is easy to see that if the flow equilibrium is disturbed either by slowing the pump speed (V_1) or, conversely, by partially closing the outflow valve on reservoir a, the effects on the small reservoir (b) will be rapid and dramatic. Reservoir b will either overflow or be emptied, while the level of reservoir a will be affected only moderately. Similarly, in biogeochemical cycles, small, actively cycled reservoirs are the most prone to disturbance by either natural or human causes.

Microbiologically mediated portions of biogeochemical cycles are essential for the growth and survival of plant and animal populations. We discussed some of the critical metabolic activities of microorganisms that directly influence plant and ani-

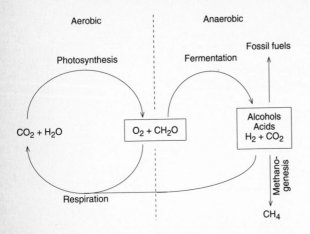

Figure 10.2

Interrelation of the biogeochemical cycles of carbon, hydrogen, and oxygen. The figure shows the involvement of oxygen and hydrogen in the aerobic and anaerobic oxidation of organic carbon and in the reduction of CO_2. The formula (CH_2O) represents organic matter on the oxidation level of carbohydrate.

mal populations in preceding chapters. It is important to recognize that the biogeochemical cycling activities of microorganisms determine, in large part, the potential productivity that can be supported within a habitat. Alterations in the biogeochemical cycling activities of microbial populations caused by human activities—by pollution, for example—can result in changes in the transfer rates of elements between reservoirs and in the sizes of reservoirs of elements in particular chemical forms within habitats. These changes alter the biochemical characteristics of a habitat and the populations that can be supported, in both quantitative and qualitative terms.

In this and the following chapter, we will discuss the microbially mediated biogeochemical cycling of various elements, treating each of the elements individually, although the biogeochemical cycles are interrelated and cannot truly be separated from each other. This applies especially to the cycling of the first three major elements, carbon, hydrogen, and oxygen, all three of which are cycled by the same two opposing processes of photosynthesis and respiration (Bolin

1970; Cloud and Gibor 1970; Krumbein and Swart 1983). Figure 10.2 makes apparent the degree to which the cycling of these elements is interwoven. This figure also illustrates the cycling of carbon between organic and inorganic forms and the processes involved in these transformations.

THE CARBON CYCLE

When examining the biogeochemical cycling of an individual element, it is useful to consider the global reservoirs of this element, the size of these reservoirs, and whether or not these reservoirs are being actively cycled. The most actively cycled reservoir of carbon is atmospheric CO_2 (0.032% of the atmosphere, or about 700 billion metric tons of carbon). The dissolved inorganic forms of carbon (CO_2, H_2CO_3, HCO_3^-, and CO_3^-) in surface seawater (500 billion tons of carbon) are in direct equilibrium with atmospheric CO_2, but a much larger portion in the deep sea (34,500 billion metric tons) equilibrates only at the slow rate of vertical seawater circulation. The living biomass in terrestrial and aquatic environments contains slightly less carbon (450 billion to 500 billion tons) than the atmosphere. Dead (but not fossil) organic matter such as humus and organic sediment contains 3,700 billion tons. All of these can be considered actively cycled carbon reservoirs. In comparison, the amounts of carbon in fossil fuels (10,000 billion tons) and carbonaceous sedimentary rock (20,000,000 billion tons) are considerably higher, but the natural turnover rates of these latter reservoirs are minute. The cited numbers (Bolin 1970) are rough estimates, but they illustrate relative reservoir sizes and serve as the basis for calculating carbon turnover rates and human's influence on them. Carbonaceous rocks, such as limestone and dolomite, may be slowly dissolved by biologically produced acid with the release of CO_2 or HCO_3^-, but compared to the bulk of carbonaceous rock, this process constitutes a negligible turnover. Some fossil fuels, in the form of coal, petroleum, natural gas, or kerogen, are recycled naturally by the biological degradation of oil and gas seepages; but again, such cycling affects only a small portion of the fossil fuels. However, since the start of the industrial revolution in the early nineteenth century, human activities have reinjected a significant percentage of fossil organic carbon into the atmosphere as CO_2 from burnt fuel.

The natural rates of carbon cycling in oceans and on land are close to a steady state; that is, the rates of movement of carbon between the atmosphere and trees or between algae and the dissolved inorganic carbon of the oceans do not change measurably from year to year and tend to balance each other (Hobbie and Melillo 1984). However, human activities have recently introduced changes in the carbon cycle that are large enough to be measured. Today, the global carbon cycling is a mixture of the natural steady-state rates and reservoirs and the changing rates and reservoirs affected by human activities. For example, the flux of carbon from algae into dissolved organic carbon in the open ocean is at steady state because human activities are not great enough to perturb that rate. In contrast, the reservoir of carbon (as CO_2) in the atmosphere is no longer in a steady state and is growing from year to year. As a result, the global carbon cycle is out of balance.

Atmospheric CO_2, because it is a relatively small carbon pool, has been measurably affected by industrial CO_2 release (Bolin et al. 1979) (Table 10.2). Between 1860 and 1980, atmospheric CO_2 rose by approximately 70 ppm from a preindustrial 270 ppm to 340 ppm (Houghton et al. 1983; Hobbie and Melillo 1984; Marche et al. 1984). The increase in atmospheric carbon dioxide is largely due to the burning of fossil fuels, with additional CO_2 contributed from forest biomass and soil humus in the course of forest clearing for agricultural land. These inputs of carbon into the atmosphere by human activities should have raised the atmospheric CO_2 concentration by substantially more than 50 ppm, but part of the input apparently has been absorbed by the sea as HCO_3^- and/or fixed into a standing crop of biomass.

There is concern today that the continued increase in atmospheric CO_2, currently at a rate of about 1 ppm per year, might create a "greenhouse effect." CO_2 is transparent to visible radiation, but

Table 10.2

Major carbon reservoirs

Reservoir	Amount (billions of metric tons of carbon)
Atmosphere before 1850	560–610
Atmosphere in 1978	692
Oceans and fresh water inorganic	35,000
Dissolved organic	1,000
Land biota	600–900
Soil organic matter	1,500
Sediments	10,000,000
Fossil fuels	10,000

Source: Bolin et al. 1979.

absorbs strongly in the infrared range. Visible sunlight striking Earth is irradiated back as longer-wavelength infrared radiation. An insulating CO_2 blanket would retain most of this radiation and thus would bring about a warming trend in the climate. The foremost effect of a doubling of the CO_2, perhaps as early as the year 2050, will be on the world climate (Hobbie and Melillo 1984). There will most likely be little direct effect on microbial activity, but if temperature increases and precipitation patterns change, there could be a strong indirect effect. Scientists generally agree that the global temperature will increase but there is less agreement as to how much and exactly how this will affect air movement and precipitation (National Research Council 1979, 1983, 1991). One computer model predicts a small change in the tropics and a large change at the poles. At 40° to 60° latitude the annual temperature increase will be 4° to 6°C, and precipitation patterns will also change. This increase in turn could reduce the size of the polar ice caps and substantially raise the level of the oceans—an unhappy prospect for densely populated coastal regions.

Contributing to the greenhouse effect is atmospheric methane, released by human activities such as drilling for oil and natural gas, landfilling of solid waste, and large-scale cattle raising. Although much lower in amount than the CO_2 produced by the burning of fossil fuels, methane traps heat four to five times as effectively as CO_2. Thus, even in relatively small amounts, it can contribute to the greenhouse effect significantly.

Although the increase in atmospheric CO_2 is well documented and will undoubtedly have some effect on our climate, the prediction of the direction and the magnitude of the effect requires modeling and many assumptions that may or may not be valid and complete. An increase in cloud cover, a possible effect of climate warming, may act as negative feedback and reduce the warming effect. Other pollutants, mainly sulfate from fossil-fuel burning, also contribute to an increased cloud cover and lessen the greenhouse effect (Charlson et al. 1992). Another factor of uncertainty is whether microbial and plant activities will dampen or amplify the effects of increased injection of CO_2 into the atmosphere, but microorganisms will probably respond to changes in CO_2 concentration and temperature. Increased fixation of CO_2 would lessen the effects; increased respiration would heighten them. The described uncertainties along with the painfulness of reducing our fossil-fuel consumption in the absence of adequate alternative energy sources have, to date, delayed meaningful policies to curb a major climate disruption by greenhouse effect. The necessary international cooperation for an effective global policy would also be difficult to achieve. Yet to delay action until unmistakable signs of a global disaster become manifest is a risky policy, since correcting the situation once it develops may prove difficult. Speculative schemes to tie up excess CO_2 in terrestrial or oceanic biomass have been advanced, but the monetary, economic, and ecological cost of such solutions may be staggering.

Carbon Transfer through Food Webs

The overall involvement of microorganisms with carbon cycling can perhaps best be discussed in the context of a theoretical food web (Figure 10.3). Every food web is based on primary producers. The net fixation of CO_2 to form organic compounds is carried out by autotrophic organisms. Among the microor-

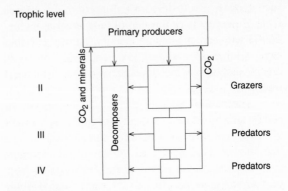

Figure 10.3

An idealized food web showing transfers between trophic levels. Organic carbon formed by primary producers is transferred to grazers and predators. Decomposers and respiration of grazers and predators return CO_2 to primary producers. The diagram shows that the supportable biomass declines at progressively higher trophic levels.

ganisms, this includes photosynthetic and chemolithotrophic organisms. The most important groups of microorganisms, in terms of their abilities to convert CO_2 to organic matter, are the algae, the cyanobacteria, and the green and purple photosynthetic bacteria (see Appendix II for a description of these organisms). Chemoautrophic microorganisms contribute to a lesser extent. The principal metabolic pathway for photosynthetic CO_2 fixation is the pentosephosphate, or Calvin, cycle. In addition, microorganisms are capable of incorporating CO_2 through the phosphoenol pyruvate carboxylase system. In the case of heterotrophic miroorganisms, exchange but no net CO_2 fixation occurs, but some chemolithotrophic microorganisms use this system either instead of or in addition to the pentose phosphate cycle for net CO_2 fixation (Wood 1989). Methanogenic archaebacteria play an important role in the anaerobic reduction of CO_2. Only a limited number of microorganisms can utilize the resulting methane (Gottschalk 1979; Haber et al. 1983). These methylotrophs are ecologically important in minimizing methane transfer to the atmosphere.

Carbon dioxide converted to organic carbon by primary producers represents the gross primary production of the community. This process is carried out predominantly by photosynthetic organisms that convert light energy to chemical energy; the chemical energy is stored within the organic compounds that are formed. The conversion of radiant energy to chemical energy in organic compounds is the essence of primary production.

A portion of the gross primary production is converted back to CO_2 by the respiration of the primary producers. The remaining organic carbon is the net primary production available to heterotrophic consumers; the heterotrophs complete the carbon cycle, ultimately converting organic compounds formed by primary producers back to CO_2 during respiration. The gross primary production and the total community respiration may or may not be in complete balance (Figure 10.4). A net gain in organic matter produced by photosynthesis and not converted back to CO_2 is known as the net community productivity. A positive net community productivity accumulates organic matter within the ecosystem. If the net community productivity is negative, then there must be an input of allochthonous organic matter or the community will dissipate the available energy and disappear. The standing biomass, or standing crop, in a habitat at any particular point in time should not be confused with productivity. Standing biomass represents stored energy. High biomass in a habitat can be due to high gross primary productivity, low respiration, or net input of organic compounds. During periods of low consumption, standing biomass may accumulate.

The rate of primary productivity can be determined by measuring fluxes of oxygen or carbon dioxide in the light and dark. Using O_2 production neglects the contribution to primary production by anaerobic photosynthetic bacteria and by chemolithotrophic bacteria. The amount of oxygen produced in the light can be used as a measure of photosynthetic net primary productivity. Oxygen consumed in the dark can be added to the measured amount of O_2 generated in the light to estimate gross primary productivity. The net assimilation of CO_2, the net community production, can be assayed by measur-

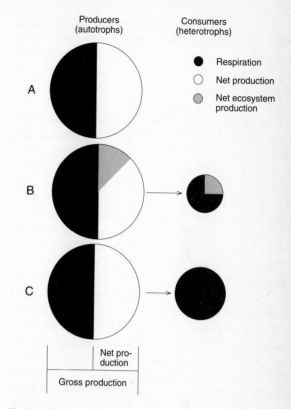

Figure 10.4

Balance of primary productivity, respiration, and community productivity. (A) Gross primary production (whole circle) and net production portion (white area) remaining after respiration of primary producers. The net primary production is available for consumers. (B) Situation where there is net ecosystem production, since heterotrophs do not consume all the available net primary production leading to the accumulation of organic matter in the ecosystem. This typically occurs during early successional stages. (C) Situation where consumers use all of the net primary production and total respiration of the community balances total primary production. No organic matter accumulates in the system. This situation is typical of climax ecosystems. (Source: Woodwell 1970, The Energy Cycle of the Biosphere. Reprinted by permission, copyright 1970 Scientific American Inc. All rights reserved.)

ing the concentration of CO_2 flowing into and out of enclosed chambers containing active biological communities; this approach measures the difference between the carbon fixed from CO_2 by primary production and the CO_2 produced by respiration. Other methods utilize $^{14}CO_2$ assimilation to measure primary productivity. Short-term $^{14}CO_2$ incorporation in light tends to measure gross photosynthetic productivity; longer incorporation periods measure net productivity (see chapter 7). The necessity of enclosing a community may alter steady-state conditions, producing inaccurate estimates of primary production (Hubel 1966).

The primary productivity of ecosystems is usually compared on a per square meter per year basis. It may be expressed as grams of carbon fixed, grams of dry organic matter (biomass) formed, or kilocalories (kcal) of energy stored. Those values are interconvertible, assuming that 50% of dry organic matter is carbon and on the average 1 g of dry organic biomass corresponds to 5 kcal of energy stored. Table 10.3 compares net primary productivity of several natural ecosystems with agricultural fields. It is important to realize that the agricultural systems (corn, rice, and sugarcane) receive nonrenewable energy subsidies derived from fossil fuels in the form of synthethic fertilizer and labor by fuel-driven farm machinery (Odum 1983). The efficiency of sun-energy use in photosynthesis is low. Less than 0.1% of the sun energy that reaches the surface of Earth is used in photosynthesis, globally producing 150 billion to 200 billion tons of dry organic matter per year. Even in intensively managed agricultural ecosystems, the efficiency of sun-energy utilization rarely exceeds 1% (Woodwell 1970).

The transfer of energy stored in organic compounds from one organism to another establishes a food chain. The transfer occurs in steps; each step constitutes a trophic level. The interrelationships of food chain steps establish the food web. Individual food chains, such as the detritus food chain, may be based on allochthonous material or on primary producers in the same ecosystem. Without the input of primary producers, however, the food web would decay. Organisms that feed directly on primary producers constitute the trophic level of grazers. Grazers are preyed upon by predators, which may in turn be preyed upon by a trophic level of larger predators.

Table 10.3

Net primary productivity of some natural and agricultural ecosystems

Description of ecosystem	Net primary productivity (g dry organic matter/m^2/year)
Tundra	400
Desert	200
Temperate grassland	up to 1,500
Temperate deciduous or evergreen forest	1,200–1,600
Tropical rainforest	up to 2,800
Cattail swamp	2,500
Freshwater ponds	950–1,500
Open ocean	100
Coastal seawater	200
Upwelling area	600
Coral reef	4,900
Corn field	1,000–6,000
Rice paddy	340–1,200
Sugarcane field	up to 9,400

Sources: Woodwell 1971; Heal and Ineson 1984.

Only 10% to 15% of the biomass from each trophic level usually is transferred to the next higher trophic level; 85% to 90% is consumed by respiration or enters the decay portion of the food web. In some extreme cases, however, values considerably higher or lower have been measured. Nevertheless, the higher a trophic level, the smaller its biomass. Many consumers feed on more than one trophic level, some primary producers (phototrophic flagellates) are also consumers (phagotrophic flagellates), and dissolved organic matter is converted partially to particulate matter by the process of "heterotrophic (secondary) production." The traditional concept of a food web as consisting of unidirectional food chains thus requires conceptual reevaluation. The depicted simple scheme, however, is adequate for our discussion of microbial involvement.

In most terrestrial and shallow-water environments, the predominant primary producers are higher plants, and the dominant grazers are invertebrate or vertebrate herbivores. Microbial primary producers are usually present, but their role is subordinate to that of higher plants. Exceptions are found in some harsh environments—such as exposed rock surfaces, polar regions, and high-salinity and thermal environments—that preclude the presence of higher plants and where, therefore, microbial producers predominate. In the limnetic zone of deep lakes and in the pelagic portion of the oceans, however, where no autochthonous higher plants exist, the entire food web is based on microbial primary producers, predominantly unicellular planktonic algae and cyanobacteria. Except in the immediate coastal zone, microorganisms are responsible for most of the ocean's primary production, which is about half of the total photosynthesis of the planet. In these environments, substantial numbers of grazers are also microbial (planktonic protozoa), though they share this role with smaller representatives of the invertebrate zooplankton (Ryther 1969).

Recent investigations indicate that extremely small planktonic forms carry out a major part of primary and secondary productivity as well as respiration in ocean waters (Sherr and Sherr 1984). Nanophytoplankton comprises most of the phytoplankton biomass in seawater, and picophytoplankton appears to be responsible for a substantial share of the total phytoplankton productivity. The nano- and picophytoplankton are better equipped to cope with the scarcity of critical mineral nutrients in the ocean, and it appears that in most areas of the ocean they are growing at near maximal rates, with primary productivity being severalfold higher than earlier estimates have suggested. A considerable percentage of the organic carbon initially fixed by phytoplankton appears to enter the pelagic food web as dissolved and nonliving particulate organic matter. This nonliving organic material is subsequently incorporated into bacterial cells that are consumed primarily by colorless (heterotrophic) phagotrophic microflagellates. Rough estimates show that one phagotrophic microflagellate occurs in one hundred bacteria, and the activity of these clears 30% to 50% of the marine-water column of bacteria per day. The phagotrophic microflagellates have growth rates equal to or greater than the bacteria on which they feed and exert sub-

stantial control over their population levels (Fenchel 1986). Surprisingly, even microflagellate members of the freshwater phytoplankton were recently reported to graze on bacteria extensively. They were estimated to consume more bacteria than the total of the various zooplankton organisms (Bird and Kalff 1986). Many colorless heterotrophic microflagellates have close morphological and taxonomic relationships to pigmented phytoflagellates. Surveying the literature Sanders and Porter (1988), concluded that numerous freshwater and marine phytoflagellates (Euglenophycophyta, Chrysophycophyta, and Pyrrophycophyta) have a potential for mixotrophic growth, that is, they are capable of photosynthesis and phagotrophy at the same time. Under the mineral nutrient limitations of most limnetic and pelagic environments, the ingested bacteria may be more critical as sources of nitrogen and phosphorus than as carbon and energy sources.

Because of the extremely small size of both primary (photosynthetic) and secondary (heterotrophic) producers in the marine environments, the phagotrophic protozoa represent an important link between these tiny producers and the higher trophic levels of the food web. It appears that earlier surveys may have underestimated the abundance and importance of this trophic level (Sherr and Sherr 1984). The motile microflagellates and microciliates are typically distributed in a patchy manner, showing strong spatial and temporal fluctuations. Their abundance in the pleuston and in association with detritus and fecal pellets is greater by several orders of magnitude than their average distribution in the water column. Pelagic sarcodina (acantharia, amoebae, foraminifera, and radiolaria) also show strong spatial and temporal fluctuations in numbers and, on the average, contribute little to the biomass of planktonic protozoa. Heterotrophic microflagellates, naked and loricate ciliates such as the tintinnids, make up the largest portion of the planktonic protozoa, in terms of both numbers and biomass. They graze on the pico- (<2 μm) and nanoplanktonic (<20 μm) primary and secondary producers that are too small to be filtered out efficiently by larger forms of zooplankton, such as the microcrustacea. The marine protozoa span a large size range between 2 and 200 μm and are well suited to graze on the smallest forms of primary producers, as well as on bacteria, the only group with some access to the large but extremely dilute dissolved organic material reservoir of the sea. The marine protozoa are of sufficient size to be grazed upon, in turn, by the larger microcrustaceae of the zooplankton, channeling biomass toward larger invertebrates and fish. Because of the number of trophic levels involved, however, the pelagic environment can support few organisms on the higher trophic levels. The length of the food chains contributes to the biological desert character of the open sea.

The decay portion of the food web is dominated by microbial forms in both aquatic and terrestrial environments. The decay portion of the food web involves the degradation of incompletely digested organic matter, such as fecal material or urea, and the decomposition of dead but not consumed plants and animals. The proportion of the biomass recycled by decay rather than by consumption varies greatly within different types of habitats (Figure 10.5), but in forest and salt-marsh habitats, decay may account for 80% to 90% of the total energy flow. On the other extreme, in pelagic and limnetic habitats, grazers rapidly consume the bulk of primary production, and relatively little biomass is channeled through the decay route (Pomeroy 1984; Sherr and Sherr 1984).

Part of the microbial biomass formed during decomposition is recycled into the food web. Both free-swimming and detritus-attached bacteria in seawater are metabolically active and are significant in the utilization of dissolved substrates and the production of biomass. Because they have access to a variety of dissolved and particulate substrates, bacteria are significant producers of biomass in the ocean. Various predators consume microorganisms as a main or supplemental source of carbon and energy. Detritus food chains are based on the consumption of microbial biomass. A crucial contribution of bacteria to food webs is their ability to assimilate dissolved organic carbon from extremely dilute solutions, and thus to convert these nutrients to biomass that can be available to other forms of life not capable of using dissolved nutrients directly. This phenomenon, known as heterotrophic, or secondary,

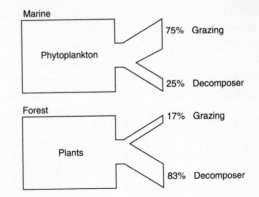

Figure 10.5
Relative proportion of energy entering grazing and decomposer food chains in marine and terrestrial habitats. Dominant producers in marine habitat are phytoplankton; in terrestrial habitat they are plants. A higher proportion of energy goes into the grazing food chain in the marine habitat; in the terrestrial habitat a higher proportion enters the decomposer food chain. (Source: Odum 1962. Reprinted courtesy of Ecological Society of Japan.)

production, is specially important in aquatic habitats with low concentrations of dissolved organic compounds. The detrital food web is intrinsically tied to secondary production including the release of soluble substances from particulates and subsequent cycling through microbial biomass.

Carbon Cycling within Habitats

The degradation and recycling of organic matter in most habitats is accomplished by heterotrophic macro- and microorganisms. Microbial activities are crucial in terms not only of the quantity, but also of the quality of their contributions. Under aerobic conditions, macro- and microorganisms share the ability to biodegrade simple organic nutrients and some biopolymers, such as starch, pectin, proteins, and so on, but microorganisms are unique in their capacity to carry out anaerobic (fermentative) degradation of organic matter. They are also responsible for the recycling of most abundant but difficult-to-digest biopolymers, such as cellulose and lignin. We discussed the importance of digestive associations of microorganisms with macroherbivores in chapter 5. The ability to degrade humic materials, hydrocarbons, and many human-made synthetics is also virtually unique to microorganisms.

The greatest range of carbon transformations occurs under aerobic conditions. Biodegradation of intact hydrocarbons and many aromatic compounds, for example, occurs in oxygenated environments only. On the other hand, certain carbon transformations, such as methanogenesis, occur exclusively under anaerobic conditions. This leads to a biogeochemical zonation of habitats. Some organic compounds can accumulate within particular habitats and be unavailable to the biological community, while in other habitats the same compounds can readily serve as sources of carbon and energy.

Respiratory metabolism yields more energy to cells than fermentative metabolism. Fermentation requires a greater consumption of organic matter to support the same biomass as respiration. Therefore, in aerobic habitats, respiration predominates over fermentation. Complete respiration results in the production of CO_2, whereas fermentation, in addition, results in the accumulation of low–molecular-weight organic alcohols and acids. At this point, several possible routes exist for further metabolism. Eventually, the anaerobic conditions may change to aerobic ones, as occurs in flooded soil that drains or dries, and thus the fermentation products are utilized further aerobically. Similarly, fermentation products may diffuse out of the anaerobic environment. As an example, the fatty acids produced in the rumen by anaerobic microorganisms are transferred to the aerobic bloodstream of the ruminant animal, where they are transformed to CO_2 by respiration. If conditions remain anaerobic and fermentation products are trapped in this environment, they may be metabolized further with the simultaneous reduction of nitrate or sulfate. In the absence of such secondary electron acceptors, further energy may be extracted from some of the fermentation products by methanogenesis (Large 1983; Zeikus et al. 1985).

Methanogenesis and Methylotrophy

Methanogens are a unique group of archaebacteria (Balch et al. 1979). Strictly anaerobic and active at redox potentials between -350 and -450 mV, they are capable of using CO_2 as electron acceptor. They reduce CO_2 using H_2 produced in the fermentation process. Since they are capable of using CO_2 as their only carbon source, they are considered chemolithotrophic. If CO_2 is considered available in carbonate form, the reaction may be represented as follows in Equation 1 (Gottschalk 1979):

(1) $\quad HCO_3^- + H^+ + 4H_2 \longrightarrow$

$$CH_4 + 3H_2O \ (\Delta G = -32.4 \text{ kcal})$$

In Equation 1, ΔG (Gibbs free energy) indicates the energy yield of the reaction in kilocalories. In exothermic (energy-yielding) reactions, ΔG is negative.

Carbon dioxide is converted to methane in a C_1 cycle involving several unusual coenzymes. In the first step, CO_2 is bound to methanofuran at the formyl reduction level and is further reduced to the methenyl, methylene, methyl, and finally methane levels while successively bound to the coenzymes tetrahydromethanopterin, 2-methylthioethanesulfonic acid, 2-mercaptoethanesulfonic acid, and a yet-unidentified coenzyme, respectively (Jones et al. 1985). The manner of CO_2 conversion to cell material in methanogens is not yet known. It does not follow the ribulose diphosphate pathway common to other chemolithotrophs.

Some methanogens, such as *Methanosarcina barkeri,* are capable of metabolizing methanol, acetate, and methylamines to methane and CO_2. Syntrophic associations between methanogens and other anaerobes, such as the ethanol-fermenting H_2-producing "S" organism associated with the methanogen *Methanobacillus omelianskii* and the short-chain fatty-acid-fermenting *Syntrophobacter* associated with some other methanogens, broaden the range of substrates suitable for methanogenesis, even though truly pure cultures of methanogenic archaebacteria have the narrow substrate range (Bryant et al. 1967). The methanogens depend upon the fermentation products of other microbes to serve as their substrates. The gaseous hydrocarbon methane is the ultimately reduced carbon compound and cannot be metabolized further without an appropriate electron sink. The single carbon unit of methane is an unusual substrate, available only to a specialized group of microorganisms, the methylotrophs (Haber et al. 1983). Many of these are obligate methylotrophs, meaning that they are restricted to the utilization of methane, methanol formate, carbon monoxide, and a few additional reduced single-carbon compounds. Facultative methylotrophs have a broader substrate range. *Methylomonas* (formerly known as *Methanomonas*) species are obligately aerobic microorganisms and utilize methane only in the presence of oxygen. Recently, however, methylotrophic microorganisms were discovered that live in marine sediments and are capable of coupling methane oxidation with the reduction of sulfate or sulfur (Hanson 1980; Large 1983).

Acetogenesis

A group of facultatively chemoautotrophic anaerobes are capable of reducing CO_2 with H_2 to acetate instead of methane. *Clostridium thermoaceticum* and *Acetobacterium woodii* carry out this reaction (Equation 2):

$$(2) \quad 2CO_2 + 4H_2 \longrightarrow$$
$$CH_3COOH + 2H_2O \ (\Delta G = -25.6 \ kcal)$$

The energy yield of the reaction is less favorable than in methanogenesis. In addition to acetogenesis from CO_2 and H_2, these organisms also ferment CO, formate, and methanol to acetate and have many metabolic features in common with methanogenic bacteria (Zeikus et al. 1985).

Carbon Monoxide Cycling

We have already discussed the role of microorganisms in the cycling of carbon between the atmosphere and the lithosphere/hydrosphere; the principal carbon form exchanged is CO_2, which is removed from the atmosphere during primary production and reintroduced principally during respiration. Microorganisms also are involved in the cycling of carbon monoxide in both a direct and an indirect manner. The global annual production is estimated at 3 billion to 4 billion metric tons per annum (bmta). A major source (1.5 bmta) of this CO is the photochemical oxidation of methane and other hydrocarbons in the atmosphere. Biologically, trace amounts of CO are formed during microbial and animal respiration by the breakdown of heme compounds. Another biological source of CO is an obscure photochemical side reaction in photosynthetic microorganisms and plants. The CO production by this mechanism is proportional to light intensity but is independent of CO_2 concentration and photosynthesis rate. Hence, CO production is not an integral part of the photosynthesis process but is part of photooxidation of cell organic carbon. Cyanobacteria and algae, along with photooxidation of dissolved organic matter, make the oceans net producers of CO. The total oceanic production of CO is around 0.1 bmta; plants and soil add another 0.1 bmta (Swinnerton et al. 1970; Weinstock and Niki 1972; Conrad 1988). The anthroprogenic contribution to CO production from the burning of biomass (wood) and fossil fuels is around 1.6 bmta. The atmospheric turnover time of CO is about 0.1–0.4 year.

The destruction of CO occurs in part by photochemical reactions in the atmosphere that convert it to CO_2. Microbial processes contribute to the substantial destruction of CO both in the ocean and in soil. Whereas the ocean is a net producer of CO, soil acts as a sink, removing an estimated 0.4 billion metric tons of CO from the atmosphere per year (Bartholomew and Alexander 1981; Conrad 1988). Carbon monoxide, though highly toxic to most aerobic organisms because of its affinity cytochromes, can be metabolized both aerobically and anaerobically by specialized microorganisms. Aerobically, the "carboxydobacteria," such as *Pseudomonas carboxidoflava* and *Pseudomonas carboxydohydrogena*, are capable of utilizing CO both as a carbon and as an energy source, although growth under such conditions is slow. The key enzyme, CO-oxidoreductase, catalyzes the reaction in Equation 3:

(3) $CO + H_2O \longrightarrow CO_2 + H_2$

In the presence of oxygen, the product H_2 is oxidized to water, yielding energy for CO_2 fixation. Growth on CO is relatively slow and inefficient; only 4% to 16% of the CO oxidized is fixed as cell carbon. Hydrogen gas is used preferentially by carboxidobacteria; growth is much more rapid and CO utilization is suppressed when hydrogen is supplied in the gas mixture. Hence, carboxidobacteria can be considered also as hydrogen bacteria that possess CO-oxidoreductase (Meyer 1989).

Anaerobically, CO can be reduced by H_2 to CH_4 by some methanogens such as *Methanosarcina barkeri* according to the reaction in Equation 4:

(4) $CO + 3H_2 \longrightarrow CH_4 + H_2O$

Alternately, CO can be reduced to acetate by acetogens such as *Clostridium thermoaceticum* (Equation 5) (Zeikus et al. 1985):

(5) $2CO + 2H_2 \longrightarrow CH_3COOH$

Limitations to Microbial Carbon Cycling

While the enzymatic ability of microorganisms to deal with naturally occurring organic substances is virtually unlimited, adverse environmental conditions, such as lack of oxygen, high acidity, and high concentrations of phenolics and tannins, can prevent the biodegradation of some types of natural substances. Such conditions are evident in muck soils (histosols), peat deposits, and some aquatic sediments. Ultimately, such accumulation of undegraded organic matter leads to fossil-fuel deposits and the removal of carbon from the biogeochemical cycling process.

The formation of humic materials from phenolic intermediates of lignin degradation and other metabolic processes represents an intermediate situation between immediate recycling and fossil-fuel deposition. Humic substances in soil are quite stable, and their average age as determined by ^{14}C dating ranges from 20 to 2,000 years. Humic compounds in peat and muck deposits persist even longer. Nevertheless, they continue to participate in the cycling process at slow rates.

Molecular structure has a major effect on cycling rates. Some organic compounds are relatively resistant to enzymatic attack; some may even be completely recalcitrant, that is, not subject to enzymatic degradation. Many synthetic chemicals, such as DDT and PCBs, discussed in chapter 13, are relatively resistant to microbial attack and therefore accumulate within the biosphere, occasionally reaching toxic levels within local habitats (Alexander 1973).

The activities of microorganisms affect the accessibility of carbon and the energy of organic compounds to the biological community. Some transformations of organic carbon—for example, the production of polymers like humic acids in soil—tend to reduce the rate of cycling or to immobilize that portion of carbon and stored energy. Other transformations, such as the anaerobic degradation of cellulose, mobilize stored carbon and energy by producing simpler organic compounds that can be more readily utilized by the biological community. Transformations that change the physical state, as the production of gaseous compounds such as CO_2 or CH_4 from liquids or solids, and transformations that alter the solubility, such as the production of glucose from cellulose, have major effects on the mobility and availability of carbon to the biological community within particular habitats.

Microbial Degradation of Polysaccharides

An important part of microbial carbon cycling is the biodegradation of plant polymers, a process that is especially critical in terrestrial environments. Plants are responsible for the principal input of organic carbon into soils, and soil microorganisms are largely responsible for the transformation of their structural polymers. As a consequence of their activity, carbon dioxide is reintroduced to the atmosphere, humic materials are formed, and simpler organic compounds are made available to other populations. Microorganisms within the gastrointestinal tract of herbivorous animals, analogously, play a major role in the degradation of plant polymers, making most of the carbon available to the animal.

Biogenic polymers recycled primarily by microbial degradation in soil include cellulose, hemicelluloses, and chitin (Figure 10.6). Cellulose, the most abundant biopolymer in the world, is a carbohydrate consisting of a linear chain of ß-1-4 linked glucose units. In soil, several varieties of fungi, including species of the genera *Aspergillus, Fusarium, Phoma,* and *Trichoderma,* and bacteria, including members of the genera *Cytophaga, Vibrio, Polyangium, Cellulomonas, Streptomyces,* and *Nocardia,* exhibit significant cellulolytic activities (Imshenetsky 1967; Ljungdahl and Eriksson 1985). Various Basidiomycota are prominent cellulose degraders in wood and litter on the soil surface. The soil pH exerts an important influence on which cellulolytic microbial populations are active. Below pH 5.5 filamentous fungi dominate, at pH 5.7–6.2 various fungi and *Cytophaga* species are involved in cellulose degradation, and at neutral-alkaline pH values *Vibrio* species and fungi predominate. Numerous additional bacteria are known to participate in cellulose degradation.

Cellulose degradation occurs under both aerobic and anaerobic conditions. Under aerobic conditions, various fungi, as well as aerobic and facultatively anaerobic bacterial populations, are involved in cellulolytic activities. The main products of cellulose degradation under aerobic conditions are carbon dioxide, water, and cell biomass. Under anaerobic conditions, members of the genus *Clostridium* appear to be the most important cellulose fermenters. Anaerobic fermentation of cellulose results in the production of low–molecular-weight fatty acids as well as carbon dioxide, water, and cell biomass. Fungi generally play a minor role in cellulose degradation under anaerobic conditions. Since fermentation yields less energy per unit of substrate consumed than does respiration, anaerobic cellulose fermenting bacteria must degrade large quantities of cellulose in order to generate cell

Figure 10.6

Composition of some structural polysaccharides. Cellulose (A) consists of β-1, 4 linked glucose (hexose) subunits (β-1, 4-glucosides). Hemicelloses are more diverse, often heteropolymeric and branched. The hemicellulose shown (B) is a β-1, 4 linked neutral xylan, consisting of xylose (pentose) subunits. Chitin (C) consists of β-1,4-linked N-acetylglucosamine (amino sugar) subunits. Microorganisms play a key role in biodegradation of these structural polymers.

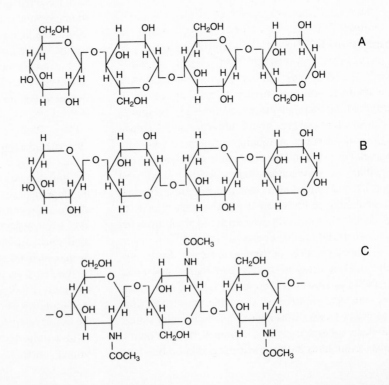

biomass. Degradation of cellulose also occurs at elevated temperatures, where it is carried out by thermophilic cellulolytic bacteria, such as *Clostridium thermocellum*. Thermophilic cellulose degradation is important in waste-composting procedures where large amounts of cellulose must be stabilized (see chapter 12).

The degradation of cellulose is catalyzed by cellulases. The enzyme system actually involves several different cellulases that catalyze various conversions of cellulose. A cellulase system involves three types of enzymes: a C_1 enzyme, a C_x or β 1-4 glucanase, and a β glucosidase. Total degradation of cellulose involves all three enzymes. The C_1 enzyme acts on native cellulose and does not appear to exhibit much action against partially degraded cellulose molecules. The C_x enzymes do not hydrolyze native cellulose but instead cleave partially degraded polymers. There are two types of cleavage exhibited by C_x enzymes. Endo β 1-4 glucanases break the chain internally, more or less at random, resulting in the formation of cellobiose and various oligomers. Exo β 1-4 glucanases attack the polymer near the end of the chain, resulting, principally, in the formation of cellobiose. The degradation of cellobiose and other relatively small oligomers is catalyzed by β glucosidase, resulting in the formation of glucose. In summary, the degradation of cellulose in soil and other habitats is a complex process involving multiple enzyme systems. Cellulolytic activities are exhibited by various microbial populations; the rates of cellulose degradation and the products formed depend on the species of microorganisms and the environmental conditions within individual habitats.

The second major class of plant constituents that enter soil habitats and that are subject to microbial degradation are the hemicelluloses (Alexander 1977). Hemicelluloses have no structural relationship to cellulose but rather are polysaccharides composed of various arrangements of pentoses such as xylose and arabinose, hexoses such as mannose, glucose and galactose, and/or uronic acids such as glucuronic and galacturonic acids. Examples of hemicellulosic compounds are xylans, mannans, and galactans. Hemicellulosic compounds are subject to degradation by various fungal and bacterial populations. Numerous fungi and bacteria, including actinomycetes and members of the genus *Bacillus,* are capable of degrading xylans. Many enzymes are involved in the degradation of hemicelluloses, including endoenzymes, which cleave bonds within the polymer somewhat at random, and exoenzymes, which normally cleave monomers or dimers from the end of the polymer. The products of hemicellulose degradation include carbon dioxide, water, cell biomass, and a variety of small carbohydrate molecules including monomers and dimers. The complexity of the hemicellulose molecules that enter soil as plant residues leads to a corresponding complexity of microbial degradation products that are formed.

Various other plant components, such as waxes, starch, and pectic substances, are also subject to degradation in various habitats. It is beyond the scope of this discussion to consider all of the molecules in plant residues that enter soil habitats and that are subject to microbial degradation. Suffice it to say that soil microorganisms through their degradative activities are effective in cycling carbon from a wide variety of plant residues, ultimately back to carbon dioxide, which can be used in primary production to complete the carbon cycle.

Chitin is another important polymer subject to microbial degradation in soil habitats. An acetylated amino sugar polymer, it is synthesized by various fungi as part of their cell walls and by various arthropods, including microcrustacea, as part of their skeletal structures. The global production of chitin in the marine and terrestrial environments amounts to many million mta (Gooday 1990), and most of this polymer is recycled by microbial degradation. Chitinases cleave the polymer from the reducing end or in a random way, eventually yielding primarily diacetylchitobiose units. This is hydrolyzed to N– acetylglucosamine monomers by acetylglucosaminidases. An alternate way of attack first deacetylates chitin to chitosan, and subsequently depolymerizes this product by chitosanase to chitobiose subunits. Glucosaminidases complete the degradation to glucosamine monomers. The ability to depolymerize and utilize chitin is quite common among bacteria, actinomycetes, and fungi. Many invertebrate and vertebrate animals produce their own chitinolytic enzymes, in contrast to cellulose digestion, but in many species intestinal microbiota make substantial contributions to chitin digestion (Kuznetsov 1970; Saunders 1970; Gooday 1990).

Agar, which is produced by numerous marine algae, is decomposed by relatively few species of marine bacteria of the genera *Agarbacterium, Flavobacterium, Bacillus, Pseudomonas*, and *Vibrio*. Many of these agar-utilizing bacteria occur as algal epiphytes. Because few microorganisms can depolymerize agar, it is especially suitable for the solidification of most microbial media.

Microbial Degradation of Lignin

Lignin is a structural plant polymer that is almost as abundant in higher plants as cellulose and hemicelluloses. Its unique structure and resistance to degradation justify a separate discussion of this structural plant polymer. In wood and other lignified structures such as grass stems, lignin occurs in intimate association with cellulose and hemicelluloses, adding structural strength and protecting the polysaccharides by its biodegradation-resistant barrier. Lignins have an aromatic structure consisting of phenylpropane subunits, linked together by carbon-carbon (C—C) or ether (C—O—C) into a highly complex three-dimensional structure (Ander and Eriksson 1978; Kirk et al. 1980; Zeikus 1981; Kirk 1984) (Figure 10.7). The biosynthesis of lignin starts from phenylalanine. Deamination, ring hydroxylation, methylation, and carboxyl reduction lead to cinnamyl alcohol precursors, which are oxidatively polymerized. Synthesis of lignin is unusual in that the polymerization does not take place on an enzyme surface. Instead, oxidases and peroxidases produce reactive quinon methid intermediates that polymerize spontaneously (Kirk 1984). Lignin and its subunits are, therefore, not optically active, and the randomness of the polymerization process makes subsequent enzymatic degradation of the product much more difficult. In this respect, there is a certain analogy between the biodegradation resistance of lignin and soil humus, although the structure of humus is much more random and, consequently, substantially more resistant to biodegradation.

Biodegradation rates for lignin are much lower than for either cellulose or hemicellulose compounds (Kirk 1984). Biodegradation of intact lignin does not occur anaerobically, although aromatic fragments produced by alkaline degradation of lignin can be metabolized anaerobically. Intact wood is attacked first by brown rot and white rot fungi, which are basidomycetes). Brown rot fungi, by a mechanism that is not yet clear, bypass the protective lignin barrier and attack the cellulose and hemicellulose components of wood directly. Logs decomposed in this manner fall apart into a brown powder consisting mainly of enzymatically liberated lignin. In contrast, white rot fungi degrade lignin preferentially, leaving a soft, fibrous, cellulosic residue. White rot and brown rot fungi are taxonomically closely related and usually work either sequentially or simultaneously, bringing about the complete biodegradation of wood.

The biodegradation of lignin is a heavily oxidative process. Like the formation process for lignin, it is indirect and random. The white rot fungus most intensively studied for its lignin-biodegradation activity, the basidiomycete *Phanerochaete chrysosporium*, appears to produce oxidizing agents such as superoxide anion (O_2^-), hydrogen peroxide (H_2O_2), hydroxyl radicals (—OH), and singlet oxygen ($^1O_2$). These break the bonds between subunits and bring about a gradual depolymerization of lignin. M. Tien and T. K. Kirk (1983) demonstrated the production of an extracellular, H_2O_2-dependent enzyme by *P. chrysosporium* that oxidatively breaks specific bonds in the lignin structure. Lignin does not serve as the sole source of carbon and energy for *P. chrysosporium*. Lignolytic activity is induced at a late growth stage and usually under nitrogen-limiting conditions. Thus, at least with this fungus, lignin biodegradation appears to be a secondary metabolic or idiophasic event.

Other fungi implicated in lignin biodegradation are *Polyporus, Poria, Fomes, Agaricus, Pleurotus, Collybia, Schizophyllum*, and *Fusarium*. Aerobic bacterial genera shown to participate in lignin biodegradation are *Arthrobacter, Flavobacterium, Micrococcus*, and *Pseudomonas*. The depolymerization of lignin

Figure 10.7

Chemical composition of the structural plant polymer lignin. The phenylpropane subunit (A) is linked by C—C and C—O—C bonds into a (B) complex three-dimensional structure that is relatively resistant to biodegradation. In the "lignocellulose complex" of wood, it also slows the biodegradation of cellulose.

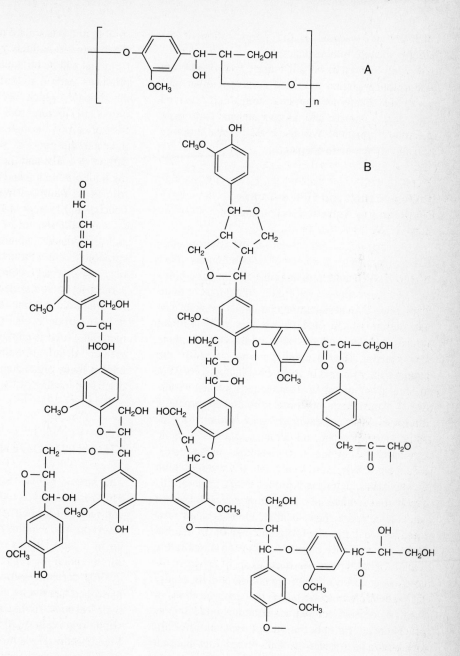

results in a variety of phenols, aromatic acids, and aromatic alcohols. Some of these are mineralized to CO_2 and H_2O, but some of the products, particularly the phenolic biodegradation intermediates, may give rise to humic compounds. This repolymerization is in part a spontaneous oxidative reaction and in part a catalysis by microbial polyphenoloxidases, peroxidases, and laccases. The apparent benefit to the microorganisms that catalyze such reactions is the detoxification of the harmful phenolic intermediates.

Only a portion of soil humus is derived from lignin biodegradation products; numerous other aromatic and nonaromatic organic compounds derived from plants, animals, and microorganisms participate in humus formation. We discussed humus structure and biodegradation in chapter 9.

Biodegradation and Heterotrophic Production in Aquatic Environments

Rivers, lakes, and coastal marine environments receive significant amounts of allochthonous plant residues that contain high amounts of structural polymers, such as cellulose and lignin. In contrast to these allochthonous materials, autochthonous organic carbon within aquatic habitats is generally formed through primary productivity of photosynthetic algae and bacteria. The qualitative composition of the organic matter formed in aquatic habitats is somewhat different from the organic compounds formed by terrestrial plants. Both the complexity and the amounts of organic matter present in the water column are generally lower than in soil habitats. Various soluble compounds, including carbohydrates, amino acids, and organic acids, are released into the water column by autochthonous primary producers. In general, this autochthonous material has a higher concentration of available nitrogen than do allochthonous plant residues. Concentrations of organic matter in natural waters are generally quite low; concentrations rarely exceed 1 mg carbon per liter in open seawater. At these low concentrations of organic carbon, the ability of aquatic bacteria to assimilate low concentrations of dissolved organic compounds becomes quite important. There appear to be significant populations of oligoheterotrophs (organisms that utilize low nutrient concentrations) in aquatic habitats. We already touched upon the microbial upgrading of detritus in chapter 5. The ability of most microorganisms to utilize inorganic nitrogen (NO_3^-, NH_4^+) is crucial to detritivores that cannot survive on nitrogen-deficient plant polymer particles alone. The utilization of both soluble and particulate organic compounds by bacteria in aquatic habitats produces organic matter in a form that can be utilized through a food web and also produces carbon dioxide, normally in the form of bicarbonate, which becomes available for primary producers. Figure 10.8 illustrates an example of the flow of carbon through an aquatic habitat. This figure indicates the pools of soluble organic and inorganic carbon in a lake and the relative transfer rates, showing both production and decomposition processes performed by microbial populations (Hutchinson 1975; Fenchel and Jorgensen 1977).

In addition to the natural input of carbon into soil and aquatic habitats, a large number of organic compounds enter these habitats through human activities. These organic compounds are, for the most part, subject to microbial degradation with the production of carbon dioxide and other compounds that enter the normal carbon cycle. The cycling of organic waste materials, such as garbage and sewage, is discussed in greater detail in chapter 12; the cycling of human-made organic chemicals, such as pesticides, is discussed in chapter 13.

THE HYDROGEN CYCLE

The largest global reservoir of hydrogen is water. This reservoir is actively cycled by photosynthesis and respiration, but because of the large size of the reservoir, the cycling rate is rather slow. Water tied up in crystal lattices of rock is not cycled actively, and the polar ice caps also remove water from active cycling for long periods of time. Substantial inert hydrogen reservoirs are liquid and gaseous fossil hydrocarbons. Living and dead organic material constitute relatively small but actively cycled reservoirs. Free gaseous H_2 is produced biologically in anaerobic fermentations and also as a side product of photosynthesis coupled with nitrogen fixation by cyanobacteria and by *Rhizobium*-legume associations (Conrad 1988). Most of the H_2 produced is further metabolized anaerobically to NH_3 or H_2S or to CH_4 as described earlier. When H_2 rises through

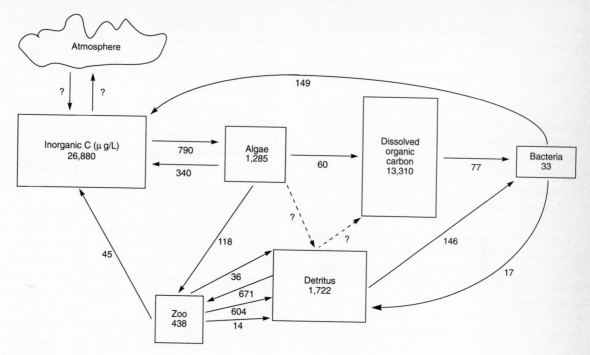

Figure 10.8

A flow diagram for carbon transfers in a lake habitat. The numbers associated with boxes indicate reservoirs of carbon; those associated with arrows indicate relative transfer rates. The figure illustrates the critical role of algae and bacteria in carbon flow through aquatic habitats. (Source: Saunders, 1970. Reprinted by permission of J. Cairns, Jr.)

oxygenated soil or sediment zones, it is metabolized oxidatively to H_2O, and only a small part—about 7 million mta—is likely to escape to the atmosphere. Although both evolution and consumption of the H_2 occur in the aquatic and the terrestrial environments, the oceans are net producers of an estimated 4 million mta of hydrogen per year, while soils serve as net hydrogen sinks. Hydrogen is produced anthropogenically by fossil fuel and biomass burning and as part of the exhaust of internal combustion engines (40 million mta). Hydrogen is also produced in the atmosphere by the photochemical decomposition of methane (40 million mta). The gravitational pull of Earth is not strong enough to prevent the loss of hydrogen to space from the upper atmosphere, where it is also produced by photodissociation of water vapor.

As discussed earlier, the present 21% atmospheric oxygen content was derived primarily by photosynthetic splitting of water, but the question inevitably arises of what happened to the double amount of hydrogen in the water molecules (Cloud and Gibor 1970; Valen 1971). The reduced carbon in living and dead organic matter and in fossil fuels is estimated to contain considerably less hydrogen than could be accounted for from the splitting of water, and there are no other substantial hydrogen sinks. The most plausible explanation appears to be that the missing amount of hydrogen may have been lost to space. Microbial processes that utilize H_2 and CH_4 and thus prevent their escape to the atmosphere minimize current losses of H_2.

The aerobic utilization of H_2 is performed by facultatively chemolithotrophic hydrogen bacteria (Schlegel 1989) in the following energy-yielding reaction shown in Equation 6:

(6) $H_2 + {}^1/_2O_2 \longrightarrow H_2O$ ($\Delta G = -56.7$ kcal/mole)

The most efficient hydrogen bacteria belong to the genus *Alcaligenes*. In addition to membrane-bound hydrogenases, these bacteria contain a soluble NAD-linked hydrogenase. A number of bacteria belonging to the genera *Pseudomonas, Paracoccus, Xanthobacter, Nocardia,* and *Azospirillum* contain membrane-bound hydrogenases only, and are capable of growing as hydrogen bacteria at slower rates. Many if not all hydrogen-activating enzymes (hydrogenases) appear to contain the trace metal nickel in their active center (Thauer et al. 1980). The overall equation of hydrogen utilization can be represented as follows in Equation 7:

(7) $6H_2 + 2O_2 + CO_2 \longrightarrow [CH_2O] + 5H_2O$

The formula in brackets represents cell material. Hydrogen bacteria fix carbon dioxide by the same mechanisms as algae and plants, that is, by carboxylation of ribulose 1,5-diphosphate and the regeneration of the above pentose through the Calvin cycle. Hydrogen bacteria can also grow on various organic substrates, or they can grow mixotrophically on a combination of H_2 and organic substrates.

The H_2 cycling activities discussed so far involve the evolution or utilization of molecular hydrogen. The most important H_2 cycling processes, photosynthesis and respiration, do not normally result in H_2 evolution or consumption, however. Instead, the electrons from H_2S (anoxyphotobacteria) or H_2O (all other photosynthetic organisms) are directly utilized in photosynthesis for reduction of CO_2. In respiration, the electrons from reduced organic compounds are passed along the respiratory chain, ultimately to reduce oxygen to water. Coupled to this process are phosphorylation reactions that satisfy the energy needs of the cell.

In dinitrogen fixation systems coupled to photosynthesis, that is, for the heterocystous cyanobacteria and the *Rhizobium*-legume symbiosis, the partial or complete uncoupling of the two systems results in the evolution of molecular hydrogen. In cyanobacteria, this seems to occur only under artificial laboratory conditions (Beneman and Weave 1974), but *Rhizobium*-legume associations seem to evolve large amounts of H_2 under agricultural field conditions. Improved *Rhizobium* strains are being constructed to eliminate this energetically wasteful process. The improved strains have high hydrogenase activity and are capable of oxidizing H_2 with the gain of energy. This energy, in turn, is used in fixation of additional nitrogen (LaFavre and Focht 1983).

A significant aspect of hydrogen cycling is the interspecies transfer of hydrogen between microorganisms with different metabolic capabilities. Interspecies hydrogen transfer, as occurs between the fermentative and methanogenic component of the synthrophic *Methanobacillus omelianskii* association (Bryant et al. 1967), is a mutually beneficial one. We will discuss competitive interspecies hydrogen transfer, as it occurs between methanogens and sulfate reducers, in chapter 11.

THE OXYGEN CYCLE

The establishment of an oxidizing atmosphere was the most profound biogeochemical transformation on our planet. Photosynthesis-derived oxygen not only transformed our atmosphere but oxidized large pools of reduced minerals such as ferrous iron and sulfides. Oxygen deposited in ferric iron and in dissolved and sedimentary sulfates exceeds the oxygen in the atmosphere by severalfold. These mineral reservoirs of oxygen, including the carbonates, participate to some extent in the oxygen cycle, but because of their large mass, their turnover rate is almost negligible. Among the more actively cycled reservoirs, atmospheric and dissolved oxygen, CO_2 and H_2O predominate. Oxygen in living and dead organic matter constitutes a relatively small but actively cycled reservoir.

The photosynthetic origin of atmospheric oxygen is generally accepted. The geological record clearly indicates that oxidized sediments such as ferric iron deposits were first formed 1.3 billion to 1.5 billion years ago. Older sediments are reduced. The oxidoreductive balance of the transition is less clear; the estimates and calculations are by necessity crude. There seems to be a shortage of reduced carbon as compared to atmospheric oxygen and oxidized minerals (Cloud and Gibor 1970; Valen 1971). As mentioned earlier, escape of hydrogen to space is one possible explanation.

Atmospheric oxygen produced in photosynthesis is removed from the atmosphere by respiration, a process that, besides producing CO_2, reconstitutes the water cleaved in photosynthesis. The presence or absence of molecular oxygen in a habitat is crucial in determining the type of metabolic activities that can occur in that habitat. Oxygen is inhibitory to strict anaerobes. Facultative anaerobes can gain more energy from organic substrates using oxygen as the terminal electron acceptor than by fermentative metabolism. For example, the aerobic metabolism of glucose yields 686 Kcal per mole; its fermentation yields only 50. Metabolic regulatory mechanisms ensure that oxygen is used preferentially as the terminal electron acceptor if it is available. Oxygen serves as the terminal electron acceptor not only in degradation of organic matter but also in the oxidation of reduced inorganic chemicals used as energy sources by chemolithotrophs.

In some habitats, the microbial utilization of oxygen during degradation of organic compounds may produce anoxic conditions. In habitats that are not in equilibrium with the atmosphere and where organic matter is undergoing rapid decomposition, oxygen is normally bound within organic compounds, and molecular oxygen becomes unavailable as an electron acceptor in respiration. Habitats such as marine sediments and flooded soils are normally depleted of molecular oxygen owing to its utilization during decomposition and the slow diffusion of molecular oxygen into these habitats.

The exhaustion of oxygen in an environment initiates the reduction of nitrate, sulfate, ferric iron, and oxidized manganese. If such electron acceptors are unavailable or become exhausted, fermentative metabolism and methanogenesis (reduction of CO_2) remain the only metabolic options.

Oxygen can be restored to an anaerobic environment by diffusion, sometimes aided by the bioturbation of worms and other burrowing animals indigenous to sediments and soils, or by photosynthetic activity. Plants, algae, and cyanobacteria produce molecular oxygen during the photosynthetic photolysis of water. Oxygen is not produced by the photosynthesis of the Anoxyphotobacteria.

The burning of fossil fuels may be expected to have an effect on atmospheric oxygen as well as on CO_2 concentrations. Indeed, alarmist articles occasionally appear in the popular press about the potential exhaustion of atmospheric oxygen by human activity. These concerns are groundless because of the sizable oxygen reservoir (21% of the atmosphere); the same processes that perturb the small CO_2 reservoir (0.03% of the atmosphere) have negligible effect on the much larger oxygen pool. The burning of all known fossil-fuel reserves would reduce the oxygen supply by only 3% (Broeker 1970).

Figure 10.9 compares the approximate turnover rates of atmospheric CO_2, O_2, and H_2O. All three materials are cycled by the balanced processes of photosynthesis and respiration, yet their differing reservoir sizes result in very different turnover rates. Based on pool sizes and utilization rates, every individual atmospheric CO_2 molecule has a chance of being assimilated by photosynthesis every 300 years or less, every atmospheric oxygen molecule will be used in respiration every 2,000 years, and every water molecule has a chance to be split by photosynthesis every 2,000,000 years. Thus, a change in the global rates of photosynthesis or respiration would have a more immediate and dramatic effect on atmospheric CO_2 concentrations than on either atmospheric O_2 or H_2O.

In the upper atmosphere, ionizing radiation transforms some of the oxygen from O_2 to O_3 (ozone). Ozone is not directly subject to biogeochemical cycling. The relatively small ozone pool in the atmosphere, however, is subject to perturbation by human activity and by alterations in the biogeochemical process of denitrification and methanogenesis. The

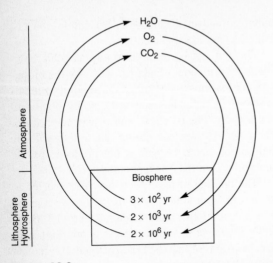

Figure 10.9

Turnover rates of CO_2, O_2, H_2O, by biogeochemical cycling showing probability of any given molecule of each being biologically processed. The smaller reservoir of CO_2 is more frequently cycled than the larger pools of O_2 and H_2O. The oceans represent the largest actively cycled H_2O reservoir. Even though all three molecules are used in the same processes of photosynthesis and respiration, their cycling times differ. Perturbation of cycling is most likely to affect the relatively small pool of CO_2. (Source: Woodwell 1970, The Energy Cycle of the Biosphere. Reprinted by permisson, copyright 1970, Scientific American Inc. All rights reserved.)

ozone layer is important for its shielding capacity against incoming UV radiation.

SUMMARY

Biogenic elements cluster in distinct areas of the periodic table and may be subdivided into major, minor, and trace elements. Biogeochemical cycling of elements with several stable valence states tends to be complex and oxidoreductive. Other elements may be cycled without valence changes by mobilization and precipitation only. A chemical form of an element represents a reservoir. Small reservoirs such as atmospheric CO_2 are turned over rapidly and are particularly sensitive to disturbances of cycling.

Carbon, hydrogen, and oxygen are cycled primarily by the two opposing processes of photosynthesis and respiration, but because of the differing reservoir sizes, the turnover rates of these three elements differ greatly. In the absence of oxygen, organic carbon may be recycled in fermentative and methanogenic processes. Reduced products of fermentation and methanogenesis often reenter the oxidative cycle. Hydrogen, evolved in some fermentation processes, may combine with various electron acceptors, including O_2, CO, and CO_2. Atmospheric oxygen is biogenic in origin. It is the preferred electron acceptor for all aerobic and facultatively anaerobic organisms.

All living organisms participate in cycling of C, H, and O. However, microorganisms dominate both the production and catabolic processes in limnetic and pelagic environments. In addition, they dominate the catabolic process in the terrestrial environment because of their unique capacity to degrade complex polymers such as cellulose, lignin, and soil humus.

REFERENCES & SUGGESTED READINGS

Alexander, M. 1973. Nonbiodegradable and other recalcitrant molecules. *Biotechnology and Bioengineering* 15:611–647.

Alexander, M. 1977, *Introduction to Soil Microbiology*. John Wiley and Sons, New York, pp. 148–202.

Ander, P., and K. E. Eriksson. 1978. Lignin decomposition. *Progress in Industrial Microbiology* 4:1–58.

Azam, F., T. Fenchel, J. G. Field, J. S. Gray, L. A. Meyer-Reil, and F. Thingstad. 1983. The ecological role of water-column microbes in the sea. *Marine Ecological Progress Series* 10:257–263.

Balch, W. E., G. E. Fox, L. J. Magrum, C. R. Woese, and R. S. Wolfe. 1979. Methanogens: Reevaluation of a unique biological group. *Microbiological Reviews* 43:260–296.

Bartholomew, G. W., and M. Alexander. 1981. Soil as a sink for atmospheric carbon monoxide. *Science* 212:1389–1391.

Beneman, J. R., and N. M. Weave. 1974. Hydrogen evolution by nitrogen-fixing *Anabaena cylindrica* cultures. *Science* 184:174–175.

Bird, D. F., and J. Kalff. 1986. Bacterial grazing by planktonic lake algae. *Science* 231:493–495.

Bolin, B. 1970. The carbon cycle. *Scientific American* 223(3):125–132.

Bolin, B., E. T. Degens, P. Duvigneaud, and S. Kempe. 1979. The global biogeochemical carbon cycle. In B. Bolin, E. T. Degens, S. Kempe, and P. Ketner (eds.). *The Global Carbon Cycle*. John Wiley and Sons, New York, pp. 1–53.

Bowien, B., and H. G. Schlegel. 1981. Physiology and biochemistry of aerobic hydrogen-oxidizing bacteria. *Annual Reviews of Microbiology* 35:405–452.

Broeker, W. S. 1970. Enough air. *Environment* 12(7):27–31.

Bryant, M. P., E. A. Wolin, M. J. Wolin, and R. S. Wolfe. 1967. *Methanobacillus omelianskii*, a symbiotic association of two species of bacteria. *Archiv für Mikrobiologie* (Berlin) 59:20–31.

Bull, A. T. 1980. Biodegradation: Some attitudes and strategies of microorganisms and microbiologists. In D. C. Ellwood, J. N. Hedger, M. J. Latham, J. M. Lynch, and J. H. Slater (eds.). *Contemporary Microbial Ecology*. Academic Press, London, pp. 107–136.

Charlson, R. J., S. E. Schwartz, J. M. Hales, R. D. Cess, J. A. Coakley, Jr., J. E. Hansen, and D. J. Hoffman. 1992. Climate forcing by anthropogenic aerosols. *Science* 255:423–430.

Cloud, P., and A. Gibor. 1970. The oxygen cycle. *Scientific American* 223(3):111–123.

Codd, G. A. (ed.). 1984. *Aspects of Microbial Metabolism and Ecology*. Academic Press, London.

Conrad, R. 1988. Biogeochemistry and ecophysiology of atmospheric CO and H_2. *Advances in Microbial Ecology* 10:231–283.

Deevey, E. S., Jr. 1970. Mineral cycles. *Scientific American* 223(3):149–158.

Eriksson, K. E., and S. Christl. 1982. Mineralisation of carbon. In R. G. Burns and J. H. Slater (eds.). *Experimental Microbial Ecology*. Blackwell Scientific Publications, Oxford, England, pp. 134–153.

Fenchel, T. 1986. The ecology of heterotrophic microflagellates. *Advances in Microbial Ecology* 9:57–97.

Fenchel, T. M., and B. B. Jorgensen. 1977. Detritus food chains of aquatic ecosystems: The role of bacteria. *Advances in Microbial Ecology* 1:1–58.

Frieden, E. 1972. The chemical elements of life. *Scientific American* 227(1):52–60.

Gooday, G. W. 1990. The ecology of chitin degradation. *Advances in Microbial Ecology* 11:387–430.

Gottschalk, G. 1979. *Bacterial Metabolism*. Springer-Verlag, New York.

Gottschalk, G. 1989. Bioenergetics of methanogenic and acetogenic bacteria. In H. G. Schlegel and B. Bowien (eds.). *Autotrophic Bacteria*. Springer-Verlag, Berlin, pp. 383–413.

Haber, C. L., L. N. Allen, S. Zhao, and R. Hanson. 1983. Methylotrophic bacteria: Biochemical diversity and genetics. *Science* 221:1147–1153.

Hanson, R. S. 1980. Ecology and diversity of methylotrophic organisms. *Advances in Applied Microbiology* 26:3–39.

Heal, O. W., and P. Ineson. 1984. Carbon and energy flow in terrestrial ecosystems: Relevance to microflora. In M. J. Klug and C. A. Reddy (eds.). *Current Perspectives in Microbial Ecology*. American Society for Microbiology, Washington, D.C., pp. 394–404.

Hobbie, J. E., and J. M. Melillo. 1984. Comparative carbon and energy flow in ecosystems. In M. J. Klug and C. A. Reddy (eds.). *Current Perspectives in Microbial Ecology*. American Society for Microbiology, Washington, D.C., pp. 389–393.

Houghton, R. A., J. E. Hobbie, J. M. Melillo, B. Moore, B. J. Peterson, G. R. Shaver, and G. M. Woodwell. 1983. Changes in the carbon content of terrestrial biota and soils between 1860 and 1980: A net release of CO_2 to the atmosphere. *Ecological Monographs* 53(3):235–262.

Hubel, H. 1966. Die ^{14}C Methode zur Bestimmung der Primärproduktion des Phytoplanktons. *Limnologica* (Berlin) 4:267–280.

Hutchinson, G. E. 1970. The biosphere. *Scientific American* 223(3):45–53.

Hutchinson, G. E. 1975. *A Treatise on Limnology. Vol. I, Geography, Physics and Chemistry*. Part 2, Chemistry of Lakes. John Wiley and Sons, New York.

Imshenetsky, A. A. 1967. Decomposition of cellulose in the soil. In T. R. G. Gray and D. Parkinson (eds.). *The Ecology of Soil Bacteria*. Liverpool University Press, Liverpool, England, pp. 256–269.

Jones, C. W. 1980. Unity and diversity in bacterial energy conservation. In D. C. Ellwood, J. N. Hedger, M. J. Latham, J. M. Lynch, and J. H. Slater (eds.). *Contemporary Microbial Ecology*. Academic Press, London, pp. 193–214.

Jones, W. J., M. I. Donnelly, and R. S. Wolfe. 1985. Evidence of a common pathway of carbon dioxide reduction to methane in methanogens. *Journal of Bacteriology* 163:126–131.

Jorgensen, B. B. 1980. Mineralization and the bacterial cycling of carbon, nitrogen, and sulphur in marine sediments. In D. C. Ellwood, J. N. Hedger, M. J. Latham, J. M. Lynch, and J. H. Slater (eds.). *Contemporary Microbial Ecology*. Academic Press, London, pp. 239–252.

Jorgensen, B. B. 1989. Biogeochemistry of chemoautotrophic bacteria. In H. G. Schlegel and B. Bowien (eds.). *Autotrophic Bacteria*. Springer-Verlag, Berlin, pp. 117–146.

Kirk, T. K. 1984. Degradation of lignin. In D. T. Gibson (ed.). *Microbial Degradation of Organic Compounds*. Marcel Dekker, New York, pp. 399–437.

Kirk, T. K., T. Higuchi, and H. M. Chang. 1980. *Lignin Biodegradation: Microbiology, Chemistry and Potential Applications*. Vols. 1 and 2. CRC Press, Boca Raton, Fla.

Kramer, P. 1981. Carbon dioxide concentration, photosynthesis, and dry matter production. *BioScience* 31:29–33.

Krumbein, W. E., and P. K. Swart. 1983. The microbial carbon cycle. In W. E. Krumbein (ed.). *Microbial Geochemistry*. Blackwell Scientific Publications, Oxford, England, pp. 5–62.

Kuznetsov, S. I. 1970. *The Microflora of Lakes and Its Geochemical Activity*. University of Texas Press, Austin.

LaFavre, J. S., and D. D. Focht. 1983. Conservation in soil of H_2 liberated from N_2 fixation by Hup⁻ nodules. *Applied and Environmental Microbiology* 46:304–311.

LaMarche, V. C., Jr., D. A. Greybill, H. C. Fritts, and M. R. Rose. 1984. Increasing atmospheric carbon dioxide: Tree ring evidence for growth enhancement in natural vegetation. *Science* 225:1019–1021.

Large, P. J. 1983. *Methylotrophy and Methanogenesis*. American Society for Microbiology, Washington, D.C.

Ljungdahl, L. G., and K. E. Eriksson. 1985. Ecology of microbial cellulose degradation. *Advances in Microbial Ecology* 8:237–299.

Manabe, S., and R. T. Wetherald. 1980. On the distribution of climate change resulting from an increase in CO_2 content in the atmosphere. *Journal of Atmospheric Science* 37:99–118.

Mertz, W. 1981. The essential trace elements. *Science* 213:1332–1358.

Meyer, O. 1989. Aerobic carbon monoxide-oxidizing bacteria. In H. G. Schlegel and B. Bowien (eds.).

Autotrophic Bacteria. Springer-Verlag, Berlin, pp. 331–350.

Morris, I. 1982. Primary production of the oceans. In R. G. Burns and J. H. Slater (eds.). *Experimental Microbial Ecology*. Blackwell Scientific Publications, Oxford, England, pp. 239–252.

National Research Council. 1979. *Carbon Dioxide and the Climate: A Scientific Assessment*. National Academy Press, Washington, D.C.

National Research Council. 1983. *Changing Climate*. National Academy Press, Washington, D.C.

National Research Council. 1991. *Global Environmental Change: The Human Dimensions*. National Academy Press, Washington, D.C.

Nedwell, D. B. 1984. The input and mineralization of organic carbon in anaerobic aquatic sediments. *Advances in Microbial Ecology* 7:93–131.

Odum, E. P. 1962. Relationships between structure and function in ecosystems. *Japanese Journal of Ecology* 12:108–118.

Odum, E. P. 1983. *Basic Ecology*. Saunders College Publishing, Philadelphia.

Pomeroy, L. R. (ed.). 1974. *Cycles of Essential Elements*. Dowden, Hutchinson and Ross, Stroudsburg, Penn.

Pomeroy, L. R. 1984. Significance of microorganisms in carbon and energy flow in marine ecosystems. In M. J. Klug and C. A. Reddy (eds.). *Current Perspectives in Microbial Ecology*. American Society for Microbiology, Washington, D.C., pp. 405–411.

Ryther, J. H. 1969. Photosynthesis and fish production in the sea. *Science* 166:72–76.

Sanders, R. W., and K. G. Porter. 1988. Phagotrophic phytoflagellates. *Advances in Microbial Ecology* 10:167–192.

Saunders, G. W. 1970. Carbon flow in the aquatic system. In J. Cairns, Jr. (ed.). *The Structure and Function of Freshwater Microbial Communities*. Research Division Monograph 3, Virginia Polytechnic Institute and State University, Blacksburg, pp. 31–46.

Schlegel, H. G. 1989. Aerobic hydrogen-oxidizing (Knallgas) bacteria. In H. G. Schlegel and B. Bowien (eds.). *Autotrophic Bacteria*. Springer-Verlag, Berlin, pp. 305–329.

Sherr, B. F., and E. B. Sherr. 1984. Role of heterotrophic protozoa in carbon and energy flow in aquatic ecosystems. In M. J. Klug and C. A. Reddy (eds.). *Current Perspectives in Microbial Ecology*. American Society for Microbiology, Washington, D.C., pp. 412–423.

Swinnerton, J. W., V. J. Linnenbom, and R. A. Lamontague. 1970. The ocean: A natural source of carbon monoxide. *Science* 167:984–986.

Thauer, R. K., G. Diekert, and P. Schönheit. 1980. Biological role of nickel. *Trends in Biochemical Sciences* 5:304–306.

Tien, M., and T. K. Kirk. 1983. Lignin-degrading enzyme from the hymenomycete *Phanerochaete chrysosporium* bonds. *Science* 221:661–663.

Underwood, E. J. 1977. *Trace Elements in Human and Animal Nutrition*. Academic Press, New York.

Valen, L. V. 1971. The history and stability of atmospheric oxygen. *Science* 171:439–443.

Weinstock, B., and H. Niki. 1972. Carbon monoxide balance in nature. *Science* 176:290–292.

Wood, H. G. 1989. Past and present utilization of CO_2. In H. G. Schlegel and B. Bowien (eds.). *Autotrophic Bacteria*. Springer-Verlag, Berlin, pp. 33–52.

Woodwell, G. M. 1970. The energy cycle of the biosphere. *Scientific American* 223(3):64–74.

Zeikus, J. G. 1977. The biology of methanogenic bacteria. *Bacteriological Reviews* 41:514–541.

Zeikus, J. G. 1981. Lignin metabolism and the carbon cycle. Polymer biosynthesis, biodegradation and environmental recalcitrance. *Advances in Microbial Ecology* 5:211–243.

Zeikus, J. G., R. Kerby, and J. A. Krzycki. 1985. Single carbon chemistry of acetogenic and methanogenic bacteria. *Science* 227:1167–1173.

11

Biogeochemical Cycling: Nitrogen, Sulfur, Phosphorus, Iron, and Other Elements

Carbon, hydrogen, and, with some exceptions, oxygen are constituents of all organic compounds. In contrast, the major elements nitrogen, sulfur, and phosphorus occur only in specific classes of biochemicals. Nitrogen is a constituent of amino acids, nucleic acids, amino sugars, and their polymers. Sulfur occurs mainly as sulfhydryl (—SH) groups in amino acids and their polymers. Phosphorus is a constituent of nucleic acids (RNA and DNA), sugar phosphates, and phosphate esters, such as the ATP/ADP/AMP system of cellular energy transfer. Minor and trace elements are required in low amounts, and not by all forms of life; they have specific restricted functions.

Carbon is assimilated by autotrophs in gaseous form (CO_2), but nitrogen, sulfur, phosphorus, and the rest of the biogenic elements are usually taken up by autotrophs and phototrophic heterotrophs in the form of mineral salts. Hence, biogenic elements other than carbon, hydrogen, and oxygen are often referred to as mineral nutrients, and their cycles as mineral cycles (Deevey 1970). Biogenic elements such as nitrogen, sulfur, iron, and manganese exist in the ecosphere in several stable valence states and tend to be cycled in a complex oxidoreductive manner. Elements such as phosphorus, calcium, and silicon exist in only one stable valence state. These elements have relatively simple cycles involving dissolution and incorporation into organic matter, balanced by mineralization and sedimentation. Some mineral deposits formed by microbial processes are unique and not duplicated by purely geochemical processes (Lowenstein 1981).

THE NITROGEN CYCLE

Nitrogen, which has stable valence states ranging from −3, as in ammonia (NH_3), to +5, as in nitrate (NO_3^-), occurs in numerous oxidation states. A large, slowly cycled reservoir for nitrogen (3.8×10^{15} metric tons) is N_2 gas of the atmosphere (79%). Large but essentially unavailable reservoirs of nitrogen are present in igneous (14×10^{15} metric tons) and sedimentary (4.0×10^{15} metric tons) rock as bound, nonexchangeable ammonia (Blackburn 1983). Physicochemical and biological weathering releases ammonia from these reservoirs so slowly that it has little influence on yearly cycling models. Geological deposits of more readily available combined nitrogen are rare. Availability of combined nitrogen is an important limiting factor for primary production in many ecosystems. The only natural accumulations of nitrate occur on some islands off the Chilean coast; these nitrate deposits are derived from the decomposition of guano deposited by seabirds. The dry climate of these islands has prevented the leaching of nitrate (Deevey 1970; Delwiche 1970).

The inorganic nitrogen salts, ammonium, nitrite, and nitrate are highly water soluble and consequently are distributed in dilute aqueous solution throughout the ecosphere; they form small, actively cycled reservoirs. Living and dead organic matter also provide relatively small, actively cycled reservoirs of nitrogen. In temperate climates, stabilized soil organic matter, or humus, forms a substantial and relatively stable nitrogen reservoir. The nitrogen of humus becomes available for uptake by living organisms only through its slow mineralization, a process measured in decades and centuries. In tropical climates, the temperature and humidity favor the rapid direct mineralization of organic matter, limiting the accumulation of litter and humus.

Plants, animals, and most microorganisms require combined forms of nitrogen for incorporation into cellular biomass (Brown and Johnson 1977), but the ability to fix atmospheric nitrogen is restricted to a limited number of bacteria and symbiotic associations. While many habitats depend on plants for a supply of organic carbon that can be used as a source of energy, all habitats depend either on the bacterial fixation of atmospheric nitrogen or on human intervention through the distribution of nitrogen fertilizers synthesized by the Haber-Bosch process, invented shortly before World War I. Plants could not continue their photosynthetic metabolism without the availability of fixed forms of nitrogen provided by microorganisms or by synthetic fertilizer.

The biogeochemical cycling of the element nitrogen is highly dependent on the activities of microorganisms. Figure 11.1 shows a generalized scheme for the biogeochemical cycling of nitrogen.

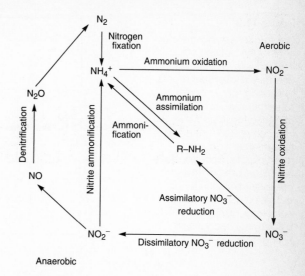

Figure 11.1

The nitrogen cycle showing chemical forms and key processes in biogeochemical cycling of nitrogen. The lower left portion of the cycle represents anaerobic, the upper right portion aerobic processes. The critical steps of nitrogen fixation, nitrification, and denitrification are all mediated by bacteria. R—NH$_2$ represents amino groups in cell protein.

Figure 11.2 shows a more detailed representation of nitrogen flow through the biosphere, indicating the magnitude of some critical transfer rates (Burns and Hardy 1975; Soderlund and Svensson 1976). Fixation on land, which includes considerable amounts from agriculturally managed legume crops, amounts to 135 million metric tons per annum. This greatly exceeds N$_2$ fixation in the much larger marine environment (40 million metric tons per annum). Anthropogenic nitrogen inputs in the form of synthetic fertilizer (30 million metric tons per annum), combustion (19 million metric tons per annum), and nitrogen fixation by leguminous and other crops (44 million metric tons per annum) approach the total of nitrogen fixation in grasslands (45 million metric tons per annum), forests (40 million metric tons per annum), other terrestrial areas (10 million metric tons per annum), and the marine environment (40 million metric tons per

Figure 11.2

General scheme for transfer of nitrogen between the atmosphere and marine and terrestrial environments indicating some transfer rates (numbers are millions of metric tons/year) and interconversions within the terrestrial and aquatic reservoirs. As minor transfer processes are omitted, a complete balance is not achieved. (Source: Burns and Hardy 1975. Reprinted by permission, copyright Springer-Verlag.)

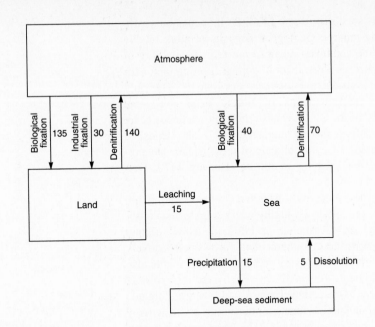

annum). Synthetic nitrogen fertilizer input is projected to rise further and may reach 100 million metric tons per annum by the year 2000. Volcanic activity, ionizing radiation, and electrical discharges supply additional combined nitrogen to the atmosphere; when washed down with precipitation, this combined nitrogen becomes available to the biosphere, but abiotically fixed nitrogen is estimated to be only 10% to 20% of the biological fixation. Microorganisms are also responsible for the return of molecular nitrogen to the atmosphere through denitrification and for the transformations that affect the mobility and accessibility of fixed nitrogen to the inhabitants of the litho- and hydroecospheres. Prior to human intervention, N_2-fixation and denitrification processes appeared to be in balance. With increased anthropogenic inputs, this may no longer be the case.

Fixation of Molecular Nitrogen

Nitrogenase is the enzyme complex responsible for nitrogen fixation (Smith 1982; Postgate 1982; Sprent

and Sprent 1990). The nitrogenase enzyme system has two coproteins, one containing molybdenum plus iron and the other containing only iron (Figure 11.3). Nitrogenase is extremely sensitive to oxygen, requiring low oxygen tensions for activity. The fixation of nitrogen needs not only nitrogenase, but also ATP, reduced ferredoxin, and perhaps other cytochromes and coenzymes (Figure 11.4). Ammonia is formed as the first detectable product of nitrogen fixation (Equation 1):

(1) $N_2 + 6e^- \longrightarrow 2NH_3$ ($\Delta G = +150$ kcal/mole)

As indicated by the positive ΔG, the reaction requires high energy input. The ammonia is assimilated into amino acids, which are subsequently polymerized into proteins (Campbell 1977; Brill 1981; Sprent and Sprent 1990).

The biological fixation of molecular nitrogen is carried out by several free-living bacterial genera, some of which may be rhizosphere associated, and by several bacterial genera that form mutualistic associations with plants. The fixation of nitrogen generally occurs in response to low or limiting concentrations

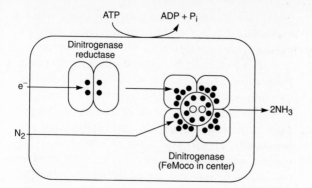

Figure 11.3

Schematic representation of the nitrogenase enzyme complex. The iron protein or dinitrogenase reductase (left) consists of two identical subunits and contains 4 iron atoms (dark). It receives electrons from highly reduced carriers (ferredoxins) from photosynthesis or from organic substrate catabolism and channels these to the dinitrogenase protein (right). The latter is an iron protein tetramer with an attached iron-molybdenum cofactor (FeMoco), containing 2 molybdenum atoms (light) per cofactor molecule. The number of the iron atoms varies between 24 and 32 with the source of the dinitrogenase. In numbers similar to the iron atoms, dinitrogenase and dinitrogenase reductase both contain reactive sulfur groups (not shown in the figure). (Source: Brill 1981; Sprent and Sprent 1990).

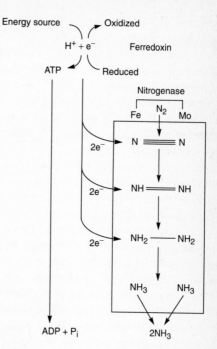

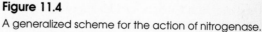

Figure 11.4

A generalized scheme for the action of nitrogenase.

of ammonia. In terrestrial habitats, the symbiotic fixation of nitrogen by rhizobia accounts for the largest contribution of combined nitrogen. The rates of nitrogen fixation by symbiotic rhizobia are often two to three orders of magnitude higher than rates exhibited by free-living nitrogen-fixing bacteria in soil. Rhizobia associated with an alfalfa field may fix up to 300 kg N/ha/year, compared to a rate of 0.5–2.5 kg N/ha/year for free-living *Azotobacter* species (Dalton 1974; Burns and Hardy 1975).

Nitrogen has no radioactive isotope with a convenient half-life; ^{13}N decays rapidly and its stable isotope (^{15}N) can be analyzed only by expensive and inconvenient mass spectrometry. Studies on the rates of nitrogen fixation were greatly enhanced by development of the acetylene reduction assay (Hardy et al. 1968). This assay is based on the fact that nitrogenase

enzymes will reduce acetylene to ethylene. The similarity of the acetylene molecule (HC≡CH) to nitrogen (N≡N) is obvious even from the written structures. The rate of formation of ethylene is a measure of nitrogenase or nitrogen-fixing activity. Ethylene can be conveniently assayed with great sensitivity using a gas chromatograph equipped with flame ionization detector. For a nitrogenase activity measurement, a small soil core, plant root system, or other environmental sample is placed in a gastight enclosure—for example, a plastic syringe or bag. The enclosure is flushed with acetylene or acetylene-oxygen mixture and incubated for several hours. Since the nitrogenase system has greater affinity to acetylene than to dinitrogen gas, meticulous removal of N_2 is not necessary. After incubation, the gas phase is assayed for the amount of ethylene formed by acetylene reduction. Procedures for chemically analyzing the disappearance of nitrogen gas from the atmosphere or the rates of formation of fixed forms of nitrogen are far less sensitive and convenient for routine use. The rate of acetylene reduction, compared with the rate of incorporation of molecular nitrogen into cell biomass, generates roughly a 3:1 conversion factor from moles of ethylene formed to moles of N_2 fixed. This conversion factor, though, is not identical in all ecosystems, necessitating a calibration against ^{15}N-incorporation measurements in each ecosystem.

Acetylene reduction and other quantitative methods, such as measuring growth on nitrogen-free media, have revealed that a number of bacterial species are capable of fixing molecular nitrogen under appropriate environmental conditions. Redox potential is especially important in determining the rates of nitrogen fixation by many bacterial species that are genetically capable of producing nitrogenase enzymes. Many aerobic nitrogen-fixing bacteria have been shown to fix nitrogen more efficiently at oxygen levels below the normal atmospheric concentration (21%). Such conditions are frequently found in subsoil and sediment habitats. In addition to *Azotobacter* and *Beijerinckia*, well-established genera of free-living nitrogen-fixing soil bacteria, more and more genera have been found over time to fix atmospheric nitrogen. They now include species of *Chromatium, Rhodopseudomonas, Rhodospirillum, Rhodomicrobium, Chlorobium, Chloropseudomonas, Desulfovibrio, Desulfotomaculum, Klebsiella, Bacillus, Clostridium, Azospirillum, Pseudomonas, Vibrio, Thiobacillus,* and *Methanobacillus* (Stewart 1973; Blackburn 1983). Several members of Actinomycetales have also been shown to fix atmospheric nitrogen, some of which are free living and others of which fix atmospheric nitrogen in association with compatible plants (see chapter 4). Overall, the sensitivity of today's methodology has revealed an increasing number of bacterial genera that can fix atmospheric nitrogen when oxygen tensions are appropriate.

Although the rates of nitrogen fixation for free-living soil bacteria are relatively low, these bacteria are widespread in soil. Rates of nitrogen fixation by free-living bacteria, such as *Azotobacter* and *Azospirillum,* are higher within the rhizosphere, because of the availability of organic compound from root exu-

dates, than in soils lacking plant roots, allowing for increased efficiency of nitrogen transfer to photosynthetic organisms (Döbereiner and Day 1974).

In aquatic habitats, cyanobacteria are the principal nitrogen fixers (Paerl 1990). Many of the filamentous nitrogen-fixing cyanobacteria, such as *Anabaena, Aphanizomenon, Nostoc, Gloeotrichia, Cylindrospermum, Calothrix, Scytonema,* and *Tolypothrix* have heterocysts. Heterocysts are thick-walled, less pigmented, and often enlarged cells occurring at more-or-less regular intervals among the regular cells. During their differentiation, heterocysts lose their oxygen-evolving photosystem II but retain their anoxygenic photosystem I. Nitrogen fixation is localized in the heterocysts, where the oxygen-sensitive nitrogenase is protected from inactivation by the photosynthetically produced oxygen. There is an active exchange between the ordinary cells and the heterocysts, the former supplying disaccharides from photosynthesis and the latter supplying fixed nitrogen, mainly as glutamine.

Some nonheterocystous cyanobacteria, such as *Oscillatoria, Trichodesmium, Microcoleus,* and *Lyngbya* have been shown to fix nitrogen. How these cyanobacteria protect their nitrogenase from oxygen inactivation is less clear, but some mechanisms have been identified that individually or in combination may explain the phenomenon. Some of the nonheterocystous cyanobacteria show a temporal separation between photosynthesis and nitrogenase activity. During daylight hours, photosynthesis and photosynthate storage take place, with little or no nitrogen fixation. During nighttime, in the absence of photosynthesis, nitrogen fixation takes place at the expense of the stored photosynthate. Other nonheterocystous nitrogen fixers form clumps or mats. In these aggregations, the outer cells photosynthesize, while the innermost cells, through a combination of shading and respiratory activity, are in a zone of reduced oxygen tension permissive of nitrogenase activity. Nutrient exchange is obviously necessary in such a situation. Apparently, nonheterocystous cyanobacteria possess mechanisms to reactivate or resynthesize their nitrogenases after oxygen exposure.

Rates of nitrogen fixation by cyanobacteria are generally one to two orders of magnitude higher than by free-living nonphotosynthetic soil bacteria. Under favorable conditions, cyanobacteria in a rice paddy may fix up to 150 kg of nitrogen/hectare/year (National Research Council 1979). Nitrogen-fixing cyanobacteria, many of which form heterocysts, are found both in marine and freshwater habitats. Some nitrogen-fixing cyanobacteria form associations with other microorganisms, as in lichens; some form symbiotic associations with plants, such as the *Azolla-Anabaena* association; others are free living.

Bacterial fixation of molecular nitrogen requires a considerable input of energy (150 kcal/mole) in the form of ATP and reduced coenzymes (Benemann and Valentine 1972; Brill 1981). The energy for carrying out nitrogen fixation may be obtained through the conversion of light energy by photoautotrophs, such as cyanobacteria, or by respiration of heterotrophs, such as *Azotobacter*. In the latter case, nitrogen fixation is limited by the availability of organic substrates, and attempts to increase nitrogen fixation by commercial *Azotobacter* inocula, such as "Nitragin," have been unsuccessful.

Nitrogen fixation within the phyllosphere and rhizosphere is ecologically important since some of the fixed nitrogen is likely to become available to the plants (see chapter 4). Exudates from the plant may supply some of the energy required for the nitrogen-fixation process. Nitrogen fixation also occurs within the digestive tracts of some animals, such as termites (see chapter 5). The fixation of nitrogen in organisms consuming primarily nitrogen-free cellulose is especially important for formation of proteins. Substantial nitrogen fixation has been observed in environments characterized by an abundance of carbon sources and a scarcity of nitrogen, such as decomposing logs or oil-contaminated soils (Coty 1967).

Azotobacter and *Beijerinckia* can fix nitrogen at normal oxygen tensions and appear to protect their nitrogenases from oxidative inactivation by a combination of compartmentalization and complex biochemical mechanisms. Other free-living nitrogen fixers, such as *Azospirillum*, fix nitrogen only at reduced oxygen tensions. Such environments, especially on the microscale, are not rare in soil and on root surfaces. Oxygen is not a problem for nitrogenases

of anoxyphototrophic bacteria such as *Chromatium, Rhodopseudomonas, Rhodospirillum, Rhodomicrobium, Chlorobium,* and *Chloropseudomonas,* or for the anaerobic heterotrophs such as *Clostridium, Desulfovibrio,* and *Desulfotomaculum.* The latter are active in anaerobic sediments and in the rhizosphere of plants growing in such sediments (see chapter 4). As in aerobic nitrogen fixation, carbon-rich nitrogen-poor substrates such as cellulose favor the process. This has been known to occur indirectly as nitrogen fixers utilized low–molecular-weight products of cellulose fermentation. More recently, direct nitrogen fixation by cellulolytic anaerobes was also documented (Leschine et al. 1988).

Ammonification

Many plants, animals, and microorganisms are capable of ammonification, a process in which organic nitrogen is converted to ammonia. Nitrogen in living and dead organic matter occurs predominately in the reduced amino form. T. H. Blackburn (1983) emphasized the importance of organic nitrogen mineralization for continued ecosystem productivity. The release of ammonia from a simple nitrogenous organic compound, urea, can be described as follows (Equation 2):

$$(2) \quad NH_2-\overset{\overset{\displaystyle O}{\displaystyle \|}}{C}-NH_2 + H_2O \xrightarrow{\text{urease}}$$

$$2NH_3 + CO_2$$

In acidic to neutral aqueous environments, ammonia exists as ammonium ions. Some of the ammonia produced by ammonification is released from alkaline environments to the atmosphere, where it is relatively inaccessible to biological systems. Ammonia and other forms of nitrogen within the atmo-ecosphere are subject to chemical and photochemical transformations and can be returned to the litho- and hydroecospheres in precipitation. The amount of ammonium nitrogen in global precipitation has been estimated at 38–85 million metric tons per annum (Soderlund and Svensson 1976).

Ammonium ions can be assimilated by numerous plants and many microorganisms, where they are incorporated into amino acids and other nitrogen-containing biochemicals. The initial incorporation of ammonia into living organic matter is often accomplished either by glutamine synthetase/glutamate synthase reactions or by direct amination of an α-keto-carboxylic acid to form an amino acid (Gottschalk 1979). The relative importance of the two assimilation pathways varies among habitats and depends on environmental factors and species composition; for example, at low NH_4^+ concentrations, aquatic bacteria primarily utilize the glutamine synthetase/glutamate synthase pathway. The incorporated amino group can be transferred through transamination to form other amino acids. The amino acids can be incorporated into proteins or transformed into other nitrogen-containing compounds that can be utilized as sources of carbon, nitrogen, and energy.

Nitrogen-containing organic compounds produced by one organism can be transferred to and assimilated by others. The transformations of organic nitrogen-containing compounds are not restricted to microorganisms. Animals, for example, produce nitrogenous wastes such as uric acid from the metabolism of nitrogen-containing organic compounds.

Nitrification

In nitrification, ammonia or ammonium ions are oxidized to nitrite ions (Equation 3) and then to nitrate ions (Equation 4).

(3) $NH_4^+ + 1\frac{1}{2}O_2 \longrightarrow$

$NO_2^- + 2H^+ + H_2O$ $(\Delta G = -66 \text{ kcal})$

(4) $NO_2^- + \frac{1}{2}O_2 \longrightarrow NO_3^-$ $(\Delta G = -17 \text{ kcal})$

The process of nitrification appears to be limited for the most part to a restricted number of autotrophic bacteria (Focht and Verstraete 1977; Hooper 1990). Different microbial populations carry out the two steps of nitrification—that is, the formation of nitrite and the formation of nitrate. Normally, however, the two processes are closely coupled and an accumulation of nitrite does not occur. The oxidation of ammonia to nitrite and the oxidation of nitrite to nitrate are both energy-yielding processes. Nitrifying bacteria are chemolithotrophs and utilize the energy derived from nitrification to assimilate CO_2. In the first reaction (Equation 3), molecular oxygen is incorporated into the nitrite molecule. The oxidation is a multistep process and involves the generation of hydroxylamine (NH_2OH) by ammonia monooxygenase. The single oxygen atom incorporated into hydroxylamine comes from atmospheric oxygen. In the next step, hydroxylamine oxidoreductase produces from hydroxylamine and water nitric acid and hydrogen, so the second oxygen atom in nitrite comes from water. The two hydrogens from water are converted back to water by a terminal oxidase using atmospheric oxygen. The production of nitric acid lowers the pH of the environment where nitrification occurs.

Though oxygen dependent, the second step of nitrification obtains the oxygen for formation of nitrate from a water molecule; the molecular oxygen serves only as an electron acceptor. Both steps of the nitrification process are aerobic. Nitrite oxidation is a single-step process and yields only low amounts of energy. Approximately 100 moles of nitrite must be oxidized for the fixation of 1 mole of CO_2, while oxidation of 35 moles of ammonia achieves the same end.

In soils, the dominant genus that is capable of oxidizing ammonia to nitrite is *Nitrosomonas,* and the dominant genus capable of oxidizing nitrite to nitrate is *Nitrobacter.* Other bacteria capable of oxidizing ammonia to nitrite are found in the genera *Nitrosospira, Nitrosococcus, Nitrosolobus,* and *Nitrosovibrio* (Bock et al. 1990). In addition to *Nitrobacter,* members of the genera *Nitrospira, Nitrospina,* and *Nitrococcus* are able to oxidize nitrite to nitrate. *Nitrobacter, Nitrospira, Nitrospina, Nitrococcus, Nitrosomonas,* and *Nitrosococcus* occur in marine habitats. *Nitrobacter, Nitrosomonas, Nitrosospira, Nitrosococcus,* and *Nitrosolobus* are found in soil habitats. Some other microorganisms, including heterotrophic bacteria and fungi, are capable of a limited oxidation of nitrogen compounds, but heterotrophic

Table 11.1

Rates of nitrification by some heterotrophic and autotrophic nitrifiers

Organism	Substrate	Product	Rate of formation (μg N/day/g dry cells)	Max. product accumulation (μg N/mL)
Arthrobacter (heterotroph)	NH_4^+	Nitrite	375–9,000	0.2–1
Arthrobacter (heterotroph)	NH_4^+	Nitrate	250–650	2–4.5
Aspergillus (heterotroph)	NH_4^+	Nitrate	1,350	75
Nitrosomonas (autotroph)	NH_4^+	Nitrite	1–30 million	2,000–4,000
Nitrobacter (autotroph)	NO_2^-	Nitrate	5–70 million	2,000–4,000

Source: Focht and Verstraete 1977.

nitrification does not appear to make a major contribution to the conversion of ammonia to nitrite and nitrate ions (Table 11.1). Since relatively few microbial genera are involved in the process of nitrification, it is not surprising that environmental stress can severely affect this process. The second step of nitrification is usually the more sensitive one, and under stress conditions, some nitrite accumulation may occur (Bollag and Kurek 1980).

The process of nitrification is especially important in soils, because the transformation of ammonium ions to nitrite and nitrate ions results in a change in charge from positive to negative. Positively charged ions tend to be bound by negatively charged clay particles in soil, while negatively charged ions freely migrate in the soil water. The process of nitrification, therefore, must be viewed as a nitrogen mobilization process within soil habitats. Ammonia in soil is normally oxidized rapidly by nitrifying bacteria. Plants readily take up nitrate ions into their roots for assimilation into organic compounds. Nitrate and nitrite ions, however, can also be readily leached from the soil column into the groundwater. This is an undesirable process, since it represents a loss of fixed forms

of nitrogen from the soil, where plants could utilize them to produce biomass.

The occurrence of nitrite in groundwater is of serious concern, since nitrite can react chemically with amino compounds to form nitrosamines, which are highly carcinogenic. Nitrate in groundwater also constitutes a health hazard. While nitrate itself is not highly toxic, it may be microbially reduced in the gastrointestinal tract to highly toxic nitrite. The normal stomach acidity of adult humans tends to prevent or minimize such reduction, but infants with lower stomach acidity are highly susceptible. Nitrite combines with the hemoglobin of the blood, causing respiratory distress, the so-called blue baby syndrome. Nitrate reduction may take place also in the rumen of livestock, resulting in animal disease or death.

Nitrate and nitrite in groundwater are problems in agricultural areas receiving heavy concentrations of synthetic nitrogen fertilizer, such as in the so-called corn belt of the Midwest, the San Joaquin Valley of California, and some areas of Long Island, New York. Wells in such areas need to be closely monitored for nitrate contamination. The deliberate use of nitrification inhibitors in combination with ammonium nitrogen can minimize the nitrate-leaching problem and at the same time lead to better plant utilization of the nitrogen fertilizer.

Nitrate Reduction and Denitrification

Nitrate ions can be incorporated by a variety of organisms into organic matter through assimilatory nitrate reduction. A heterogeneous group of microorganisms, including many bacterial, fungal, and algal species, is capable of assimilatory nitrate reduction. The process of assimilatory nitrate reduction involves several enzyme systems, including nitrate and nitrite reductases, to form ammonia, which can be subsequently incorporated into amino acids (Gottschalk 1979). The enzyme systems appear to involve soluble metalloproteins and require reduced cofactors, including reduced nicotinamide adenine dinucleotide phosphate (NADPH). Assimilating nitrate enzyme systems that have been examined in bacteria, algae, and fungi are repressed or suppressed by the presence of ammonia or reduced nitrogenous organic metabolites in the growth environment. Normal atmospheric concentrations of oxygen do not appear to inhibit assimilatory nitrate reductase enzyme systems. Assimilatory nitrate reduction does not result in the accumulation of high concentrations of extracellular ammonium ions, since ammonia is incorporated relatively rapidly into organic nitrogen. Excess ammonium would act through feedback inhibition to shut off nitrate reduction.

In the absence of oxygen, nitrate ions can act as terminal electron acceptors. As reviewed by D. D. Focht and W. Verstraete (1977), the process is known as nitrate respiration, or dissimilatory nitrate reduction. During dissimilatory nitrate reduction, nitrate is converted to a variety of reduced products, while organic matter is simultaneously oxidized. In the absence of oxygen, dissimilatory nitrate reduction allows utilization of organic compounds with a much higher energy yield than fermentation processes.

There are two types of dissimilatory nitrate reduction. A variety of facultatively anaerobic bacteria, including *Alcaligenes, Escherichia, Aeromonas, Enterobacter, Bacillus, Flavobacterium, Nocardia,*

Spirillum, Staphylococcus, and *Vibrio,* reduce nitrate under anaerobic conditions to nitrite. The nitrite produced by these species is excreted, or, under appropriate conditions, some of these organisms will reduce nitrite via hydroxylamine to ammonium (nitrate ammonification). These organisms do not produce gaseous nitrogen products, that is, they do not denitrify. Nitrate ammonification plays an important role in stagnant water, sewage plants, and some sediments (Koike and Hattori 1978). Unlike assimilatory nitrate reduction, dissimilatory nitrate reductase is not inhibited by ammonia; thus, ammonium ions can be excreted in relatively high concentrations. As compared to denitrification, however, nitrate ammonification is an environmentally less significant process for the reductive removal of nitrate and nitrite ions; its importance appears to be limited by the number of reducing equivalents that must be consumed in the system.

Denitrifying nitrate reducers such as *Paracoccus denitrificans, Thiobacillus denitrificans,* and various pseudomonads have a more complete reduction pathway, converting nitrate through nitrite to nitric oxide (NO), and nitrous oxide (N_2O) to molecular nitrogen. The denitrification sequence is as follows (Equation 5):

$$(5) \quad NO_3^- \longrightarrow NO_2^- \longrightarrow NO \longrightarrow N_2O \longrightarrow N_2$$

In soil, the primary denitrifying genera are *Pseudomonas* and *Alcaligenes;* many additional genera, such as *Azospirillum, Rhizobium, Rhodopseudomonas,* and *Propionibacterium,* have also been reported to denitrify under some conditions (Focht 1982). Usually, a mixture of nitrous oxide and nitrogen is evolved, depleting the environment of combined nitrogen. The proportion of the denitrification products is dependent both on the denitrifying microorganisms and on environmental conditions. The lower the pH of the habitat, the greater the proportions of nitrous oxide formed. Formation of molecular nitrogen is favored by an adequate supply or an oversupply of reducing equivalents.

Simultaneously with denitrification, organic matter is oxidized. Utilization of glucose through nitrate reduction by *Pseudomonas denitrificans* can be described as follows (Equation 6):

(6) $C_6H_{12}O_6 + 4NO_3 \longrightarrow 6CO_2 + 6H_2O + 2N_2$

The enzymes involved in these processes are the dissimilatory nitrate and nitrite reductase systems. Dissimilatory nitrate reductases are particle bound, competitively inhibited by oxygen, and not inhibited by ammonia; this is in contrast to assimilatory nitrate reductases, which are soluble, inhibited by ammonia, and not substantially inhibited by oxygen.

Denitrification occurs under strictly anaerobic conditions or under conditions of reduced oxygen tension. Some denitrification may occur in generally aerobic environments if these contain anoxic microhabitats (Hutchinson and Mosier 1979). Denitrification is more common in standing waters than in running streams and rivers. Denitrification rates typically are higher in the hypolimnion of eutrophic lakes during summer and winter stratification than during fall and spring turnover.

The measurement of denitrification rates was hampered by analytical difficulties mentioned in connection with N fixation. The evolution of nitrogenous oxides can be measured by gas chromatography (Payne 1971, 1973), but the evolution of molecular nitrogen is difficult to quantify against a background of omnipresent atmospheric nitrogen. Consequently, the discovery that denitrification could be arrested at the nitrous oxide level by the acetylene blockage of the N_2O reductase (Balderston et al. 1976; Hallmark and Terry 1985) was an important breakthrough. The addition of acetylene gas at 0.01 atm to the incubation atmosphere allows the gas chromatographic measurement of the denitrification process in the form of a single product, N_2O. Measurement of N_2O evolution with time under these conditions gives a good estimate of denitrification rates. The method is applicable to pure cultures as well as to environmental samples and in situ measurements, but is limited in sensitivity by the low nitrate availability in many habitats.

Various steps within the nitrogen cycle are subject to repression. For example, ammonium ions normally inhibit the fixation of atmospheric nitrogen. Nitrification occurs readily in neutral, well-drained soils but is inhibited in anaerobic or highly acidic soils. Assimilatory nitrate reductase activity is repressed by ammonia; dissimilatory nitrate reductase is repressed by oxygen.

Because various environmental conditions favor specific nitrogen cycling processes, there is a spatial zonation of cycling processes. Fixation of nitrogen occurs in both surface and subsurface habitats. Nitrification occurs exclusively in aerobic habitats. Denitrification predominates in waterlogged soils and in anaerobic aquatic sediments. The cycling of nitrogen within a given habitat also exhibits seasonal fluctuations; during spring and fall blooms of cyanobacteria, for example, rates of nitrogen fixation in aquatic habitats usually are high, reflecting population fluctuations and availability of needed energy and mineral nutrients for fixation of molecular nitrogen.

THE SULFUR CYCLE

Sulfur, a reactive element with stable valence states from −2 to +6 (Table 11.2), is among the ten most abundant elements in the crust of Earth. At an average concentration of 520 ppm, it rarely becomes a limiting nutrient (Ehrlich 1981; Jorgensen 1983). At least some elemental sulfur deposits and some sulfide ores appear to be of biogenic origin (Ivanov 1968). Eruptive and postvolcanic activity continue to introduce additional sulfur into the ecosphere at relatively low rates. With the notable exceptions of ferric and calcium sulfates, most sulfate salts are readily soluble in water. Sulfate is the second most abundant anion in seawater, and the SO_4^{2-} of the marine environment represents a large, slowly cycled sulfur reservoir. Living and dead organic matter comprise a smaller, more rapidly cycled sulfur reservoir. Largely inert sulfur reservoirs are metal sulfides in rock, elemental sulfur deposits, and fossil fuels. Human activities, including strip mining and the burning of fossil fuels, have mobilized a part of these inert sulfur reservoirs with destructive pollution consequences. We will discuss some of these consequences later in this chapter.

Plants, algae, and many heterotrophic microorganisms assimilate sulfur in the form of sulfate. For incorporation into cysteine, methionine, and coenzymes in the form of sulfhydryl (—SH) groups, sulfate

Table 11.2

Oxidation states of sulfur in various compounds

Form	Example	Oxidation state
S^{2-}	Sulfides, mercaptans	−2
$S°$	Elemental sulfur	0
S_2O_4	Hyposulfite	+2
SO_3^{2-}	Sulfite	+4
SO_4^{2-}	Sulfate	+6

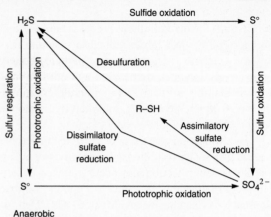

Figure 11.5

Representation of the sulfur cycle showing biogeochemical transformations of oxidized and reduced forms of sulfur. R—SH represents sulfhydryl groups of cell protein. Note the similarities between the sulfur cycle and the nitrogen cycle (Figure 11.1).

needs to be reduced to the sulfide level by assimilatory sulfate reduction. A direct uptake as sulfide is not feasible for most microorganisms because of the high toxicity of H_2S. In assimilatory sulfate reduction, toxicity is avoided by immediately reacting the reduced sulfur with an acceptor—for example, serine—to yield cysteine.

Organosulfur decomposition in soils and sediments yields mercaptans and H_2S (Bremner and Steele 1978). Analogous to ammonification, this process is referred to as desulfuration. Volatile mercaptans and H_2S are the offensive odor components of rotten eggs and cabbage. The action of cysteine desulfhydrase can be described as shown in Equation 7:

$$(7) \quad SH—CH_2-CH-COOH + H_2O \xrightarrow{\text{cysteine}}$$
$$\qquad\qquad\qquad |$$
$$\qquad\qquad\quad NH_2$$

$$OH—CH_2-CH-COOH + H_2S$$
$$\qquad\qquad\quad |$$
$$\qquad\qquad\quad NH_2$$

In the marine environment, a major decomposition product of organosulfur is dimethylsulfide (DMS). This product originates from dimethylsulfoniopropionate (DMSP), a major metabolite of marine algae that may have a role in osmoregulation. Dimethylsulfide is released during zooplankton grazing on phytoplankton and also during decay processes (Dacey and Wakeham 1986). The volatile DMS

escapes the oceans; according to some estimates, 90% of the total sulfur flux from the marine environment to the atmosphere occurs in the form of DMS. Another major product is H_2S. Once they escape to the atmosphere, DMS, H_2S, and mercaptans are subject to photooxidative reactions that ultimately yield sulfate.

If H_2S does not escape to the atmosphere, it may be subject to microbial oxidation under aerobic conditions, or it may be phototrophically oxidized under anaerobic conditions. Under anaerobic conditions, sulfate as well as elemental sulfur may serve as electron acceptors, while organic substrates are oxidized. Figure 11.5 summarizes the microbial cycling of sulfur.

Oxidative Sulfur Transformations

In the presence of oxygen, reduced sulfur compounds are capable of supporting chemolithotrophic microbial metabolism. *Beggiatoa, Thioploca, Thiothrix* (Nelson 1990), and the thermophilic *Thermothrix*

Figure 11.6

Chemolithotrophic cultivation of *Beggiatoa*, a typical gradient organism, in a sulfide-oxygen gradient. At right is a culture tube with sulfide agar, overlayered with initially sulfide-free soft mineral agar. The airspace in the closed tube is the source of oxygen. Stab-inoculated *Beggiatoa* grows in a narrowly defined gradient of H_2S and oxygen, as shown. Source: Nelson 1990.

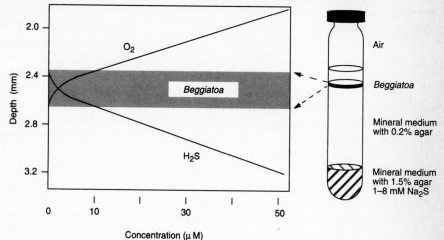

(Caldwell et al. 1976), are filamentous, microaerophilic bacteria capable of oxidizing H_2S according to Equation 8.

$$(8) \quad H_2S + \frac{1}{2}O_2 \longrightarrow S° + H_2O$$

$$(\Delta G = -50.1 \text{ kcal/mole})$$

Sulfur globules are deposited within the cells. In the absence of H_2S, these sulfur globules are slowly oxidized further to sulfate. These typical gradient organisms position themselves on the interface of an anaerobic environment, the sediment, and the partially oxygenated water in contact with the sediment (Jorgensen and Revsbeck 1983).

The large size and observable sulfur globules of *Beggiatoa* allowed Sergei Winogradsky to develop the concept of microbial chemolithotrophy almost a hundred years ago. Although *Beggiatoa* clearly derived energy from H_2S and $S°$ oxidation, until recently it was impossible to grow *Beggiatoa* in pure culture under strictly chemolithotrophic conditions, and it was widely assumed that Winogradsky developed the right concept on the wrong microorganism. By a close control of H_2S—O_2 gradients, however, it was possible to demonstrate clearly that at least some *Beggiatoa* strains are capable of chemolithotrophic

metabolism (Nelson and Jannasch 1983; Nelson 1990) (Figure 11.6).

Some species of *Thiobacillus* (*T. thioparus, T. novellus*) also oxidize H_2S and other reduced sulfur compounds and, because they have a low acid tolerance, deposit elemental sulfur rather than generate sulfuric acid by further oxidation. The filamentous sulfur bacteria and these *Thiobacillus* species are facultatively chemolithotrophic (Kuenen et al. 1985). Other members of the genus *Thiobacillus* species produce sulfate from the oxidation of elemental sulfur and other inorganic sulfur compounds. Elemental sulfur is oxidized according to Equation 9:

$$(9) \quad S° + 1\frac{1}{2}O_2 + H_2O \longrightarrow H_2SO_4$$

$$(\Delta G = -149.8 \text{ kcal/mole})$$

These *Thiobacillus* species are acidophilic, grow well at pH 2–3, and are obligate chemolithotrophs, obtaining their energy exclusively from the oxidation of inorganic sulfur and their carbon from the reduction of carbon dioxide. Most *Thiobacillus* species are obligate aerobes requiring molecular oxygen for the oxidation of the inorganic sulfur compounds. *Thiobacillus denitrificans,* however, can utilize nitrate ions as terminal electron acceptors in the oxidation of inorganic sulfur compounds, as shown for elemental sulfur in Equation 10:

$$(10) \quad 4NO_3^- + 3S° \longrightarrow 3SO_4^{2-} + 2N_2$$

This organism, though, is not capable of assimilatory nitrate reduction and requires ammonium as a nitrogen source. Members of the genus *Sulfolobus* (archaebacteria) oxidize elemental sulfur in hot acidic habitats to generate their required energy. Chemoautotrophic sulfur-oxidizing bacteria are widely distributed; they are quite active in soils and aquatic habitats. A variety of other heterotrophic microorganisms oxidize inorganic sulfur to sulfate or thiosulfate but do not appear to derive energy from this transformation.

Hydrogen sulfide is also subject to phototrophic oxidation in anaerobic environments (Pfennig 1990; Trüper 1990). Photosynthetic sulfur bacteria, the Chromatiaceae, Ectothiorhodospiraceae, and Chlorobiaceae, are capable of photoreducing CO_2 while oxidizing H_2S to $S°$ (Equation 11), in striking analogy to the photosynthesis of eukaryotes (Equation 12). The formula $[CH_2O]$ symbolizes photosynthate.

$$(11) \quad CO_2 + H_2S \longrightarrow [CH_2O] + S$$

$$(12) \quad CO_2 + H_2O \longrightarrow [CH_2O] + O_2$$

Most Chromatiaceae store sulfur globules intracellularly, whereas Ectothiorhodospiraceae and Chlorobiaceae excrete sulfur globules. All have a limited capacity to oxidize sulfur further to sulfate, and they may contribute to biological sulfur deposition. Some cyanobacteria are capable of both oxygenic and anoxygenic photosynthesis; thus, they may participate in the phototrophic oxidation of H_2S. Microbial oxidation of reduced sulfur is essential for continued availability of this element in nontoxic forms, but the chemoautotrophic fixation of CO_2 connected with this activity contributes only minimally to the carbon cycling in most ecosystems.

Notable exceptions to the above statement are the recently discovered deep-sea hydrothermal vent ecosystems (Karl et al. 1980). Rifts caused by the spreading of the seafloor from midocean ridges allow seawater to percolate deeply into the crust and to react with hot basaltic rock. This hot anoxic hydrothermal fluid, containing geothermally reduced H_2S, $S°$, H_2, NH_4^+, NO_2^-, Fe^{2+}, and Mn^{2+}, as well as some CH_4 and CO, moves upward and through the vents and enters the cold, oxygenated water of the deep sea (Figure 11.7). Depending on the degree of mixing with seawater during upflow, the temperature of the hydrothermal vents ranges from warm (5°C–23°C) to extremely hot (350°C–400°C). Because of the prevailing high hydrostatic pressures, the water does not boil at these high temperatures, but on cooling, the precipitation of elemental sulfur and of metal sulfides forms plumes called white or black smokers, respectively (Jannasch and Mottl 1985).

The vent communities, located at a depth of 800–1,000 m, receive no sunlight and minimal organic nutrient input from the low-productivity surface water above, yet their biomass exceeds that of the surrounding seafloor by orders of magnitude. Microbial mats cover all available surfaces on and near the vents, and high densities of unique clams, mussels, vestimentiferan worms, and other invertebrates cluster in the vicinity (Grassle 1985). Some of these graze or filter-feed on microorganisms; others are directly symbiotic with microorganisms and exhibit chemoautotrophic activity. Energetically, the whole vent community is supported by the chemoautotrophic oxidation of reduced sulfur, primarily by *Beggiatoa, Thiomicrospira*, and additional sulfide or sulfur oxidizers of great morphological diversity (Jannasch and Wirsen 1981; Jannasch 1990). Oxidation of H_2, CO, NH_4^+, NO_2^-, Fe^{2+}, and Mn^{2+} are assumed to contribute to chemoautotrophic production, though measurements of these processes in the vent environment are yet to be accomplished (Jannasch and Mottl 1985). Methane, derived from reduction of CO_2 with geothermally produced hydrogen by extremely thermophilic *Methanococcus* species detected in the anoxic hydrothermal fluid, is also oxidized by methanotrophic bacteria and provides additional carbon and energy input for the vent ecosystem.

In environments other than the deep-sea thermal vents, generation of reduced minerals used in chemolithotrophic production is directly tied to oxidation of photosynthetically produced organic matter. There-

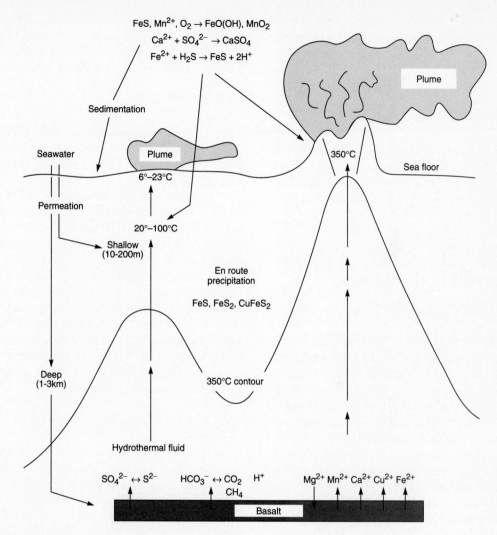

$$FeS, Mn^{2+}, O_2 \rightarrow FeO(OH), MnO_2$$
$$Ca^{2+} + SO_4^{2-} \rightarrow CaSO_4$$
$$Fe^{2+} + H_2S \rightarrow FeS + 2H^+$$

Sedimentation

Plume

Seawater

Plume

350°C

Sea floor

6°–23°C

Permeation

20°–100°C

Shallow
(10-200m)

En route
precipitation

$FeS, FeS_2, CuFeS_2$

Deep
(1-3km)

350°C contour

Hydrothermal fluid

$SO_4^{2-} \leftrightarrow S^{2-}$ $HCO_3^- \leftrightarrow CO_2$ H^+ Mg^{2+} Mn^{2+} Ca^{2+} Cu^{2+} Fe^{2+}
CH_4

Basalt

Figure 11.7

Schematic diagram showing inorganic chemical processes occurring at warm- and hot-water vent sites. Deeply circulating seawater is heated to 350°C to 400°C and reacts with crustal basalts, leaching various species into solution. The hot water rises, reaching the seafloor directly in some places and mixing first with cold, down-seeping seawater in others. On mixing, iron-copper-zinc sulfide minerals and anhydrite (a form of calcium sulfate) precipitate. (Source: Jannasch and Mottl 1985.)

fore, sustained primary production without input of sun energy is unthinkable even by chemolithotrophs. The described deep-sea hydrothermal vent communi-

ties are unique by being independent of solar energy input and by their direct use of geothermal energy instead. It needs to be pointed out, however, that the reduced minerals represent an energy source only in the oxidized environment of the seawater, a condition originally created by oxygenic photosynthesis powered by sun energy.

Reductive Sulfur Transformations

The analogy between H_2O and H_2S in oxygenic and anaerobic phototrophy, respectively, was pointed out in the preceding section. It stands to reason, then, that elemental sulfur should assume a role similar to oxygen in respiratory processes, though this role of elemental sulfur has been demonstrated only recently (Pfennig and Biebl 1981). *Desulfuromonas acetoxidans* grows on acetate, anaerobically reducing stoichiometric amounts of $S°$ to H_2S (Equation 13).

(13) $CH_3COOH + 2H_2O + 4S° \longrightarrow 2CO_2 + 4H_2S$

The free energy yield ($\Delta G = 5.7$ kcal/mole) of the above reaction is rather low, yet no other sources of carbon and energy are required for growth. *Desulfuromonas* is unable to reduce sulfate or live by fermentative metabolism and uses the substrate acetate that is not metabolized by most sulfate reducers. *Desulfuromonas* occurs in anaerobic sediments rich in sulfide and elemental sulfur. It also lives syntrophically with the phototrophic green sulfur bacteria (Chlorobiaceae) that photooxidize H_2S to $S°$ and excrete elemental sulfur extracellularly. *Desulfuromonas* regenerates H_2S by sulfur respiration, using, at least in part, organic matter leaked by *Chlorobium* cells (Kuenen et al. 1985).

From submarine hydrothermal vent environments, extremely thermophilic anaerobic archaebacteria were isolated that are capable of sulfur respiration with hydrogen gas (Fischer et al. 1983; Stetter and Gaag 1983; Stetter 1990). Several species of *Thermoproteus, Pyrobaculum,* and *Pyrodictium* were described to carry out this chemolithotrophic reaction. *Pyrodictium occultum* is the most thermophilic member of the group, having an optimal growth temperature of 105°C and a maximal growth temperature of 110°C. It is probably an advantage for these extreme thermo-

philes that sulfur is present at these temperatures in a molten state, while at room temperature it is in a hydrophobic solid. Sulfur respiration with molecular hydrogen can be described as shown in Equation 14:

(14) $H_2 + S° \longrightarrow H_2S$

Some of these archaebacteria are facultative chemolithotrophs and are capable of heterotrophic or mixotrophic growth during sulfur respiration (Parameswaran et al. 1987).

Martinus Beijerinck described the use of sulfate (SO_4^{2-}) as terminal electron acceptor in anaerobic respiration much earlier. When obligately anaerobic bacteria carry out dissimilatory sulfate reduction, they are referred to as sulfate reducers (Postgate 1984; Thauer 1990). The traditional sulfate-reducing genera *Desulfovibrio* and *Desulfotomaculum* were recently joined by several newly described morphological and physiological types, such as *Desulfobacter, Desulfobulbus, Desulfococcus, Desulfonema,* and *Desulfosarcina* (Pfennig et al. 1981). The reduction of sulfate results in production of hydrogen sulfide according to the following equation (Equation 15):

(15) $H_2 + SO_2^{2-} \longrightarrow H_2S + 2H_2O + 2OH^-$

In addition to anaerobic sulfate-reducing bacteria, some species of *Bacillus, Pseudomonas,* and *Saccharomyces* have been found to liberate hydrogen sulfide from sulfate, but these additional genera do not appear to play a major role in the dissimilatory reduction of sulfate. Sulfate reduction can occur over a wide range of pH, pressure, temperature, and salinity conditions. Only relatively few compounds can serve as electron donors for sulfate reduction. Although hydrogen and sulfate can serve as their only source of energy for growth, sulfate reducers are not chemolithotrophs. They lack the enzyme systems to assimilate CO_2 and require organic carbon sources. The most common electron donors are pyruvate, lactate, and molecular hydrogen. Sulfate reduction is inhibited by the presence of oxygen, nitrate, or ferric ions. The rate of sulfate reduction is often carbon limited. The addition of organic compounds to marine sediments can result in greatly accelerated rates of dissimilatory sulfate reduction.

The production of even small amounts of hydrogen sulfide by sulfate reducers can have a marked effect on populations within a habitat. Hydrogen sulfide is extremely toxic to aerobic organisms because it reacts with the heavy metal groups of the cytochrome systems. Production of hydrogen sulfide by *Desulfovibrio* in waterlogged soils can kill nematodes and other animal populations. Hydrogen sulfide also has antimicrobial activity and can adversely affect microbial populations in soil. Heavy metals are extremely reactive with hydrogen sulfide, resulting in the precipitation of metallic sulfides. Black metal sulfides give reduced sediments their characteristic dark color. Hydrogen sulfide is also highly toxic to the cytochrome systems of plant roots and can kill the plants.

In contrast to the specialized dissimilatory sulfate reducers, many organisms are capable of assimilatory sulfate reduction. Assimilatory sulfate reduction produces low concentrations of hydrogen sulfide, which are immediately incorporated into organic compounds. Many microorganisms and plants can utilize sulfate ions as the source of sulfur required for incorporation into proteins and other sulfur-containing biochemicals. Plant roots readily take up sulfate from soils, incorporating it into organic matter.

The assimilation of inorganic sulfate (Ehrlich 1981) involves a series of transfer reactions initiated by the reaction of sulfate with ATP to form adenosine-5′-phosphosulfate (APS) and pyrophosphate (PP) (Equation 16). A second reaction between ATP and APS produces 3′-phosphoadenosine--5-phosphosulfate (PAPS) and ADP (Equation 17).

$$(16) \quad SO_4^{2-}ATP \xrightarrow[\text{ATP sulfurylase}]{} APS + PP_i$$

$$(17) \quad APS + ATP \xrightarrow[\text{APS kinase}]{} APS + ADP$$

The active sulfate of PAPS is subsequently reduced, as shown in Equation 18, to yield sulfite and adenosine 3′, 5′-diphosphate (PAP). A second reduction

step (Equation 19) yields sulfide that is immediately incorporated into an amino acid Equation 20).

$$(18) \quad PAPS + 2e^- \xrightarrow[\substack{\text{NADH:PAPS} \\ \text{reductase}}]{} SO_3^{2-} + PAP$$

$$(19) \quad SO_3^{2-} + 6H^+ + 6e^- \xrightarrow[\substack{\text{NADH:SO}_3^{2-} \\ \text{reductase}}]{} SO_3^{2-} + PAP$$

$$(20) \quad S^{2-} + serine \longrightarrow cysteine + H_2O$$

The biochemical mechanism involved in dissimilatory sulfate reduction is similar to the one described above, but the generated H_2S is released to the environment. In sulfate-rich marine environments, most of the H_2S originates from dissimilatory sulfate reductions, but in rich organic sediments, H_2S may accumulate also from the decomposition of organosulfur compounds.

An interesting aspect of the microbial metabolism of sulfur compounds is the ability of microorganisms to fractionate sulfur isotopes. Sulfate-reducing bacteria exhibit a slight preference for ^{32}S over ^{34}S. In contrast, the oxidation of hydrogen sulfide does not exhibit any preferential utilization of sulfur isotopes. The hydrogen sulfide formed by the action of sulfate-reducing bacteria has a higher proportion of ^{32}S than the original sulfate. The alteration of the relative abundances of ^{32}S and ^{34}S allows for the differentiation of biologically generated sulfide and sulfide from strictly geochemical processes (Ivanov 1968). Sulfide readily reacts with heavy metals, and, because of the crustal abundance of iron, it occurs in geological deposits primarily as ferrous sulfide (FeS).

Sulfate reduction is probably also involved in the formation of some elemental sulfur deposits. Hydrogen sulfide generated by sulfate reduction may be photooxidized to elemental sulfur under anaerobic conditions by Chromatiaceae, Ectothiorhodospiraceae, or Chlorobiaceae. Under aerobic conditions, especially when oxygen is limiting, hydrogen sulfide may be oxidized to sulfur by the chemical reaction with oxygen or by the activities of the *Beggiatoa-Thiothrix*

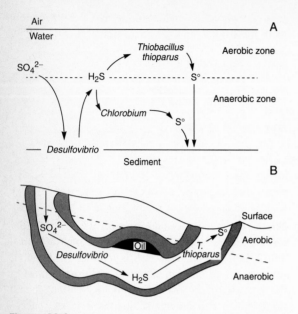

Figure 11.8

The biological deposition of sulfur (A) in a lake and (B) in geological strata. Sulfate is converted to H_2S in anaerobic zones. The H_2S is oxidized by *Thiobacillus thioparus* in both cases and in the lake habitat, also by *Chlorobium,* forming and leading to the deposition of elemental sulfur. In geological strata, oil or gas deposits supply organic matter for sulfate reduction.

sulfur deposits in Russia and Texas were formed by similar mechanisms.

Sulfate reduction contributes to the atmospheric cycling of sulfur. Until recently, it was believed that most of the biogenic sulfur transfer to the atmosphere (an estimated 142 million metric tons per annum when calculated as sulfur) takes place in the form of H_2S (Kellog et al. 1972). This was most likely an overestimate, and H_2S represents less than half of a 65–125 metric tons per annum biogenic sulfur transfer to the atmosphere (Kelly and Smith 1990). The balance of the atmospheric transfer takes place in form of volatile organosulfur compounds, principally dimethyl sulfide (DMS), with minor contributions of carbon disulfide (CS_2) and carbonyl sulfide (COS). The major source of all volatile sulfur is the ocean, with minor contributions from swamps and lakes. Although volatile sulfur is produced also in soil, this environment is a net sink rather than a net producer of volatile sulfur.   Various thiobacilli rapidly oxidize hydrogen sulfide and other reduced volatile sulfur compounds to sulfate in aerobic soils and sediments. Hydrogen sulfide, dimethyl sulfide, and other reduced volatile sulfur compounds, when they reach the atmosphere, are subject to oxidation and photooxidation reactions, ultimately converting them to sulfate.

Some Practical Implications of the Sulfur Cycle

Sulfur oxidation produces substantial amounts of strong mineral acid. Within soils, this can lead to solubilization and mobilization of phosphorus and other mineral nutrients with a generally beneficial effect on both microorganisms and plants. The activity of *Thiobacillus thiooxidans* may be used for adjusting soil pH. *Thiobacillus thiooxidans* and *T. ferrooxidans* are used in microbial mining operations. When mining activities, especially strip mining, uncover large amounts of reduced sulfide rock, the activities of the same thiobacilli give rise to acid mine drainage, a destructive pollution phenomenon.

Fossil fuels, especially coal and some heating oils, contain substantial amounts of sulfur. Much of

group. Acid-intolerant thiobacilli (*Thiobacillus thioparus*) may also be involved in the formation of sulfur deposits. Figure 11.8 depicts some scenarios for the biogenic deposition of elemental sulfur in a lake and geological formations.

Present-day sulfur biogenesis has been observed in some Libyan lakes fed by artesian springs containing up to 100 mg/L H_2S (Postgate 1984). Chromatiaceae and Chlorobiaceae photooxidize the H_2S to elemental sulfur. Some H_2S is oxidized to sulfate, but this sulfate is again reduced to H_2S by *Desulfovibrio,* which utilizes the photosynthetically produced biomass. The H_2S produced by *Desulfovibrio* is again available for photooxidation to $S°$. The elemental sulfur precipitates and is collected on the commercial scale by the local population. It is very likely that the large fossil elemental

this sulfur is present as pyrite (FeS_2) and originates from H_2S produced by sulfate reducers. Some organosulfur is also present (Altschuler et al. 1983). When burned, most of this sulfur is converted to SO_2, which combines with atmospheric moisture to form sulfurous acid (H_2SO_3) (Kellog et al. 1972). Photooxidation also converts some SO_2 to SO_3, resulting in sulfuric acid. Atmospheric inversions in urban areas during the winter heating season can lead to the formation of highly irritating and unhealthy acid smog. This condition differs from the urban smog that arises in warmer weather predominantly from automobile exhaust, in which ozone and nitrogen oxides are the main irritants.

On the larger scale, the burning of fossil fuels gives rise to the formation of acid rain. Rainwater, which normally has a pH just below neutrality due to the weak acidity of H_2CO_3, becomes quite acidic (pH 3.5–4) from sulfurous and sulfuric acids. Acid rain corrodes buildings and monuments, especially those made of limestone and marble; it may also damage plant leaves. It has little detrimental effect on well-buffered soils and hard waters high in carbonate, but it becomes highly destructive in ecosystems that are not sufficiently buffered—for example, in glacier-scoured portions of Scandinavia, Canada, and New England. These regions have characteristically thin, weakly buffered soils, soft-water lakes, and rivers on hard granite bedrock. Here acid rain causes a substantial decrease in soil and water pH, leading to the elimination of important microbial, plant, and animal species and a general decline in ecosystem diversity and productivity. Strongly industrialized regions are not the only areas affected by acid rain. Wind patterns distribute the air pollutants; much of the acid rain problem in Scandinavia appears to originate from the industrialized regions of England and the German Ruhr Valley. The New England problem seems to originate from the industrialized regions of the Atlantic seaboard and the Great Lakes. The Ohio Valley in the United States contributes to the Canadian problem.

Because of the acid rain problem, air pollution standards limit sulfur dioxide emissions, and there is pressure to further tighten these standards. The option to burn clean fuels, like natural gas and low-sulfur oil,

is becoming increasingly expensive. Scrubbing systems, designed to remove SO_2 from stack gases when high-sulfur fuels are being used, are expensive and have a variety of operational problems.

Microorganisms show some potential for reducing the acid rain problem by removing sulfur from fossil fuels prior to burning. Various chemical approaches have been considered for the removal of sulfur from coal prior to its use. The activity of sulfur-oxidizing bacteria could theoretically be used for the same purpose and perhaps at lower cost. The physical state of coal, which occurs as large chunks, is not conducive to microbial activities that could remove the sulfur. However, if coal distribution systems are developed in which the coal is ground and transported as a slurry through existing oil pipeline systems, the removal of sulfur from the slurried coal by the activity of the thiobacilli appears to be an option well worth exploring.

An important practical implication of the sulfur cycle is the anaerobic corrosion of steel and iron structures set in sulfate-containing soils and sediments. This type of corrosion can severely damage or destroy pipes and pilings and has posed an unexpected engineering problem—it was believed that corrosion in anaerobic environments was minimal. The process consists of spontaneous chemical and microbially mediated steps, but the bacterial contribution is essential for driving the whole process (Postgate 1984). The surface of metallic iron spontaneously reacts with water, forming a thin double layer of ferrous hydroxide and hydrogen (Equation 21). The process would tend to stop here except for the activity of *Desulfovibrio desulfuricans*, which removes the protective H_2 layer and forms H_2S (Equation 22). The H_2S attacks iron in a spontaneous chemical reaction, forming ferrous sulfide and hydrogen (Equation 23). As the sum of the above reactions (Equation 24), metallic iron is rapidly converted to ferrous hydroxide and ferrous sulfide. The reaction steps are:

(21) $Fe° + 2H_2O \longrightarrow Fe(OH)_2 + H_2$

(22) $4H_2 + CaSO_4 \longrightarrow H_2S + Ca(OH)_2 + 2H_2O$

$$(23) \quad 2H_2S + Fe^{2+} \longrightarrow FeS + H_2$$

$$(24) \quad 4Fe° + CaSO_4 + 4H_2O \longrightarrow$$

$$FeS + 3Fe(OH)_2 + Ca(OH)_2$$

THE PHOSPHORUS CYCLE

Phosphorus is not an abundant component of the eco-sphere (Cosgrove 1977; Ehrlich 1981). Its availability is further restricted by its tendency to precipitate in the presence of bivalent metals (Ca^{2+}, Mg^{2+}) and fer-ric (Fe^{3+}) ions at neutral to alkaline pH. Large, slowly cycled reservoirs of phosphate occur in marine and other aquatic sediments. Small, actively cycled reser-voirs of phosphate are dissolved phosphate in soils and waters and phosphate in living and dead organic matter. A largely inert reservoir has been phosphate rock, such as apatite ($3Ca_3[PO_4]_2$•$Ca[FeCl]_2$), but this reservoir is being increasingly tapped by the fertilizer industry. Much of this phosphate is eventually lost to marine sediments, raising concerns about future avail-ability of an important fertilizer. In primary phos-phates, such as NaH_2PO_4, only one of the three PO_4^{3-} valences is linked to a metal. Primary phosphates have high water solubilities. Secondary and tertiary phosphates have only one or no hydrogen atoms in their molecule, respectively, and their water solubili-ties are progressively lower. In the manufacture of phosphate fertilizer, the tertiary phosphate of apatite is converted by acid treatment to secondary and pri-mary "superphosphate."

Phosphorus is an essential element in all living systems. Within biological systems, the most abun-dant forms of phosphorus are phosphate esters and nucleic acids. Phosphate diester bonds form the links within nucleic acid molecules. Phosphate also forms an essential portion of the ATP molecule. The hydrolysis of phosphate from ATP to ADP forms the basis for most energy transfer reactions within bio-logical systems. Phospholipids, which contain hydrophilic phosphate groups, are essential compo-nents of cell membranes.

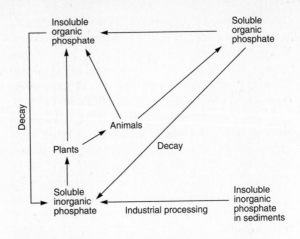

Figure 11.9

The phosphorus cycle, showing various transfers, none of which alters the oxidation state of phosphate.

The microbial cycling of phosphorus for the most part does not alter the oxidation state of phosphorus (Stewart and McKercher 1982). Most phosphorus trans-formations mediated by microorganisms can be viewed as transfers of inorganic to organic phosphate or as trans-fers of phosphate from insoluble, immobilized forms to soluble or mobile compounds (Figure 11.9).

Although phosphate is normally not reduced by microorganisms, it appears that some soil and sediment microorganisms may be capable of utilizing phosphate as a terminal electron acceptor under appropriate envi-ronmental conditions. Reliable and recent documenta-tion for this transformation is not available. Phosphate is likely to serve as a terminal electron acceptor in the absence of sulfate, nitrate, and oxygen. The final prod-uct of phosphate reduction would be phosphine (PH_3). Phosphines are volatile and spontaneously ignite on contact with oxygen, producing a green glow. The pro-duction of phosphines is sometimes observed near burial sites and swamps where there is extensive decomposition of organic matter. Phosphines can also ignite the methane produced in similar environments, giving rise to "ghostly" light phenomena. A microbial involvement in phosphine production is likely but requires experimental confirmation.

Within many habitats, phosphates are combined with calcium, rendering them insoluble and unavailable to plants and many microorganisms. Some heterotrophic microorganisms are capable of solubilizing phosphates from such sources. Phosphate is assimilated by these microorganisms, which also solubilize a large proportion of the insoluble inorganic phosphate, releasing it for use by other organisms. The mechanism of phosphate solubilization is normally by production of organic acids. Some chemolithotrophic microorganisms, such as *Nitrosomonas* and *Thiobacillus,* mobilize inorganic phosphates by producing nitric and sulfuric acids, respectively (Tiessen and Stewart 1985).

Within soils, phosphate also exists as insoluble iron, magnesium, and aluminum salts. The mobilization of insoluble ferric phosphates may occur when microorganisms reduce ferric ions to ferrous ions under anaerobic conditions. Flooding of soils enhances release of phosphate by this mechanism.

Plants and microorganisms can readily take up soluble forms of inorganic phosphates and assimilate them into organic phosphates. As an example, inorganic phosphate reacts with ADP to generate ATP when coupled with a sufficiently exothermic reaction. Inorganic phosphate also reacts with carbohydrates such as glucose at the initiation of glycolysis.

The reverse process, mineralization of organic phosphates, also occurs; this process is catalyzed by phosphatase enzymes. Many microorganisms produce phosphatase enzymes. Some bacteria and fungi produce phytase, which releases soluble inorganic phosphates from inositol hexaphosphate (phytic acid) (Equation 25).

$$(25) \quad \text{Inositol hexaphosphate} + 6H_2O \xrightarrow{\text{phytase}}$$

$$\text{inositol} + 6PO_4^{3-}$$

Microbial activities may also immobilize phosphorus, rendering it inaccessible to the biological community. The assimilation of phosphorus into microbial cell constituents, such as membranes, removes phosphate from the available pool. In some situations, microorganisms aid plants by mobilizing phosphates; in others, they may compete with plants for available phosphate resources.

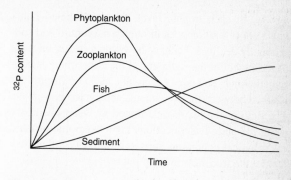

Figure 11.10

Idealized uptake and distribution with time of $^{32}PO_4^{3-}$ in an aquatic ecosystem. The phosphate is taken up directly by the phytoplankton and reaches the higher trophic levels more slowly through feeding. A large portion of the added phosphate is eventually sequestered in the sediment.

Productivity in many habitats is phosphate limited. In aquatic environments, phosphate concentrations exhibit seasonal fluctuations that are associated with algal and cyanobacterial blooms. The precipitation of phosphorus, especially in marine habitats, greatly limits primary productivity. In aquatic habitats, phosphorus may exist in soluble or particulate form. These forms exhibit differential reactivity and availability to the biological community. Phosphate concentrations in freshwater habitats are closely correlated with eutrophication. Phosphate-limited lakes normally are oligotrophic. The addition of phosphates from sources such as detergent fillers causes these lakes to become eutrophic (Chapra and Robertson 1977).

Phosphorus, being an essential and often also a limiting mineral nutrient with a convenient radioisotope (^{32}P), is particularly suitable for demonstrating the movement and distribution of a nutrient element through food webs. If introduced as $^{32}PO_4^{3-}$ into a natural or model aquatic ecosystem (Figure 11.10), primary producers (phytoplankton) are rapidly labeled. The zooplankton does not take up the phosphate directly, but via ingested phytoplankton, and is therefore labeled more slowly. Higher trophic levels (fish and macro-invertebrates) are labeled even more slowly. In each trophic level, the radiolabeled phosphate concentration goes through a maximum and subsequently declines. There is, however, a steady increase of radiolabel in the sediment. If the experiment is continued long enough, eventually an equilibrium is reached between the precipitated and cycled phosphate. The ratio of the two is determined by water chemistry (high pH and bivalent cations favor precipitation) and hydrological conditions such as stratification that may restrict the free movement of phosphate.

THE IRON CYCLE

Iron is the fourth most abundant element in the crust of Earth, but only a small portion of this iron is available for biogeochemical cycling (Ehrlich 1981; Nealson 1983a). The cycling of iron consists largely of oxidation-reduction reactions that reduce ferric iron to ferrous iron and oxidize ferrous iron to ferric iron (Figure 11.11). These oxidation-reduction reactions are important in both organic iron-containing compounds and in inorganic iron compounds.

Ferric (Fe^{3+}) and ferrous (Fe^{2+}) ions have very different solubility properties. Ferric iron precipitates in alkaline environments as ferric hydroxide. Ferric iron may be reduced under anaerobic conditions to the more soluble ferrous form. Under some anaerobic conditions, however, sufficient H_2S may be evolved to precipitate iron as ferrous sulfide. Flooding of soil, which creates anaerobic conditions, favors the accumulation of ferrous iron. In aerobic habitats, such as well-drained soil, most of the iron exists in the ferric state.

In organic compounds, iron is often attached to organic ligands by chelation. Chelated iron can undergo oxidation-reduction transformations, which are utilized in electron transport processes. Cytochromes of electron transport chains contain iron that undergoes oxidation-reduction transformations during the transfer of electrons.

The low solubility of ferric iron necessitates relatively elaborate uptake and transport mechanisms for this otherwise abundant element. Several groups of bacteria utilize siderophores, special iron-chelating

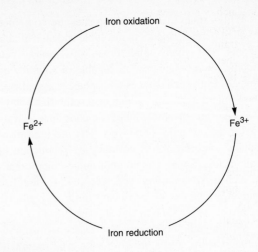

Iron oxidation

Fe^{2+}　　　　　　Fe^{3+}

Iron reduction

Figure 11.11

The iron cycle, showing interconversion of ferrous and ferric iron.

agents, to facilitate the solubilization and uptake of iron. The excretion of siderophores is induced by iron deficiency. To date, two groups of siderophores have been studied extensively. Phenol-catechol derivatives are synthesized by enterobacteria and bear such common names as enterochelin and enterobactin. Streptomycetes and some other bacterial groups synthesize derivatives of hydroxamic acids referred to as ferrioxamines. In each case, the ferric iron is chelated by multiple hydroxyl or carbonyl groups as the ferric iron is enclosed in the molecular cage of the chelator. Siderophores have molecular weights between 500 and 1,000 and are not transported back through the cell membrane, but rather pass on the iron molecule to siderophore receptor proteins in the membrane. The siderophore receptor shuttles the iron molecule through the membrane and releases the siderophore to chelate additional molecules of ferric iron (Lewin 1984).

Under alkaline to neutral conditions, ferrous iron is inherently unstable in the presence of O_2 and is oxidized spontaneously to ferric iron. Under such conditions, microorganisms have little chance to extract energy from the oxidation process. Under acidic reaction conditions in oxygenated environments, however, ferrous iron is relatively stable, and under such conditions, acidophiles such as *Thiobacillus ferrooxidans*, *Leptospirillum ferrooxidans*, and some strains of *Sulfolobus acidocaldarius* are capable of the chemolithotrophic oxidation of ferrous iron as shown in Equation 26.

(26) $2Fe^{2+} + \frac{1}{2}O_2 + 2H^+ \longrightarrow$

$2Fe^{3+} + H_2O \ (\Delta G = -6.5 \ kcal/mole)$

With the exception of *L. ferrooxidans*, the same microorganisms can also oxidize reduced sulfur compounds. Nonacidophilic iron bacteria, such as the prosthecate *Hyphomicrobium*, *Pedomicrobium*, and *Planctomyces*, and the filamentous *Sphaerotilus* and *Leptothrix*, were at times described as being chemolithotrophs or at least as deriving energy from ferrous iron oxidation. A number of additional "iron bacteria," such as *Gallionella*, *Metallogenium*, *Seliberia*, *Ochrobium*, *Siderocapsa*, *Naumanniella*, and *Siderococcus*, were also described from enrichments but were never obtained as stable axenic cultures (Jones 1986; Kuenen and Bos 1990). Although these organisms appear to catalyze ferrous iron oxidation to some extent and become encrusted with the ferric precipitate, the chemolithotrophic or energy-yielding nature of the process has never been demonstrated convincingly. Cell walls may simply act as catalytic surfaces for ferric iron precipitation, or the cells may use ferrous iron oxidation as a sink for harmful excess oxygen.

The activity of iron-oxidizing bacteria can lead to substantial iron deposits. Typically, groundwater seeping through sand formations dissolves ferrous salts. Underground, a lack of oxygen usually prevents iron oxidation. When groundwater seeps to the surface, usually in a swampy area, iron oxidizers convert the ferrous (Fe^{2+}) ions to ferric (Fe^{3+}), which precipitates as ferric hydroxide and forms bog-iron deposits. Such easily accessible surface deposits were mined extensively and smelted in the early industrial age.

Since oxygenic photosynthesis changed the atmosphere of our planet to an oxidizing one, most iron in the biosphere is kept in the oxidized (ferric) state. Where limited oxygen diffusion and vigorous heterotrophic microbial activity create anaerobic conditions, as occurs in the hypolimnion of stratified lakes, waterlogged soils, and aquatic sediments, ferric iron may act as an electron sink and be reduced to Fe^{2+}. A large and heterogeneous group of heterotrophic bacteria, including *Bacillus*, *Pseudomonas*, *Proteus*, *Alcaligenes*, clostridia, and enterobacteria, appear to be involved in iron reduction.

The mechanism of iron reduction remains largely unexplored. The facts that nitrate inhibits Fe^{3+} reduction and that nitrate reductase negative mutants lose their ability to reduce Fe^{3+} link iron reduction to the nitrate reductase system at least in some microorganisms (Nealson 1983a). Some iron reduction may occur nonenzymatically when reduced products of microbial metabolism, such as formate or H_2S, react chemically with Fe^{3+}. In soils, iron reduction is linked to a condition called gleying. Anoxic conditions due to waterlogging or high clay content give rise to the formation of reduced ferrous iron, which gives the soil a greenish grey color and a sticky consistency. The predominant iron reducers within gleyed soils appear to be *Bacillus* and *Pseudomonas* species.

THE MANGANESE CYCLE

Manganese is an essential trace element for plants, animals, and many microorganisms. It is also cycled by microorganisms between its oxidized and reduced states much like iron, except manganese occurs in the ecosphere either in the reduced manganous (Mn^{2+}) or in the oxidized manganic (Mn^{4+}) state (Ehrlich 1981; Nealson 1983b; Kuenen and Bos 1990). The manganous ion is stable under aerobic conditions at pH values of less than 5.5, but it is also stable at higher pH values under anaerobic conditions. In the presence of oxygen, at pH values greater than 8, the manganous ion is spontaneously oxidized to the tetravalent manganic ion. The manganic ion forms a dioxide (MnO_2) that is insoluble in water. Manganic oxide is not readily assimilated by plants. In some marine and freshwater habitats, the precipitation of manganese forms characteristic manganese nodules (Figure 11.12). The

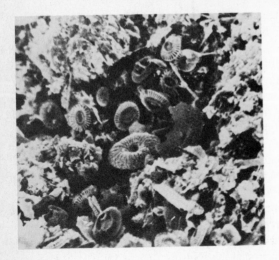

Figure 11.12

Scanning electron micrographs showing microorganisms on the surfaces of ferromanganese nodules. (Source: LaRock and Ehrlich 1975. Reprinted by permission, copyright Springer-Verlag.)

manganese for these nodules originates in anaerobic sediments and is oxidized and precipitated, at least in part with the aid of bacteria, when it enters aerobic habitats, forming the nodules. Manganese is a relatively rare and strategically important metal, and the mining of deep-sea manganese nodules is under consideration.

Various soil and aquatic bacteria and fungi have been reported to catalyze the oxidation of Mn^{2+} to Mn^{4+} as shown in Equation 27.

$$(27) \quad Mn^{2+} + \frac{1}{2}O_2 + H_2O \longrightarrow$$

$$MnO_2 + 2H^+ \ (\Delta G = -7.0 \ kcal/mole)$$

Gallionella, Metallogenium, Sphaerotilus, Leptothrix, Bacillus, Pseudomonas, and *Arthrobacter* strains were reported to carry out Mn^{2+} oxidation either in an inducible or in a constitutive manner. Manganese-based chemolithotrophy has been repeatedly suggested but difficult to prove. Manganese oxidation may be catalyzed by oxidases or catalases, and in some instances Mn^{2+} oxidation appeared to be linked

to ATP synthesis (Ehrlich 1985). Working in continuous culture, P. E. Kepkay and K. H. Nealson (1987) presented solid evidence for chemoautotrophic growth and CO_2 fixation of a Mn^2-oxidizing marine *Pseudomonas*, but further studies will be needed to confirm and clarify this report. In many previously reported cases, simple surface catalysis on the sheath or cell wall of the bacteria promoted manganese oxidation, without an energy benefit to the microorganism.

Metabolism in anaerobic environments by a broad and heterogeneous group of bacteria results in Mn^{4+} reduction, increasing the solubility and mobility of the resulting Mn^{2+}. Reduction of Mn^{4+} might occur enzymatically, but if so, the process is still obscure. As in the case of ferric iron, reduced products of microbial metabolism also react chemically with Mn^{4+}, reducing it to Mn^{2+}.

CALCIUM CYCLING

As a bivalent cation, calcium is an important solute of the cytoplasmic environment and is required for the activity of numerous enzymes. It also stabilizes structural components of the bacterial cell wall. From the geochemical point of view, however, biological precipitation and dissolution in the form of carbonate ($CaCO_3$) and bicarbonate ($Ca[HCO_3]_2$), respectively, are of the highest significance (Ehrlich 1981). Carbonate precipitation or dissolution may be an incidental consequence of metabolic processes that affect pH. Carbonate precipitation is involved also in formation of exoskeleta of microorganisms and invertebrates. Vertebrate animals deposit carbonates in bones and teeth.

Calcium bicarbonate has a high water solubility, calcium carbonate a much lower one. The equilibrium between HCO^{3-} and CO_3^{2-} is influenced by CO_2, which dissolves in water as carbonic acid (H_2CO_3). The prevailing pH strongly influences the formation of H_2CO_3, a very weak acid, and its salts. Increasing the hydrogen ion concentration encourages dissolution of carbonate; a decrease in hydrogen ion concentration encourages its precipitation.

In well-buffered alkaline to neutral environments rich in Ca^{2+}, CO_2 from aerobic or anaerobic microbial oxidations precipitates, at least in part, in the form of $CaCO_3$. The processes of ammonification, nitrate reduction, and sulfate reduction increase alkalinity and under appropriate conditions can contribute to $CaCO_3$ precipitation. The most significant process contributing to biological carbonate precipitation, however, is photosynthesis. In seawater, the principal dissolved form of calcium is bicarbonate, which is in equilibrium with carbonate and CO_2 (Equation 28).

$$(28) \quad Ca(HCO_3)_2 \underset{\longleftarrow}{\overset{\longrightarrow}{\quad}} CaCO_3 + H_2O + CO_2$$

If photosynthesis removes CO_2 from this equilibrium by assimilation, calcium shifts from the bicarbonate to the carbonate form. The latter, being less soluble, precipitates. Cyanobacterial photosynthesis has contributed to the formation of calcified stromatolites by prokaryotic microbial mats (Knoll and Awramik 1983). As endosymbionts, cyanobacteria and algae contribute to exoskeleton formation in Foraminifera and various invertebrate animals, most significantly in corals. Calcium carbonate deposition by corals forms reefs and islands in tropical waters (Figure 11.13). The shells of Foraminifera accumulate as calcareous oozes and eventually turn into limestone deposits. The famous White Cliffs of Dover on the British channel coast were formed from biologically precipitated calcium carbonate, primarily shells of Foraminifera. Magnesium is a bivalent cation that behaves much like calcium and is abundant in seawater. However, $MgCO_3$ has considerably higher solubility in seawater than $CaCO_3$; therefore, calcium is preferentially used in the exoskeletal structures of marine microorganisms.

Microbial metabolic processes that produce organic acids (fermentation) or inorganic acids (nitrification or sulfur oxidation) contribute to the dissolution and mobilization of carbonates. Calcium readily reacts with phosphate anions, forming insoluble tertiary phosphate, which is unavailable for uptake (Equation 29).

$$(29) \quad 3Ca^{2+} + 2PO_4^{3-} \longrightarrow Ca_3(PO_4)_2$$

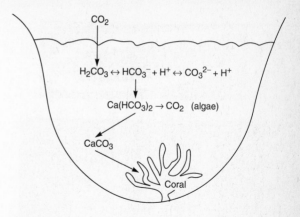

Figure 11.13
The reactions of calcium that lead to the formation of corals in seawater.

Production of organic and inorganic acids by microorganisms solubilizes this precipitated phosphate, a process that effects phosphorus mobilization in soils and sediments.

SILICON CYCLING

Silicon is the second most abundant element in the crust of Earth, 28% by weight (Ehrlich 1981). It occurs primarily as silicon dioxide (SiO_2) and as silicates, the salts of silicic acid (H_4SiO_4). Silicon dioxide may be considered as the anhydride of silicic acid. Water solubility of silicic acid is low, ranging in natural waters from a few µg/L to a maximum of 20 µg/L (Krumbein and Werner 1983). Although silicon as an element is closely related to carbon and is capable of polymerization as siloxanes ($HO-Si^{2+}-O-Si^{2+}-O-Si^{2+}-OH$), its biological role appears to be restricted to structural purposes in many grassy plants and a few invertebrates, such as siliceous sponges and the radula of some molluscs. It also forms the exoskeleton of important microbial groups such as the diatoms, radiolaria, and silicoflagellates. The form of silicon used for this purpose is amorphous and hydrated silica ($SiO_2 \cdot nH_2O$; opal).

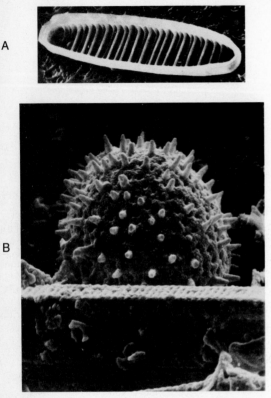

Figure 11.14

Scanning of electron micrographs showing silicon-containing structures. (A) The shell of a diatom (*Surirella*) showing the inside of one of the valves. (B) A Radiolaria. (Source: C. Versfelt.)

Fungi, cyanobacteria, and lichens living on and within siliceous rocks in harsh environments actively dissolve silica through the excretion of carboxylic acids. In particular, 2-ketogluconic, citric, and oxalic acids have been implicated in dissolution of siliceous rock. The solubilization of siliceous rock by the chelating action of these organic acids contributes to the rock-weathering and soil-formation process.

Diatoms play the most important role in the precipitation of dissolved silica. Up to 90% of the siliceous oozes accumulating as pelagic sediment consist of diatom frustules, with Radiolaria contributing most

of the rest. These oozes give rise to geological deposits known as Fuller's Earth, which is used as a filtering aid in the laboratory and in the manufacture of dynamite from nitroglycerine. Dissolved silicic acid is an essential and sometimes limiting nutrient for diatoms and is the major cause of seasonal diatom successions in some lakes (Hurley et al. 1985). The uptake and ordered deposition of silica involves formation of the organic silica complexes having Si—C and Si—O—C type bonds. The silica is deposited in a preformed organic matrix in a highly ordered fashion (Figure 11.14).

Purely chemical precipitation of silica may encrust and preserve microbial cells. Originally, calcareous stromatolites become impregnated and preserved by siliceous minerals (Knoll and Awramik 1983). Contemporary siliceous stromatolites are being formed in some hot springs (Walter et al. 1972). The prokaryotic mats in these hot springs passively become encrusted with silica as the mineral precipitates during the cooling of the thermal waters. Some fossil siliceous stromatolites may have been formed in a similar fashion.

Various heavy metals exhibit biogeochemical cycles. For example, mercury can exist in a variety of inorganic and organic forms. Various microorganisms are capable of forming methyl mercury and may also form methylate arsenic, selenium, tin, and perhaps lead. We will discuss these methylation processes and the role of microorganisms in concentrating heavy metals and radionuclides in chapter 13.

INTERRELATIONS BETWEEN THE CYCLING OF INDIVIDUAL ELEMENTS

In chapters 10 and 11, we have discussed the cycling of each element separately. It needs to be emphasized, however, that in reality the cycle of each element is dependent on, or at least influenced by, the cycling of other elements. This is true not only for C, H, and O, which are cycled by the same two processes of photosynthesis and respiration, but also for cycles that are driven by different biochemical processes and performed by distinct microorganisms.

The study of biogeochemical cycle interactions is a technically challenging task for the investigator. In a complex natural system with a wide variety of microorganisms, substrates, and electron sinks, it is difficult to determine which group of microorganisms degrades a certain substrate and at what rate. This field of investigation has gained much from the use of selective inhibitors of certain cycling activities (Oremland and Capone 1988). As examples, 2-bromoethanesulfonate can be used to selectively inhibit methanogenesis in sediments, molybdate to suppress sulfate reduction, nitrapyrin to shut down nitrification, and acetylene to block denitrification on the nitrous oxide level. While the effectivity and selectivity of these inhibitors needs to be carefully established case by case, an ability to selectively inhibit a metabolic group within a complex system provides an excellent experimental tool for defining the normal contribution of that group to the system.

The reductive portions of the N, S, Fe, and Mn cycles are driven by chemical energy fixed in organic substances during photosynthesis. The chemolithotrophic reoxidation of N, S, Fe, and perhaps Mn are, in turn, linked to the conversion of CO_2 into cell material, again involving the cycling of C, H, and O. Solubilization, uptake, and precipitation of Ca and Si are directly or at least energetically linked to photosynthetic and respiratory cycling of C, H, and O. Acids from nitrification and sulfur oxidation help to mobilize phosphorus; photosynthesis or respiration are required for its uptake and conversion into high-energy phosphates. Sulfur is oxidized with reduction of nitrate by *Thiobacillus denitrificans,* and some extremely thermophilic methanogens can transfer hydrogen not only to CO_2 but also to $S°$. Those are just some of the most obvious examples showing the interdependence of biogeochemical cycles; many other examples could be found. An instructive and ecologically important example of biogeochemical cycle interactions is the sequential use of electron acceptors during the oxi-

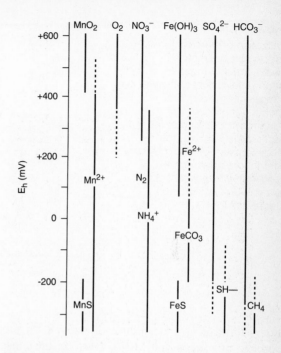

Figure 11.15

Redox potential (E_h) ranges for microbial utilization of potential electron acceptors. A microbial community will preferentially transfer electrons from an organic substrate to the most oxidizing electron acceptor available in their environment. This "choice" by the community, brought about by a combination of metabolic regulation and competition between populations, maximizes energy yield for the community as a whole. (Source: Nedwell 1984.)

dation of organic substrates (Jorgensen 1980; Nedwell 1984).

From the pool of potential electron acceptors, the microbial community selects the one that maximizes energy yield from the available substrate. This seemingly intelligent decision is in part due to metabolic regulation within a single population, and in part due to the inevitable outcome of competition between populations with diverse metabolic capabilities. Through various regulatory mechanisms, facultatively anaerobic microorganisms shut off their less efficient fermentative or dissimilatory nitrate reduc-

Figure 11.16

Utilization of electron acceptors in a marine sediment, showing the strong interactions between biogeochemcial cycles and the stratification of electron acceptor use. The dotted line indicates the sediment surface. Sediment depths are approximate and vary with sediment and season. (Based in part on Jorgensen 1980.)

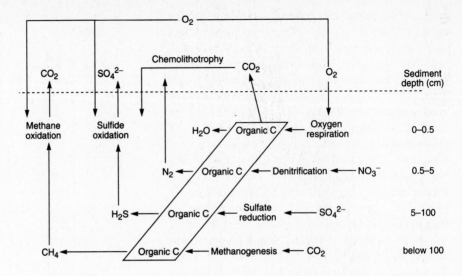

tion pathways in the presence of oxygen. In the absence of oxygen and MnO_2, NO_3^- is the most oxidizing electron acceptor (Figure 11.15). From a substrate equally utilizable by nitrate and sulfate reducers, nitrate reducers will obtain the higher energy yield. Therefore, they will obtain the higher biomass per unit substrate utilization and will effectively outcompete the sulfate reducers. Nitrate and iron, usually scarce in aquatic sediments, are rapidly depleted, leaving sulfate as the most oxidizing electron acceptor. When competing for a common substrate (H_2), methanogens have a lower utilization efficiency and a higher threshold for hydrogen uptake than sulfate reducers (Lovley 1985). Consequently, methanogens cannot effectively compete with sulfate reducers until all or most of the sulfate is depleted. In low-sulfate freshwater sediments, this occurs fairly rapidly; in sulfate-rich marine sediments, it occurs much more slowly.

In aquatic sediments, the sequence of electron acceptor utilization can be observed spatially in horizontal layers of increasing depth. In a typical littoral marine sediment, only the first few millimeters of the sediment are oxygenated, though bioturbation by invertebrates may extend this oxygenated

zone downward. For a few centimeters under the oxygenated zone, nitrate serves as the electron acceptor. Below this, for several meters, sulfate is the principal electron acceptor. Methanogenesis is usually confined to the sulfate-depleted deeper sediment layers, though the generated methane may diffuse upward into the zone of sulfate reduction. Figure 11.16 illustrates the described sequential utilization of electron acceptors in aquatic sediments.

The described classical sequence of electron acceptor utilization is further illuminated but not altered by recently recognized patterns of interspecies hydrogen transfer (Wolin and Miller 1982; Thiele and Zeikus 1988). Hydrogenogens, that is, hydrogen-producing fermentative microorganisms, are at a thermodynamical disadvantage if hydrogen accumulates. They live syntrophically with hydrogentrophs, that is, hydrogen consumers, such as sulfate reducers, methanogens, and acetogens. The syntrophic relationship allows the fermentation to proceed and at the same time supplies the sulfate reducer or methanogen with the hydrogen necessary for SO_4^{2-} or CO_2 reduction, respectively. Recent measurements on anaerobic sewage sludge and lake sediment showed that most of the H_2-dependent methanogenesis in these ecosystems occurs via interspecies hydrogen transfer, rather than by the utilization of hydrogen dissolved in water (Conrad et al. 1985). Coculture experiments with a methanogen and a sulfate reducer on the substrates methanol and acetate, not utilizable by the sulfate reducer (Phelps et al. 1985), showed that the sulfate reducer competed effectively for the hydrogen generated by the methanogen from these substrates. The methanogen produced 10 μmoles of H_2 per L headspace when growing alone, while H_2 was less than 2 μmoles per L headspace when the sulfate reducer was present. The evolution of methane was reduced in coculture. Although the environmental relevance of this coculture experiment remains to be established, it shows that interspecies hydrogen transfer may also be competitive rather than syntrophic, and it confirms the competitive advantage of sulfate reducers over methanogens as long as sulfate is present.

SUMMARY

The major elements nitrogen and sulfur are cycled in a complex, oxidoreductive fashion. Their reduced forms support chemolithotrophic metabolism. Their oxidized forms are used as electron sinks in anaerobic environments. Denitrification is balanced by the agriculturally important process of dinitrogen fixation. Phosphorus is a major nutrient element in short supply. It is cycled by dissolution as primary phosphate and by precipitation as secondary or tertiary phosphate. In their reduced forms, iron and manganese may support chemolithotrophic metabolism. Their oxidized states serve as electron sinks. Calcium and silicon have important roles in the formation of microbial exoskeleta and form large deposits of biological origin. Biogeochemical cycles interact with each other extensively in space and time. Under each given set of circumstances, the process that allows for the maximal energy flow is selected.

REFERENCES & SUGGESTED READINGS

Aleem, M. I. H. 1977. Coupling of energy with electron transfer reactions in chemolithotrophic bacteria. In B. A. Haddock and W. A. Hamilton (eds.). *Microbial Energetics.* Cambridge University Press, Cambridge, England, pp. 351–381.

Altschuler, Z. S., M. M. Schneppe, C. C. Silber, and F. O. Simon. 1983. Sulfur diagenesis in Everglades peat and origin of pyrite in coal. *Science* 221:221–227.

Balderston, W. L., B. Sherr, and W. J. Payne. 1976. Blockage by acetylene of nitrous oxide reduction in *Pseudomonas perfectomarinus. Applied and Environmental Microbiology* 31:504–508.

Belser, L. W. 1979. Population ecology of nitrifying bacteria. *Annual Reviews of Microbiology* 33:309–333.

Belser, L. W., and E. L. Schmidt. 1978. Nitrification in soils. In D. Schlesinger (ed.). *Microbiology 1978.* American Society for Microbiology, Washington, D.C., pp. 348–351.

Benemann, J. R., and R. C. Valentine. 1972. The pathways of nitrogen fixation. *Advances in Microbial Physiology* 8:59–104.

Blackburn, T. H. 1983. The microbial nitrogen cycle. In C. W. E. Krumbein (ed.). *Microbial Geochemistry.* Blackwell Scientific Publications, Oxford, England, pp. 63–89.

Bock, E., H. P. Koops, and H. Harms. 1990. Nitrifying bacteria. In H. G. Schlegel and B. Bowien (eds.) *Autotrophic Bacteria.* Springer-Verlag, Berlin, pp. 81–96.

Bollag, J. M., and E. J. Kurek. 1980. Nitrite and nitrous oxide accumulation during denitrification in presence of pesticide derivatives. *Applied and Environmental Microbiology* 39:845–849.

Bremner, J. M., and C. G. Steele. 1978. Role of microorganisms in the atmospheric sulfur cycle. *Advances in Microbial Ecology* 2:155–201.

Brill, W. 1975. Regulation and genetics of bacterial nitrogen fixation. *Annual Review of Microbiology* 29:109–129.

Brill, W. 1979. Nitrogen fixation: Basic to applied. *American Scientist* 67:458–466.

Brill, W. 1980. Biochemical genetics of nitrogen fixation. *Microbiological Reviews* 44:449–467.

Brill, W. 1981. Biological nitrogen fixation. *Scientific American* 245(3):68–81.

Brown, C. M. 1982. Nitrogen mineralisation in soils and sediments. In R. G. Burns and J. H. Slater (eds.). *Experimental Microbial Ecology*. Blackwell Scientific Publications, Oxford, England, pp. 154–163.

Brown, C. M., and B. Johnson. 1977. Inorganic nitrogen assimilation in aquatic microorganisms. *Advances in Aquatic Microbiololgy* 1:49–114.

Burns, R. C., and R. W. F. Hardy. 1975. *Nitrogen Fixation in Bacteria and Higher Plants*. Springer-Verlag, New York.

Caldwell, D. E., S. J. Caldwell, and J. P. Laycock. 1976. *Thermothrix thiopara* gen. et sp. nov., a facultatively anaerobic facultative chemolithotroph living at neutral pH and high temperature. *Canadian Journal of Microbiology* 22:1509–1517.

Campbell, R. 1977. *Microbial Ecology*. Blackwell Scientific Publications, Oxford, England.

Chapra, S. C., and A. Robertson. 1977. Great Lakes eutrophication: The effect of point source control on total phosphorus. *Science* 196:1448–1449.

Conrad, R., T. J. Phelps, and J. G. Zeikus. 1985. Gas metabolism evidence in support of the juxtaposition of hydrogen-producing and methanogenic bacteria in sewage sludge and lake sediments. *Applied and Environmental Microbiology* 50:595–601.

Cosgrove, D. J. 1977. Microbial transformations in the phosphorous cycle. *Advances in Microbial Ecology* 1:95–134.

Coty, V. F. 1967. Atmospheric nitrogen fixation by hydrocarbon-oxidizing bacteria. *Biotechnology and Bioengineering* 9:25–32.

Dacey, J. W. H., and S. G. Wakeham. 1986. Oceanic dimethylsulfide: Production during zooplankton grazing on phytoplankton. *Science* 233:1314-1316.

Dalton, H. 1974. Fixation of dinitrogen by free-living microorganisms. *CRC Critical Reviews in Microbiology* 3:183–220.

Dalton, H., and L. E. Mortenson. 1972. Dinitrogen (N_2) fixation with a biochemical emphasis. *Biological Reviews* 36:231–260.

Deevey, E. S., Jr. 1970. Mineral cycles. *Scientific American* 223(5):148–158.

Delwiche, C. C. 1970. The nitrogen cycle. *Scientific American* 223(3):137–146.

Delwiche, C. C., and B. A. Bryan. 1976. Denitrification.

Annual Reviews of Microbiology 30:241–262.

Döbereiner, J. 1974. Nitrogen-fixing bacteria in the rhizosphere. In A. Quispel (ed.). *The Biology of Nitrogen Fixation*. North-Holland Publishing Co., Amsterdam, The Netherlands, pp. 86–120.

Döbereiner, J., and J. M. Day. 1974. Nitrogen fixation in the rhizosphere of tropical grasses. In W. D. P. Stewart (ed.). *Nitrogen Fixation by Free-living Bacteria*. Cambridge University Press, Cambridge, England.

Ehrlich, H. L. 1981. *Geomicrobiology*. Marcel Dekker, New York.

Ehrlich, H. L. 1985. Mesophilic manganese oxidizing bacteria from hydrothermal discharge areas at 21° North on the East Pacific Rise. In D. C. Caldwell, J. A. Brierley, and C. L. Brierley (eds.). *Planetary Ecology*. Van Nostrand Reinhold Co., New York, pp. 186–194.

Fenchel, T., and T. H. Blackburn. 1979. *Bacteria and Mineral Cycling*. Academic Press, London.

Fischer, F., W. Zillig, K. O. Stetter, and G. Schreiber. 1983. Chemolithoautotrophic metabolism of anaerobic extremely thermophilic archaebacteria. *Nature* (London) 301:511–513.

Focht, D. D. 1982. Denitrification. In R. G. Burns and J. H. Slater (eds.). *Experimental Microbial Ecology*. Blackwell Scientific Publications, Oxford, England, pp. 194–211.

Focht, D. D., and W. Verstraete. 1977. Biochemical ecology of nitrification and denitrification. *Advances in Microbial Ecology* 1:135–214.

Gersberg, R., K. Krohn, N. Peele, and C. R. Goldman. 1976. Denitrification studies with ^{13}N-labeled nitrate. *Science* 192:1229–1231.

Gibson, A. H. (ed.). 1977. *A Treatise on Dinitrogen Fixation*. Sec. IV: Agronomy and Ecology. John Wiley and Sons, New York.

Gottschalk, G. 1979. *Bacterial Metabolism*. Springer-Verlag, New York.

Grassle, J. F. 1985. Hydrothermal vent animals: Distribution and biology. *Science* 229:713–716.

Hall, J. B. 1978. Nitrogen-reducing bacteria. In D. Schlesinger (ed.). *Microbiology 1978*. American Society for Microbiology, Washington, D.C., pp. 296–298.

Hallmark, S. L., and R. E. Terry. 1985. Field measurement of denitrification in irrigated soils. *Soil Science* 140:35–44.

Halstead, R. L., and R. B. McKercher. 1975. Biochemistry and cycling of phosphorus. In E. A. Paul and A. D. McLaren (eds.). *Soil Biochemistry*. Vol. 4. Marcel Dekker, New York, pp. 31–63.

Hardy, R. W. F., R. P. Holsten, E. K. Jackson, and R. C. Burns. 1968. The acetylene ethylene assay for N_2-fixation: Laboratory and field evaluation. *Plant Physiology* 43:1185–1207.

Hooper, A. B. 1990. Biochemistry of the nitrifying lithoautotrophic bacteria. In H. G. Schlegel and B. Bowien (eds.). *Autotrophic Bacteria*. Springer-Verlag, Berlin, pp. 239–265.

Hurley, J. P., D. E. Armstrong, G. J. Kenoyer, and C. J. Bowser. 1985. Ground water as silica source for diatom production in a precipitation dominated lake. *Science* 227:1576–1578.

Hutchinson, G. L., and A. R. Mosier. 1979. Nitrous oxide emissions from an irrigated cornfield. *Science* 205:1125–1126.

Ingraham, J. L. 1980. Microbiology and genetics of denitrifiers. In C. C. Delwiche (ed.). *Denitrification, Nitrification and Atmospheric Nitrous Oxide*. John Wiley and Sons, New York, pp. 45–65.

Ivanov, M. V. 1968. *Microbiological Processes in the Formation of Sulfur Deposits*. Translated from Russian. Israel Program of Scientific Translations, Ltd., U.S. Department of Commerce, Springfield, Va.

Jannasch, H. W. 1990. Chemosynthetically sustained ecosystems in the deep sea. In H. G. Schlegel and B. Bowien (eds.). *Autotrophic Bacteria*. Springer-Verlag, Berlin, pp. 147–166.

Jannasch, H. W., and C. O. Wirsen. 1981. Morphological survey of microbial mats near deep-sea thermal vents. *Applied and Environmental Microbiology* 41:528–538.

Jannasch, H. W., and M. J. Mottl. 1985. Geomicrobiology of deep-sea hydrothermal vents. *Science* 229:717–725.

Jones, J. G. 1986. Iron transformations by freshwater bacteria. *Advances in Microbial Ecology* 9:149–185.

Jorgensen, B. B. 1980. Mineralization and the bacterial cycling of carbon, nitrogen and sulphate in marine sediments. In D. C. Ellwood, M. J. Latham, J. N. Hedger, J. M. Lynch, and J. H. Slater (eds.). *Contemporary Microbial Ecology*. Academic Press, New York, pp. 239–251.

Jorgensen, B. B. 1983. The microbial sulfur cycle. In W. E. Krumbein (ed.). *Microbial Geochemistry*. Blackwell Scientific Publications, Oxford, England, pp. 91–214.

Jorgensen, B. B., and N. P. Revsbeck. 1983. Colorless sulfur bacteria *Beggiatoa* spp. and *Thiovulum* spp. in O_2 and H_2S microgradients. *Applied and Environmental Microbiology* 45:1261–1270.

Karl, D. M., C. O. Wirsen, and H. W. Jannasch. 1980. Deep-sea primary production at the Galapagos hydrothermal vents. *Science* 207:1345–1347.

Kellog, W. W., R. D. Cadle, E. R. Allen, A. L. Lazrus, and E. A. Martell. 1972. The sulfur cycle. *Science* 175:587–596.

Kelly, D. P., and N. A. Smith. 1990. Organic sulfur compounds in the environment, biochemistry, microbiology and ecological aspects. *Advances in Microbial Ecology* 11:345–385.

Kepkay, P. E., and K. H. Nealson. 1987. Growth of a manganese oxidizing *Pseudomonas* sp. in continuous culture. *Archives for Microbiology* 148:63–67.

Knoll, A. H., and S. M. Awramik. 1983. Ancient microbial ecosystems. In W. E. Krumbein (ed.). *Microbial Geochemistry*. Blackwell Scientific Publications, Oxford, England, pp. 287–315.

Koike, I., and A. Hattori. 1978. Denitrification and ammonia formation in anaerobic coastal sediments. *Applied and Environmental Microbiology* 35:278–282.

Krumbein, W. E., and D. Werner. 1983. The microbial silica cycle. In W. E. Krumbein (ed.). *Microbial Geochemistry*. Blackwell Scientific Publications, Oxford, England, pp. 125–157.

Kuenen, J. G., L. A. Robertson, and H. V. Gemerden. 1985. Microbial interactions among aerobic and anaerobic sulfur-oxidizing bacteria. *Advances in Microbial Ecology* 8:1–59.

Kuenen, J. G., and P. Bos. 1990. Habitats and ecological niches of chemolitho(auto)trophic bacteria. In H. G. Schlegel and B. Bowien (eds.). *Autotrophic Bacteria*. Springer-Verlag, Berlin, pp. 53–80.

LaRock, P. A., and H. L. Ehrlich. 1975. Observations of bacterial microcolonies on the surface of ferromanganese nodules from Blake Plateau by scanning electron microscopy. *Microbial Ecology* 2:84–96.

Leschine, S. B., K. Howell, and E. Canale-Parola. 1988. Nitrogen fixation by anaerobic cellulolytic bacteria. *Science* 242:1157–1159.

Lewin, R. 1984. How microorganisms transport iron. *Science* 225:401–402.

Lovley, D. R. 1985. Minimum threshold for hydrogen metabolism in methanogenic bacteria. *Applied and Environmental Microbiology* 49:1530–1531.

Lowenstein, H. A. 1981. Minerals formed by organisms. *Science* 211:1126–1131.

National Research Council. 1979. *Microbial Processes: Promising Technologies for Developing Countries*. National Academy of Sciences, Washington D.C., pp. 59–79.

Nealson, K. H. 1983a. The microbial iron cycle. In W. E. Krumbein (ed.). *Microbial Geochemistry*. Blackwell Scientific Publications, Oxford, England, pp. 159–190.

Nealson, K. H. 1983b. The microbial manganese cycle. In W. E. Krumbein (ed.). *Microbial Geochemistry*. Blackwell Scientific Publications, Oxford, England, pp. 191–221.

Nedwell, D. B. 1984. The input and mineralization of organic carbon in anaerobic aquatic sediments. *Advances in Microbial Ecology* 7:93–131.

Nelson, D. C. 1990. Physiology and biochemistry of filamentous sulfur bacteria. In H. G. Schlegel and B. Bowien (eds.). *Autotrophic Bacteria*, Springer-Verlag, Berlin, pp. 219-238.

Nelson, D. C., and H. W. Jannasch. 1983. Chemoautotrophic growth of a marine *Beggiatoa* in sulfide gradient cultures. *Archives for Microbiology* 136:262–269.

Oremland, R. S., and D. C. Capone. 1988. Use of specific inhibitors in biogeochemistry and microbial ecology. *Advances in Microbial Ecology* 10:285–383.

Paerl, H. W. 1990. Physiological ecology and regulation of N_2 fixation in natural waters. *Advances in Microbial Ecology* 11:305–344.

Parameswaran, A. K., C. N. Provan, F. J. Sturmand, and R. M. Kelly. 1987. Sulfur reduction by the extremely thermophilic arachaebacterium *Pyrodictium occultum*. *Applied and Environmental Microbiology* 53:1690–1693.

Payne, W. J. 1971. Gas chromatographic analysis of denitrification by marine organisms. In L. H. Stevenson and R. R. Colwell (eds.). *Estuarine Microbial Ecology*. University of South Carolina Press, Columbia, pp. 53–71.

Payne, W. J. 1973. Reduction of nitrogenous oxides by microorganisms. *Bacteriological Reviews* 37:409–452.

Pfennig, N. 1990. Ecology of phototrophic purple and green sulfur bacteria. In H. G. Schlegel and B. Bowien (eds.). *Autotrophic Bacteria*. Springer-Verlag, Berlin, pp. 97–116.

Pfennig, N., and H. Biebl. 1981. The dissimilatory sulfur-reducing bacteria. In M. P. Starr, H. Stolp, H. G. Trüper, A. Balows, and H. G. Schlegel (eds.). *The Prokaryotes*. Springer-Verlag, Berlin, pp. 941–947.

Pfennig, N., F. Widdel, and H. G. Trüper. 1981. The dis-

similatory sulfate-reducing bacteria. In M. P. Starr, H. Stolp, H. G. Trüper, A. Balows, and H. G. Schlegel (eds.). *The Prokaryotes*. Springer-Verlag, Berlin, pp. 926–940.

Phelps, T. J., R. Conrad, and J. G. Zeikus. 1985. Sulfate-dependent interspecies H_2 transfer between *Methanosarcina barkeri* and *Desulfovibrio vulgaris* during coculture metabolism of acetate or methanol. *Applied and Environmental Microbiology* 50:589–594.

Pomeroy, L. R. (ed.). 1974. *Cycles of Essential Elements*. Dowden, Hutchinson and Ross. Stroudsburg, Penn.

Postgate, J. R. 1982. *The Fundamentals of Nitrogen Fixation*. Cambridge University Press, Cambridge, England.

Postgate, J. R. 1984. *The Sulphate-reducing Bacteria*. Cambridge University Press, Cambridge, England.

Prosser, J. I., and D. J. Cox. 1982. Nitrification. In R. G. Burns and J. H. Slater (eds.). *Experimental Microbial Ecology*. Blackwell Scientific Publications, Oxford, England, pp. 178–193.

Setter, K. O. 1990. Extremely thermophilic chemolithotrophic archaebacteria. In H. G. Schlegel and B. Bowien. *Autotrophic Bacteria*. Springer-Verlag, Berlin, pp 167–176.

Silverman, M. P., and H. L. Ehrlich. 1964. Microbial formation and degradation of minerals. *Advances in Applied Microbiology* 6:153–206.

Smith, D. W. 1982. Nitrogen fixation. In R. G. Burns and J. H. Slater (eds.). *Experimental Microbial Ecology*. Blackwell Scientific Publications, Oxford, England, pp. 212–220.

Sorderlund, R., and B. H. Svensson. 1976. The global nitrogen cycle. Scope Report 7, Global Cycles. *Bulletin of the Ecological Research Commission* (Stockholm) 22:23–73.

Sprent, J. I., and P. Sprent. 1990. *Nitrogen Fixing Organisms: Pure and Applied Aspects*. Chapman Hall, London.

Stetter, K. O., and G. Gaag. 1983. Reduction of molecular sulphur by methanogenic bacteria. *Nature* (London) 305:301–310.

Stewart, J. W. B., and R. B. McKercher. 1982. Phosphorus cycle. In R. G. Burns and J. H. Slater (eds.). *Experimental Microbial Ecology*. Blackwell Scientific Publications, Oxford, England, pp. 229–238.

Stewart, W. D. P. 1973. Nitrogen fixation by photosynthetic microorganisms. *Annual Reviews of Microbiology* 27:283–316.

Thauer, R. K. 1990. Energy metabolism of sulfate-reducing bacteria. In H. G. Schlegel and B. Bowien (eds.). *Autotrophic Bacteria.* Springer-Verlag, Berlin, pp. 397–413.

Thiele, J. H., and G. Zeikus. 1988. Control of interspecies electron flow during anaerobic digestion: Significance of formate transfer during syntrophic mehanogenesis in flocs. *Applied and Environmental Microbiology* 54:20–29.

Tiessen, H., and J. W. B. Stewart. 1985. The biogeochemistry of soil phosphorus. In D. C. Caldwell, J. A. Brierley, and C. L. Brierley (eds.). *Planetary Ecology.* Van Nostrand Reinhold Co., New York, pp. 463–472.

Trüper, H. G. 1990. Physiology and biochemistry of phototropohic bacteria. In H. G. Schlegel and B. Bowien (eds.). *Autotrophic Bacteria.* Springer-Verlag, Berlin, pp. 267–281.

van Berkum, P., and B. B. Bohlool. 1980. Evaluation of nitrogen fixation by bacteria in association with roots of tropical grasses. *Microbiological Reviews* 44:491–517.

Wallace, W., and D. J. D. Nicholas. 1969. The biochemistry of nitrifying organisms. *Biological Reviews* 44:359–391.

Walter, M. R., J. Bauld, and T. D. Brode. 1972. Siliceous algal and bacterial stromatolites in hot spring and geyser effluents of Yellowstone National Park. *Science* 178:402–405.

Winter, H. C., and R. H. Burris. 1976. Nitrogenase. *Annual Review of Biochemistry* 45:409–426.

Wolin, M. J., and T. L. Miller. 1982. Interspecies hydrogen transfer: 15 years later. *ASM News* 48:561–565.

PART FIVE

Biotechnological Aspects of Microbial Ecology

12

Ecological Aspects of Biodeterioration Control: Soil, Waste, and Water Management

Microbial ecology is not an abstract science. The presence and functioning of microbial communities affect our everyday lives in many ways. The application of the principles and methods outlined in the previous chapters allows us to understand phenomena that influence our economic well-being, the general public health, and global environmental quality. It provides scientific explanations for traditional practices that have often evolved empirically. More importantly, an understanding of microbial ecology allows us to reach acceptable solutions to current problems in agriculture, resource recovery, public health, and pollution control by a planned, scientific—rather than a haphazard—approach. Microbial ecology is not only intriguing, but also a highly relevant scientific discipline with many practical implications and applications.

Deterioration of foods and materials, healthful drinking water, fertile agricultural soil, and the acceptable disposal of liquid and solid waste materials are long-standing problems of human society, and some solutions to these problems had evolved long before there was any appreciation of their microbiological aspects. The trends toward increased world population, urbanization, and industrialization, though, have created a situation where traditional solutions have become largely inadequate. The revision and optimization of these processes require an understanding of the underlying microbial mechanisms from both biochemical and ecological points of view.

CONTROL OF BIODETERIORATION

All foods and many traditional materials used by human society are potential substrates for microbial growth. In most cases, their biochemical transformation by microorganisms results in spoilage or deterioration. Food preservation and prevention of biodeterioration are major areas of concern to food and industrial microbiologists and to microbial ecologists. The emphasis of this brief discussion is on principles used to control biodeterioration rather than on procedural details that are available in textbooks (Desrosier

1970; Frazier and Westhoff 1978) and reviews (Hurst and Collins-Thompson 1979). The control methods that are employed represent applications of Liebig's and Shelford's laws, that is, the adjustment of environmental conditions to exceed the growth requirements and tolerance ranges of microbes to prevent potentially destructive microbial growth.

To control undesirable microbial activity, one or more of the following basic approaches and combinations are used:

1. All undesirable microorganisms are destroyed or removed by physical or chemical means, and recontamination is prevented by a physical barrier.

2. Food or other materials subject to biodeterioration are kept under environmental conditions that preclude or minimize microbial activity.

3. Food or material is modified, by processing or additives, to reduce its availability as a microbial substrate.

The first approach, sterilization, is used in canning, the preparation of surgical utensils and microbiological media, and so forth. Initial sterilization is achieved by heat treatment, radiation sterilization, or chemosterilization. We discussed the sensitivity of microorganisms to these factors and the death curves of microbial populations previously. The commonly used physical processes are exposure to wet heat, dry heat, gamma radiation (usually from a ^{60}Co source), and UV radiation. Because of its low penetration, sterilization by UV radiation is largely restricted to gases or surfaces. Gamma irradiation is used on some heat-sensitive materials, such as sutures and plastics, and some foods. The most frequently used chemosterilants are ethylene oxide, propylene oxide, chlorine, and ozone. Because of residual toxicity and taste problems, chemosterilization is rarely applied to foods, but it is useful for surgical and other medical implements, microbiological plasticware, and so on. Ethylene oxide is commonly used in sterilization of such materials. Chlorine, chloramines, and ozone are employed in the treatment of drinking water. These chemicals are strong oxidizing agents that effectively kill microorganisms.

All of the sterilization treatments tend to affect adversely the material to be preserved, and often only partial sterilization is used in order to minimize damage to the product. Pasteurization of milk and other products, such as beer, is aimed at the destruction of heat-sensitive human pathogens and an overall reduction in microbial numbers; the complete sterilization of milk and beer by heat would render them unpalatable.

In the canning industry, the most dangerous and highly heat-resistant spoilage organism is *Clostridium botulinum*. Minute amounts of the botulin neurotoxin produced by this spore-forming anaerobe can cause rapid fatality by respiratory paralysis. Canned food products that may serve as substrates for *Clostridium* and other endospore-forming bacteria need extensive heat processing to eliminate all viable endospores. Foods that are by nature acidic (fruit preserves, tomatoes, and so on) are satisfactorily preserved with a milder heat process, since *C. botulinum* fails to grow and produce toxin at low pH values. Significantly, when plant-breeding programs developed low-acid tomato varieties, clostridial spoilage problems occurred with several outbreaks of botulism, and the canning process had to be revised.

Indicator microorganisms are used in safety and quality control procedures whenever a potential problem organism cannot be detected with ease and reliability. For safety, an indicator organism should be at least as resistant or persistent as the problem organism. Its presence and survival should indicate the potential presence of viable problem organisms. For easy detection, indicator organisms should be more numerous than the problem organisms, and they should be easy to grow and to identify. In sterilization quality control, each sterilization batch includes at least one sample intentionally contaminated with an organism that is highly resistant to the treatment. These contaminated samples are incubated after the sterilization process under appropriate conditions; lack of growth indicates the success of the sterilization process. Because of the high resistance of its endospores and its easy cultivation, *Bacillus pumilus* serves most frequently as the indicator organism in sterilization procedures. The use of an indicator organism is especially critical in cases where contamination would become obvious only after the appearance of serious health consequences; therefore, it is used with disposable syringes, intravenous tubing, sutures, bandages, and similar medical implements.

Microorganisms can be removed by filtration from solutions or gases that cannot be sterilized by any of the previously discussed approaches. Filtration is used for some beverages, for some pharmaceutical products, and for the sterilization of air. Filtration causes minimal sterilization damage to the product.

Following each of the previously described procedures for killing or removing contaminating microorganisms, the product is physically separated from the environment to prevent recontamination. Generally a barrier, such as a metal can, glass bottle, or plastic-sealed wrapper, is used to block exchange with the nonsterile surroundings. A break in this protective barrier, such as a puncture of a can, allows recontamination and negates the value of the sterilization procedure. If the integrity of the barrier remains intact, however, and the product has been truly sterilized, storage can be at room temperature, and the storage time without biodeterioration (shelf life) can be virtually indefinite.

The second approach, for control of microbial growth, manipulates environmental conditions to restrict microbial activity but does not attempt to eliminate or exclude microorganisms. The environmental parameters most frequently controlled for suppression of microbial activity are water activity and temperature. To reduce water activity, the heat of the sun or fire is used to evaporate water. Meat, fish, vegetables, biscuits, cereals, and so on are preserved from spoilage in this manner. Freeze-drying is a modern modification of this approach that minimizes the deleterious effects of drying on some foods. An alternate way of lowering water activity is by adding high concentrations of salt or sugar.

Salted meat and fish products, candied fruit, and refined sugar itself are examples of products preserved by their low water activity. Most bacteria will not grow at a water activity below 0.95, although halophilic bacteria occasionally grow on salted fish and meat products. On the other hand, some common fungi will grow at water activities as low as 0.65–0.70, as evidenced by the appearance of molds on various foods stored under damp conditions. Growth of some molds on some grains and nuts stored under damp conditions can result in the synthesis of powerful mycotoxins, even when the overall appearance of the produce is only minimally affected. Aflatoxins, produced by certain strains of *Aspergillus flavus,* have high acute toxicity and are also potent carcinogens.

Wood, leather, and cellulosic fibers, such as cotton, hemp, and jute, are natural products quite susceptible to biodeterioration. A general approach to prevent these materials from biodeteriorating is to keep them dry. Addition of preservatives is an alternative that will be discussed later. Wooden structures are built to minimize exposure to moisture by the use of stone or concrete foundations and roofing. The roofs of the picturesque covered bridges served to protect the wooden bridge structure, not its users, from the elements. The primary function of paint, stain, and varnish is to prevent wood from soaking up moisture. In all of these examples, low water activities limit biodeterioration.

Because they affect the quality and taste of food only minimally or not at all, low temperatures (freezing and refrigeration) are popular food preservation methods in developed countries. Freezing arrests biodeterioration (although not chemical deterioration) indefinitely, and at refrigerator temperatures—that is, below 5°C—growth of most spoilage organisms is slow. Before the ready availability of home refrigeration units, this mode of food preservation was seasonally and regionally restricted; it is still unaffordable for the great majority of the world's population.

The third approach, for control of microbial growth, modifies the availability of the substrate to microorganisms through processing or additives. In situations where microorganisms cannot be excluded and environmental parameters (moisture or temperature, for example) cannot be readily controlled, this is the only remaining alternative. Additives designed to suppress microbial activity are not necessarily innocuous to higher organisms. Even traditional additives, such as smoke and nitrate, are controversial, because they probably contribute carcinogenic benzpyrenes and nitrosamines, respectively, to the human diet.

Nitrite derived from the added nitrate, however, is considered essential for control of *Clostridium botulinum* in sausage-type meat products.

Some common modern spoilage retardants used in food products are acetic, lactic, propionic, citric, and sorbic acids or their salts. Fatty acids have a general antibacterial activity, the mechanism of which is complex and not well understood. Microorganisms exhibit great variations in sensitivity to these agents. Sodium or calcium propionate is extensively used to inhibit the development of molds in bread and other bakery products, while the same agents do not interfere with the activity of yeasts prior to the baking process. Sodium benzoate and the methyl and propyl esters of *p*-hydroxybenzoic acid (methylparaben and propylparaben) are broad-spectrum microbial inhibitors used in various food products. Salicyclic acid and salicylates are used as food preservatives in many countries but not in the United States. Benzoates and salicylates interfere with a number of enzymatic processes, but their exact mode of action is not well defined. In sugar manufacturing, molasses and sugar solutions are protected from rapid biodeterioration by osmotolerant bacteria and yeasts by the use of formaldehyde, sodium metabisulfite, or quaternary ammonium compounds.

Ecologically most interesting are the food preservation methods that rely on microbial transformations of one kind to prevent other undesirable microbial transformations. In effect, microorganisms capable of preemptive colonization are used to prevent an ecological succession that would lead to spoiled food products. Lactic acid fermentations often are used in this manner to produce fermented milk products (yogurt, sour milk, and sour cream), various pickles, sauerkraut, and silage for animal feed. In each case, profuse acid production by lactobacilli prevents the activity of other decomposing bacteria. The food product is preserved, at least as long as anaerobic conditions prevail. Somewhat similar stabilization of milk products occurs during the formation of various cheeses by propionibacteria. Acetic acid (vinegar) produced from ethanol by *Acetobacter aceti* is utilized in stabilizing food products such as pickled herring. Wine, beer, and other alcoholic beverages are preserved by their ethanol content produced by yeasts during the fermentation process.

For some uses, wood and natural fibers cannot be kept from prolonged moisture exposure, and the addition of antibacterial preservatives is needed to prevent rapid biodeterioration. Wood that must be buried for such uses as fence posts, telephone poles, or pilings is commonly impregnated with coal tar or creosote. Both substrates consist of highly condensed and partially oxygenated petroleum fractions. They are antimicrobial, hydrophobic, and highly resistant to decomposition. Other wood preservatives are copper naphthalene and ammoniated or chromated copper arsenate. These products rely on heavy metal toxicity. Tar from the dry distillation of wood or from petroleum is used on rope and sailcloth. Some common modern commercial biocides used on tent canvas, camping gear, and other field equipment are chlorophenols, dithiocarbamates, acrolein, and dibromonitrilopropionamide. Phenylmercury biocides are used in the pulp industry to retard the development of slime-forming microorganisms. This practice and the dressing of seeds with organomecurials prior to planting have contributed substantially to environmental mercury pollution and are gradually being replaced by other biocides.

The tanning of leather is essentially a stabilizing measure. Tannic acid traditionally was used to establish extensive cross-linkage of protein strands, resulting in substantial resistance to proteolytic enzymes. Chromic salts are used for the same purpose in modern tanning processes.

Fouling Biofilms

The fouling of submerged surfaces, especially in the marine environment, poses a special deterioration problem. Surface fouling is a complex phenomenon initiated by microorganisms but completed by invertebrate animals, such as shipworms (*Teredo*) and other molluscs, barnacles, polychaetes, brachiopods, sponges, and bryozoa. Under euphotic conditions, microalgae and macroalgae also participate in the surface-fouling phenomenon (Sieburth 1975).

Surface fouling results in the direct deterioration of materials such as wood, rubber, many plastics, and insulation materials. Indirect damage occurs from the increased weight and hydrodynamic resistance of cables and other structures, leading to their mechanical failure. Corrosion of ferrous metals is enhanced by damage to the paint coat. Fouling increases the drag of boat and ship hulls, resulting in decreased speed and fuel inefficiency. Surface fouling reduces the efficiency of heat exchangers and is a big problem in cooling systems that use fresh or saltwater.

The ecological succession of surface fouling is initiated by the permanent attachment of heterotrophic marine microorganisms (DiSalvo and Daniels 1975). These microorganisms are attracted even to inert surfaces, such as glass, since they concentrate by adsorption the dilute organic nutrients in seawater. The colonizing microorganisms secrete adhesive mucopolysaccharides and establish a primary surface film (Corpe et al. 1976). Figures 12.1–12.3 show, via scanning electron micrography, examples of this primary surface film, which appears to be critical for the subsequent colonization by larvae of various invertebrate animals.

Marine surface fouling can be retarded or prevented by shielding with materials containing toxic

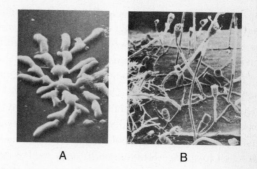

A   B

Figure 12.2

Microbial surface fouling. (A) Microorganisms attached to a fouling slide immersed in a polluted harbor in the Bahamas for six days. (B) Microbial colony developing on a glass slide immersed in San Francisco Bay water for five days. (Source: Di Salvo and Daniels 1975. Reprinted by permission, copyright Springer-Verlag.)

Figure 12.3

Microbial surface fouling. (A) Fungi on wood surface from Point Judith, Rhode Island, showing mycelia and spores of *Zalerion maritimum*. (B) Microorganisms colonizing nylon fishing line immersed for four days at Pigeon Key, Florida, showing attachment of pennate diatoms. (Source: Sieburth 1975. Reprinted by permission, copyright University Park Press.)

Figure 12.1

Photomicrographs showing microbial surface fouling. (A) Estuarine bacteria cultured on a clean glass slide for forty-eight hours in an aquarium containing San Francisco Bay water. (B) Colony and lawn of an estuarine bacterial isolate grown on clean glass slide. (Source: Di Salvo and Daniels 1975. Reprinted by permission, copyright Springer-Verlag.)

A   B

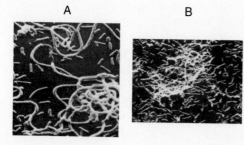

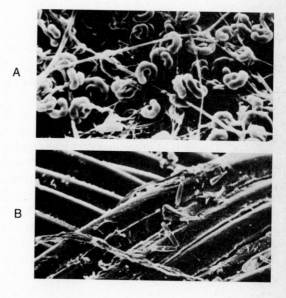

substances—for example, substances containing heavy metals, such as copper and lead. Wood is often impregnated with tar, creosote, or other wood preservatives. Submerged metal and other surfaces usually are protected by paint containing toxic heavy metals, such as mercury, lead, tin, chromate, copper, and so on. A disadvantage of the latter approach is that the toxic heavy metals gradually leach into the marine environment and create a pollution problem. A novel approach to surface-fouling control may evolve from recent investigations on microbial chemotaxis. It may be possible to retard surface fouling by the use of specific repellents rather than by using indiscriminately toxic heavy-metal biocides. Acrylamide, tannic acid, and benzoic acid exhibit promising repellent characteristics and these or similar substances may be of use in future antifouling paints (Chet and Mitchell 1976).

Fouling of surfaces is also a problem in highly humid terrestrial environments. The walls of shower stalls, bathrooms, indoor pools, cold rooms, basements, and the like may be colonized by microorganisms, especially by molds that are able to grow at low water activities. Paints that incorporate antimicrobial substances and periodic scrubbing with disinfectants are the common remedies to this problem.

MANAGEMENT OF AGRICULTURAL SOILS

Soil management practices vary according to crop and soil characteristics but share certain common features. With the exception of rice grown in paddies, all major crops require aerobic soil conditions. Since the aeration status of soils is controlled principally by their water saturation, it is of prime importance to provide adequate drainage for agricultural land. Prolonged anaerobic soil conditions are injurious to plant roots, not only by direct oxygen deprivation but also because of the microbial use of secondary electron acceptors, such as nitrate, sulfate, and ferric iron. The resulting denitrification causes loss of vital nitrogen; toxic H_2S is produced; and sticky greenish ferrous iron is deposited in gley soils.

The soil pH regulates the solubility of plant nutrients as well as the bioavailability of potentially toxic heavy metals; thus, the management of soil pH is often critical in agricultural production. Biodegradation of plant material and some mineral cycling activities, such as sulfur oxidation and nitrification, result in acid production; in some soils, it is necessary to maintain the pH close to neutrality by liming. On the other hand, the pH may deliberately be lowered for the control of soil-borne plant pathogens. Figure 12.4 shows the pH dependence of *Streptomyces scabies* in culture solution and the corresponding occurrence of the potato scab disease in relation to soil pH (Waksman 1952). Since potatoes grow well in acidic soil, it is simple to control this plant disease by acidifying the soil. As it would be dangerous and uneconomic to apply acid directly, soil

Figure 12.4

Growth of *Streptomyces scabies* in culture solution and the occurrence of potato scab disease in the field as related to prevalent pH. Soil acidification is an effective control measure for this plant pathogen. (Source: Waksman 1952. Reprinted by permission of John Wiley and Sons.)

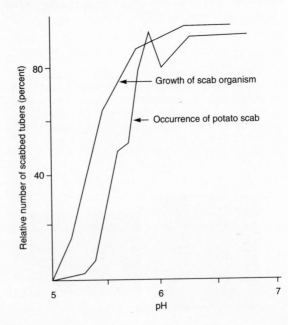

acidification is conveniently accomplished by applying powdered sulfur to the field. The activity of *Thiobacillus thiooxidans* converts sulfur to H_2SO_4, accomplishing the desired acidification.

Before fertilizer application is undertaken, careful consideration must be given to the biogeochemical cycling activities of microorganisms. Solubility and leaching characteristics of the fertilizer form used need to be considered, too. In order to avoid losses by leaching and denitrification, nitrogen fertilizer is commonly applied as an ammonium salt, free ammonia, or urea, even though plants prefer nitrogen in the form of nitrate. Nitrification that yields this form of nitrogen may be too fast in some agricultural soils, leading to nitrogen loss and groundwater contamination by nitrate. In such cases, nitrification inhibitors, such as nitrapyrin, are applied along with the nitrogen fertilizer. Nitrification inhibitors can increase crop yield 10%–15% for the same amount of nitrogen applied, while decreasing the problem of groundwater pollution by nitrate.

Leguminous crops planted in rotation with others are able, through their symbiotic nitrogen fixation, to reduce the requirement for expensive nitrogen fertilizer. Nitrogen fixation can be improved by inoculation of legume seeds with appropriate *Rhizobium* strains. In some molybdenum-deficient soils, a dramatic improvement in nitrogen fixation can also be achieved by application of small amounts of molybdenum, an elemental constituent of the nitrogenase enzyme complex.

Organic matter (humus) is an important constituent of soils. It acts as a nutrient reserve, increases ion exchange capacity, and loosens the structure of the soil. When virgin lands are put to agricultural use, their humus content decreases for the next forty to fifty years, eventually equilibrating at a much lower value. The probable causes for this phenomenon are increased aeration of the soil through tilling and removal of most of the produced organic matter with the harvest of the crop. Use of heavy farm equipment causes compaction, which reduces aeration and water infiltration in the affected soils. Fields denuded from crop cover are subject to erosion of the topsoil by wind and water, especially if the land is sloping. For the year 1977, the Soil Conservation Service estimated that U.S. farmlands

lost 3 billion metric tons of topsoil by erosion. Such losses not only endanger continued agricultural productivity but also constitute a major cause of pollution and silting in waterways and reservoirs.

A relatively novel practice that can provide a solution to the combined problems of humus loss, topsoil erosion, and soil compaction is "no-tillage farming," or "conservation till farming." In this type of soil management, ploughing is eliminated or reduced to a minimum. Crop residues are left on the field as mulch cover. Planting is done with drill seeding machines, and herbicides rather than cultivation are used to control weeds. In areas especially vulnerable to erosion, grass or other cover is planted for the winter and then killed with herbicides just before the next year's crop is planted. A five-year monitoring period demonstrated that on land with a 9% slope, conventional farming resulted in an annual topsoil loss of 1,761 kg/ha. On the same land, no-tillage farming reduced topsoil loss to 27 kg/ha. Additional benefits of no-tillage farming are moisture conservation achieved by the accumulating mulch layer, an increased soil humus content, improved soil structure, fuel conservation, and savings on heavy farm equipment. These benefits are balanced in part by an increased need for pesticides and care in management. Nevertheless, in the majority of the agricultural situations, no-tillage farming offers not only soil conservation but also tangible economic advantages in the forms of lower production costs and higher yields. About 15% of the farmland in the United States is under such management. This figure is expected to rise to 45%–65% by the year 2000. This management practice holds great promise for other parts of the world that are threatened by erosion and desertification of agricultural land (Phillips et al. 1980; Gebhardt et al. 1985).

TREATMENT OF SOLID WASTE

Urban solid waste production in the United States is estimated to amount to roughly 150 million tons per year. Part of this material is inert, composed of glass, metals, plastics, and so on, but the rest is decomposable solid organic waste, such as kitchen scraps, paper, and other garbage. Other large sources of waste are sewage sludge derived from treatment of liquid wastes and animal waste from cattle feed lots and large-scale poultry and swine farms.

In traditional small-scale farm operations, most organic solid waste was composted and recycled to the land as manure fertilizer. In societies characterized by urbanization and large-scale agriculture, the disposal of organic waste becomes a difficult and expensive problem. Following is a discussion of the options available for dealing with modern solid waste problems, with an emphasis on the disposal methods that rely on microbial activity.

Landfills

The simplest way to handle solid waste disposal at the lowest direct cost is in landfills. In this procedure, solid wastes, both organic and inorganic, are deposited in low-lying and hence low-value land. Exposed waste causes various esthetic and public health problems, attracts insects and rodents, and poses a fire hazard. The "sanitary landfill" (Fig. 12.5) is an improvement over this in which each day's waste deposit is covered with a layer of soil (U.S. Department of Health, Education and Welfare 1970). After completion of the landfill, the site becomes usable for recreation and eventually for construction. This simple and inexpensive disposal technique, however, has several disadvantages. The limited number of suitable disposal sites available in urban areas are rapidly becoming filled, necessitating longer hauling of the solid waste to more distant sites. The organic content of the landfill undergoes slow, anaerobic decomposition over a period of thirty to fifty years. During this period, the landfill slowly subsides, and methane is produced. Premature construction on the landfill site may result in structural damage to the buildings and an explosion hazard due to methane seeping into basements and cellars. Methane seepage may also damage plantings on the disposal site.

Most importantly, anaerobic decomposition products, heavy metals, and a variety of hazardous

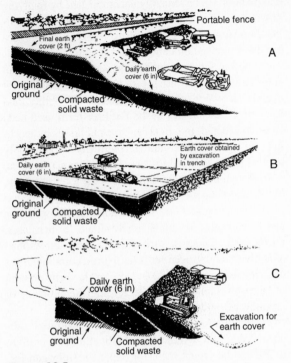

Figure 12.5

The sanitary landfill method. (A) The area method. The bulldozer spreads and compacts solid wastes. The scraper (foreground) is used to haul the cover material at the end of the day's operations. Note the portable fence that catches any blowing debris. This is used with any landfill method. (B) The trench method. The waste collection truck deposits its load into the trench where the bulldozer spreads and compacts it. At the end of the day, the dragline excavates soil from the future trench; this soil is used as the daily cover material. Trenches can also be excavated with a front-end loader, a bulldozer, or a scraper. (C) The ramp variation. Solid wastes are spread and compacted on a slope. The daily cell may be covered with earth scraped from the base of the ramp. This variation is used with either the area or the trench method. (Source: U.S. Department of Health, Education and Welfare 1970.)

pollutants may seep from the landfill site into underground aquifers, polluting much-needed urban water resources. Groundwater contamination arising from landfills, leaky gasoline and solvent tanks, and improper industrial and domestic waste disposal practices is a serious problem in the United States (Pye and Patrick 1983). Once pollutants reach an aquifer, their presence is characterized by slow movement and high persistence. Self-purification processes in underground aquifers, in contrast to those in surface waters, are minimal. Low numbers of microorganisms, oxygen and nutrient limitations, and frequently the inherently recalcitrant nature of the pollutants all contribute to the persistence of groundwater contaminants. Once polluted, an aquifer tends to stay that way for many decades. Remedial actions, such as pumping out the aquifer, followed by conventional treatment of the contaminated water or an *in situ* treatment of the aquifer with oxygenated and nutrient-supplemented water, with or without a microbial inoculum, tend to be both inadequate and prohibitively expensive. For these reasons, prevention of groundwater contamination is by far the most effective remedy. Increasingly, restrictions are being placed on the siting and operation of landfills, and alternatives to the described technique must be found.

The energy shortage of the 1970s prompted some proposals and pilot projects to recover the methane gas released from completed landfills. Such recovery requires a sizable investment in the form of a collection pipe and pumping system and for sealing of the landfill to prevent escape of the produced gas. Gas collected from landfills contains high amounts of N_2, CO_2, and H_2S impurities, along with methane, and so requires purification prior to use. Methane production starts several months to a year after construction of the landfill, then goes through a peak period of production, followed by a gradual decline over the next five to ten years. The economics of methane recovery from landfill sites are not favorable at current and projected natural gas prices.

Incineration or pyrolysis of solid waste requires a sizable capital investment in facilities and is also beset by substantial operational problems and expenses because of the necessity of controlling air pollution. Only a small part of these expenses can be recovered in terms of heat energy or fuel gas from incineration or pyrolysis, respectively. Furthermore, residual ash produced still presents a disposal problem (Pavoni et al. 1975).

Composting

Composting of organic waste appears to offer an attractive alternative. Both composting and the incineration-pyrolysis approach require some initial sorting of the solid waste into organic and inorganic portions. This can be accomplished either at the source by the separate collection of garbage (organic waste) and trash (inorganic waste) or at the receiving facility. Many communities have now started "at the source" separation programs for recyclables such as aluminum, glass, metal cans, newspaper, cardboard, and certain plastics. Alternately, waste may be sorted at the receiving facility. Ferrous metals can be removed by magnetic separators. Mechanical separators based on air flotation or on inertial energy content have been used to sort solid waste with fair success. Glass, aluminum, scrap iron, newspaper, and some plastic materials can be recycled, with some of the sorting expense recovered while the total disposal problem is reduced. The remaining waste, largely organic, can then be ground, mixed with sewage sludge and/or bulking agents such as shredded newspaper or wood chips, and composted (Harteinstein 1981).

Composting is a microbial process that converts putrefyable organic waste materials into a stable, sanitary, humuslike product that is reduced in bulk and can be used for soil improvement. Composting is accomplished in static piles (windrows), aerated piles, or continuous feed reactors. The static pile process is simple but relatively slow, typically requiring many months for stabilization. Odor and insect problems can be controlled by covering the piles with a layer of soil, finished compost, or wood chips. Unless turned several times, the finished compost is rather uneven in quality. Under favorable conditions, self-heating in static piles typically raises the temperature inside a compost pile to 55°C–60°C or above in two to three days. After a few days at peak temperature, there is a gradual temperature decline. Turning of a static compost pile at this time may cause a secondary temperature rise brought about by the replenishment of the exhausted oxygen supply. Turning also helps to make the compost more uniform, since otherwise the thermophilic processes are restricted to the core of the compost pile. Following the thermophilic phase are several months of "curing" at mesophilic temperatures. During this period, the thermophilic populations decline and are replaced by mesophiles that survived the thermophilic period. Because of the slowness of the process, large amounts of land are required, a disadvantage in densely populated urban areas.

The aerated pile process achieves substantially faster composting rates through improved aeration (Epstein et al. 1976). The so-called Beltsville process involves suction of air through perforated pipes buried inside the compost pile. This design achieves at least partial oxygenation of the pile, but temperature control is inadequate. Inside the pile, temperatures rise to self-limiting levels of 70°C–80°C. The improved Rutgers process (Finstein et al. 1983) reverses the airflow from suction to injection. Thermostats placed inside the pile control blower operation, starting when the tem-perature exceeds 60°C. The injection of air not only oxygenates the pile but cools it sufficiently to avoid a self-limiting rise in the temperature. The heat generated by the biodegradation process is effectively used in evaporating water and results in a dryer and more sta-ble compost. The aerated pile process goes to comple-tion in about three weeks. Wood chips, if used as bulking agents, are removed from the final product by screening and are reused.

Figure 12.6 shows a continuous feed compost-ing reactor. This process is even faster and forms a more uniform product, but it also requires a high ini-tial investment. Composting in the reactor is accom-plished in two to four days. A part or all of the reactor is maintained at thermophilic temperatures, using the heat produced in the composting process. After pro-cessing in the reactor, the product requires "curing" for about a month prior to packaging and shipment (Pavoni et al. 1975).

Figure 12.6

Continuous process of composting. (Source: Pavoni et al. 1975. Reprinted by permission, copyright Van Nostrand Reinhold Co.)

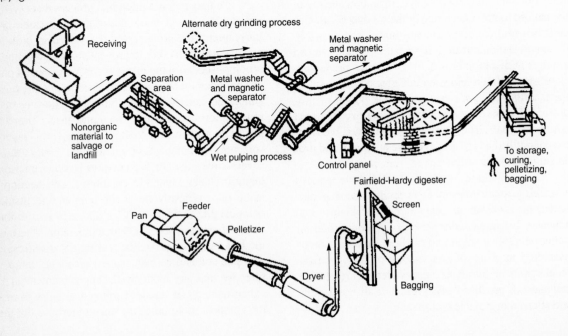

Regardless of the process design, conducting the composting process in the thermophilic temperature range is desirable, because this speeds the process and destroys pathogens that may be present in fecal matter and in sewage sludge. The aerobic oxidation reactions catalyzed by microorganisms produce heat. Under appropriate circumstances, which include sufficient mass and low heat conductivity of the composted material, this self-heating may raise temperatures inside a compost pile to 76°C–78°C. Temperatures this high are actually inhibitory to biodegradation. Maximal thermophilic activity occurs between 52°C and 63°C. Aeration or turning may be adjusted to prevent excessive self-heating.

The composting process is initiated by mesophilic heterotrophs. As the temperature rises, these are replaced by thermophilic forms. Thermophilic bacteria prominent in the composting process are *Bacillus stearothermophilus, Thermomonospora, Thermoactinomyces,* and *Clostridium thermocellum.* Important fungi in the thermophilic composting process are *Geotrichum candidum, Aspergillus fumigatus, Mucor pusillus, Chaetomium thermophile, Thermoascus auranticus,* and *Torula thermophila* (Finstein and Morris 1975).

For optimal composting, several conditions are critical. Adequate moisture (50%–60% water content) must be present, but excess moisture (70% or above) should be avoided, since it interferes with aeration and lowers self-heating because of its large heat capacity. Carbon-to-nitrogen ratios should not be greater than 40:1. Lower nitrogen content does not permit the formation of a sufficient microbial biomass. Excessive nitrogen (C:N = 25:1 or narrower) leads to volatilization of ammonia, causing odor problems, and lowers the fertilizer value of the resulting compost.

The economic viability of composting is often mistakenly judged only by the sale value of the generated compost. The primary purpose of composting, however, is to dispose of noxious wastes in an environmentally acceptable manner. Any costs recovered from the sale of compost reduce the cost of waste disposal operation but should not be expected to render the waste disposal operation self-supporting. Landfill operations may have lower direct costs than composting, but the long-range environmental costs in terms of groundwater contamination favor the composting process. Because of their water pollution potential, raw or anaerobically digested sewage sludges and other unstable organic wastes are acceptable by federal and state regulations only in a few designated landfill operations. In addition, compost is useful in soil improvement in certain situations.

Compost is a good soil conditioner and supplies some plant nutrients but cannot compete with synthetic fertilizers in agricultural production. If sewage sludge is a major component of the compost mixture, the finished compost may contain relatively high concentrations of potentially toxic heavy metals, such as cadmium, chromium, and thallium. Little is known about the behavior of these metals in agricultural soils, and because of the possibility that some of these heavy metals may contaminate agricultural products, the large-scale use of sewage sludge or of sewage-sludge-derived compost in agriculture needs careful monitoring. Only low-metal sludges may be used in this manner, and only for limited periods of time. Compost finds unrestricted application in parks and gardens for ornamental plants and in land reclamation, such as for strip-mining reclamation and highway beautification projects. A relatively new advancement in biological solid waste treatment developed in Belgium is anaerobic composting (Baere et al. 1986; Wilde and Baere 1989). The DRANCO (DRy ANaerobic COnversion) process is carried out at thermophilic temperatures (50°C–55°C) in specially constructed airtight reactors. The reactor is charged from the top with organic waste (35%–40% solids), is preheated to 50°C by steam, and moves passively through the vertical cylinder of the reactor with a residence time of eighteen to twenty-one days. Gastight loading and extraction systems prevent loss of biogas, and insulation prevents excessive heat loss from the reactor. About 30% of the generated biogas (163 m^3 per metric ton of waste) is used in running the process; 70% is used for generating electricity. The discharged and stabilized waste is dewatered on a filter press and dried further using the waste heat of the generator. The stable peat-moss-like end product (70% solids, 300–350 kg per ton of wet waste) is marketed under the trade name Humotex for gardening purposes. The thermophilic nature of the process assures destruction of pathogens, and the production of electricity and Humotex reduces operating costs, making this a promising system for composting the organic portion of urban solid waste. Compared to alternative disposal methods, composting has considerable environmental advantages. It is likely to be practiced in the future on an increased scale.

TREATMENT OF LIQUID WASTES

Liquid wastes are produced by everyday human activities (domestic sewage) and by various agricultural and industrial operations. Following natural drainage patterns or sewers, liquid waste discharges enter natural bodies of surface water, such as rivers, lakes, and oceans. At much slower rates, they may also percolate to the groundwater table, especially if it is high or if fissures are present in the unsaturated (vadose) soil layer. People use these same water bodies in alternative ways: as sources of drinking, household, industrial, and irrigation water; for fish and shellfish production; and for swimming and other recreational activities. Therefore, it is crucial to maintain the quality of these natural waters to the best of our ability. Water quality is a broad concept. Its maintenance means that natural waters cannot be overloaded with organic or inorganic nutrients or with toxic, noxious, or esthetically unacceptable substances. They should not become vehicles of disease transmission from fecal contamination, nor should their oxygenation, temperature, salinity, turbidity, or pH be altered significantly.

Natural waters have an inherent self-purification capacity. Organic nutrients are utilized and mineralized by heterotrophic aquatic microorganisms. Ammonia is nitrified and, along with other inorganic nutrients, utilized and immobilized by algae and higher aquatic plants. Allochthonous populations of enteric and other pathogens are reduced in numbers and eventually eliminated by competition and predation pressures exerted by the autochthonous aquatic populations. Thus, natural waters can accept sufficiently low amounts of raw sewage without significant deterioration of water quality. Dense human populations, community living patterns, and large-scale agricultural and industrial activities typically produce liquid wastes on a scale that overwhelms the homeostasis of aquatic communities and causes unacceptable deterioration of water quality (Dart and Stretton 1977; LaRiviere 1977).

Historically, the first human concern with water quality was prompted by destructive epidemics of cholera, typhoid fever, dysentery, and other diseases spread by waterborne pathogens. Waste treatment, protection of drinking water sources, and disinfection of drinking water and sewage, gradually introduced in the early years of this century, largely eliminated the spread of waterborne epidemics in developed countries; concern remained, however, for the health, environmental, and esthetic problems caused by an overload of organic nutrients in sewage-polluted waters. Even as treatment methods were developed that alleviated most of these problems, total sewage volume increased further, and the inorganic nutrient content of sewage started to command additional attention because of "cultural eutrophication" problems caused by the excessive mineral nutrient enrichment of natural waters (National Academy of Sciences 1969; Keeney et al. 1971). Human-made synthetic organic molecules, not readily subject to microbial degradation during sewage treatment, are the subject of additional contemporary concerns.

Biological Oxygen Demand

In contemporary liquid waste treatment, the usual order of operations is the reduction of the biological oxygen demand (BOD) associated with suspended and dissolved organics, occasionally followed by the removal of inorganic nutrients and recalcitrant organics prior to the discharge of the effluent (Hawkes 1963; Mitchell 1974). Biological oxygen demand is a measure of oxygen consumption required by the microbial oxidation of readily degradable organics and ammonia in sewage. The five-day BOD expresses the oxygen consumed by a properly diluted sewage sample during five days of incubation at 20°C. The sewage sample is diluted in oxygen-saturated water and enclosed without air space in a BOD bottle. After incubation, the decrease in dissolved oxygen is measured, usually by an oxygen electrode. Since some organics are not readily oxidized by microorganisms, and since microbial biomass is also formed, BOD is lower than the chemical oxygen demand (COD) measured using strong chemical oxidants.

Because its solubility is low, dissolved oxygen in natural waters seldom exceeds 8 mg/L, and because of heterotrophic microbial activity, it is often considerably lower than that. Replenishment of dissolved oxygen from the atmosphere (reaeration) and/or by photosynthetic O_2 evolution can be considerably slower than the utilization rate of oxygen by heterotrophic microorganisms in the presence of abundant organic substrates. Consequently, the principal impact of sewage overload on natural waters is the lowering or exhaustion of their dissolved oxygen (DO) content. Once oxygen is exhausted, self-purification processes slow drastically. Fermentation products and the reduction of secondary electron sinks, such as nitrate and sulfate, give rise to noxious odors, tastes, and colors; the water becomes anaerobic or septic. Oxygen deprivation kills obligately aerobic organisms, including some microbial forms, fishes, and invertebrate animals. The decomposition of these dead organisms constitutes an additional oxygen demand. Turbidity and toxic metabolic products, such as H_2S, also interfere with photosynthetic oxygen regeneration, thus further slowing the recovery process.

Input of a single dose of sewage into a lentic (stagnant, nonflowing) environment causes a septic period of oxygen depletion with an associated reduction in biological diversity (Hynes 1960). After a prolonged septic period, oxygen diffusion eventually reaerates the system and allows the mineralization of the accumulated fermentation products. The mineral nutrients liberated from the degraded organic matter may then give rise to an algal bloom. If there is no further disturbance, secondary succession will eventually restore the aquatic community to its former state. When a lotic (flowing river) environment receives a steady input of raw sewage, the previously described events may be observed as steady state conditions at various distances downstream from the sewage outfall. Figure 12.7 illustrates the relative

changes in some environmental parameters and populations in a river receiving sewage. Depending on the rate of sewage discharge, flow rate, water temperature, and other environmental factors, water quality may return close to the original state at distances from a few up to several dozen kilometers downstream from the sewage outfall (Hynes 1960; Warren 1971).

Figure 12.7

Diagrammatic presentation of the effects of an organic effluent on a river and the changes as one passes downstream from the outfall. (A and B) Physical and chemical changes. (C) Changes in microorganisms. (D) Changes in invertebrate animals. Sewage fungus is predominantly *Sphaerotilus natans*. (Source: Warren 1971. Based on Hynes 1960. Reprinted by permission of W. B. Saunders Co.)

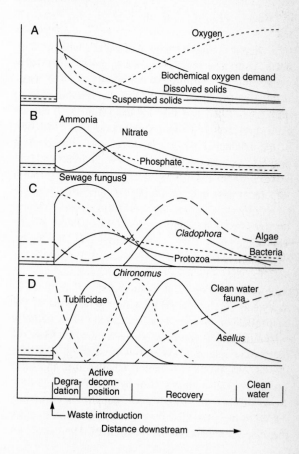

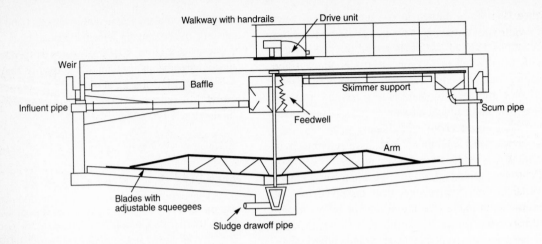

Figure 12.8
Typical clarifier cross section of a modern settling tank used for primary treatment of sewage. Grit and large debris are screened out from the sewage before it enters this settling tank. (Source: Schroeder 1977. Reprinted by permission, copyright McGraw-Hill.)

Septic conditions in natural waters, whether they occur temporarily or as a steady state, are clearly undesirable. To avoid these conditions, sewage treatment aims to reduce the BOD of the sewage prior to discharge. Typically, this is achieved in three stages, each stage bringing about a more complete reduction of the original BOD. These stages are referred to as primary, secondary, and tertiary sewage treatment.

Sewage enters the treatment plant through screens, traps, and skimming devices that remove larger objects, grit, floating scum, and grease. Primary sewage treatment removes only suspended solids. This removal is achieved in settling tanks or basins (Figure 12.8) where the solids are drawn off from the bottom. They may be subjected to anaerobic digestion and/or composting prior to final disposal in landfills or as soil conditioner. Only a low percentage of the suspended or dissolved organic material is actually mineralized during liquid waste treatment. Most of the organic material is removed by settling; in a way, the disposal problem is merely displaced to the solid waste area rather than solved by complete recycling. Nevertheless, the fragility of aquatic ecosystems, owing to their low dissolved oxygen content, renders this displacement essential (Warren 1971).

The liquid portion of the sewage containing dissolved organic matter is subjected to further treatment or is discharged after primary treatment only. Liquid wastes vary in composition, and if they contain mainly solids and little dissolved organic matter, primary treatment may remove 70%–80% of the BOD and may be considered adequate. For typical domestic sewage (Table 12.1), however, primary treatment removes only 30%–40% of the BOD, and secondary treatment is necessary for acceptable BOD reduction (Schroeder 1977).

In secondary sewage treatment, a smaller portion of the dissolved organic matter is mineralized and a larger portion is converted from a dissolved state to removable solids. The combination of primary and secondary treatments reduces the original BOD of the sewage by 80%–90%. The secondary sewage treatment step relies on microbial activity. The treatment may be aerobic or anaerobic and may be conducted in a large variety of devices. In some of these treatments, the microorganisms are associated with surface films; in others, they are homogeneously suspended. A correctly designed and operated secondary treatment unit should produce effluents with BOD and/or suspended solids less than 20 mg/L.

Table 12.1

Characteristics of typical municipal wastewater

	Concentration, mg/L
Solids, total	700
Dissolved	500
Fixed	300
Volatile	200
Suspended	200
Fixed	50
Volatile	150
Ultimate biochemical oxygen demand (BOD)	300
Total organic carbon (TOC)	200
Chemical oxygen demand (COD)	400
Nitrogen (as N)	40
Organic	15
Free ammonia	25
Nitrites	0
Nitrates	0
Phosphorus (as P)	10
Organic	3
Inorganic	7
Grease	100

Source: Schroeder 1977.

Aerobic Treatments

A simple and relatively inexpensive film-flow-type aerobic sewage treatment installation is the trickling filter (Figure 12.9). The sewage is distributed by a boom-type sprinkler revolving over a bed of porous material. It slowly percolates through this porous bed, and the effluent is collected at the bottom. Dense, slimy bacterial growth coats the porous material of the filter bed. *Zooglea ramigera* and similar bacteria play a principal role in generating the slime matrix (Figure 12.10), which accommodates a heterogeneous microbial community including bacteria, fungi, protozoa, nematodes, and rotifers (Mack et al. 1975; Hawkes 1977). This microbial community absorbs and mineralizes the dis-

Figure 12.9

Trickling filter for the secondary treatment of sewage. Enlarged section shows construction details. (Source: Warren 1971. Reprinted by permission, copyright W. B. Saunders Co.)

solved organic nutrients in the sewage, thus reducing the BOD of the effluent (Fig. 12.11). Aeration is provided passively by the porous nature of the bed. Insects, mainly fly larvae, consume the excess microbial biomass generated but constitute a nuisance to

Figure 12.10

Zooglea ramigera, a sewage bacterium that produces large amounts of extracellular polysaccharide. It plays an important role in surface slime and floc formation during sewage treatment. (Courtesy of P. Dugan, Ohio State University.)

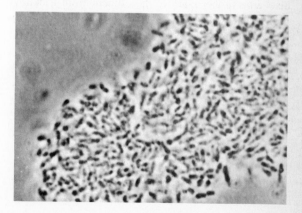

Figure 12.11

Main pathways of energy and materials transfer during treatment of dissolved organic wastes by means of trickling filters. Respiration at each level leads to mineralization of organic material and to release of heat energy. (Source: Hawkes 1963. Reprinted by permission of Pergamon Press Ltd.)

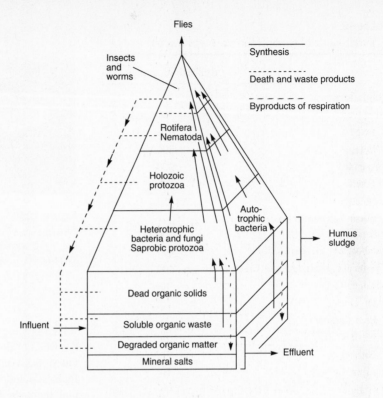

nearby residential areas. They can be controlled by the continuous rather than intermittent operation of the trickling filters, because continuous spreading suppresses the fly larvae, which successfully feed on the exposed microbial biomass only when the sprinkling is shut off.

Allowing sloughed-off biomass to settle prior to discharge further clarifies the effluent from trickling filters. The sewage may be passed through two or more trickling filters or recirculated several times through the same filter. A drawback of this otherwise simple and inexpensive treatment system is that a nutrient overload may lead to excess microbial slime, reducing aeration and percolation rates and necessitating a renewal of the trickling filter bed. Cold winter temperatures strongly reduce the effectiveness of these outdoor treatment facilities.

A more advanced aerobic film-flow-type treatment system is the so-called rotating biological contactor or biodisc system. Closely spaced discs, usually manufactured from plastic material (Figure 12.12), are rotated in a trough containing the sewage effluent. The partially submerged discs become coated with a microbial slime similar to that described in the case of trickling filters. The continuous rotation keeps the slime well aerated and in contact with the sewage. The thickness of the microbial slime layer in all film-flow processes is governed by the diffusion of nutrients through the film. When the film grows to a thickness that prevents nutrients from reaching the innermost microbial cells, these will die, autolyze, and cause the detachment of the slime layer. The sloughed-off microbial biomass can easily be removed by settling (Howell 1977; Schroeder 1977). Biodisc systems require less space than trickling filters, are more efficient and stable in operation, and produce no aerosols, but they require a higher initial investment. They have been used successfully in treatment of both domestic and industrial sewage effluents.

Figure 12.12

Two views of a rotating biological contactor, or "biodisc unit." (Source: Schroeder 1977. Reprinted by permission, copyright McGraw-Hill.)

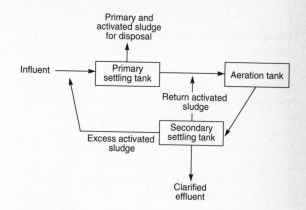

Figure 12.13

Flow diagram of an activated sludge secondary sewage treatment system. A portion of the sludge is recycled as inoculum for the incoming sewage. (Source: Imhoff and Fair 1956. Reprinted by permission of John Wiley and Sons.)

rial populations. Gram-negative rods predominate, with coliforms (*Escherichia*), *Enterobacter,* pseudomonads, *Achromobacter, Flavobacterium,* and *Zooglea* being most frequently isolated. *Micrococcus, Arthrobacter,* various coryneforms, and mycobacteria also occur,

Figure 12.14

Aeration basin of an activated sludge sewage treatment plant. (Source: Casida 1968. Reprinted by permission, copyright John Wiley and Sons.)

A popular aerobic suspension type of liquid waste treatment system is the activated sludge process (Figures 12.13 and 12.14). After primary settling, the sewage, containing dissolved organic compounds, is introduced into an aeration tank. Aeration is provided by air injection and/or mechanical stirring. Microbial activity is maintained at high levels by reintroduction of most of the settled activated sludge, hence the name of the process. During the holding period in the aeration tank, vigorous development of heterotrophic microorganisms takes place.

The heterogeneous nature of the substrate allows the development of diverse heterotrophic bacte-

along with *Sphaerotilus* and other large filamentous bacteria. Filamentous fungi and yeasts normally occur in low numbers and play a subordinate role in the activated sludge process. The protozoa are mainly represented by the ciliates. Along with rotifers, these protozoa are important predators of bacteria. The bacteria occur individually in free suspension and also aggregate as floc. The floc consists predominantly of microbial biomass cemented by bacterial slimes, such as produced by *Zooglea ramigera* and similar organisms. Most of the ciliate protozoa, such as *Vorticella*, are of the attached filter-feeding type; they adhere to the floc but feed predominantly on the bacteria in suspension. The floc is too large to be ingested by the ciliates and rotifers and thus may be considered a defense mechanism against predation. In the raw sewage, suspended bacteria predominate, but during the holding time in the aeration tank, the numbers of suspended bacteria decrease. At the same time, those associated with floc greatly increase in numbers (Casida 1968; Pike and Curds 1971). The diversity and density of bacteria in activated sewage sludge renders this material a popular inoculum for various enrichment cultures. Table 12.2 shows some typical total and viable bacterial counts at various stages of sewage treatment.

During the holding period in the aeration tank, a portion of the dissolved organic substrates is mineralized. Another portion is converted to microbial biomass. In the advanced stage of aeration, most of the microbial biomass is associated with floc that can be removed from suspension by settling. The settling characteristics of the sewage sludge floc are critical for their efficient removal. Poor settling characteristics are associated with the "bulking" of the sewage sludge, a problem caused by proliferation of filamentous bacteria, such as *Sphaerotilus, Beggiatoa, Thiothrix,* and *Bacillus,* and with filamentous fungi, such as *Geotrichum, Cephalosporium, Cladosporium,* and *Penicillium.* The causes for bulking are not always understood, but this condition is frequently associated with high C:N and C:P ratios and/or low dissolved oxygen concentrations.

A portion of the settled sewage sludge is recycled for inoculation of the incoming raw sewage; the excess sludge requires either incineration, additional treatment by anaerobic digestion and/or composting, or disposal in landfills. A past practice of ocean dumping is no longer considered acceptable. Combined with primary settling, the activated sludge process tends to reduce the BOD of the effluent to 5%–15% of the raw sewage.

Table 12.2

Numbers of total and viable bacteria in samples from different stages of sewage treatment and in the suspended biomass

	Bacterial count				
	In samples (number/mL)		In biomass (number/g)		Percent of bacteria viable
Stage of treatment	Total	Viable	Total	Viable	
Settled sewage	6.8×10^8	1.4×10^7	3.2×10^{12}	6.6×10^{10}	2.0
Activated sludge mixed liquor	6.6×10^9	5.6×10^7	1.4×10^{12}	1.2×10^{10}	0.85
Filter slimes	6.2×10^{10}	1.5×10^9	1.3×10^{12}	3.2×10^{10}	2.5
Secondary effluents	5.2×10^7	5.7×10^5	4.3×10^{12}	4.7×10^{10}	1.1
Tertiary effluents	3.4×10^7	4.1×10^4	3.4×10^{12}	4.1×10^9	0.12

Source: Pike and Curds 1971.

The treatment also drastically reduces the numbers of intestinal pathogens in the sewage, even prior to final disinfection by chlorination. The combined effects of competition, adsorption, predation, and settling accomplish this reduction. Predation by ciliates, rotifers, and *Bdellovibrio* is probably indiscriminate and affects pathogens as well as nonpathogenic heterotrophs. Pathogens, however, tend to grow poorly or not at all under the conditions prevailing in the aeration tank, while the nonpathogenic heterotrophs proliferate vigorously. Thus, nonpathogenic heterotrophs compensate for their removal by predation, while the pathogens are decimated continuously. Settling of the floc removes additional pathogens. Typically, numbers of *Salmonella, Shigella,* and *Escherichia coli* are 90%–99% lower in the effluent of the activated sludge treatment process than in the incoming raw sewage. Enteroviruses are removed to a similar degree; here the main removal mechanism appears to be adsorption of the virus particles to the settling sewage sludge floc.

The activated sludge process is essentially a continuous culture process, suitable for mathematical and computer modeling (Pike and Curds 1971). In a steady-state condition, the growth of the sludge bacteria (those associated with floc) must be equal to the sludge wastage (those removed). In Figure 12.15, the steady-state populations of microorganisms are modeled at various dilution and sludge wastage rates. Model A assumes the presence of attached ciliates, a condition that most closely resembles the real-life situation. Simulation B assumes only free-swimming ciliates, and simulation C assumes the absence of ciliates. The role of the attached ciliates in controlling the dispersed bacteria is clearly evident. Thus, the presence of the attached ciliates is critical for removal of the portion of the BOD consisting of dispersed bacteria that would be otherwise discharged with the effluent. In fact, the types and numbers of protozoa associated with the activated sludge floc can be used as an indicator of sludge condition and, thus, treatment performance. The sludge is in poor condition when few ciliates and many flagellates are observed; ciliates predominate in "good" sludge. The role of the ciliates was also experimentally demonstrated in bench-scale treatment units inoculated with controlled bacterial and protozoan populations, as Table 12.3 shows.

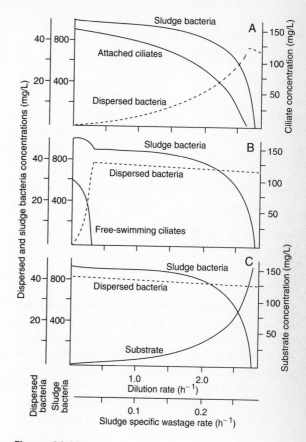

Figure 12.15

Steady-state populations of microorganisms at various dilution and sludge wastage rates by computer modeling. (A) Model assuming the presence of attached ciliates. (B) Model assuming the presence of free-swimming ciliates. (C) Model assuming the absence of ciliates. Sludge wastage rate is kept at 1/10 the dilution rate. The substrate concentration curve shown in C is identical for A and B, but is omitted from these plots. The simulations clearly show the strong effect of the attached ciliates on the dispersed bacteria. The reduction of the latter helps to achieve a low BOD discharge. (Source: Pike and Curds 1971. Reprinted by permission, copyright Academic Press, London.)

The activated sludge treatment system is efficient and flexible and is able to withstand considerable variations in sewage flow rate and concentration.

Table 12.3

Effect of ciliated protozoa on the effluent quality of bench-scale, activated sludge plants

Effluent analysis	Without ciliates	With ciliates
BOD (mg/L)	53–70	7–24
COD (mg/L)	198–250	124–142
Permanganate value (mg/L)	83–106	52–70
Organic nitrogen (mg/L)	14–21	7–10
Suspended solids (mg/L)	86–118	26–34
Optical density at 620 nm	0.95–1.42	0.23–0.34
Viable bacteria count (10^6/mL)	106–160	1–9

Source: Pike and Curds 1971.

It is widely used for the treatment of domestic sewage and industrial effluents. Figures 12.16 and 12.17 show some interesting variations of the process developed for treatment of industrial wastes. Both modifications are designed to improve the oxygen transfer rate and thus achieve a higher BOD reduction rate per unit volume of treatment system. The UNOX process (Figure 12.16), developed by Union Carbide, uses oxygen instead of air for aeration. Closed tanks and a specially designed stirring system prevent the wastage of oxygen. The deep shaft process of ICI (Figure 12.17) uses air injection and achieves high oxygen dissolution by increasing hydrostatic pressure through an ingenious circulation pattern (Howell 1977).

Oxidation ditches and lagoons are low-cost substitutes for the previously described treatment systems. They tend to be inefficient and require large holding capacities and long retention times. As oxygenation is usually achieved by diffusion and by the photosynthetic activity of algae, these systems need to be shallow. Oxygenation is usually incomplete, with consequent odor problems. Performance is strongly influenced by seasonal temperature fluctuations, and usefulness, therefore, is largely restricted to warmer climates.

Most frequently, treated sewage is discharged into surface waters, but in some arid areas, sufficient surface water flow may not exist. In addition, it may be necessary to regenerate the groundwater removed from wells. In such areas, partially treated sewage is used for groundwater recharge. Typically, the sewage is subjected to primary settling and is subsequently channeled through a series of holding ponds. The first holding pond tends to reaerate the oxygen-depleted water. After settling of most of the algal and bacterial

Figure 12.16

The "UNOX" wastewater treatment system, developed by Union Carbide. This system achieves higher biological oxidation rates per volume using oxygen instead of air in its aeration tank. Higher oxygen partial pressures result in more rapid oxygen transfer rates. (Source: Howell 1977. Reprinted by permission, copyright The Institute of Petroleum, London.)

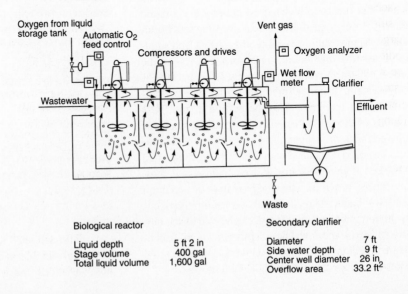

Biological reactor	
Liquid depth	5 ft 2 in
Stage volume	400 gal
Total liquid volume	1,600 gal

Secondary clarifier	
Diameter	7 ft
Side water depth	9 ft
Center well diameter	26 in
Overflow area	33.2 ft^2

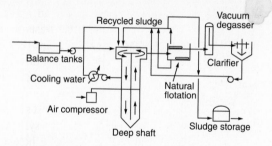

Figure 12.17

The ICI deep shaft process for wastewater treatment. This process achieves high oxygen partial pressures and transfer rates through increased hydrostatic pressure. The shaft is 20–30 m deep and is separated by a central divider. Circulation is initially established by air injection into the rising half of the shaft. After circulation is established, air injection is switched to the downcomer side, as shown in the drawing. Air bubbles are retained for a long time period, and the increasing hydrostatic pressure facilitates O_2 dissolution. (Source: Howell 1977. Reprinted by permission, copyright The Institute of Petroleum, London.)

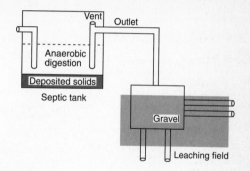

Figure 12.18

Diagram of a septic tank. (Source: Mitchell 1974. Reprinted by permission, copyright Prentice-Hall.)

biomass, the water is transferred to large, shallow infiltration ponds. From these infiltration ponds, the water flows through natural sand and soil layers and slowly returns to the underground aquifer. Clogging of the infiltration ponds by undegraded microbial polysaccharides and by accumulation of reduced ferrous sulfide is a recurrent problem in groundwater recharge operations. The problem is countered by periodic rest periods for individual infiltration basins. These interruptions allow degradation of the excess polysaccharides and reaerate the sediment of the infiltration ponds with the concurrent oxidation of the clogging ferrous sulfide.

Anaerobic Treatments

Anaerobic wastewater treatment methods are generally slower but save the energy necessary for forced aeration. Some anaerobic treatment systems can also salvage a part of the chemical energy content of the wastewater by generating biogas, a useful fuel (Cillie

et al. 1969; Toerien and Hatting 1969; Zinder 1984). The simplest anaerobic treatment system is the septic tank (Figure 12.18), popular in rural areas without sewer systems. The organics in the wastewater undergo limited anaerobic digestion. Residual solids settle to the bottom of the tank. The clarified effluent is distributed over a leaching field, where dissolved organics undergo oxidative biodegradation. Septic tank treatment does not reliably destroy intestinal pathogens, and it is important that the leaching field not be in the proximity of water wells. The Imhoff tank (Figure 12.19) has an improved design that maintains anaerobic conditions more strictly, produces some utilizable biogas, and facilitates the settling of solids but requires expert maintenance (Imhoff and Fair 1956; Mitchell 1974).

Large-scale anaerobic digestors (Figure 12.20) are used for further processing of sewage sludge produced by primary and secondary treatments. While anaerobic digestors can and have been used directly for the treatment of sewage, the economics of the situation favor aerobic processes for such relatively dilute wastes. Large-scale anaerobic digestors, therefore, are used primarily for processing of settled sewage sludge and for treatment of some very high BOD industrial effluents.

Conventional anaerobic digestors are large fermentation tanks designed for continuous operation under anaerobic conditions. Provisions for mechani-

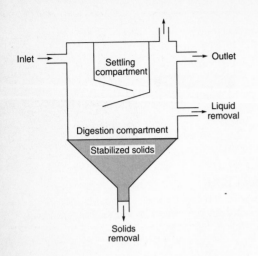

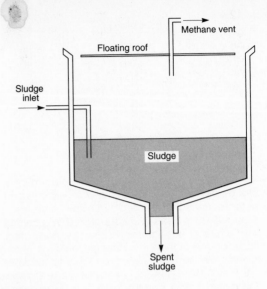

Figure 12.19

Diagram of an Imhoff tank. A separate settling compartment is baffled in a way so as to prevent gas bubbles from carrying solids back up into the tank. The flow passes along this baffled compartment perpendicularly to the plane of the drawing. Stabilized solids are removed from the bottom of the tank. Gases are vented and may be utilized as fuel. (Source: Mitchell 1974. Reprinted by permission, copyright Prentice-Hall.)

Figure 12.20

Greatly simplified diagram of an anaerobic digestor. (Source: Mitchell 1974. Reprinted by permission, copyright Prentice-Hall.)

cal mixing, heating, gas collection, sludge addition, and draw off of stabilized sludge are incorporated. The digestor contains high amounts of suspended organic matter; values between 20 and 100 g/L are considered favorable. A considerable part of this suspended material is bacterial biomass, viable counts of which can be as high as 10^9–10^{10} bacteria per ml. Anaerobic counts are typically two to three orders of magnitude higher than aerobic ones. Fungi and protozoa are present in low numbers only and are not considered to play a significant role in anaerobic digestion.

The anaerobic digestion of wastes can be considered as a two-step process, even though it really is a coupled sequence of microbiological interactions. First, complex organic materials, including microbial biomass, are depolymerized and converted to fatty acids, CO_2, and H_2. A large variety of nonmethanogenic obligately or facultatively anaerobic bacteria par-

ticipate in these processes. In the next step, methane is generated, either by the direct reduction of methyl groups to methane or by the reduction of CO_2 to methane by molecular hydrogen or by other reduced fermentation products, such as fatty acid, methanol, or even carbon monoxide (Balch et al. 1979). Table 12.4 lists some of the methanogenic bacteria and the mechanisms of their methane production.

The operation of anaerobic digestors requires the close control of several parameters, such as retention time, temperature, pH, and C:N and C:P ratios. The reactors are usually heated to 35°C–37°C for optimal performance. Control is necessary for pH to stay within the 6.0–8.0 values, with pH 7.0 being optimal. Extremes of pH, and the influx of heavy metals solvents or other toxic materials, can easily upset the operation of the anaerobic digestor. In a "stuck" or "sour" digestor, methane production is interrupted and fatty acids and other fermentation products accumulate. Restoring a stuck digestor to

Table 12.4

The origin of biogenic methane and some representative methanogenic bacteria

Source of CH_4–carbon	Electron donor	Methanogenic bacteria
CO_2	H_2	*Methanobacterium bryantii*
CO_2	H_2	*Methanobacterium thermoautotrophicum*
CO_2	H_2	*Methanomicrobium mobile*
CO_2	H_2, formate	*Methanococcus vannielii*
CO_2	H_2, formate	*Methanobrevibacter ruminantium*
CO_2	H_2, CO, formate	*Methanobacterium formicicum*
CO_2, methanol, methylamine, di- and tri- methylamine, acetate	H_2	*Methanosarcina barkeri*

normal operation is difficult. The reactor must usually be cleaned and charged with large volumes of anaerobic sludge or appropriate methanogenic bacteria from an operational unit.

A number of advanced process designs have been developed recently for anaerobic sewage treatment (Switzenbaum 1983). The relatively high cost of anaerobic digestors has previously limited them to processing high BOD liquids, primarily sewage sludge. Some advanced designs make anaerobic digestors more efficient and economical in handling more dilute wastes, such as primary-treated sewage. The anaerobic activated sludge process closely parallels its aerobic counterpart: Settled biomass is continuously recycled to maximize the rate of digestion per reactor volume. Other designs involve sludge beds, anaerobic filters, or anaerobic expanded fluidized beds. Each design attempts to assure that the dilute sewage passes through a dense stationary anaerobic biomass. This passage minimizes reactor volume and fluid retention time while maximizing the efficiency of the digestion process. Bacteria involved in the anaerobic digestion process, like their aerobic counterparts, tend to form granules, or aggregates. Complex metabolic interactions, including interspecies electron transfer, take place between fermenters, acetogens, sulfate reducers, and methanogens within these granules (Thiele and Zeikus 1988; Wu et al. 1991). The intimate contact between the bacteria within the anaerobic granules appears to be essential for efficient anaerobic digestion. We discussed the nature of the interactions in more detail in chapter 3.

A normally operating anaerobic digestor yields a reduced volume of sludge. This material still causes odor and water pollution problems and can be directly disposed of only at a limited number of landfill sites. Aerobic composting further consolidates the sludge and renders it suitable for disposal in any landfill site or for use as a soil conditioner. The biogas produced consists of methane, CO_2, small amounts of other gases, such as N_2 and H_2S, and, occasionally, traces of H_2. This gas either is burned directly in the treatment plant to drive the pumps and maintain the temperature of the digestor at the desired level or, after purification, may be added to municipal gas distribution systems (Finstein and Morris 1979).

Anaerobic digestors are relatively expensive to construct and require expert maintenance, though the production of methane gas offsets some of the cost. The use of anaerobic lagoons as an inexpensive alternative is seldom feasible because of the noxious odors associated with such an operation. A special case in which the anaerobic lagoon performs in a relatively acceptable fashion is in treatment of meat-packing-plant effluents. Fatty materials and animal stomach contents form a thick, insulating scum layer on the surface of such anaerobic lagoons that substantially reduces the odor problem and renders such facilities acceptable if they are a sufficient distance from residential areas.

Tertiary Treatments

The previously discussed aerobic and anaerobic biological liquid waste treatment processes are designed to reduce the BOD represented by dissolved, biodegradable organic substrates; all are classified as secondary liquid waste treatment. Tertiary treatment is any practice beyond secondary. Tertiary liquid waste treatments are aimed at the removal of nonbiodegradable organic pollutants and mineral nutrients, especially nitrogen and phosphorus salts. The removal of nonbiodegradable organic pollutants, such as chlorophenols, polychlorinated biphenyls, and other synthetic pollutants (see chapter 13) is necessary because of the potential toxicity of these compounds. Activated carbon filters are used to remove such materials from secondary-treated industrial effluents. Previous secondary treatment is necessary in order to avoid overloading this expensive treatment stage with biodegradable materials that can be removed in more economic ways.

Concern about nitrogen salts and phosphate in sewage effluents is relatively recent and is connected with the phenomenon of cultural eutrophication of natural waters (Keeney et al. 1971; Godzing 1972). Eutrophication, the process of becoming rich in nutrients, is a natural sequence in the geological history of lakes and ponds. Initially, deep, cold, nutrient-poor (oligotrophic) waters become shallower, warmer, and more eutrophic as silt and organic matter accumulates. Eventually, the lake is filled in. Cultural eutrophication, frequently referred to only as eutrophication with the same intended meaning, is the result of human activities and causes dramatic changes in lake character within years or decades instead of geological ages. Sudden nutrient enrichment through sewage discharge or agricultural runoff triggers explosive algal blooms. Owing to various known or unknown causes, such as a period of cloudy weather that intensifies mutual shading, exhaustion of micronutrients, toxic products, or disease, the algal population eventually crashes. The decomposition of the dead algal biomass by heterotrophic microorganisms exhausts the dissolved oxygen in the water, precipitating extensive fish kills and septic conditions. Even if eutrophication does not go to this extreme, algal mats, turbidity, discoloration, and shifts of the fish population from valuable species to more tolerant but less valued forms represent undesirable eutrophication changes.

Eutrophication is a complex phenomenon that can be understood in the context of Liebig's Law of the Minimum. Many natural oligotrophic lakes are limited by phosphorus or by nitrogen, but other nutrient limitation possibilities, such as dissolved CO_2 or sulfur, may also exist. If the sewage effluent supplies large amounts of the inorganic nutrient that happens to be limiting in the receiving water body, a eutrophication response is likely to occur. If this is the case and the sewage discharge cannot be stopped or diverted, the obvious remedy is to remove the causative inorganic nutrients from the sewage discharge.

The phosphate filler of many detergent formulations, which is added to precipitate bivalent ions such as Ca^{2+} and Mg^{2+} in hard water, is most frequently blamed in eutrophication cases. In Lake Washington, located adjacent to Seattle and surrounding communities, the pattern of eutrophication and the improvement following sewage diversion from Lake Washington to Puget Sound correlated best with phosphate concentrations (Edmondson 1970). Work on Wisconsin lakes indicated that phosphate concentrations exceeding $10\mu g$ P/L at the beginning of the growing season are likely to result in destructive algal blooms. In these and many other cases, there is little doubt that phosphate has been the main offender, but in the case of nitrogen-limited waters, a treatment to remove phosphate alone is clearly useless. In practice, if eutrophication of the receiving waters is a concern, an effort is made to remove both nitrogen and phosphorus from the effluent (Godzing 1972).

Phosphate is commonly removed by precipitation as calcium or iron phosphate. This can be accomplished as an integral part of primary or secondary settling or in a separate facility with recycling of the precipitating agent. Nitrogen, present mainly as ammonia, can be removed by "stripping," that is, by volatilization as NH_3 at a high pH. Some of the ammonia eliminated from the sewage in this manner, however, may return to the watershed in the form of precipitation and still cause eutrophication problems.

Another alternative for ammonia nitrogen removal is breakpoint chlorination. Addition of hypochlorous acid in 1:1 molar ratio results in the formation of monochloramine as in Equation 1:

$$(1) \quad HOCl + NH_3 \longrightarrow NH_2Cl + H_2O$$

Further addition of hypochlorous acid results in the formation of dichloramine (Equation 2):

$$(2) \quad HOCl + NH_2Cl \longrightarrow NHCl_2 + H_2O$$

Breakpoint chlorination occurs at the approximate molar ratio of two moles of chlorine to one mole of ammonia (7.6 mg Cl to 1 mg ammonium-N) and results in the near-quantitative conversion of ammonia to molecular nitrogen according to the simplified reaction in Equation 3:

$$(3) \quad 2NHCl_2 + H_2O \longrightarrow N_2 + HOCl + 3H^+ + 3Cl^-$$

As chlorination of the sewage effluent is commonly practiced for disinfection purposes, chlorination to this "breakpoint" can accomplish the removal of ammonium nitrogen in the same process. The removal of ammonium nitrogen also lowers the BOD of the effluent, since the nitrification process would consume oxygen dissolved in the receiving water.

Biological processes for removal of nitrogen and, to a lesser degree, phosphorus have been proposed, and some of these are gaining acceptance. In-plant nitrification can solve the problem of BOD associated with ammonium nitrogen. Vigorous and prolonged aeration is necessary to convert ammonium nitrogen to nitrate by biological nitrification. After conversion to nitrate, it is also possible to remove most of the nitrogen from effluent by denitrification. To achieve this result, aeration is discontinued and organic matter added. The anaerobic oxidation of organic matter converts the nitrate to gaseous denitrification products (see chapter 11). Methanol has been used as an additional carbon source, but cost considerations favor the use of the dissolved carbon substrates of the raw sewage itself to drive the denitrification process. This usage can be accomplished by intermittent aeration or by an admixture of raw sewage to extensively aerated and thus nitrified sewage.

Bacterial uptake and storage of phosphorus in the form of polyphosphates may be encouraged by treatment process modifications (Toerien et al. 1990). Enhanced phosphate removal from sludge appears to be associated with polyphosphate (poly-P) storage by the *Acinetobacter-Moraxella-Mima* group of bacteria. Under anoxic conditions, these microorganisms incorporate large amounts of fatty acids and store them in the form of poly-β-hydroxybutyrate (PHB). Phosphate is released into soluble form during the anoxic period. When the sewage is subsequently re-aerated, the poly-P bacteria rapidly oxidize their intracellular PHB reserves while engaging in "luxury" phosphate uptake, far in excess of their growth requirements. The incorporated phosphate is stored as energy-rich poly-phosphate within the cell, thus removing phosphorus from the liquid effluent.

Several biological phosphorus removal systems have been designed and incorporated into the activated sludge treatment system. All involve one or more alternate anoxic-oxic sewage storage cycles. It is important to have little or no nitrate present during the anoxic phase. Nitrate is used as alternate electron sink by the poly-P bacteria, and its presence interferes with PHB storage that, in turn, primes the poly-P bacteria for their phosphate uptake binge. Biological phosphate removal saves cost and avoids the impact of precipitating chemicals.

Other schemes of biological N and P removal involve the uptake of nitrogen and phosphorus into algal biomass in shallow holding ponds. Problems with harvesting and utilizing the algal biomass in animal feed, for example, need to be solved before this approach can be practiced on a large scale. Higher aquatic plants, such as water hyacinths and cattails, also have potential for the removal of nitrogen and phosphorus from sewage effluents and might prove to be easier to harvest if appropriate uses for such biomass could be found. One should be aware, however, of the practical difficulty of devoting large land areas to such operations near urban centers.

A highly advanced tertiary treatment system is currently operating in South Tahoe, California; the description of this system (Figure 12.21) serves well to illustrate the integration of various previously men-

Figure 12.21

(A) Diagram of the advanced tertiary waste-water treatment system in South Tahoe, California. Phosphate, nitrogen, and recalcitrant organic compounds are removed. (B) Photograph of the South Tahoe sewage treatment system. (A. Source: Slechta and Culp 1967. Reprinted by permission of Water Pollution Control Federation. B. Source: Warren 1971. Reprinted by permission of W. B. Saunders Co.)

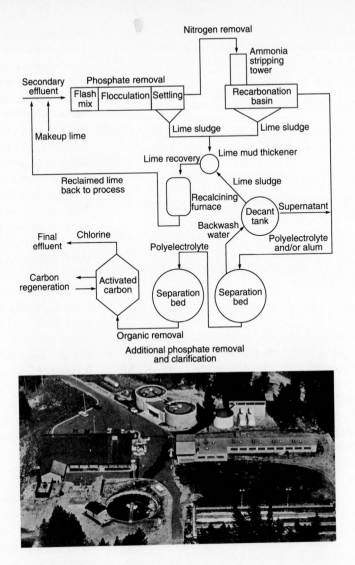

tioned approaches (Slechta and Culp 1967). This advanced treatment system became necessary to prevent the eutrophic deterioration of scenic Lake Tahoe, an oligotrophic crater lake.

After conventional primary and secondary treatments, phosphate is precipitated by liming, and ammonia is removed by stripping the high pH effluent. This step is accomplished in a stripping tower at elevated temperature and with vigorous aeration. After ammonia stripping, the pH is adjusted to neutrality. Following additional settling and clarification, aided by polyelectrolyte addition, nonbiodegradable organics are removed by filtration through activated carbon. The effluent is disinfected by chlorination. The lime sludge from the phosphate precipitation step is recycled by conversion to CaO in a recalcinating furnace. The quality of the treated effluent is high (Table 12.5).

Table 12.5

Quality of purified reclaimed water after tertiary treatment at South Tahoe, California

Parameter	Wastewater (mg/L)	Reclaimed water (mg/L)
BOD	200–400	1
COD	400–600	3.25
Total organic carbon	0	1–7.5
Phosphate	25–30	0.2–1.0
Organic nitrogen	10–15	0.3–2.0
Ammonia	25–35	0.3–1.5
Nitrate and nitrite	0	0

Source: Mitchell 1974.

Disinfection

The final step in sewage treatment is disinfection, designed to kill enteropathogenic bacteria or viruses that were not eliminated during the previous steps of sewage treatment. Disinfection is commonly accomplished by chlorination, using either chlorine gas (Cl_2) or hypochlorite ($Ca[OCl]_2$ or $NaOCl$). Chlorine gas reacts with water to yield hypochlorous and hydrochloric acids (Equation 4):

$$(4) \quad Cl_2 + H_2O \longrightarrow HOCl + HCl$$

Hypochlorite also reacts with water to form hypochlorous acid (Equation 5):

$$(5) \quad Ca(OCl)_2 + 2H_2O \longrightarrow 2HOCl + Ca(OH)_2$$

or (Equation 6):

$$(6) \quad NaOCl + H_2O \longrightarrow HOCl + NaOH$$

The hypochlorous acid or the hypochlorite (ClO^-) ion is the actual disinfectant. The total concentration of hypochlorous acid and hypochlorite is designated as free residual chlorine. Hypochlorite is a strong oxidant which is the basis of its antibacterial action. As an oxidant, it also reacts with residual dissolved or suspended organic matter, ammonia, reduced iron, manganese, and sulfur compounds. The oxidation of such compounds competes for HOCl and reduces its disinfectant action, making necessary chlorination to a point that these reactions are satisfied with several mg/L excess free residual chlorine remaining in solution. This process requires high amounts of hypochlorite, resulting in high salt concentrations in the effluent. Therefore, it is desirable to remove nitrogen and other contaminants by alternate means and use chlorination for disinfection only.

A disadvantage of disinfection by chlorination is that more resistant types of organic molecules, including some lipids and hydrocarbons, are not oxidized completely but become, instead, partially chlorinated. Chlorinated hydrocarbons tend to be toxic and are difficult to mineralize (see chapter 13). Since alternative means of disinfection are more expensive, however, chlorination remains the principal means of sewage disinfection.

TREATMENT AND SAFETY OF WATER SUPPLIES

Closely related to the safe disposal of liquid wastes is the problem of safe and healthful water for drinking and other uses. Increasing population densities and urbanization during the eighteenth and nineteenth centuries initially were not accompanied by adequate sanitation practices. This situation created conditions for devastating epidemics caused by enteropathogenic microorganisms such as *Vibrio cholerae* (cholera), *Salmonella typhi* (typhoid fever), various other *Salmonella* and *Shigella* strains (gastrointestinal infections of varying severity), and *Entamoeba histolytica* (amoebic dysentery). The common feature of these diseases is that the infectious organisms are shed in the feces of sick individuals and clinically asymptomatic carriers. Fecal contamination, through untreated or inadequately treated sewage effluents entering lakes, rivers, or groundwaters that in turn serve as municipal water supplies, creates conditions for rapid dissemination of the pathogens. The primary route of infection is ingestion of drinking water, but fruits, vegetables, and eating utensils washed with contaminated water are other possible carriers. The obvious remedy to this situation is to disrupt the transmission of enteropathogenic organisms through fecal contamination of water supplies. This is done to a certain degree during sewage treatment, but the adequate treatment of all sewage discharges is by no means assured. A major sanitation effort, therefore, is necessary to treat and safely distribute public water supplies. Such sanitation practices have led to the virtual elimination of waterborne infections in developed countries, but these infections continue to be major causes of sickness and death in undeveloped regions (Sobsey and Olson 1983).

Public water supplies originate from ground water through wells or from surface waters such as rivers, lakes, or reservoirs. In the case of groundwater, disinfection is often all that is required. Surface waters and some groundwaters have to be treated by settling and filtration. Dissolved ferrous iron and divalent manganese are oxidized prior to settling by addition of permanganate. Filtration is carried out through gravel and fine sand beds. Microbial films on the surface of the bed particles enhance the effectiveness of the filtration. The filter beds have to be backwashed frequently to prevent clogging.

Disinfection of municipal water supplies has traditionally been accomplished by chlorination, as described for the disinfection of sewage discharges (Equations 4-6). This treatment is relatively inexpensive, and the free residual chlorine content of the treated water is a built-in safety factor against pathogens surviving the actual treatment period and against recontamination. The disadvantage of chlorination is the creation of trace amounts of trihalomethane (THM) compounds. The fact that THMs were formed in virtually all municipal water supplies that used chlorination for disinfection linked the formation of these contaminants to the chlorination process. Since some of the THMs are suspected carcinogens, the U.S. Environmental Protection Agency (EPA) established in 1979 a maximal THM limit in drinking water of 100 µg/L. To keep the levels within this limit using traditional chlorine disinfection, organic compounds were removed from the water meticulously by sand filtration and other techniques. This method is often both impractical and too expensive. Fortunately, disinfection by monochloramine is effective but produces much lower amounts of THMs. As an example, traditional chlorination of Ohio River water produced 160 µg THM/L, while chloramine treatment produced THM levels consistently below 20 µg/L. Monochloramine may be generated right in the water to be disinfected by adding ammonium prior to or simultaneously with chlorine or hypochlorite. Once hypochlorous acid is formed according to Equations 4–6, it reacts with the added ammonia according to Equation 1, generating monochloramine. The practice of using chloramines as drinking water disinfectants, the least expensive way to reduce THM formation, is spreading rapidly (Wolfe et al. 1984).

Ozone (O_3) has been used for water disinfection with good results both in Europe and in the United States. Ozone treatment kills pathogens reliably and does not result in the synthesis of undesirable THM contaminants. Since ozone is an unstable gas, however, water treated with ozone does not have any residual antimicrobial activity and is prone to chance recontamination. Ozone has to be generated from air on site in ozone reactors, using electrical corona discharge. Since only about 10% of the electricity actually generates ozone and the rest is lost as heat, disinfection by ozone is considerably more expensive than by chlorination (Mitchell 1974).

Water Quality Testing

The public health importance of clean drinking water requires an objective test methodology to evaluate the effectiveness of treatment procedures and to establish drinking water safety standards (Bonde 1977; Hoadley and Putka 1977). The detection of actual enteropathogens such as *Salmonella* or *Shigella* in routine monitoring studies would be a difficult and uncertain undertaking. Instead, the bacteriological tests of drinking water simply establish the fact and the degree of fecal contamination of the water sample by demonstrating the presence of so-called indicator organisms. The most frequently used indicator organism for fecal contamination is the normally nonpathogenic *Escherichia coli*. Positive tests for *E. coli* do not prove the presence of enteropathogenic organisms but establish the possibility of such a presence. Because *E. coli* is plentiful and easier to grow than the enteropathogens, the test has a built-in safety factor for detecting potentially dangerous fecal contamination.

Ideally, an indicator organism should be present whenever the pathogens concerned are present, should be present only when there is a real danger of pathogens being present, must occur in greater numbers than the pathogens to provide a safety margin, must survive in the environment as long as the potential pathogens, and must be easy to detect with a high reliability of correctly identifying the indicator organism regardless of what other organisms are present in the sample. *Escherichia coli* meets many of these criteria, but there are limitations to its use as an indicator organism, and various other species have been proposed as additional or replacement indicators of water safety. There probably is no universal indicator organism for determining water quality. Under different conditions, different populations may be better indicators than others (Wolf 1972).

Prior to 1989, the EPA had certified only two techniques for the detection of coliform bacteria in water: the multiple tube fermentation technique and the membrane filtration test. These procedures take several days to complete. In 1991, the EPA eliminated the requirement for the enumeration of coliform bacteria in water samples, instituting regulations based only on the presence or absence of coliform bacteria. This was done in response to studies that demonstrated that the level of coliform bacteria was not quantitatively related to the potential for an outbreak of waterborne disease, while the presence-absence of coliform bacteria provided adequate water quality information. Culture-requiring methods for the detection of coliform bacteria, however, are limited in their ability to detect viable but nonculturable bacteria. In oligotrophic situations, like drinking water distribution systems, a proportion of the total population of coliform bacteria may be unrecoverable while in the pseudosenescent state associated with bacteria adapted to low-nutrient situations.

The classic test for the detection of fecal contamination involves, in the first step, inoculation of lactose broth with undiluted or appropriately diluted water samples. Gas formation detected in small inverted test tubes (Durham tubes) gives positive presumptive evidence of contamination by fecal coliforms; this is the presumptive test. Gas formation in lactose broth at elevated incubation temperature is a characteristic of fecal *Salmonella, Shigella,* and *E. coli* strains but also of the nonfecal coliform *Enterobacter aerogenes*. In this second stage, the confirmed test, positive lactose broth cultures are streaked out on a lactose-peptone agar medium containing sodium sulfite and basic fuchsin (m-Endo agar). Fecal coliform colonies on this medium acquire a characteristically greenish metallic sheen, while *Enterobacter* forms reddish colonies and nonlactose fermenters form colorless colonies. Subculture in lactose broth at 35°C should yield gas formation and completes a positive test for fecal coliforms.

It is possible to simplify this three-stage test. In the Eijkman test, suitable dilutions are incubated in lactose broth at 44.5°C. At this temperature, fecal coliforms still grow, but nonfecal coliforms are inhibited. Gas formation in lactose broth at 44.5°C constitutes a one-step positive test, but precise temperature control is mandatory, since temperatures only a few degrees higher inhibit or kill fecal coliforms. Another technique is filtration of known volumes of diluted or undiluted water samples through 0.45 μm pore-size Millipore filters and incubation of these filters directly on Endo medium. Colonies of fecal coliforms show up with a characteristic metallic sheen and can be counted (Figure 12.22). The dried filters can be filed for a permanent record.

Several alternate procedures are able to provide the information necessary for determining drinking water quality in the presence-absence test format. These tests can be completed in twenty-four hours or less. One test protocol uses defined substrate technology to determine enzymatic activities that are diagnostic of coliform bacteria and the fecal coliform *E. coli*. A medium containing isopropyl β-D-thiogalactopyranoside (IPTG) is used to detect β-galactosidase, which is diagnostic of total coliform bacteria, and a medium containing *o*-nitrophenol-β-D-galactopyranoside-4-methylumbelliferyl-β-D-glucuronide (MUG) is

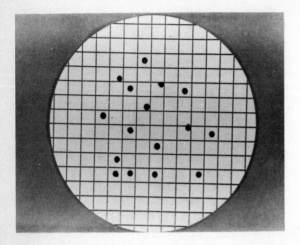

Figure 12.22
Filter used in filtration of fecally contaminated water showing typical colonies of *Escherichia coli* after incubation on Endo-agar.

used to detect β-glucuronidase, which is diagnostic of *E. coli*. The defined substrate test requires only a day to complete (Edberg et al. 1988, 1989a, 1989b). Gene probes and other molecular approaches may also be used to detect coliform bacteria and *E. coli*. A. K. Bej et al. (1990, 1991a, 1991b) developed a method using the polymerase chain reaction (PCR) to detect total coliform bacteria based upon the *lac*Z gene and *E. coli* based upon the *uid*A gene; these genes code for the enzymes detected in the defined substrate tests.

The drinking water standard does not absolutely exclude the possibility of ingesting enteropathogens but seems to reduce this possibility to a statistically tolerable limit. Enteropathogens are likely to be present in much lower numbers than fecal coliforms, and a few infective bacteria are usually unable to overcome body defenses. A minimum infectious dose from several hundred to several thousand bacteria is necessary for various diseases to establish an actual infection. Drinking water supplies meeting the 1 per 100 mL coliform standard have never been demonstrated to be the source of waterborne bacterial infections (Olson and Nagy 1984).

Fecal coliform counts are also used to establish the safety of water for shellfish harvesting and recreational uses. Shellfish tend to concentrate bacteria and other particles through their filter-feeding activity. Some shellfish are consumed raw and thus can transmit infection by waterborne pathogens. This is reflected by the relatively stringent standard (70/100 mL) for waters in use for shellfishing. Clinical evidence for infection by enteropathogenic coliforms through recreational use of waters for wading or swimming is unconvincing, but as a precaution, beaches are usually closed when fecal coliform counts exceed the recreational standard of 1,000 per 100 mL. Some regional standards require that disinfected sewage discharges not exceed this limit.

We discussed the fate of nonpathogenic and pathogenic fecal coliforms in natural waters earlier in the context of allochthonous organisms and competitive interactions. In general, fecal coliforms are poor competitors at the low substrate concentrations that prevail in natural waters and so tend to be eliminated by competition and predation. Low water temperatures, sediment adsorption, and anoxic conditions occasionally contribute to the prolonged survival of fecal coliforms in natural waters.

Enteroviruses and the cysts of some pathogenic protozoa are somewhat more resistant to disinfection by chlorine chloramine or ozone than bacteria, and occasionally active virus particles or live cysts are recovered from water treated to meet fecal coliform standards (Sobsey and Olson 1983; Olson and Nagy 1984). Thus, water treated according to accepted standards may occasionally become a source of viral or protozoan infection. As many as one hundred different viral types have been demonstrated to be shed in human feces, but practical concern has been mainly with the causative viruses of infectious hepatitis, polio, and viral gastroenteritis. There is evidence that infectious hepatitis was, on occasion, spread by water supplies, though a much more prevalent way of infection is the consumption of raw shellfish from fecally contaminated waters. Spread of polio infection through water supplies and/or recreational use of beaches has been suspected in some cases, but conclusive proof is lacking (Shuval and Katznelson 1972;

Vaughan et al. 1979). The situation with viral gastro-enteritis is similar. It seems fair to say that while the possibility of the occasional spread of viral infection through drinking water adequately treated by bacteriological standards cannot be excluded, there is no supporting evidence for actual epidemics caused by such water. Similarly, it is unclear whether properly disinfected water has ever spread protozoan pathogens such as the flagellate *Giardia lamblia* or the amoeba *Negleria fowleri* (Sobsey and Olson 1983). There is ample evidence for destructive epidemics by enteroviruses and protozoa caused by untreated drinking water in various underdeveloped countries.

The considerable uncertainty about the hazards posed by enteroviruses in water supplies and recreational waters is due to the difficulty of detecting and enumerating viruses in such waters. For detection, viruses have to be grown in suitable tissue culture cell lines and/or detected by immunological procedures. Virus concentrations are generally low, 0.1–1 particle per ml treated

sewage effluent and as low as 0.1–1 per L in receiving natural waters. Therefore, the first problem is to concentrate the viruses sufficiently for detection and enumeration. Table 12.6 lists and compares some of these methods.

The oldest procedure is the gauze pad method. Gauze pads are suspended in water for about twenty-four hours. They concentrate viruses by absorption. Water expressed from the retrieved gauze pad contains higher numbers of viruses than directly collected water samples. Sample incorporation relies on concentrated media that allow the introduction of relatively large (10–50 ml volume) water samples.

Ultracentrifugation concentrates viruses by sedimentation. It requires expensive instrumentation and is time-consuming. Filters do not actually filter out viruses but retain them by adsorption. Subsequently, viruses are eluted from the filter, or the filter is dissolved. Hydro-extraction is actually a dialysis process. Water is extracted through the dialysis tubing by

Table 12.6

A comparison of various methods for detecting viruses in water

Method	Sample volume (liters)	Concentration factor	Recovery efficiency (%)
Gauze pad		10–50	0.5–5
Sample incorporation	0.06–0.15	10–20	95
Ultracentrifugation	1–10	30–50	20–100
Membrane filter adsorption	10–1,000	10^3	5–38
Soluble ultrafilters	10	10^3	35–100
Hydro extraction	1	10^2	24–64
Phase separation (two steps)	5	10^3	35–100
Electrophoresis	0.3	10^2	100
Adsorption to particulate materials			
calcium phosphate	3.0	10^3	50–100
cobalt chloride	1	10^3	100
aluminum hydroxide	3.8–19	10^3	50–100
ion exchange resins	0.1	50–100	20–50
passive hemagglutination			80–100
insoluble polyelectrolytes	1,000	10^3–10^5	35–80

Source: Shuval and Katznelson 1972.

concentrated polyethylene glycol. Virus particles remain with other large molecules in the tubing and are concentrated. Phase separation involves partitioning between two liquid phases. Electrophoresis takes advantage of the negative charge on virus particles. Adsorption to and elution from various particulate materials has also been used with varying success.

Considerable research effort is currently directed at elucidating the duration and mechanisms of enterovirus survival in natural waters. Generalizations are difficult at this time, but enteroviruses appear to be eliminated from natural waters more slowly than fecal coliforms. Adsorption to sediment particles appears to offer considerable protection against inactivation and, under some circumstances, may promote dispersion of the viruses (Smith et al. 1978). Changing environmental conditions, such as pH salinity, may cause the desorption of the still-infective viral particles. Recent surveys sporadically identified human enteroviruses in both freshwater and saltwater environments, but the numbers of enteroviruses showed little correlation to the number of fecal coliforms. Whether this is due to differential survival or to nonidentical contamination sources is not clear at this time. Virus recovery, cultivation, and enumeration from natural waters clearly needs further scientific progress before standards and routine monitoring processes can be established.

SUMMARY

Practical applications of microbial ecology are evident in the practices of biodeterioration control and of soil, water, and waste management. Sterilization and aseptic containment prevent spoilage of food and perishable materials. Alternatively, unfavorable physicochemical conditions are used to retard biodeterioration. Agricultural soils are managed to encourage aerobic and discourage anaerobic cycling processes. In addition, fertilizer technology and soil pH control take into account and often utilize microbial cycling activities. The composting of solid waste and the biological treatment of sewage are complex microbial processes that recycle and/or stabilize organic wastes and protect the environment. Disinfective treatment of municipal water supplies guards against the spread of waterborne epidemics. The effectiveness of such disinfective treatment is monitored through counts of indicator organisms.

REFERENCES & SUGGESTED READINGS

Baere, L. D., O. Verdonck, and W. Verstraete. 1986. High rate dry anaerobic composting process for the organic fraction of solid wastes. *Proceedings, Biotechnology and Bioengineering Symposium No. 15.* Wiley, New York, pp. 321–330.

Balch, W. E., G. E. Fox, L. J. Magrum, C. R. Woese, and R. S. Wolfe. 1979. Methanogens: Reevaluation of a unique biological group. *Bacteriological Review* 43:260–296.

Bej, A. K., R. J. Steffan, J. L. DiCesare, L. Haff, and R. M. Atlas. 1990. Detection of coliform bacteria in water by using polymerase chain reaction and gene probes. *Applied and Environmental Microbiology* 56:307–314.

Bej, A. K., J. L. DiCesare, L. Haff, and R. M. Atlas. 1991a. Detection of *Escherichia coli* and *Shigella* spp. in water by using polymerase chain reaction (PCR) and gene probes for *uid*. *Applied and Environmental Microbiology* 57:1013–1017.

Bej, A. K., M. H. Mahbubani, J. L. DiCesare, and R. M. Atlas. 1991b. Polymerase chain reaction-gene probe detection of microorganisms by using filter-concentrated samples. *Applied and Environmental Microbiology* 57:3529–3534.

Bonde, G. J. 1977. Bacterial indicators of water pollution. *Advances in Aquatic Microbiology* 1:273–364.

Casida, L. E., Jr. 1968. *Industrial Microbiology.* John Wiley and Sons, New York.

Chet, I., and R. Mitchell. 1976. Control of marine fouling by chemical repellents. In J. M. Sharpley and A. M. Kaplan (eds.). *Proceedings of the Third International Bio-degradation Symposium.* Applied Science Publishers, Essex, England, pp. 515–521.

Cillie, G. G., M. R. Henzen, G. J. Stander, and R. D. Baillie. 1969. The application of the process in waste purification. Part IV, Anaerobic digestion. *Water Research* 3:623–643.

Corpe, W. A., L. Matsuuchi, and B. Armbruster. 1976. Secretion of adhesive polymers and attachment of marine bacteria to surfaces. In J. M. Sharpley and A. M. Kaplan (eds.). *Proceedings of the Third International Biodegradation Symposium.* Applied Science Publishers, Essex, England, pp. 433–442.

Dart, R. K., and R. J. Stretton. 1977. *Microbiological Aspects of Pollution Control.* Elsevier Publishing Company, Amsterdam, The Netherlands.

Desrosier, N. W. 1970. *The Technology of Food Preservation.* 3d ed. The AVI Publishing Co., Westport, Conn.

Di Salvo, L. H., and G. W. Daniels. 1975. Observations on estuarine microfouling using the scanning electron microscope. *Microbial Ecology* 2:234–240.

Edberg, S. C., M. J. Allen, D. B. Smith, and the National Collaborative Study. 1988. Enumeration of total coliforms and *Escherichia coli* from drinking water: Comparison with the standard multiple tube fermentation method. *Applied and Environmental Microbiology* 54:1595–1601.

Edberg, S. C., M. J. Allen, D. B. Smith, and the National Collaborative Study. 1989a. Defined substrate method for the simultaneous detection of total coliforms and *Escherichia coli* from drinking water: Comparison with presence- absence techniques. *Applied and Environmental Microbiology* 55:1003–1008.

Edberg, S. C., M. J. Allen, D. B. Smith, and N. J. Kriz. 1989b. Enumeration of total coliforms and *Escherichia coli* from source water by the defined substrate technology. *Applied and Environmental Microbiology* 56:366–368.

Edmondson, W. T. 1970. Phosphorus, nitrogen and algae in Lake Washington after diversion of sewage. *Science* l69:690–691.

Epstein, E., G. B. Wilson, W. D. Burge, D. C. Mullen, and N. K. Enkiri. 1976. A forced aeration system for composting wastewater sludge. *Journal (Water Pollution Control Federation)* 48:655–694.

Finstein, M. S., and M. L. Morris. 1975. Microbiology of municipal solid waste composting. *Advances in Applied Microbiology* 19:113–151.

Finstein, M. S., and M. L. Morris. 1979. Anaerobic digestion and composting: Microbiological alternatives for sewage sludge treatment. *ASM News* 45:43–48.

Finstein, M. S., F. C. Miller, P. F. Strom, S. T. MacGregor, and K. M. Psarianos. 1983. Composting ecosystem management for waste treatment. *Biological Technology* 1:347–353.

Flint, K. P. 1982. Microbial ecology of domestic wastes. In R. G. Burns and J. H. Slater (eds.). *Experimental Microbial Ecology.* Blackwell Scientific Publications, Oxford, England. pp. 575–590.

Frazier, W. D., and D. C. Westhoff. 1978. *Food Microbiology.* McGraw-Hill Book Co., New York.

Gebhardt, M. R., T. C. Daniel, E. E. Schweizer, and R. R. Allmaras. 1985. Conservation tillage. *Science* 230:625–630.

Godzing, T. J. 1972. The role of nitrogen in eutrophication processes. In R. Mitchell (ed.). *Water Pollution Microbiology.* Wiley-Interscience, New York, pp. 43–68.

Harteinstein, R. 1981. Sludge decomposition and stabilization. *Science* 212:743–749.

Hawkes, H. A. 1963. *The Ecology of Waste Water Treatment.* Pergamon Press, Oxford, England.

Hawkes, H. A. 1977. The ecology of activated sludge. In K. W. A. Chater and H. J. Somerville (eds.). *The Oil Industry and Microbial Ecosystems.* Institute of Petroleum, London, pp. 217–233.

Hoadley, A. W., and B. J. Putka (eds.). 1977. *Bacterial Indicators/Health Hazards Associated with Water.* ASTM Technical Publication 635. American Society for Testing Materials, Philadelphia.

Howell, J. A. 1977. Alternative approaches to activated sludge and trickling filters. In K. W. A. Chater and J. H. Somerville (eds.). *The Oil Industry and Microbial Ecosystems.* Institute of Petroleum, London, pp. 199–216.

Hurst, A., and D. L. Collins-Thompson. 1979. Food as bacterial habitat. *Advances in Microbial Ecology* 3:79–134.

Hynes, H. B. N. 1960. *The Biology of Polluted Waters.* Liverpool University Press, Liverpool, England.

Imhoff, K., and G. M. Fair. 1956. *Sewage Treatment.* 2d ed. John Wiley and Sons, New York.

Keeney, D. R., R. A. Herbert, and A. J. Holding. 1971. Microbiological aspects of the pollution of fresh waters with inorganic nutrients. In G. Sykes and F. A. Skinner (eds.). *Microbial Aspects of Pollution.* Academic Press, London, pp. 181–200.

LaRiviere, J. W. M. 1977. Microbial ecology of liquid waste treatment. *Advances in Microbial Ecology* 1:215–259.

Mack, W. N., J. P. Mack, and O. Ackerson. 1975. Microbial film development in a trickling filter. *Microbial Ecology* 2:215–226.

Mitchell, R. 1974. *Introduction to Environmental Microbiology*. Prentice-Hall, Englewood Cliffs, N.J.

National Academy of Sciences. 1969. *Eutrophication: Causes, Consequences, Correctives*. National Academy of Sciences, Washington, D.C.

Olson, B. H., and L. A. Nagy. 1984. Microbiology of potable water. *Advances in Applied Microbiology* 30:73–132.

Pavoni, J. L., J. E. Heer, Jr., and D. J. Hagerty. 1975. *Handbook of Solid Waste Disposal, Materials, and Energy Recovery*. Van Nostrand Reinhold Co., New York.

Phillips, R. E., R. L. Blevins, G. W. Thomas, W. W. Frye, and S. H. Phillips. 1980. No-tillage agriculture. *Science* 208:1108–1113.

Pike, E. B., and C. R. Curds. 1971. The microbial ecology of the activated sludge process. In G. Sykes and F. A. Skinner (eds.). *Microbial Aspects of Pollution*. Academic Press, London, pp. 123–147.

Pye, V. I., and R. Patrick. 1983. Ground water contamination in the United States. *Science* 221:713–718.

Schroeder, E. D. 1977. *Water and Wastewater Treatment*. McGraw-Hill, New York.

Shuval, H. J., and E. Katznelson. 1972. The detection of enteric viruses in the water environment. In R. Mitchell (ed.). *Water Pollution Microbiology*. Wiley-Interscience, New York, pp. 347–361.

Sieburth, J. M. 1975. *Microbial Seascapes*. University Park Press, Baltimore.

Slechta, A. F., and G. L. Culp. 1967. Water reclamation studies at the South Tahoe public utility district. *Journal (Water Pollution Control Federation)* 39:787–814.

Smith, E. M., C. P. Gerba, and J. L. Melnick. 1978. Role of sediment in persistence of enteroviruses in the estuarine environment. *Applied and Environmental Microbiology* 35:685–689.

Sobsey, M. D., and B. H. Olson. 1983. Microbial agents of water-borne disease. In P. S. Berger and Y. Argaman (eds.). *Assessment of Microbiology and Turbidity Standards of Drinking Water*. U.S. Environmental Protection Agency, Washington, D.C., pp. 1–69.

Switzenbaum, M. S. 1983. Anaerobic treatment of wastewater: Recent developments. *ASM News* 49:532–536.

Thiele, J. H., and G. Zeikus. 1988. Control of interspecies electron flow during anaerobic digestion: Significance of formate transfer versus hydrogen transfer during syntrophic methanogenesis in flocs. *Applied and Environmental Microbiology* 54:20–29.

Toerien, D. F., and W. H. J. Hatting. 1969. Anaerobic digestion. Part I, The microbiology of anaerobic digestion. *Water Research* 3:385–416.

Toerien, D. F., A. Gerber, L. H. Hötter, and T. E. Cloete. 1990. Enhanced biological phosphorus removal in activated sludge systems. *Advances in Microbial Ecology* 11:173–230.

U.S. Department of Health, Education and Welfare. 1970. *Sanitary Landfill Facts*. SW 41s. Government Printing Office, Washington D.C.

Vaughan, J. M., E. F. Landry, M. Z. Thomas, F. J. Vicale, and W. F. Penello. 1979. Survey of human enterovirus occurrence in fresh and marine surface waters on Long Island. *Applied and Environmental Microbiology* 38:290–296.

Waksman, S. A. 1952. *Soil Microbiology*. John Wiley and Sons, New York.

Warren, C. E. 1971. *Biology and Water Pollution Control*. W. B. Saunders, Philadelphia.

Wilde, D. B., and D. L. Baere. 1989. Experiences on the recycling of municipal solid waste to energy and compost. Paper presented at the *Fifth International Conference on Solid Waste Management and Secondary Materials*, 5–8 December, Philadelphia.

Wolf, H. W. 1972. The coliform count as a measure of water quality. In R. Mitchell (ed.). *Water Pollution Microbiology*. Wiley-Interscience, New York, pp. 333–345.

Wolfe, R. L., N. R. Ward, and B. H. Olson. 1984. Inorganic chloramines as drinking water disinfectants: A review. *Journal of the American Water Works Association* 76:74–88.

Wu, W. M., R. F. Hickey, and J. G. Zeikus. 1991. Characterization of metabolic performance of methanogenic granules treating brewery wastewater: Role of sulfate-reducing bacteria. *Applied and Environmental Microbiology* 57:3438–3449.

Zinder, S. H. 1984. Microbiology of anaerobic conversion of organic wastes to methane: Recent developments. *ASM News* 50:294–298.

13

Microbial Interactions with Xenobiotic and Inorganic Pollutants

During the second half of the twentieth century, pollution problems have become more acute than ever before. Pollution and its control have ceased to be the concerns of only sanitary engineers; they have become slogans for mass movements and the focal topic of noisy scientific, public, and legislative debates. The current environmental predicament of our planet has two recognizable components: conventional pollution, which now occurs on a vastly increased scale; and novel pollution problems, largely unknown before World War II. The population explosion, combined with an international trend toward urbanization, has increased the intensity of conventional pollution to a level where the old sanitation principle, "the solution to pollution is dilution," is clearly an inadequate approach.

To stave off unacceptable deterioration of the environment, industrialized societies are being forced to pay heavily for the abatement of wastes and pollutants. The scientific background of these conventional pollution problems is well understood and effective control is largely a question of technology and finances. In contrast, many novel pollution problems, caused by chemicals to which the environment had not been exposed previously, have caught both scientists and lay people unprepared. Of necessity, regulatory decisions with far-reaching economic and environmental consequences are being made on the basis of limited evidence.

The majority of the novel pollution phenomena involve organic chemicals called xenobiotics, chemicals synthesized by humans that have no close natural counterparts. These xenobiotic chemicals include pesticides, plastics, and other synthetics, many of which persist because transport mechanisms and catabolic pathways for them have not evolved. Others are converted to even less desirable residues so that they cause untoward effects. Acid mine drainage, oil, and heavy-metal pollution do not involve synthetics, but human activities are responsible for introducing these materials into the ecosphere at problem levels. The principles that govern the distribution and fate of these chemical pollutants in the ecosphere show similarities regardless of their origin. Whereas the impact of conventional pollution is typically felt only within definable geographic boundaries, such as within a watershed, the pollution problems discussed in this chapter often have global implications as well.

PERSISTENCE AND BIOMAGNIFICATION OF XENOBIOTIC MOLECULES

Given favorable environmental conditions, all natural organic compounds degrade. While this hypothesis has never been put to a rigorous and exhaustive test, its truth appears self-evident from the fact that we do not observe any large-scale accumulations of natural organic substances. If any organic compound produced in the ecosphere were inherently resistant to recycling, huge deposits of this material would have accumulated throughout the geological ages. This clearly has not been the case; substantial organic deposits, such as fossil fuels, accumulate only under conditions that are adverse to biodegradation.

M. Alexander (1965) formalized this general understanding of the biodegradative capacity of microorganisms as the principle of microbial infallibility, expressing the empirical observation that no natural organic compound is totally resistant to biodegradation provided that environmental conditions are favorable. This fact should hardly be surprising. The evolution of various biopolymers was slow and gradual, allowing time for a parallel evolution of microbial catabolism appropriate for every newly available substrate. The time scale of this evolution was measured in billions of years. In contrast, the explosive development of synthetic organic chemistry during the last century has led to the large-scale production of a bewildering variety of organic compounds that, either by design or by accident, eventually wind up in the environment. Once there, their fates vary. Most synthetic compounds have natural counterparts or are sufficiently similar to naturally occurring compounds to be subject to microbial metabolism (Dagley 1975). Others are xenobiotic, having molecular structures and chemical bond sequences not recognized by existing degradative enzymes. These resist biodegradation or are metabolized incompletely, with the result that some xenobiotic compounds accumulate in the environment. Human ingenuity outpaced microbial evolution, and the fallibility of the taken-for-granted "biological incinerators" was exposed.

One can imagine many reasons for a xenobiotic organic compound proving recalcitrant (totally resistant) to biodegradation (Alexander 1981). Unusual substitutions (such as with chlorine and other halogens), unusual bonds or bond sequences (such as in tertiary and quaternary carbon atoms), highly condensed aromatic rings, and excessive molecular size (in the case of polyethylene and other plastics) are some of the common reasons for recalcitrance. Other, more subtle reasons may be the failure of the compound to induce the synthesis of degrading enzymes, even though it may be susceptible to their action. A failure of the compound to enter the microbial cell for lack of suitable permeases, unavailability of the compound due to insolubility or adsorption phenomena, and excessive toxicity of the parent compound or its metabolic products all could contribute to lack of degradation.

Unfortunately, the term degradation has been used to describe transformations of every type, including those that yield products more complex than the starting material, as well as those responsible for the complete oxidation of organic compounds to CO_2, H_2O, NO_3^-, and other inorganic components. On occasion, microbial transformations are synthetic rather than degradative. They result in residues that are more stable than the parent compound, yet this phenomenon is sometimes called degradation, since the parent compound disappears. This ambiguous terminology has proven confusing to legislators and interested lay people alike, who, for example, would hear one expert testify that the insecticide DDT is degradable, since it is converted in part to DDD, DDE, and other closely related compounds (Wedemeyer 1967) (Figure 13.1). Another expert would insist that DDT is persistent, because the basic carbon skeleton remains unaltered. To preclude misunderstanding, the term mineralization has been proposed for describing the ultimate degradation and recycling of an organic molecule to its mineral constituents.

What happens if a persistent organic compound is introduced into the ecosphere? Particle size and solubility characteristics, among other factors, will influence its mode and rate of travel, but chances are good that it will be distributed to places far removed

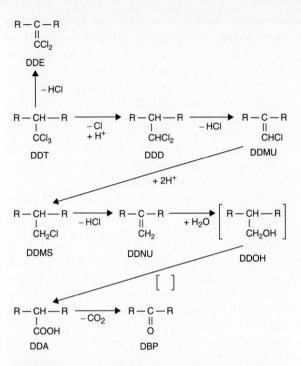

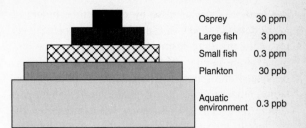

Osprey	30 ppm
Large fish	3 ppm
Small fish	0.3 ppm
Plankton	30 ppb
Aquatic environment	0.3 ppb

Figure 13.2

Biomagnification of DDT. Minute concentrations of dissolved DDT move by physical partitioning into plankton. Further concentration of the pollutant takes place through the shrinkage of biomass between successive trophic levels. DDT concentrations are idealized and, while typical, do not represent actual analytical data.

Figure 13.1

Transformations of the insecticide DDT by *Enterobacter* (*Aerobacter*) *aerogenes*. All transformations are restricted to the –CCl₃ moiety. Anaerobic conditions favor dechlorination steps. The final product DBP is still recalcitrant and biologically active. R = a *p*-chlorophenyl moiety;
DDT = 1,1,1-trichloro-2,2-*bis*(*p*-chlorophenyl) ethane;
DDE = 1,1-dichloro-2,2,-*bis*(*p*-chlorophenyl) ethylene;
DDD = 1,1-dichloro-2,2-*bis*(*p*-chlorophenyl) ethane;
DDMU = 1-chloro-2,2-*bis*(*p*-chlorophenyl) ethylene;
DDMS = 1-chloro-2,2-*bis*(*p*-chlorophenyl) ethane;
DDNU = *unsym-bis*(*p*-chlorophenyl) ethylene;
DDOH = 2,2-*bis*(*p*-chlorophenyl) ethanol;
DDA = 2,2-*bis*(*p*-chlorophenyl) acetate;
DBP = 4,4´-dichlorobenzophenone. (Source: Wedemeyer 1967. Reprinted by permission of American Society for Microbiology.)

similar; they have been detected in remote and even Arctic regions, thousands of miles removed from the nearest possible application site (Cade et al. 1971). It is important, therefore, to keep in mind that assessing the effect of a persistent organic chemical on the environment to which it is directly applied is insufficient. The chemical may be dispersed and elicit unexpected detrimental effects in a more fragile ecosystem far removed from the original application site.

The production and release of persistent organic chemicals into the environment have a relatively short history. Their distribution, however, is widespread (National Research Council 1972; Higgins and Burns 1975). Distribution tends to dilute; organo-chlorines in environments that were only indirectly exposed often are present in the low parts per billion (ppb) range. Why should such low concentrations still cause concern? They do so because of a phenomenon called biological magnification, or biomagnification for short. To be subject to this phenomenon, the pollutant must be both persistent and lipophilic (Gosset et al. 1983). Because of their lipophilic character, the minute dissolved amounts are partitioned from the surrounding water into the lipids of both prokaryotic and eukaryotic microorganisms (Figure 13.2). Concentrations in their cells, as compared to the surround-

from its original application site. The fallout from atmospheric nuclear tests has dramatized the capacity of atmospheric forces to distribute chemicals from point sources globally (Hanson 1967). The experience with chlorinated hydrocarbon insecticides has been

ing medium, may increase by one to three orders of magnitude. Members of the next higher trophic level ingest the microorganisms. Only 10%–15% of the biomass is transferred to the higher trophic level, the rest is dissipated in respiration. The persistent lipophilic pollutant, however, is neither degraded nor excreted to a significant extent, and so is preserved practically without loss in the smaller biomass of the second trophic level. Consequently, its concentration increases by almost an order of magnitude. The same thing occurs at the successively higher trophic levels. The top trophic level, composed of birds of prey, mammalian carnivores, and large predatory fish, may carry a body burden of the environmental pollutant that exceeds the environmental concentration by a factor of 10^4–10^6.

If the pollutant is a biologically active substance, such as a pesticide, at such levels it may cause death or serious debilitation of the affected organism. Chlorinated hydrocarbons, including DDT, were implicated in the death or reproductive failure of various birds of prey (Cade et al. 1971). In the case of local application, even a relatively short food chain such as the sprayed vegetation $\longrightarrow$ earthworm $\longrightarrow$ robin can be destructive to the bird. We humans derive our food from various trophic levels and are in a less exposed position than a top-level carnivore. Nevertheless, at the time of unrestricted DDT use, the average individual in the United States, with no occupational exposure, carried a body burden of 4–6 ppm DDT and its derivatives. Although this amount was not considered acutely dangerous, the trend for an increasing contamination of the higher trophic levels of the ecosphere became sufficiently clear and led to the ban on the use of DDT in the United States and most other developed countries for all but emergency situations.

Recalcitrant Halocarbons

Many xenobiotic pollutants, such as DDT, are halocarbons that have proven recalcitrant to microbial attack. The carbon-halogen bond is highly stable. Cleavage of this bond is not exothermic but rather requires a substantial energy input; it is an endothermic reaction. As a consequence, halocarbons are chemically and biologically stable. Stability is desirable in pesticides, but the same property causes persistence and is undesirable as an environmental contaminant. Limited photochemical and biochemical mechanisms in the biosphere detoxify and recycle them. Furthermore, the energetics of their degradation are such that there is little prospect for microorganisms ever to evolve a capability for utilization of extensively halogenated carbon compounds as growth substrates.

Important groups of halocarbons include halocarbon propellants, solvents and refrigerants, certain organochlorine insecticides, polychlorinated or polybrominated biphenyls and triphenyls, chlorodibenzodioxins, and chlorodibenzofurans.

Halocarbon Propellants and Solvents   Halocarbon propellants and solvents are C_1–C_2 alkanes in which all or nearly all hydrogen atoms have been replaced by fluorine-chlorine combinations. They serve as solvents and aerosol propellants in spray cans for such things as cosmetics, paint, and insecticides and as the working fluid in the condensor units of air conditioners and refrigerators. Most commonly used are CCl_3F and CCl_2F_2 freons, designated by the codes F-11 and F-12, respectively. Spilled halocarbon propellants and solvents have seriously contaminated groundwater. Being inert and quite volatile, some of these components rise to the stratosphere upon release, depleting Earth's protective ozone layer by photochemical interactions (Molina and Rowland 1974; National Research Council 1976). Partial destruction of the ozone layer would result in increased UV radiation on the surface of Earth, causing an increased incidence of skin cancer and mutagenesis. While volatile halocarbons represent only one of several ozone depletion mechanisms, legislation has been enacted in the United States and other developed countries to phase out halocarbons as aerosol propellants.   Intensive research efforts are under way to find a suitable substitute for halocarbons as refrigerants, but their use as coolants will continue for some time to come.   Some effort is being made to extract and re-use the refrigerant fluids from defunct cooling units.

Dehalogenation of organic compounds is thermodynamically favored under anaerobic conditions;

microorganisms have been found that produce deha-logenases and carry out reductive dehalogenation (Suflita et al. 1982). Sulfate-reducing bacteria trans-form tetrachloroethene to trichloroethene and *cis*-1,2-dichloroethene by anaerobic dehalogenation (Bagley and Gossett 1990). The fully chlorinated but unsatur-ated tetrachloroethylene ($Cl_2C=CCl_2$) was subject to stepwise dechlorination and partial conversion to CO_2. Tetrachloroethylene degradation has been dem-onstrated for methanogenic bacterial consortium growing on acetate in an anaerobic reactor (Vogel and McCarty 1985; Galli and McCarty 1989). Aerobi-cally, dichloromethane was shown to serve as the only carbon source for a pseudomonad, and dichloroethane served as substrate for a microbial consortia (Stucki et al. 1981). Extensive aerobic degradation of trichlo-roethylene (TCE), a widely distributed halocarbon pollutant, by a methane-utilizing microbial consor-tium has been demonstrated (Fogel et al. 1986; Little et al. 1988). The low specificity of methane monoox-ygenase allows the conversion of TCE to TCE epoxide, which subsequently spontaneously hydro-lyzes to polar (formic acid, glyoxylic acid) products utilizable by microorganisms. *Methylococcus capsula-tus* was reported to convert chloro- and bromomethane to formaldehyde, dichloromethane to CO, and trichlo-romethane to CO_2 while growing on methane (Dalton and Stirling 1982). Thus, methanotrophic bacteria show some promise for bioremediation of halocarbon-polluted aquifers. Trichloroethylene is also degraded aerobically by *Alcaligenes* via alternate pathways (Harker and Kim 1990).

Pesticides The chlorinated pesticides are rela-tively resistant to microbial attack. In general, the more extensive the chlorine substitution, the more persistent the pesticide (Ghosal et al. 1985; Kennedy et al. 1990; Chaudhry and Chapalamadugu 1991). Mirex ($C_{10}Cl_{12}$) and Kepone ($C_{10}Cl_{10}O$) are exten-sively chlorinated insecticidal compounds (Figure 13.3). Mirex has been used in the United States in the fire ant control program. Kepone was manufactured principally for export and as a precursor for other pes-ticides. Kepone came to national attention in the United States when it poisoned a substantial number

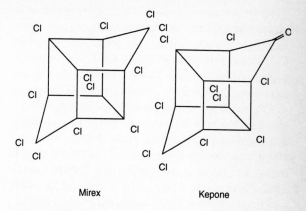

Figure 13.3

Structural formulae of the insecticides Mirex and Kepone. Extensive chlorination renders these compounds extremely resistant to biodegradation.

of workers in a manufacturing plant at Hopewell, Vir-ginia. The subsequent investigation revealed exten-sive environmental contamination around the plant. The Kepone incident necessitated the prolonged clos-ing for fishing of the contaminated section of the James River. No biological mineralization was dem-onstrated for either Kepone or Mirex, although both compounds are subject to limited dechlorination by anaerobic microbial consortia (Orndorff and Colwell 1980) and photodegradation (National Research Council 1978b).

The much less extensively chlorinated pesticide DDT is subject to some biochemical transformations both *in vitro* and in the environment, but the basic car-bon skeleton persists in nature for excessively long periods. The white rot fungus *Phanerochaete chrysos-porium* is able to attack DDT (Bumpus and Aust 1987). F. K. Pfaender and M. Alexander (1972) succeeded in demonstrating the *in vitro* mineralization of DDT ring-carbon by sequentially using cell-free extracts of a hydrogen-oxidizing bacterium, whole cells of *Arthro-bacter*, the addition of cosubstrates, and a regimen of alternating anaerobic and aerobic conditions. Neverthe-less, they came to the conclusion that biochemical DDT mineralization in the environment either does not occur or occurs at exceedingly slow rates. Other than the unfavorable energetics of the dechlorination steps, no satisfactory explanation could be found for this fact. Chlorinated pesticides, including DDT, are subject to reductive dehalogenation, that is, removal of the chlorine substituents by anaerobic microorganisms (Genthner et al. 1989a, 1989b). A nitrogen-heterocyclic herbicide, for example, has been shown to be attacked by reductive dehalogenation (Adrian and Suflita 1990). Chlorobenzoates and chlorobenzenes are also degraded by this mechanism (Dolfing and Tiedje 1987; Fathepure et al. 1988; Stevens and Tiedje 1988; Stevens et al. 1988; Linkfield et al. 1989; Mohn and Tiedje 1990a, 1990b).

Trichlorophenoxyacetic acid (2,4,5-T) is relatively resistant to biodegradation. A. M. Chakrabarty and coworkers (Kellogg et al. 1981; Chatterjee et al. 1982; Kilbane et al. 1982) were able to isolate a recombinant 2,4,5-T-degrading *Pseudomonas cepacia* only following molecular breeding in which a mixture of plasmid-bearing strains were maintained in a chemostat in the presence of 2,4,5-T. In contrast to 2,4,5-T, dichlorophenoxyacetic acid (2,4-D) is readily degraded by many aerobic soil microorganisms. The genes involved in the 2,4-D degradation pathway have been identified, and most have been cloned (Harker et al. 1989).

Polychlorinated Biphenyls and Dioxins

Polychlorinated biphenyls (PCBs) are mixtures of biphenyls with one to ten chlorine atoms per molecule (Figure 13.4). They are oily fluids with high boiling points, great chemical resistance, low electrical conductivity, and high light-breaking index. Because of these properties, they have been used as plasticizers in polyvinyl polymers, as insulators-coolants in transformers, and as heat-exchange fluids in general. Minor uses involve inks, paints, sealants, insulators, flame retardants, and, as microbiologists should know, microscope immersion oils. The chemical inertness of PCBs is paralleled by their great biological stability, which increases with the degree of their chlorination.

The structure of PCBs is similar in some ways to that of DDT, and so is their behavior in the environ-

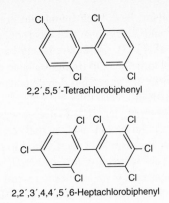

2,2´,5,5´-Tetrachlorobiphenyl

2,2´,3´,4,4´,5´,6-Heptachlorobiphenyl

Figure 13.4
Polychlorinated biphenyls (PCBs). These compounds are only two examples of the 210 different chloro-substitution possibilities. Resistance to biodegradation increases with the number of chloro-substitutions.

ment in regard to persistence and potential for biological magnification. Whereas most of the enumerated uses of PCBs would not seem to be conducive to widespread environmental contamination, PCB residues have in fact been detected in a large percentage of random environmental samples, and they accumulate in higher-trophic-level animals (Gustafson 1970; National Research Council 1979).

Concentrations of PCB over 1 ppm have been detected in one-third of the sampled U.S. population with no known occupational exposure. In some freshwater habitats, older individuals of predatory fish, such as trout and salmon, accumulate 20–30 ppm of PCBs, and predatory and fish-eating birds were found to contain, on occasion, several hundred ppm of PCBs. In 1968, PCB-contaminated cooking oil, caused by a leaky heat exchanger, poisoned nearly a thousand people in Japan; the main symptoms were liver damage and a severe skin condition called chloracne. In this case, the victims had consumed several grams of the substance (Fujiwara 1975). Low PCB intakes from general environmental contamination are not known to harm humans or most other mammals, but along with chlorinated hydrocarbon insecticides,

PCBs were implicated in eggshell thinning and reproductive failure of some predatory and fish-eating birds. Reproductive failure in cultured mink was caused by the approximately 5 ppm PCB content of their fish-based feed, and some species of fish, shrimp, and algae were affected by PCB concentrations in the low ppb range. Besides having acute effects, PCBs—and some other chlorinated hydrocarbons, including DDT—are under suspicion of being potential human carcinogens, although this point is still controversial.

The various degrees of chlorination and the high number of isomers render the positive identification and quantitative measurement of PCBs a difficult analytical task; in the past, PCB residues were frequently confused with those of chlorinated hydrocarbon insecticides. Photochemical conversion of DDT to PCBs may contribute to the widespread occurrence of these environmental contaminants (Maugh 1973).

Biphenyl is able to serve some microorganisms as their only source of carbon and energy, and mono-, di-, tri-, and tetrachlorobiphenyls are, to some degree, subject to biodegradation, as evidenced by formation of hydroxylated metabolites (Ahmed and Focht 1973; Massé et al. 1984; Safe 1984). In the more extensively chlorinated biphenyls, the substituents prevent ring hydroxylation; these highly chlorinated analogs cause the most concern with respect to persistence and environmental contamination. Voluntary curbs by manufacturers coupled with government regulations have removed PCBs from food-processing equipment and food-packaging material. Disposable items using PCBs as plasticizers have been phased out in an effort to reduce the amount of contaminating PCBs. As of 1978, the use and discharge of PCBs in the United States came under a complete government ban. Continuing sources of PCB pollution, however, are previously contaminated sediments, landfills, and older electric transformers.

Although relatively resistant to biodegradation, a number of microorganisms have been isolated that transform PCBs (Chaudhry and Chapalamadugu 1991). Degradation of PCBs is carried out by the white rot fungus *Phanerochaete* (Bumpus 1989), aerobically by *Acinetobacter* (Adriaens and Focht 1990),

and *Alcaligenes* (Bedard et al. 1987), and anaerobically by reductive dehalogenation (Quensen et al. 1988). Degradation of PCBs typically is by cometabolism and is enhanced by the addition of less chlorinated analogs such as dichlorobiphenyl (Adriaens et al. 1989; Novick and Alexander 1985). Extensive degradation of some PCB congeners has been found in soils and aquatic waters and sediments (Bedard et al. 1987; Brown et al. 1987; Novick and Alexander 1985) (Figure 13.5). The specific congeners are differentially degraded, and various PCB products, according to their composition, exhibit different degrees of susceptibility to biodegradative transformations.

Frequent contaminants of PCBs and of the herbicide 2,4,5-T, to be discussed later in this chapter, are 2,3,7,8-tetrachlorodibenzodioxin (TCDD), related chlorodibenzodioxins, and chlorodibenzofurans (Figure 13.6). Some of these contaminants arise during manufacture, others are formed through thermal degradation of the chlorophenols and chlorobiphenyls. Not only high recalcitrance, but also an unusually high degree of acute and chronic toxicity characterize TCDD and related compounds. Contamination as low as in the ppb range is considered hazardous to human health. Human settlements that became contaminated with TCDD and related substances—as occurred in Times Beach, Missouri, in the United States, through spreading of contaminated waste oil and in Soveso in Italy, by the explosion of a chemical reactor—were evacuated and closed indefinitely (Tucker et al. 1983). J. A. Bumpus et al. (1985) reported oxidation of TCDD and some other chlorinated hydrocarbon pollutants by the lignin-degrading basidiomycete *Phanerochaete chrysosporium*. The amounts of $^{14}CO_2$ evolved from the radiolabeled test substrates ranged only between 1%-4% and may not have been in excess of the radiochemical impurities of the test substances. Therefore, the biodegradability of TCDD continues to be doubtful.

Polybrominated biphenyls (PBBs), which are used as flame retardants, and polychlorinated terphenyls (PCTs), which have three attached phenyl rings and are employed as plasticizers (Figure 13.7), have pollution potentials similar to PCBs. To date, they have been manufactured on a much smaller scale than

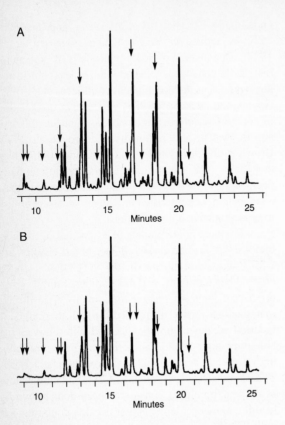

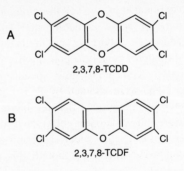

Figure 13.6

(A) 2,3,7,8-tetrachlorodibenzodioxin (TCDD). (B) 2,3,7,8-tetrachlorodibenzofuran, contaminants and/or thermal degradation products of PCBs and certain phenoxyalkanoic acid herbicides. These and related chlorodibenzodioxins and chlorodibenzofurans are extremely toxic and highly resistant to biodegradation.

Figure 13.5

Biodegradation of arochlor PCBs. Each peak in the gas chromatographic tracing represents a specific PCB congener. The height of each peak represents the concentration of that congener. The arrows point to congeners that have been biodegraded. (Source: Kohler et al. 1988.)

Figure 13.7

Examples of polychlorinated terphenyls and polybrominated biphenyls. Commercial mixtures contain numerous partially halogenated homologues. The extensive halogenation of these compounds results in high resistance to biodegradation.

PCBs. Their behavior as environmental contaminants received little attention until an unfortunate incident in 1976, when in a chemical plant that manufactured both PBB flame retardants and cattle feed additives, large amounts of cattle feed were inadvertently mixed with PBBs and were fed to dairy cows in Michigan. The incident necessitated the destruction of many dairy herds, causing great economic losses. Although no human fatalities could be proven, large numbers of people were contaminated with PBBs through the dairy products before the cause was traced and elimi-

nated. While the incident was clearly a case of human error, it has served as a reminder of the long residence time and high biomagnification potential of PBBs as environmental pollutants.

Synthetic Polymers

Synthetic polymers are easily molded into complex shapes, have high chemical resistance, and are more-or-less elastic. Some can be formed into fibers or thin, transparent films. These properties have made them popular in the manufacture of garments, durable and disposable goods, and packaging materials. About 50 million metric tons per year of plastics are produced in the United States (Thayer 1990). Disposable goods and packaging material, about one-third of the total plastic production, have the largest environmental impact. More than 90% of the plastic material in municipal garbage consists of polyethylene, polyvinyl chloride, and polystyrene, in roughly equal proportions (Figure 13.8).

Typically, these materials have molecular weights ranging from several thousand to 150,000 and appear to resist biodegradation indefinitely. Older reports on plastic biodegradation involve, on closer scrutiny, only the degradation of the plasticizers, the additives designed to render the plastic pliable. Frequently, plasticizers are esters of long-chain fatty acids and alcohols, such as dioctyl adipate, or esteus of phthalic acid, such as dioctyl phthalate. Their biodegradation renders the material brittle, but the polymer structure is not affected. The contention that the basic polymer structures are not subject to biodegradation was originally based on relatively crude weight-loss measurements but recently has been confirmed by highly sensitive measurements using ^{14}C-labeled polymers. Resistance to biodegradation seems to be associated here with excessive molecular size. Biodegradation of long chain C_{30} and higher n-alkanes declines with increasing molecular weight, and n-alkanes in excess of a molecular weight of 550–600 become refractory to biodegradation. If the molecular size of polyethylene is reduced to a molecular weight under 500, by pyrolysis, for example, the fragments are susceptible to biodegradation.

Plastic in the environment has been regarded as more of an aesthetic nuisance than a hazard, since the material is biologically inert. Reports about large numbers of floating plastic spherules in the ocean and their ingestion by small fish with consequent intestinal blockage, however, indicate that even inert materials are not

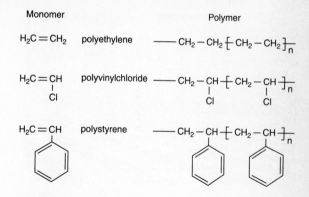

Figure 13.8

Chemical structure of some plastic materials of importance in environmental contamination. Left: Monomeric starting material. Right: Polymer chain with repeating unit shown in brackets. Excessive molecular weight and a lack of depolymerizing exoenzymes make these polymers virtually nonbiodegradable. Monomers or short polymer chain fragments of the same materials tend to be biodegradable.

necessarily innocuous (Carpenter 1972). Remains of plastic nets and ties also ensnare and destroy wildlife.

A new agricultural practice involving plastics, the so-called film-mulching technique (Figure 13.9), lends special urgency to the development of biodegradable plastic film material. This practice involves covering fields with plastic film in order to control weeds and conserve moisture. It is impractical to reuse or even to collect the film at the end of the growing season; a large accumulation of plastic fragments in the field is equally undesirable. Since polyethylene, polybutene, and other polyolefins are susceptible to photochemical degradation, it is possible to devise materials that sustain sufficient photochemical damage during a growing season to become susceptible to subsequent microbial degradation (Reich and Bartha 1977) (Figure 13.10). The resistance of the material to various irradiation levels can be adjusted by appropriate additions of antioxidants. Tests have shown that the native polymer was resistant to biodegradation, but the irradiated material became susceptible to it.

Figure 13.9

Photograph of film-mulched agricultural field. (Source: B. L. Pollack; New Jersey Agricultural Experiment Station.)

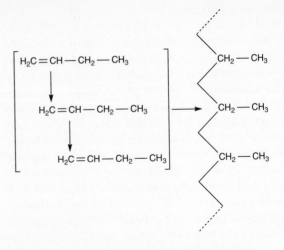

Figure 13.10

Chemical structure of a polybutene film-mulch polymer, which is degradable by a combined photodegradation-biodegradation process. Left: Monomeric precursor. Right: Polybutene polymer. Photochemical reactions break the polymer chain at the susceptible tertiary carbon allowing a subsequent biodegradative attack on the resulting chain fragments. (Source: Reich and Bartha 1977. Reprinted by permission, copyright Williams and Wilkins.)

Responding to pressure from environmental groups and legislators, several manufacturers started to sell products stamped "environmentally friendly" or "degradable." A lack of clear standards and definitions made it difficult to challenge these somewhat misleading claims. Some of these materials contain photosensitizers or pro-oxidants (oxidation catalysts), alone or in combination with 3%–15% modified starch (Evans and Sikdar 1990; Thayer 1990). Prolonged sun irradiation (several weeks to several months) causes breaks in the large polymer molecules, resulting in brittleness and disintegration. Extensive photodamage may prepare the molecules for limited microbial degradation (Lee and Levy 1991). Without photodamage, only the accessible starch granules are degraded, leading to some porosity and brittleness but no further degradation (Lee and Levy 1991; Krupp and Jewell 1992).

Truly biodegradable poly-β-hydroxyalkanoates are produced by *Alcaligenes eutrophus, Bacillus cereus,* and various pseudomonads (Brandl et al. 1988; Ramsay et al. 1990). These intracellular storage products have thermoplastic properties and can be molded. Their properties can be influenced by the substrate fed to the bacteria. A β-hydroxybutyric and β-hydroxyvaleric acid copolymer gives the best prop-

erties. Up to 50% of the cell dry weight is β-hydroxyalkanoate under the proper growth conditions. Nevertheless, the price of this product is five to seven times higher than that of petrochemical-based polyethylene, restricting its use to speciality products. Recently, some shampoo bottles and disposable diaper liners were also manufactured from this material. Properly modified, starch and other natural polymers may yet yield useful and biodegradable plastics. The synthetic caprolactone polyester that contains hydrolyzable bonds is also a good candidate as biodegradable plastic.

Alkyl Benzyl Sulfonates

Alkyl benzyl sulfonates (ABS) are the major components of anionic detergents. Like other surface active agents, the molecule has a polar (sulfonate) and a non-polar (alkyl) end. Emulsification of fatty substances, and hence cleaning, occurs when ABS molecules form a monolayer around lipophilic droplets or particles. These molecules orient with their nonpolar end toward the lipophilic substance and their sulfonate end toward the surrounding water. Nonlinear ABS (Figure 13.11) is resistant to biodegradation and causes extensive foaming of rivers that receive ABS-containing wastes. Nonlinear ABS is easier to manufacture and has slightly superior detergent properties, but the methyl branching of the alkyl chain interferes with biodegradation, since the tertiary carbon atoms block the normal β-oxidation sequence (Larson 1990). The problem of restricted biodegradation of nonlinear ABS was largely alleviated when the detergent industry switched to linear ABS that was free of this blockage and was consequently more easily biodegraded. The ABS story is a remarkable one, since it was one of the first instances when a synthetic molecule was redesigned to remove obstacles to biodegradation while essentially preserving the other useful characteristics of the compound (Brenner 1969).

Petroleum Hydrocarbons

Petroleum is, in one sense, a natural product, resulting from the anaerobic conversion of biomass under high temperature and pressure. It has always entered the biosphere by natural seepage, but at rates much slower than the forced recovery of petroleum by drilling, currently estimated to be about 2 billion metric tons per year. Petroleum, or at least most of its components, is subject to biodegradation, but at relatively slow rates. The production, transportation, refining, and ultimate disposal are introducing, by conservative estimate, 3.2 million metric tons of oil annually into the oceans (National Research Council 1985). This load is heavily concentrated around offshore production sites, major shipping routes, and refineries and frequently exceeds the self-purification capacity of the receiving waters. Oil floating on water is technically difficult to contain and collect. It is destructive to birds and various forms of marine life and, when driven ashore, causes heavy economic and esthetic

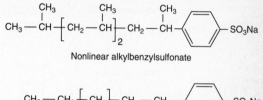

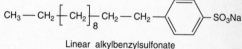

Figure 13.11

Chemical structures of some nonlinear and linear alkyl benzyl sulfonates (ABS). The methyl branches in the former group of compounds interfere with rapid biodegradation and cause foaming problems in contaminated water. Linear ABS compounds are free of this problem.

damage. Cleanup costs are high, often amounting to ten to fifteen dollars per gallon of spilled oil. Prevailing weather conditions often completely thwart the containment and cleanup of oil slicks. Dispersing of the oil with the use of detergents or sinking the oil with chalk or siliconized sand are essentially cosmetic measures. They remove the oil from the surface but increase the exposure of marine life to the pollutant and are, therefore, best avoided.

In addition to killing birds, fish, shellfish, and other invertebrates, oil pollution appears also to have more subtle effects on marine life. Dissolved aromatic components of petroleum disrupt, even at a low ppb concentration, the chemoreception of some marine organisms. Since feeding and mating responses largely depend on chemoreception, such disruption can lead to elimination of many species from the polluted area even when the pollutant concentration is far below the lethal level as defined in the conventional sense. Another disturbing possibility is that some condensed polynuclear components of petroleum that are relatively resistant to biodegradation are carcinogenic and may move up marine food chains and taint fish or shellfish.

Petroleum is a complex mixture of aliphatic, alicyclic, and aromatic hydrocarbons (Figure 13.12) and a smaller proportion of nonhydrocarbon com-

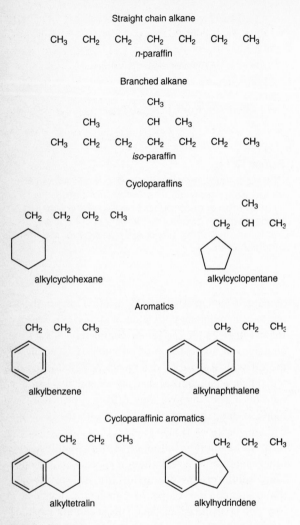

Straight chain alkane

CH_3 CH_2 CH_2 CH_2 CH_2 CH_2 CH_3

n-paraffin

Branched alkane

iso-paraffin

Cycloparaffins

alkylcyclohexane   alkylcyclopentane

Aromatics

alkylbenzene   alkylnaphthalene

Cycloparaffinic aromatics

alkyltetralin   alkylhydrindene

Figure 13.12

Structures of hydrocarbons in crude oil. These include aromatic, paraffinic, cycloparaffinic, and cycloparaffinic aromatics. Homologues, isomers, and combinations result in hundreds of individual hydrocarbon compounds in crude oil samples.

tion of each crude oil varies with its origin. Most xenobiotic pollutants can be regarded as substituted or modified hydrocarbons; therefore, it is useful to review here briefly the biodegradation pathways of various hydrocarbon compounds.

Susceptibility to biodegradation varies with the type and size of the hydrocarbon molecule (Atlas 1984; National Research Council 1985). *n*-Alkanes of intermediate chain length (C_{10}–C_{24}) are degraded most rapidly. Short-chain alkanes are toxic to many microorganisms, but they generally evaporate from oil slicks rapidly. Very long chain alkanes become increasingly resistant to biodegradation. As the chain length increases and the alkanes exceed a molecular weight of 500, the alkanes cease to serve as carbon sources. Branching, in general, reduces the rate of biodegradation, since tertiary and quaternary carbon atoms interfere with degradation mechanisms or block degradation altogether. Aromatic compounds, especially of the condensed polynuclear type, are degraded more slowly than alkanes. Alicyclic compounds are frequently unable to serve as the sole carbon source for microbial growth unless they have a sufficiently long aliphatic side chain, but they can be degraded via cometabolism by two or more cooperating microbial strains with complementary metabolic capabilities (Atlas 1981).

The initial attack on alkanes occurs by enzymes that have a strict requirement for molecular oxygen, that is, monooxygenases (mixed function oxidases) or dioxygenases (Britton 1984; Singer and Finnerty 1984). In the first case (Equation 1), one atom of O_2 is incorporated into the alkane, yielding a primary alcohol. The other is reduced to H_2O, with the reduced form of nicotinamide dinucleotide phosphate ($NADPH_2$) serving as electron donor.

(1) $R\ CH_2\ CH_3 + O_2 + NADPH_2 \longrightarrow$

$R\ CH_2\ CH_2\ OH + NADP + H_2O$

In the second case (Equation 2), both atoms of O_2 are transferred to the alkane, yielding a labile hydroperoxide intermediate that is subsequently reduced by $NADPH_2$ to an alcohol and H_2O (Equation 3).

pounds, such as naphthenic acids, phenols, thiols, heterocyclic nitrogen, and sulfur compounds, as well as metalloporphyrins. There are several hundred individual components in every crude oil, and the composi-

(2) $R-CH_2-CH_3 + O_2 \longrightarrow$

$R-CH_2-CH_2-OOH$

(3) $R-CH_2-CH_2-OOH + NADPH_2 \longrightarrow$

$R-CH_2-CH_2-OH + NADP + H_2O$

Most frequently, the initial attack is directed at the terminal methyl group, forming a primary alcohol that, in turn, is further oxidized to an aldehyde and fatty acid. Occasionally, both terminal methyl groups are oxidized in this manner, resulting in the formation of a dicarboxylic acid. This variation, described as diterminal or ω-oxidation, is one of several ways to bypass a block to β-oxidation due to branching of the carbon chain.

Once a fatty acid is formed, further catabolism occurs by the β-oxidation sequence. The long-chain fatty acid is converted to its acyl coenzyme A form (Equation 4) and is acted upon by a series of enzymes, with the result that an acetyl CoA group is cleaved off and the fatty acid is shortened by a two-carbon unit (Equations 4-9). This sequence is then repeated. The acetyl CoA units are converted to CO_2 through the tricarboxylic acid cycle. The end products of hydrocarbon mineralization, thus, are CO_2 and H_2O.

(4) $R-CH_2-CH_2-CH_2-COOH \xrightarrow{+CoA}$

(5) $R-CH_2-CH_2-CH_2-\overset{\displaystyle O}{\overset{\displaystyle \|}{C}}-CoA \xrightarrow{-2H^+}$

(6) $R-CH_2-CH=CH-\overset{\displaystyle O}{\overset{\displaystyle \|}{C}}-CoA \xrightarrow{+H_2O}$

(7) $R-CH_2-\underset{\displaystyle OH}{\overset{\displaystyle }{CH}}-CH_2-\overset{\displaystyle O}{\overset{\displaystyle \|}{C}}-CoA \xrightarrow{-2H^+}$

(8) $R-CH_2-\underset{\displaystyle O}{\overset{\displaystyle \|}{C}}-CH_2-\overset{\displaystyle O}{\overset{\displaystyle \|}{C}}-CoA \xrightarrow{+CoA}$

(9) $R-CH_2-\underset{\displaystyle O}{\overset{\displaystyle \|}{C}}-CoA + CH_3-\overset{\displaystyle O}{\overset{\displaystyle \|}{C}}-CoA$

The β-oxidation sequence does not necessarily require the presence of molecular oxygen, and after the initial oxygenation, fatty acid biodegradation may proceed under anaerobic conditions. A direct dehydrogenation of an intact hydrocarbon (Equation 10)—leading through a 1-alkene (Equation 11), alcohol (Equation 12), and aldehyde (Equation 13) to a fatty acid (Equation 14)—has been suggested as a potential anoxic biodegradation mechanism.

(10) $R-CH_2-CH_3 \xrightarrow{-2H^+}$

(11) $R-CH=CH_2 \xrightarrow{+H_2O}$

(12) $R-\underset{\displaystyle OH}{\overset{\displaystyle }{CH}}-CH_3 \xrightarrow{-2H^+}$

(13) $R-CH_2-CHO \xrightarrow{-2H^+, +H_2O}$

(14) $R-CH_2-COOH$

This sequence would allow the anaerobic metabolism of an intact hydrocarbon in the presence of an electron acceptor such as nitrate. The evidence for such a sequence at this time is tenuous, and significant metabolism of intact aliphatic hydrocarbons in a strictly anaerobic environment is yet to be demonstrated. Geologic and environmental experience indicate that in the absence of molecular oxygen, hydrocarbons persist.

Some microorganisms attack alkanes subterminally; that is, oxygen is inserted on a carbon atom within the chain instead of at its end (Figure 13.13). In this manner, a secondary alcohol is formed first, which then is further oxidized to a ketone and finally to an ester. The ester bond is cleaved, yielding a primary alcohol and a fatty acid. The sum of the carbon atoms in the two fragments is equal to that of the parent hydrocarbon. The alcohol fragment is oxidized through the aldehyde to the fatty acid analog, and both fragments are metabolized further by the β-oxidation sequence (Britton 1984; Singer and Finnerty 1984).

Alicyclic hydrocarbons having no terminal methyl groups are biodegraded by a mechanism simi-

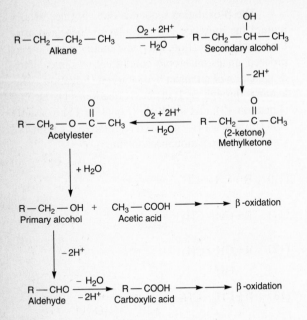

Figure 13.13

Pathway through which subterminal oxidation of alkanes yields two fatty acid moieties, which are metabolized further by β-oxidation.

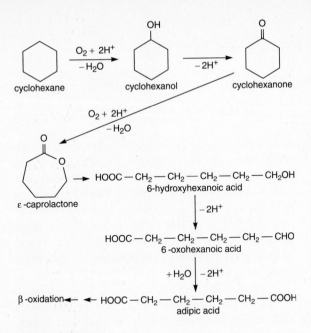

Figure 13.14

Microbial oxidation of cyclohexane as an example for metabolism of alicyclic hydrocarbons.

lar to the subterminal oxidation shown in Figure 13.13. In the case of cyclohexane (Figure 13.14), hydroxylation by a monooxygenase leads to an alicyclic alcohol. Dehydrogenation leads to the ketone. Further oxidation inserts an oxygen into the ring and a lactone is formed. The hydroxyl group is oxidized, in sequence, to an aldehyde and carboxyl group. The resulting dicarboxylic acid is further metabolized by β-oxidation. Microorganisms have been found that can grow on cyclohexane (Trower at al. 1985) and hence must be capable of carrying out the whole degradation sequence. More frequently, however, organisms capable of converting cyclohexane to cyclohexanone are unable to lactonize and open the ring and vice versa. Consequently, commensalism and cometabolism play an important role in the biodegradation of alicyclic hydrocarbons (Perry 1984; Trudgill 1984).

Aromatic hydrocarbons are oxidized by dioxygenases to labile *cis, cis*-dihhydrodiole that spontane-

ously convert to catechols (Figure 13.15). The dihydroxylated aromatic ring is opened by oxidative "ortho cleavage," resulting in *cis, cis*-muconic acid. This is metabolized further to β-ketoadipic acid, which is oxidatively cleaved to the common tricarboxylic cycle intermediates, succinic acid, and acetyl-CoA. Alternatively, the catechol ring may be opened by meta cleavage, adjacent to rather than between the hydroxyl groups, yielding 2-hydroxy-*cis,cis*-muconic semialdehyde. Further metabolism leads to formic acid, pyruvic acid, and acetaldehyde. Condensed aromatic ring structures, if degradable, are also attacked by dihydroxylation and the opening of one of the rings (Figure 13.16). The opened ring is degraded to pyruvic acid and CO_2, and a second ring is attacked in the same fashion. Many condensed polynuclear aromatic compounds, however, are degraded only with difficulty or not at all (Cerniglia 1984; Gibson and Subramanian 1984). One reason

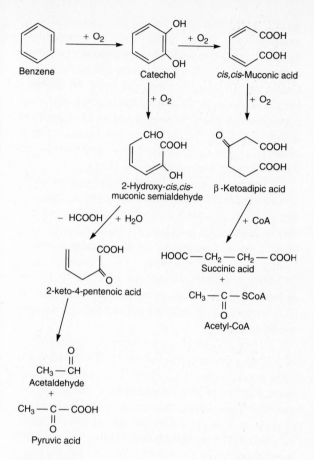

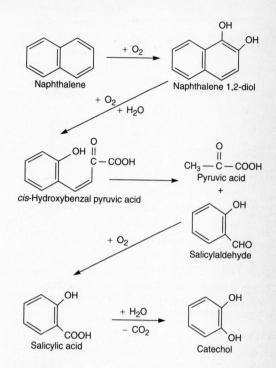

Figure 13.15

Microbial metabolism of the aromatic ring (simplified) by *meta* or *ortho* cleavage, as shown for benzene.

Figure 13.16

Microbial metabolism of a condensed aromatic ring structure (simplified) as shown for naphthalene. The resulting catechol is metabolized further by *ortho* or *meta* cleavage, as shown in Figure 13.15.

for resistance to biodegradation is that induction of the enzymes responsible for polynuclear aromatic hydrocarbon degradation depends upon the presence of lower–molecular-weight aromatics (Heitkamp and Cerniglia 1988).

Importantly, eukaryotic microorganisms (fungi and algae)—like mammalian liver systems—produce *trans* diols, whereas most bacteria oxidize aromatic hydrocarbons to *cis* diols (Cerniglia 1984; Gibson and Subramanian 1984) (Figure 13.17). *Trans* diols of various polynuclear aromatic hydrocarbons are carcinogenic, whereas *cis* diols are not.

The generally accepted view that hydrocarbons without hydroxyl-, carbonyl- or carboxyl-substituents are not degraded under strictly anaerobic conditions needed revision in case of the simple aromatics benzene and toluene. D. Grbic-Galic and T. M. Vogel (1987) clearly demonstrated that these compounds were metabolized under anaerobic methanogenic conditions. The oxygen used in the ring hydroxylation came from H_2O (Vogel and Grbic-Galic 1986). Although the biodegradation of aromatics is definitely much slower under anaerobic conditions than under aerobic ones, the rates of anaerobic toluene and *m*-xylene biodegradation were considered significant enough for use in aquifer bioremediation (Zeyer et al. 1986, 1990).

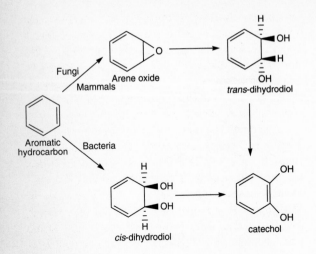

Figure 13.17
Fungi and other eukaryotic cells form *trans*-diols, whereas most bacteria form *cis*-diols when they oxidize aromatic hydrocarbons.

Crude Oil Biodegradation Crude oils are composed of complex mixtures of paraffinic, alicyclic, and aromatic hydrocarbons. A marine oil spill left to its natural fate is gradually degraded by the just-described biological and also by some nonbiological mechanisms. Autooxidation in the absence of light plays a minor role because the low temperatures of the marine environment provide no opportunity for activation. Photochemical oxidation, however, may contribute substantially to the self-purification of the marine environment. Laboratory experiments suggest that in eight hours of effective sunshine, as much as 0.2 metric tons of oil per km^2 may be destroyed by photooxidation, though the actual figures are probably substantially lower. The contribution of photodegradation is made especially important by the fact that this process preferentially attacks the same tertiary carbon atoms that tend to block biodegradation.

Communities exposed to hydrocarbons become adapted, exhibiting selective enrichment and genetic changes resulting in increased proportions of hydrocarbon-degrading bacteria and bacterial plasmids encoding hydrocarbon catabolic genes (Leahy

and Colwell 1990). Because adapted microbial communities have higher proportions of hydrocarbon degraders, they can respond to the presence of hydrocarbon pollutants within hours. In the case of the Amoco Cadiz and Tanio spills along the coast of Brittany, France, for example, the adapted hydrocarbon-degrading populations increased by several orders of magnitude within a day of the spillages, and biodegradation occurred as fast as or faster than evaporation in the days following these spills (Atlas 1981).

The measurement of biodegradation rates under favorable laboratory conditions has led to the estimate that as much as 0.5–60 g oil per m^3 seawater per day may be mineralized to CO_2 in this manner, but this estimate is far too optimistic. *In situ* measurements of hydrocarbon biodegradation in seawater using ^{14}C-labeled hexadecane indicate rates of 1–30 mg hydrocarbon per m^3 seawater per day converted to CO_2, depending on temperature and mineral nutrient conditions.

The principal forces limiting the biodegradation of polluting petroleum in the sea are the resistant and toxic components of oil itself; low water temperatures; scarcity of mineral nutrients, especially nitrogen and phosphorus; the exhaustion of dissolved oxygen; and, in previously unpolluted pelagic areas, the scarcity of hydrocarbon-degrading microorganisms (Atlas 1981).

Low winter temperatures can limit rates of hydrocarbon biodegradation, increasing residence time of oil pollutants (Bodennec et al. 1987). In a recent diesel oil spill from the Bahia Paraiso in the Antarctic, M. C. Kennicutt (1990) reported turnover rates, measured as ^{14}C-hexadecane mineralization, in excess of two years. This was considerably slower than the rates previously measured in Arctic zones.

Microbial degradation of oil has been shown to occur by attack on the aliphatic or light aromatic fractions of the oil. Although some studies have reported their removal at high rates under optimal conditions (Rontani et al. 1985; Shiaris 1989), high–molecular-weight aromatics, resins, and asphaltenes are generally considered to be recalcitrant or exhibit only low rates of biodegradation. In aquatic ecosystems, dispersion and emulsification of oil in slicks appear to be prerequisites for rapid biodegradation; large masses of mousse, tar balls, or high concentrations of oil in quiescent environments tend to persist because of the limited surface areas available for microbial activity. Petroleum spilled on or applied to soil is largely adsorbed to particulate matter, decreasing its toxicity but possibly also contributing to its persistence. In the short run, petroleum and petroleum fractions containing asphaltic components are not degraded quantitatively. The residues, along with polymerization products formed from reactions of free radical degradation intermediates with each other, form tar globules. The tar, a partially oxygenated high–molecular-weight material, is quite resistant to further degradation, and floating tar globules are encountered in the marine environment in increasing quantities (Butler et al. 1973).

As the water solubility of many solid hydrocarbons is extremely low, transport limitation can cause an apparent recalcitrance of otherwise degradable hydrocarbons. This was demonstrated by packaging highly recalcitrant n-hexatriacontane (C_{36}) into liposomes (Miller and Bartha 1989). Hexatriacontane transported in this manner into the cells of a *Pseudomonas* strain originally isolated on n-octadecane (C_{18}) utilized the C_{36} hydrocarbon at the same rate as the water-soluble substrate succinate. R. A. Efroymson and M. Alexander (1991) have also shown that some bacteria can utilize hydrocarbons directly from the nonaqueous phase.

An ability to isolate high numbers of certain oil-degrading microorganisms from an environment is commonly taken as evidence that those microorganisms are the most active oil degraders of that environment. This projection is questionable, but *in situ* measurements of hydrocarbon biodegradation activity are few. Using selective inhibitors, H. Song et al. (1986) determined that in a field soil with no hydrocarbon spill history, 80% of added hexadecane was degraded by bacteria and only 20% by fungi. In the same soil, bacterial and fungal population segments shared glucose degradation evenly.

Accidental oil spills are generally easier to contain and clean on land than on water, though gasoline and other low-viscosity distillation products may seep into subsoils and persist there because of prevailing anoxic conditions. They may also pollute groundwater and pose an explosion hazard in nearby basements and sewers. Oil spills are destructive to vegetation not only because of contact toxicity, but because hydrocarbon biodegradation in the soil renders plant root zones anoxic. The lack of oxygen and accompanying H_2S evolution kills the roots of most plants, including those of large, well-established trees (Bossert and Bartha 1984).

Natural gas seepages from underground pipe systems or from former landfills have a similar deleterious effect on vegetation. Here, the anoxic conditions are created by a combination of the physical displacement of soil air by natural gas and the activity of *Methylomonas* and other bacteria capable of oxidizing gaseous alkanes. Destruction of ornamental and shade trees occurred on a large scale in some communities that replaced coal-derived generator gas with natural gas. The former is generated by blowing a mixture of steam and air through hot coals, giving rise to a combustible gas mixture containing hydrogen, carbon monoxide, nitrogen, and various impurities. This gas mixture has a high moisture content. When natural gas that consisted mainly of methane and was quite dry was distributed through the same old municipal pipe networks, pipe joint gaskets dried out, shrank, and became leaky. The underground gas seepage caused widespread tree kills through anoxic soil conditions, though the direct toxicity of methane and other gaseous alkanes to plants is quite low (Hoeks 1972).

Pesticides

Monocultures are inherently unstable ecosystems, and of necessity a large part of any agricultural production effort goes toward the control of competing weeds, destructive insects, and other pests. The cost of manual labor and the increasing scale of agricultural operations in developed countries both have worked in favor of chemical pest control. By 1975, 1,170 pesticides were registered for use in the United States—including 425 herbicides, 410 fungicides, and 335 insecticides. The total annual production of synthetic organic pesticides in the same year was 725 million kg. In terms of effectiveness, economy, and quality control, pesticides have been an unqualified success;

the dangers inherent in our ignorance of the ultimate fate and unintended side effects of pesticides were recognized only gradually. Rachel Carson's controversial *Silent Spring* (1962) played a large role in alerting legislators and the general public to the proven and potential dangers of unwise pesticide use. In the subsequent, highly emotional debate, groups with conflicting interests and philosophies made sweeping endorsements or condemnations of pesticides. In reality, a sudden discontinuation of pesticide use would surely precipitate an economic and public health crisis, while uncontrolled pesticide use would, in the long term, cause irreversible environmental damage. A responsible pesticide policy clearly demands careful risk-benefit analysis for each compound and its various uses.

Most of the currently used organic pesticides are subject to extensive mineralization within the time span of one growing season or less. Biochemical processes, alone or in combination with purely chemical reactions, are responsible for their disappearance. Synthetic pesticides show a bewildering variety of chemical structures, yet most can be traced to relatively simple hydrocarbon skeletons that bear a variety of substituent groups, such as halogen, amino, nitro, hydroxyl, and others. Aliphatic carbon chains are degraded by the β-oxidation sequence. The resulting C_2 fragments are further metabolized via the tricarboxylic acid cycle. Aromatic ring structures are metabolized by dihydroxylation and ring cleavage. Prior to these transformations, chemical groups attached to the aromatic ring are completely or partially removed. Substituent groups uncommon in natural compounds, such as halogens and nitro- and sulfonate groups, if situated so as to impede the oxygenations, will frequently cause recalcitrance, whereas methyl, methoxy, carboxyl, and carbonyl groups can often be removed metabolically from blocking positions. If the dihydroxylation can occur on a substituted aromatic compound, the previously described degradation pathways will be followed with minor modification according to the position and nature of the substituents. Saturated ring systems, such as cyclohexane and decalin, are more refractory than their aromatic analogs. They rarely support

growth of any single microorganism, but some are degraded in the presence of other substrates and by the cooperative effort of several microbial strains. This phenomenon is observed also with other difficult-to-degrade compounds and relates to the concept of cometabolism (Horvath 1972).

Cometabolism is the phenomenon that occurs when a compound is transformed by a microorganism, yet the organism is unable to grow on the compound and does not derive energy, carbon, or any other nutrient from the transformation. Cometabolic transformation occurs when an enzyme of a microorganism, growing on compound A, recognizes as substrate compound B and transforms it to a product. The transformation is limited, because the next enzyme of the organism that should attack in sequence has a higher specificity and does not recognize the product of B as a substrate. In the case of a pure culture, cometabolism is a dead-end transformation without benefit to the organism. In a mixed culture or in the environment proper, however, such an initial cometabolic transformation may pave the way for subsequent attack by another organism. In this manner, cometabolic and substrate-utilization synergistic-type transformations may eventually lead to the recycling of relatively recalcitrant compounds that do not support the growth of any microbial culture.

As organic molecules to be degraded become more complex and consist of aliphatic as well as aromatic, alicyclic, or heterocyclic portions, few generalizations can be made about their degradation pattern. If the moieties of the molecule are connected by ester, amide, or ether bonds that can be cleaved by microbial enzymes, the initial attack usually takes this form, and the resulting compounds are subsequently metabolized as outlined before. If such an attack cannot occur, degradation will commonly be initiated at the aliphatic end of the molecule. If this end is blocked by extensive branching or by other substituents, however, the attack may start from the aromatic end. The site and mode of the initial attack is determined not only by molecular structure, but also by the enzymatic capabilities of the microorganisms involved, as well as by the prevailing environmental conditions, such as the redox potential, pH, and ionic environment. These factors will modify not only the rate, but also the pathway and the ultimate products of the degradation (Hill and Wright 1978).

The chemical structures of some biodegradable and some recalcitrant pesticides are compared in Figure 13.18. 2,4-Dichlorophenoxyacetic acid (2,4-D) is biodegraded within days; 2,4,5-trichlorophenoxyacetic acid (2,4,5-T) differs only by one additional chlorine substitution in the *meta*-position, yet this compound persists for many months. The additional substitution interferes with the hydroxylation and cleavage of the aromatic ring. The primary amine group in isopropyl-N-phenyl-carbamate (propham) is cleaved by microbial amidases so rapidly that for some applications the addition of amidase inhibitors becomes necessary. The secondary amine group of N-isopropyl-2-chloroacetanilide (propachlor) is not subject to attack by such amidases, and this compound persists considerably longer. 1-Naphthyl-N-methylcarbamate (carbaryl) is biodegraded quite rapidly, while hexachloro-octahydro-dimethanonaphthalene (aldrin) persists or undergoes only a minor change by epoxidation. 1,1,1-Trichloro-2,2-*bis*(*p*-methoxyphenyl)-ethane (methoxychlor) is less persistent than 1,1,1-trichloro-*bis*(*p*-chlorophenyl)-ethane (DDT), since the *p*-methoxy groups are subject to dealkylation, while the *p*-chloro-substitution endows DDT with great biological and chemical stability.

Similarly to halogen substituents, nitro-substitution of aromatic rings has a strong negative effect on biodegradability. Both halogen and nitro-substituents are electron-withdrawing and tend to deactivate the aromatic group. Compounds with several nitro-substituents tend to be recalcitrant (Hallas and Alexander 1983). Nitroaromatics are common military explosives (such as trinitrotoluene, or TNT), solvents (such as nitrobenzene), and pesticides (such as nitrophenols). Their manufacture and disposal left many sites polluted. The toxicity and mutagenicity of these residues are high. While oxidative transformations of haloaromatics are not favored thermodynamically (they burn poorly or not at all), nitroaromatics have high chemical energy potentials, exemplified by TNT. However, to date no metabolic pathways have evolved to release this potential for microbial growth.

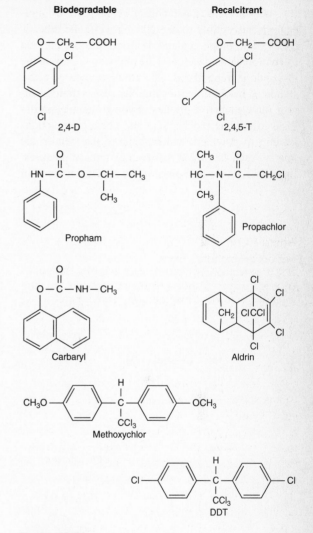

Figure 13.18

Molecular structures of some biodegradable and some recalcitrant pesticides. The upper four compounds are herbicides, the lower four insecticides. The terms *biodegradable* and *recalcitrant* are used here in a relative sense and imply neither instant biodegradation nor indefinite resistance. The pairs of compounds (left and right) were selected for overall structural similarity in order to help pinpoint molecular features that render the compounds in the right column recalcitrant. These features are the 5-chloro-substitution in the case of 2,4,5-T, the N-alkyl-substitution in the case of propachlor, the multiple of chloro-substitutions in the case of aldrin, and the two *p*-chloro-substitutions in the case of DDT.

In some cases, one portion of the pesticide molecule is susceptible to degradation while the other is recalcitrant. Some acylanilide herbicides are cleaved by microbial amidases, and the aliphatic moiety of the molecule is mineralized. The aromatic moiety, being stabilized by chlorine substitutions, resists mineralization, but the reactive primary amino group may participate in various biochemical and chemical reactions leading to polymers and complexes that render the fate of such herbicide residues extremely complex. Figure 13.19 shows some of the transformations of

N-(3,4-dichlorophenyl)-propionamide (propanil) (Bartha and Pramer 1970). Microbial acylamidases cleave the propionate moiety, which is subsequently mineralized. A portion of the released 3,4-dichloroaniline (DCA) is acted upon by microbial oxidases and peroxidases, with the result that they dimerize and polymerize to form highly stable residues such as 3,3´,4,4´tetrachloroazobenzene (TCAB) and related azo compounds. p-Bromoaniline, liberated during the biodegradation of the phenylurea herbicide 3-(p-bromophenyl)-1-methoxy-1-methylurea (bromuron), undergoes acetylation by soil fungi (Figure 13.20).

Figure 13.19

Microbial metabolism of the herbicide N-(3,4-dichlorophenyl)-propionamide (propanil). The aliphatic portion of the molecule is degraded, but the aromatic portion is dimerized and polymerized to persistent residues.

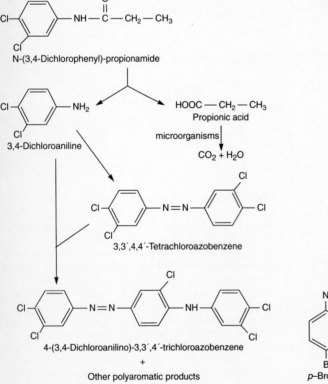

Figure 13.20

Metabolism of the phenylurea herbicide 3-(p-bromophenyl)-1 methoxy-1-methylurea by stepwise dealkylation and conversion to p-bromoaniline. Instead of further biodegradation, p-bromoaniline is acetylated to p-bromoacetanilide.

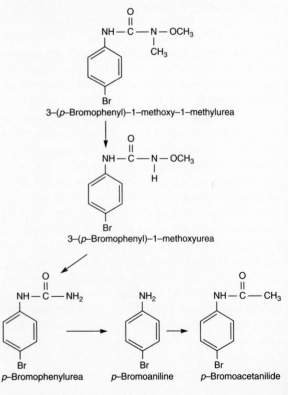

The reasons for such transformations are still somewhat obscure. They may occur by chance when a microbial enzyme having another synthetic function recognizes and acts on the human-made residue. In some cases, the reaction seems to detoxify the residue from the microbe's point of view, but the overall persistence and environmental impact of the pesticide is increased by such synthetic transformations (Bartha and Pramer 1970).

Only a small portion of the total halogenated aniline residues derived from the degradation of various phenylurea, phenylcarbamate and acylanilide herbicides, and some other pesticidal compounds undergoes transformation as previously described. The bulk of the liberated anilines disappears, that is, becomes unextractable, without evidence of rapid conversion to CO_2. The major mechanism of this phenomenon is the chemical attachment of the anilines to humic substances present in both soils and surface waters. There is increasing evidence that residues of other pesticides may become similarly attached to humic compounds. Humification is part of the natural carbon cycle, and it should not have been unexpected that aromatic residues of pesticides may be subject to it. At least some pesticide residues preserve their xenobiotic character when bound to humus and when subsequently liberated by the biodegradation of the surrounding humic matrix (Figure 13.21). Thus, humus-bound pesticide

Figure 13.21

Some attachment mechanisms of pesticide residues to humus. Subsequent microbial mobilization of the residues may result in contamination of organisms and agricultural products. White arrows represent chemical processes; black arrows represent enzymatic processes. The xenobiotic residues are separated from the type structure of natural humus with a dashed line.

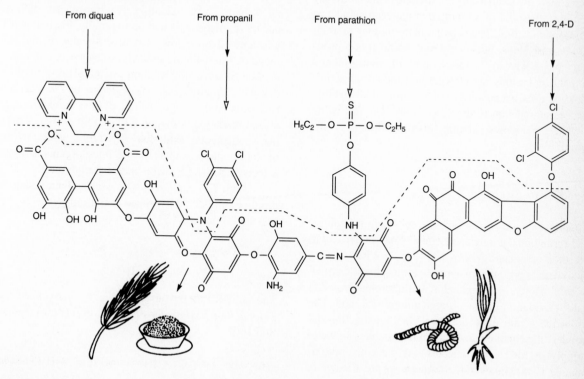

residues may contaminate crops and biota that were never directly exposed to the pesticide. Humus-bound residues may explain some puzzling instances of low-level crop contamination by a "wrong" pesticide, but the levels involved are so low that any concerns about human health and effects on soil fertility seem unjustified (Bartha 1980).

MICROBIAL INTERACTIONS WITH SOME INORGANIC POLLUTANTS

The pollution problems discussed thus far in this chapter mostly arise from the inability of microorganisms to mineralize certain organic molecules efficiently. In contrast, the pollution problems to be discussed in the following section are a consequence of the normal biogeochemical cycling activities of microorganisms. Pollution problems arise when the activities of industrialized societies inadvertently increase the pool of an inorganic material subject to microbial cycling. The original material may be relatively innocuous, but its microbial conversion products may not be so. To foresee the full environmental impact of various agricultural and industrial activities requires a knowledge of the microbial cycling processes relevant to those activities. The following selected examples should serve to illustrate this point.

Acid Mine Drainage

Coal and various metal ores are enclosed in geological formations of a reduced nature. Coal, in particular, is often associated with pyrite (FeS_2). When mining activities expose this material to atmospheric oxygen, a combination of autoxidation and microbial iron and sulfur oxidation produces large amounts of acid. The iron-rich acidic mine drainage kills aquatic life and renders the contaminated stream unsuitable as a water supply or for recreational use. At present, approximately 10,000 miles of U.S. waterways are affected in

this manner, predominantly in the states of Pennsylvania, Virginia, Ohio, Kentucky, and Indiana.

Some acid mine drainage originates from subsurface mining because of water flowing through the mine itself and runoff from mounds of mine tailings or gob piles, rock and low-grade coal separated from the high-grade coal before shipment. This problem is limited and relatively easily controlled. In subsurface mining, as coal is removed, the empty seam is either filled in with tailings or allowed to collapse. In this procedure, only a limited amount of rock is exposed to oxidative action at any one time. In contrast, strip mining removes the overburden and leaves it with the tailings as porous rubble, exposed to oxygen and percolating water. As a result of iron and sulfur oxidation, the pH drops rapidly and prevents the establishment of vegetation and a stable soil cover that would, eventually, seal the rubble from oxygen. The strip-mined region continues to give rise to acid mine drainage until most of the sulfide is oxidized and leached out. The recovery of the land may take from 50 to 150 years.

The mechanism of pyrite oxidation is complex. At a neutral pH, oxidation by atmospheric oxygen occurs spontaneously and quite rapidly, but below pH 4.5, autoxidation slows down drastically. In the pH range of 3.5 to 4.5 the stalked iron bacterium *Metallogenium* catalyzes the reaction. As the pH drops below 3.5, the acidophilic bacteria of the genus *Thiobacillus* take over. At this stage, the rate of the microbially catalyzed oxidation is several hundred times higher than the spontaneous oxidation. Therefore, while pyrite oxidation starts spontaneously, microorganisms have a decisive role in its maintenance at high rates.

Both *Thiobacillus thiooxidans* and autoxidation produce sulfate and hydrogen ions from pyrite, in the first step (Equation 15):

$$(15) \quad 2FeS_2 + 7O_2 + 2H_2O \longrightarrow$$
$$2Fe^{2+} + 4SO_4^{2-} + 4H^+$$

The solubilized ferrous iron (Fe^{2+}) is oxidized by *Thiobacillus ferrooxidans* to ferric iron (Fe^{3+}) (Equation 16):

$$(16) \quad 2Fe^{2+} + {}^1/_2O_2 + 2H^+ \longrightarrow 2Fe^{3+} + H_2O$$

The second reaction (Equation 16) is the rate-limiting and, therefore, most critical step in the oxidation sequence. The ferric iron produced is able to oxidize the remaining sulfide to sulfate in a nonbiological reaction (Equation 17).

(17) $8Fe^{3+} + S^{2-} + 4H_2O \longrightarrow$
$$8Fe^{2+} + SO_4^{2-} + 8H^+$$

As seen in Equation 17, the ferric iron is reduced back to ferrous iron, which is available again for oxidation in the reaction of Equation 16. Alternately, the ferric iron produced in reaction (16) may precipitate as ferric hydroxide (Equation 18), releasing the hydrogen ions consumed in reaction (16) plus four additional hydrogen ions.

(18) $2Fe^{3+} + 6H_2O \longrightarrow 2Fe(OH)_3 + 6H^+$

The overall reaction for the oxidation of pyrite can be summarized as shown in Equation 19:

(19) $2FeS_2 + 71/2O_2 + 7H_2O \longrightarrow$
$$2Fe(OH)_3 + 4H_2SO_4$$

The sulfuric acid produced accounts for the high acidity and the precipitated ferric hydroxide for the deep brown color of the effluent.

Conventional water treatment techniques were designed for organic pollution and are ineffective against pollution by acid mine drainage. The best way to deal with the problem is to prevent it at the source. It is often feasible to seal abandoned subsurface mines, thus preventing or restricting the availability of oxygen for pyrite oxidation. In the case of strip mining, acid mine drainage can be effectively controlled by prompt reclamation of the land. This involves spreading topsoil over the rubble and establishing a vegetation cover. The same technique is effective on mounds of mine tailings. Of course, this involves extra cost, but after prolonged controversy, federal laws were enacted in the United States that require the prompt and complete reclamation of strip-mined land areas.

If the sealing off of pyritic material from oxygen cannot be accomplished, acid mine drainage can still be curbed, at least in theory, by suppressing the activity of the iron- and sulfur-oxidizing bacteria. Broad-spectrum antimicrobial agents would be dangerous pollutants themselves and so cannot be considered for this purpose. In the laboratory, good control of iron and sulfur oxidizers has been achieved with low concentrations of some carboxylic and alpha-keto acids. These compounds are not known to be hazardous to other organisms.

The feasibility of a novel treatment technique for acid mine effluents using the activity of *Desulfovibrio* and *Desulfotomaculum* was demonstrated on the laboratory scale. The mine effluent is combined with large amounts of organic waste materials, such as sawdust. The activity of aerobic and facultatively anaerobic cellulolytic microorganisms lowers the redox potential and produces degradation intermediates that can be utilized by the sulfate reducers. The hydrogen sulfide generated first reduces the ferric iron to ferrous and precipitates the latter as FeS. The process restores neutral pH and removes the iron and sulfur from the effluent, but the economic and environmental feasibility of the proposed process is yet to be explored (Higgins and Burns 1975).

Microbial Conversions of Nitrate

Nitrate reduction and denitrification are normal parts of the nitrogen cycle and balance the nitrification and nitrogen-fixation processes, respectively. Heavy use of nitrogen fertilizers in agriculture can greatly increase the nitrate pool. Nitrate itself is relatively innocuous, but its reductive microbial conversion products cause both local and global pollution problems. For this reason, nitrate in drinking water should not exceed 10 ppm. Nitrate can be reduced to nitrite in anaerobic soils, sediments, and aquatic environments.

Nitrate reduction in foods and feeds is of special concern. Production of nitrite in damp forage, which is high in nitrate, has caused poisoning incidents in cattle. Spoilage of high-nitrate vegetable foods, such as spinach, may cause similar nitrite poisoning in humans. Cured meat products are preserved, in part, by addition of nitrate or nitrite. In the former case, some of the added nitrate is converted by microbial action to nitrite, which is the active preserving and curing agent. Nitrite, through its reaction with myo-globin, lends a pleasing red color to cured meat. Unfortunately, the same affinity to hemoglobin causes its toxicity, and residual nitrite in cured meat is limited to 200 mg per kg. Nitrite, by a still obscure mechanism, inhibits *Clostridium botulinum,* the most dangerous food-poisoning agent. Therefore, even though there is valid concern about the safety of nitrate and nitrite in food products, an abrupt phaseout of this curing agent would probably have grave health consequences.

Besides its direct toxicity, nitrite may react in the environment or in foods with secondary amines to form N-nitrosamines. This reaction may occur spontaneously in some cases or may be mediated by microbial enzymes. For instance, N-nitrosodimethylamine ($[CH_3]_2N$—NO) has been reported in various food products. N-nitrosamines are potent carcinogens, and their ingestion in food and water has proven to be a health hazard. Whether N-nitrosamine-forming reactions can proceed to a significant extent in the intestinal tract of a healthy individual is uncertain at this time. N-nitrosamines are relatively labile compounds and are subject to chemical as well as microbial degradation. Nevertheless, because of their high carcinogenic potential, even low and occasional exposure may be hazardous. Considerable research effort was devoted to a better understanding of the formation and effects of these pollutants (Wolff and Wasserman 1972).

Further microbial reduction of nitrite in anaerobic soils and sediments leads either to ammonium (NH_4^+) or, through nitric oxide (NO), to nitrous oxide (N_2O) and/or to elemental nitrogen. Nitrous oxide and N_2 are both products of denitrification, and at a low pH, their release predominates. Nitrous oxide may rise to the stratosphere and through its photodecomposition contribute to the depletion of the protective ozone layer of the atmosphere. In this manner, excessive denitrification, resulting from an elevated nitrate pool caused by synthetic fertilizer input, becomes a global pollution problem.

Ozone formation in the stratosphere is initiated by photodissociation when light energy (hv) hits molecular oxygen as seen in Equation 20:

$$(20) \quad O_2 + hv \longrightarrow O + O$$

Some of the singlet oxygen reacts with molecular oxygen to form ozone (Equation 21):

$$(21) \quad O + O_2 \longrightarrow O_3$$

In reality, the reactions are more complex, some of them contributing to ozone formation, others depleting it. The equilibrium concentration of ozone strongly depends on the concentration of singlet oxygen.

Ozone is an effective absorber of UV radiation. Without a protective ozone layer, high amounts of UV radiation would reach the surface of Earth with deleterious consequences for living organisms. Even a modest depletion of the ozone layer would be expected to increase the incidence of skin cancer.

Nitrous oxide depletes ozone through a series of photochemical and chemical reactions. The photodissociation of N_2O (Equation 22) produces N_2 and singlet-state oxygen in which one of the electrons has been raised to an excited state (O^*). The excited singlet oxygen then may react with additional N_2O to form NO (Equation 23).

$$(22) \quad N_2O + hv \longrightarrow N_2 + O^*$$

$$(23) \quad N_2O + O^* \longrightarrow 2NO$$

Nitric oxide (NO) reacts with ozone according to Equations 24 and 25.

$$(24) \quad NO + O_3 \longrightarrow NO_2 + O_2$$

$$(25) \quad NO_2 + O \longrightarrow NO + O_2$$

The pair of catalytic reactions (Equations 24 and 25) deplete the stratospheric ozone and also the singlet oxygen that is involved in the formation of ozone, without changing the concentrations of either NO or NO_2. This is a highly efficient catalytic sequence for the conversion of $O_3 + O$ to O_2. Figure 13.22 shows a summary of the natural ozone production and removal processes.

As to the extent and rate of ozone depletion by the above mechanism, calculations and predictions contain many speculative elements and vary widely. There is little doubt, however, that increased denitrifi-

Figure 13.22

A summary of natural ozone production and removal processes in the atmosphere. (Source: National Research Council 1976.)

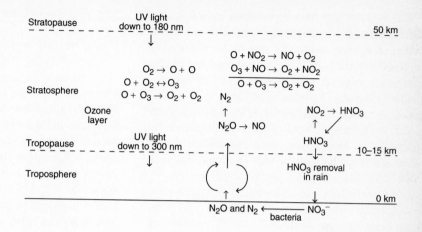

cation, caused by intensive nitrogen fertilizer use, contributes to ozone depletion. The fact that fluorocarbons and oxides of nitrogen from internal combustion and jet engines also contribute to ozone depletion makes the problem even more acute. More prudent and modest use of nitrogen fertilizer could control the problem, but this would also result in decreased agricultural yields. An increased reliance on symbiotic nitrogen fixation would also diminish the problem. An interesting alternative approach is the use of selective nitrification inhibitors. If the fertilizer is applied as ammonia or urea, denitrification cannot occur until the fertilizer is oxidized to nitrate. Very low concentrations of nitrification inhibitors, such as nitrapyrin (Figure 13.23), inhibit the activity of the nitrifying

Figure 13.23

2-chloro-6-(trichloromethyl)-pyridine (nitrapyrin), a highly effective nitrification inhibitor. When added to ammonium fertilizer, it ensures better fertilizer utilization by crop plants and decreases nitrate contamination of groundwater.

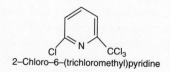

2–Chloro–6–(trichloromethyl)pyridine

bacteria of the *Nitrosomonas* group. This leads to better plant utilization of the nitrogen fertilizer and a decreased nitrate pool. The latter, in turn, decreases the problems connected with nitrate in groundwater and the deleterious effects of nitrate reduction-denitrification products (Huber et al. 1977). Better timing of fertilizer addition and altered tillage practices can also reduce the problem of nitrate leaching.

Microbial Methylations

Microorganisms are capable of transferring methyl groups onto various heavy metals and some metalloids such as selenium and arsenic. Methylation processes result in increased toxicity and biomagnification of these elements. Modern technology uses heavy metals extensively, both in their elemental and combined forms. Microbial interactions with some of these heavy metals result in the mobilization and potentiation of heavy metals as environmental toxicants. Mercury, which has perhaps received the most attention in recent years as a heavy-metal pollutant, serves well to illustrate a process that is not restricted to this metal alone (Summers and Silver 1978).

Metallic mercury is extensively used in the electrical industry, instrument manufacturing, electrolytic processes, and chemical catalysis. Mercury salts and phenylmercury compounds show strong antimicrobial activity and find applications as fungicides, disinfectants, and spoilage retardants. The worldwide annual production and use of mercury is estimated to be 10,000 metric tons. Only a small portion of this material is recycled; most of it winds up as environmental pollutant. The burning of fossil fuels is estimated to release an additional 3,000 metric tons of mercury. The natural input of mercury from rock by weathering is only a fraction of the amount that is released through human activities (National Research Council 1978a).

Mercury salts and phenylmercury compounds, though fairly toxic, are excreted efficiently; therefore, environmental contamination by trace amounts of such compounds were initially not regarded with alarm. A tragic occurrence in Japan dramatized the danger of mercury in the environment and called attention to an insidious microbial process—the methylation of mercury in anaerobic sediments.

In the early 1950s, many inhabitants of a small fishing village at Minamata Bay came down with severe disturbances of the central nervous system. These included tremors and impairment of vision, speech, and coordination. Death or permanent impairment occurred in more than one hundred cases. Most severely affected were the families that consumed, as a major part of their diet, fish and shellfish taken from the bay. Effluents of a chemical plant, containing various mercury compounds, including some methylmercury, were discharged into the same bay. Methylmercury is lipophilic and highly neurotoxic. It was taken up directly, as well as via the food chain by the fish and shellfish in the bay. The inorganic mercury wastes, once incorporated into the anaerobic sediment of the bay, may have been methylated by microorganisms and may have added to the methylmercury contamination, though the contribution of biological methylation versus direct methylmercury discharge was not quantified, and the contribution from microbial methylation probably was minor in this incident.

In contrast to the inorganic and phenylmercury compounds, alkyl mercury is excreted in humans very slowly, with a half-life of thirty days. Consequently, methylmercury concentrations built up in persons consuming tainted seafood that contained as much as 50 ppm of the pollutant, causing the described symptoms, now often referred to as Minamata disease. A similar incident occurred at another Japanese fishing village at Niiagata Bay (D'Itri 1972).

Fortunately, environmentally mediated methylmercury poisoning on a similar scale did not occur elsewhere, but methylmercury did build up to alarming levels in lakes and freshwater fish in Scandinavia because of the use of organomercury fungicides on seeds. A hazardous situation also developed in some of the Great Lakes in the United States, caused by industrial discharges, which forced large areas to be closed to fishing. Ocean fish of higher trophic levels and older age such as tuna and swordfish contain lower, but significant, amounts of mercury, and occasionally, some catches had to be condemned. There is no evidence that the mercury in pelagic fish is a consequence of industrial pollution; comparable amounts have been found in museum specimens dating back as far as one hundred years. The phenomenon illustrates the general tendency of aquatic food chains to accumulate heavy metals and indicates that the margin of safety for increasing the mercury concentration of the marine environment is narrow indeed.

The environmental methylation of mercury and some other heavy metals was attributed to the system responsible for the anaerobic generation of methane (Wood 1971; Ridley et al. 1977; Oremland et al. 1991). Methylcobalamine (methyl-vitamin B_{12}) is able to transfer its methyl group to Hg^{2+} ions, yielding reduced vitamin B_{12} (B_{12}-r). The process may be enzymatically mediated, but it also proceeds spontaneously. Under environmental conditions, the predominant product is monomethylmercury (Hg^+CH_3), but in a second step, smaller amounts of dimethylmercury (CH_3HgCH_3) may also be formed (Figure 13.24). The biomethylation of mercury may be incidental or may represent a detoxification mechanism for the microbe. The synthesis and persistence of methylmercury is favored under anaerobic conditions, while a high redox potential promotes the microbial degradation of methylmercury (Olson and Cooper 1974, 1976; Compeau and Bartha 1984). There appears to be a dynamic equilibrium between generation of methylmercury in anaerobic sediments and its

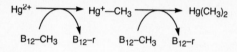

Figure 13.24

Methylation of mercuric ions to monomethyl- and dimethyl-mercury. The donor of the methyl groups here is methyl-cobalamine (the methylated form of vitamin B_{12}). (Source: Wood 1971.)

oxidation in aerated water layers. The resulting minute methylmercury concentrations in the water are sufficient to contaminate aquatic food chains (Figure 13.25).

Phenylmercury compounds are also subject to microbial degradation, and the liberated inorganic mercury, if incorporated into the anaerobic sediment, may be converted to methylmercury. Dimethylmercury is highly volatile and tends to escape to the atmosphere. It is also chemically unstable and decomposes to monomethylmercury under low pH conditions. These factors combine with its low synthesis rate to render dimethylmercury relatively unimportant as an environmental pollutant. The main concern is with monomethylmercury, taken up by fish and shellfish directly through the gills, as well as through the food chain. Additional mercury methylation may result from the activity of anaerobic microorganisms in the intestinal tract.

The involvement of methanogenes and some other bacteria in the environmental mercury methylation process was inferred from *in vitro* experiments using isolated microorganisms and laboratory media (Robinson and Tuovinen 1984). Use of selective inhibitors of methanogenesis and sulfate reduction (Oremland and Capone 1988) directly in anaerobic aquatic sediments yielded the surprising result that more than 95% of mercury methylation in these sediments was performed by sulfate-reducing bacteria, a group not previously suspected to be involved in the biomethylation process (Compeau and Bartha 1985). In a mercury-methylating *Desulfovibrio desulfuricans* strain, the methyl group transferred to mercury originated from C_3 of serine. The methyl transfer agent was found to be vitamin B_{12}, a cobalt corrinoid (Berman et al. 1990; Choi et al. 1991).

Our knowledge of the microbial cycling of other heavy metals and metalloids in nature is rudimentary. Extracts of *Veillonella alcaligenes* (*Micrococcus lactilyticus*) were shown to perform reductive transformations of many metal and metalloid compounds, including selenium, lead, tin, cobalt, and thallium, but we know little about the rate or the significance of these transformations in nature.

Selenium is an essential micronutrient for animals and bacteria that in other than minute concentrations becomes highly toxic (Doran 1982). Chemically,

Figure 13.25

The cycling of mercury and the biological magnification of methylmercury in the aquatic food chain.

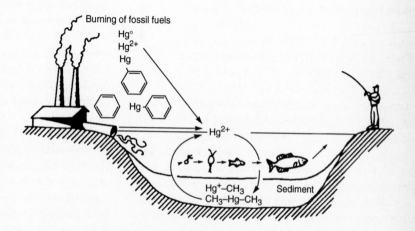

it is closely related to sulfur. Its reduced form, selenide (Se^{2-}), is oxidized by microorganisms to Se^0, SeO_3^{2-} and SeO_4^{2-}, possibly supporting chemolithotrophic growth, though the evidence on this point is yet inadequate. Like sulfate, oxidized forms of selenium serve as electron acceptors under anaerobic conditons for oxidation of hydrogen or organic hydrogen donors. In addition, inorganic selenium can be methylated primarily to dimethylselenide, a volatile product. In contrast to mercury, methylation of selenium decreases rather than increases its toxicity and probably represents a detoxification reaction that occurs readily in aerobic soils and sediments. Under anaerobic conditions, methylation of inorganic selenium was low, but dimethylselenide was released from organoselenium compounds. For methylation of inorganic selenium, both methylcobalamine and S-adenosylmethionine function as methyl donors.

Arsenic is a highly toxic, nonbiogenic element. Its toxicity is at least in part due to the fact that arsenate may replace phosphate in biochemical reactions, thus interfering with the synthesis of essential high-energy phosphate esters. Arsenate can be methylated by some filamentous fungi such as *Scopulariopsis brevicaulis* to mono-, di-, and trimethylarsines (Figure 13.26). The methyl group is donated here either by S-adenosylmethionine or methyl-cobalamine. The products are volatile and highly toxic compounds. Some poisonings occurred when fungi growing on damp wallpaper converted and volatilized arsenate in paint in this manner and the residents inhaled the resulting methylarsines.

Microbial Accumulation of Heavy Metals and Radionuclides

Various heavy metals other than mercury, including tin, cobalt, chromium, nickel, cadmium, and thallium, are used in metal alloys or as catalysts. Their mining, smelting, and ultimate disposal cause heavy-metal pollution problems. All these metals are substantially toxic to plants, animals, and many microorganisms.

Radionuclides as environmental pollutants originate from atmospheric testing of thermonuclear

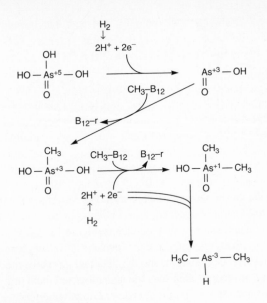

Figure 13.26

The methylation of arsenic to mono-, di-, and trimethylarsines by filamentous fungi. (Source: Wood 1971. Reprinted by permission, copyright John Wiley and Sons.)

weapons, uranium mining and processing, disposal of nuclear wastes, the routine operation of nuclear power plants, and accidents at such installations. To date, atmospheric nuclear testing has been the most significant radionuclide pollution source, but the accumulation of radioactive wastes poses an increasing and not yet satisfactorily solved problem.

Microorganisms, owing to their large surface-to-volume ratio and high metabolic activity, are important vectors in introducing heavy metal and radionuclide pollutants into food webs. At neutral to alkaline pH, heavy metals in soils and sediments tend to be immobilized by precipitation and/or adsorption to cation exchange sites of clay minerals. Microbial production of acid and chelating agents can reverse this adsorption and mobilize the toxic metals. Microbial metabolic products that can chelate metals include dicarboxylic and tricarboxylic acids, pyrocatechol, aromatic hydroxy acids, polyols, and some specific chelators such as the enterochelins and ferrioxamines (see chapter 11).

Mobilization is often followed by uptake and intracellular accumulation of the heavy metals, both by microorganisms and by plant roots. It is not entirely clear why some of these toxic metals are taken up and stored by microorganisms, but intracellular sequestering seems to confer heavy-metal resistance on at least some bacteria. Filamentous fungi were shown to transport heavy metals and radionuclides along their hyphae. This has some implications for the potential role of mycorrhizal fungi in transmitting such pollutants into higher plants. Direct root uptake of heavy metals mobilized by microbial acid production or chelation is an alternative possibility.

The heavy metal cadmium is of special concern in this respect. Cadmium is highly toxic and tends to accumulate with even very low exposures because it is excreted extremely slowly. Its approximate half-life in humans is ten years. Cadmium in humans causes, at low chronic exposure, hypertension and kidney damage. Higher exposures in Japan through rice grown on industrially contaminated fields have caused the painful and crippling itai-itai (ouch-ouch) bone-and-joint disease. Unfortunately, cadmium occurs in low concentrations in some phosphate rocks and thus is present also in phosphate fertilizer spread on agricultural fields. Another potential source of cadmium and other heavy metals is sewage sludge. Such sludge is being considered for use as soil conditioner. Since a substantial portion of sewage sludge is derived from microbial biomass, the relatively high concentration of heavy metals in this material also reflects the ability of microorganisms to concentrate these pollutants. Obviously, the overall mobility of heavy metals in agricultural soils, their uptake by crop plants, and the role of soil microorganisms in these processes should be clarified prior to any large-scale use of sewage sludge on agricultural fields.

Microbial accumulation of radionuclides, with clear human health implications, has been demonstrated in some Arctic areas (Hanson 1967). Lichens are extremely effective in concentrating the radionuclides such as ^{90}Sr and ^{137}Cs from atmospheric fallout (Figure 13.27). During periods of snow cover, when the lichens are shielded from direct fallout, concentrations of radionuclides in the lichens decrease.

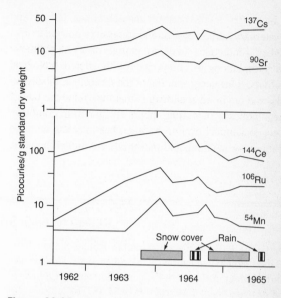

Figure 13.27

Concentrations of ^{54}Mn, ^{90}Sr, ^{106}Ru, ^{137}Cs, and ^{144}Ce in *Cladonia letiaria* lichens in Anaktuvuk Pass, Alaska, during 1962–1965. (Source: Hanson 1967. Reprinted by permission, copyright Pergamon Press Ltd.)

During periods of rain, concentrations of radionuclides increase. Since lichens serve as the primary producers in a food chain of lichen ⟶ caribou ⟶ humans, there can be an efficient transfer of such concentrated elements to the highest member of the food chain. These radionuclides, being deposited in bone tissue, may affect blood cell synthesis in the bone marrow and cause leukemia.

SUMMARY

Industrial societies with mechanized agricultural production are now beset by a number of pollution problems that were largely unknown prior to World War II. Prominent among these problems is environmental pollution by xenobiotic synthetic chemicals that fail to be recycled by microorganisms and may be biomagnified with destructive effects. Reasons for recalcitrance often involve carbon-halogen bonds, tertiary and quaternary carbon atoms, and excessive molecular size. Other xenobiotics undergo partial or even synthetic transformations to compounds with altered biological activity. Our understanding of the molecular, biochemical, and environmental factors that determine the fate of xenobiotics is increasing, but is still far from complete. Biogeochemical cycling by microbes interacts with the mining, agricultural, and industrial activities of humans, resulting in local and global pollution problems. Acid mine drainage originates from strip-mining regions and involves the activity of sulfide and iron-oxidizing bacteria. Excess nitrogen fertilizer may result in global ozone depletion due to the activity of nitrate reducers. Microbial methylation of some heavy metals, accumulation of the same, and additional metals and radionuclides in microorganisms facilitate the entry of these pollutants into food webs and, ultimately, into the human diet.

REFERENCES & SUGGESTED READINGS

Adriaens, P. H., P. E. Kohler, and D. Kohler-Staub. 1989. Bacterial dehalogenation of chlorobenzoates and coculture biodegradation of 4,4'–dichlorobiphenyl. *Applied and Environmental Microbiology* 55:887–892.

Adriaens, P., and D. D. Focht. 1990. Continuous coculture degradation of selected polychlorinated biphenyl congeners by *Acinetobacter* spp. in an aerobic reactor system. *Environmental Science and Technology* 24:1042–1049.

Adrian, N. R., and J. M. Suflita. 1990. Reductive dehalogenation of a nitrogen heterocyclic herbicide in anoxic aquifer slurries. *Applied and Environmental Microbiology* 56:292–294.

Ahmed, M., and D. D. Focht. 1973. Degradation of polychlorinated biphenyls by two species of *Achromobacter*. *Canadian Journal of Microbiology* 19:47–52.

Alexander, M. 1965. Biodegradation: Problems of molecular recalcitrance and microbial fallibility. *Advances in Applied Microbiology* 7:35–80.

Alexander, M. 1981. Biodegradation of chemicals of environmental concern. *Science* 211:132–138.

Atlas, R. M. 1981. Microbial degradation of petroleum hydrocarbons: An environmental perspective. *Microbiological Reviews* 45:180–209.

Atlas, R. M. (ed.). 1984. *Petroleum Microbiology*. Macmillan, New York.

Bagley, D. M., and J. M. Gossett. 1990. Tetrachloroethene transformation to trichloroethene and *cis*–1, 2–dichloroethene by sulfate-reducing enrichment cultures. *Applied and Environmental Microbiology* 56:2511–2516.

Bartha, R. 1980. Pesticide residues in humus. *ASM News* 46:356–360.

Bartha, R., and D. Pramer. 1970. Metabolism of acylanilide herbicides. *Advances in Applied Microbiology* 13:317–341.

Bedard, D. L., R. E. Wagner, and M. J. Brennan. 1987. Extensive degradation of arochlors and environmentally transformed polychlorinated biphenyls by *Alcaligenes eutrophus* H850. *Applied and Environmental Microbiology* 53:1094–1102.

Berman, M., T. Chase, Jr., and R. Bartha. 1990. Carbon flow in mercury biomethylation by *Desulfovibrio desulfuricans*. *Applied and Environmental Microbiology* 56:298–300.

Bodennec, G. J., P. Desmarquest, B. Jensen, and R. Kantin. 1987. Evolution of hydrocarbons and the activity of bacteria in marine sediments contaminated with discharges of petroleum. *International Journal of Environmental and Analytical Chemistry* 29:153–178.

Bossert, I., and R. Bartha. 1984. The fate of petroleum in soil ecosystems. In R. M. Atlas (ed.). *Petroleum Microbiology*. Macmillan, New York, pp. 435–473.

Brandl, H., R. A. Gross, and R. W. Lenz. 1988. *Pseudomonas oleovorans* as a source of poly (ß–hydroxyalkanoates) for potential applications as biodegradable polyesters. *Applied and Environmental Microbiology* 54:1977–1982.

Brenner, T. E. 1969. Biodegradable detergents and water pollution. *Advances in Environmental Science Technology* 1:147–196.

Britton, L. N. 1984. Microbial degradation of aliphatic hydrocarbons. In D. T. Gibson (ed.). *Microbial Degradation of Organic Compounds*. Marcel Dekker, New York, pp. 89–129.

Brown, J. F., D. L. Bedard, and M. J. Brennan. 1987. Polychlorinated biphenyl dechlorination in aquatic sediments (river sediments). *Science* 236:709–712.

Bumpus, J. A. 1989. Biodegradation of polycyclic aromatic hydrocarbons by *Phanerochaete chrysosporium.* *Applied and Environmental Microbiology* 55:154–158.

Bumpus, J. A., M. Tien, D. Wright, and S. D. Aust. 1985. Oxidation of persistent environmental pollutants by a white rot fungus. *Science* 228:1434–1436.

Bumpus, J. A., and S. D. Aust. 1987. Biodegradation of DDT [1,1,1–trichloro–2,2–bis(4–chlorophenyl) ethane] by the white rot fungus *Phanerochaete chrysosporium.* *Applied and Environmental Microbiology* 53:2001–2028.

Butler, J. N., B. F. Morris, and J. Sass. 1973. Pelagic tar from Bermuda and the Sargasso Sea. *Bermuda Biological Station Special Publication No. 10.* Bermuda.

Cade, T. J., J. L. Lincer, C. M. White, D. G. Roseneau, and L. G. Swartz. 1971. DDE residues and eggshell changes in Alaskan falcons and hawks. *Science* 172:955–957.

Carpenter, E. J. 1972. Polystyrene spherules in coastal waters. *Science* 178:749–750.

Carson, R. 1962. *Silent Spring.* Houghton-Mifflin, New York.

Cerniglia, C. E. 1984. Microbial transformation of aromatic hydrocarbons. In R. M. Atlas (ed.). *Petroleum Microbiology.* Macmillan, New York, pp. 99–128.

Chatterjee, D. K., J. J. Kilbane, and A. M. Chakrabarty. 1982. Biodegradation of 2,4,5-trichlorophenoxyacetic acid in soil by a pure culture of *Pseudomonas cepacia.* *Applied and Environmental Microbiology* 44:514–516.

Chaudhry, G. R., and S. Chapalamadugu. 1991. Biodegradation of halogenated organic compounds. *Microbiological Reviews* 55:59–79.

Choi, S. C., M. Berman, and R. Bartha. 1991. Evidence for a novel Hg2-methylating corrinoid in *Desulfovibrio desulfuricans.* *91st Annual Meeting, American Society for Microbiology,* 5–9 May, Abstract Q261, Dallas TX.

Colwell, E. B. (ed.). 1971. *The Ecological Effects of Oil Pollution on Littoral Communities.* Applied Science Publishers, London.

Compeau, G., and R. Bartha. 1984. Methylation and demethylation of mercury under controlled redox, pH and salinity conditions. *Applied and Environmental Microbiology* 48:1203–1207.

Compeau, G., and R. Bartha. 1985. Sulfate-reducing bacteria: Principal methylators of mercury in anoxic estuarine sediment. *Applied and Environmental Microbiology* 50:498–502.

Dagley, S. 1975. A biochemical approach to some problems of environmental pollution. In P. N. Campbell and W. N. Aldridge (eds.). *Essays in Biochemistry.* Vol. 2. Academic Press, London, pp. 81–130.

Dalton, H., and D. I. Stirling. 1982. Co-metabolism. *Philosophical Transactions of the Royal Society* (London), Ser. B 297:481–491.

D'Itri, F. M. 1972. *The Environmental Mercury Problem.* CRC Press, Cleveland, Ohio.

Dolfing, J., and J. M. Tiedje. 1987. Growth yield increase linked to reductive dechlorination in a defined 3-chlorobenzoate degrading methanogenic coculture. *Archives of Microbiology* 149:102–105.

Doran, J. W. 1982. Microorganisms and the biological cycling of selenium. *Advances in Microbial Ecology* 6:1–32.

Drinkwine, A., S. Spurlin, J. Van Emon, and V. Lopez-Avila. 1991. Immuno-based personal exposure monitors. In *Field Screening Methods for Hazardous Wastes and Toxic Chemicals.* U.S. Environmental Protection Agency and U.S. Army Toxic and Hazardous Materials Agency, Las Vegas, Nev. pp. 449–459.

Efroymson, R. A., and M. Alexander. 1991. Biodegradation by an *Arthrobacter* species of hydrocarbons partitioned into an organic solvent. *Applied and Environmental Microbiology* 57:1441–1447.

Evans, J. D., and S. K. Sikdar. 1990. Biodegradable plastics: An idea whose time has come. *Chemtech* 20:38–42.

Fathepure, B. Z., J. M. Tiedje, and S. A. Boyd. 1988. Reductive dechlorination of hexachlorobenzene to tri- and dichlorobenzenes in anaerobic sewage sludge. *Applied and Environmental Microbiology* 54:327–330.

Fogel, M. M., A. R. Taddeo, and S. Fogel. 1986. Biodegradation of chlorinated ethanes by a methane-utilizing mixed culture. *Applied and Environmental Microbiology* 51:720–724.

Fujiwara, K. 1975. Environmental and food contamination with PCBs in Japan. *Science of the Total Environment* 4:219–247.

Galli, R., and P. L. McCarty. 1989. Biotransformation of 1,1,1–trichloroethane, trichloromethane, and tetrachloromethane by a *Clostridium* sp. *Applied and Environmental Microbiology* 55:837–844.

Genthner, B. R. S., W. A. Price, and P. H. Pritchard. 1989a. Anaerobic degradation of chloroaromatic compounds in aquatic sediments under a variety of enrichment conditions. *Applied and Environmental Microbiology* 55:1466–1471.

Genthner, B. R. S., W. A. Price, and P. H. Pritchard. 1989b. Characterization of anaerobic dechlorinating consortia derived from aquatic sediments. *Applied and Environmental Microbiology* 55:1472–1476.

Ghosal, D., I. S. You., D. K. Chatterjee, and A. M. Chakrabarty. 1985. Microbial degradation of halogenated compounds. *Science* 228:135–142.

Gibson, D. T., and V. Subramanian. 1984. Microbial degradation of aromatic hydrocarbons. In D. T. Gibson (ed.). *Microbial Degradation of Organic Compounds*. Plenum Press, New York, pp. 181–252.

Gosset, R. W., D. A. Brown, and D. R. Young. 1983. Predicting the bioaccumulation of organic compounds in marine oganisms using octanol/water partition coefficients. *Marine Pollution Bulletin* 14:387–392.

Grbic-Galic, D., and T. M. Vogel. 1987. Transformation of toluene and benzene by mixed methanogenic cultures. *Applied and Environmental Microbiology* 53:254–260.

Greaves, M. P., H. A. Davies, J. A. P. Marsh, and G. I. Wingfield. 1977. Herbicides and soil microorganisms. *CRC Critical Reviews of Microbiology* 5:1–38.

Gustafson, C. G. 1970. PCBs: Prevalent and persistent. *Environmental Science and Technology* 4:814–819.

Hallas, L. E., and M. Alexander. 1983. Microbial transformation of nitroaromatic compounds in sewage effluent. *Applied and Environmental Microbiology* 45:1234–1241.

Hanson, W. C. 1967. Cesium-137 in Alaskan lichens, caribou and Eskimos. *Health Physics* 13:383–389.

Harker, A. R., R. H. Olsen, and R. J. Seidler. 1989. Phenoxyacetic acid degradation by the 2,4–dichlorophenoxyacetic acid (TFD) pathway of plasmid pJP4: Mapping and characterization of the TFD regulatory gene, *tfd*R. *Journal of Bacteriology* 171:314–320.

Harker, A. R., and Y. Kim. 1990. Trichloroethylene degradation by two independent aromatic-degrading pathways in *Alcaligenes eutrophus* JMP134. *Applied and Environmental Microbiology* 56:1179–1181

Heitkamp, M. A., and C. E. Cerniglia. 1988. Mineralization of polycyclic aromatic hydrocarbons by a bacterium isolated from sediment below an oil field. *Applied and Environmental Microbiology* 54:1612–1614.

Higgins, I. J., and R. G. Burns. 1975. *The Chemistry and Microbiology of Pollution*. Academic Press, London.

Hill, I. R., and S. J. L. Wright (eds.). 1978. *Pesticide Microbiology*. Academic Press, London.

Hoeks, J. 1972. Changes in composition of soil near air leaks in natural gas mains. *Soil Science* 113:46–54.

Horvath, R. S. 1972. Microbial co-metabolism and the degradation of organic compounds in nature. *Bacteriological Reviews* 36:146–155.

Huber, D. M., H. L. Warren, D. W. Nelson, and C. Y. Tsai. 1977. Nitrification inhibitors—New tools for food production. *BioScience* 27:523–529.

Johnson, L. M., C. S. McDowell, and M. Krupha. 1985. Microbiology in pollution control: From bugs to biotechnology. *Developments in Industrial Microbiology* 26:365–376.

Kellogg, S. T., D. K. Chatterjee, and A. M. Chakrabarty. 1981. Plasmid assisted molecular breeding—New technique for enhanced biodegradation of persistent toxic chemicals. *Science* 214:1133–1135.

Kennedy, D. W., S. D. Aust, and J. A. Bumpus. 1990. Comparative biodegradation of alkyl halide insecticides by the white rot fungus, *Phanerochaete chrysosporium* (BKM–F–1767). *Applied and Environmental Microbiology* 56:2347–2353.

Kennicutt, M. C. 1990. Oil spillage in Antartica. *Environmental Science and Technology* 24:620–624.

Kilbane, J. J., D. K. Chatterjee, J. S. Karns, S. T. Kellog, and A. M. Chakrabarty. 1982. Biodegradation of 2,4,5-trichlorophenoxyacetic acid by a pure culture of *Pseudomonas cepacia*. *Applied and Environmental Microbiology* 44:72–78.

Kohler, H. -P. E., D. Kohler-Staub, and D. D. Focht. 1988. Cometabolism of polychlorinated biphenyls: Enhanced transformation of Aroclor 1254 by growing bacterial cells. *Applied and Environmental Microbiology* 54:1940–1945.

Krupp, L. R., and W. J. Jewell. 1992. Biodegradability of modified plastic films in controlled biological environments. *Environmental Science and Technology* 26:193–198.

Larson, R. J. 1990. Structure-activity relationships for biodegradation of linear alkylbenzenesulfonates. *Environmental Science and Technology* 24:1241–1246.

Leahy, J. G., and R. R. Colwell. 1990. Microbial degradation of hydrocarbons in the environment. *Microbiological Reviews* 54:305–315 .

Lee, K., and E. M. Levy. 1991. Bioremediation: Waxy crude oils stranded on low-energy shorelines. In *Proceedings of the 1991 International Oil Spill Conference*. American Petroleum Institute, Washington, D.C., pp. 541–547.

Linkfield, T. G., J. M. Suflita, and J. M. Tiedje.1989. Characterization of the acclimation period before anaerobic dehalogenation of halobenzoates. *Applied and Environmental Microbiology* 55:2773–2778.

Little, C. D., A. V. Palumbo, and S. E. Herbes. 1988. Trichloroethylene biodegradation by a methane-oxidizing bacterium. *Applied and Environmental Microbiology* 54:951–956.

Massé, R., F. Messier, L. Peloquin, C. Ayotte, and M. Sylvestre. 1984. Microbial degradation of 4-chlorobiphenyl, a model compound of chlorinated biphenyls. *Applied and Environmental Microbiology* 47:947–951.

Maugh, T. H. 1973. DDT: An unrecognized source of polychlorinated biphenyls. *Science* 180:578–579.

Miller, R. M., and R. Bartha. 1989. Evidence from liposome encapsulation for transport-limited microbial metabolism of solid alkanes. *Applied and Environmental Microbiology* 55:269–274.

Mohn, M. M., and J. M. Tiedje. 1990a. Catabolite thiosulfate disproportionation and carbon dioxide reduction in strain DCB-1, a reductively dechlorinating anaerobe. *Journal of Bacteriology* 172:2065–2070.

Mohn, M. M., and J. M. Tiedje. 1990b. Strain DCB-1 conserves energy for growth from reductive dechlorination coupled to formate oxidation. *Archives of Microbiology* 153:267–271.

Molina, M. J., and F. S. Rowland. 1974. Stratospheric sink for chlorofluoromethanes—chlorine atom-catalysed destruction of ozone. *Nature* (London) 249:810–812.

National Research Council. 1972. *Degradation of Synthetic Organic Molecules in the Biosphere.* National Academy of Sciences, Washington, D.C.

National Research Council. 1976. *Halocarbons: Effects on Stratospheric Ozone.* National Academy of Sciences, Washington, D.C.

National Research Council. 1978a. *An Assessment of Mercury in the Environment.* National Academy of Sciences, Washington, D.C.

National Research Council. 1978b. *Kepone, Mirex, Hexachlorocyclopentadiene: An Environmental Assessment.* National Academy of Sciences, Washington, D.C.

National Research Council. 1979. *Polychlorinated Biphenyls.* National Academy of Sciences, Washington, D.C.

National Research Council. 1985. *Oil in the Sea: Inputs, Fates and Effects.* National Academy Press, Washington, D.C.

Novick, N. J., and M. Alexander. 1985. Cometabolism of low concentrations of propachlor, alachlor, and cycloate in sewage and lake water. *Applied and Environmental Microbiology* 49:737–743.

Olson, B. H., and R. C. Cooper. 1974. *In situ* methylation of mercury in estuarine sediment. *Nature* (London) 252:682–683.

Olson, B. H., and R. C. Cooper. 1976. Comparison of aerobic and anaerobic methylation of mercuric chloride in San Francisco Bay sediments. *Water Research* 10:113–116.

Oremland, R. S., and D. G. Capone. 1988. Use of "specific" inhibitors in biogeochemistry and microbial ecology. *Advances in Microbial Ecology* 10:285–383.

Oremland, R. S., C. W. Culbertson, and M. R. Winfrey. 1991. Methylmercury decomposition in sediments and bacterial cultures: Involvement of methanogens and sulfate reducers in oxidative demethylation. *Applied and Environmental Microbiology* 57:130–137.

Orndorff, S. A., and R. R. Colwell. 1980. Microbial transformation of Kepone. *Applied and Environmental Microbiology* 39:398–406.

Perry, J. J. 1984. Microbial metabolism of cyclic alkanes. In R. M. Atlas (ed.). *Petroleum Microbiology.* Macmillan, New York, pp. 61–97.

Pfaender, F. K., and M. Alexander. 1972. Extensive microbial degradation of DDT *in vitro* and DDT metabolism by natural communities. *Journal of Agricultural Food Chemistry* 20:842–846.

Quensen, J. F., J. M. Tiedje, and S. A. Boyd. 1988. Reductive dechlorination of polychlorinated biphenyls by anaerobic microorganisms from sediments. *Science* 242:752–754.

Ramsey, B. A., K. Lomaliza, C. Chavarie, B. Dube, P. Bataille, and J. Ramsay. 1990. Production of poly-(β-hydroxybutyric-co-β-hydroxyvaleric) acids. *Applied and Environmental Microbiology* 56:2093–2098.

Reich, M., and R. Bartha. 1977. Degradation and mineralization of a polybutene film-mulch by the synergistic action of sunlight and soil microbes. *Soil Science* 124:177–180.

Ridley, W. P., L. J. Dizikes, and J. M. Wood. 1977. Biomethylation of toxic elements in the environment. *Science* 197:329–332.

Robinson, J. B., and O. H. Tuovinen. 1984. Mechanism of microbial resistance and detoxification of mercury and organomercury compounds: Physiological, biochemical and genetic analyses. *Microbiological Reviews* 48:95–124.

Rontani, J. F., F. Bosser-Joulak, E. Rambeloarisoa, J. E. Bertrand, G. Giusti, and R. Faure. 1985. Analytical study of Asthart crude oil asphaltenes biodegradation. *Chemosphere* 14:1413–1422.

Safe, S. H. 1984. Microbial degradation of polychlorinated biphenyls. In D. T. Gibson (ed.). *Microbial Degradation of Organic Compounds*. Marcel Dekker, New York, pp. 361–369.

Shiaris, M. P. 1989. Seasonal biotransformation of naphthalene, phenanthrene, and benzo[a]pyrene in surficial estuarine sediments. *Applied and Environmental Microbiology* 55:1391–1399.

Singer, M. E., and W. R. Finnerty. 1984. Microbial metabolism of straight-chain and branched alkanes. In R. M. Atlas (ed.). *Petroleum Microbiology*. Macmillan, New York, pp. 1–59.

Song, H. G., T. A. Pedersen, and R. Bartha. 1986. Hydrocarbon mineralization in soil: Relative bacterial and fungal contribution. *Soil Biology Biochemistry* 18:109–111.

Stevens, T.O., and J. M. Tiedje. 1988. Carbon dioxide fixation and mixotrophic metabolism by strain DCB-1, a dehalogenating anaerobic bacterium. *Applied and Environmental Microbiology* 54:2944–2948.

Stevens, T.O., T. G. Linkfield, and J. M. Tiedje. 1988. Physiological characterization of strain DCB-1, a unique dehalogentaing sulfidogenic bacterium. *Applied and Environmental Microbiology* 54:2938–2943.

Stucki, G., W. Brunner, D. Staub, and T. Leisinger. 1981. Microbial degradation of chlorinated C1 and C2 hydrocarbons. In T. Leisinger, A. M. Cook, R. Hütter, and J. Nüesch (eds.). *Microbial Degradation of Xenobiotics and Recalcitrant Compounds*. Academic Press, London, pp. 131–137.

Suflita, J. M., A. Horowitz, D. R. Shelton, and J. M. Tiedje. 1982. Dehalogenation: A novel pathway for the anaerobic biodegradation of haloaromatic compounds. *Science* 214:1115–1117.

Summers, A. O., and S. Silver. 1978. Microbial transformations of metals. *Annual Reviews of Microbiology* 32:637–672.

Thayer, A. M. 1990. Degradable plastics create controversy in solid waste issues. *Chemical Engineering News* 68(26):7–14.

Trower, M. K., R. M. Buckland, R. Higgins, and M. Griffin. 1985. Isolation and characterization of a cyclohexane-metabolizing *Xanthobacter* sp. *Applied and Environmental Microbiology* 49:1282–1289.

Trudgill, P. W. 1984. Microbial degradation of the alicyclic ring. In D. T. Gibson (ed.). *Microbial Degradation of Organic Compounds*. Marcel Dekker, New York, pp. 131–180.

Tucker, R. E., A. L. Young, and A. P. Gray (eds.). 1983. *Human and Environmental Risks of Chlorinated Dioxins and Related Compounds*. Plenum Press, New York.

Vogel, T. M., and P. L. McCarty. 1985. Biotransformation of tetrachloroethylene to trichloroethylene, dichloroethylene, vinyl chloride and carbon dioxide under methanogenic conditions. *Applied and Environmental Microbiology* 49:1080–1083.

Vogel, T. M., and D. Grbic-Galic. 1986. Incorporation of oxygen from water into toluene and benzene during anaerobic fermentative transformation. *Applied and Environmental Microbiology* 52:200–202.

Wedemeyer, G. 1967. Dechlorination of l, l, l-trichloro-2,2,-bis (p-chlorophenyl) ethane by *Aerobacter aerogenes*. *Applied Microbiology* 15:569–574.

Wolff, I. A., and A. E. Wasserman. 1972. Nitrates, nitrites and nitrosamines. *Science* 177:15–19.

Wood, J. M. 1971. Environmental pollution by mercury. *Advances in Environmental Science Technology* 2:39–56.

Zeyer, J., E. P. Kuhn, and R. P. Schwarzenbach. 1986. Rapid microbial mineralization of toluene and 1,3-dimethylbenzene in the absence of molecular oxygen. *Applied and Environmental Microbiology* 52:944–947.

Zeyer, J., P. Eicher, J. Dolfing, and P. R. Schwarzenbach. 1990. Anaerobic degradation of aromatic hydrocarbons. In D. Kamely, A. Chakrabarty, and G. S. Omenn (eds.). *Biotechnology and Biodegradation*. Gulf Pub. Co., Houston, Tex., pp. 33–40.

14

Biodegradability Testing and Monitoring the Bioremediation of Xenobiotic Pollutants

In response to the pollution and biomagnification caused by xenobiotics that we outlined in chapter 13, the United States and other developed nations took a number of actions to deal with these problems. In this chapter, we describe these actions and policies. In some cases, regulatory agencies such as the U. S. Environmental Protection Agency (EPA) prohibited the manufacture and/or domestic use of the most offending substances, such as DDT, PCBs, and similar agents, especially when less hazardous substitutes for them could be found. In addition, the manufacture and sale of newly developed chemical agents became contingent on their passing certain environmental safety tests. Extensive environmental monitoring by federal and state agencies was initiated to detect and correct chemical pollution problems before ecosystems were destroyed and humans poisoned. Finally, the complex and difficult task of hazardous chemical waste disposal and the restoration of polluted sites was gradually confronted.

Greatly complicating the present task is the legacy of a less responsible past that has left a multitude of haphazard chemical-waste dump sites around the world that now pollute groundwater and constitute an imminent health hazard to nearby communities. The cleanup process will require a variety of different technologies. Some toxic chemical wastes will require high-temperature incineration or storage in secure lined and capped chemical landfills. Other hazardous wastes, however, may be biodegraded, detoxified, or immobilized more cost-effectively by a variety of novel microbial treatment techniques. Increasingly, such measures are being referred to as bioremediation.

BIODEGRADABILITY AND ECOLOGICAL SIDE EFFECT TESTING

Testing for Biodegadability and Biomagnification

As described in the previous sections, microorganisms have a finite capacity to recycle synthetic organic molecules. Because recalcitrance of an organic substance introduced into the biosphere on a large scale may cause problems, the EPA has increased its demands for experimental proof from manufacturers that their product will be recycled to harmless compounds. In practice, this usually requires proof of biodegradation within a reasonable period of time and, in addition, a demonstration that essential processes in the environment are not disrupted. Such procedures add substantially to the development costs of new products affected by such regulations, and the consumer eventually bears these costs. In the long run, however, the haphazard introduction of new materials into the biosphere would prove even costlier.

The U. S. government's authority to regulate the manufacture and distribution of potentially dangerous chemicals was vastly broadened by the 1976 Toxic Substances Control Act (TOSCA). This legislation requires the producer to show that a newly introduced chemical will not have undue toxic environmental side effects. Permits for the large-scale manufacture and sale of new chemicals are contingent on the outcome of the tests. The legislation is based on the experience that, regardless of intended use, any recalcitrant and toxic chemical manufactured on a large scale is likely to cause environmental pollution problems, as occurred with DDT, PCBs, and similar compounds. Test procedures must be reasonably representative, accurate, and broadly comparable, yet simple and routine enough not to impose unacceptable delays and financial burdens on the introduction of new products. Pesticides have been subject to testing requirements for some time, and these TOSCA test procedures are similar, at least in part, to test procedures currently applied to pesticides.

The classical approach to the demonstration of biodegradability has been the enrichment culture method, with the test substance serving as substrate. This technique is still useful for the isolation and purification of the responsible microorganisms, but conditions are far removed from those prevailing in the natural environment, such as in the soil, and the results are not representative. Substances that cannot serve as sole carbon sources may be degraded by a complex microbial community in the presence of other sub-

stances. The organism isolated in elective culture may not be the one that is the most active in its native environment. For this reason, the currently accepted test procedure is to introduce the test substance into a representative environment, for example, a biologically active soil or water sample, at a normal rate and, for safety, also at an excessive rate ten times higher, and to incubate the treated samples under representative environmental conditions (Pramer and Bartha 1972). Biodegradation of the test substance is monitored by respirometric or CO_2-evolution techniques. In aquatic systems, BOD measurement using unacclimated or acclimated sewage sludge as inoculum is useful. Acclimation or adaptation is accomplished by exposing sewage sludge to increasing concentrations of the test substance over a period of time while feeding the sewage sludge microbial community with raw or synthetic sewage. If specific analytical techniques are available, the disappearance of the test substance from the soil or aquatic sample can be monitored by periodic extraction of a replicate sample followed by gas chromatography or other suitable microanalytical process.

Because degradation products cannot always be anticipated, it is of great advantage to use radiolabeled test substances, such as ^{14}C. Conversion of the bulk of the test substance to $^{14}CO_2$ is the most clear-cut proof of mineralization. The ^{14}C label, if it is not uniform, must be carefully placed. Biodegradation may affect only a few carbon atoms in the molecule of the test substance, but if these happen to be the labeled ones, the mistaken conclusion would be that the whole molecule was mineralized (Figure 14.1). If the ^{14}C label is uniform or is placed in the difficult-to-degrade portion of the molecule, it can be of great help in the identification of degradation products. These may have biological activity substantially different from the parent substance. The custom labeling of test substances, however, involves considerable expense and, is therefore not always used.

Biodegradation experiments in environmental samples incubated under controlled laboratory conditions are probably the best routine tests for recalcitrance. While they are certainly superior to experiments with pure cultures or even mixed enrichments, they do not duplicate certain field conditions. Photodegra-

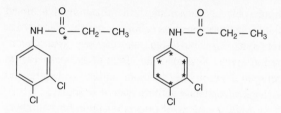

Figure 14.1

Significance of the position of ^{14}C label in bio-degradation tests. Asterisks indicate label positions in the herbicide N-(3,4-dichlorophenyl)-propionamide (propanil). Tests with propanil labeled in the acyl moiety (A) indicated more than 90% mineralization in two weeks. Labeling in the 3,4-dichloroaniline ring (B) resulted in less than 5% mineralization during the same time period.

dation, evaporation, and leaching are excluded or restricted in these laboratory experiments. Some co-metabolic pesticide transformations seem to be greatly stimulated in the plant rhizosphere by the presence of root exudates (Hsu and Bartha 1979), yet to test for such effects on a routine basis appears to be too laborious. Laboratory tests require final verification by field experience.

If biodegradation tests indicate a high degree of recalcitrance and the residue has a strongly lipophilic character, there is a good chance that the substance will be subject to biomagnification. The lipophilic character and, hence, the potential for biomagnification can be predicted to some degree from the partition coefficient of the pollutant between water and a less polar solvent—*n*-octanol, for example (Gosset et al. 1983). The stronger the lipophilic character, the greater the potential for biomagnification. However, this simple partitioning test fails to predict changes that render the pollutant more polar and facilitate its excretion.

To use the environment as the testing laboratory, as occurred with DDT, PCBs, and other persistent pollutants, is too costly in terms of ecological damage. R. L. Metcalf and associates (1971) designed a useful model ecosystem for testing biomagnification potential (Figure 14.2). The model ecosystem is in a small aquarium that contains a sand slope and is par-

tially filled with water. Sorghum is planted on the sand slope, and the radiolabeled test compound, such as DDT or an analog, is applied in a small volume of acetone to the plant leaves. Salt-marsh caterpillar (*Estigmene*) larvae are introduced and feed on the sorghum leaves. Their droppings enter the aquatic environment containing a mixed bacterial and planktonic inoculum from a pond. Other added organisms are a filamentous green alga (*Oedogonium*), an aquatic snail (*Physa*), mosquito larvae (*Culex*), and the mosquito fish (*Gambusia*). The ecosystem is maintained for thirty days after introduction of the radiolabeled test compound. After this time, the macroscopic organisms are sorted, the planktonic organisms are filtered out, and radioactivity is determined in each group of organisms and also in the sand and water. A compound like DDT typically shows 10^5 to 10^6-fold biomagnification in *Gambusia,* which represents the highest trophic level in this model ecosystem. Less persistent and/or more polar compounds show proportionally less biomagnification. This model ecosystem is a flexible one and can be modified with relative ease to model other food chains. Improvements on the

Figure 14.2

Model ecosystem developed by R. L. Metcalf and associates for testing the biomagnification tendency of pesticides. (Source: Metcalf et al. 1971, *Environmental Science and Technology.* Reprinted by permission, copyright American Chemical Society.)

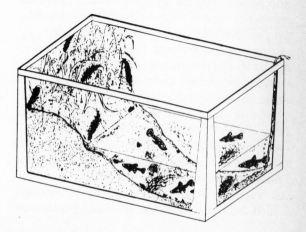

system also include an enclosed atmosphere and appropriate scrubbers to account for evolution of $^{14}CO_2$ and other volatiles. The predictive value of tests in such model systems appears to be good, and their cost is not excessive.

Related test systems for the evaluation of pollutant behavior are microcosms (Pritchard and Bourquin 1984). These differ from model ecosystems in that they are not synthetic but attempt to bring a minimally disturbed subsample of the real environment into the laboratory. Such systems are of value because they give not only qualitative but also reasonably good quantitative results about the behavior of a pollutant in the sampled environment. Biodegradation of a pollutant in an appropriate microcosm closely resembles the rate measured in the corresponding environment. A simple microcosm is the "eco-core," which models estuarine and other shallow-water environments. A sediment core is placed in a glass column of corresponding diameter and overlayered with a simultaneously collected water sample. The water column is aerated, but the anaerobic sediment is left undisturbed. The effluent air is scrubbed through an alkali trap, and the $^{14}CO_2$ evolution rate from radiolabeled pollutants is monitored. Periodic analysis of replicate eco-cores for nonvolatile residues complements $^{14}CO_2$ analysis.

Testing for Effects on Nontarget Microorganisms

In the case of biologically active substances that are introduced into the environment in large quantities, such as pesticides, it should be ascertained that nontarget organisms are not unduly affected. Such testing may involve toxicity tests for wildlife or fish, but as applied to the natural microbiological activity in soil, it should be demonstrated that the microbial processes essential to soil fertility are not seriously disrupted (Greaves et al. 1977). Test procedures in this area are in a state of evolution. Of the various approaches, the most sound and relevant ones appear to be those that measure direct effects on biogeochemical cycling in environmental samples (Johnen and Drew 1977; Atlas et al. 1978). For example, the effect of a pesticide on

the aerobic and anaerobic mineralization of a model plant constituent, such as cellulose, may be measured at field application and at ten times the suggested concentration levels. Similarly, the effect of the pesticide on nitrogen fixation, nitrification, sulfur oxidation, and perhaps phosphatase activity may be tested.

Measurements of effects on microbial numbers appear to be less useful, but effects on ATP or on the adenylate energy charge may prove to be relevant toxicity indicators. Least useful are tests of pesticide toxicity *in vitro* on a selected microbial strain, such as *Azotobacter* or *Nitrobacter*. Such tests ignore the diversity and homeostasis of natural microbial communities, as well as the detoxification of the pesticide in the natural environment by soil absorption and microbial degradation. No valid extrapolations are possible from such *in vitro* pure culture experiments to effects on biogeochemical cycling in the environment, and there seems to be little scientific or practical justification for such tests.

One xenobiotic may influence the biodegradation of another; for example, some carbamate insecticides inhibit acylamidases that cleave acylanilide and phenylcarbamate herbicides. The great diversity of xenobiotics seems to preclude routine testing for such possible interactions. At best, the effect of a newly developed xenobiotic might be tested on the mineralization of some other radiolabeled compounds selected to represent some prominent classes of xenobiotic compounds.

BIOSENSOR DETECTION OF POLLUTANTS

Traditionally, the monitoring of environmental pollutants and the contaminants they produce has involved physicochemical analytical techniques. Spectrometry (UV, visible, IR) and chromatography—such as high-pressure liquid chromatography (HPLC) or gas chromatography (GC), often coupled with mass spectrometry (MS)—represent increasingly powerful analytical tools that are capable of detecting, quantifying, and identifying xenobiotic pollutants in parts per million (ppm) and even in the parts per billion (ppb) range. These powerful analytical techniques are now being supplemented by biosensors, which are biological systems based upon reporter genes or immunoassays that can detect specific pollutants. Biosensors are just beginning to be commercialized for a few chemicals; for the most part though, they are still in experimental development for environmental monitoring (Van Emon 1989; Van Emon et al. 1989, 1991).

Reporter Genes

Reporter genes can be readily detected when they are expressed. These genes can be joined through recombinant DNA technology to the genes that code for specific metabolic pathways that are inducible. When the pathway is induced, so is the reporter gene. Thus, a reporter gene that codes for production of light, such as the *lux* gene that codes for luminescence, can be coupled with various inducible genes. If light is emitted, the pathway has been induced, indicating the presence of a specific inducer substance. If that inducer is a pollutant, the reporter gene can be used as a sensitive indicator of that pollutant. Biosensors based upon the *lux* gene have been genetically engineered to detect naphthalene; in the presence of naphthalene, light is emitted at an intensity proportional to the concentration of naphthalene (Burlage et al. 1990).

Immunoassay Biosensors

Immunoasssay biosensors have been developed to detect a variety of organic molecules, including benzopyrene and parathion (Vanderlaan et al. 1988; Van Emon et al. 1989; Reynolds and Yacynych 1991; Schultz 1991). These assays are based upon the specificities of immunological reactions. Because antibodies react only with specific antigens, immunoassays can be used to detect substances that are both pollutants and antigenic, that is, capable of reacting with a specific diagnostic antibody. Monoclonal antibodies, which react only with a single specific antigen, are particularly useful as diagnostic biosensors for the immunological detection of environmental pollutants.

Although still a novel application of this technology, immunoassays of various types have been developed for environmental pollutants (Lukens et al. 1977; Chamerlik-Cooper et al. 1991; Dohrman 1991). These immunoassays use different labels for detection, including radioactive labels in radioimmunoassays (RIAs), enzymatic reactions that produce colored products in enzyme-linked immunosorbant assays (ELISAs), and fluorescent dyes in fluorescent immunoassays (FIAs). Radioimmunoassays have been developed for detection of dieldrin, aldrin, warfarin, benomyl, and parathion; ELISAs for molinate, metalaxyl, chlorsulfuron, paraquat, triadimefon, diflubenzuron, and terbutryn; and FIAs for diclofop methyl. Portable, field-test immunoassay kits for polychlorinated biphenyls (PCBs) and pentachlorophenol (PCP) are now commercially available. Field-test kits have also been developed for the toxic aromatic hydrocarbons benzene, toluene, and xylene.

Present research in immunochemical biosensors is moving toward the development of innovative assays for air samples (Drinkwine et al. 1991; Stopa et al. 1991). Efforts in this area of environmental detection, previously thought incompatible with immunological techniques, are evidence of the growing realization of the potential value of these methods in environmental measurement.

BIOREMEDIATION

The biodegradation of pollutants in the environment is a complex process, whose quantitative and qualitative aspects depend on the nature and amount of the pollutant present, the ambient and seasonal environmental conditions, and the composition of the indigenous microbial community (Atlas 1981; Leahy and Colwell 1990; Hinchee and Olfenbuttel 1991a, 1991b). Although in practice the separation of causes is often difficult, in theory one may assume that the failure of chemical-waste biodegradation may be due to unfavorable environmental conditions, to the absence of appropriate catabolic pathways, or to both. Established or emerging biotechnological approaches can correct such situations and protect the environment from xenobiotic pollutants. With the recognition that rates of pollutant biodegradation are often limited by environmental constraints or possibly by the lack of suitable microbial populations, it is possible to carry out bioremediation programs to overcome these limitations (Bluestone 1986; Zitrides 1990).

Bioremediation involves the use of microorganisms to remove pollutants (Atlas and Pramer 1990). The two general approaches to bioremediation are environmental modification, such as through nutrient application and aeration, and the addition of appropriate xenobiotic degraders by seeding. The end products of effective bioremediation, such as water and carbon dioxide, are nontoxic and can be accommodated without harm to the environment and living organisms. Using bioremediation to remove pollutants is inexpensive compared to physical methods for decontaminating the environment, which can be extraordinarily expensive. As an example, more than $1 million a day was spent in an only partially successful attempt to clean the oiled rocks of Prince William Sound, Alaska, using water washing and other physical means, after the Exxon Valdez tanker ran aground there in 1989. Neither the government nor private industry can afford the cost to clean up the nation's chemically contaminated sites by physical means. Therefore, a renewed interest in bioremediation developed when this approach gave positive results in the Exxon Valdez incident (Beardsley 1989). While conventional technologies call for moving large quantities of toxic-waste-contaminated soil to incinerators, bioremediation typically can be performed on-site and requires only simple equipment. Bioremediation, though, is not the solution for all environmental pollution problems. Like other technologies, bioremediation is limited in the materials that it can treat, and by conditions at the treatment site and the time available for treatment.

Bioremediation has great potential for destroying environmental pollutants, especially in case of fuel or creosote-contaminated soil (Mueller et al. 1989, 1991; Song et al. 1990). Several hundred contaminated sites have been identified where bioremediation is being tested as a possible clean-up technology. Petroleum and creosote are most often the

pollutants of concern, comprising about 60% of the sites where bioremediation is being used in field demonstrations or for full-scale operations (Hinchee and Olfenbuttel 1991a, 1991b).

Bioremediation Efficacy Testing

To demonstrate that a bioremediation technology is potentially useful, it is important to document enhanced biodegradation of the pollutant under controlled conditions. This generally cannot be accomplished *in situ* and thus must be accomplished in laboratory experiments. Laboratory experiments demonstrate the potential a particular treatment may have to stimulate the removal of a xenobiotic from a contaminated site (Bailey et al. 1973). Laboratory experiments that closely model real environmental conditions are most likely to produce relevant results (Bertrand et al. 1983). In many cases, this involves using samples collected in the field and containing indigenous microbial populations. In such experiments, it is important to include appropriate controls, such as sterile treatments, to separate the effects of the abiotic weathering of the pollutant from actual biodegradation. Such experiments do not replace the need for field demonstrations but are critical for establishing the scientific credibility of specific bioremediation strategies. They are also useful for the screening of potential bioremediation treatments.

The parameters typically measured in laboratory tests of bioremediation efficacy include enumeration of microbial populations (Song and Bartha 1990), measurement of rates of microbial respiration (oxygen consumption and/or carbon dioxide production), and determination of degradation rates (disappearance of individual and/or total pollutants) as compared to untreated controls. The methodologies employed in these measurements are critical.

Undoubtedly, the most direct measure of bioremediation efficacy is the monitoring of disappearance rates of the pollutant. When using this approach, the appropriate controls and the proper choice of analytical techniques become especially critical. The "disappearance" of pollutants may occur not only by biodegradation but also by evaporation, photodegradation, and leaching.

In order to promote biotechnology and bioremediation, the EPA and the University of Pittsburgh Trust have entered into a multiyear cooperative agreement to establish the National Environmental Technology Applications Corporation (NETAC). The purpose of NETAC is to facilitate the commercialization of technologies being developed by the government and the private sector that will positively affect the nation's most pressing environmental problems. The corporation's efforts encompass encouraging new technologies with promising commercial potential, as well as innovations aimed solely at modifying and improving existing technologies or processes (USEPA 1990). Under the agreement with the EPA, NETAC convenes a panel of experts to review proposals for products that have been developed for use as bioremediation. The panel recommends those products that offer the most promise for success in the field. The recommended products are subjected to laboratory testing at the EPA laboratory in Cincinnati, Ohio. The objective of the laboratory protocol is to determine if commercial bioremediation products can enhance the biodegradation of a pollutant to a degree significantly better than that achievable by simple environmental modification. The laboratory tests consist of electrolytic respirometers set up to measure oxygen uptake over time and shake flask microcosms to measure degradation and microbial growth (Venosa et al. 1991). Pollutant constituents remaining are analyzed. Products that demonstrate potential efficacy are screened for toxicities, and products passing the toxicity tests are then considered for field testing.

The evaluation of pollutant biodegradation *in situ* is far more difficult than in laboratory studies. Analyses that require enclosure, such as respiration measurements, typically are precluded from field evaluations. Field evaluations, therefore, have relied upon the enumeration of pollutant-degrading microorganisms and the recovery and analysis of residual pollutants. This is especially complicated, since the distribution of pollutants in the environment typically is patchy; therefore, a high number of replicate samples must be obtained for results to be statistically valid.

Side Effects Testing

In addition to demonstrating efficacy, bioremediation treatments must not produce any untoward ecological effects (Colwell 1971; O'Brien and Dixon 1976; Doe and Wells 1978). The focus of ecological effects testing of bioremediation has been on the direct toxicity of chemical additives, such as fertilizers, to indigenous organisms. Standardized toxicological tests are used to determine the acute toxicities of chemicals. Chronic toxicities and sublethal effects may also be determined. Generally, toxicity tests are run using the microcrustacean *Daphnia,* bivalves (such as oyster larvae) and fish (such as rainbow trout). Sometimes regionally important species, such as salmon or herring, are included. Additionally, tests are run to determine effects on algal growth rates. These tests are aimed at determining levels of fertilizer application that will stimulate biodegradation without causing eutrophication.

To exclude the possibility that some unsuspected toxic or mutagenic residue of pollutant degradation remains undetected, it is sometimes desirable to complement residue analysis with bioassays. As a rapid and convenient measure of acute toxicity, the reduction of light emission by *Photobacterium phosphoreum* (Microtox assay) has been used in the verification of landtreatment efficacy for fuel-contaminated soil (Wang and Bartha 1990; Wang et al. 1990). These measurements showed that the moderate acute toxicity of the intact fuel increased in the early phase of biodegradation, but upon completion of the biodegradation process, it returned to the background level of uncontaminated soil. Bioremediation accelerated both the transient increase and the ultimate disappearance of toxicity. The Microtox assay has been used also to determine treatability, such as a need to dilute the soil prior to bioremediation when high concentrations of toxic residues interfered with biodegradation activity (Matthews and Hastings 1987).

Mutagenic residues, especially those of polycyclic aromatic hydrocarbons, are assayed for by the Ames test (Maron and Ames 1983). In a case of a soil contaminated by diesel fuel, the initially moderate mutagenic activity of the residue increased sharply with the onset of microbial degradation, especially when the assay was run without activating microsomal enzymes (S-9 mix). However, as biodegradation progressed, the mutagenic activity returned to the background level of uncontaminated soil (Wang et al. 1990).

In the case of soil contaminated by oily sludge or fuel spills, the progress of bioremediation and the decrease in toxicity was sometimes documented also by seed germination and plant growth bioassays (Dibble and Bartha 1979c; Bossert and Bartha 1985; Wang and Bartha 1990). As compared to microbial tests, seed and plant bioassays have moderate sensitivities and usually show effects only above 1.0% hydrocarbon of the soil dry weight. Their main function is to determine when the revegetation of an oil-inundated site can be attempted.

APPROACHES TO BIOREMEDIATION

Environmental Modification for Bioremediation

Some common environmental limitations to biodegradation of hazardous chemical wastes are excessively high waste concentrations, lack of oxygen, unfavorable pH, lack of mineral nutrients, lack of moisture, and unfavorable temperature. Once the limitations by environmental conditions are corrected, the ubiquitous distribution of microorganisms allows, in most cases, for a spontaneous enrichment of the appropriate microorganisms. In the great majority of cases, an inoculation with specific microorganisms is neither necessary nor useful. Exceptions exist when the biodegrading microorganisms are poor competitors and fail to maintain themselves in the environment or when the chemical waste is only cometabolized and thus fails to provide a selective advantage to the catabolic organism(s). A massive accidental spill of a toxic chemical in a previously unexposed environment constitutes another situation where inoculation with preadapted microbial cultures may hasten biodegradative cleanup.

However, inoculation should always be combined with efforts to provide the inoculum with reasonable growth conditions in the polluted environment. As a minimum, suitable growth temperature, adequate water potential, suitable pH, suitable nutrient balance, and, for aerobic processes, adequate oxygen supply need to be assured. If the pollutant to be eliminated does not support microbial growth, the addition of a suitable growth substrate and/or repeated massive inoculations is necessary. Inoculation in absence of the appropriate ecological considerations rarely attains the desired improvement in biodegradation.

An approach originally developed for the isolation of cometabolizing microorganisms shows promise for promoting the biodegradation of some recalcitrant chemical wastes. If a xenobiotic organochlorine compound fails to support microbial growth, a microorganism may be enriched for on a nonchlorinated or less chlorinated structural analog. If the organism has sufficiently broad-spectrum enzymes, it may cometabolize the xenobiotic chemical even though the organism cannot grow on the chemical. It stands to reason that soil contaminated with a recalcitrant organochlorine compound may be detoxified if a biodegradable structural analog is provided. This was in fact demonstrated in the case of PCBs and 3,4-dichloroaniline. In the former case, biphenyl aniline was used as the stimulating substrate; in the latter, aniline was used (You and Bartha 1982; Brunner et al. 1985). The disappearance of the xenobiotics was accelerated with this approach by an order of one magnitude or more.

The availability of oxygen in soils, sediments, and aquifers is often limiting and dependent on the type of soil and on whether the soil is waterlogged (Jamison et al. 1975; Huddleston and Cresswell 1976; Bossert and Bartha 1984; von Wedel et al. 1988; Lee and Levy 1991). The microbial degradation of petroleum contaminants in some groundwater and soil environments is severely limited by oxygen availability. In surface soil, and in on-site bioremediation, oxygenation is best assured by providing adequate drainage. Air-filled pore spaces in the soil facilitate diffusion of oxygen to hydrocarbon-utilizing microorganisms, while in waterlogged soil, oxygen diffusion is extremely slow and cannot keep up with the demand of heterotrophic decomposition processes. Substantial concentrations of decomposable hydrocarbons create a high oxygen demand in soil, and the rate of diffusion is inadequate to satisfy it even in well-drained and light-textured soils. Cultivation (ploughing and rototilling, for example) has been used to turn the soil and assure its maximal access to atmospheric oxygen.

In many cases, hydrocarbons have contaminated soil and groundwater from leaking underground storage tanks. Microorganisms in such subsurface environments degrade hydrocarbons and other xenobiotics only extremely slowly, but these rates can be substantially enhanced (Johnson et al. 1985; Aelion et al. 1987; Aelion and Bradley 1991; Frederickson et al. 1991). If digging up and surface spreading of the contaminated soil (on-site bioremediation) is not feasible, *in situ* treatment of the undisturbed soil is performed. The most troublesome problem to overcome in such cases is the oxygen limitation in groundwater and in deep soil layers. Because oxygen solubility in water is low, this can best be achieved by pumping down dilute solutions of hydrogen peroxide in appropriate and stabilized formulations (Brown et al. 1984 1985; Yaniga and Smith 1984; American Petroleum Institute 1987; Thomas et al. 1987; Berwanger and Barker 1988). The decomposition of hydrogen peroxide releases oxygen which can support aerobic microbial metabolism. To avoid formation of gas pockets and microbial toxicity, the practical concentration of hydrogen peroxide in injected water is kept around 100 ppm (Brown et al. 1984; Yaniga and Smith 1984). As an example, D. J. Berwanger and J. F. Barker (1988) investigated *in situ* bioremediation of aromatic hydrocarbons in an originally anaerobic, methane-saturated groundwater situation using hydrogen peroxide as an oxygen source. Batch biodegradation experiments were conducted with groundwater and core samples obtained from a Canadian landfill. Hydrogen peroxide, added at a nontoxic level, provided oxygen, which promoted the rapid biodegradation of benzene, toluene, ethyl benzene, and *o-, m-,* and *p*-xylene.

Thermodynamic considerations, as well as practical experience, indicate that dechlorination of halo-carbon compounds is favored under anaerobic conditions (Suflita et al. 1982). Unfortunately, these same conditions usually prevent the oxidative cleavage of aromatic rings. One would expect, however, that sequential anaerobic-aerobic conditions would promote the biodegradation of halocarbons. S. Fogel et al. (1983) demonstrated this for the insecticide methoxychlor in soil. As measured by $^{14}CO_2$ evolution from radiolabeled methoxychlor, the insecticide was mineralized slowly under continuous aerobic conditions, and no $^{14}CO_2$ evolved during anaerobic incubation. When anaerobic incubations were switched to aerobic, however, a rapid burst of $^{14}CO_2$ evolution was observed. Redox potential can be easily manipulated in waste streams, and the same can be accomplished in contaminated soils through periodic flooding. Therefore, the demonstrated principle of sequential treatment may well find practical application.

Several major oil spills have focused attention on the problem of hydrocarbon contamination in marine and estuarine environments and the potential use of bioremediation through nutrient addition to remove petroleum pollutants. Since microorganisms require nitrogen and phosphorus for incorporation into biomass, the availability of these nutrients within the same area as the hydrocarbons is critical. Under conditions where nutrient deficiencies limit the rate of petroleum biodegradation, the beneficial effect of fertilization with nitrogen and phosphorus has been conclusively demonstrated and offers great promise as a countermeasure for combating oil spills (Pritchard and Costa 1991). R. M. Atlas and R. Bartha (1973) developed an oleophilic nitrogen and phosphorus fertilizer that places the nitrogen and phosphorus at the oil-water interface, the site of active oil biodegradation. Because the fertilizer is oleophilic, it remains with the oil and is not rapidly diluted from the site where it is effective. In the aftermath of the Amoco Cadiz oil spill of 1978 in France, a commercial oleophilic fertilizer was developed by Elf Aquitaine (Tramier and Sirvins 1983; LaDousse et al. 1987; Sveum and LaDousse 1989; LaDousse and Tramier 1991). The product, called Inipol EAP 22, is a microemulsion that contains urea as a nitrogen source, lauryl phosphate as a phosphate source, and oleic acid as a carbon source to boost the populations of hydrocarbon-degrading microorganisms.

Microbial Seeding and Bioengineering Approaches to the Bioremediation of Pollutants

Microbial pollution control products, including dried or liquid microbial inocula with or without nutrient additives, currently have annual sales of $7 million to $10 million, and the potential market may be as high as $200 million. L. M. Johnson et al. (1985) describe several successful examples of pollution abatement by specifically designed microbial agents. An inoculation program with hydrocarbon-degrading bacteria significantly improved the performance of a petrochemical waste-treatment system. The removal of tertiary butyl alcohol from a waste stream was improved by inoculation with specifically adapted microorganisms, and a large accidental formaldehyde spill was successfully cleaned up by biodecontamination. Unfortunately, rigorous controls are not available in such case studies, and it is often difficult to determine how much of the improvement can be ascribed to bioremediation treatment.

Since bioremediation relies upon the biodegradation capacity of the microorganisms in contact with the pollutants, some have proposed seeding with pollutant-degrading bacteria. Seeding involves the introduction of microorganisms into the natural environment for the purpose of increasing the rate or extent, or both, of biodegradation of pollutants. The rationale for this approach is that indigenous microbial populations may not be capable of degrading xenobiotics or the wide range of potential substrates present in complex pollutant mixtures. Several commercial enterprises began to market microorganism preparations for removing petroleum pollutants. Commercial mixtures of microorganisms are being marketed for use in degrading oil in pollutants in contained treatment bioreactors as well as for *in situ* applications (Applied Biotreatment Association 1989, 1990).

Many different microorganisms are being marketed or considered for future bioremediation applications. Pure cultures as well as undefined mixtures of microorganisms are used for bioremediation. *Pseudomonas* species are often selected because of their ability to degrade a wide range of pollutants, including the low–molecular-weight hydrocarbons associated with many fuel spills. The white rot fungus *Phanerochaete* is being proposed for bioremediation of many more complex pollutant compounds (Bumpus et al. 1985). *Phanerochaete chrysosporium* produces laccases and peroxidases that normally are involved in the degradation of lignin, which is a complex polyaromatic substance. These enzymes are also able to attack a diverse range of compounds, and *Phanerochaete* has been shown to biodegrade pesticides such as DDT (Bumpus and Aust 1987); munitions, such as TNT (Fernando et al. 1990); high–molecular-weight polynuclear aromatics, such as benzpyrene (Bumpus 1989); and plastics, such as polyethylene (Lee et al. 1991).

The absence of a catabolic pathway for a xenobiotic compound appears to be the ultimate obstacle to its biodegradative cleanup. But this obstacle is no longer an absolute one. Biochemical pathways are under constant evolution, and plasmid-mediated genetic information exchange between microbial strains can greatly accelerate this process. In each case, selective pressure is applied to evolve the desired characteristic, such as utilization of a recalcitrant xenobiotic compound. Typically, an enrichment culture is started in a chemostat that is fed small concentrations of the xenobiotic compound and higher concentrations of a related but utilizable substrate. During the enrichment procedure, which may take weeks or even months, the concentration of the utilizable substrate is gradually decreased, while the concentration of the xenobiotic compound is increased. Spontaneous mutants with increased ability to utilize the xenobiotic compound are selected for under these circumstances. Spontaneous mutation may be supplemented by treating a part or all of the chemostat culture with UV light or chemical mutagens. Recombinant DNA technology holds great promise for developing microbial strains that can aid in the environmental removal of toxic chemicals (Kilbane 1986).

One approach for providing the enzymatic capability to degrade diverse pollutants is to use genetic engineering to create microorganisms with the capacity to degrade a wide range of hydrocarbons. A hydrocarbon-degrading pseudomonad engineered by A. M. Chakrabarty was the first organism that the Supreme Court of the United States ruled, in a landmark decision, could be patented. However, considerable controversy surrounds the release of such genetically engineered microorganisms into the environment, and field testing of these organisms must be delayed until the issues of safety, containment, and potential for ecological damage are resolved (Sussman et al. 1988). Given the current regulatory framework for the deliberate release of specifically genetically engineered microorganisms, it is unlikely that any genetically engineered microorganism would gain the necessary regulatory approval in time to be of much use in treating the oil spill for which it was created.

Advances in the directed evolution of microorganisms have been aided by the recognition that many genes coding for the biodegradation of xenobiotics are located on transposable chromosomal elements (transposons) or on plasmids (Eaton and Timmis 1984). Evidence indicates that the transfer and recombination of such movable genetic elements play important roles in the evolution of antibiotic and heavy-metal resistance and utilization of novel substrates in natural environments (Hardy 1981). It is possible to augment the evolution of new degradative pathways by feeding into the chemostat enrichment microorganisms known to harbor plasmids, coding for portions of the desired biodegradative pathway. Exchange, recombination, and amplification of genetic information under selective pressure, along with spontaneous and induced mutation, can greatly accelerate an evolutionary process. Combinations of the described techniques yielded specially constructed *Pseudomonas* strains capable of degrading a wide array of chlorobenzoates and chlorophenols (Hartmann et al. 1979; Reineke and Knackmuss 1979). After the introduction of these strains into the waste stream of a chemical manufacturing plant, the removal of haloaromatic xenobiotics was greatly improved (Knackmuss 1983, 1984). Plasmid-assisted molecular breeding also produced a *Pseudomonas cepacia* strain with the capacity to grow on the herbi-

cide 2,4,5-T, though no 2,4,5-T-degrading microorganism could be isolated previously from the environment (Kellogg et al. 1981). Subsequently, it was demonstrated that the treatment of a heavily contaminated soil sample with this engineered *P. cepacia* effectively eliminated the 2,4,5-T contaminant while growing on it (Ghosal et al. 1985) (Figure 14.3). The soil detoxified in this manner was subsequently able to support the growth of herbicide-sensitive plants.

The introduction of genetically engineered microorganisms into the environment, for whatever purpose, has its own hazards. It requires scientific forethought and governmental regulation (Halverson et al. 1985). It can be argued, however, that the procedures previously described simply speed up evolutionary processes that would eventually take place naturally. They do not involve *in vitro* splicing and manipulation of DNA.

Figure 14.3

Disappearance of 2,4,5-T and simultaneous growth of *Pseudomonas cepacia* AC1100 in soil. (Source: Kellogg et al. 1981.)

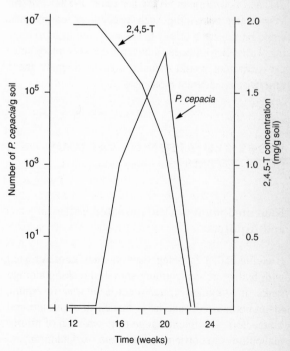

Therefore, the environmental release of *in vivo*-engineered organisms for cleanup of xenobiotic residues is not expected to be highly controversial.

Microbial exoenzymes have been demonstrated to be effective in detoxifying pesticide residues in soils, in used pesticide containers, and in manufacturing wastes (Munnecke 1981). The lignin-degrading enzyme complex of the basidiomycete *Phanerochaete chrysosporium* was effective in oxidizing and immobilizing a number of highly recalcitrant halocarbon xenobiotics, including DDT and TCDD (Bumpus et al. 1985).

Finally, microbial treatment methods may complement other disposal techniques. As an example, photodegradation is quite effective in dehalogenating halocarbon compounds and hydroxylating condensed polycyclic aromatics, but the complete photodegradative destruction of these compounds would be technically difficult and prohibitively expensive. On the other hand, a short photodegradative treatment can be sufficient to prime a recalcitrant halocarbon for biodegradation. A moderate dose of UV radiation was demonstrated to prime the biodegradation of such problem xenobiotics as 2,4-dichlorophenol, 2,4,5-trichlorophenol, and benzo(a)pyrene (Miller et al. 1988a, 1988b). Wastewater containing such recalcitrant compounds could be subjected briefly to UV radiation prior to mixing with other effluents. Conventional secondary sewage treatment would then be able to degrade these otherwise recalcitrant molecules.

BIOREMEDIATION OF VARIOUS ECOSYSTEMS

Bioremediation of Contaminated Soils and Aquifers

Bioremediation is being used at sites contaminated with hydrocarbons from underground leaking storage tanks. In some cases, treatment is *in situ;* in others, the contaminated material is treated in above-ground bioreactors. Treatments range from addition of microbial cultures to nutrient addition and oxygenation.

Effluent treatment by refineries and petrochemical plants generates large amounts of oily sludges, including gravity (API) separator sludges, air flotation sludges, centrifugation residues, filter cakes, and biotreatment sludges. Tank bottoms, sludges from cleaning operations, and, occasionally, used lubricating and crankcase oils that are considered nonrecyclable are additional sources of oily wastes (CONCAWE 1980; Bartha and Bossert 1984). From the available disposal alternatives, the use of soil bioremediation is the most economical. The process is referred to as "landtreatment," or "landfarming," and constitutes a deliberate disposal process in which place, time, and rates can be controlled. In landtreatment, the chosen site has to meet certain criteria and needs to undergo preparation to assure that floods, runoff, and leaching will not spread the hydrocarbon contamination in an uncontrolled manner (Bartha and Bossert 1984). Oily sludges are applied to soil at rates to achieve approximately 5% hydrocarbon concentration in the upper 15- to 20-cm layer of soil. Hydrocarbon concentrations above 10% definitely inhibit the biodegradation process. This limit translates to approximately 100,000 L of hydrocarbon per ha, usually in three to four times as high sludge volume (Dibble and Bartha 1979a). The soil pH is adjusted to a value between 7 and 8 or to the nearest practical value, using agricultural limestone. Nitrogen and phosphorus fertilizers are applied in ratios of hydrocarbon:N = 200: 1 and hydrocarbon:P = 800:1. Adequate drainage is essential, but irrigation is necessary only in very arid and hot climates.

Undegraded hydrocarbons do not leach readily into the groundwater from landtreatment sites (Dibble and Bartha 1979b), and the environmental impact of properly operated sites appears to be minimal (Arora et al. 1982). Currently, open-air landtreatment is practiced in the United States (American Petroleum Institute 1980; Arora et al. 1982), in various European countries (CONCAWE 1980; Shailubhai 1986), in Brazil (Amaral 1987; Tesan and Barbosa 1987), and most likely in many other countries. As some volatiles are inevitably lost to the atmosphere in this type of treatment, a variation of the procedure may be performed in a temporary plastic foil enclosure with treatment of the exhaust air. Typically, a portion of the waste organic chemical is mineralized during treatment; another portion is, after partial biodegradation, incorporated into soil humus, bringing about a high degree of detoxification and immobilization. In some cases, ascertainment that such bound residues will not be remobilized at a later time is necessary.

Essentially the same process as landtreatment has been used for bioremediation of surface soils contaminated by crude oil or by refined hydrocarbon fuels due to accidental spills and leaks. Because of their accidental nature, the location and rate of these contamination events cannot be controlled, but it is important to know whether or not biodegradation can be applied as an effective cleanup measure at such sites and how to optimize the process. Extensive site characterization is needed to establish the levels of contaminants and the direction of subsurface movement of any pollutant plumes (Figure 14.4).

If fuel and site characteristics are favorable, the decision for "on site" bioremediation is economically and environmentally sound. The contaminated soil is dug out and spread out in a 15- to 20-cm-thick layer. While bioremediation may be slow and has its limitations, the alternatives are more expensive and often are environmentally inferior (Jones and Greenfield 1991).

Investigating the bioremediation potential of common fuels, H. G. Song et al. (1990) concluded that the process is most suitable for the medium fuel distillates such as jet fuel, diesel oil, and No. 2 fuel oil. While gasoline responds to bioremediation, in surface soils, biodegradation cannot keep up with evaporation rates, and most of the product is lost to the atmosphere. In laboratory experiments, H. G. Song et al. (1990) had limited success with No. 6 (residual) fuel oil. However, M. Jones and J. H. Greenfield (1991) reported quite promising results in an on-site bioremediation effort in Florida. Soil contaminated by an average of 10,000 ppm of No. 6 fuel oil was treated with fertilizer and a commercial microbial inoculum. The soil was turned, cultivated, and kept moist. In three hundred days, about 90% of the contaminant was eliminated, leaving approximately 1,000 ppm residue, which included multiring PAHs. The inoculum without fertilizer was ineffective, but in combination

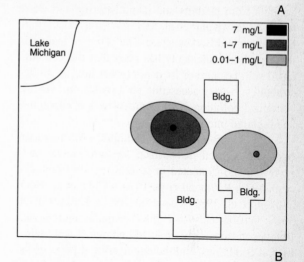

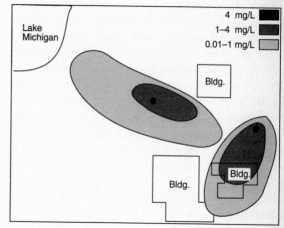

Figure 14.4

Site characterization following a pollutant spill, showing the plume of contaminants. (A) Trichloroethylene (B) Vinyl chloride. (Source: McCarty et al. 1991.)

with fertilizer, it significantly improved degradation rates. Climate factors and photodegradation may have contributed to the relative success of their bioremediation effort with a difficult product.

Aromatic solvents such as benzene and alkylbenzenes, toluene and xylene (BTX), chlorinated aromatics, and haloethanes and halomethanes leak into aquifers from landfills and manufacturing and waste storage sites (Kuhn et al. 1985; Wilson et al. 1986; Berwanger and Barker 1988). The common feature in these pollution incidents has been that the contaminated soil, because of its depth or because of surface structures, was inaccessible to excavation, yet the continued pollution of the aquifers was not tolerable and required intervention.

In situ bioremediation techniques for hydrocarbon-contaminated subsurface systems have been reviewed and evaluated in several reports (Vanlooke et al. 1975; Raymond et al. 1976; Brown et al. 1985; Wilson et al. 1986; Lee et al. 1987; Thomas et al. 1987; Wilson and Ward 1987). Aquifer bioremediation depends strongly on local geological and hydrological conditions, and the engineering aspects of the process are beyond the scope of this text. In broad outline, an attempt is usually made to isolate the contamination plume from wells and other sensitive areas. This may be accomplished by using physical barriers, such as cement, bentonite, or grout injection, or dynamically by pumping the polluted portion of the aquifer. In the latter case, groundwater flow is redirected toward rather than away from the contaminated plume, thus preventing the spreading of contamination.

Besides containment, the pumping process recovers free-flowing hydrocarbon and contaminants dissolved in water. In theory, prolonged pumping could eventually flush out all the contaminant, but the solubility characteristics of hydrocarbons make reliance on physical flushing alone prohibitively slow and expensive. Hydrocarbons have much higher affinities than water to soil particles (Verstraete et al. 1976). To clean an aquifer by simple water flushing may take fifteen to twenty years and several thousand times the volume of water in the contaminated portion of the aquifer. Physical flushing may be facilitated by the addition of dispersants and emulsifiers through injection wells around the periphery of the contamination plume, while water is continuously withdrawn from the center of the plume. However, the use of combined biodegradation and emulsification by hydrocarbon-utilizing microorganisms is usually more efficient and economic (Brown et al. 1985; Thomas et al. 1987; Beraud et al. 1989).

Bioremediation of aquifers contaminated with halogenated aromatics, haloethanes, and halomethanes presents additional complex problems (Kuhn et al. 1985; Wilson et al. 1986; Berwanger and Barker 1988). While some of these materials are dehalogenated anaerobically, others cannot serve as substrates under either aerobic or anaerobic conditions and are attacked only cometabolically in the presence of methane or toluene, and only under oxidative conditions.

Since methylotrophs through their production of methane monooxygenase (MMO) are able to degrade trichloroethylene (TCE), dichloroethylene (DCE), and vinylchloride by cometabolism (Fogel et al. 1986), several investigators have considered using methylotrophs for the bioremediation of sites contaminated with these halogenated compounds. P. L. McCarty et al. (1991) found that they could stimulate indigenous methylotrophic populations. They developed a model *in situ* bioremediation treatment that would require 5,200 kg of methane and 19,200 kg of oxygen in order to convert 1,375 kg of chlorinated hydrocarbons from an aquifer of 480,000 m^3 containing a contaminant load of 1,617 kg of halogenated compounds. A. V. Palumbo et al. (1991) found, however, that the presence of perchlormethene (PCE) inhibited the methylotrophs and suggested that anaerobic PCE removal would be necessary prior to stimulating methylotrophs to remove TCE. L. G. Senprini et al. (1991) found that carbon tetrachloride-, trichloroethane-, and Freon-contaminated sites could be bioremediated by stimulating indigenous denitrifying populations through the addition of acetate.

Bioremediation of Marine Oil Pollutants

The Exxon Valdez spill formed the basis for a major study on bioremediation and the largest application of this emerging technology. The initial approach to the cleanup of the oil spilled from the Exxon Valdez was physical. Washing of oiled shorelines with high-pressure water was expensive, and cleaned shorelines became re-oiled forcing recleaning. Bioremediation, therefore, was considered as a method to augment other cleanup procedures. The EPA and Exxon entered into an agreement to jointly explore the feasiility of using bioremediation. The project focused on determining whether nutrient augmentation could stimulate rates of biodegradation by the indigenous natural populations.

The application of the oleophilic fertilizer produced dramatic results, stimulating biodegradation so that the surfaces of the oil-blackened rocks on the shoreline turned white and were essentially oil free within 10 days after treatment (Pritchard and Costa 1991) (Figure 14.5). The striking results strongly supported the idea that oil degradation in Prince William Sound was limited by the amounts of available nutrients and that fertilizer application was a useful bioremediation strategy. The use of Inipol and a slow-release fertilizer was approved for shoreline treatment and was used as a major part of the cleanup effort. Monitoring tests demonstrated that fertilizer application sustained higher numbers of oil-degrading micro-

organisms in oiled shorelines and that rates of biodegradation were enhanced, as evidenced by the chemical changes detected in recovered oil from treated and untreated reference sites (Prince et al. 1990; Chianelli et al. 1991). Owing to its effectiveness, bioremediation became the major treatment method for removing oil pollutants from the impacted shorelines of Prince William Sound.

Bioremediation of Air Pollutants

Air emissions from various industrial or waste-treatment processes release into the atmosphere substances that may be noxious or hazardous to humans or may contribute to smog formation or to depletion of the atmospheric ozone layer. Air emissions are controlled by various physicochemical means, such as

Figure 14.5

Photograph showing results of Inipol application to a shoreline in Prince William Sound, Alaska. The "white window" square in the photograph was treated with the oleophilic fertilizer ten days earlier. While the surrounding untreated area remained black and oil covered, the treated area was relatively free of oil. Tests indicated that simple physical removal was not responsible and that biological degradation of oil was occurring. (Source; Russ Chianelli, EXXON Corporate Research, Annondale, N. J.)

chemical scrubbing or washing, condensation, filtration, adsorption, and flaring (incineration). All of these processes tend to be cumbersome and costly. Most volatile organic emissions tend to be potential microbial substrates, and the use of microorganisms for air pollution control in some cases represents a more cost-effective alternative. Microorganisms are used for air emission control in three types of devices: biofilters, trickling filters, and bioscrubbers (Ottengraf 1986; Leson and Winer 1991).

Biofilters have been used to control odors and volatile organic chemicals (VOCs) in contaminated air (Ottengraf and Van den Oever 1983; Leson and Winer 1991). Packed beds and soil beds have been used at sewage treatment plants to control odors associated with hydrogen sulfide, terpenes, mercaptans, and other compounds. The odor-forming compounds are sorbed onto a biofilm and transformed by the microorganisms in that biofilm to their mineral constituents. The operation of such biofilters for the treatment of contaminated air is analogous to wastewater treatment in biofilm-bearing devices. In both applications, bacteria are immobilized on a solid support as a biofilm and the microorganisms form products that are less toxic and generally environmentally acceptable. The classical biofilter technology uses peat, compost, bark, or soils as filter media and includes initial addition of nutrients and buffers. During operation, only lost moisture is replenished. Table 14.1 gives a summary for the removal of specific odor compounds.

Limitations exist in the current biofilter technology. Key limitations are that residual biomass accumulates in the filter media and that acids produced from the degradation of chlorinated organics and sulfur compounds consume the buffer capacity in the bed. Thus the process, at high organic loadings in air, experiences increasing pressure loss across the media and decreasing process pH. The time of operation before media replacement depends upon the pollutant loadings in the air. Low pollutant loadings result in process runs of more than one year. In trickling-filter-type devices, the biofilm is associated with a porous solid, like in the case of a biofilter, but in addition to air that moves through this medium, liquid is also cir-

Table 14.1

Removal of pollutants from air by biofilters

Substance	Percent removal
Hydrogen sulfide	>99
Dimethyl sulfide	>91
Terpene	>98
Organo-sulfur gases	>95
Ethyl benzene	>92
Tetrachloroethylene	>86
Chlorobenzene	>69

culated, either in the same direction or countercurrent to the air. This type of device offers more process control as the circulating liquid can be periodically amended with nutrients or can be neutralized or diluted if salts accumulate. This type of device is essential if VOC metabolism produces acidity, alkalinity, or salts, as for instance in the case of chlorinated solvent vapors. It is essential that the VOC compound be capable of supporting growth of a properly adapted microbial consortium. It may be possible to remove biologically some nonsubstrate type of compounds like trichloroethylene (TCE) cometabolically, but in such cases, low amounts of an appropriate and growth-inducing substrate like methane or toluene need to be added to the airstream. Biomass buildup can also become a problem in trickling air filters, especially at high loadings. In the EPA's Risk Reduction Engineering Laboratory in Cincinnati, Ohio, high pollutant loadings (greater than 200 ppm) of organic compounds in air were fed to a trickling-filter-type biotower. The biotower removed essentially all of the organic pollutants from the air in two minutes of retention time but exhibited relatively rapid accumulation of residual biomass and tower flooding in several months.

In bioscrubbers, air moves countercurrent to a fine spray of microbial suspension that washes out water-soluble vapors from the air. The actual degradation takes place in a stirred reactor into which the collected spray is channeled. This device contains no solid phase. It is effective only for highly water-soluble VOCs. The use of microorganisms for control of air emissions is a relatively new technique. Nevertheless, in Europe, some large units with 75,000 m^3 per h capacity are operating with good results, and this biotechnology is likely to evolve further (Leson and Winer 1991).

SUMMARY

Pollution problems caused by xenobiotics have necessitated an elaborate environmental safety testing program for newly introduced chemicals. In addition to routine toxicity tests, evaluations of biodegradability, biomagnification, and environmental fate are performed in microcosms prior to licensing and release.

The environmental monitoring of xenobiotic pollutants relies on sophisticated analytical instrumentation and, increasingly, also on microbial biosensors. Remedies to existing pollution problems by xenobiotics include redesigning the molecules of synthetic products, testing the fate and effects of newly introduced chemicals in model ecosystems and microcosms, optimizing environmental conditions for biodegradation, and engineering new microorganisms with broader biodegradative abilities. The destruction of the stratospheric ozone layer, acid mine drainage, heavy-metal and metalloid biomethylation, and radionuclide pollution are all processes that are either mediated or exacerbated by microbial processes. Amelioration of these environmental pollution problems depends, in part, on a better understanding and eventual control of the microbial processes involved. A new biotechnological solution to pollution problems—bioremediation—has emerged.

REFERENCES & SUGGESTED READINGS

Aelion, C. M., C. M. Swindoll, and F. K. Pfaender. 1987. Adaptation to and biodegradation of xenobiotic compounds by microbial communities from a pristine aquifer. *Applied and Environmental Microbiology* 53:2212–2217.

Aelion, C. M., and P. M. Bradley. 1991. Aerobic biodegradation potential of subsurface microorganisms from a jet fuel-contaminated aquifer. *Applied and Environmental Microbiology* 57:57–63.

Amaral, S. P. 1987. Landfarming of oily wastes: Design and operation. *Water Science Technology* 19:75–86.

American Petroleum Institute. 1980. *Manual on Disposal of Petroleum Wastes.* American Petroleum Institute, Washington, D.C.

American Petroleum Institute. 1987. *Field Study of Enhanced Subsurface Biodegradation of Hydrocarbons Using Hydrogen Peroxide as an Oxygen Source.* American Petroleum Institute Pub. 4448. American Petroleum Institute, Washington, D.C.

Applied Biotreatment Association. 1989. *Case History Compendium.* Applied Biotreatment Association, Washington, D.C.

Applied Biotreatment Association. 1990. *The Role of Biotreatment of Oil Spills.* Applied Biotreatment Association, Washington, D.C.

Arora, H. S., R. R. Cantor, and J. C. Nemeth. 1982. Land treatment: A viable and successful method of treating petroleum industry wastes. *Environment International* 7:285–292.

Atlas, R. M. 1981. Microbial degradation of petroleum hydrocarbons: An environmental perspective. *Microbiological Reviews* 45:180–209.

Atlas, R. M., and R. Bartha. 1973. Stimulated biodegradation of oil slicks using oleophilic fertilizers. *Environmental Science and Technology* 7:538–541.

Atlas, R. M., D. Pramer, and R. Bartha. 1978. Assessment of pesticide effects on non-target soil microorganisms. *Soil Biology Biochemistry* 10:231–239.

Atlas, R. M., and D. Pramer. 1990. Focus on bioremediation. *ASM News* 56:7.

Atlas, R. M., and R. Bartha. 1992. Hydrocarbon biodegradation and oil spill bioremediation. *Advances in Microbial Ecology*. In press.

Bailey, N. J. L., A. M. Jobson, and M. A. Rogers. 1973. Bacterial degradation of crude oil: Comparison of field and experimental data. *Chemical Geology* 11:203–221.

Bartha, R. 1986. Biotechnology of petroleum pollutant biodegradation. *Microbial Ecology* 12:155–172.

Bartha, R., and I. Bossert. 1984. The treatment and disposal of petroleum wastes. In R. M. Atlas (ed.). *Petroleum Microbiology*. Macmillan, New York, pp. 553–577.

Beardsley, T. 1989. No slick fix: Oil spill research is suddenly back in favor. *Scientific American* 261(3):43.

Beraud, J. F., J. D. Ducreux, and C. Gatellier. 1989. Use of soil-aquifer treatment in oil pollution control of underground waters. In *Proceedings of the 1989 Oil Spill Conference*. American Petroleum Institute, Washington, D.C., pp. 53–59.

Bertrand, J. C., E. Rambeloarisoa, J. F. Rontani, G. Giusti, and G. Mattei. 1983. Microbial degradation of crude oil in sea water in continuous culture. *Biotechnology Letters* 5:567–572.

Berwanger, D. J., and J. F. Barker. 1988. Aerobic biodegradation of aromatic and chlorinated hydrocarbons commonly detected in landfill leachate. *Water Pollution Research Journal of Canada* 23(3):460–475.

Bluestone, M. 1986. Microbes to the rescue. *Chemical Week* 139(17):34–35.

Bossert, I., W. M. Kachel, and R. Bartha. 1984. Fate of hydrocarbons during oily sludge disposal in soil. *Applied and Environmental Microbiology* 47:763–767.

Bossert, I., and R. Bartha. 1984. The fate of petroleum in soil ecosystems. In R. M. Atlas (ed.). *Petroleum Microbiology*. Macmillan, New York, pp. 435–473.

Bossert, I., and R. Bartha. 1985. Plant growth in soils with a history of oily sludge disposal. *Soil Science* 140:75–77.

Brown, R. A., R. D. Norris, and R. L. Raymond. 1984. Oxygen transport in contaminated aquifers. In *Proceedings of the Conference on Petroleum Hydrocarbons and Organic Chemicals in Ground Water—Prevention, Detection, and Restoration*. National Water Well Association, Worthington, Ohio, pp. 441–450.

Brown, R. A., R. D. Norris, and G. R. Brubaker. 1985. Aquifer restoration with enhanced bioreclamation. *Pollution Engineering* 17:25–28.

Brunner, W., S. H. Southerland, and D. D. Focht. 1985. Enhanced biodegradation of polychlorinated biphenyls in soil by analog enrichment and bacterial inoculation. *Journal of Environmental Quality* 14:324–328.

Bull, A. T. 1980. Biodegradation: Some attitudes and strategies of microorganisms and microbiologists. In D. C. Ellwood, J. N. Hedger, M. J. Latham, and J. M. Lynch (eds.). *Contemporary Microbial Ecology*. Academic Press, New York, pp. 107–136.

Bull, A. T., C. R. Ratledge, and D. C. Ellwood (eds.). 1979. *Microbial Technology: Current State and Future Prospects*, Twenty-Ninth Symposium of the Society for General Microbiology. Cambridge University Press, Cambridge, England.

Bumpus, J. A. 1989. Biodegradation of polycyclic aromatic hydrocarbons by *Phanerochaete chrysosporium*. *Applied and Environmental Microbiology* 55:154–158.

Bumpus, J. A., M. Tien, D. Wright, and S. D. Aust. 1985. Oxidation of persistent environmental pollutants by a white rot fungus. *Science* 228:1434–1436.

Bumpus, J. A., and S. D. Aust. 1987. Biodegradation of DDT [1,1,1–trichloro–2,2–bis(4–chlorophenyl) ethane] by the white rot fungus *Phanerochaete chrysosporium*. *Applied and Environmental Microbiology* 53:2001–2028.

Burlage, R. S., G. S. Sayler, and F. Larimer. 1990. Monitoring of naphthalene catabolism by bioluminescence with *nah–lux* transcriptional fusions. *Journal of Bacteriology* 172:4749–4757.

Chamerlik-Cooper, M., R. E. Carlson, and R. O. Harrison. 1991. Determination of PCBs by enzyme immunoassay. In *Field Screening Methods for Hazardous Wastes and Toxic Chemicals*. U.S. Environmental Protection Agency and U.S. Army Toxic and Hazardous Materials Agency, Las Vegas, Nev., pp. 625–628.

Chianelli, R. R., T. Aczel, R. E. Bare, G. N. George, M. W. Genowitz, M. J. Grossman, C. E. Haith, F. J. Kaiser, R. R. Lessard, R. Liotta, R. L. Mastracchio, V. Minak–Bernero, R. C. Prince, W. K. Robbins, E. I. Stiefel, J. B. Wilkinson, S. M. Hinton, J. R. Bragg, S. J. McMillan, and R. M. Atlas. 1991. Bioremediation technology development and application to the Alaskan spill. In *Proceedings of the 1991 International Oil Spill Conference*. American Petroleum Institute, Washington, D.C., pp. 549–558.

Colwell, E. B. (ed.). 1971. *The Ecological Effects of Oil Pollution on Littoral Communities*. Applied Science Publishers, London.

CONCAWE. 1980. *Sludge Farming: A Technique for the Disposal of Oily Refinery Wastes*. Rep. 3/80. CONCAWE, The Hague, The Netherlands.

Dibble, J. T., and R. Bartha. 1979a. Effect of environmental parameters on the biodegradation of oil sludge. *Applied and Environmental Microbiology* 37:729–739.

Dibble, J. T., and R. Bartha. 1979b. Leaching aspects of oil sludge biodegradation in soil. *Soil Science* 127:365–370.

Dibble, J. T., and R. Bartha. 1979c. Rehabilitation of oil-inundated agricultural land: A case history. *Soil Science* 128:56–60.

Doe, K. G., and P. G. Wells. 1978. Acute toxicity and dispersing effectiveness of oil spill dispersants: Results of a Canadian oil dispersant testing program (1973 to 1977). In L. T. McCarthy, Jr., G. P. Lindblom, and H. F. Walter (eds.). *Chemical Dispersants for the Control of Oil Spills.* American Society for Testing and Materials, Philadelphia, pp. 50–65.

Dohrman, L. 1991. Immunoassays for rapid environmental contaminant monitoring. *American Laboratory* Oct:29–30.

Drinkwine, A., S. Spurlin, J. Van Emon, and V. Lopez-Avila. 1991. Immuno-based personal exposure monitors. In *Field Screening Methods for Hazardous Wastes and Toxic Chemicals.* U.S. Environmental Protection Agency and U.S. Army Toxic and Hazardous Materials Agency, Las Vegas, Nev., pp. 449–459.

Eaton, R. W., and K. N. Timmis. 1984. Genetics of xenobiotic degradation. In M. J. Klug and C. A. Reddy (eds.). *Current Perspectives in Microbial Ecology.* American Society for Microbiology, Washington, D.C., pp. 694–703.

Fernando, T., J. A. Bumpus, and S. D. Aust. 1990. Biodegradation of TNT (2,4,6–trinitrotoluene) by *Phanerochaete chrysosporium. Applied and Environmental Microbiology* 56:1666–1671.

Fogel, M. M., A. R. Taddeo, and S. Fogel. 1986. Biodegradation of chlorinated ethanes by a methane-utilizing mixed culture. *Applied and Environmental Microbiology* 51:720–724.

Fogel, S., R. L. Lancione, and A. E. Sewal. 1982. Enhanced biodegradation of methoxychlor in soil under sequential environmental conditions. *Applied and Environmental Microbiology* 44:113–120.

Frankenberger, W. T., Jr., K. D. Emerson, and D. W. Turner. 1989. *In situ* bioremediation of an underground diesel fuel spill: A case history. *Environment Management* 13(3):325–332.

Fredrickson, J. K., F. J. Brockman, and D. J. Workman. 1991. Isolation and characterization of a subsurface bacterium capable of growth on toluene, naphthalene, and other aromatic compounds. *Applied and Environmental Microbiology* 57:796–803.

Ghosal, D., I. S. You., D. K. Chatterjee, and A. M. Chakrabarty. 1985. Microbial degradation of halogenated compounds. *Science* 228:135–142.

Gosset, R. W., D. A. Brown, and D. R. Young. 1983. Predicting the bioaccumulation of organic compounds in marine organisms using octanol/water partition coefficients. *Marine Pollution Bulletin.* 14:387–392.

Greaves, M. P., H. A. Davies, J. A. P. Marsh, and G. I. Wingfield. 1977. Herbicides and soil microorganisms. *CRC Critical Reviews of Microbiology* 5:1–38.

Halvorson, H. O., D. Pramer, and M. Rogul (eds.). 1985. *Engineered Organisms in the Environment: Scientific Issues.* American Society for Microbiology, Washington, D.C.

Hardy, K. 1981. *Bacterial Plasmids.* Aspects of Microbiology Series, No. 4. American Society for Microbiology, Washington, D.C.

Hartmann, J., W. Reineke, and H. J. Knackmuss. 1979. Metabolism of 3-chloro, 4-chloro-, and 3,5-dichlorobenzoate by a pseudomonad. *Applied and Environmental Microbiology* 37:421–428.

Hinchee, R. E., and R. F. Olfenbuttel (eds.). 1991a. *In Situ Bioreclamation: Applications and Investigations for Hydrocarbon and Contaminated Site Remediation.* Butterworth-Heinemann, Boston.

Hinchee, R. E., and R. F. Olfenbuttel (eds.). 1991b. *On-Site Bioreclamation: Processes for Xenobiotic and Hydrocarbon Treatment.* Butterworth-Heinemann, Boston.

Hsu, T. S., and R. Bartha. 1979. Accelerated mineralization of two organophosphate insecticides in the rhizosphere. *Applied and Environmental Microbiology* 37:36–41.

Huddleston, R. L., and L. W. Cresswell. 1976. Environmental and nutritional constraints of microbial hydrocarbon utilization in the soil. In *Proceedings of the 1975 Engineering Foundation Conference: The Role of Microorganisms in the Recovery of Oil.* NSF/RANN, Washington, D.C., pp. 71–72.

Jain, R. K., and G. S. Sayler. 1987. Problems and potential for *in situ* treatment of environmental pollutants by engineered microorganisms. *Microbiological Sciences* 4:59–63.

Jamison, V. M., R. L. Raymond, and J. O. Hudson, Jr. 1975. Biodegradation of high-octane gasoline in groundwater. *Developments in Industrial Microbiology* 16:305–312.

Johnen, B. G., and E. A. Drew. 1977. Ecological effects of pesticides on soil microorganisms. *Soil Science* 123:319–324.

Johnson, L. M., C. S. McDowell, and M. Krupha. 1985. Microbiology in pollution control: From bugs to biotechnology. *Developments in Industrial Microbiology* 26:365–376.

Jones, M., and J. H. Greenfield. 1991. *In situ* comparison of bioremediation methods for a Number 6 residual fuel oil spill in Lee County, Florida. In *Proceedings of the 1991 International Oil Spill Conference*. American Petroleum Institute, Washington, D.C., pp. 533–540.

Kellogg, S. T., D. K. Chatterjee, and A. M. Chakrabarty. 1981. Plasmid assisted molecular breeding—new technique for enhanced biodegradation of persistent toxic chemicals. *Science* 214:1133–1135.

Kilbane, J. J. 1986. Genetic aspects of toxic chemical degradation. *Microbial Ecology* 12:135–146.

Kilbane, J. J., D. K. Chatterjee, and A. M. Chakrabarty. 1983. Detoxification of 2,4,5-trichlorophenoxyacetic acid from contaminated soil by *Pseudomonas cepacia*. *Applied and Environmental Microbiology* 45:1697–1700.

Knackmuss, H. J. 1983. Xenobiotic degradation in industrial sewage: Haloaromatics as target substances. In C. F. Phelps and P. H. Clarke (eds.). *Biotechnology*. The Biochemical Society, London, pp. 173–190.

Knackmuss, H. J. 1984. Biochemistry and practical implications of organohalide degradation. In M. J. Klug and C. A. Reddy (eds.). *Current Perspectives in Microbial Ecology*. American Society for Microbiology, Washington, D.C., pp. 687–693.

Kuhn, E. P., P. J. Colberg, J. L. Schnoor, O. Wanner, A. J. B. Zehnder, and R. P. Schwartzenbach. 1985. Microbial transformations of substituted benzenes during infiltration of river water to groundwater: Laboratory column studies. *Environmental Science and Technology* 19:961–968.

LaDousse, A., C. Tallec, and B. Tramier. 1987. Progress in enhanced oil degradation. In *Proceedings of the 1987 Oil Spill Conference*. American Petroleum Institute, Washington, D.C., Abstract 142.

LaDousse, A., and B. Tramier. 1991. Results of 12 years of research in spilled oil bioremediation: Inipol EAP 22. In *Proceedings of the 1991 International Oil Spill Conference*. American Petroleum Institute, Washington, D.C., pp. 577–581.

Leahy, J. G., and R. R. Colwell. 1990. Microbial degradation of hydrocarbons in the environment. *Microbiological Reviews* 54:305–315 .

Lee, B., A. L. Pometto, and A. Fratzke. 1991. Biodegradation of degradable plastic polyethylene by *Phanerochaete* and *Streptomyces* species. *Applied and Environmental Microbiology* 57:678–685.

Lee, K., and E. M. Levy. 1991. Bioremediation: Waxy crude oils stranded on low-energy shorelines. In *Proceedings of the 1991 International Oil Spill Conference*. American Petroleum Institute, Washington, D.C., pp. 541–547.

Lee, M. D., J. T. Wilson, and C. H. Ward. 1987. *In situ* restoration techniques for aquifers contaminated with hazardous wastes. *Journal of Hazardous Material* 14:71–82.

Leson, G., and A. M. Winer. 1991. Biofiltration: An innovative air pollution control technology for VOC emissions. *Journal of Air and Waste Management Association* 41:1045–1054.

Lukens, H. R., C. B. Williams, S. A. Levinson, W. B. Dandliker, D. Murayama, and R. L. Baron. 1977. Fluorescence immunoassay technique for detecting organic environmental contaminants. *Environmental Science and Technology* 11:292–297.

McCarty, P. L., L. Semprini, M. E. Dolan, T. C. Harmon, C. Tiedeman, and S. M. Gorelick. 1991. *In situ* methanotrophic bioremediation for contaminated groundwater at St. Joseph, Michigan. In R. E. Hinchee and R. F. Olfenbuttel (eds.). *On-site Bioreclamation: Processes for Xenobiotic and Hydrocarbon Treatment*. Butterworth-Heinemann, Boston, pp. 16–40.

Maron, D. M., and B. N. Ames. 1983. Revised methods for the *Salmonella* mutagenicity test. *Mutation Research* 113:173–215.

Matthews, E., and L. Hastings. 1987. Evaluation of toxicity test procedure for screening treatability potential of waste in soil. *Toxicity Assessment* 2:265–281.

Means, A. J. 1991. Observations of an oil spill bioremediation activity in Galveston Bay, Texas. U.S. Dept. of Commerce, National Oceanic and Atmospheric Administration, National Ocean Service, Seattle, Wash.

Metcalf, R. L., G. K. Sangha, and I. P. Kapoor. 1971. Model ecosystems for evaluation of pesticide biodegradability and ecological magnification. *Environmental Science and Technology* 5:709–713.

Miller, R., G. M. Singer, J. D. Rosen, and R. Bartha. 1988a. Photolysis primes the biodegradation of benzo-(a)pyrene. *Applied and Environmental Microbiology* 54:1724–1730.

Miller, R., G. M. Singer, J. D. Rosen, and R. Bartha. 1988b. Sequential degradation of chlorophenols by photolytic and microbial treatment. *Environmental Science and Technology* 22:1215–1219.

Moore, A. T., A. Vira, and S. Fogel. 1989. Biodegradation of trans-1,2-dichloroethylene by methane-utilizing bacteria in an aquifer simulator. *Environmental Science and Technology* 23:403–406.

Morgan, P., and R. J. Watkinson. 1989. Hydrocarbon biodegradation in soils and methods for soil biotreatment. *CRC Critical Reviews in Biotechnology* 8(4):305–333.

Mueller, J. G., P. J. Chapman, and P. H. Pritchard. 1989. Creosote-contaminated sites: Their potential for bioremediation. *Environmental Science and Technology* 23:1197–1201.

Mueller, J. G., D. P. Middaugh, and S. E. Lantz. 1991. Biodegradation of creosote and pentachlorophenol in contaminated groundwater: Chemical and biological assessment. *Applied and Environmental Microbiology* 57:1277–1285.

Munnecke, D. M. 1981. The use of microbial enzymes for pesticide detoxification. In T. Leisinger, A. M. Cook, R. Huffer, and J. Nüesch (eds.). *Microbial Degradation of Xenobiotics and Recalcitrant Compounds.* Academic Press, New York, pp. 251–270.

O'Brien, P. Y., and P. S. Dixon. 1976. The effects of oil and oil components on algae: A review. *British Phycology Journal* 11:115–142.

Office of Technology Assessment. 1991. *Bioremediation for Marine Oil Spills.* U.S. Congress, Washington, D.C.

Ottengraf, S. P. P. 1986. Exhaust gas purification. In W. Schønborn (ed.). *Biotechnology.* Vol. 8. VHC Verlagsgellschaft, Weinheim, Germany, pp. 425–452.

Ottengraf, S. P. P., and A. H. C. Van den Oever. 1983. Kinetics of organic compound removal from waste gases with a biological filter. *Biotechnology and Bioengineering* 25:3089–3102.

Palumbo, A. V., W. Eng, P. A. Boerman, G. W. Strandberg, T. L. Donaldson, and S. E. Herbes. 1991. Effects of diverse organic contaminants on trichloroethylene degradation by methanotrophic bacteria and methane-utilizing consortia. In R. E. Hinchee and R. F. Olfenbuttel (eds.). *On-site Bioreclamation: Processes for Xenobiotic and Hydrocarbon Treatment.* Butterworth-Heinemann, Boston, pp. 77–91.

Pramer, D., and R. Bartha. 1972. Preparation and processing of soil samples for biodegradation studies. *Environmental Letters* 2:217–224.

Prince, R., J. R. Clark, and J. E. Lindstrom. 1990. *Bioremediation Monitoring Program.* Joint Report of EXXON, the U.S. Environmental Protection Agency, and the Alaskan Department. of Environmental Conservation, Anchorage, Alaska.

Pritchard, P. H., and A. W. Bourquin. 1984. The use of microcosms for evaluation of interactions between pollutants and microorganisms. *Advances in Microbial Ecology* 7:133–215.

Pritchard, P. H., and C. F. Costa. 1991. EPA's Alaska oil spill bioremediation project. *Environmental Science and Technology* 25:372–379.

Raymond, R. L., V. W. Jamison, and J. O. Hudson. 1976. Beneficial stimulation of bacterial activity in ground waters containing petroleum products. In *Water—1976.* American Institute of Chemical Engineers, New York, pp. 319–327.

Reineke, W., and H. J. Knackmuss. 1979. Construction of haloaromatic utilizing bacteria. *Nature* (London) 277:385–386.

Reynolds, E. R., and A. M. Yacynych. 1991. Miniaturized electrochemical biosensors. *American Laboratory* March:19–28.

Schultz, J. S. 1991. Biosensors. *Scientific American* 265(2):64–69.

Semprini, L., G. D. Hopkins, P. V. Roberts, and P. L. McCarty. 1991. *In situ* biotransformation of carbon tetrachloride, Freon-113, Freon-11 and 1,1,1-TCA under anoxic conditions. In R. E. Hinchee and R. F. Olfenbuttel (eds.). *On-site Bioreclamation: Processes for Xenobiotic and Hydrocarbon Treatment.* Butterworth-Heinemann, Boston, pp. 41–58.

Shailubhai, K. 1986. Treatment of petroleum industry oil sludge in soil. *Trends in Biotechnology* 4:202–206.

Song, H. G., X. Wang, and R. Bartha. 1990. Bioremediation potential of terrestrial fuel spills. *Applied and Environmental Microbiology* 56:652–656.

Song, H. G., and R. Bartha. 1990. Effect of jet fuel spills on the microbial community of soil. *Applied and Environmental Microbiology* 56:641–651.

Stopa, P.J., M.T. Goode, A.W. Zulich, D. W. Sickenberger, E. W. Sarver, and R. A. Mackay. 1991. Real time detection of biological aerosols. In *Field Screening Methods for Hazardous Wastes and Toxic Chemicals.* U.S. Environmental Protection Agency and U.S. Army Toxic and Hazardous Materials Agency, Las Vegas, Nev., pp. 793–795.

Suflita, J. M., A. Horowitz, D. R. Shelton, and J. M. Tiedje. 1983. Dehalogenation: A novel pathway for the anaerobic biodegradation of haloaromatic compounds. *Science* 214:1115–1117.

Sussman, M., C. H. Collins, F. A. Skinner, and D. E. Stewart-Tull (eds.). 1988. *Release of Genetically-engineered Microorganisms.* Academic Press, London.

Sveum, P., and A. LaDousse. 1989. Biodegradation of oil in the Arctic: Enhancement by oil-soluble fertilizer application. In *Proceedings of the 1989 Oil Spill Conference*. American Petroleum Institute, Washington, D.C., pp. 439–446.

Tesan, G., and D. Barbosa. 1987. Degradation of oil by land disposal. *Water Science Technology* 19:99–106.

Thomas, J. M., M. D. Lee, P. B. Bedient, R. C. Borden, L. W. Carter, and C. H. Ward. 1987. *Leaking Underground Storage Tanks: Remediation with Emphasis on in situ Bioreclamation*. EPA/600/S2–87/008. U.S. Environmental Protection Agency, Ada, Okla.

Tramier, B., and A. Sirvins. 1983. Enhanced oil biodegradation: A new operational tool to control oil spills. In *Proceedings of the 1983 Oil Spill Conference*. American Petroleum Institute, Washington, D.C.

United States Environmental Protection Agency. 1990. *ORD/NETAC:Bringing Innovative Technologies to the Market, EPA Alaskan Oil Spill Bioremediation Project Update*. Office of Research and Development, Washington, D.C.

Vanderlaan, M., E. B. Watkins, and L. Stanker. 1988. Environmental monitoring by immunoassay. *Environmental Science and Technology* 22:247–288.

Van Emon, J. M. 1989. Selected references addressing the development and utilization of immunoassay. In U. M. Cowgill and L. R. Williams (eds.). *Aquatic Toxicology and Hazard Assessment*. ASTM STP 1027. American Society for Testing and Materials, Philadelphia, pp. 427–431.

Van Emon, J. M., J. N. Seiber, and B. D. Hammock. 1989. Immunoassay techniques for pesticide analysis. In G. Zweig (ed.). *Analytical Methods for Pesticides and Plant Growth Regulators*, Vol. XVII. Academic Press, New York, pp. 217–263.

Van Emon, J. M., R. W. Gerlach, R. J. White, and M. E. Silverstein. 1991. U.S. EPA evaluation of two pentachlorophenol immunoassay systems. In *Field Screening Methods for Hazardous Wastes and Toxic Chemicals*. U.S. Environmental Protection Agency and U.S. Army Toxic and Hazardous Materials Agency, Las Vegas, Nev., pp. 815–818.

Vanloocke, R., R. DeBorger, J. P. Voets, and W. Verstraete. 1975. Soil and groundwater contamination by oil spills: Problems and remedies. *International Journal of Environmental Studies* 8:99–111.

Venosa, A. D., J. R. Haines, W. Nisamaneepong, R. Goving, S. Pradhan, and B. Siddique. 1991. Screening of commercial inocula for efficacy in enhancing oil biodegradation in closed laboratory system. *Journal of Hazardous Materials* 28:131–144.

Verstraete, W., R. Vanlooke, R. deBorger, and A. Verlinde. 1976. Modelling of the breakdown and the mobilization of hydrocarbons in unsaturated soil layers. In J. M. Sharpley and A. M. Kaplan (eds.). *Proceedings of the Third International Biodegradation Symposium*. Applied Science Publishers, London, pp. 98–112.

von Wedel, R. J., J. F. Mosquera, C. D. Goldsmith, G. R. Hater, A. Wong, T. A. Fox, W. T. Hunt, M. S. Paules, J. M. Quiros, and J. W. Wiegand. 1988. Bacterial biodegradation and bioreclamation with enrichment isolates in California. *Water Science Technology* 20:501–503.

Wang, X., X. Yu, and R. Bartha. 1990. Effect of bioremediation on polycyclic aromatic hydrocarbon residues in soil. *Environmental Science and Technology* 24:1086–1089.

Wang, X., and R. Bartha. 1990. Effects of bioremediation on residues: Activity and toxicity in soil contaminated by fuel spills. *Soil Biology and Biochemistry* 22:501–506.

Wilson, B. H., G. B. Smith, and J. F. Rees. 1986. Biotransformations of selected alkylbenzenes and halogenated aliphatic hydrocarbons in methanogenic aquifer material: A microcosm study. *Environmental Science and Technology* 20:997–1002.

Wilson, J. 1991. Performance evaluations of *in situ* bioreclamation of fuel spills at Traverse City, Michigan. In *Proceedings of the In Situ and On-Site Bioreclamation: An International Symposium*. Butterworth, Stoneham, Mass.

Wilson, J. T., and C. H. Ward. 1987. Opportunities for bioreclamation of aquifers contaminated with petroleum hydrocarbons. *Developments in Industrial Microbiology* 27:109–116.

Yaniga, P. M., and W. Smith. 1984. Aquifer restoration via accelerated *in situ* biodegradation of organic contaminants. In *Proceedings of the Conference on Petroleum Hydrocarbons and Organic Chemicals in Ground Water—Prevention, Detection, and Restoration*. National Water Well Association, Worthington, Ohio, pp. 451–470.

You, I. S., and R. Bartha. 1982. Stimulation of 3,4-dichloroaniline mineralization by aniline. *Applied and Environmental Microbiology* 44:678–681.

Zitrides, T. G. 1990. Bioremediation comes of age. *Pollution Engineering* 12:59–60.

15

Microorganisms in Mineral and Energy Recovery and Fuel and Biomass Production

RECOVERY OF METALS

As the Earth's high-grade deposits of petroleum and certain metals are depleted, there is a need to find innovative and economical procedures to recover oil and metals from low-grade deposits that, for technical or economic reasons, have been beyond the reach of current recovery techniques. Ores with low metal content are not suitable for direct smelting, but, at least in the case of some sulfide or sulfide-containing ores, it is possible to extract metal from low-grade ore economically using the activity of sulfur-oxidizing bacteria, especially *Thiobacillus ferrooxidans* (Zimmerley et al. 1958; Kuznetsov et al. 1963; Beck 1967; Tuovinen and Kelly 1974; Kelly 1976; Karavaiko et al. 1977; Torma 1977; Brierley 1978, 1982; Murr et al. 1978; Kelly et al. 1979; Lundgren and Silver 1980; Norris and Kelly 1982; Karavaiko 1985; Ehrlich and Brierley 1990). This microbial recovery of metals is sometimes referred to as microbial mining, or biohydrometallurgy (Mersou 1992).

The process is currently applied on a commercial scale to low-grade copper and uranium ores (Brierley 1978, 1982). Laboratory-scale experiments indicate that the process also has promise for recovering nickel, zinc, cobalt, tin, cadmium, molybdenum, lead, antimony, arsenic, and selenium from low-grade sulfide-containing ores. The general metal recovery process can be represented by Equation 1:

(1) $\quad MS + 2O_2 \xrightarrow[\textit{T. ferrooxidans}]{} MSO_4$

where M represents a bivalent metal that is insoluble as a sulfide but soluble, and thus leachable, as a sulfate. *Thiobacillus ferrooxidans* is a chemolithotrophic bacterium that derives energy through the oxidation of either a reduced sulfur compound or ferrous iron. It exerts its bioleaching action directly by oxidizing the metal sulfide and/or indirectly by oxidizing the ferrous iron content of the ore to ferric iron; the ferric iron, in turn, chemically oxidizes the metal to be recovered by leaching.

When microbes are employed to bioleach metals from low-grade ores, it is usually necessary to mine and break up the ore and heap it in piles on a water-impermeable formation or on a specially con-

Figure 15.1

Heap leaching of low-grade ore. The ore is mined, crushed, and heaped in the form of a truncated cone on a suitable asphalt pad. The leaching liquor is pumped to the top of the heap and percolates through the ore. The leachate is collected, processed, and recycled. (Source: Zajic 1969. Reprinted by permission, copyright Academic Press.)

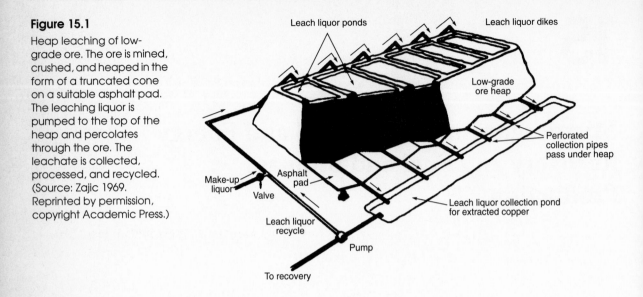

structed apron (Brierley 1982) (Figure 15.1). Water is then pumped to the top of the ore heap and allowed to trickle down through the ore to the apron. Figure 15.2 shows a continuous reactor-type leaching operation for recovery of copper from low-grade sulfide ore. The leachate is collected and processed for the recovery of the metal. Most commonly, the leached metal is partitioned into an organic solvent and subsequently recovered by "stripping" (evaporating) the solvent. Both the leaching liquor and the solvent are recycled. Figure 15.3 shows the flowchart of a typical operation.

Geological formations play a key role in determining the suitability of ore deposits for recovery by *in situ* bioleaching. Rarely, under favorable geological conditions, microorganisms may be induced to release the metals into solution without any mechanical mining of the metal-bearing rocks. If the ore formation is sufficiently porous and overlies water-impermeable strata, a suitable pattern of boreholes is established. Some of the holes are utilized for the injection of the leaching liquor; others are used for the recovery of the leachate (Figure 15.4).

The mineralogy of the ore is significant in determining its susceptibility to bioleaching. In some cases, the chemical form of the element within the ore

Figure 15.2

Diagram of apparatus for extracting copper from low-grade ore by a continuous leaching process. The oxidation of sulfide and ferrous iron is carried out by *Thiobacillus ferrooxidans* generating the acid for leaching. Copper is precipitated by exchange, using scrap iron. This process was the basis of U.S. patent No. 2,829,964, assigned to Kennecott Copper Company. (Source: Zimmerley et al. 1958.)

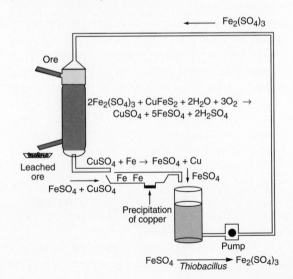

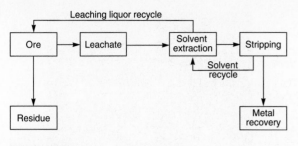

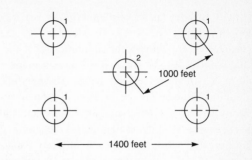

Figure 15.3

Flowchart of metal recovery from low-grade ores by bioleaching using the activity of *Thiobacillus ferrooxidans*. (Source: Torma 1977.)

is resistant to microbial attack. In other cases, microorganisms can attack the ore initially, but toxic products are produced that preclude further bioleaching activities (Tuttle and Dugan 1976). In still other cases, the ore minerals exist in chemical forms that are readily subject to bioleaching. The size of mineral particles to be leached is critical in determining the rate of leaching; increasing the surface area, which can be achieved by crushing or grinding, greatly enhances the rate of bioleaching.

A variety of ecological factors affects the efficiency of bioleaching. Factors that influence the activities of the most important leaching bacterium, *T. ferrooxidans*, include temperature, generally considered optimal between 30°C and 50°C; acidity, optimally at pH 2.3–2.5; iron supply, optimally 2–4 g/L of leach liquor; oxygen; and the availability of other nutrients required for growth (Brierley 1978; Summers and Silver 1978). *Thiobacillus ferrooxidans* is generally unable to initiate growth on ferrous iron above pH 3.0. Ammonium nitrogen, phosphorus, sulfate, and magnesium are essential for the growth of *T. ferrooxidans*. These nutrients may be limiting in some leaching operations and need to be added. Availability of sufficient water can also be an important consideration in bioleaching processes. In open-air leaching operations, light has an inhibitory effect on some species of *Thiobacillus*, with the greatest inhibition occurring at the shorter wavelengths. Suspended particles of ferric iron offer a degree of protection against inhibitory light intensities.

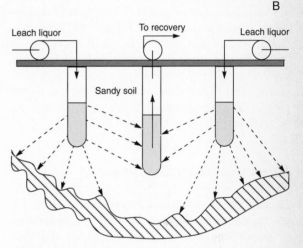

Figure 15.4

Hole-to-hole leaching of low-grade ores, an *in situ* process for metal recovery by bioleaching using the activity of thiobacilli. The process may be practiced in relatively porous ore overlying impermeable bedrock. Leaching is from injection wells to a collection well. (A) Overhead view of layout of wells. 1 = leach liquor shafts; 2 = recovery shaft. (B) Side view of wells. (Source: Zajic 1969. Reprinted by permission, copyright Academic Press.)

Temperatures in some mineral deposits can be significantly increased by the oxidative activities of the thiobacilli and may exceed the tolerance limits of a *Thiobacillus* species, leading to decreased bioleaching activity. Because of the high temperatures associated with some leaching operations, thermophilic sulfur-oxidizing microorganisms may play an important role in the bioleaching process (Brierley and

Murr 1972; Brierley and Lockwood 1977; Brierley and Brierley 1978; Brock 1978). Members of the genus *Sulfolobus* are obligate thermophilic archaebacteria that oxidize ferrous iron and sulfur in a manner similar to the members of the genus *Thiobacillus*. Thermophilic *Thiobacillus* species have also been isolated from various hot sulfur-rich environments. The abilities of these acid-tolerant thermophilic bacteria to oxidize inorganic substrates make them likely candidates for use in bioleaching metal sulfides. *Sulfolobus* has been used especially in the bioleaching of molybdenite (molybdenum sulfide) (Figure 15.5). *Thiobacillus* species are unable to leach molybdenite efficiently because of the toxicity of molybdenum to this bacterium; *Sulfolobus* is less sensitive to molybdenum. Still other microorganisms, perhaps in mixed populations, are potential candidates for use in bioleaching processes (Norris and Kelly 1982).

Microbial Assimilation of Metals

Microorganisms have been considered not only for use in bioleaching of metals but also as accumulators of metals from dilute solutions (Charley and Bull 1979; Norris and Kelly 1979, 1982; Brierley 1982; Ehrlich and Brierley 1990). Large numbers of bacteria, yeasts, and algae are capable of accumulating metal ions in their cells to concentrations several orders of magnitude higher than the background concentration of these metals. The mechanism of accumulation may involve intracellular uptake and storage via active cation transport systems, surface binding, or some undefined mechanisms. These processes have a potential use in extracting rare metals from dilute solution or removing toxic metals from industrial effluents. The selectivity and gradient of accumulation in general do not, however, compete favorably with conventional chemical extraction and partition processes, and it is doubtful whether such microbial accumulations will ever form the basis of a technological process. Bioaccumulation of toxic heavy metals has a definite practical significance, however, in terms of an increased but undesirable introduction of these pollutants into food webs.

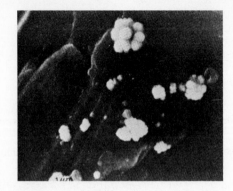

Figure 15.5

Scanning electron micrograph of thermophilic, acidophilic, and chemoautotrophic bacteria capable of oxidizing reduced iron and sulfur compounds. The bacteria are adhering to ore particles. They appear to be related to *Sulfolobus* and are potentially useful for leaching low-grade chalcopyrite and molybdenite ores. (Source: Brierley and Murr 1972. Reprinted from *Science*, copyright American Association for the Advancement of Science.)

Copper Bioleaching

Bioleaching with thiobacilli and related iron-oxidizing bacteria contributes to the commercially important recovery of copper (Groudev and Genchev 1978; Rangachari et al. 1978; Groudev 1979). Copper is a metal generally in short supply and is in high demand by the electrical industry and for various metal alloys. In the bioleaching of copper, the action of *Thiobacillus* involves both the direct oxidation of CuS and the indirect oxidation of CuS via generation of ferric (Fe^{3+}) ions from ferrous sulfide (Tuovinen 1990) (Table 15.1; Figure 15.6). Ferrous sulfide is present in the economically most important copper ores, such as chalcopyrite ($CuFeS_2$); a typical low-grade ore contains 0.1%–0.4% copper.

Copper is recovered from the leaching solution either by solvent partitioning or by the use of scrap iron; in the latter method, copper replaces iron according to Equation 2:

(2) $CuSO4 + Fe° \longrightarrow Cu° + FeSO_4$

Table 15.1

Reactions involved in microbial bioleaching of copper

Direct oxidation of monovalent copper in chalcocite (Cu_2S) by *Thiobacillus ferrooxidans*:

$$2Cu_2S + O_2 + 4H^+ \longrightarrow 2CuS + 2Cu^{2+} + 2H_2O$$
$$\text{Chalcocite} \qquad\qquad \text{Covellite}$$
$$(4Cu^+ \longrightarrow 4Cu^{2+} + 4e^-)$$

No change in valence of sulfur. Electrons from Cu^+ oxidation probably used as energy source by the bacteria.

Indirect oxidation of covellite by *T. ferrooxidans* via elemental sulfur:

$$CuS + 1/2O_2 + 2H^+ \longrightarrow Cu^{2+} + H_2O + S^\circ \text{ (spontaneous)}$$

Accumulation of a film of elemental sulfur on the mineral causes the reaction to cease.

$$S^\circ + 1.5O_2 + H_2O \longrightarrow 2H^+ + SO_4^{2-} \text{ (bacterial)}$$

Removal of protective film by bacteria keeps the reaction going:

$$\text{Overall: } CuS + 2O_2 \longrightarrow Cu^{2+} + SO_4^{2-}$$

Indirect oxidation of covellite (also other ores) with ferric iron (Fe^{3+}) and regeneration of ferric iron from ferrous by bacterial oxidation:

$$CuS + 8Fe^{3+} + 4H_2O \longrightarrow Cu^{2+} + 8Fe^{2+} + SO_4^{2-} + 8H^+ \text{ (chemical)}$$
$$8Fe^{2+} + 2O_2 + 8H^+ \longrightarrow 8Fe^{3+} + 4H_2 \text{ (bacterial)}$$
$$\text{Overall: } CuS + 2O_2 \longrightarrow Cu^{2+} + SO_4^{2-}$$

Recovery of elemental copper from copper ions by reaction with scrap iron:

$$Cu^{2+} + Fe^\circ \longrightarrow Fe^{2+} + Cu^\circ \text{ (chemical)}$$

Sources: Brock and Gustafson 1976; Torma 1976; Brock 1978; Summers and Silver 1978.

Figure 15.6

Two biological leaching mechanisms for pyrite (FeS_2)/chalcopyrite ($CuFeS_2$). (A) Direct leaching with transfer of electron (e^-) to molecular oxygen. (B) indirect leaching with Fe^{3+} as the primary electron acceptor. (Source: Tuovinen 1990.)

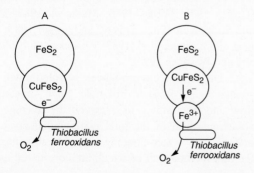

The latter method is more advantageous for copper recovery in the bioleaching process. When solvent partitioning is used, residues in the leaching liquor, unless carefully removed by activated carbon, may inhibit the activity of *T. ferrooxidans*. In laboratory studies, as much as 97% of the copper in low-grade ores has been recovered by bioleaching, but such high yields are seldom achieved in actual biomining operations. Nevertheless, even a 50%–70% recovery of copper by bioleaching from an ore that would otherwise be useless is an important achievement. In case of low-grade ores, bioleaching costs only one-half to one-third as much as direct smelting (Mersou 1992). The principal disadvantage of bioleaching is the relative slowness of the process. It may take decades to recover metals from a deposit by bioleaching that could be exploited by mining and smelting in relatively only a few years.

Uranium Bioleaching

Uranium is another metal that can be recovered by microbial leaching (Brierley 1982; McCready and Gould 1990). Uranium is used as a fuel in nuclear power generation, and the microbial recovery of uranium from otherwise useless low-grade ores can be considered a contribution to energy production. Bioleaching can have a direct bearing on the economics of nuclear facilities by providing a mechanism for commercial utilization of low-grade uranium deposits and for the recovery of uranium from low-grade nuclear wastes.

Tetravalent uranium oxide (UO_2) occurs in low-grade ores and is insoluble. It can be converted to the leachable exhalant form by oxidation with ferric (Fe^{3+}) ions. Carbonate-rich uranium deposits are leached with alkaline bicarbonate-carbonate solutions. Pyrite-containing uranium deposits are leached with dilute sulfuric acid solutions. The ferrous ions (Fe^{2+}) produced during uranium oxidation are converted back to Fe^{3+} by chemical oxidants such as chlorate (ClO_3^-), manganese dioxide (MnO_2), or hydrogen peroxide (H_2O_2). The chemical oxidants add substantial cost to the leaching process, but in pyrite-containing uranium ores, the Fe^{3+} oxidant may be produced and regenerated at lower cost by the action of *T. ferrooxidans* (Guay et al. 1977; Torma 1985). *Thiobacillus ferrooxidans* oxidizes the ferrous iron in pyrite (FeS), which often accompanies uranium ores, to ferric iron, which in turn acts as an oxidant to convert UO_2 chemically to the leachable UO_2SO_4 (Table 15.2).

The feasibility of utilizing *Thiobacillus* for the recovery of uranium depends primarily on the composition of the mineral deposit (Tomizuka and Yagisawa 1978). Pyritic uranium oxide ores, for example, are suitable for bioleaching, but uranium ores that lack iron-containing minerals or are high in carbonates are not. In general, bacterial leaching of uranium is feasible in geological strata where the ore is in the tetravalent state and is associated with reduced sulfur and

Table 15.2

Reactions involved in microbial bioleaching of uranium

Indirect oxidation of uranium ore with ferric iron:

UO_2^{2-} tetravalent uranium, insoluble oxide

$UO_2SO_4{}^{2-}$ hexavalent uranium (uranyl ion, UO_2^{2-}), soluble sulfate

$$UO_2 + 2Fe^{3+} + SO_4^{2-} \longrightarrow UO_2SO_4 + 2Fe^{2+}$$
$$(U^{4+} + 2Fe^{3+} \longrightarrow U^{6+} + 2Fe^{2+})$$

Fe^{2+} is reoxidized by *T. ferrooxidans*.

iron minerals. In an acidic medium, those energy sources provide a suitable environment for the growth of *T. ferrooxidans*.

RECOVERY OF PETROLEUM

Petroleum and natural gas are relatively clean and convenient energy sources, and the United States relies on these fuels for about 70% of its energy needs. Domestic production supplies less than two-thirds of the U.S. oil demand. High petroleum imports and unpredictable supplies from politically volatile countries have led to an unfavorable trade balance and economic vulnerability. Limited prospects exist for finding new high-quality oil reserves, but microbial processes—called microbially enhanced oil recovery (MEOR)—may contribute to improved petroleum recovery from existing old wells, as well as from formations that for technical or economic reasons have not been tapped to date (Geffen 1976; Forbes 1980; Matthews 1982; Moses and Springham 1982; Finnerty and Singer 1983; Hitzman 1983; Singer et al. 1983; Westlake 1984; McInerney and Westlake 1990).

Under ideal circumstances, petroleum will gush spontaneously from a drill hole. Such spontaneous flow and/or pumping recovers on the average only about one-third of the total petroleum deposit. Gas pressurizing, water flooding, miscible flooding, and thermal methods, all of which are considered as secondary recovery techniques, have been used for additional recovery of petroleum. Gas pressurizing and water flooding involve the deliberate injection of gas or water into drill holes in order to dislodge and push petroleum toward producing wells. Injected under high pressure, carbon dioxide becomes an effective lipophilic solvent that decreases the viscosity of heavy crude oils but separates from the petroleum spontaneously after the reduction of pressure. For these reasons, miscible flooding with pressurized CO_2 is an especially promising technique. Thermal methods involve steam injection or controlled underground combustion designed to increase the flow of heavy, viscous petroleum. Tertiary oil recovery techniques, which include the use of solvents, surfactants, and polymers designed to dislodge oil from geological formations, have the potential for recovering an additional 60 billion to 120 billion barrels (8.5 billion to 17 billion metric tons) of petroleum from U.S. deposits alone (Figure 15.7). Some promising surfactants and polymers, such as the

xanthan gums (Figure 15.8), are produced by the bacterium *Xanthomonas campestris* (Sandvik and Maerker 1977; Cooper and Zajic 1980; Cooper 1983; Gutnick et al. 1983; Westlake 1984; Zajic and Mahomedy 1984). The critical characteristic of these polymers is their substantial viscosity combined with flow characteristics that allow them to pass through small pore spaces. These materials are produced by conventional fermentation processes and are injected as additives in water-flooding operations.

Another possible though less promising approach involves the use of microorganisms *in situ* for dislodging oil (Moses and Springham 1982; Finnerty and Singer 1983). Claude E. ZoBell (1947) first promoted this idea. Microbial growth on rock pore surfaces dislodges oil directly by physical displacement or indirectly by synthesis of surface active metabolites and gases such as H_2, CO_2, CH_4, and H_2S. In addition, the viscosity of the oil may be altered by partial microbial degradation. ZoBell (1947) considered *Desulfovibrio* strains as promising microbial agents for increasing petroleum recovery by *in situ* action, and J. W. M. La Riviere (1955a, 1955b) experimentally demonstrated enhanced oil release by microbial action in the laboratory.

In *in situ* tests, microbial suspensions injected in combination with a carbon source such as molasses or milk whey into an oil formation were shown to enhance oil recovery from wells with previously low rates of production (Finnerty and Singer 1983). In these tests, the wells were sealed and time was allowed for microbial action, after which production was resumed; oil flow was increased when the wells were reopened. In spite of these preliminary results, difficulties with the *in situ* approach are numerous, and the practicality of this approach remains in doubt. Growth conditions, salinity, pH, temperature, and redox potential are quite unfavorable; anaerobic utilization of petroleum is marginal at best; introduction of air can cause undesirable changes in petroleum quality; and, perhaps most importantly, microbial growth may actually close pore spaces and thus decrease the flow of oil (Updegraff 1983).

In addition to oil wells, tar sands and oil shales form extensive fossil hydrocarbon reserves. Tar sands

Figure 15.7

The U.S. crude oil "barrel" as of 1975, showing target of tertiary oil recovery. The oil shortage and world energy crisis beginning in the late 1970s made secondary and tertiary processes for the recovery of oil essential and economical. (Source: Geffen 1976.)

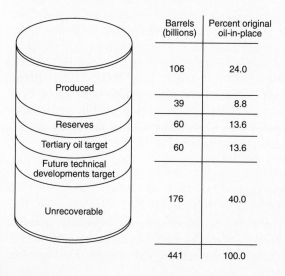

	Barrels (billions)	Percent original oil-in-place
Produced	106	24.0
	39	8.8
Reserves	60	13.6
Tertiary oil target	60	13.6
Future technical developments target		
Unrecoverable	176	40.0
	441	100.0

Figure 15.8

Structure of xanthan gum produced by *Xanthomonas campestris*. The gum is used in the tertiary recovery of oil. As an additive in water flooding operations, it helps to push crude oil toward production wells. (Source: Sutherland and Ellwood 1979. Reprinted by permission; copyright Cambridge University Press.)

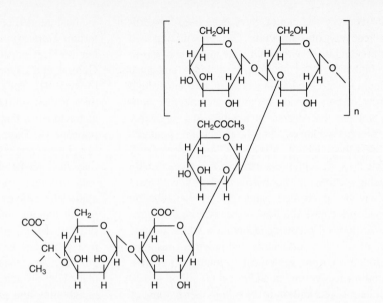

contain mainly bitumen, a highly viscous hydrocarbon mixture soluble in aromatic solvents; oil shales are impermeable, fine-grained, sedimentary rock formations that contain some bitumen and considerable amounts of kerogen, a highly cross-linked and condensed hydrocarbon complex that is insoluble in benzene. Tar sands and oil shales may contain as much as 50%–60% hydrocarbon material, but 20%–30% is a more common figure for deposits of economic importance. The current technology of oil production from tar sands and oil shales uses heat treatment to recover and process the hydrocarbons into a usable product. The costs of processing these oil reserves are too high at the present time, and technologies using microorganisms are being considered to improve the economics of recovering oil from these reserves (Westlake 1984).

Biosurfactants may be useful for releasing oil from tar sands (Zajic and Gerson 1978; Singer et al. 1983; Westlake 1984). In the case of oil shales that occur in association with substantial amounts of carbonates and some pyrite, the carbonates can be dissolved by acid, increasing porosity and reducing the bulk of the shale so that the oil can be more efficiently recovered (Figure 15.9). The acid can be produced most economically by using the thiobacilli as in metal

recovery (Meyer and Yen 1976; Yen 1976a, 1976b; Westlake 1984). In laboratory experiments, bioleaching of oil shale has increased oil yield by approximately 30% (Figure 15.10).

PRODUCTION OF FUELS

Microorganisms used in enhanced oil recovery have the potential of contributing to fuel production in several additional ways. Microorganisms may be used to convert waste products, plants, or microbial biomass into liquid or gaseous fuels (Schlegel and Barnea 1976; Keenan 1979; National Research Council 1979; Bungay 1981) and can also be used to convert solar energy into biomass that can be fermented to yield fuels (Bennemann and Weissman 1976).

Ethanol

The ability of microorganisms, especially yeasts, to produce ethanol by the fermentation of carbohydrate-containing materials is well known. The microbial

Figure 15.9

Flowchart of bioleaching of oil shale with recycling of sulfur. Kerogen is liberated from the leached shale and is converted subsequently to refinery stock by hydrogenation. The leaching process uses the activity of thiobacilli; sulfur is recycled using *Desulfovibrio* and organic wastes as substrates. (Source: Yen 1976a.)

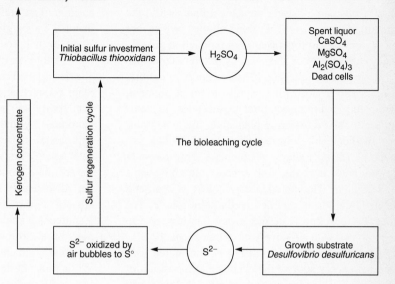

Figure 15.10

Effect of bioleaching on oil shale. Oxidizability of the kerogen by permanganate is used as a measure of liberation from the shale showing accessibility of hydrocarbon. Raw unleached shale uses little permanganate; bioleached shale shows permanganate use about four times as high, indicating the effectiveness of bioleaching. (Source: Yen, 1976a.)

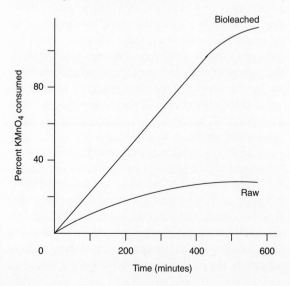

formation of ethanol is used in the production of many beverages; ethanol can also serve as a valuable fuel resource (National Research Council 1979; Venkata-subramanian and Keim 1981; Eveleigh 1984). The addition of ethanol to gasoline can greatly extend this petroleum-based fuel (Anderson 1978). Gasoline-ethanol mixtures are commonly known as gasohol. The normal ratio of gasoline to ethanol is 9:1. Gasohol can be used directly in present internal combustion automobile engines without any engineering modifications. Gasohol is an efficient fuel, lowering the release of atmospheric hydrocarbon pollutants compared to petroleum. Brazil, which has suitable land and climate for large-scale growing of sugarcane that can be fermented to ethanol, has embarked on an ambitious and generally successful program to replace gasoline with ethanol (Jackson 1976; Lindeman and Rocchiccioli 1979).

The extensive production of ethanol requires a large supply of carbohydrate substrates. The production of substrates suitable for conversion to fuels such as ethanol and methane involves the conversion of solar energy to plant or microbial biomass and/or the utilization of organic wastes. The use of photosynthetic activities for the production of suitable fermentation substrates raises some conflict with the agricultural production of food supplies, which are also becoming limiting because of population growth. It is necessary to seek increasingly efficient processes for the primary conversion of solar energy to cellular biomass and to utilize the energy stored in biomass. Fortunately, many agricultural wastes such as stems and leaves of fruit-bearing plants are, in theory, suitable candidates as substrates for microbial fuel production. These sources of biomass are normally plowed back into the soil, where microorganisms degrade them. The stored energy is lost as heat, and the fixed carbon is returned to the atmosphere. The materials recycling is desirable, but this process can be combined with a recovery of chemical energy.

In situations where ethanol is considered as a potential energy source, the chemical composition of the substrate is critical in determining whether it can be converted to ethanol efficiently. Ethanol is normally produced from simple carbohydrates by yeasts. The yeasts used in commercial production are unable to attack complex polymers, and it is usually necessary initially to degrade plant polymers with enzymes produced by the plant itself or by microorganisms other than yeasts. In the brewing industry, for example, the starches in the grain substrates are initially subjected to amylase attack, producing simple carbohydrates from starches. If cellulosic materials are considered for the production of ethanol, it will be necessary initially to catalyze the breakdown of this polymer to simple carbohydrates before yeasts or genetically engineered microorganisms with the necessary enzymes can be utilized for ethanol production. Because of their heat tolerance and rapid metabolism, several thermophilic microorganisms appear to be promising prospects for future ethanol production from cellulosic wastes (Eveleigh 1984).

The production of agricultural crops, both as food sources and as potential substrates for microbial fuel production, is limited by a number of factors, including the availability of arable land and the availability of fixed forms of the nitrogen required for plant and microbial growth. Attempts to establish additional nitrogen-fixing symbiotic relationships between plants and bacteria, through genetic manipulation, may have an important influence on whether sufficient plant material can be grown cheaply for utilization as a substrate for fuel production.

Although microorganisms may be capable of alleviating the scarcity of fixed forms of nitrogen for plant production, it is still likely that land resource availability will severely limit the expansion of conventional agricultural practices. Therefore, novel approaches to terrestrial biomass production from solar energy conversion may have to be considered for adequate production of substrates that can be converted to usable forms of chemical energy, such as ethanol. Forest biomass represents a largely unused resource that could be tapped for fuel production. Biomass production in terrestrial ecosystems ranges from less than 1 to more than 700 tons per ha per year. While much of the biomass produced in forest ecosystems will not be available for microbial fuel production because it is inaccessible or dedicated to other uses, forest residuals such as branches, wood chips, and sawdust are good potential substrates (Steinbeck 1976). Intensive forest management can greatly increase the production of plant biomass for microbial conversion, but the slow metabolism of lignocellulosic material is a technical obstacle to the use of forest biomass. A novel "steam explosion" technology, which first exposes wood to high-pressure steam and subsequently explosively releases the pressure, shredding the wood and achieving partial breakdown of the lignocellulose complexes, may overcome this difficulty and convert wood into a fermentable substrate. This, together with recent progress in understanding the biochemistry and genetics of lignin biodegradation (Paterson et al. 1984), may make the microbial conversion of lignocellulosic material into a liquid fuel a technical and economic reality.

Algae, cyanobacteria, and submerged or semi-submerged higher plants growing in aquatic, marine, and freshwater habitats can make a major contribution to the production of biomass through the conversion of solar energy. At present, these habitats are not intensively managed for the production of biomass. There are several advantages to considering aquatic habitats in the production of substrates for microbial energy production. Algae and cyanobacteria are not grown on a large scale as food resource, thus eliminating the conflict of resource utilization between food production and production of substrates for energy. The rapid growth rates of algae and cyanobacteria permit efficient substrate production. Microalgae may reach actual yields of 55 tons per hectare per year. Compared to higher plants, however, microalgae and cyanobacteria are difficult to harvest. Productivity in aquatic habitats is also naturally limited by the availability of nitrogen and phosphorus. Cyanobacteria are potentially attractive sources of substrates for microbial energy production, since many cyanobacteria are capable of nitrogen fixation. In enclosed basins, it is possible to add fertilizers and maintain effective nutrient recycling within the system to promote biomass production. In open ocean systems, it might become possible to take advantage of natural nutrient upwellings or to create artificial upwellings and thus to enhance biomass production for conversion to fuel (Roels et al. 1976; Soeder 1976).

Methane

Methane, produced through methanogenesis by bacteria, is an extremely important potential fuel source (Pfeffer 1976; National Academy of Sciences 1977; National Research Council 1979). Methane can be used in the generation of mechanical, electrical, and heat energy. It can be used as a fuel source for homes and industry by transmission through natural gas pipelines and converted by microbial action or chemical means to methanol, which can be used as fuel in internal combustion engines.

Large amounts of methane are produced during the anaerobic decomposition of organic materials, but this energy resource normally is reoxidized, as in many aquatic habitats, or lost to the atmosphere, as in most terrestrial habitats. Methane production can be based on the decomposition of waste materials or on the conversion of biomass produced as a step in the conversion of solar energy to usable chemical energy. The production and utilization of methane as an energy source can close an important loop in the bio-geochemical cycling of carbon, allowing for a recycling of energy that is initially trapped in organic compounds through the conversion of solar energy.

Methane can be produced from the biomass of primary producers, such as marine seaweeds. The growth of marine algae for the production of methane by methanogenic bacteria is an attractive way of coupling solar energy with the production of a gaseous hydrocarbon. Anaerobic photoorganotrophic bacteria can be grown on animal wastes in the presence of light to produce high-protein biomass and various fatty acids (Ensign 1976). The use of such photoheterotrophic bacterial populations permits the direct coupling of solar energy conversion and recycling of organic wastes to produce a higher-grade product that can be used either as an animal feed or in the generation of methane.

One of the most promising ways of producing methane is through the bioconversion of waste materials (Table 15.3). Anaerobic digestors can be used for the production of methane, and indeed, methane is a

Table 15.3

Gas yields from anaerobic digestion of wastes

Nature of solid waste	Total gas yield (m^3/kg)
Municipal sewage sludge	0.43
Municipal garbage only	0.61
Municipal paper only	0.23
Municipal refuse combined	0.28
Dairy wastes, sludge	0.98
Yeast wastes, sludge	0.49
Brewery wastes, hops	0.43
Stable manure, with straw	0.29
Horse manure	0.40
Cattle manure	0.24
Pig manure	0.26
Beet leaves	0.46
Maize tops	0.49
Grass	0.50

Source: Imhoff et al. 1971.

normal by-product of anaerobic sewage treatment facilities. In some sewage treatment plants, the methane generated in anaerobic digesters is used as an energy source to power the plant. The energy generated by sewage treatment facilities is also used to supply power for some small municipalities. Landfills containing organic wastes generate methane for years; pilot projects have been conducted to trap and utilize this gas. It is theoretically possible to place anaerobic digestors in private residences and small industries to permit the recycling of available energy resources with maximal retention and utilization within closed systems. Municipal and agricultural wastes could be supplied to central facilities for the large-scale production of methane.

Unlike the production of ethanol, which is carried out by pure cultures of yeasts, the production of methane is carried out by a mixed microbial community. Some microbial populations are involved in the breakdown of various organic materials, including complex polymers and simple fatty acids, hydrogen, carbon dioxide, and alcohols. Methanogenic bacteria utilize these fermentation products for the production of methane. Methanogenesis is commonly carried out by microbial populations involved in synergistic or mutualistic relationships. The microorganisms responsible for the conversion of organic matter to methane represent a balanced microbial community. This mixture presents some problems for bioengineers, who design fermenters for industrial applications that optimize conditions for defined microbial populations. Effective designs for the conversion of organic matter to methane require relatively compact fermenters for the ready trapping of gaseous methane (Switzenbaum 1983).

The evolution of methane from anaerobic digestors and other bioconversion processes occurs simultaneously with the evolution of carbon dioxide. The ratios of methane to carbon dioxide depend on the chemical composition of the substrate and the environmental conditions under which the bioconversion is carried out. Biotechnological processes can, and must, be adjusted to maximize the proportions of methane in the evolved gases. The methane must be trapped and separated from other gases to be a useful

energy resource that can supplement and/or replace natural gas as a fuel.

Hydrocarbons Other Than Methane

To date, little consideration has been given to the development of processes for the microbial production of hydrocarbons other than methane. Microorganisms are believed to play a role in the formation of petroleum deposits. The formation of petroleum, however, appears to be the result of physicochemical as well as microbial processes and occurs over long periods of time. Some microorganisms are known to produce hydrocarbons. Hydrocarbon biosynthesis occurs in various algae, fungi, and bacteria, such as the bacterium *Micrococcus luteus* (Tornabene 1976).

Some bacteria, fungi, and algae synthesize significant levels of hydrocarbons, suggesting the possibility that microorganisms that could produce hydrocarbons on a commercial scale could be found. Studies have shown that a tenfold increase in hydrocarbon production can be achieved with *Micrococcus* through genetic transformation. The successful utilization of hydrocarbon production by microorganisms apparently may depend on their genetic modification and selection of efficient hydrocarbon-producing strains. In theory, the algae and cyanobacteria that produce hydrocarbons may permit the conversion of solar energy to chemical forms of energy that could be utilized to power conventional combustion engines. This possibility deserves further consideration but, at this time, faces considerable technical and economic obstacles.

Hydrogen

Hydrogen, not presently a major fuel, can be produced by various microorganisms and could become an important fuel resource (Reeves, et al. 1976; National Research Council 1979). Hydrogen is produced as an end product from the fermentation of carbohydrates by bacteria that carry out a mixed acid fermentation. Rapid hydrogen removal favors the fermentation process. The theoretical efficiency of the conversion of carbohydrates to energy stored in hydrogen is approximately 33%, compared to 85% for methane formation from organic matter. It is evident, therefore, that methane formation from organic compounds, rather than hydrogen formation, should be considered as the preferred process for microbial energy production via fermentation.

In contrast to hydrogen production from fermentation processes, phototrophic microorganisms may be capable of producing significant amounts of molecular hydrogen (Bennemann and Weissman 1976). Various microorganisms are capable of biophotolysis, that is, the production of hydrogen from water, using solar energy. Hydrogen production occurs in various photosynthetic bacteria, including purple and green sulfur bacteria, and in some cyanobacteria. The photosystems of cyanobacteria, algae, and plants are all capable of splitting water, releasing free oxygen and using the hydrogen to reduce carbon dioxide to carbohydrate. Theoretically, it should be possible to uncouple the process from CO_2 reduction and directly evolve hydrogen gas as the result of photosynthesis. In reality, the photoproduction of hydrogen has substantial obstacles. Algae rarely possess the enzyme hydrogenase needed for hydrogen release; this enzyme must be added to algal chloroplast preparations from bacterial sources. These hydrogenases are extremely sensitive to oxygen, yet the photolysis of water inevitably produces this gas. Photoproduction of hydrogen by biological systems is an intriguing and important concept, but at present it is still far from the stage of practical utilization.

Various environmental conditions affect the rates of evolution of hydrogen gas through biophotolysis. In some nitrogen-fixing cyanobacteria, it is possible to release hydrogen effectively through biophotolysis when the organisms are nitrogen starved. Photoheterotrophic bacteria, such as *Rhodospirillum* or *Rhodopseudomonas,* can be used for hydrogen production from waste materials. As with other potential sources of future energy supplies, technological and economic considerations play a large part in determining which fuel resources are developed and utilized. A lack of knowledge about many areas of biophotolysis prevents assessment of the feasibility of producing and utilizing hydrogen generated by microorganisms on a large scale. The prospects for developing and employing biophotolysis as a viable future energy source, therefore, are uncertain. It can be expected, however, that efforts will be made in the future to determine the feasibility of utilizing photosynthetic microorganisms in the conversion of solar energy to hydrogen gas.

PRODUCTION OF MICROBIAL BIOMASS

Microbial biomass may be produced not only as a starting material for fuel production but also for human and animal consumption as food or feed (Mateles and Tannenbaum 1968; Gaden and Humphrey 1977; Litchfield 1977; Rose 1979; Bhattacharjee 1980; Linton and Drozd 1982; Shennan 1984). Microbial biomass is a potentially attractive food additive because it usually contains a large percentage of high-grade protein. Microbial biomass production for food or feed is frequently referred to as single-cell protein production. Human diets in many underdeveloped and developing countries are essentially grain based and deficient in protein in general, and in some essential amino acids in particular. As an example, corn-based diets are deficient in lysine, and such deficiencies can lead to malnutrition and deficiency diseases, especially in children, even if caloric intake is otherwise adequate.

Because microorganisms can convert relatively inexpensive materials, such as molasses and possibly even cellulosic wastes, into protein with high efficiency, the protein supplement from microbial sources should be more affordable in countries with low per capita income than other protein-rich food products, such as fish, meat, milk, and eggs. To illustrate this point, the conversion of 100 kg of carbohydrate can yield up to 65 kg of yeast, but only 10–20 kg of meat products.

The disadvantages of single-cell protein production include the high nucleic acid content and the digestion-resistant cell walls of some microorganisms, as well as the need for the industrial facilities and technical sophistication required to safely and efficiently produce single-cell protein. The main barriers to the use of single-cell protein as a food supplement, however, are sociological rather than technical. Food habits are deeply ingrained in cultural and ethnic heritages and are not easily modified. The general public gives little consideration to the fact that microbes are present in some traditional food products, such as lactobacilli in yogurt, yet most people consider a microbial source of protein to be a distasteful or disquieting notion. The acute food shortages in Europe during and following World War II prompted some experimentation with microbial biomass in human nutrition. For instance, a highly nutritious and reasonably palatable liverwurst substitute was concocted from a combination of compressed bakers' yeast, *Saccharomyces cerevisiae,* and soybean meal, but this and similar substitutes were promptly discontinued when traditional food products became plentiful. It is somewhat paradoxical that industrialized societies, capable of producing and perhaps more ready to accept microbial food products, normally have no need for them, whereas undernourished societies that lack manufacturing and distribution capabilities are also culturally less ready to accept novel food products. Consequently, for the time being, the major nutritional role of microbial biomass is in animal feed supplements.

An objective rather than cultural-sociological problem with microbial single-cell protein is the high DNA-RNA content of bacterial and yeast cells. The metabolism of nucleic acids yields excessive amounts of uric acid that can cause kidney stones and gout. Two grams of nucleic acids per day from microbial single-cell supplements appears to be the upper limit for human consumption. If, therefore, such microbial single-cell proteins should become a major part of the human diet, efforts will have to be made to reduce the nucleic acid content of the microbial biomass. This reduction can be accomplished with acid precipitation, with acid or alkaline hydrolysis, or by nuclease enzymes, but the treatment increases the cost of the single-cell protein product. Digestibility of microbial cell walls varies, and in some cases, pretreatment to increase digestibility will be necessary. Some bacterial strains produce toxic or indigestible by-products.

Poly-β-hydroxybutyric acid, a common microbial storage product, though not toxic, is apparently not digestible by mammals. Mice fed on biomass of the hydrogen-oxidizing microorganisms developed intestinal blockages due to this storage product. Selection of appropriate microbial strains or mutants, however, can eliminate this type of problem (Schlegel and Oeding 1971).

The group of microorganisms most extensively grown for single-cell protein are yeasts (Rose 1979). Their high nutritional value (50%–60% protein), easy cultivation, and lack of toxic by-products all contribute to their usefulness. A common substrate for cultivation of various *Saccharomyces* species is molasses from sugar refining. An interesting and novel process grows yeast biomass on starch in a single step, employing a mixed culture of the amylase-producing yeast *Endomycopsis fibuliger* and *Candida utilis* (Figure 15.11); the former organism overproduces amylase, and the bulk of the sugar produced from the starch is utilized by *C. utilis*, which constitutes 85%–90% of the final yeast biomass (Skogman 1976).

Prior to 1970, when petroleum was considered inexpensive, hydrocarbons were thought to be suitable substrates for single-cell protein production, and several commercial processes were developed (De Pontanel 1972; Shennan 1984). Some yeasts, such as *Candida lipolytica*, that readily utilize normal alkanes of intermediate chain length (C_{12}-C_{16}) were grown as feed yeast by the British Petroleum Company and various other commercial enterprises. When the price of petroleum increased, however, these commercial operations became uneconomic. Changing organisms and substrates to natural gas-derived methanol, for example, may make single-cell protein production economically viable once again.

From the point of view of cost, the use of waste materials for single-cell protein production seems particularly attractive (Bellamy 1974; Birch et al. 1976; Skogman 1976; Litchfield 1977). Various schemes have been proposed to convert animal waste from cattle feed lots or hog farms into feed protein. Technical and sanitary difficulties have thus far thwarted the use of animal waste in this manner, but the approach has merits since it not only creates an important commod-

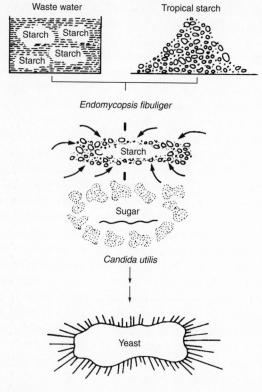

Figure 15.11

One-step production of yeast biomass from starch-containing wastewaters of food processing plants or from starch especially produced for this purpose, such as from cassava. *Endomycopsis fibuliger* produces amylase, an uncommon metabolic feature in yeasts, and supplies sugar for the more rapidly growing *Candida utilis.* The final yeast biomass consists of 85%–90% *C. utilis* and 10%–15% *E. fibuliger.* (Source: Skogman 1976. Reprinted by permission of Applied Science Publications, Inc.)

ity but also provides partial solution to a waste disposal problem. The waste sulfite liquor from wood-pulp processing has been used commercially with good success for production of yeasts for animal feed.

Algae and cyanobacteria have the advantages as potential sources of single-cell protein; only inorganic nutrients and sunlight are required for their production (Bennemann et al. 1976, 1977). The cyanobacte-

rium *Spirulina,* which forms large surface blooms on Lake Chad, Africa, has been traditionally used as a food supplement by the local population and has high potential as a food or feed supplement. The historic record indicates that a similar utilization of *Spirulina* was traditional in Mexico at the time of the Spanish Conquest (Ciferri 1983). The cyanobacterial bloom that floats to the surface is simply scooped up and usually dried for storage and distribution. It is eaten raw or added to various dishes. Of the dry weight, 60%–70% is high-grade protein. Total nucleic acids are less than 5%, which is considerably less than in most bacteria and yeasts. *Spirulina maxima* is commercially produced at Lake Texcoco, Mexico, at a rate approaching 2 metric tons per day. The *Spirulina maxima* biomass is currently used as animal feed, and uses in human food are under consideration. The *Spirulina* biomass is a by-product of a large solar evaporator for production of soda lime. *Spirulina* is harvested from the external portion of the helical evaporator, where the brine is most dilute. Harvest is by filtration. After homogenization and pasteurization, the product is spray-dried (Ciferri and Tiboni 1985).

In some Pacific islands, certain species of marine macroalgae are traditional dietary supplements. *Ulva, Caulerpa* (Chlorophycophyta), *Laminaria* (Phaeophycophyta), and *Porphyra* (Rhodophycophyta) are used in this manner. They are consumed fresh as salad, dried as a snack, or are added as supplements to other dishes. These algae add important vitamins and minerals to the rice- and fish-based diets of these regions. The unicellular freshwater alga *Chlorella* can be grown in simple media and has a good potential as food supplement. In Japan, cultivated *Chlorella* has been used as a protein and vitamin supplement in yogurt, ice cream, bread, and other food products. The acceptability problem of microbial biomass as food can be solved by its conversion to fish or shellfish protein as practiced in aquaculture. Food value is lost in the process, but easier harvestability and better palatability of the product compensate for this. According to one scheme, atolls surrounded by deep ocean could be used in combined energy and biomass production (Pinchot 1970; Othmer and Roels 1973). The heat energy of warm tropical surface seawater (25°C–

30°C) could be used to evaporate a low-boiling working fluid and to drive low-pressure turbines. The vapors are condensed by piping to the surface cold seawater (5°C–6°C) from the depth of 900–1,000 m. After its use for cooling, the nutrient-rich and now-warm deep seawater is piped into the shallow lagoon of the atoll and used in production of algal biomass. The latter supports the production of fish, shrimp, or shellfish. Serious technical problems remain to be surmounted, especially in respect to marine surface fouling by bulky heat exchangers, but the soundness of the concept has been verified in small-scale demonstration projects.

The discussion of microbial biomass as food would not be complete without mentioning mushroom cultivation, although mushrooms are grown more for their flavor than for their nutritional value (Gray 1970; Chang and Hayes 1978; National Research Council 1979). The Basidiomycota cultivated in this manner are grown essentially on waste materials, and their fruiting bodies are marketed fresh, dried, or canned. The annual consumption of mushrooms in the United States currently exceeds 1,000 metric tons. *Agaricus bisporus* is the mushroom commercially grown in Europe and in the United States. Composted horse manure is used as the growth substrate; cultivation takes place in caves, cellars, or specially constructed mushroom houses. The substrate is placed in trays or beds and is inoculated with spawn, a mycelial mass grown from spores under controlled and sterile conditions. Temperature and humidity conditions are carefully controlled during the next fifteen to thirty days of mycelial proliferation. At the end of this period, the compost is covered with a thin layer of soil or sand casing. Ventilation needs to be carefully controlled in the following period, since elevated CO_2 concentrations inhibit fruiting body formation. One to three months after application of the casing, the mushroom caps appear and are harvested. Production continues in periodic cycles, called breaks. Commonly the compost is replaced after four breaks, allowing two six-month production cycles per year.

In various countries of Indochina, the straw mushroom *Volvariella volvacea* is cultivated on a commercial scale. Rice straw, tied in bundles and soaked in water, is made into beds approximately 3 by 1 m and placed either in open fields or in sheds. Occasionally, small supplements of rice bran and distillery residues are added. Inoculation is from spent straw beds. Mushrooms are produced three to four weeks after inoculation, and production continues in about ten-day intervals for several cycles. By this simple cultivation process, *V. volvacea* is grown by farmers for home use and local markets, and also by large-scale operators for canning.

The Shiitake mushroom *Lentinus edodes* is commercially cultivated in Japan on deciduous logs, mainly oak. The wood used for this purpose, boles 10–30 cm in diameter, has little or no commercial value. The logs are soaked, the bark is pounded and lacerated, and the sapwood is inoculated with a spore suspension of *L. edodes*. The logs are placed for five to eight months in near-horizontal position in shady outdoor "laying yards," where mycelial proliferation takes place. At the end of this period the logs are transferred to a vertical position and a lower-humidity environment in "raising yards," where the mushroom caps develop. Logs continue to produce for three to six years and, aside from occasional watering and harvesting, require little care.

SUMMARY

Microbial activities can be used to aid in the recovery of metals from low-grade ores. *Thiobacillus* and other iron- and sulfur-oxidizing bacteria are currently employed in some mining operations to increase the rate of recovery of copper and uranium. The recovery process generally is based upon the transformation of iron sulfide (pyrite) associated with the mineral resource. By either direct or indirect microbial oxidation, the insoluble metal sulfide is transformed to water-soluble metal sulfate, which can be leached from the mineral deposit and recovered. The use of bioleaching for mineral recovery is governed by the economics of the process and the value of the recovered mineral resource. Bioaccumulation of metals by microorganisms may provide another mechanism for enhanced metal recovery.

Microbial fuel production can contribute to meeting world energy requirements. Microorganisms can produce various gaseous and liquid fuels. The brightest prospects for employing microorganisms in the production of fuel resources involve a conversion of solar energy to biomass and the subsequent formation of alcohols, hydrogen, and hydrocarbons, particularly methane, by the microbial fermentation of this biomass. Obviously, the described solar energy conversion schemes will need to compete with the more direct photovoltaic use of sun energy. However, the periodicity of sunlight and the difficulty of storing electricity for vehicular propulsion favor production of conventional fuels by solar energy. Several microbially produced fuels could partially replace petroleum and natural gas as energy sources in conventional engines. In addition to being utilized for the production of conventional fuels, microorganisms can be utilized in bioleaching processes to enhance the recovery of petroleum hydrocarbons and nuclear fuels, such as uranium. Microbial recovery processes can extend our available energy resources. The production of energy will undoubtedly be a major concern of scientists, including microbial ecologists, until ways are found to convert solar energy to usable fuels that are safe and that can be produced on a scale that will permit a high worldwide standard of living.

Traditional uses of microbial biomass for human food are limited to yeasts, mushrooms, and certain algae. An expanded future use of microbial biomass for human food or animal feed requires suitable and inexpensive substrates that will make the process economical and provide a nutritious product acceptable to consumers. Hydrocarbons were once considered as prime substrates for single-cell protein production, but the increased cost of petroleum made these processes uneconomic. New processes are being developed based upon natural-gas-derived methanol and upon utilization of cellulose and various waste materials.

References & Suggested Readings

Anderson, E. V. 1978. Gasohol: Energy mountain or molehill? *Chemical and Engineering News* 56:8–15.

Beck, J. V. 1967. The role of bacteria in copper mining operations. *Biotechnology and Bioengineering* 9:487–497.

Bellamy, W. D. 1974. Single cell proteins from cellulosic wastes. *Biotechnology and Bioengineering* 16:869–890.

Bennemann, J. R., B. Koopman, J. Weissman, and W. J. Oswald. 1976. Biomass production and waste recycling with blue-green algae. In H. C. Schlegel and J. Barnea (eds.). *Microbial Energy Conversion*. Report of the UNITAR/BMFT Göttingen Seminar. Pergamon Press, New York, pp. 399–412.

Bennemann, J. R., and J. C. Weissman. 1976. Biophotolysis: Problems and prospects. In H. C. Schelgel and J. Barnea (eds.). *Microbial Energy Conversion*. Report of the UNITAR/BMFT Göttingen Seminar. Pergamon Press, New York, pp. 413–426.

Bennemann, J. R., B. Koopman, J. Weissman, and W. J. Oswald. 1977. Energy production by microbial photosynthesis. *Nature* (London) 268:19–23.

Bhattacharjee, J. K. 1980. Microorganisms as potential sources of food. *Advances in Applied Microbiology* 13:139–161.

Birch, G. G., K. J. Parker, and J. T. Morgan (eds.). 1976. *Food from Waste*. Applied Science Publishers, London.

Brierley, C. L. 1978. Bacterial leaching. *CRC Critical Reviews in Microbiology* 6:207–262.

Brierley, C. L., and L. E. Murr. 1972. Leaching: Use of a thermophilic and chemoautotrophic microbe. *Science* 179:488–489.

Brierley, J. A., and S. J. Lockwood. 1977. The occurrence of thermophilic iron-oxidizing bacteria in a copper leaching system. *FEMS Microbiology Letters* 2:163–165.

Brierley, J. A., and C. L. Brierley. 1978. Microbial leaching of copper at ambient and elevated temperatures. In L. E. Murr, A. E. Torma, and J. A. Brierley (eds.). *Metallurgical Applications of Bacterial Leaching and Related Microbiological Phenomena*. Academic Press, New York, pp. 477–490.

Brierley, C. L. 1982. Microbiological mining. *Scientific American* 247(2):44–53.

Brierley, C. L. 1990. Metal immobilization using bacteria. In H. L. Ehrlich and C. L. Brierley (eds.). *Microbial Mineral Recovery*. McGraw-Hill, New York, pp. 303–324.

Brock, T. D. 1978. *Thermophilic Microorganisms and Life at High Temperatures*. Springer-Verlag, New York.

Brock, T. D., and J. Gustafson. 1976. Ferric iron reduction by sulfur- and iron-oxidizing bacteria. *Applied and Environmental Microbiology* 32:567–571.

Bungay, H. R. 1981. *Energy, the Biomass Options*. John Wiley and Sons, New York.

Chang, S. T., and W. A. Hayes. 1978. *The Biology and Cultivation of Edible Mushrooms*. Academic Press, New York.

Charley, R. C., and A. T. Bull. 1979. Bioaccumulation of silver by a multi-species community of bacteria. *Archives of Microbiology* 123:239–244.

Ciferri, O. 1983. *Spirulina,* the edible microorganism. *Microbiological Reviews* 47:551–578.

Ciferri, O., and O. Tiboni. 1985. The biochemistry and industrial potential of *Spirulina*. *Annual Review of Microbiology* 39:503–526.

Cooper, D. G. 1983. Biosurfactants and enhanced oil recovery. In S. B. Clark and E. S. Donaldson (eds.). *Proceedings of 1982 International Conference on the Microbial Enhancement of Oil Recovery*. Technology Transfer Branch, Bartlesville Technology Center, Bartlesville, Okla., pp. 112–114.

Cooper, D. G., and J. E. Zajic. 1980. Surface active compounds from microorganisms. *Advances in Applied Microbiology* 26:229–253.

De Pontanel, H. G. (ed.). 1972. *Proteins from Hydrocarbons*. Proceedings of Symposium Aix-en-Provence. Academic Press, London.

Ehrlich, H. L., and C. L. Brierley (eds.). 1990. *Microbial Mineral Recovery*. McGraw-Hill, New York.

Ensign, J. C. 1976. Biomass production from animal wastes by photosynthetic bacteria. In H. G. Schlegel and J. Barnea (eds.). *Microbial Energy Conversion*. Report of the UNITAR/BMFT Göttingen Seminar. Pergamon Press, New York, pp. 455–482.

Eveleigh, D. E. 1984. Biofuels and oxychemicals from natural polymers: A perspective. In M. J. Klug and C. A. Reddy (eds.). *Current Perspectives in Microbial Ecology*. American Society for Microbiology, Washington, D.C., pp. 553–557.

Finnerty, W. R., and M. E. Singer. 1983. Microbial enhancement of oil recovery. *Biotechnology* 1:47–54.

Forbes, A. D. 1980. Microorganisms in oil recovery. In D. E. F. Harrison, I. J. Higgins, and R. Watkinson (eds.). *Hydrocarbons in Biotechnology*. Heyden and Son, London, pp. 169–180.

Gaden, E. L., and A. E. Humphrey. 1977. Single cell protein from renewable and non-renewable resources. *Biotechnology and Bioengineering Symposium 7*. John Wiley and Sons, New York.

Geffen, T. M. 1976. Present technology for oil recovery. In *The Role of Microorganisms in the Recovery of Oil*. Proceedings 1976 Engineering Foundation Conferences, NSF/RA-770201. Government Printing Office, Washington, D.C., pp. 23–32.

Gray, W. D. 1970. *The Use of Fungi as Food and in Food Processing*. CRC Press, Cleveland, Ohio.

Groudev, S. 1979. Mechanism of bacterial oxidation of pyrite. *Mikrobiologija Acta Biologica Iugoslavica* (Belgrade) 16:75–87.

Groudev, S. N., and F. N. Genchev. 1978. Mechanisms of bacterial oxidation of chalcopyrite. *Mikrobiologija Acta Biologica Iugoslavica* (Belgrade) 15:139–152.

Guay, P., M. Silver, and A. E. Torma. 1977. Ferrous iron oxidation and uranium extraction by *Thiobacillus ferrooxidans*. *Biotechnology and Bioengineering* 19:727–740.

Gutnick, D. L., Z. Zosim, and E. Rosenberg. 1983. The interaction of emulsan with hydrocarbons. In S. B. Clark and E. S. Donaldson (eds.). *Proceedings of 1982 International Conference on the Microbial Enhancement of Oil Recovery*. Technology Transfer Branch, Bartlesville Energy Technology Center, Bartlesville, Okla., pp. 6–11.

Hitzman, D. O. 1983. Petroleum microbiology and the history of its role in enhanced oil recovery. In S. B. Clark and E. S. Donaldson (eds.). *Proceedings of 1982 International Conference on the Microbial Enhancement of Oil Recovery*. Technology Transfer Branch, Bartlesville Energy Technology Center, Bartlesville, Okla., pp. 162–218.

Imhoff, K., W. J. Müller, and D. K. B. Thistlethwayte. 1971. *Disposal of Sewage and Other Water-borne Wastes*. Ann Arbor Science Publishers, Ann Arbor, Mich.

Jackson, E. A. 1976. Brazil's national alcohol program. *Process Biochemistry* 11:29–30.

Jernelov, A., and A. L. Martin. 1975. Ecological implications of metal metabolism by microorganisms. *Annual Reviews of Microbiology* 29:61–78.

Karavaiko, G. I. 1985. *Microbiological Processes for the Leaching of Metals from Ores*. UNEP, Center of International Projects GKNT, Moscow.

Karavaiko, G. I., S. I. Kuznetsov, and A. I. Golomzik. 1977. *The Bacterial Leaching of Metals from Ores*. Technicopy, Stonehouse, England.

Keenan, J. D. 1979. Review of biomass to fuels. *Process Biochemistry* 5:9–15.

Kelly, D. P. 1976. Extraction of metals from ores by bacterial leaching: Present status and future prospects. In J. Barnea and H. G. Schlegel (eds.). *Microbial Energy Conversion*. Report of the UNITAR/BMFT Göttingen Seminar. Pergamon Press, New York, pp. 329–338.

Kelly, D. P., P. R. Norris, and C. L. Brierley. 1979. Microbiological methods for the extraction and recovery of metals. In A. T. Bull, D. C. Ellwood, and C. Ratledge (eds.). *Microbial Technology: Current State, Future Prospects*. Cambridge University Press, Cambridge, England, pp. 263–308.

Kuznetsov, S. I., M. V. Ivanov, and N. N. Lyalikova. 1963. *Introduction to Geological Microbiology*. McGraw-Hill, New York.

La Riviere, J. W. M. 1955a. The production of surface-active compounds by microorganisms and its possible significance in oil recovery. Part I. Some general observations on the change of surface tension in microbial cultures. *Antonie van Leeuwenhoek* 21:1–8.

La Riviere, J. W. M. 1955b. The production of surface-active compounds by microorganisms and its possible significance in oil recovery. Part II. On the release of oil from oil-sand mixtures with the aid of sulfate reducing bacteria. *Antonie van Leeuwenhoek* 21:8–27.

Lindeman, L. R., and C. Rocchiccioli. 1979. Ethanol in Brazil: Brief summary of the state of Industry in 1977. *Biotechnology and Bioengineering* 21:1107–1119.

Linton, J. D., and J. W. Drozd. 1982. Microbial interactions and communities in biotechnology. In A. T. Bull and J. H. Slater (eds.). *Microbial Interactions and Communities*. Academic Press, London, pp. 357–406.

Litchfield, J. H. 1977. Comparative technical and economic aspects of single-cell protein processes. *Advances in Applied Microbiology* 22:267–305.

Lundgren, D. G., and M. Silver. 1980. Ore leaching by bacteria. *Annual Review of Microbiology* 34:263–283.

McCready, R. G. L., and W. D. Gould. 1990. Bioleaching of uranium. In H. L. Ehrlich and C. L. Brierley (eds.). *Microbial Mineral Recovery*. McGraw-Hill, New York, pp. 107–126.

McInerney, M. J., and D. W. Westlake. 1990. Microbial

enhanced oil recovery. In H. L. Ehrlich and C. L. Brierley (eds.). *Microbial Mineral Recovery*. McGraw-Hill, New York, pp. 409–445.

Mateles, R. I., and S. R. Tannenbaum. 1968. *Single Cell Protein*. MIT Press, Cambridge, Mass.

Matthews, D. 1982. Here's how enhanced oil recovery works. *EXXON USA* 21(4):16–23.

Mersou, J. 1992. Mining with microbes. *New Scientist* 133(1802):17–19.

Meyer, W. C., and T. F. Yen. 1976. Enhanced dissolution of shale oil by bioleaching with *Thiobacilli*. *Applied and Environmental Microbiology* 32:310–316.

Moses, V., and D. G. Springham. 1982. *Bacteria and the Enhancement of Oil Recovery*. Applied Science Publishers, London.

Murr, L. E., A. E. Torma, and J. A. Brierley (eds.). 1978. *Metallurgical Applications of Bacterial Leaching and Related Microbiological Phenomena*. Academic Press, London.

National Academy of Sciences. 1977. *Methane Generation from Human, Animal, and Agricultural Wastes*. National Academy of Sciences, Washington, D.C.

National Research Council. 1979. *Microbial Processes: Promising Technologies for Developing Countries*. National Academy of Sciences, Washington, D.C.

Norris, P. R., and D. P. Kelly. 1979. Accumulation of metals by bacteria and yeasts. *Developments in Industrial Microbiology* 20:299–308.

Norris, P. R., and D. P. Kelly. 1982. The use of mixed microbial cultures in metal recovery. In A. T. Bull and J. H. Slater (eds.). *Microbial Interactions and Communities*. Academic Press, London, pp. 443–474.

Othmer, D. F., and O. A. Roels. 1973. Power, freshwater, and food from cold deep seawater. *Science* 182:121–125.

Paterson, A., A. J. McCarthy, and P. Broda. 1984. The application of molecular biology to lignin degradation. In J. M. Graingor and J. M. Lynch (eds.). *Microbiological Methods for Environmental Biotechnology*. Academic Press, New York, pp. 33–68.

Pfeffer, J. T. 1976. Methane from urban wastes: Process requirements. In H. C. Schlegel and J. Barnea (eds.). *Microbial Energy Conversion*. Report of the UNITAR/BMFT Göttingen Seminar. Pergamon Press, New York, pp. 139–156.

Pinchot, G. F. 1970. Marine farming. *Scientific American* 223 (6):14–21.

Rangachari, P. N., V. S. Krishnamachar, S. G. Pail, M. N. Sainani, and H. Balakrishnan. 1978. Bacterial leaching of copper sulfide ores. In L. E. Murr, A. E. Torma, and J. A. Brierley (eds.). *Metallurgical Applications of Bacterial Leaching and Related Microbiological Phenomena*. Academic Press, London, pp. 427–439.

Reeves, S. G., K. K. Rao, L. Rosa, and D. D. Hall. 1976. Biocatalytic production of hydrogen. In H. G. Schlegel and J. Barnea (eds.). *Microbial Energy Conversion*. Report of the UNITAR/BMFT Göttingen Seminar. Pergamon Press, New York, pp. 235–243.

Roels, O. A., S. Laurence, M. W. Farmer, and L. van-Hemelryck. 1976. Organic production potential at artificial upwelling marine culture. In H. G. Schlegel and J. Barnea (eds.). *Microbial Energy Conversion*. Report of the UNITAR/BMFT Göttingen Seminar. Pergamon Press, New York, pp. 69–81.

Rose, A. H. (ed.). 1979. Microbial Biomass. *Economic Microbiology*. Vol. 4. Academic Press, London.

Sandvik, E. I., and J. M. Maerker. 1977. Application of xanthan gum for enhanced oil-recovery. In R. F. Gould (ed.). *Extracellular Polysaccharides*. American Chemical Society Symposium Ser. 48, Washington, D.C., pp. 242–264.

Schlegel, H. G., and V. Oeding. 1971. Selection of mutants not accumulating storage materials. In *Radiation and Radioisotopes for Industrial Microorganisms*. International Atomic Energy Agency, Vienna, Austria, pp. 223–231.

Schlegel, H. G., and J. Barnea (eds.). 1976. *Microbial Energy Conversion*. Report of the UNITAR/BMFT Göttingen Seminar. Pergamon Press, New York.

Shennan, J. L. 1984. Hydrocarbons as substrates in industrial fermentation. In R. M. Atlas (ed.). *Petroleum Microbiology*. Macmillan, New York, pp. 643–683.

Singer, M. E., W. R. Finnerty, P. Bolden, and A. D. King. 1983. Microbial processes in the recovery of heavy petroleum. In S. B. Clark and E. S. Donaldson (eds.). *Proceedings of 1982 International Conference on the Microbial Enhancement of Oil Recovery*. Technology Transfer Branch, Bartlesville Energy Technology Center, Bartlesville, Okla., pp. 94–101.

Skogman, H. 1976. Production of symba-yeast from potato wastes. In G. G. Bach, K. J. Parker, and J. T. Woregan (eds.). *Food from Waste*. Applied Science Publishers, Essex, England, pp. 167–176.

Soeder, C. J. 1976. Primary production of biomass in fresh water with respect to microbial energy conversion. In H. G. Schlegel and J. Barnea (eds.). *Microbial Energy*

Conversion. Report of the UNITAR/BMFT Göttingen Seminar. Pergamon Press, New York, pp. 59–68.

Steinbeck, K. 1976. Biomass production of intensively managed forest ecosystems. In H. G. Schlegel and J. Barnea (eds.). *Microbial Energy Conversion*. Report of the UNITAR/BMFT Göttingen Seminar. Pergamon Press, New York, pp. 35–44.

Summers, A. O., and S. Silver. 1978. Microbial transformations of metals. *Annual Review of Microbiology* 32:637–672.

Sutherland, I. W., and D. C. Ellwood. 1979. Microbial exopolysaccharides-industrial polymers of current and future potential. In A. T. Bull, D. C. Ellwood, and C. Ratledge (eds.). *Microbial Technology: Current State, Future Prospects*. Cambridge University Press, Cambridge, England, pp. 107–150.

Switzenbaum, M. S. 1983. Anaerobic treatment of wastewater: Recent developments. *ASM News* 49:532–536.

Tomizuka, N., and M. Yagisawa. 1978. Optimum conditions for leaching of uranium and oxidation of lead sulfide with *Thiobacillus ferrooxidans* and recovery of metals from bacterial leaching solution with sulfate-reducing bacteria. In L. E. Murr, A. E. Torma, and J. A. Brierley (eds.). *Metallurgical Applications of Bacterial Leaching and Related Microbiological Phenomena*. Academic Press, London, pp. 321–344.

Torma, A. E. 1976. Biodegradation of chalcopyrite. In J. M. Sharpley and A. M. Kaplan (eds.). *Proceedings of the 3rd International Biodegradation Symposium*. Applied Science Publishers, Essex, England, pp. 937–946.

Torma, A. E. 1977. The role of *Thiobacillus ferrooxidans* in hydrometallurgical processes. In N. Blakebrough, A. Fiechter, and T. K. Ghose (eds.). *Advances in Biochemical Engineering*. Vol. 6. Springer-Verlag, Berlin, pp. 1–37.

Torma, A. E. 1985. Scientific fundamentals of technology of tank leaching of uranium. In G. I. Karavaiko and S. N. Groudev (eds.). *Biotechnology of Metals*. UNEP, Centre of International Projects GKNT, Moscow, pp. 266–274.

Tornabene, T. G. 1976. Microbial formation of hydrocarbons. In H. G. Schlegel and J. Barnea (eds.). *Microbial Energy Conversion*. Report of the UNITAR/BMFT Göttingen Seminar. Pergamon Press, New York, pp. 281–299.

Tuovinen, O. H. 1990. Biological fundamentals of mineral leaching processes. In H. L. Ehrlich and C. L. Brierley (eds.). *Microbial Mineral Recovery*. McGraw-Hill, New York, pp. 55–78.

Tuovinen, O. H., and D. P. Kelly. 1974. Use of microorganisms for the recovery of metals. *International Metallurgical Reviews* 19:21–71.

Tuttle, J. H., and P. R. Dugan. 1976. Inhibition of growth, iron and sulfur oxidation in *Thiobacillus ferrooxidans* by simple organic compounds. *Canadian Journal of Microbiology* 22:719–730.

Updegraff, D. M. 1983. Plugging and penetration of petroleum reservoir rock by microorganisms. In S. B. Clark and E. S. Donaldson (eds.). *Proceedings of 1982 International Conference on the Microbial Enhancement of Oil Recovery.* Technology Transfer Branch, Bartlesville Energy Technology Center, Bartlesville, Okla., pp. 81–85.

Venkatasubramanian, K., and C. R. Keim. 1981. Gasohol: A commercial perspective. *Proceedings of the New York Academy of Sciences* 369:187–204.

Westlake, D. W. S. 1984. Heavy crude oils and oil shales: Tertiary recovery of petroleum from oil-bearing formations. In R. M. Atlas (ed.). *Petroleum Microbiology.* Macmillan, New York, pp. 537–552.

Yen, T. F. 1976a. Current status of microbial shale oil recovery. In *The Role of Microorganisms in the Recovery of Oil.* Proceedings 1976 Engineering Foundation Conferences. NSF/RA-770201. Government Printing Office, Washington, D.C., pp. 81–115.

Yen, T. F. 1976b. Recovery of hydrocarbons from microbial attack on oil-bearing shales. In *The Genesis of Petroleum and Microbiological Means for Its Recovery.* Institute of Petroleum, London, pp. 22–46.

Zajic, J. E. 1969. *Microbial Biogeochemistry.* Academic Press, New York.

Zajic, J. E., and D. F. Gerson. 1978. Microbial extraction of bitumen from Athabasca oil sand. In E. V. Lown and O. P. Strausz (eds.). *Oil Sands and Oil Shale Chemistry.* Verlag-Chemie, Weinheim, Germany, pp. 145–161.

Zajic, J. E., and A. Y. Mahomedy. 1984. Biosurfactants: Intermediates in the biosynthesis of amphipathic molecules in microbes. In R. M. Atlas (ed.). *Petroleum Microbiology.* Macmillan, New York, pp. 221–297.

Zimmerley, S., D. Wilson, and J. Prater. 1958. *Cyclic leaching process employing iron-oxidizing bacteria.* USA Patent No. 2,829,964, assigned to Kennecott Copper Corp.

ZoBell, C. E. 1947. Bacterial release of oil from oil-bearing materials. Vols. I, II. *World Oil* 126(13):36–44:127(1):35–40.

16

Ecological Control of Pests and Disease-causing Populations

A highly interesting and economically important area of applied microbial ecology is the ecological control of pests and disease-causing organisms. Biological control using microorganisms to initiate disease in pest populations has the potential of reducing agricultural reliance on chemical pesticides. Historically, the negative interactions (amensalism, predation, and parasitism) among microbial populations and between microbes and higher organisms formed the basis for the biological control of pests and pathogens. Biological methods to control populations of disease-causing organisms and pests are based on modification of host and/or vector populations, modification of reservoirs of pathogens, and the direct use of microbial pathogens and predators (National Academy of Sciences 1968; Wilson 1969; Baker and Cook 1974; Woods 1974; Henis and Chet 1975; Huffaker and Messenger 1976; Simmonds et al. 1976; Coppel and Mertins 1977; Burges 1981; Batra 1982; Arntzen and Ryan 1987; Chet 1987; Hedrin et al. 1988; Harman and Lumsden 1990). Biotechnology may improve approaches for controlling pest and disease-causing populations (Nakas and Hagedorn 1990).

MODIFICATION OF HOST POPULATIONS

There are several important methods utilized for controlling pests and pathogens based on modification of the host population. Host populations have natural defense mechanisms that defend against infection by pathogens and attack by pests, and it is possible to enhance these natural defense mechanisms to control disease. In agriculture, selective breeding has been extensively utilized to develop plant populations that are genetically resistant to disease-causing pathogens. Many new plant strains have been developed on this basis. Selective breeding is successful because of the specificity of host-pathogen interactions. Unfortunately, pest populations are subject to evolution, selection, and geographical spreading. Thus, there is a need in agriculture for a continuous breeding program in order to develop new strains of resistant plants and to replace strains of plants that are, or are becoming, susceptible to diseases caused by pathogens and pests (Beck and Maxwell 1976).

Resistance to disease is the inherent capacity of a plant that enables it to prevent or restrict the entry or subsequent activities of a pathogen to which the plant is exposed under conditions that could lead to disease. The basis for resistance in plant populations may be anatomical or biochemical. Insects or nematodes, some of which are vectors for pathogens, may be unable to penetrate plants with thicker cuticles or cork layers. For example, the resistance of some species of barberry to penetration by basidiospores of *Puccinia graminis* has been attributed to the thickness of the cuticle on the epidermis of the leaves. Strains of plants capable of closing their stomata when conditions are favorable for infection also possess a mechanism for disease resistance (Snyder et al. 1976). Infected plants may respond with morphological and/or biochemical changes that render the plant more resistant to the pathogen.

Agricultural management practices are frequently directed at avoiding plant disease (National Academy of Sciences 1968; Batra 1982). Because many plant pathogens are transmitted from one infected individual to another, increasing the spacing between individual plants decreases the likelihood that a pathogenic microorganism will be successfully transmitted from infected to uninfected plants. In cases where pathogens are not spread from plant to plant, high-density planting produces good crops with only minor losses due to disease. For example, dense seeding of cotton reduces losses due to *Verticillium* wilt. In contrast, dense plantings in hemlock result in increased losses due to twig rust, because this pathogen spreads from plant to plant. Density of host plant populations also results in environmental modifications that affect the development of plant pathogens. Dense planting of tomatoes, for example, results in retention of humidity, which favors development of blight due to *Botrytis*.

Environmental conditions at the time of planting are important in controlling pathogenic infections of plants. At suboptimal temperatures, plants often increase the exudation of amino acids and carbohydrates by roots. Such exudates act as nutrients that favor the rapid development of microbial populations in the vicinity of the plant. Development of *Pythium*

and *Rhizoctonia* populations on exudates often results in fungal infection of the developing seedlings. Other suboptimal abiotic factors, such as lack of oxygen because of excessive moisture content, also result in increased production of exudates by plants, which, in turn, favors the development of pathogenic microbial populations.

Considerations analogous to those discussed for plant populations can also be applied to domesticated animal populations. Management of domesticated animals includes practices that decrease the probability of widespread disease among the animal population. Some animal populations have been bred specifically for their resistance to disease (Davis et al. 1975).

Immunization

In vertebrate animals, including humans, disease control is often based on the immune response, in which exposure to antigens results in the formation of antibodies. An antigen can be any biochemical that elicits antibody formation and reacts with antibodies. Antibodies are blood serum proteins. Antibodies react with antigens, and if the antigens are toxins, these reactions can neutralize the toxins and prevent onset of disease. If the antigens are on the surface of a pathogenic microorganism, reaction with antibodies can result in agglutination and loss of infectivity. Reaction with antibodies can also result in the lysis of infectious microorganisms. In addition to forming antibodies, the immune response involves the appearance of specifically altered lymphoid cells, such as lymphocytes and macrophages, that phagocytize infecting microorganisms.

The stimulation of antibody formation by intentional exposure to antigens is known as immunization. When the antigens are from infective pathogens, the term vaccination is applied. When a vertebrate animal is exposed to a foreign antigen for the first time, there is a primary immune response. The primary immune response is characterized by a relatively long lag period before the onset of antibody formation and by a relatively low production of antibody. Subsequent exposure to the same antigen results in a secondary, or memory, response, characterized by a much shorter lag period, a much higher level of antibodies, and greater specificity of the antigen-antibody reaction. The secondary response normally is triggered by a lower threshold dose of antigen than the primary response. The capacity to produce a secondary response can provide long-lasting immunity against infection.

The introduction of a pathogen with its associated antigens into an immune animal results in the rapid production of antibodies that react with the antigens to prevent the onset of disease. The ability to produce a secondary response varies with the particular antigens and with the physiological state of the host animal. Some antigens, such as the toxin produced by *Clostridium tetani,* evoke secondary responses even decades after primary exposure. Under conditions of physiological and metabolic stress, the immune response capability of the animal may be impaired, allowing invasion by pathogens and the development of disease.

Vaccination is used as a preventive measure against several human diseases including smallpox, diphtheria, tetanus, polio, measles, mumps, German measles, yellow fever, typhus fever, typhoid fever, paratyphoid fever, cholera, plague, influenza, pertussis, and tuberculosis. It is used also to prevent animal diseases, including rabies and distemper in dogs and cholera in hogs. Vaccines normally contain antigens as killed noninfectious pathogens or as live pathogens of attenuated virulence. Vaccination results in a modification of the host population by establishing the capability of the secondary or memory immune response, greatly increasing the resistance to the particular disease-causing microorganisms. If the pathogen has a narrow host range and all potentially exposed hosts can be rendered disease resistant by vaccination, the disease can be effectively eradicated. A well-coordinated campaign by the World Health Organization succeeded in the worldwide eradication of smallpox, eliminating the need for continued immunization.

In response to infecting pathogens, plants produce a variety of polycyclic and polyaromatic compounds, called phytoalexins, that have antimicrobial activities. Plants that have previously been exposed to invading microorganisms and have survived the infection retain such chemicals for a period of time, during which the plants have an increased resistance to infection by pathogenic microorganisms. Exposure to attenuated pathogens that cause only mild symptoms of disease can also induce such acquired systemic resistance (Deacon 1983).

In plants, a more specific resistance mechanism against infection by closely related viruses is cross protection. A plant systemically infected by one type of virus will not develop additional disease symptoms when challenged with a closely related second virus. This protection is always reciprocal among two related viruses. The phenomenon allows plant protection against some virulent viruses by an approach reminiscent of immunization, although the mechanism of protection differs from the animal immune response. Tobacco mosaic virus (TMV) causes, on the average, 15% loss of greenhouse-grown tomatoes. Deliberate infection of the tomato seedlings with an attenuated TMV mutant that causes an essentially symptomless infection provides protection against attack by virulent TMV. The inoculation process is fairly simple. Tomato plants infected with the mild TMV mutant are ground up, and the virus is enriched with differential centrifugation and sprayed in water suspension on the tomato seedlings. The spray also contains carborundum particles as abrasive agent, inflicting on the leaves small wounds through which the virus can enter.

A similarly successful protection of citrus trees against the tristeza virus was recently instituted in Brazil and Peru. The citrus trees were inoculated with an attenuated virus mutant that arose naturally and was isolated from healthy-looking trees in badly affected orchards. The virus is systemic and easily transmitted through the grafting propagation of the infected trees. It provides satisfactory protection even in environments where wild strains of the tristeza virus and their aphid vectors pose a severe challenge.

The described successes have sparked additional research on cross protection of plants against viral infections. The mechanism of cross protection is not yet fully understood. A probable theory is that the single-stranded RNA genome of TMV and similar viruses acts in the early infection stage as messenger RNA and is the template for production of a replicase subunit. This replicase subunit combines with a host protein to form a functional replicase. The specificity of replicases makes it likely that the replicase produced by the mild virus will attach to the RNA of the infecting wild strain because of the similarity of the two binding sites. However, the mutant-specific replicase is unable to replicate the wild-type virus RNA, thus aborting the infection process (Deacon 1983). Other theories on the mechanism of cross protection assume some type of interaction between the plant and coat proteins of the inducing and the pathogenic viruses. This theory gains experimental support from the fact that tobacco plants transformed to incorporate genetic information encoding for the coat protein of TMV acquired at least partial resistance against infection with virulent TMV (Abel et al. 1986). This type of genetic manipulation may become a powerful tool of plant protection against viral diseases.

MODIFICATION OF RESERVOIRS OF PATHOGENS

Various control measures are utilized to eliminate or diminish reservoirs of pathogenic microorganisms. Many such procedures can be called sanitation practices and include sewage treatment and disinfection of drinking water supplies. These procedures reduce the reservoir populations of potential animal pathogens. Many practices involved in food processing and preparation are carried out to reduce the pathogens in food products. These practices include heat and filter sterilization and pasteurization. Disinfection of utensils used in food processing also removes a potential reservoir of pathogens. Similarly, instruments and materials used in medical and dental procedures are normally sterilized to remove possible pathogenic microorganisms. In horticultural practices, soils are sometimes fumigated or heat-sterilized to remove plant pathogens. Since many plant pathogens contaminate plant seeds, disinfection of seeds by chemical or hot-water treatments can greatly reduce the incidence of disease within a plant population.

Removal of infected plant or animal tissues is also used to control the spread of disease because it effectively reduces the reservoir of pathogens. In agricultural practice, this is often accomplished by crop rotation. The removal of the infected crop also removes the pathogens contained in infected tissues. The specificity of host-pathogen interactions precludes the establishment of infection by that pathogen in other crops used in a system of rotation. Given sufficient time, the infected tissues are eliminated from the soil and the pathogen population is greatly reduced. The reintroduction of the original plant species can then be accomplished without undue loss. When the reservoir of pathogens increases, the crops are again rotated (Batra 1982).

For some diseases, control is achieved by eliminating the populations of animals or plants that act as alternate hosts for the pathogens. To this end, rodent control programs are aimed at reducing the reservoir of human pathogens associated with rodent populations. Elimination of rat populations by means as drastic as burning whole villages has been used historically to control outbreaks of bubonic plague, for which the rat population acts as a reservoir. The apparent association of histoplasmosis with infected bird roosts has been used as justification for blackbird eradication campaigns. Other control methods include the vaccination and population control of canines that may harbor the rabies virus transmittable to humans. Elimination of alternate hosts can also be used to control plant diseases. The economically important rust fungus, *Puccinia graminis,* which infects wheat, uses the barberry shrub, *Berberis vulgaris,* as an alternate host; eliminating or reducing the population size of this alternate host can curtail the transmission of disease caused by this rust fungus (Moore-Landecker 1972).

Another procedure for controlling reservoirs of pathogenic microbial populations is the containment of infected organisms. Hobbyists who maintain tropical fish aquariums, as well as commercial plant and animal breeders, know well that infected organisms must be separated lest the pathogens or parasites spread from infected individuals to other members of the population. Quarantine regulations for human travelers, as well as for plant and animal shipments, are designed to prevent the unwitting introduction of pathogens and pests into areas that are normally free from such organisms.

MODIFICATION OF VECTOR POPULATIONS

Programs designed to eliminate or reduce animal (usually insect) populations that transmit microbial pathogens are often used to control diseases that are spread in this manner (Laird 1978). A number of human diseases are transmitted by mosquito vectors; by controlling the population levels of mosquitoes, the incidence of these diseases can be greatly reduced. Mosquito populations can be controlled by using chemical insecticides, by biological control methods, and by habitat destruction. The tsetse fly, vector for the sleeping sickness caused by the flagellate *Trypanosoma,* requires moist jungle soil for larval development. Deforestation is an extreme but effective measure for controlling this vector. Less drastic actions, such as the elimination of stagnant water bodies required by mosquito larvae or the introduction of larva-eating fish species, decrease the reproduction and thus the population density of mosquito vectors.

Control of fleas and ticks is often achieved simultaneously with the control of animal reservoirs of pathogenic microbial populations. For example, elimination of rat populations also eliminates or greatly reduces populations of rat fleas. Some animals, such as monkeys, have developed grooming behavior that reduces the populations of tick and flea vectors of microbial pathogens. Humans use appropriate sanitary procedures to eliminate populations of body lice, which also may act as vectors for pathogenic microorganisms.

Insects and nematodes act as vectors for viral, bacterial, and fungal plant pathogens. Outbreaks of plant diseases that are transmitted by vectors can be avoided by appropriate vector control. Control of aphids is especially important, because these organisms often act as vectors for various viral plant diseases.

MICROBIAL AMENSALISM AND PARASITISM TO CONTROL MICROBIAL PATHOGENS

Attempts to utilize antagonistic relationships between microbial populations have been considered mainly to control plant pathogens. Although many plant diseases are caused by fungal and bacterial populations, microbial plant pathogens constitute only a small proportion of the microbes growing in the vicinity of plants. Many saprophytic microorganisms that occur in the rhizosphere and phyllosphere protect plants against pathogens, and this phenomenon may be utilized in plant protection.

Microbial antagonism toward animal pathogens has been clearly demonstrated but does not offer the same possibilities for disease control as with plants. Autochthonous microbial populations on the skin surface and in the gastrointestinal tract are responsible for preventing infections by many potential pathogens. Germfree animals are much more susceptible to infectious disease than are animals with associated indigenous microbial populations. The widespread use of antibiotics in disease treatment threatens to diminish this natural antagonism of indigenous microorganisms against invading populations, including pathogens. Disruption of the normal microbiota can increase the host's susceptibility to invasion by opportunistic pathogens. Assuring the maintenance of the normal indigenous microbial populations can reduce opportunity for infection.

Antifungal Amensalism and Parasitism

Amensalism is used to control plant pathogenic fungi. As an example, papaya (*Carica papaya*) grown in Hawaii, mainly on lava rock, is highly susceptible to root rot by *Phytophthora palmivora,* particularly in its seedling state. The porous and practically humus-free volcanic rubble has a rudimentary microbial community that offers no hindrance to the spread of *P. palmivora*. Losses from the root rot can be effectively controlled, however, by drilling small depressions, filling these with microbe-rich topsoil, and planting papaya seeds in the depressions. Antagonistic and competitive effects of the established microbial community toward *P. palmivora* slow down the spread of the pathogen and allow the seedlings to become established and disease resistant (Schroth and Hancock 1981). By stimulating the saprophytic segment of the microbial community, organic supplements added to the rhizosphere or to the soil in general often suppress the spread and infectivity of fungal plant pathogens. In so-called disease-suppressive soils, a poorly understood combination of physicochemical and biological factors limits the survival, spread, and infectivity of soilborne fungal plant pathogens (Baker 1980; Schroth and Hancock 1982). This phenomenon is called soil fungistasis. Soil fungistasis is believed to be one of the many factors that contribute to the suppression of plant pathogens in some soils (Lockwood 1979).

Some bacterial species produce antifungal substances. Addition of some *Bacillus* and *Streptomyces* strains to soil or seeds has been shown to control plant disease caused by the fungus *Rhizoctonia solani*, which causes damping off disease, and several other plant diseases in cucumbers, peas, and lettuce. *Bacillus subtilis* also has been found to control stem rot and wilt of carnations caused by a *Fusarium* species. (Schroth and Hancock 1981; Schroth et al. 1984).

Mycoparasitism of *Trichoderma harzianum* and *T. hamatum* on the soilborne plant pathogens *Botrytis cinerea*, *Sclerotium rolfsii*, and *Rhizoctonia solani* offers a possibility to control several economically important plant diseases caused by the latter fungi. The *T. mycelium* winds around the mycelial strands of the plant pathogen (Figure 16.1) and, by lytic action, penetrates their cell wall. The cell contents of the plant pathogens are lysed and utilized by *Trichoderma*. It seems, however, that *Trichoderma* does not attack young, vigorously growing hyphae of the pathogens, but only stressed or senescent ones. This fact limits the effectiveness of the biological control (Henis 1983;

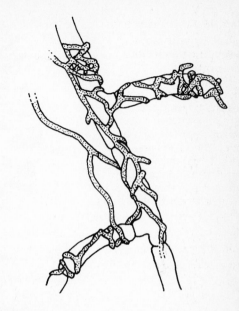

Figure 16.1

Trichoderma can coil its hyphae around the wider hyphae of the pathogenic fungus *Rhizoctonia solani*. (Source: Deacon 1983.)

Dubos 1984). Additional mycoparasitic relationships are being investigated for biological control of plant pathogens. The basidiomycete *Laetisaria arvalis* shows control potential against the plant diseases caused by *Pythium ultimum* and *R. solani* (Schroth and Hancock 1981). The mechanism of control was recently identified as the production of 8-hydroxylinoleic acid (laetisaric acid) by *L. arvalis* (Bowers et al. 1986). By a mechanism not yet clearly identified, this compound induces rapid lysis of *P. ultimum*, *R. solani*, *Fusarium solani*, and several additional plant pathogenic fungi. The simple chemical structure of this antibiotic lends itself to synthetic production and structural optimization for maximal potency.

Heterobasidion annosum (formerly *Fomes annosus*) is a basidiomycete that causes serious damage to managed forests, especially to pine trees, by causing butt and root rot (Deacon 1983). The infection occurs on the stumps of cut trees (Figure 16.2). Treatment of stumps with chemical agents did not succeed in elimi-

Figure 16.2

Peniophora gigantea prevents airborne spores of the pathogen *Heterobasidion annosum* from colonizing stumps of trees and traveling into the root zone. Once a population of *Peniophora* is established in the tree stump, it can restrict the spread of the pathogen from infected areas to uninfected roots. (Source: Deacon 1983.)

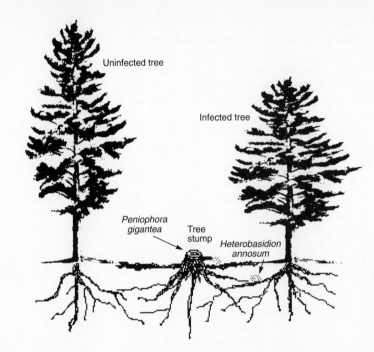

nating the problem. *Peniophora gigantea,* a mildly parasitic basidiomycete, sometimes colonizes pine stumps naturally, and in such cases excludes *H. annosum*. This antagonistic action has been used to control *H. annosum* infection with spectacular success. A *P. gigantea* spore suspension, which is readily preserved in a concentrated sucrose solution, is painted on stumps of cut trees. It rapidly colonizes the cut stump and excludes *H. annosum. Peniophora gigantea* is not an aggressive pathogen and does not spread to the roots of healthy standing trees. The control measure is elegant, highly effective, and very economic. The mechanism of *H. annosum* exclusion by *P. gigantea* is termed hyphal interference. On making contact with *P. gigantea* hyphae, *H. annosum* stops growing, shows impaired membrane integrity, and degenerates. The effect is strictly localized to contact areas, and its biochemical basis is yet unknown.

Pruning of fruit trees often creates a portal of entry for fungal plant pathogens. Canker of apple trees, caused by *Nectria galligena,* and silver leaf disease of plum and other fruit trees, caused by *Stereum purpureum,* can be controlled by painting the pruning

wounds with a spore suspension of *Trichoderma viride,* a highly antagonistic saprophyte. The dieback and gummosis disease of apricots, caused by the fungus *Eutypa armeniaceae,* can be controlled by painting pruning wounds with *Fusarium lateritium* spores. These control techniques are in various stages of development and have not yet been practiced extensively (Deacon 1983).

Antibacterial Amensalism and Parasitism

Amensalism also plays a role in protecting plants against bacterial diseases. Agrocins, highly specific antibiotic-like substances with killing effect toward *Agrobacterium* strains closely related to the agrocin producer, are highly effective in controlling crown gall infections (Kerr and Tate 1984; Whipps and Lynch 1986). Agrocin 84 is a plasmid-encoded adenine nucleotide antimetabolite produced by certain strains of *A. radiobacter*. It kills selectively the pathogenic strains of *A. tumefaciens*. Its highly selective action depends on a permease encoded on the same *Ti* plasmid that is responsible for the pathogenic property of *A. tumefaciens*. Nearly 100% biological control of crown gall infection could be achieved by simply dipping seeds, cuttings, or roots of young plants into a suspension of an Agrocin 84-producing strain of *A. radiobacter*. Agrocin 84 is effective only against certain strains of *A. tumefaciens*, but, fortunately, the controlled strains are the most frequent pathogens. Other types of agrocins with different specificities have been found and are under investigation.

MICROBIAL PATHOGENS AND PREDATORS FOR CONTROLLING PEST POPULATIONS OF PLANTS AND ANIMALS

The use of living organisms to control plant and animal diseases is called biological control. Biological control methods can augment the use of chemical pesticides in controlling pest populations. Populations of pathogenic or predatory microorganisms that are antagonistic toward particular pest populations provide a natural means of controlling population levels of pest animals and weeds. Preparations of such antagonistic microbial populations are called microbial pesticides.

The ideal microbial pesticide should not be subject to attack by hyperparasites; should be virulent, causing disease in the pest population when applied at the recommended concentration; should not be sensitive to moderate environmental variations; should survive following application until infection within the pest population has been established; should rapidly establish disease in the pest populations so as to minimize destruction by the pest populations; should be rather specific for the pest population; and should not cause disease in nontarget populations.

The most spectacular successes in biological control of pests have been achieved when a pest has been accidentally introduced to a new environment—for example, a new continent (Batra 1982). In such accidental introductions, pests have been frequently separated from their parasite or predator control agents. In the new environment devoid of such control agents, their effect was often devastating. The identification of the control agents that kept them in balance at their place of origin and the intentional introduction of those agents to the problem area often achieved a quick and economic solution to the pest problem. Several examples of such successes will be presented.

The biological control of a pest that is not a recent introduction and that has evolved in balance with its own parasites and predators is considerably more difficult. Such organisms become intolerable pests, not because of an absolute absence of parasites or predators, but because of environmental changes introduced by man. Changes such as monoculture may favor the pest and, at the same time, put its parasites and predators at disadvantage. In such cases, one needs to modify the environmental conditions and agricultural production techniques rather than simply to identify and introduce a biological control agent. Another important consideration in biological control is timing. Natural biological control mechanisms often eliminate pest populations only after they have reached very high densities and have inflicted unacceptable crop damage. The aim may be to activate biological control at a lower pest population level by hastening the outbreak of a disease before extensive damage to the crop occurs (Batra 1982).

Microbial Control of Insect Pests

Microbial control methods have been developed for the suppression of arthropod pests, especially insects (Falcon 1971; Lipa 1971; McLaughlin 1971; Stairs 1971; David 1975; Weiser et al. 1976; Ferron 1978; Longworth and Kalmakoff 1978; Deacon 1983; Aronson et al. 1986). Several commercial microbial insecticides have been developed and marketed. Microbial suppression of insect populations is aimed at those that cause crop and other plant damage and at those that act as vectors of disease-causing microorganisms. Potentially, many viral, bacterial, fungal, and protozoan populations can be used in the control of insect pests and vectors.

Viral Pesticides Insect pathogenic viruses have the potential to become useful pesticidal agents (Krieg 1971; Stairs 1971; David 1975; Tinsley 1979; Deacon 1983). More than 450 viruses have been described from approximately 500 arthropod species. Insect pathogenic viruses frequently cause natural epizootics, the analog of epidemics applied to animals. Viruses pathogenic for insects are found in the families Baculoviridae, Poxviridae, Reoviridae, Iridoviridae, Parvoviridae, Picornaviridae, and Rhabdoviridae. Some of these are nuclear polyhedrosis viruses (NPV), cytoplasmic polyhedrosis viruses (CPV), and granulosis viruses (GV). Nuclear polyhedrosis viruses develop in the host cell nuclei; the virions are occluded singly or in groups in polyhedral inclusion bodies. Cytoplasmic polyhedrosis viruses develop only in the cytoplasm of host midgut epithelial cells; the virions are occluded singly in polyhedral inclusion bodies. Granulosis viruses develop in either the nucleus or the cytoplasm of host fat, tracheal, or epidermal cells; the virions are occluded singly or, rarely, in pairs in small occlusion bodies called capsules.

Baculoviruses are perhaps the most studied insect viruses. They include nuclear polyhedrosis and granulosis viruses. Pathogenic baculoviruses have been found principally for Lepidoptera, Hymenoptera, and Diptera. Infection is often transmitted by ingestion of contaminated food. Cell invasion probably begins in the midgut. Several nuclear polyhedrosis viruses are being produced on a large scale in the United States for control of insect pests. Lepidoptera and Hymenoptera larvae that feed on plant leaves are important pests that cause great economic damage. Inoculation of leaves with polyhedrosis viruses can initiate epizootics, resulting in decreases in pest populations. Many nuclear polyhedrosis viruses kill host larvae, releasing the polyhedra over the plant. The polyhedra remain infective for a long time. Nuclear polyhedrosis viruses have been used extensively in controlling pests of forest trees (Stairs 1972). Gypsy moths, tent caterpillars, and spruce budworms are subject to epizootics caused by nuclear polyhedrosis viruses.

Nuclear polyhedrosis viruses cause disease in sawflies. The accidental introduction of the European sawfly into North America in the twentieth century threatened the spruce forests of North America. Introduction of nuclear polyhedrosis viruses into the European sawfly population caused a spectacular epizootic that reduced the sawfly populations and saved many spruce forests. Similarly, the European pine sawfly, which was causing serious damage to pines in New Jersey, Ohio, and Michigan, has been controlled by introducing insects containing nuclear polyhedrosis viruses into the population, resulting in epizootics. Other sawfly populations have been subjected to similar controls.

Viruses have been used in attempts to control outbreaks of a variety of pests, including gypsy moths, Douglas fir tussock moths, pine processionary caterpillars, red-banded leaf rollers (pest of apples, walnuts, and other deciduous fruits), Great Basin tent caterpillars, alfalfa caterpillars, white butterflies, cabbage loopers, cotton bollworms, corn earworms, tobacco budworms, tomato worms, army worms, wattle bagworms, and others. A few viruses cause diseases in mites (Lipa 1971; David 1975) (Figure 16.3). Mites cause extensive crop damage, for example, to European citrus fruits; control of these mite populations is of great economic importance (Gustafsson 1971).

Bacterial Pesticides There are several bacterial pathogens of insects that currently are used as insecticides or that have potential for such use in the future (Aronson et al. 1986). They include *Rickettsiella popillae, Bacillus popilliae, B. thuringiensis, B. lentimorbus, B. sphaericus, Clostridium malacosome, Pseudomonas aeruginosa,* and *Xenorhabdus nematophilus. Bacillus thuringiensis* has many subspecies that differ in the number and type of plasmids they contain. The genetic information coding for the insecticidal toxins of these strains is borne on these plasmids.

Bacillus thuringiensis, or BT as it is commonly known, is an important biological control agent used since the late 1960s for the control of lepidopteran target pests (Aronson et al. 1986). *Bacillus thuringiensis* is called a crystalliferous bacterium because, in addition to endospores, it produces discrete parasporal bodies within the cell (Figure 16.4). Several toxic substances have been isolated from *B. thuringiensis* and are designated as either exotoxins or endotoxins,

Figure 16.3

Photograph of a mite infected with a virus showing characteristic crystal inclusions (white spots). The crystals apparently result from metabolic disturbance and indicate the presence of infecting virus. (Source: Lipa 1971. Reprinted by permission, copyright Academic Press Ltd., London.)

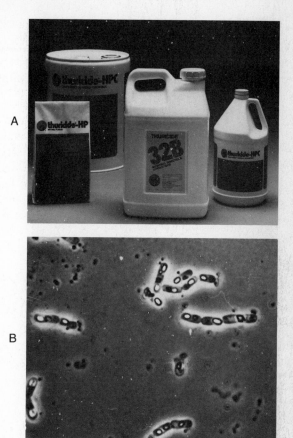

Figure 16.4

(A) Various commercial products containing *Bacillus thuringiensis*. (B) A micrograph of *B. thuringiensis* showing spores and parasporal inclusions of this broadly used, toxin-producing microbial pesticide. (A. Courtesy of Sandoz, Inc., San Diego. B. Courtesy of B. N. Herbert, Shell Research Ltd., Kent, England.)

the latter being responsible for most of the insecticidal activity (Hurley et al. 1987; Hofte and Whiteley 1989). The endotoxins comprise the paracrystalline inclusion body. In most cases, this is located outside the exosporium; in a few strains, it is associated with the exosporium. The main component of the paracrystalline inclusions are 130- to 140-kilodalton (kd) size polypeptides. These molecules are termed protoxins. They are solubilized in the alkaline midgut of susceptible insect larvae, releasing an active toxin estimated to have a size of 30–80 kd. The ω-endotoxins are a family of polypeptides showing some conserved (identical) and some variable portions. The variations determine specific activity against particular insect species.

Recent novel *B. thuringiensis* isolates have extended its utility to include the control of certain dipteran and coleopteran pest species. At least twelve manufacturers in five countries have commercially produced *B. thuringiensis* preparations for control of pest insects. There are currently 410 preparations of

Table 16.1

Some registered uses for *Bacillus thuringiensis* products in the U.S.A.

Pest	Plant
Alfalfa caterpillar *(Colias eurytheme)*	Alfalfa
Artichoke plume moth *(Platyptilia carduidactyla)*	Artichokes
Bollworm *(Heliothis zea)*	Cotton
Cabbage looper *(Trichoplusia ni)*	Beans, broccoli, cabbage, cauliflower, celery, collards, cotton, cucumbers, kale, lettuce, melons, potatoes, spinach, tobacco
Diamondback moth *(Plutella maculipennis)*	Cabbage
European corn borer *(Ostrinia nubilalis)*	Sweet corn
Imported cabbageworm *(Picris rapae)*	Broccoli, cabbage, cauliflower, collards, kale
Tobacco budworm *(Heliothis virescens)*	Tobacco
Tobacco hornworm *(Manduca sexta)*	Tobacco
Tomato hornworm *(Manduca quinquemaculata)*	Tomatoes
Fruit tree leaf roller *(Archips argyrospilus)*	Oranges
Orange dog *(Papilio cresphontes)*	Oranges
Grape leaf folder *(Desmia funeralis)*	Grapes
Great Basin tent caterpillar *(Malacosoma fragile)*	Apples
California oakworm *(Phryganidia californica)*	Oaks
Fall webworm *(Hyphantria cunea)*	Ailanthus, ash, roses, wisteria
Fall cankerworm *(Alsophila pometaria)*	Maple, elm
Gypsy moth *(Lymantria [Porthetria] dispar)*	Pine
Linden looper *(Erannis tiliaria)*	Linden
Spring cankerworm *(Paleacrita vernata)*	Maple, elm

B. thuringiensis registered in the United States for use on many agricultural crops, forest trees, and ornamentals for control of various insect pests (Table 16.1), and it has been tested successfully against more than 140 insect species, primarily members of Lepidoptera and Diptera.

Molecular genetic analyses of insecticidal activity has shown that the genes encoding *B. thuringiensis* insecticidal crystal proteins are most often carried by plasmids, certain of which have the ability to transfer via conjugation. Many toxin genes have been isolated and characterized with respect to their DNA sequence. These analyses have yielded important information regarding the diversity of *B. thuringiensis*

toxin genes and set the stage for future gene structure and function studies.

Various preparations of *B. thuringiensis* effect commercially acceptable suppression of various insect parasites of plants (Table 16.2). Additional targets of *B. thuringiensis* include cabbage worms and cabbage loopers, which impact vegetable crops, and tent caterpillars, bagworms, and cankerworms, which are pests of forest trees. Suppression of gypsy moths and spruce budworm can be achieved, but only when high application rates are used and uniform foliage coverage is achieved. Some strains of *B. thuringiensis* (subspecies *kurstaki* and *israelensis*) and *B. sphaericus* produce endotoxins that are effective against mosquitoes,

Table 16.2

Commercial *Bacillus thuringiensis* (BT) products

BT strain	Product	Target insect
Bacillus thuringiensis var. *aizawai*	Certan	Wax moth larvae
Bacillus thuringiensis var. *israelensis*	Vectobac-AS Skeetal Teknar	Mosquito and blackfly larvae
Bacillus thuringiensis var. *kurstaki*	Dipel Bactospeine Thuricide Javelin	Lepidopteran larvae
Bacillus thuringiensis var. *san diego*	M-One	Colorado potato beetle larvae

including some that are malaria vectors. Since the resistance of malaria vectors against conventional insecticides is rapidly rising, an alternate strategy to control these mosquitoes is urgently needed.

Attempts are under way to improve *B. thuringiensis*-based bioinsecticides by changing the conventional formulation of *B. thuringiensis* crystal protein (Rowe and Margaritis 1987; Currier and Gawron-Burke 1990). This strategy involves altering the active ingredient, either the *B. thuringiensis* strain or the crystal protein, and/or changing aspects of the formulation itself. The ability to improve *Bacillus thuringiensis* strains has been greatly aided by molecular genetics used to alter the genes responsible for insecticidal activity. The diversity of *B. thuringiensis*-toxin-encoding genes, especially as related to differences in insecticidal activity, coupled with the ability of certain toxin-encoding plasmids to conjugally transfer, provides the basis for the directed development of *B. thuringiensis* strains. Novel combinations of δ-endotoxin genes can be derived without the use of recombinant DNA techniques via selective plasmid curing and conjugal transfer. Selective plasmid curing to remove plasmids carrying toxin genes of low activity can yield strains that can be used as recipients in conjugal transfer. Conjugal transfer of toxin plasmids into partially cured derivatives can produce *B. thur-*

ingiensis strains with elevated activities against specific insect pests as well as strains with a broader spectrum of insecticidal activity.

The other strategy is to use recombinant DNA technology to incorporate the genes encoding the toxins of *B. thuringiensis* directly into plant species or rhizosphere microorganisms to afford them protection against insect parasites.

Bacillus popilliae and *B. lentimorbus* control Coleoptera such as the scarabid Japanese beetles, causing the "milky disease" of their larvae. The Japanese beetle feeds voraciously on some 300 species of plants and has been responsible for large economic losses. In Japan, the beetle encounters natural antagonists and is only a minor pest. In the United States, where the beetle does not have antagonists, good success in suppressing this pest has been obtained using bacteria that produce milky disease. For many years, spore dust containing *B. popilliae* and *B. lentimorbus* has been marketed under the trade name Doom (Figure 16.5). *Bacillus lentimorbus* does not produce a parasporal body and infects mainly first and second instar grubs. Some strains of *B. popilliae* produce parasporal bodies and infect third instar grubs. The use of these *Bacillus* species to produce milky disease in these grubs has been largely responsible for the control of Japanese beetle populations. Whereas in the past there have been major infestations of Japanese beetles in the United States, today there are relatively few. *Bacillus popilliae* fails to form spores on artificial media; for production of spore powders, infected insects have to be ground up. Obtaining grubs, inoculating them, and subsequently processing the infected grubs is a cumbersome and expensive process. However, *B. popilliae* spores survive in soil for years and provide long-lasting control. *Bacillus popilliae* is ineffective against the adult beetle; thus, in severe infestations, short-term control of adults by chemical insecticides needs to be combined with long-term biological control of the larvae.

Among the nonsporulating bacterial pathogens of insects, *Pseudomonas aeruginosa* has received consideration, but problems have been associated with its use as an insecticide because it is an opportunistic pathogen capable of producing disease in

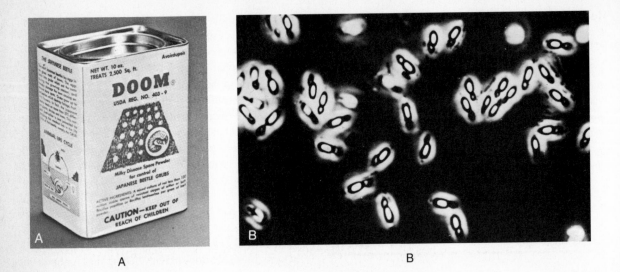

A   B

Figure 16.5

(A) A can of "Doom." (B) A micrograph of *Bacillus popilliae*, one of the two active components in this microbial pesticide that is effective in controlling Japanese beetles. (A. Source: Fairfax Laboratories, Clinton Corners, New York. B. Source: Michael Klein, U.S. Department of Agriculture, Wooster, Ohio.)

humans. It also tends to infect only stressed or injured insects and has a very limited shelf life and survival on exposed dry surfaces such as leaves and stems of plants. Other bacterial pathogens of arthropods occur among the Rickettsiae; rickettsial infections cause diseases in Coleoptera, Diptera, and Orthoptera, cause the blue disease of Japanese beetles, and cause diseases of grubs of the European chafer. Problems similar to those discussed in connection with *P. aeruginosa* have prevented their use in biological control.

Protozoan Pesticides   Many protozoa are path–ogens of arthropods, but for the most part, protozoa are not suited for use as short-term, quick-acting microbial insecticides; few act with sufficient speed to prevent severe crop damage. Protozoa must be applied before outbreaks of disease to effect proper control. The relatively slow development of protozoan infections of pest animal populations—and difficulties with storage and environmental stability—limit the prospects of their use in biological control.

There have been few attempts to utilize protozoa as a practical measure of pest suppression. There does appear to be some potential for use of protozoa on grasshoppers, mosquitoes, and boll weevils to augment other methods of controlling these pest populations. Some attempts have also been made to control lepidopteran pests of fruit trees using sporozoan protozoa (Pramer and Al-Rabiai 1973).

Fungal Pesticides   Fungi are potentially important in the control of pest populations (Roberts and Yendol 1971; Ferron 1978; Deacon 1983; Howe 1990). Most studies on entomogenous fungi have been concerned with members of the fungal genera *Beauveria, Metarrhizium, Entomophthora,* and *Coelomomyces. Beauveria bassiana* was used extensively in the Soviet Union to control the eastward spread of the Colorado beetle and to control the codling moth. It also infects the corn earworm, a major pest in the United States (Figure 16.6). Fungi of the genus *Aschersonia* were used to control pests of

Figure 16.6

Scanning electron micrographs showing interaction of *Beauveria bassinana* and corn earworm. (A) First instar corn earworm larva (*Heliothis zea*) with many setae (1) and spiracles (2). (B) Hyphae of poorly pathogenic mutant of *B. bassiana* growing errantly over the larval surface with little or no penetration of the larval integument. (C) Germinating conidia (c) and germ tubes (g) of *B. bassiana* penetrating the integument of a corn earworm larva. With highly pathogenic mutants, penetration occurs soon after germination. (D) Enzymatic degradation (arrow) of the larval integument by a penetrating germ tube of *B. bassiana*. (Source: E. Grula.)

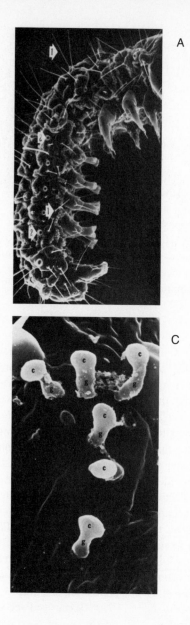

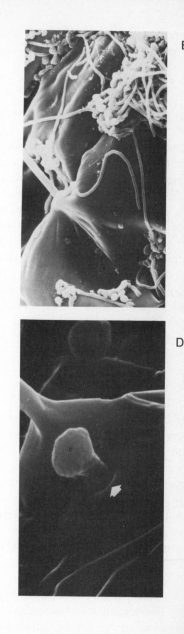

citrus trees in the Soviet Union near the Black Sea. In Florida, *Hirsutella* has been used against the citrus rust mite. The fungus *Metarrhizium* is used in Brazil to control populations of leaf hoppers and frog hoppers. Members of the genus *Entomophthora* show promise as pathogens of aphids. Some members of the genus *Coelomomyces* are pathogenic to the larvae of the mosquito populations of *Anopheles, Opifex, and Aedes.* Because these mosquitoes are important vectors for microbial pathogens of humans, control of these pests by fungal pathogens could become an important contribution to disease control. Similarly, several fungal species have been shown to produce diseases in mites, which also are important vectors for microbial diseases of humans.

Microbial Control of Other Animal Pests

Microbial control of invertebrate animal populations other than insects has received less attention to date but may become important in the future. Some aquatic snails, for example, harbor important human pathogens, such as the causative agent of schistosomiasis. Some bacterial and protozoan populations are parasites of molluscs and may provide a mechanism for controlling these invertebrates (Steinhaus 1965).

Predaceous nematode-trapping fungi have been suggested for the control of pest nematode populations in soil (Pramer 1964; Pramer and Al-Rabiai 1973; Eren and Pramer 1978) (Figure 16.7). Nematode-trapping fungi have been used in attempts to control root knot disease of pineapples in Hawaii. Successful control was achieved but depended on supplementing the soil with organic matter. It is not clear whether control of the nematodes was due to the predaceous fungi or to an alteration in food web relationships caused by the addition of organic supplements. Subsequent attempts to demonstrate that nematode-trapping fungi, rather than soil supplements were primarily responsible for controlling nematode populations in soil had ambiguous results. Under appropriate conditions, nematode-trapping fungi, together with supplemental organic matter, appear to have the potential for diminishing the effects of nematodes on susceptible plant populations. Other possibilities for controlling pest nematodes include infection with parasitic fungi, protozoa, bacteria, and viruses.

Microorganisms have also been used in attempts to control pest populations of mammals. Pasteur demonstrated that *Pasteurella multocida* was capable of killing wild rabbits when they ate contaminated cut alfalfa. For lack of natural enemies, the imported European rabbit was becoming a serious pest, and Pasteur sent this organism to Australia to be used in a campaign to control rabbit populations, but the program was never initiated.

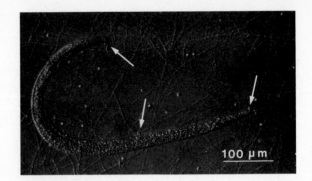

Figure 16.7

Nomarski interference micrograph of the nematode-trapping fungus *Dactylaria candida* showing the nematode captured by several adhesive knobs of the fungus, indicated by arrows. (Courtesy of S. Olson and B. Norbring-Herz, University of Lund.)

Rabbits were also significant agricultural pests in some regions of Europe. After a chance observation in Uruguay that imported European rabbits (*Oryctolagus cuniculus*) were wiped out by a disease caused by the myxoma virus (Poxviridae), to which the native South American rabbit (*Sylvilagus brasiliensis*) was resistant (due to coevolution), the disease was deliberately imported to a walled estate in France. In spite of precautions, the disease escaped containment and in a few years swept through Europe, severely decimating rabbit populations. Transmission of the virus depends on insect vectors; in Europe, the rabbit flea served as the principal vector. By the late 1950s, however, the evolution of attenuated virus strains and increasing immunity allowed a gradual recovery of the rabbit population in Europe.

Because of the severity of the rabbit problem, myxomatosis was deliberately introduced also in Australia in the early 1950s. Here the disease was spread primarily by mosquito vectors, and mortality of infected rabbits ranged around 99.4%–99.8%. For unknown reasons, control in Australia was more persistent than in Europe, and in 1970 the rabbit population was only 1 percent of the 1950 level (Deacon 1983).

Microbial Control of Weeds and Cyanobacterial Blooms

Although much of the work concerning the use of microorganisms as pesticides has been aimed at insect and other animal pest populations, some studies have been aimed at the potential use of plant pathogenic microorganisms to control weeds (Charudattan 1975; Andres et al. 1976). Such pathogenic microbial populations can be called microbial herbicides. Studies to develop microbial herbicides for important agricultural weeds have lagged behind studies on microbial insecticides. In part, this may have been due to the experiences of plant pathologists with accidental introduction of pathogens that destroyed important crops, such as the chestnut blight in the United States and the potato blight in Ireland. There have been understandable safety concerns and a reluctance on the part of investigators to develop microbial herbicides lest they introduce pathogens that might cause catastrophic economical or ecological losses.

There are, however, some precedents in which the accidental or intentional introduction of microbial pathogens resulted in the biological control of terrestrial weeds. In Australia, bathurst burr was successfully attacked by *Colletotrichum xanthii*. Crofton weed, also in Australia, has been attacked by a leaf spot disease. It appears that the pathogen *Cercospora eupatorii* was accidentally introduced with a gallfly imported from Hawaii for the control of this weed. Crofton weed is now held in check by the leaf spot disease and by a native beetle. The rust fungus *Puccinia chondrillina* has been introduced into Australia to control brush skeleton weed; the fungus has spread rapidly and appears to have been effective (Hasan 1974). The choice of *P. chondrillina* was based on the observation that in the western Mediterranean this fungus is more effective in reducing skeleton weed than associated arthropod grazers. Similar studies have demonstrated that powdery mildews can also be used to control skeleton weed. In the Soviet Union, *Alternaria cuscutacidae* was used to control weeds in alfalfa fields. In Europe, *Uromyces rumicis* has been used to control curly dock. On the experimental scale, the indigenous rust fungus *P. caniculata* has been

shown to effectively control the yellow nutsedge, which causes a serious weed problem in the United States (Phatak et al. 1983). The control system has good potential for herbicidal use but needs further development and safety testing. Several aquatic weeds, such as water hyacinth, hydrilla, alligator weed, and water milfoil, pose important problems in the southeastern United States. These weeds are subject to infections by pathogenic microorganisms. Various fungal, bacterial, and viral populations have been examined for their abilities to control aquatic weeds (Wilson 1969; Zettler and Freeman 1972).

In freshwater ponds and lakes, cyanobacteria are frequently the dominant component of destructive algal blooms. In addition to their basic sequence of bloom, die-off, and oxygen depletion, which results in septic odors and fish kills, some cyanobacteria, such as *Anabaena flos-aquae,* form potent neurotoxins (Carmichael et al. 1975). These may cause respiratory arrest and rapid death in livestock and waterfowl that drink the water during the cyanobacterial bloom. Cyanobacteria are subject to viral and fungal infections and also to bacterial lysis; cyanophages are similar to other bacteriophages both in their morphology and in their life cycle and are believed to be an important factor in the natural control of cyanobacterial population densities. Various chitrids (Phycomycetes) are parasitic on cyanobacteria; aquatic myxobacteria are also capable of lysing cyanobacteria. All microbial antagonists to cyanobacteria, especially the cyanophages, are receiving considerable attention as potential control agents of cyanobacterial blooms (Stewart and Daff 1977).

GENETIC ENGINEERING IN BIOLOGICAL CONTROL

Genetics has always played an important role in biological control. Plants and animals have been bred and selected for disease resistance, pathogenic viruses have been mutated to obtain attenuated strains for cross protection or immunization, and biological control agents have been mutated and selected for enhanced virulence and toxin production. We presented examples of such manipulations in the preceding sections. *In vitro* genetic engineering has broadened the possibilities for custom designing biological control agents but has also unleashed unprecedented controversy over whether these possibilities should be utilized. There are legitimate ecological concerns about unforeseen and perhaps unforeseeable side effects the environmental release of such organisms might cause (Halvorson et al. 1985). Some groups have also raised objections to genetic engineering and the release of engineered organisms on philosophical, ethical, and religious grounds, but, for broader appeal, these are often presented as farfetched safety concerns. It is important, therefore, to differentiate in this debate legitimate ecological concerns from deliberate obstructionism. Following, we describe some examples of engineered biological control agents that, except for the described controversy, would be ready for field testing. The EPA and other regulatory authorities have implemented extensive review processes to ensure the safety of a deliberate release of recombinant microorganisms (Milewski 1990).

Frost Protection

Freezing weather in spring and fall can cause extensive damage to nonhardy agricultural crops. Damage is not from the low temperature per se, but from formation of ice crystals that disrupt cell membranes. In the absence of ice-nucleating agents, water can be cooled several degrees below 0°C without freezing. The most important ice-nucleating agents on leaf surfaces are cells of *Pseudomonas syringae,* a pseudomonad pathogenic to lilacs but a harmless commensal epiphyte for most other plants, except for its ice-nucleating activity in freezing weather (Lindow 1983, 1988a).

The ice-nucleating factor in this bacterium appears to be a membrane-bound proteinaceous substance. The DNA sequences coding for the ice-nucleating agent have been identified and partially characterized by cloning procedures. Deletion (Ice$^-$) mutants were prepared by site-directed *in vitro* exci-

sion and ligation of the DNA (Lindow 1985). They appear to correspond in every respect to the wild strain, except for a lack of ice-nucleating activity.

Treating leaves to reduce the population sizes of ice-nucleating bacteria reduces the incidence of frost injury to plants (Lindow 1985, 1987, 1988b, 1990). Bactericides that reduce population sizes of *P. syringae* enhance protection against frost damage. Non-ice-nucleating bacteria also reduce the incidence of freezing injury (Lindow 1990). Non-ice-nucleating bacteria (Ice$^-$ bacteria) can act as biological control agents to prevent frost injury to plants. The Ice$^-$ bacteria prevent the growth of wild *P. syringae* strains on plants but do not kill or displace these strains once they are established on plants. Competitive exclusion of Ice$^+$ *P. syringae* strains has been reported on plants only when the population size of Ice$^-$ strains was at least 10^5 cells per gram (Lindow 1990).

It has been shown in laboratory and greenhouse-scale experiments that spraying young plants with suspensions of Ice$^-$ *P. syringae* mutant preempts subsequent colonization by the wild strain, preventing ice nucleation and crop damage in subfreezing weather. The mutant shows no detectable differences from the wild strain except for its lack of ice-nucleating activity. After great controversy and repeated delays, field trials using this engineered microorganism were finally approved.

Initial field trials of Ice$^-$ *P. syringae* strains applied to strawberries and potatoes were carefully designed and intensively monitored to restrict and to detect the dissemination of recombinant strains. Limited dispersal of recombinant Ice$^-$ *P. syringae* strains was detected in field trials (Lindow and Panopoulos 1988). Ice$^-$ *P. syringae* strains inoculated onto potato plants comprised a significant part of the total epiphytic microflora for four to six weeks. The population size of Ice$^+$ *P. syringae* strains on plants colonized by Ice$^-$ *P. syringae* strains was significantly decreased compared to that on uninoculated control plants (Lindow 1990). The incidence of frost injury was significantly lower to potato plants inoculated with Ice$^-$ *P. syringae* strains than to uninoculated control plants in several natural field frost events that occurred in the field in a northern California test site.

Bacillus thuringiensis Pesticides

The potency of δ-endotoxin in *Bacillus thuringiensis* against lepidopteran insects was described earlier. *Bacillus thuringiensis* toxin genes have been introduced via recombinant DNA technology into the genome of plants or plant-associated microorganisms (Adang et al. 1987; Currier and Gawron-Burke 1990). When specific promoters are used that allow toxin gene expression in the plant or plant-associated microorganism, insect control can be achieved in laboratory tests. *Bacillus thuringiensis* toxin genes have been introduced into tobacco plants, tomato plants, cotton plants, and a root-colonizing pseudomonad.

A group of scientists at Monsanto Agricultural Products Company in St. Louis, Missouri, undertook to transfer the genetic information coding for δ-endotoxin formation from *B. thuringiensis* variety *kurstaki* to a bacterium that would readily colonize the rhizosphere and could be used as seed inoculant (Watrud et al. 1985). If expressed, the δ-endotoxin would confer resistance of the crop plants against lepidopteran pests. By *in vitro* genetic engineering techniques, the portion of the *B. thuringiensis* plasmid coding for formation of δ-endotoxin was first ligated into an *Escherichia coli* plasmid. Subsequently, the δ-endotoxin genes were transferred to *Pseudomonas fluorescens*. The genes were expressed, and δ-endotoxin was formed. In feeding trials, the efficacy and action spectrum of the δ-endotoxin formed by the engineered *P. fluorescens* was identical to that of *B. thuringiensis* subspecies *kurstaki*. Growth chamber and greenhouse experiments demonstrated the capacity of the engineered *P. fluorescens* to protect, as seed inoculant, the plants from lepidopteran target pests with no side effects on nontarget insects such as bees. This engineered biological control system also appears to be ready for field testing.

Other Applications

In addition to the two described examples, genetic engineering could make other useful contributions to biological control. Moving microbial genes coding for the biodegradation of a herbicide into a crop plant could be used to make the plant resistant to a herbicide (Goodman and Newell 1985). In this way, an inherently nonselective herbicide could be made into a selective one. Recombinant virus vaccines may be more effective, more easily produced, and more easily applied than traditional ones (Moss and Butler 1985). There is little doubt that genetically engineered biocontrol systems will contribute substantially to protection of human health crops and livestock and will decrease our overdependence on chemical pesticides.

ADDITIONAL PRACTICAL CONSIDERATIONS

It should be pointed out that the development and use of microorganisms in controlling pest populations requires considerations that go well beyond simply finding a pathogen for a particular weed or animal pest population (Falcon 1976). There are questions of economics, production, quality control, application, side effects, and safety. In many cases, the economics of developing, producing, and applying microbial pesticides would not permit the commercial development of an otherwise promising biological control system. Many pathogens known to produce epizootics cannot presently be grown in the laboratory or do not produce infective stages when grown outside of host populations. Many protozoa, for example, do not produce infective spores or cysts when cultured on artificial media.

Large, stable batches of potential microbial pesticides must be produced and stockpiled for field application. Quality control must ensure that different batches of microbial pesticides have the same virulence, so that standard application rates may be utilized. The microorganisms must remain viable long enough to contact and establish widespread disease within the pest population. This is most readily accomplished using resistant stages, such as spores or cysts.

The effectiveness of the microbial pathogens must be carefully evaluated. It is difficult to achieve repeatable results with some candidate microbial pathogens. Environmental conditions must be taken into account, as these affect the virulence of microbial pesticides. Persistence must also be considered. Some candidate microorganisms have short survival times, whereas others persist for long periods of time. Persistence is required for long-term control of pest populations.

Ideally, microbial pesticides should exhibit a high degree of host specificity. They should not cause diseases in nontarget populations. The host specificity, though, should not be so narrow as to preclude effectiveness against genetic variations within a pest population. In many cases, it is difficult to predict whether a microbial pesticide could establish disease in nontarget populations. It is impossible to test infectivity against all possible nontarget populations. Clearly, microbial pesticides should be harmless to humans and valued plant and animal populations (Kurstak 1978). Microbial pesticides are probably best used in an integrated program that employs management practices for agricultural crops and domestic animals that minimize opportunities for infection or interaction, along with limited applications of chemical pesticides, carefully timed for maximum effect (Worthy 1973). The use of chemical insecticides can be further minimized and their action can be made more specific when used in combination with insect pheromones. Pheromones are highly specific intraspecies chemical attractants that can be used to attract the target insect to the insecticide rather than to blanket the environment with the insecticidal compound (Silverstein 1981). Integrated pest control programs incorporate the described features and combine maximum effectiveness with minimal environmental disruption. The use of this approach is on the increase. Biological control systems of the traditional and of the genetically engineered types will undoubtedly play major roles in integrated pest control programs (Batra 1982).

SUMMARY

Plant and animal populations can be bred for resistance or rendered more resistant to infection by previous exposure to microbial pathogens. Both cases represent an ecological change in that the ability of a pathogenic population to establish an antagonistic relationship with host populations is reduced. Some outbreaks of plant and animal diseases can be controlled by reducing or removing the reservoirs of pathogenic microorganisms. These practices are aimed at reducing the probability that healthy individuals will contact disease-causing microorganisms. A variety of disinfection and isolation procedures is used to reduce the reservoirs of pathogenic microorganisms. Control of vector populations can be used with great effectiveness to prevent the transmission of microbial populations to susceptible hosts.

Microorganisms, including viruses, bacteria, and fungi, have great potential for use in controlling populations of pest animals and weeds. To date, relatively little of this potential has been exploited on a commercial scale. Care must be taken to maximize effectiveness and minimize undesirable side effects of such control measures. Additional research is needed to further the practical application of biological control of pests.

REFERENCES & SUGGESTED READINGS

Abel, P. P., R. S. Nelson, B. De, N. Hoffmann, S. G. Rogers, R. T. Fraley, and R. N. Beachy. 1986. Delay of disease development in transgenic plants that express the tobacco mosaic virus coat protein gene. *Science* 232:738–743.

Adang, M. J., E. Firoozabady, J. Klein, D. DeBaker, V. Sekar, J. D. Kemp, E. Murray, T. A. Rocheleau, K.

Rashka, G. Staffeld, C. Stock, D. Sutton, and D. J. Merlo. 1987. Expression of a *Bacillus thuringiensis* insecticidal crystal protein gene in tobacco plants. In C. J. Arntzen and C. Ryan (eds.). *Molecular Strategies for Crop Protection.* Alan R. Liss, Pub., New York, pp. 345–353.

Andres, L. A., C. J. Davis, P. Harris, and A. J. Wapsphere. 1976. Biological control of weeds. In C. B. Huffaker and

P. S. Messenger (eds.). *Theory and Practice of Biological Control*. Academic Press, New York, pp. 481–500.

Arntzen, C. J., and C. Ryan (eds.). 1987. *Molecular Strategies for Crop Protection*. Alan R. Liss, Pub., New York.

Aronson, A. I., W. Beckman, and P. Dunn. 1986. *Bacillus thuringiensis* and related insect pathogens. *Microbiological Reviews* 50:1–24.

Baker, K. F. 1980. Microbial antagonism: The potential for biological control. In D. C. Ellwood, J. N. Hedger, M. J. Latham, J. M. Lynch, and J. H. Slater (eds.). *Contemporary Microbial Ecology*. Academic Press, New York, pp. 327–347.

Baker, K. F., and R. J. Cook. 1974. *Biological Control of Plant Pathogens*. W. H. Freeman, San Francisco.

Batra, S. W. T. 1982. Biological control in agroecosystems. *Science* 215:134–139.

Beck, S. D., and F. G. Maxwell. 1976. Use of plant resistance. In C. B. Huffaker and P. S. Messenger (eds.). *Theory and Practice of Biological Control*. Academic Press, New York, pp. 615–636.

Bowers, W. S., H. C. Hock, P. H. Evans, and M. Katayama. 1986. Thallophytic allelopathy: Isolation and identification of laetisaric acid. *Science* 232:105–106.

Burges, H. D. (ed.). 1981. *Microbial Control of Pests and Plant Diseases 1970–1980*. Academic Press, New York.

Carlton, C. B. 1988. Development of genetically improved strains of *Bacillus thuringiensis*. In P. A. Hedrin, J. J. Menn, and R. M. Hollingworth (eds.). *Biotechnology of Plant Protection*. American Chemical Society, Washington, D.C., pp. 261–279.

Carmichael, W. W., D. F. Biggs, and P. R. Gorham. 1975. Toxicology and pharmacological action of *Anabaena flos-aquae* toxin. *Science* 187:542–544.

Charudattan, R. 1975. Use of plant pathogens for control of aquatic weeds. In A. W. Bourquin, D. G. Ahearn, and S. P. Myers (eds.). *Impact of the Use of Microorganisms on the Aquatic Environment*. EPA 660-3-75-001. U.S. Environmental Protection Agency, Corvallis, Ore., pp. 127–146.

Chet, I. (ed.). 1987. *Innovative Approaches to Plant Disease Control*. Wiley, New York.

Coppel, H. C., and J. W. Mertins. 1977. *Biological Insect Pest Suppression*. Springer-Verlag, Berlin.

Currier, T. C., and C. Gawron-Burke. 1990. Commercial development of *Bacillus thuringiensis* bioinsecticide products. In J. P. Nakas and C. Hagedorn (eds.). *Biotechnology of Plant-Microbe Interactions*. McGraw-Hill, New York., pp. 111–143.

David, W. A. L. 1975. The status of viruses pathogenic for insects and mites. *Annual Reviews of Entomology* 20:97–117.

Davis, D. E., K. Myers, and J. B. Hoy. 1975. Biological control among vertebrates. In C. B. Huffaker and P. S. Messenger (eds.). *Theory and Practice of Biological Control*. Academic Press, New York, pp. 501–520.

Deacon, J. W. 1983. *Microbial Control of Plant Pests and Disease*. Aspects of Microbiology No. 7. American Society for Microbiology, Washington, D.C.

Dubos, B. 1984. Biocontrol of *Botrytis cinerea* on grapevines by an antagonist strain of *Trichoderma harzianum*. In M. J. Klug and C. A. Reddy (eds.). *Current Perspectives in Microbial Ecology*. American Society for Microbiology, Washington, D.C., pp. 370–373.

Eren, J., and D. Pramer. 1978. Growth and activity of the nematode-trapping fungus *Arthrobotrys conoides* in soil. In M. W. Loutit and J. A. R. Miles (eds.). *Microbial Ecology*. Springer-Verlag, Berlin, pp. 121–127.

Falcon, L. A. 1971. Use of bacteria for microbial control of insects. In H. D. Burges and N. W. Hussey (eds.). *Microbial Control of Insects and Mites*. Academic Press, London, pp. 67–96.

Falcon, L. A. 1976. Problems associated with the use of arthropod viruses in pest control. *Annual Reviews of Entomology* 21:305–324.

Ferron, P. 1978. Biological control of insect pests by entomogenous fungi. *Annual Reviews of Entomology* 23:409–442.

Goodman, R. M., and N. Newell. 1985. Genetic engineering of plants for herbicide resistance: Status and prospects. In H. O. Halvorson, D. Pramer, and M. Rogul (eds.). *Engineered Organisms in the Environment: Scientific Issues*. American Society for Microbiology, Washington, D.C., pp. 47–53.

Gustafsson, M. 1971. Microbial control of aphids and scale insects. In H. D. Burges and N. W. Hussey (eds.). *Microbial Control of Insects and Mites*. Academic Press, London, pp. 375–384.

Halvorson, H. O., D. Pramer, and M. Rogul (eds.). 1985. *Engineered Organisms in the Environment: Scientific Issues*. American Society for Microbiology, Washington, D.C.

Harman, G. E., and R. D. Lumsden. 1990. Biological disease control. In J. M. Lynch (ed.). *The Rhizosphere*. Wiley, New York, pp. 259–280.

Hasan, S. 1974. First introduction of a rust fungus in Australia for the biological control of skeleton weed. *Phytopathology* 64:253–254.

Hedrin, P. A., J. J. Menn, and R. M. Hollingworth (eds.). 1988. *Biotechnology of Plant Protection*. American Chemical Society, Washington, D.C.

Henis, Y. 1983. Ecological principles of biocontrol of soilborne plant pathogens: *Trichoderma* model. In M. J. Klug and C. A. Reddy (eds.). *Current Perspectives in Microbial Ecology*. American Society for Microbiology, Washington, D.C., pp. 353–361.

Henis, Y., and I. Chet. 1975. Microbial control of plant pathogens. *Advances in Applied Microbiology* 19:85–111.

Hofte, H., and H. Whiteley. 1989. Insecticidal crystal proteins of *Bacillus thuringiensis*. *Microbiological Reviews* 53:242–255.

Howell, C. R. 1990. Fungi as biological control agents. In J. P. Nakas and C. Hagedorn (eds.). *Biotechnology of Plant-Microbe Interactions*. McGraw-Hill, New York, pp. 257–317.

Huffaker, C. B., and P. S. Messenger (eds.). 1976. *Theory and Practice of Biological Control*. Academic Press, New York.

Hurley, J. M., L. A. Bulla, Jr., and R. E. Andrews, Jr. 1987. Purification of the mosquitocidal and cytolytic proteins of *Bacillus thuringiensis* subsp. *israelensis*. *Applied and Environmental Microbiology* 53:1316–1321.

Kerr, A., and M. E. Tate. 1984. Agrocins and the biological control of crown gall. *Microbiological Sciences* 1:1–4.

Krieg, A. 1971. Possible use of Rickettsiae for microbial control of insects. In H. D. Burges and N. W. Hussey (eds.). *Microbial Control of Insects and Mites*. Academic Press, London, pp. 173–180.

Kurstak, E. 1978. Viral and bacterial insecticides: Safety considerations. In M. W. Loutit and J. A. R. Miles (eds.). *Microbial Ecology*. Springer-Verlag, Berlin, pp. 265–268.

Laird, M. 1978. Microbial and integrated control of vectors of medical importance. In M. W. Loutit and J. A. R. Miles (eds.). *Microbial Ecology*. Springer-Verlag, Berlin, pp. 272–277.

Lindemann, J., and T. V. Suslow. 1987. Competition between ice nucleation active wild-type and ice nucleation deficient deletion mutant strains of *Pseudomonas syringae* and *P. fluorescens* biovar I and biological control of frost injury on strawberry blossoms. *Phytopathology* 77:882–886.

Lindow, S. E. 1983. The role of bacterial ice nucleation in frost injury to plants. *Annual Reviews of Phytopathology* 21:363–384.

Lindow, S. E. 1985. Ecology of *Pseudomonas syringae* relevant to the field use of Ice⁻ deletion mutants constructed *in vitro* for plant frost control. In H. O. Halvor-son, D. Pramer, and M. Rogul (eds.). *Engineered Organisms in the Environment: Scientific Issues*. American Society for Microbiology, Washington, D.C., pp. 23–35.

Lindow, S. E. 1987. Competitive exclusion of epiphytic bacteria by Ice⁻ mutants of *Pseudomonas syringae*. *Applied and Environmental Microbiology* 53:2520–2527.

Lindow, S. E. 1988a. Construction of isogenic Ice⁻ strains of *Pseduomonas syringae* for evaluation of specificity of competition on leaf surfaces. In F. Megusar and M. Gantar (eds.). *Microbial Ecology*. Slovene Society for Microbiology, Ljuvljana, Yugoslavia, pp. 509–515.

Lindow, S. E. 1988b. Lack of correlation of antibiosis in antagonism of ice nucleation active bacteria on leaf surfaces by non-ice nucleation active bacteria. *Phytopathology* 78:445–450.

Lindow, S. E. 1990. Use of genetically altered bacteria to achieve plant frost control. In J. P. Nakas and C. Hagedorn (eds.). *Biotechnology of Plant-Microbe Interactions*. McGraw-Hill, New York, pp. 85–110.

Lindow, S. E., and N. J. Panopoulos. 1988. Field tests of recombinant Ice⁻ *Pseudomonas syringae* for biological frost control in potato. In M. Sussman, C. H. Collins, and F. A. Skinner (eds.). *Proceedings of the First International Conference on Release of Genetically Engineered Microorganisms*. Academic Press, London, pp. 121–138.

Lipa, J. J. 1971. Microbial control of mites and ticks. In H. S. Burges and N. W. Hussey (eds.). *Microbial Control of Insects and Mites*. Academic Press, London, pp. 357–373.

Lockwood, J. L. 1979. Soil mycostasis: Concluding remarks. In B. Schippers and W. Gams (eds.). *Soil-borne Plant Pathogens*. Academic Press, London, pp. 121–129.

Longworth, J. F., and J. Kalmakoff. 1978. An ecological approach to the use of insect pathogens for pest control. In M. W. Loutit and J.A.R. Miles (eds.). *Microbial Ecology*. Springer-Verlag, Berlin, pp. 269–271.

McLaughlin, R. E. 1971. Use of protozoans for microbial control of insects. In H. D. Burges and N. W. Hussey (eds.). *Microbial Control of Insects and Mites*. Academic Press, London, pp. 151–172.

Milewski, E. 1990. EPA regulations governing release of genetically engineered microorganisms. In J. P. Nakas and C. Hagedorn (eds.). *Biotechnology of Plant-Microbe Interactions*. McGraw-Hill, New York, pp. 319–340.

Moore-Landecker, E. 1972. *Fundamentals of the Fungi*. Prentice-Hall, Englewood Cliffs, N.J., pp. 103–105.

Moss, B., and M. L. Butler. 1985. *Vaccinia* virus vectors:

Potential use as live recombinant virus vaccines. In H. O. Halvorson, D. Pramer, and M. Rogul (eds.). *Engineered Organisms in the Environment: Scientific Issues*. American Society for Microbiology, Washington, D.C., pp. 36–39.

Nakas, J. P., and C. Hagedorn (eds.). 1990. *Biotechnology of Plant-Microbe Interactions*. McGraw-Hill, New York.

National Academy of Sciences. 1968. Plant-disease development and control. *Principles of Plant and Animal Pest Control*. Vol. I, Publication 1596. National Academy of Sciences, Washington, D.C.

Phatak, S. C., D. R. Sumner, H. D. Wells, D. K. Bell, and N. C. Glaze. 1983. Biological control of yellow nutsedge with the indigeneous rust fungus *Puccinea canaliculata*. *Science* 219:1446–1447.

Pramer, D. 1964. Nematode-trapping fungi. *Science* 144:382–388.

Pramer, D., and S. Al-Rabiai. 1973. Regulation of insect populations by protozoa and nematodes. *Annals of the New York Academy of Sciences* 217:85–92.

Roberts, D. W., and W. G. Yendol. 1971. Use of fungi for microbial control of insects. In H. D. Burges and N. W. Hussey (eds.). *Microbial Control of Insects and Mites*. Academic Press, London, pp. 125–150.

Rowe, G. E., and A. Margaritis. 1987. Bioprocess developments in the production of bioinsecticides of *Bacillus thuringiensis*. *CRC Critical Reviews in Biotechnology* 6:87–127.

Schroth, M. N., and J. G. Hancock. 1981. Selected topics in biological control. *Annual Reviews of Microbiology* 35:453–476.

Schroth, M. N., and J. G. Hancock. 1982. Disease-suppressive soil and root-colonizing bacteria. *Science* 216:1376–1381.

Schroth, M. N., J. E. Loper, and D. C. Hildebrand. 1984. Bacteria as biocontrol agents of plant disease. In M. J. Klug and C. A. Reddy (eds.). *Current Perspectives in Microbial Ecology*. American Society for Microbiology, Washington, D.C., pp. 362–369.

Silverstein, R. M. 1981. Pheromones: Background and potential for use in insect pest control. *Science* 213:1326–1332.

Simmonds, F. J., J. M. Franz, and R. I. Sailer. 1976. History of biological control. In C. B. Huffaker and P. S. Messenger (eds.). *Theory and Practice of Biological Control*. Academic Press, New York, pp. 17–41.

Snyder, W. C., G. W. Wallis, and S. N. Smith. 1976. Biological control of plant pathogens. In C. B. Huffaker and P. S. Messenger (eds.). *Theory and Practice of Biological Control*. Academic Press, New York, pp. 521–542.

Stairs, G. R. 1971. Use of viruses for microbial control of insects. In H. D. Burges and N. W. Hussey (eds.). *Microbial Control of Insects and Mites*. Academic Press, London, pp. 97–124.

Stairs, G. R. 1972. Pathogenic microorganisms in the regulation of forest insect populations. *Annual Reviews of Entomology* 17:355–372.

Steinhaus, E. A. 1965. Diseases of invertebrates other than insects. *Bacteriological Reviews* 29:388–396.

Stewart, W. D. P., and M. J. Daft. 1977. Microbial pathogens of cyanophycean blooms. *Advances in Aquatic Microbiology* 1:177–218.

Tinsley, T. W. 1979. The potential of insect pathogenic viruses as pesticidal agents. *Annual Reviews of Entomology* 24:63–87.

Watrud, L. S., F. J. Perlak, M. T. Tran, K. Kusano, E. J. Mayer, M. A. Miller-Wiedeman, M. G. Obukowicz, D. R. Nelson, J. P. Kreitinger, and R. J. Kaufman. 1985. Cloning of *Bacillus thuringiensis* subsp. *kurstaki* deltaendotoxin gene into *Pseudomonas fluorescens*: Molecular biology and ecology of an engineered microbial pesticide. In H. O. Halvorson, D. Pramer, and M. Rogul (eds.). *Engineered Organisms in the Environment: Scientific Issues*. American Society for Microbiology, Washington, D.C., pp. 40–46.

Weiser, J., G. E. Bucher, and G. O. Poinas, Jr. 1976. Host relationships and utility of pathogens. In C. B. Huffaker and P. S. Messenger (eds.). *Theory and Practices of Biological Control*. Academic Press, New York, pp. 169–188.

Whipps, J. M., and J. M. Lynch. 1986. The influence of the rhizosphere on crop productivity. *Advances in Microbial Ecology* 9:187–244.

Wilson, C. L. 1969. Use of plant pathogens in weed control. *Annual Reviews of Phytopathology* 7:411–434.

Woods, A. 1974. *Pest Control: A Survey*. John Wiley and Sons, New York.

Worthy, W. 1973. Integrated insect control may alter pesticide use patterns. *Chemical Engineering News* 51:13–19.

Zettler, F. W., and T. E. Freeman. 1972. Plant pathogens as biocontrols of aquatic weeds. *Annual Reviews of Phytopathology* 10:455–470.

PART SIX

Appendices

Appendix 1: Statistics in Microbial Ecology

The use of statistical analyses is an essential part of many scientific studies. This is especially true in the field of microbial ecology because of the high degree of variability in natural ecosystems and the heterogeneity of distribution of microorganisms within habitats. Statistics are used to describe experimental results numerically in ways that can be readily used to communicate the informational content and limitations of those results. Proper use of statistics allows scientists to characterize their results so that large amounts of data are reduced to a limited set of statistical descriptors that summarize the data in a manner that can be viewed and comprehended. Statistics also allow the comparison of experimental data so that the degree of confidence in making particular interpretations of the data can be assessed; that is, statistics determine whether the data really support a particular conclusion or not. It is important always to remember that numbers are only the representatives of facts and not facts themselves. Statistical analyses and the degree of confidence that can be placed in the conclusions made by the investigator depend upon the adequacy of the data collected. Thus, when designing an experiment, microbial ecologists must consider the types of statistical analyses that will be necessary to interpret the results and to establish the validity of that interpretation. Inadequate experimental design leads to speculation rather than to scientifically valid conclusions.

DATA COLLECTION AND EXPERIMENTAL DESIGN

Independent and Dependent Variables

The type of data to be obtained is an important consideration in designing an experiment. Some data may be considered as independent variables, which, by strict statistical definition, are under the control of the experimenter; other data, which are not under the control of the experimenter, are considered by statisticians to be dependent variables. For example, in a laboratory experiment designed to determine the influence of temperature on numbers of viable microorganisms, the experimenter may use a series of incubators with carefully regulated temperatures and viable plate counts to determine numbers of viable microorganisms. In this case, temperature is the independent variable because it is under the control of the experimenter, and the number of viable microorganisms enumerated, which is not controlled, is the dependent variable. Indeed, the purpose of many such experiments is to determine the effect of an independent variable, such as temperature, on a dependent variable, such as growth rate.

Microbial ecologists who collect field data, however, often encounter a problem in statistical analyses because they are unable to control any of the variables they measure. In such observational studies, they often still want to infer relationships among variables. They may, for example, measure temperature and determine its effect on numbers of viable microorganisms in an experiment analogous to the laboratory experiment just described. In the field, however, they have no control over the temperature, and by the statistician's definition, both temperature and numbers of microorganisms are thus dependent variables. The independent variables in such field experiments may be time and place of sample collection, which are the only factors under the control of the experimenter. In such cases, microbial ecologists have to employ judgment as biologists and to decide which variables should be treated as independent. Such decisions are based on experimental evidence gained in carefully controlled laboratory experiments, and few would contest, for example, that growth rate is dependent on temperature and that temperature measured in the field should be "properly" considered an independent variable when studying the relationship of temperature and growth rate. In other cases, insufficient laboratory evidence about the relationship between two variables makes such judgments difficult. For these reasons, it is often unwise to rely completely on either laboratory or field experiments alone; laboratory experiments must be put to the test of field verification, and apparent causal relationships observed in the field must be confirmed in carefully controlled laboratory situations.

The establishment of proper controls is important for all experimental designs and field studies. Ideally, the researcher is able to manipulate a single independent variable and observe its effects on one or more dependent variables. In reality, some variable that is not being measured may affect the variables that are being measured. Statistical analysis should reveal if the variability of results is due to factors that have not been measured. Proper testing should also reveal if the interactions of multiple independent variables influence a measured dependent variable; for example, cell viability may depend on the interactive effects of temperature and moisture.

Bias

An important consideration in data collection is that of bias, the sources of which should be identified. There are several sources of bias in the analytical procedures used for data collection, sample collection, and statistical method. For many statistical procedures it is necessary to randomize some aspects of experimental design and data collection to minimize bias. For example, in a study to determine the effects of a pesticide on rates of denitrification, the results would be invalid if the sites of the experimental plots were placed together in a well-drained portion of a field and the control plots were located in a poorly drained area.

It is also important to run some experiments in the blind, that is, to prevent bias on the part of the researcher by shielding knowledge of which tests are the controls and which are the experimentals. This prevents researchers from unconsciously viewing the results in the way that they expect or want them to come out. The inherent bias of certain analytical procedures must also be understood, and proper limitations must be placed on the interpretation of results. For example, the plate count procedure for enumeration of fungi is biased toward enumeration of propagules, such as spores, and against filamentous forms. Total fungal biomass cannot be estimated by such a procedure that has a major inherent bias, but such results can and must be interpreted within the limitations of the populations enumerated.

ESTIMATION OF CENTRAL TENDENCY

Experimental replication is essential for statistical analyses, but presenting all of the data collected is unwieldy. In summarizing data, it is often desirable to find a convenient means of comparison; therefore, the data routinely are reduced to representative values. This reduction is accomplished in part by describing the central tendency of the data, that is, the center around which values tend to fall, by calculating the median, mode, and/or mean of the data. The median is defined as that point in the distribution of data points above and below which half of the data values fall; the mode is the data point value that occurs most frequently; the arithmetic mean (also known as the average value or simply as the mean) is the sum of the data point values divided by the number of data points (Equation 1).

(1) $\bar{x} = \dfrac{\sum x}{n}$

where

$\bar{x}$ = the mean

$\sum$ = the sum of

x = the numerical value of a data point

n = the number of data points (sample size)

The choice of which measure of central tendency best describes the data, that is, which measure gives the truest picture of where values most commonly occur, depends upon the distribution of data. If the data has a normal distribution, the mean, median, and mode are identical, but if the distribution of the data points is skewed, these measures of central tendency differ (Figure A1.1).

The mean is the most widely used as the measure of central tendency in microbial ecology. Different values, though, can have the same mean; for example, the mean of the values 1, 2, 3, 6, and 13 is 5, as is the mean of the values 4, 5, and 6. The mean is particularly sensitive to extreme values if these values are not balanced on the other side of the mean; thus, if all estimates of the viable microorganisms are of the order of 10^5, except for a single measurement that is

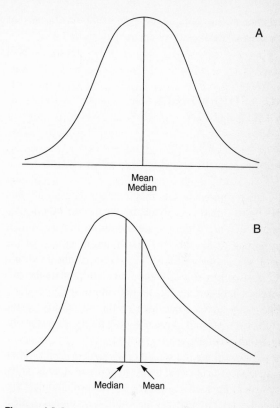

Figure A1.1

The locations of the means and medians for (A) normal and (B) skewed distributions of data points.

of the order of 10^7, the exceptionally high value will have a marked effect on the mean. For this reason, the mean must be viewed with caution and subjected to further statistical analyses before it can be used for comparison of control and experimental data sets.

If the data set has a particularly skewed distribution, then the mean of that data after it has been transformed by a mathematical operation, such as transformation by logarithms, may be a better representative value of central tendency. For example, counts of microorganisms that are expressed as multiples of powers of ten have a skewed distribution. In this case the data can be transformed into logarithms and, because the distribution of the logarithmic values are not as skewed as the original data, the mean in the log scale could then be used as the reasonable measure of central tendency.

MEASUREMENTS OF VARIABILITY

Range, Variance, and Standard Deviation

Because some data sets may contain data points that are tightly clustered about the mean and others data points that are widely dispersed about the mean, it is important to describe the dispersion or spread of the data. The simplest description of the variability within a set of collected data is the range, defined as the difference between the largest and smallest data values. The range gives some idea of the variability of the collected data but does not effectively describe the distribution of a given set of data.

Describing the spread of the data is critical when statistical tests are used to test hypotheses, that is, when statistics are used to determine the probabilities that statements or hypotheses are valid. A quantitative measure of the spread of the data, defined as the sum of the squared deviations from the mean divided by the sample size minus one, is called the variance (Equation 2).

$$(2) \quad s^2 = \frac{\sum (x - \bar{x})^2}{n - 1}$$

where

s^2 = variance of the data

$\sum$ = the sum of

x = the numerical value of a data point

$\bar{x}$ = the mean of the data

n = the number of data points (sample size)

One less than the number of data points represents the degrees of freedom, which is defined as the number of values that are free to vary after certain restrictions have been placed on the data. The concept of degrees of freedom is based on the premise that once certain values are defined, others cannot vary. As a trivial example of this concept, if it is known that there are five different microorganisms in a sample and which organisms these are, once the first four are identified, the identity of the fifth is automatically known even without observing it. In other words, the fact that the first four have been specified also specifies the fifth and, in this case, the degrees of freedom are $n-1$, or 4 degrees of freedom.

Often the positive square root of the variance, known as the standard deviation (Equation 3), is used to describe the variability of a set of collected data.

$$(3) \quad SD = \sqrt{s^2}$$

where

SD = the standard deviation

s^2 = the variance

The standard deviation quantitatively describes the spread of the data.

Estimates of the variance are based on a limited sample size collected from an overall population. For example, if one wants to know distribution by size of bacteria in a lake, one collects only a limited number of samples and makes size measurements on a portion of the bacteria contained in the samples; one does not actually measure the size of all the bacteria within the population of the lake.

If samples are repeatedly collected from a population and the means calculated for each individual set of collected data, there will be a distribution about the true population mean. The standard error of the mean is calculated to describe the distribution of such individually calculated means (Equation 4).

$$(4) \quad SE = \frac{SD}{\sqrt{n}}$$

where

SE = the standard error of the mean

SD = the standard deviation

n = the number of data points (sample size)

The standard error describes the variability of the means about the true mean of the population.

The distribution of sample means drawn from an arbitrarily distributed population tends to be bell

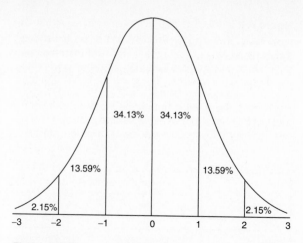

Figure A1.2

The normal distribution showing the proportions of area underlying different regions of the curve. The units along the x axis are standard deviation units from the mean (shown by 0). The proportions of area shown indicate the probability of a data point occurring at a specified standard deviation from the mean. Thus, 68.26% of the data points will occur within one deviation unit from the mean in this distribution. These area proportions can be used as probabilities in hypothesis testing that assume a normal distribution.

shaped, that is, to form a normal distribution. This is described by the central limit theorem, which states that if random samples of a fixed sample size are repeatedly drawn from a population, the resulting distribution of means will be normally distributed with a mean equal to the true population mean.

HYPOTHESIS TESTING

Significance and Hypothesis Testing

It is possible to relate deviations from the means to probability values in terms of the area under a normal distribution curve (Figure A1.2). This allows one to develop hypotheses and to test that the hypotheses are correct by comparison with the expected normal distri-

bution. Hypothesis testing is based on the ability to compare an observed result with an expected result. In testing a hypothesis, an attempt is made to determine whether the hypothesis is true for the entire population by estimating statistical parameters based on a limited sample drawn from the population. The intent of the statistical testing is to permit educated guesses or inferences about the populations themselves.

Most frequently, the microbial ecologist wants to determine whether an observed result is significantly different from an expected result. The ability to make such a determination often involves determining whether the mean of an experimental group differs significantly from the mean of a control group, or whether the mean of a population representing one habitat differs significantly from the mean of a population of another habitat. In this process, it is necessary to define significance based on the probability that the observed result could be due to the natural variability of a population. It is necessary for the microbial ecologist to use the term significant in a statistical sense and to define the level of significance or confidence coefficient that is being used.

In statistical tests of significance, the hypothesis is stated in the null form (H_0); that is, for example, there is no difference between the groups being compared. Thus, if one is trying to determine whether streptomycin has an effect on a bacterial population, the null hypothesis will state that there is no difference between the mean of the control population and the mean of the experimental population exposed to streptomycin. One also must state an alternate hypothesis that would be established by the rejection of the null hypothesis. The alternate hypothesis is usually that there is a difference between the means of the groups being compared.

In order to determine whether or not to reject the null hypothesis, it is necessary to establish levels of significance based on the idea that chance alone caused the observed results. The level of significance (α value) is the probability that an error will be made in rejecting the null hypothesis when it is true. Generally, the acceptable levels of significance for biological experiments are the $\alpha = 0.05$ and $\alpha = 0.01$ levels. At the 0.05 level of significance, one accepts that 5% of the time a null hypothesis that indeed was true will be rejected. Phrased another way, at the $\alpha = 0.05$

level, the investigator is 95% confident that a true null hypothesis will not be rejected; at the 0.01 level, the investigator is 99% confident.

Two types of errors may be made in testing hypotheses, known as type I and type II errors (Figure A1.3). In a type I error, a correct null hypothesis is mistakenly rejected. In a type II error, there is a failure to reject a false null hypothesis. The α value equals the probability of making a type I error; the β value equals the probability of making a type II error. By setting the α level at 0.01, the probability of making a type I error decreases compared to $\alpha = 0.05$, but at the same time the probability of making a type II error increases.

Since many experiments in microbiology are aimed at concluding a significant difference between experimental and control populations, that is, rejecting the null hypothesis, there is greater concern about making a type I error, concluding that there was an effect when actually there was none, than about making a type II error. For example, when testing the effects of various concentrations of a disinfectant on microbial populations in potable water, it is critical to establish with a high level of confidence that the particular concentration of that disinfectant had the desired effect. To mistakenly reject a true hypothesis in this case could result in the use of an ineffective disinfectant.

An important aspect of statistical tests for testing hypotheses is power. The power of a particular test can be defined as the ability of the test to detect false null hypotheses. As the power of the test increases, fewer type II errors will be made. Generally, the power of statistical tests increases as the sample size increases. Power may be defined as $1-\beta$, that is, one minus the probability of making a type II error.

Parametric Tests of Significance

Various statistical tests are employed to assess hypotheses. These tests differ in the underlying assumptions about the data and its distribution. Statistical tests of significance differ in their applicability to particular data sets and the ability of their power. A parametric test of significance generally assumes a normal distribution of the population, whereas a nonpara-

Figure A1.3

Probability of making type I and type II errors. Top shows that a type I error occurs when a true null hypothesis is rejected and that a type II error occurs when a false null hypothesis is not rejected. The lower section shows the region of rejection for different levels of α. As shown, decreasing the α value decreases the probability of making a type I error but increases the probability of making a type II error (b has a lower α value than a).

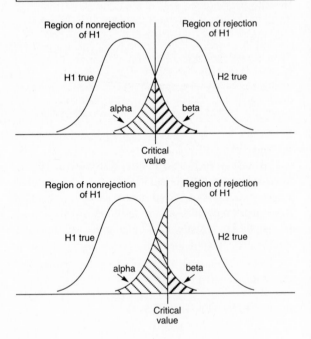

Decision	State of affairs in the population	
	H_1 true	H_1 false H_2 true
Reject H_1	Type 1 error (alpha)	No error
Do not reject H_1	No error	Type 2 error (beta)

metric test of significance does not make that assumption about the distribution of the data. Accordingly, a nonparametric test is referred to as a distribution-free test of significance. For a given sample size, a parametric test has more power than a nonparametric test if its assumptions are satisfied. For any given sample size, the parametric test of significance, which typically assumes normally distributed populations with the same variance, entails less risk of making a type II error than its nonparametric counterparts at a given level of significance. The parametric tests are more likely to reject the null hypothesis when the null hypothesis is indeed false. Thus, given a choice between nonparametric and parametric tests of significance, the parametric test should be employed as long as the necessary assumptions of that test are fulfilled; if the conditions for using parametric test procedures are not met, however, nonparametric tests should be employed.

The Student t Test A widely used statistical test for determining the validity of hypotheses is the Student t test. The t test allows one to test the null hypothesis that the means of two groups are the same. The t test is based on probabilities obtained from t distributions. For the t statistic, there is a family of distributions that vary as a function of degrees of freedom. The t distributions are symmetrical about a mean of 0, as are normal distributions. The t distributions, however, are more spread out than on the normal curve. Consequently, the proportion of area beyond the specific value of t is greater than the proportion of area beyond the corresponding value obtained from the normal curve. The greater the degrees of freedom, the more the t distribution resembles the normal curve; for large sample sizes with high degrees of freedom, the t distributions approach the normal curve. There are tables of critical values for t, which are values of the t distributions that correspond to varying levels of significance; that is, there are published tables that allow one to compare a calculated t value from one's own data with a t value determined by the level of significance. By calculating a t value for one's data and comparing it with a critical value based on a t distribution with the degrees of freedom determined from the sample sizes, one is able to decide whether or not to reject the null hypothesis.

The t value is calculated by dividing the difference of two means by the standard error of the difference of the means. The exact formula used for this calculation depends on whether the variances in the populations from which the means are derived are identical. Usually,

the t test is employed with the assumption that the variances within the populations are identical (Equation 5).

$$(5) \quad t = \frac{(\bar{x}_A - \bar{x}_B)}{\sqrt{\left(\dfrac{\sum (x_a - \bar{x}_A)^2 + \sum (x_b - \bar{x}_B)^2}{(n_A + n_B - 2)}\right)\left(\dfrac{1}{n_A} + \dfrac{1}{n_B}\right)}}$$

where

$$
\begin{aligned}
t &= \text{student } t \text{ value} \\
\bar{x}_A &= \text{mean of group } A \\
\bar{x}_B &= \text{mean of group } B \\
\sum (x_a - \bar{x}_A)^2 &= \text{sum of squares for group } A \\
\sum (x_b - \bar{x}_B)^2 &= \text{sum of squares for group } B \\
n_A &= \text{number of data points in group } A \\
n_B &= \text{number of data points in group } B
\end{aligned}
$$

The mean obtained from one group can be called the control mean and the mean obtained from the other group called the experimental group mean. In such cases, the null hypothesis would state that there was no experimental effect, that is, that the population means of the experimental and control groups were not significantly different. As an example, in a simple experiment aimed at determining the effects of temperature on the degradation rates of a particular pesticide, the pesticide would be added to aliquots of soil and incubated at two different temperatures, after which the extent of degradation would be assessed by recovering any untransformed pesticide from replicate aliquots of soil incubated at the two different temperatures. The variance within the population would be assumed to be the same because the same soil with the same microbial populations was used in all cases. The calculated value would be compared with the critical values obtained from standard tables and a decision made whether to reject or accept the null hypothesis that there was no difference between the means of the degradation rates obtained at different temperatures.

There are several important restrictions on the use of the t test. The t test is used for the comparison of two means; it does not permit the direct comparison of all the means obtained from more than two groups. In the above example, one could not directly compare the means obtained over a gradient of temperatures, and the *t* test could not be directly used to test hypotheses concerning the effects of a gradient of concentrations or any other environmental factor. When comparing the means of more than two groups, it is wrong to use a *t* statistic initially, even to compare repeatedly the means of two groups. This is because the critical values of the *t* statistic are chosen to minimize the probability of making a type I error. When a large number of *t* tests are performed, there is a greater likelihood that a type I error will be made within all the comparisons. Thus, even though paired *t* tests are often used, it is better to use the analysis of variance procedure with multiple range tests, described in the following section, to make such multigroup comparisons.

Analysis of Variance For hypothesis testing involving the comparison of multiple groups, the appropriate procedure is an analysis of variance (ANOVA). In an analysis of variance procedure an *F* statistic is calculated that can be compared with the values derived from an *F* distribution for the comparison of multiple groups. An analysis of variance permits an unambiguous assessment of differences when more than one comparison is made. To perform an analysis of variance, it is necessary to group the data within each group, called a cell, containing the replicates of a particular treatment. Normally, there will be a control group and various experimental groups. We should point out, however, that the analysis of variance does not indicate the direction or the magnitude of the actual effect, but only determine whether the observed results could have arisen by chance at a particular probability level.

An analysis of variance is achieved by obtaining two independent estimates of variance, one based on variability between groups (between-group variance) and the other based on variability within groups (within-group variance). The *F* ratio is obtained by dividing the between-group variance by the within-group variance (Equation 6).

$$(6) \quad F = \frac{\sum n_i \left((\bar{x}_i - \bar{x})^2 / (k-1) \right)}{\sum \sum (x_{ij} - x_i)^2 / (n-k)}$$

where

$$F = \text{the } F \text{ statistic}$$
$$x_{ij} = \text{the } j\text{th observation in Group } i$$
$$\bar{x}_i = \text{the mean of Group } i$$
$$\bar{x} = \text{the overall mean}$$
$$k = \text{the total number of groups}$$
$$n = \text{the number of observations}$$
$$n_i = \text{the number of observations in Group } i$$

In this ratio are two different degrees of freedom, one for the within-group variance and the other for the between-group variance. The degrees of freedom of between groups is simply the number of groups minus one. The number of degrees of freedom of the within groups is the total sample size minus the number of groups. The *F* distribution is determined by the two values for the degrees of freedom. Each combination of degrees of freedom for between and within groups determines a separate *F* distribution. If the differences among means are large relative to the within-group differences, the *F* ratio is large. Conversely, if the between-group variance is small relative to the within-group variance, the *F* ratio is small. A table is generated for an analysis of variance that includes the calculated *F* ratio (Table A1.1).

There are tables of critical values for *F* distributions of these combinations of degrees of freedom that can be used in hypothesis testing. The use of the F ratio assumes that the population from which the samples are drawn is normally distributed. In the event of a significant *F* ratio, that is, an *F* ratio greater than the critical value, the null hypothesis can be rejected. Then it can be concluded that the means were not drawn from the same population, or there appears to be a significant difference among means. An actual probability value, *p*, also can be determined, permitting a precise statement of the probability of obtaining a deviation as large as the observed one or larger.

The simplest analyses of variance (when there are multiple treatments) occur when there are different combinations of only two treatments or factors. In this case, a two-way analysis of variance is performed. If both temperature and water availability were varied in an experi-

Table A1.1

Formulae used in calculating an analysis of variance table

Source of Variation	Sum of squares	df	Mean squares	F ratio
Between	$SS_b = \sum n_i (\bar{x}_i - \bar{x})^2$	$k-1$	$MS_b = \dfrac{SS_b}{k-1}$	$F_{(k-1),\Sigma n_i - k} = \dfrac{MS_b}{MS_w}$
Within	$SS_w = SS_t - SS_b$	$\sum n_i - k$	$MS_w = \dfrac{SS_w}{\sum n_i - k}$	
Total	$SS_t = \sum\limits_{i=1}^{k} \sum\limits_{j=1}^{n} [\,(x_{ij} - \bar{x})^2\,]$	$\sum n_i - 1$		

ment and the rate of respiration measured as the dependent variable, there would be two treatment factors. The two-way analysis of variance determines not only whether there is a significant overall effect, but also whether individual factors could have produced the observed results or whether the observed results represent an interaction between the two factors. The multifactor univariate analysis of variance allows for partitioning of the variance into a segment due to factor 1, a segment due to factor 2, a segment due to the interaction of factors 1 and 2, and a residual or error variance.

Table A1.2 shows an example of the results of a two-way analysis of variance. In this particular experiment, the effects of a fungicide on soil respiration were being tested using different soils. The two factors were soil type, including the inherent biological, chemical, and physical properties of the different soils, and fungicide concentration. Although there were only two independent factors in this experiment, each factor was represented by multiple levels or groups, that is, more than two soils were used and more than two concentrations of the fungicide were applied. There were two degrees of freedom for the fungicide factor and eight degrees of freedom for the soil factor. The calculated F statistics indicated that there was a significant effect of the fungicide on soil respiration and that there were sig-

Table A1.2

Two-way analysis of variance testing the effects of fungicides on soil microorganisms

Source of variation	Results of a two-way ANOVA		
	df	F	Significance of F
Main effects	10	107.10	0.000
Soil	8	133.40	0.000
Fungicide	2	5.10	0.000
Two-way interaction (soil x fungicide)	16	0.180	1.000
Explained	26	41.30	0.000

nificant differences in respiration rates in the different soils. This was shown by a significant α value of less than 0.01 for the two main effects. There was, however, no significant interaction between the two factors; that is, fungicide application had the same effect on soil respiration regardless of soil type.

Even more complex analyses of variance can be performed using more than two factors. A multivariate analysis of variance (MANOVA) is used with one or more independent and two or more dependent variables. When factors interact, however, it is often difficult to interpret the results of multivariate analyses of variance in terms of understandable biological concepts. The experimenter is frequently restricted to making the statement that the effects were interactive without being able to specify the contributions of the individual factors to the observed results. The microbial ecologist must interpret the statistical significance of a given result in a biologically meaningful manner. For example, the application of fungicide to a soil may have a statistically significant effect at the $\alpha = 0.01$ probability level, as previously discussed; however, the alteration in soil respiration may be less than 5%. The microbial ecologist must judge whether a 5% change in soil respiration is an ecologically significant change, and not just a statistically significant change. Microbial ecologists cannot ignore the fact that they are biologists and so must make biologically sound conclusions. They cannot simply replace the need to make biological judgments of significance with their ability to estimate statistical significance.

Multiple Range Tests In a simple case where there are only two groups, the F ratio yields probability values identical to those obtained with the student t ratio. Indeed, in the one degree of freedom situation, t equals the square root of F. Rejection of the null hypothesis using the F statistic with three or more groups indicates that there is a significant difference between one or more groups and the others. It does not indicate, however, which groups are significantly different from the others.

Having established that the F ratio is significant, it is then possible to compare the individual groups. This is accomplished using a variety of multiple range or multiple comparison tests. These multiple comparison tests include the Student Newman-Keuls, Duncan, Tukey, and Scheffé procedures. Each of these procedures makes somewhat different assumptions with respect to how it protects against making a type I error. In the Duncan procedure, there is a protection level of α for the collection of comparisons rather than an α level for each individual comparison. The various multiple comparison tests should give the same results when there is a major difference between means. When there are marginal differences, that is, differences close to the accepted α level, different results will be obtained using the different multiple comparison tests. With increasing sample size, the results of the various multiple comparison tests approach each other.

Comparing the Student Newman-Keuls and the Duncan test procedures, the Duncan procedure has the smaller critical value. Thus, on average, a larger difference between two means is required for statistical significance under the Newman-Keuls procedure. The Newman-Keuls procedure is more conservative than the Duncan procedure. The Tukey test is more conservative than either the Newman-Keuls or Duncan approaches; that is, fewer significant differences will be obtained with the Tukey procedure. Still more conservative is the Scheffé method. In the Scheffé procedure, the type I error is at a maximum set α value for any of the possible comparisons.

Yet another approach is the Dunnett procedure. The Dunnett procedure permits comparison of all experimental treatments against a single control group. Using the Dunnett procedure, the overall probability of making a type I error will be at the set α level.

The importance of the analysis of variance procedure coupled with the multiple comparison tests is that it allows one to set an appropriate probability level for rejecting the null hypothesis for the overall analysis. These procedures allow one to analyze data statistically from multiple groups, while still retaining the capability to determine which groups are significantly different from each other. The analysis of variance procedure without the coupled use of multiple comparison tests would not permit determination of which groups were different from each other.

Nonparametric Tests of Significance

In order to use the parametric tests discussed here, it is necessary to make assumptions about the distribution of the population. In many cases examined by microbial ecologists, the assumption of a normal distribution is valid, and these tests can be utilized. Even if the data are not normally distributed, it may be possible to transform the data to obtain a normal distribution that can be subjected to these statistical analyses. For example, it may be possible to use a logarithmic transformation of the data and subject the transformed data to analysis of variance. In cases where it is known that the data do not fit the necessary assumptions or where it is desirable to be conservative and not to make assumptions about the distribution of the data, it is necessary to use less powerful, nonparametric tests.

Chi Square   A common nonparametric statistical test is the χ^2 (chi square) test (Equation 7).

$$(7) \quad \chi^2 = \sum \frac{(O - E)^2}{E}$$

where

$$\chi^2 = \text{Chi square}$$
$$O = \text{the observed frequencies}$$
$$E = \text{the expected frequencies}$$

The χ^2 test of the independence of categorical variables is valuable when attempting to analyze the apportionment of a characteristic within a population. For example, the χ^2 test is useful in examining whether an observed frequency distribution of a feature, such as the ability to grow at 43°C, within a population is statistically different from the expected distribution of this feature. The raw data for a χ^2 analysis should be nominal scale data, that is, on a scale in which data are placed into discrete groups, such as bacteria versus fungi.

Mann Whitney U Test   The Mann Whitney U test is one of the more powerful nonparametric statistical tests. The Mann Whitney U test may be employed as an alternative to the student t ratio when the data are on an ordinal scale. The Mann Whitney U test is used with two independent groups. The null hypothesis in the Mann Whitney U test is that both samples are drawn from populations with the same distribution. The alternative hypothesis is that the parent populations from which the samples are drawn have different medians. The Mann Whitney U test assumes that the distributions have the same form, but have different medians. In using the Mann Whitney test, the concern is with the sampling distribution of the statistic U. In order to find U, the scores must be ranked from the lowest to the highest, while still retaining the identity of the group from which they came, that is, the control group or experimental group (Equation 8).

$$(8) \quad U = n_1 n_2 + \frac{n_2 (n_2 + 1)}{2} - R_2$$

where

$$U = \text{the Mann Whitney statistic}$$
$$R_2 = \text{the sum of ranks assigned to Group 2}$$
$$n_1 = \text{the number of data points in Group 1}$$
$$n_2 = \text{the number of data points in Group 2}$$

U represents the number of times that the n_1 value precedes the n_2 value. U is large if the n_1 population is located below the n_2 population.

It is apparent that the microbial ecologist has a variety of statistical tests available for analyzing experimental data. The choice of the test procedure depends on the nature of the data, including the scale in which the data are gathered; on the experimental design, including how many factors and how many groups are to be considered in the analysis; and on the assumptions that each statistical test makes about the distribution of the data. The microbial ecologist should choose the most powerful statistical test that can be applied to the data. An appropriate null hypothesis should be stated and the appropriate statistical test decided upon before beginning the experiments. The microbial ecologist must understand the assumptions being made when using statistical tests and must determine the confidence intervals that he or she wishes to accept.

CORRELATION AND REGRESSION

Many studies in microbial ecology attempt to determine whether there is a relationship between two variables. Toward this end, it is possible to perform correlation and regression analyses. There are several types of correlation coefficients. A correlation coefficient can be calculated to express quantitatively the extent to which two variables are related. The choice as to the correlation coefficient appropriate to use depends on the scale of measurement in which each variable is expressed, whether the distribution of the data is continuous or discrete, and whether there is a linear or a nonlinear relationship.

Perhaps the most common correlation coefficient used in microbial ecology is the Pearson product moment correlation coefficient (r) (Equation 9). It is necessary to pair the data so that there are matching measurements of each of the variables being examined in order to calculate this correlation coefficient .

$$(9) \quad r = \frac{\sum (x - \bar{x})\,(y - \bar{y})}{\sqrt{\left(\sum (x - \bar{x})^2\right)\left(\sum (y - \bar{y})^2\right)}}$$

where

r = Pearson's correlation coefficient

x = an x value for a given measurement

$\bar{x}$ = the mean of the x values

y = the corresponding paired y value

$\bar{y}$ = the mean of y values

The values of the correlation coefficient vary between −1 and +1. A correlation coefficient that approaches −1 expresses a negative linear relationship; a coefficient that approximates +1 shows a positive linear relationship. A positive relationship is one in which both variables increase in magnitude. A negative relationship is one in which one variable increases, while the other variable decreases. A correlation coefficient of 0 indicates the lack of a linear relationship between the two variables.

The results of correlation analyses may be represented simply by indicating the correlation coefficient.

The data can also be represented graphically as a scatter diagram (Figure A1.4). A scatter diagram plots all of the data that have been collected, not just a measure of central tendency. The limitations of interpreting the results of correlation analysis should be considered.

The Pearson r coefficient is employed with interval or ratio scale data. The Pearson r correlation coefficient assumes a linear relationship between the two variables. A lack of evidence of a relationship, that is, a Pearson r value that approximates 0, may arise because the variables are indeed unrelated or may arise when the variables are related in a nonlinear fashion. In this latter instance, the use of the Pearson r coefficient is inappropriate. There are other correlation coefficients that permit estimation of correlation according to nonlinear functions. In some cases, it may be possible to transform the data in order to obtain a linear relationship where the Pearson r coefficient can be used. For example, in a growing bacterial culture, there is a nonlinear relationship between viable cell number and time, but there is a linear relationship between log number of viable microorganisms and time. This is the reason that growth rate for bacterial cultures is normally plotted as the log number versus time.

The statistical significance of a calculated correlation coefficient can be determined with the t test.

Figure A1.4

Scatter diagrams showing varying degrees of correlation between two variables. The correlation coefficient r varies from −1 to +1. If $r = 1$, there is a perfect positive correlation; $r = -1$ is a perfect negative correlation; $r = 0$ if there is no correlation.

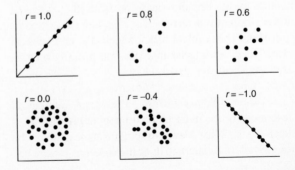

In this case, the interest is in deciding whether there is a true relationship between two variables. Therefore, one establishes a null hypothesis that the true correlation coefficient in the overall population equals 0, that is, that there is no correlation between the data variables measured within the population. One then calculates a t value. A critical α value is chosen as a criterion for determining whether or not to reject the null hypothesis. If the null hypothesis is rejected, the alternate hypothesis states that the correlation at the calculated r value between the two variables is significant.

In addition to determining whether there is a relationship between two variables, it is often desirable to define a mathematical relationship between the two variables that permits prediction of one variable if the value of the other is known. This is accomplished by performing a regression analysis. There is a distinction between correlation and regression analyses. In regression analysis, there should be a true independent variable, that is, a variable under the control of the experimenter. Regression analysis establishes a relationship that permits prediction of the dependent variable (predicted variable) for a given value of an independent variable (predictor variable). However, when there is no true independent variable, such as when a microbial ecologist measures numbers of phytoplankton and chlorophyll concentrations in a lake (neither variable is controlled and thus neither is a true independent variable), statisticians often will perform only correlation analyses and not regression analyses on the data. Biologists, though, sometimes apply regression analyses to field data when a relationship is known to exist.

In regression analysis, a relationship of best fit is used to describe the data. The experimenter must decide the type of relationship that the data fit—for example, a linear relationship. The data would then be fit to a regression line that best describes the data, assuming that a linear relationship exists. In other cases, where a particular nonlinear relationship is assumed, the data would be fit to a line of best fit described by the mathematical formula for that relationship. The regression line is constructed so as to minimize the variance of the data about the line.

The slope of the regression line is known as the regression coefficient. In constructing the regression line, it is necessary to define the slope of the line and the intercept of an axis. Although regression analysis minimizes the variance, a residual variance remains. As with correlation coefficients, it is possible to test the significance of the regression coefficient using the student t test. In this case, the null hypothesis states that there is no difference between the calculated regression coefficient and a true population regression coefficient of 0; that is, the population regression coefficient indicates that no prediction of y can be made from x, nor of x from y.

Certain procedures in microbial ecology are largely based on regression analysis. For example, the determination of heterotrophic potential uses regression analysis to determine the maximal rate of uptake of a substrate and the turnover time of a substrate. The commonly used determination of heterotrophic potentials assumes a linear relationship between the inverse of velocity of substrate uptake and the inverse of substrate concentration. The line of best fit to the collected data must be constructed in heterotrophic potential analyses in order to determine the intercepts and the slope that can be converted to V_{max}, turnover, and substrate affinity values. Similarly, biochemically oriented microbial ecologists utilize regression analyses when studying the kinetics of enzyme reactions.

We should give a word of caution about the interpretation of correlation and regression analyses. A positive correlation does not necessarily establish a cause-effect relationship. Two factors may show a significant correlation because they are both similarly affected by some other variable. For a cause-and-effect relationship, however, there should also be a significant correlation. For example, temperature affects bacterial activity; there is a direct cause–and-effect relationship between temperature and bacterial activity. One would, therefore, assume that there would be a significant correlation between the prevailing temperatures and rates of bacterial metabolism. Similarly, temperature affects protozoan metabolism, and one would expect a significant correlation between temperature and protozoan activity. If one were measuring rates of protozoan and bacterial metabolism without carefully controlling and measuring temperature, and one found a significant correlation between bacterial activity and protozoan activity, it would be wrong to conclude that bacterial activity affected protozoan activity, or vice versa. This is a rather clear-cut example in which one variable, temperature, caused the two measured variables, protozoan activity and bacterial activity, to exhibit a significant correlation. The existence of a hidden or ignored variable is less obvious in many cases, and the possibility of postulating the false cause-and-effect relationships based on correlation analyses is very real.

CLUSTER ANALYSIS

Cluster analysis is an extension of correlation analysis that has assumed an important role in microbial ecology. Cluster analysis methods permit grouping of variables according to the magnitudes and interrelationships of their correlation or similarity coefficients. The methods of cluster analysis form the basis of numerical taxonomic studies, which has added a great deal to our understanding of the distribution of microbial populations in natural habitats.

In order to perform a cluster analysis, it is first necessary to establish a correlation or similarity matrix that shows the relationship between individuals (Figure A1.5). In numerical taxonomy, the individuals are normally separate microbial strains; in other studies, the individuals might be considered as samples from various habitats or particular features of habitats or organisms. Many coefficients can be used to describe the relationship or correlation between individuals; the choice of a particular coefficient depends on the scale in which the data are recorded and the particular purpose for performing the cluster analysis. It is possible to use the Pearson product moment correlation coefficient in some cases, but frequently the scale of the data requires the use of different association coefficients.

Two of the association coefficients most commonly used by microbial ecologists in numerical taxonomic studies are the simple matching coefficient (Equation 10) and the Jaccard coefficient (Equation 11).

$$(10)\quad S_M = \frac{(aa) + (bb)}{(aa) + (ab) + (ba) + (bb)}$$

where

S_M = the simple matching coefficient

aa = the number of positive matches

bb = the number of negative matches

$ab + ba$ = the number of mismatches

$$(11)\quad S_J = \frac{(aa)}{(aa) + (ab) + (ba)}$$

where

S_J = the simple matching coefficient

aa = the number of positive matches

$ab + ba$ = the number of mismatches

These coefficients are suitable for use with binary data; the data for calculating either of these coefficients is generated by determining whether a microbial strain is positive or negative for a given feature. The similarity is then calculated according to how well the organisms match in their positive and negative features.

The difference between the simple matching and Jaccard coefficients rests in how negative matches (that is, both organisms lacking a particular feature) are considered; in a simple matching coefficient, negative matches are used in the calculation, whereas the Jaccard coefficient does not consider negative matches. The question of whether or not to utilize negative matches depends on the appropriateness of the particular test feature. The term appropriate refers to whether a positive score was possible for a given test on the organisms being examined. For example, the feature "weight greater than 1 ton" might well be appropriate for examining a population of elephants, but would hardly permit a positive response when examining a bacterial population.

Several sources of errors can be identified in calculating similarity coefficients. These include sampling errors and experimental or observational errors. In microbiological testing, a certain error rate is unavoid-

Figure A1.5

Flow diagram of procedures used in cluster analysis for numerical taxonomy.

General process	Example
Gather data and form data matrix	5 strains were tested for 5 features (aerobiosis, polar flagella, pigment production, Gram-negative and rod shape)

		Strain				
		1	2	3	4	5
T	1	+	+	+	+	+
e	2	–	–	–	–	–
s	3	+	+	–	+	+
	4	+	–	+	+	+
	5	–	–	+	–	–

General process	Example
Calculate similarity index and arrange similarity matrix	The similarities were calculated using the simple matching coefficient

		Strain				
		1	2	3	4	5
T	1	1.0				
e	2	0.0	1.0			
s	3	0.8	0.2	1.0		
	4	1.0	0.0	0.8	1.0	
	5	0.2	0.8	0.0	0.2	1.0

General process	Example
Perform clustering and form rearranged similarity matrix	Similarity matrix rearranged using single linkage

		Strain				
		1	4	3	5	2
S	1	1.0				
t	4	0.8	1.0			
r	3	1.0	0.8	1.0		
a	5	0.2	0.2	0.0	1.0	
	2	0.0	0.0	0.2	0.8	1.0

General process	Example
Present graphic cluster analysis	Results presented as cluster triangle showing that 2 clusters (80% similarity)

		Strain			
		1	4	3	5
S	4	+			
t	3	x	x		
r	5	0	0	0	
a	2	0	0	0	x

+ = 81–100% similarity
x = 60–80% similarity
0 = <60% similarity

able. Some common taxonomic tests, such as hydrogen sulfide production, are particularly variable and subject to error. A proportion p of the test results can be assumed to be erroneous. By employing repetitive testing and analyses of variance, it is possible to estimate p. The effect of these errors is to shift the observed similarity coefficient from the true similarity coefficient. The variance produced by experimental error can be described mathematically if p is known. The effect of experimental error becomes serious when p is greater than 0.1. In most microbiological studies, p has been observed to be less than or equal to 0.05, which does not introduce a serious error into the similarity measurement.

Having selected and calculated similarity or correlation coefficients, the coefficients are arranged in a matrix to show the relationships between all the possible combinations of pairs of individuals. In cluster analysis, the matrix of correlation or similarity coefficients then is sorted so that individuals with the greatest similarity are proximally located and those with low similarities are distantly located. This provides the basis for graphically showing the relative similarities among large numbers of organisms.

There are several algorithms and strategies employed in cluster analysis. The result of clustering is to produce one or more divisions of the correlation matrix that can be recognized as unit clusters of related individuals. Some cluster methods are hierarchical and retain the information of how each pair of individuals is related; other methods are nonhierarchical and do not retain the ranking as subsidiary clusters become more inclusive. Some clustering techniques are agglomerative. These techniques begin with all the separate entities and establish larger and larger groups until eventually there is a single set containing all the entities. In contrast, divisive techniques begin with one large set and subdivide the set into finer and finer divided subsets. Some cluster techniques are overlapping methods, in which inclusion in one group does not preclude inclusion in another; other methods are nonoverlapping, in which case groups are mutually exclusive.

Within the algorithm used for clustering procedures it is possible to weight certain types of relationships over others. Weighted relationships can be considered more important than others. Weighting can be achieved by altering the scale of distance between individuals.

Three major techniques of clustering are used in cluster analysis. These are the single linkage, complete linkage, and average linkage methods. These methods differ in how new clusters are formed. In the single linkage, or nearest neighbor, technique, an individual is added to a cluster based on its similarity to any individual within the cluster. The highest similarity between individuals determines the similarity relationship to the entire cluster. Clusters are related to each other by a single linkage of individuals of highest similarity. The complete linkage, or farthest neighbor, technique is the antithesis of the single linkage technique. An individual that is a candidate for admission to an existing cluster is considered to have a similarity to that cluster equal to the similarity of the least similar or farthest member within the cluster. Clusters join each other at a similarity existing between the least similar individuals within each cluster, that is, the farthest pairs. Single linkage clustering often leads to the formation of heterogeneous clusters, that is, clusters with a high variance. Complete linkage clustering leads to the formation of discrete clusters of low variance. Complete linkage tends to force clusters apart; single linkage tends to force clusters together.

Average linkage clustering techniques have been developed to avoid the extremes that occur in either single linkage or complete linkage clusters. In average linkage clustering, an individual is considered for entry into an existing group at its average similarity to all members in the group. Clusters are linked by the average similarity of all of the members within the respective clusters. Average linkage clustering can be based on the arithmetic average or on centroid clustering. In arithmetic average clustering, one computes the arithmetic average of the similarity coefficients between an individual being considered for admission into a cluster and the members of the existing cluster. In centroid clustering, one finds the centroid of the existing cluster and determines the dissimilarity of any individual candidate or other existing cluster from this point. The centroid or average organism can be determined by the mean of the members in the cluster. It is also possible to use a median, or hypothetical

Figure A1.6

Dendrogram representation with individual strains arranged along one axis and percent similarity shown on the other axis. Strains are linked together by lines that show the similarity level. The hierarchial structure of taxonomic groupings, which shows decreasing similarity from strain to species to genus, is seen in this dendrogram.

modal organism, which represents the commonest state as the centroid of a cluster.

Once the relationships between individuals and clusters have been established and sorted, the results of the cluster analysis can be graphically represented. The most common representations of cluster analyses used by microbiologists are the cluster triangle and the dendrogram (Figure A1.6); each graphically shows the percent similarity of individuals to each other within clusters and the percent similarity of clusters to each other.

Statistical methods in the techniques of cluster analysis can be extended to develop probabilistic identification matrices. The development of such matrices has greatly aided the identification of microorganisms in clinical situations. Microbial ecologists are similarly attempting to develop probabilistic identification matrices for microorganisms from various habitats. Probabilistic identification matrices contain information on the natural variance of selected features within a given population. The features of an unknown microorganism are compared with the probabilistic occurrence of features in the population of a known microorganism. A

similarity coefficient or likelihood score is generated by this comparison.

Unlike similarity coefficients generated when directly comparing individual strains, the likelihood score normally cannot have a maximal value of one. The best likelihood, that is, the highest similarity score possible, is the product of the probabilities entered for each of the features within the matrix for the known population. Normalization of likelihood scores following comparison of an unknown with multiple knowns within the probabilistic matrix allows for a ranking of the best matches as well as the calculation of a normalized identification score.

FACTOR ANALYSIS

Unlike cluster analysis, where there is no directly expressed understanding of why variables cluster together, the correlation or similarity matrix is analyzed in factor analysis to express covariation in terms of the underlying factors that explain a large part of the variance and covariance. The number of factors in a study is generally much lower than the number of variables. Factor analysis appears to be an extremely powerful statistical tool for use in microbial ecology since it allows for resolving complex relationships into the interaction of fewer and simpler factors. These factors may be physiological and environmental variables underlying the community structure in a particular ecosystem. The aim of factor analysis is often to identify the causal factors behind the ecological correlations.

Figure A1.7 shows the major steps in principal component analysis, a form of factor analysis. The first step is to create a correlation matrix from the basic data matrix. The principal components of the correlation matrix are then computed. This involves the computation of the characteristic roots of the characteristic equation that describes the matrix. A new matrix is generated in this procedure. The roots of such an equation are known as eigenvalues, which are a set of positive scalar quantities; there is an equal number of associated orthogonal vectors called eigenvectors. The eigenvectors themselves are known as

Figure A1.7

Flow diagram showing the major steps in performing a principal component analysis. (Source: Rosswall and Kvillner 1978. Reprinted by permission of Advances in Microbial Ecology, copyright Plenum Press.)

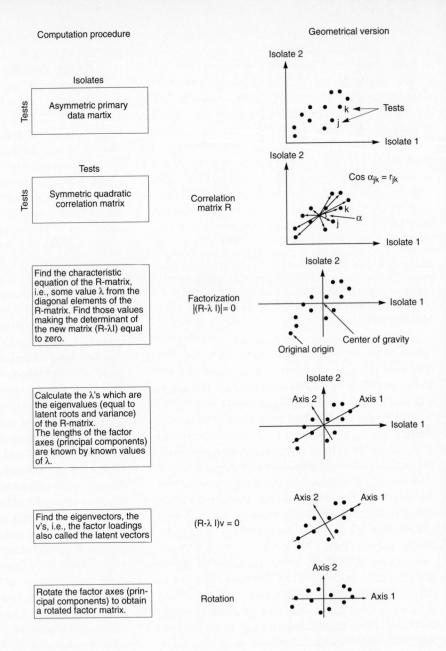

Computation procedure

Geometrical version

Isolates

Tests

Asymmetric primary data martix

Isolate 2

k ← Tests

j

Isolate 1

Tests

Tests

Symmetric quadratic correlation matrix

Correlation matrix R

Isolate 2

Cos $\alpha_{jk} = r_{jk}$

k

α

j

Isolate 1

Find the characteristic equation of the R-matrix, i.e., some value λ from the diagonal elements of the R-matrix. Find those values making the determinant of the new matrix (R-λI) equal to zero.

Factorization $|(R-\lambda \ I)| = 0$

Isolate 2

Isolate 1

Original origin

Center of gravity

Calculate the λ's which are the eigenvalues (equal to latent roots and variance) of the R-matrix. The lengths of the factor axes (principal components) are known by known values of λ.

Isolate 2

Axis 2   Axis 1

Isolate 1

Find the eigenvectors, the v's, i.e., the factor loadings also called the latent vectors

$(R-\lambda \ I)v = 0$

Axis 2   Axis 1

Rotate the factor axes (principal components) to obtain a rotated factor matrix.

Rotation

Axis 2

Axis 1

factor loadings. The sum of the squared factor loadings of all features is equal to the eigenvalue of each individual factor.

Principal component analysis results in the establishment of normalized vectors that give directions to a set of orthogonal axes known as the principal axes. The coordinates of the principal axes are linear combinations of the original variables and summarize the major dimensions of variation. In order to achieve a simplified structure of the factor matrix that is generated, the axes (factors) are rotated. In principal component analysis, the rotation is orthogonal. The result is the production of meaningful axes. The first principal axis corresponds to the largest eigenvalue and accounts for the greatest amount of variance within a sample. The second principal axis accounts for the second largest amount of variance, and so forth, for the other eigenvalues. Often only three principal axes will account for most of the variance. This can then be represented as a three-dimensional projection. Such projections can replace dendrograms to represent taxonomic relationships between microorganisms. Even more important in microbial ecology, ecological relationships among microorganisms, their activities, and their environment can be expressed in terms of the principal components.

Principal factor analysis can also be used to analyze taxonomic and ecological data. There are some differences between principal component analysis and principal factor analysis. In principal component analysis, the diagonals of the correlation matrix are equal to one, whereas in principal factor analysis, diagonals are reduced to percentages of variation due to common factors. In principal component analysis, the axes that are obtained are orthogonal, that is, they are uncorrelated, and are arranged in decreasing order of magnitude; in principal factor analysis, some features are assumed to be correlated with others. When only the two largest eigenvalues are being considered, such as occurs in most numerical taxonomic studies, the differences between principal component analysis and principal factor analysis are usually not significant. Another type of factor analysis is multiple factor analysis. In such analyses, when the factor axes are rotated, they can depart from an orthogonal relationship. This allows angular relationships to describe correlation.

In some cases, the factors can be assigned biological meaning. In other cases, the factors are complex and it is difficult to determine their precise biological or ecological meaning. Factor analyses have been applied to various studies in microbial ecology. These have included the natural variations in bacterial populations in soil and fresh water and the response of bacterial populations to environmental pollutants. Perhaps the real value of factor analysis is its ability to quantitate relationships that previously could be described only qualitatively. As the science of microbial ecology continues to develop, it has become increasingly necessary to quantitate ecological relationships.

COMPUTERS—THE PRAGMATIC APPROACH TO STATISTICS

Computers have provided microbial ecologists with the capacity to apply sophisticated statistical treatments to their data. Without computers, the time necessary to perform the required calculations for statistical treatments such as cluster and factor analyses would be overwhelming and preclude any practical use by the microbial ecologist. Even less sophisticated statistical methods, such as analysis of variance and t tests, require extensive mathematical calculations for large data sets and could not be used without devoting large amounts of time to this task. Microbial ecologists would have to become statisticians and know the nuances of the mathematical formulas used in the various statistical treatments. What would take days without computers takes only seconds with computers. As long as microbial ecologists understand the principles of the various statistical methods so that they select the right analytical methods and restrict their interpretive conclusions to the limitations of those methods, they can use computers for rapid and automated analysis of data.

It is also fortunate that the microbial ecologist does not have to become a computer programmer. Many statistical computer program packages are available for use, some of which run on microcomputers; even some programmable pocket calculators are supported by software for performing statistical analyses. Larger computers available at most academic institutions, government research facilities, and industrial concerns have statistical programs that can be used for analyzing microbial and ecological data. Common packages of statistical programs are SAS (Statistical Analysis System), SPSS (Statistical Package for the Social Sciences), BMD (Biomedical Computer Programs), JMP, and SuperANOVA, among others. These packages of statistical programs have the capability of performing normal statistical tests, including correlation analyses, regression analyses, analyses of variance, and, to a limited extent, cluster analysis and factor analysis. These packages also provide programs for forming tables and graphic displays of the data and the results of statistical analyses. Each package of programs is supported by a published manual that describes each of the statistical analytical methods

available and the format that the data must be in for use with the programs. An elementary knowledge of computer formatting, which can be obtained in a few hours, is all that is required for entering the data in the appropriate format for analysis. Additional specialized computer programs are available from commercial, academic, and government sources for performing more sophisticated cluster and factor analyses.

REFERENCES & SUGGESTED READINGS

Barr, A. J., J. H. Goodnight, J. P. Sall, and J. A. Helwig. 1976. *Statistical Analysis System.* SAS Institute, Raleigh, N.C.

Colwell, R. R. 1973. Genetic and phenetic classification of bacteria. *Advances in Applied Microbiology* 16:137–176.

Daniel, W. W. 1983. *Biostatistics: A Foundation for Analysis in the Health Sciences.* John Wiley and Sons, New York.

Daniel, W. W. 1990. *Applied Nonparametric Statistics.* PWS-Kent, Boston.

Dixon, W. J., and M. B. Brown (eds.). 1977. *Biomedical Computer Programs.* University of California Press, Berkeley.

Everitt, B. S., and G. Dunn. 1983. *Advanced Methods of Data Exploration and Modelling.* Gower, London.

Finn, J. D. 1974. *A General Model for Multivariate Analysis.* Holt, Rinehart and Winston, New York.

Gyllenberg, H. G. 1965. Character correlations in certain taxonomic and ecologic groups of bacteria: A study based on factor analysis. *Annals of Society of Experimental Biology of Finland* 43:82–90.

Kachigan, S. K. 1982. *Multivariate Statistical Analysis: A Conceptual Approach.* Radius Press, New York.

Koopmans, L. 1985. *An Introduction to Contemporary Statistics.* PWS-Kent, Boston.

Kotz, S., N. L. Johnson, and C. B. Read (eds.). 1985. *Encyclopedia of Statistical Sciences.* Wiley, New York.

Kuzma, J. W. 1992. *Basic Statistics for the Health Sciences.* Mayfield Publishing Company, Palo Alto, Calif.

Milton, J. S. 1992. *Statistical Methods in the Biological and Health Sciences.* McGraw-Hill, New York.

Myers, R. 1990. *Classical and Modern Regression with Applications.* PWS-Kent, Boston.

Nie, N. H., C. H. Hull, J. G. Jenkins, K. Steinbrenner, and D. H. Bent. 1975. *Statistical Package for the Social Sciences.* McGraw-Hill, New York.

Ott, L. 1984. *An Introduction to Statistical Methods and Data Analysis.* Duxbury Press, Boston.

Philips, D. S. 1978. *Basic Statistics for Health Science Students.* W. H. Freeman, San Francisco.

Quadling, C., and J. W. Hopkins. 1967. Evaluation of tests and grouping of cultures by a two-stage principal component method. *Canadian Journal of Microbiology* 13:1379–1400.

Remington, R. D., and M. A. Schork. 1985. *Statistics with Applications to the Biological and Health Sciences.* Prentice Hall, Englewood Cliffs, N. J.

Robinson, J. A. 1985. Determining microbial kinetic parameters using nonlinear regression analysis. *Advances in Microbial Ecology* 8:61–114.

Rosner, B. 1986. *Fundamentals of Biostatistics.* PWS Publishers, Boston.

Rosswall, T., and E. Kvillner. 1978. Principal-components and factor analysis for the description of microbial populations. *Advances in Microbial Ecology* 2:1–48.

Runyon, R. P., and A. Haber. 1967. *Fundamentals of Behavioral Statistics.* Addison-Wesley, Reading, Mass.

Schefler, W. C. 1979. *Statistics for the Biological Sciences.* Addison-Wesley, Reading, Mass.

Shott, S. 1990. *Statistics for Health Professionals.* W. B. Saunders, Philadelphia.

Sneath, P. H. A., and R. R. Sokal. 1973. *Numerical Taxonomy: The Principles and Practice of Numerical Classification.* W. H. Freeman, San Francisco.

Sokal, R. R., and F. J. Rohlf. 1973. *Introduction to Bio-statistics*. W. H. Freeman, San Francisco.

Sokal, R. R., and F. J. Rohlf. 1981. *Biometry*. W. H. Free-man, San Francisco.

Soumare, S., J. Losfeld, and R. Blondeau. 1973. Apports de la taxonomie numerique a l'etude de spectre bacteria de la microflore des sols du nord de la France. *Annals of the Microbiology Institute Pasteur* 124B:81–94.

Steel, R. G. D., and J. H. Torrie. 1980. *Principles and Procedure of Statistics: A Biometrical Approach*. McGraw-Hill Book Co., New York.

Sundman, V. 1970. Four bacterial soil populations char-acterized and compared by a factor analytical method. *Canadian Journal of Microbiology* 16:455–464.

Sundman, V. 1973. Description and comparison of micro-bial populations on ecological studies with the aid of fac-tor analysis. *Bulletin of Ecological Resources Commission* (Stockholm) 17:135–141.

Sundman, V., and H. G. Gyllenberg. 1967. Application of factor analysis in microbiology. Part I. General aspects on the use of factor analysis in microbiology. *Annals of the Academy of Sciences of Finland* Ser. A IV 112:1–32.

Winer, B. J. 1971. *Statistical Principles in Experimental Design*. McGraw-Hill Book Co., New York.

Appendix 2: Survey of Microorganisms

ACELLULAR MICROORGANISMS

Prions

Prions are specific infectious protein molecules that contain the information that codes for their own replication (Prusiner 1982, 1984). They seem to be an exception to what appeared to be a universal characteristic of living systems—genetic information in nucleic acid molecules. All other organisms store their genetic information in nucleic acids, in DNA (deoxyribonucleic acid) or, less commonly, in RNA (ribonucleic acid). It is not yet clear how a protein can direct its own replication, and thus we do not understand how prions replicate.

Viroids

Whereas prions are protein molecules, viroids are composed exclusively of RNA (Diener 1981). Viroids, like prions, are simply molecules that contain the information for their own reproduction; they have no other structures, and their RNA genomes are quite small. Inside a host cell, a viroid is capable of initiating its own replication. The replication of viroids sometimes manifests itself as disease symptoms in the host organism. In the early 1970s, viroids were shown to cause potato spindle tuber disease, chrysanthemum stunt, and citrus exocortis.

Viruses

Viruses are often classified into large groups according to the type of host cell they infect (Wildy 1971; Fenner 1976; Luria et al. 1978; Fraenkel-Conrat and Kimball 1982). Formal classification systems are largely based on the nature of the nucleic acid molecule and on the arrangement of the capsid. Viruses can be separated into groups on the basis of the type and form of the nucleic acid genome and of the size, shape, structure, and mode of replication of the virus particle. Critical features used in classifying viruses include the host cell in which the virus reproduces, the

Table A2.1

Descriptions of the animal viruses

Family name	Nucleic acid type	Nucleic acid strands	Nucleic acid weight (millions)	Capsid symmetry	Virion envelopment	Site of envelopment
Parvoviridae	DNA	Single	1.4	Cubic	Naked	—
Papovaviridae	DNA	Double	3–5	Cubic	Naked	—
Adenoviridae	DNA	Double	23	Cubic	Naked	—
Iridoviridae	DNA	Double	130	Cubic	Enveloped	Cytoplasm
Herpesviridae	DNA	Double	54–92	Cubic	Enveloped	Nucleus
Poxviridae	DNA	Double	160	Complex	Complex	Cytoplasm
Picornaviridae	RNA	Single	2.5–2.8	Cubic	Naked	—
Calciviridae	RNA	Single	2	Cubic	Naked	—
Reoviridae	RNA	Double	12–15	Cubic	Naked	—
Togaviridae	RNA	Single	3–4	Cubic	Enveloped	Cytoplasm
Orthomyxoviridae	RNA	Single	2–4	Helical	Enveloped	Cytoplasm
Paramyxoviridae	RNA	Single	4–8	Helical	Enveloped	Cytoplasm
Rhabdoviridae	RNA	Single	3–4	Helical	Enveloped	Cytoplasm
Retroviridae	RNA	Single	10–12	Helical	Enveloped	Cytoplasm
Bunyaviridae	RNA	Single	6	Helical	Enveloped	Organelle
Arenaviridae	RNA	Single	3.2	—	Enveloped	Cytoplasm
Coronaviridae	RNA	Single	—	—	Enveloped	Organelle

nature of the genetic material (DNA or RNA—single or double stranded), the molecular weight of the nucleic acid, the presence or absence of an envelope, the symmetry of the capsid (protein coat surrounding nucleic acid), the number of capsomeres (subunits of capsid), and the site of capsid assembly (Table A2.1).

Animal Viruses The animal viruses, which encompass a wide variety of morphological types, traditionally refer to those viruses that infect only vertebrate animals. Viruses that infect invertebrate animals and microorganisms other than bacteria are usually classified separately (Dalton and Haguenau 1973; Fenner et al. 1974; Maramorosch 1977; Fraenkel-Conrat and Wagner 1978; Palmer and Martin 1982). The DNA animal viruses are divided into six major families and the RNA animal viruses into eleven

major families. Most of the animal viruses are recognized also by the names of the diseases they cause—for example, rabies virus, yellow fever virus, and so forth. Infections of animal populations by specific viruses are ecologically important in controlling population levels.

Plant Viruses The formal systematics of plant viruses has attempted to retain information concerning the infected host cells, so plant viruses are grouped primarily according to the type of disease they cause (Fraenkel-Conrat and Wagner 1977; Maramorosch 1977). The names of the viral groups themselves are key to the types of diseases caused by the viruses in that group, including the nature of the host plant and the symptoms of the diseases. Table A2.2 lists some properties of the major plant virus groups.

Table A2.1

Descriptions of the animal viruses; Continued

Family name	Reaction to solvents	Number of capsomeres	Diameter of helix (nm)	Diameter of virion (nm)	pH 3 stability	Site of capsid assembly
Parvoviridae	Resistant	32	—	18–22	—	Nucleus
Papovaviridae	Resistant	72	—	45–55	—	Nucleus
Adenoviridae	Resistant	252	—	70–90	—	Nucleus
Iridoviridae	Sensitive	1500	—	130	—	Cytoplasm
Herpesviridae	Sensitive	162	—	100	—	Nucleus
Poxviridae	Resistant	—	—	230–300	—	Cytoplasm
Picornaviridae	Resistant	32	—	20–30	Resistant	Cytoplasm
Calciviridae	Resistant	32	—	40	Sensitive	Cytoplasm
Reoviridae	Resistant	32 or 92	—	60–80	Variable	Cytoplasm
Togaviridae	Sensitive	32 or 42	—	40–60	Sensitive	Cytoplasm
Orthomyxoviridae	Sensitive	—	6–9	80–120	Sensitive	Cytoplasm
Paramyxoviridae	Sensitive	—	18	100–300	Sensitive	Cytoplasm
Rhabdoviridae	Sensitive	—	2–6	60–220	Sensitive	Cytoplasm
Retroviridae	Sensitive	—	16	100	Sensitive	Cytoplasm
Bunyaviridae	Sensitive	—	2–10	100	Sensitive	Cytoplasm
Arenaviridae	Sensitive	—	—	50–300	Sensitive	Cytoplasm
Coronaviridae	Sensitive	—	—	70–120	Sensitive	Cytoplasm

Viruses of Microorganisms The viruses that multiply within the host cells of microorganisms are divided initially according to the group of microorganisms within whose cells they can reproduce, into bacteriophage (viruses that infect bacteria), cyanophage (viruses that infect cyanobacteria), mycoviruses (viruses that infect fungi), phycoviruses (viruses that infect algae), and viruses of protozoa (Dalton and Haguenau 1973; Lemke and Nash 1974; Hollings 1978; Luria et al. 1978). The viral groups are largely defined and separated based on molecular considerations. The viruses of microorganisms exhibit great host-cell specificity that may in part be due to the nature of the receptor sites that are required for viral adsorption. Bacteriophage, therefore, are often named according to the species of bacteria they infect; for example, coliphage are viruses that infect *E. coli*.

PROKARYOTES

Extremely diverse groups of organisms exhibiting widely differing morphological, ecological, and physiological properties are included among the bacteria (Skerman 1967; Buchanan and Gibbons 1974; Laskin and Lechevalier 1977; Starr et al. 1981; Holt and Krieg 1984). The unifying feature of the bacteria is the fact that they all are prokaryotic cells.

Archaebacteria (Archaea)

The archaebacteria, or Archaea, as they have recently been renamed, have been shown to be phylogenetically related; they have been separated from other prokaryotes based on analysis of their 16S ribosomal

Table A2.2

Description of the major groups of plant viruses

Virus	Description
Brome mosaic virus	Small, icosahedral RNA virus
Cauliflower mosaic virus	Double-stranded DNA, reproduce in cytoplasm
Cucumber mosaic virus	Naked, icosahedral RNA viruses
Barley yellow dwarf virus	Small, isometric virus, RNA genome
Potato virus X	Flexuous rods, 480–580 nm, RNA genome
Potato virus Y	Flexuous, rod-shaped, helical symmetry, single-stranded RNA
Tobacco mosaic virus	Rod-shaped, helical symmetry, single-stranded RNA
Tobacco necrosis virus	Isometric RNA virus
Tobacco rattle virus	Rod-shaped, nematode-transmitted, positive-stranded RNA virus, segmented genome
Tobacco ringspot virus	Polyhedral, nematode-transmitted RNA virus
Tomato bushy stunt virus	Small RNA virus, cubic symmetry, resistant to elevated temperatures and organic solvents
Turnip yellow mosaic virus	Icosahedral virus, RNA genome, transmitted by flea beetles
Watermelon mosaic virus	Flexuous rods, 700–950 nm, RNA genome

RNA molecules and also have been shown to have several morphological and physiological features that make them distinct from other bacteria, including the lack of peptidoglycan in their cell walls and the unusual ether linkage that occurs in their membrane lipid molecules (Woese 1981; Woese et al. 1990). The archaebacteria include three distinct groups of prokaryotes that evolved along different lineages: the extreme halophiles, which grow at salt concentrations of 8%–32%; the strict anaerobes, which produce methane; and the sulfur-dependent archaebacteria, which are also called the thermoacidophiles because they grow at elevated temperatures and low pH (Table A2.3).

The methane-producing, or methanogenic, bacteria are a highly specialized physiological group of archaebacteria able to form methane by the reduction of carbon dioxide (Zeikus 1977; Balch et al. 1979; Mah and Smith 1981). The methanogenic bacteria are very strict obligate anaerobes; they utilize electrons generated in the oxidation of hydrogen or simple organic compounds, such as acetate and methanol, to produce methane and are unable to use carbohydrates, proteins, or other complex organic substrates. Based upon genetic analyses of the mole% G + C, the individual species of methanogens appear to be as greatly separated as the larger groups within the eubacteria. The methanogens often form consortia in association with other microorganisms. The microorganisms associated with the methanogens maintain the low oxygen tensions and provide the carbon dioxide and fatty acids required by the methanogenic bacteria. Such associations are extremely important in the rumen of animals, such as cows.

Genera in the family Halobacteriaceae have the characteristic properties of the archaebacteria but are aerobic (Larsen 1981). *Halobacterium,* a genus of this group, has an unusual bacteriorhodopsin-mediated phototrophic metabolism. All members of this family are obligate halophiles, growing only in media containing at least 15% sodium chloride (NaCl). Members of this family are found in ecosystems that have

Table A2.3

Description of the major genera of Archaea

Genus	Properties
	Extreme halophiles
Halobacterium	Extremely halophilic; some metabolize carbohydrates; aerobic and anaerobic respiration; can use S° as electron acceptor; photoheterotrophic; Gram-negative rods
Halococcus	Extremely halophilic; Gram-negative cocci
Natronobacterium	Extremely halophilic; extremely alkalinophilic; optimal growth at low Mg^{2+}; Gram-negative rods
Natronococcus	Extremely halophilic; extremely alkalinophilic; Gram-negative cocci
	Methanogens
Methanobacterium	Methane from formate and $H_2 + CO_2$; Gram-variable rods
Methanobrevibacter	Methane from formate and $H_2 + CO_2$; Gram-positive rods
Methanothermus	Methane from $H_2 + CO_2$; Gram-positive rods
Methanococcus	Methane from formate and $H_2 + CO_2$; Gram-negative cocci; protein wall polymer
Methanomicrobium	Methane from formate and $H_2 + CO_2$; Gram-negative rods; protein wall polymer
Methanogenium	Methane from formate and $H_2 + CO_2$; Gram-negative cocci; glycoprotein wall polymer
Methanospirillum	Methane from formate and $H_2 + CO_2$; Gram-negative spirals
Methanosarcina	Methane from formate, methylamines, methanol, acetate, and $H_2 + CO_2$; Gram-positive cocci in tetrads
Methanococcoides	Methane from methanol and methylamines; Gram-negative cocci
Methanothrix	Methane from acetate; Gram-negative rods–filaments
Arcahaeoglobus	Organic compound + SO_4^{2-} yields $H_2S + CO_2 + CH_4$; $H_2 + SO_4^{2-}$ yields $H_2S + H_2O$
	Sulfur-dependent
Sulfolobus	Organic compound + O_2 yields $H_2S + CO_2$; $FeS_2 + O_2 + H_2O$ yields $Fe_2(SO_4)_3 + H_2SO_4$; $S + O_2 + H_2O$ yields H_2SO_4; lobed cocci; optimum growth 77°C, pH 3
Acidianus	$H_2 + S°$ yields H_2S; $S° + O_2 + H_2O$ yields H_2SO_4; cocci; optimum growth 87°C, pH 2
Thermoproteus	Organic compound + S° yields $H_2S + CO_2$; $H_2 + S°$ yields H_2S; rods; optimum growth 88°C, pH 6
Thermofilum	Organic compound + S°yields $H_2S + CO_2$; rods; optimum growth 88°C, pH 5.5
Desulfurococcus	Organic compound + S° yields $H_2S + CO_2$; cocci; optimum growth 92°C, pH 6
Pyrodictium	$H_2 + S°$ yields H_2S; discs-filaments; optimum temperature 105°C; optimum pH 6
Pyrococcus	Organic compound + S° yields $H_2S + CO_2$; cocci; optimum growth 100°C, pH 7
Staphylothermus	Organic compound + S° yields $H_2S + CO_2$; cocci in clusters; optimum growth 92°C, pH 6.5
Thermococcus	Organic compound + S° yields $H_2S + CO_2$; cocci; optimum growth 88°C, pH 6
Thermodiscus	Organic compound + S°yields $H_2S + CO_2$; discs; optimum growth 90°C, pH 5.5

extremely high NaCl concentrations, such as some salt lakes, the Dead Sea, and foods preserved by salting.

The sulfur-oxidizing genus *Sulfolobus* share other common properties with the archaebacteria (Brock 1981). Members of this genus are thermophiles, with an optimum growth temperature of 70°C–75°C. *Sulfolobus* species occur in hot, acidic environments. Association with unusual habitats appears to be another common characteristic of many archaebacteria.

Eubacteria (Bacteria)

Phototrophic Bacteria   The phototrophic bacteria are distinguished from other bacterial groups by their ability to use light energy to drive the synthesis of ATP (Stanier et al. 1981) (Table A2.4). Most of the organisms included in this group are autotrophs capable of using carbon dioxide as the source of cellular carbon; some are photoheterotrophs, using light to generate ATP but requiring organic compounds for cellular synthesis. Some of the phototrophic bacteria (Oxyphotobacteria) use water as an electron donor and liberate oxygen. The remainder of the photobacteria do not evolve oxygen and, with one exception, can be classified as belonging to Anoxyphotobacteria. The

exception, *Halobacterium*, belongs to archaebacteria and has a unique mode of phototrophic metabolism.

The Cyanobacteriales and Prochlorales are Oxyphotobacteria, occupying intermediary positions between the other phototrophic bacteria and the eukaryotic algae (Stanier and Cohen-Bazire 1977; Lewin 1981; Stanier et al. 1981). The primary photosynthetic pigment in both cases is chlorophyll a, but the prochlorophytes also possess chlorophyll b, making them similar to the green algae. Presumably, the prochlorobacteria are more closely related to the green algae than to the cyanobacteria. Some cyanobacteria, on the other hand, are capable of using H_2S as an electron donor for anoxygenic photosynthesis (Padau 1979), closely relating them to the Anoxyphotobacteria. Clearly, there is a phylogenetic relationship among the photosynthetic organisms, with the Oxyphotobacteria occupying an intermediate position between the Anoxyphotobacteria and the algae.

The cyanobacteria, or blue-green bacteria, are the most diverse and widely distributed photosynthetic bacteria, with more than a thousand species of cyanobacteria having been reported (Stanier and Cohen-Bazire 1977; Stanier et al. 1981). The primary photosynthetic pigment of the cyanobacteria is chlorophyll a; the outer surfaces of the photosynthetic

Table A2.4

Characteristics of the major groups of phototrophic bacteria

Taxonomic groups	Metabolism	Photosynthetic pigments	Electron donors	Carbon source
Cyanobacteria	Oxygenic photosynthesis	Chlorophyll a, phycobiliproteins	H_2O, (H_2S)*	CO_2
Green bacteria	Anoxygenic photosynthesis	Bacteriochlorophyll a or b, carotenoids	H_2, H_2S, S	CO_2
Halobacterium	Anoxygenic purple membrane mediated	Bacteriorhodopsin	—	Organic C
Prochlorobacteria	Oxygenic photosynthesis	Chlorophyll a + b, β carotenes	H_2O	CO_2
Purple bacteria	Anoxygenic photosynthesis	Bacteriochlorophyll a or b, carotenoids	H_2, H_2S, S	Organic C or CO_2

* Under some conditions, photosynthesis is anoxygenic and H_2S serves as the electron donor.

Table A2.5

The subgroups of the cyanobacteria

Group	Description
Chroococcaean	Unicellular rods or cocci; reproduce by binary fission or budding
Pleurocapsalean	Single cells enclosed in a fibrous layer; reproduce by multiple fission, producing baeocytes
Oscillatorian	Cells form trichomes but do not form heterocysts
Heterocystous	Cells form trichomes with both vegetative cells and heterocysts

membranes have associated granules known as phycobilisomes, which are composed of auxiliary photosynthetic pigments.

There are four major subgroups of cyanobacteria (Table A2.5). The chroococcacean cyanobacteria are unicellular rods or cocci. They reproduce either by binary fission (family Chroococcaceae) or by budding (family Chamesiphonaceae). *Synechococcus, Synechocystis,* and *Chamaesiphon* are representative genera of chroococcacean cyanobacteria.

The pleurocapsalean cyanobacteria are distinguished from the chroococcacean cyanobacteria by the fact that they exhibit multiple fission to produce small coccoid reproductive cells. In the phycological literature, these reproductive cells are referred to as endospores, but to avoid confusion with endospore-forming bacteria, it has been proposed that the term baeocyte be used to describe the reproductive cells of the pleurocapsalean cyanobacteria. Because binary fission does not result in complete separation of the cells, the pleurocapsalean cyanobacteria form multicellular aggregates.

The oscillatorian cyanobacteria form filamentous structures exclusively composed of vegetative cells, known as trichomes. *Spirulina, Oscillatoria,* and *Pseudanabaena* are representative genera of oscillatorian cyanobacteria.

The heterocystous cyanobacteria form differentiated cells known as heterocysts when growing in the absence of fixed forms of nitrogen. Heterocysts are nonreproductive cells that are distinguished from the adjoining vegetative cells by the presence of refractory polar granules and a thick outer wall. The ability to form heterocysts is associated with the physiological capability of fixing atmospheric nitrogen, and the physiologically specialized heterocyst cells appear to be the anatomical sites of nitrogen fixation in heterocystous cyanobacteria. Being able to carry out both oxygen-yielding photosynthesis and nitrogen fixation is a unique characteristic of cyanobacteria principally found among the heterocystous cyanobacteria. The heterocystous cyanobacteria are ecologically important because they can form both organic carbon and fixed forms of nitrogen that can support the nutritional requirements of other organisms. *Nostoc* and *Anabaena* are probably the best-known genera of heterocystous cyanobacteria.

The prochlorales are similar to the cyanobacteria except that they lack phycobilin pigments and also synthesize chlorophyll b (Lewin 1981). The only known genus, *Prochloron,* occurs as single-celled extracellular symbionts of marine invertebrates, appearing bright green on the surfaces of the animals with which they are associated. Various species of *Prochloron* have been recognized in field studies, but until the organisms are grown in pure culture, the validity of these species remains ambiguous.

The Anoxyphotobacteria require an electron donor other than water; they do not evolve oxygen, and they carry out photosynthesis anaerobically. The anaerobic photosynthetic bacteria typically occur in aquatic habitats, often growing at the sediment-water interface of shallow lakes where there is sufficient light penetration to permit photosynthetic activity, where anaerobic conditions are sufficient to permit the existence of these organisms, and where there is a source of reduced sulfur or organic compounds to act as electron donors for the generation of reduced coenzymes. The phototrophic bacteria include the Rhodospirillaceae (purple nonsulfur bacteria), Chromatiaceae (purple sulfur bacteria), Chlorobiaceae (green sulfur bacteria), and Cloroflexaceae (green flexibacteria) (Pfennig 1977; Trüper and Pfennig 1981) (Table A2.6). The green and purple sulfur bacteria utilize

Table A2.6

Descriptions of the anoxygenic phototrophic bacteria

Family	Description
Rhodospirillaceae (purple nonsulfur bacteria)	Cells photoassimilate simple organic substrates; most species unable to grow with sulfide as the sole electron donor; cells contain bacteriochlorophyll a or b; representative genera are *Rhodospirillum, Rhodopseudomonas,* and *Rhodomicrobium*
Chromatiaceae (purple sulfur bacteria)	Cells able to grow with sulfide and sulfur as the sole electron donor; sulfur deposited inside or outside of cell; cells contain bacteriochlorophyll a or b; representative genera are *Chromatium, Thiocystis, Thiosarcina, Thiospirillum, Thiocapsa, Lamprocystis, Thiodictyon, Thiopedia, Amoebobacter,* and *Ectothiorhodospira*
Chlorobiaceae (green sulfur bacteria)	Cells able to grow with sulfide and sulfur as the sole electron donor; sulfur deposited only outside of cell; cells contain bacteriochlorophyll a, b, or c; representative genera are *Chlorobium, Prosthecochloris, Chloropseudomonas, Pelodictyon,* and *Clathrochloris*
Chloroflexaceae (green flexibacteria)	Cells have flexible walls, gliding motility, form filaments, utilize organic C sources; cells contain bacteriochlorophyll a, b, or c; representative genera are *Chloroflexus, Chloronema,* and *Oscillochloris*

reduced sulfur compounds, such as hydrogen sulfide, as electron donors for generating reducing power. Most of the purple nonsulfur bacteria and green flexibacteria are unable to use reduced sulfur compounds, but rather these organisms utilize organic compounds to support photosynthetic growth.

The purple sulfur bacteria produce carotenoid pigments and may appear orange-brown, red-brown, purple-red, or purple-violet. They are potentially mixotrophic (capable of both photoautotrophic and heterotrophic growth), and all strains are capable of photoassimilating simple organic substrates such as acetate (Pfennig and Trüper 1981). Some genera, such as *Thiodictyon* and *Thiopedia,* contain gas vacuoles that permit an adjustment of cell buoyancy in a water column to a depth that is appropriate for light penetration and oxygen concentration, making anaerobic photosynthetic metabolism possible.

The green sulfur bacteria produce green or green-brown carotenoid pigments; they assimilate carbon dioxide, utilizing sulfide or elemental sulfur as electron donors, and they deposit sulfur granules

extracellularly (Pfennig and Trüper 1981). Some genera of Chlorobiaceae, such as *Pelodictyon,* produce gas vacuoles, but others, such as *Chlorobium* and *Chloropseudomonas,* do not contain gas vacuoles. These bacteria often occur in similar ecological situations as the purple sulfur bacteria.

The purple nonsulfur bacteria generally produce red-purple carotenoid pigments. Their photosynthetic development depends on the ability of the cells to photoassimilate simple organic compounds (Biebl and Pfennig 1981). Because they generally require preformed organic matter for growth and are able to utilize light energy for generating ATP, the type of metabolism carried out by these organisms is sometimes referred to as photoheterotrophic or photoorganotrophic metabolism. The basic metabolic pathways of the Rhodospirillaceae are the same as those of other autotrophic microorganisms, but their ability to assimilate organic compounds and the requirement of many members of the Rhodospirillaceae for organic compounds establish a resemblance of these organisms to heterotrophs.

The Chloroflexaceae, a relatively newly discovered family of anaerobic phototrophic bacteria, have flexible walls, form filaments, and exhibit gliding motility (Castenholz and Pierson 1981). Gliding motility and formation of filaments were previously thought to be restricted among phototrophic bacteria to the cyanobacteria. Photosynthesis is anoxygenic, and some organic compounds are needed to achieve optimal growth. *Chloroflexus* resembles a green sulfur bacterium in cell ultrastructure and photosynthetic pigments but resembles a nonsulfur purple bacterium in its photosynthetic and catabolic metabolism. *Choroflexus aurantiacus* has been isolated from alkaline hot springs in various parts of the world.

Gliding Bacteria   The Myxobacterales (fruiting myxobacteria) and the Cytophagales are grouped together based on their gliding motility on solid surfaces (Reichenbach and Dworkin 1981).

The myxobacteria are small rods normally embedded in a slime layer. They lack flagella but are capable of gliding movement. A unique feature of the myxobacteria is that under appropriate conditions they aggregate to form fruiting bodies. Frequently, the fruiting bodies of myxobacteria occur on decaying plant material, on the bark of living trees, or on animal dung, appearing as highly colored slimy growths that may extend above the surface of the substrate. Most of the myxobacteria produce a variety of hydrolytic enzymes, such as cellulases, and many are capable of lysing other microorganisms.

In contrast to the Myxobacterales, the Cytophagales do not produce fruiting bodies; they are unified only by the presence of gliding motion and lack of fruiting-body formation. Some of the Cytophagales are chemolithotrophs. For example, *Beggiatoa* forms filaments, oxidizes hydrogen sulfide, and deposits sulfur intracellularly when growing on hydrogen sulfide. Cells of *Cytophaga* contain deep yellow-orange or red pigments and hydrolyze agar, cellulose, and chitin. As a consequence of their hydrolytic activities, these gliding bacteria play an important ecological role in the decomposition of organic matter.

Sheathed Bacteria   The sheathed bacteria comprise those bacteria whose cells occur within a filamentous structure known as a sheath (van Veen et al. 1978; Mulder and Deinema 1981). The formation of a sheath enables these bacteria to attach themselves to solid surfaces. This is important to the ecology of these bacteria because many sheathed bacteria live in low-nutrient aquatic habitats. By absorbing nutrients from the water that flows by the attached cells, these bacteria are able to conserve their limited energy resources. Additionally, the sheaths afford protection against predators and parasites. In some cases, the sheaths may be covered with metal oxides. For example, in the genus *Leptothrix,* sheaths are encrusted with iron or manganese oxides (van Veen et al. 1978). In the genus *Sphaerotilus,* the sheath is sometimes encrusted with iron oxides; *Sphaerotilus natans,* often referred to as "sewage fungus," occurs in polluted flowing waters, such as sewage effluents, where it may be present in high concentrations just below sewage outfalls.

Budding and/or Appendaged Bacteria   The budding and/or appendaged bacteria have in common the formation of extensions or protrusions from the cell (Poindexter 1981; Staley et al. 1981). The cell appendages of the bacteria in this group, known as prosthecae, afford the cell greater efficiency in concentrating available nutrients. Many of the appendaged bacteria grow well at low nutrient concentrations. Many of the bacteria in this group primarily occur in aquatic habitats where concentrations of organic matter typically are low. *Caulobacter,* for example, is able to grow in dilute concentrations of organic matter in lakes and is even able to grow in distilled water. The appendages of *Caulobacter* are referred to as stalks. In some cases, the stalks of individual cells provide a holdfast by which the organisms can attach to a substrate.

Spirochetes   Spirochetes are helically coiled rods, with the cell wound around one or more central axial fibrils (Johnson 1976, 1981; Holt 1978). In addition to their characteristic morphology, the spirochetes exhibit a unique mode of motility; these bacteria move by a flexing motion of the cell, exhibiting greatest velocities in viscous solutions, where motility by bacteria with external flagella is slowest.

Spiral and Curved Bacteria Members of the spiral and curved bacteria group are helically curved rods that may have less than one complete turn (comma shaped) or many turns (helical), but unlike the spirochetes, the cells are not wound around a central axial filament (Krieg 1976; Raj 1981; Stolp 1981). *Bdellovibrio,* a genus of uncertain affiliation within this group, has the outstanding characteristic of being able to penetrate and reproduce within the periplasmic space of prokaryotic cells. All naturally occurring strains of *Bdellovibrio* have been found to be bacterial parasites. In the periplasmic space, *Bdellovibrio* grows into a sausage-shaped body that subdivides into usually less than ten polarly flagellated cells. Subsequently, the host cell lyses, releasing the *Bdellovibrio* progeny.

Gram-Negative Aerobic Rods and Cocci Several major families of bacteria are grouped together as Gram-negative aerobic rods and cocci (Table A2.7). The metabolism of members of the family Pseudomonadaceae is respiratory, but some strains are able to carry out anaerobic respiration (Palleroni 1981). The Pseudomonadaceae are unable to fix atmospheric nitrogen. Many *Pseudomonas* species are nutritionally versatile and are capable of degrading many natural and synthetic organic compounds. *Pseudomonas* species are widely distributed in soil and aquatic ecosystems, occurring as free-living bacteria or in association with plants and animals.

The family Azotobacteraceae is characterized by its capacity to fix molecular nitrogen (Becking 1981; Gordon 1981). The genera *Azotobacter* and *Bei-*

jerinckia are particularly important free-living nitrogen-fixing bacteria. The Rhizobiaceae are also capable of fixing atmospheric nitrogen. *Rhizobium* and *Bradyrhizobium* can fix atmospheric nitrogen in symbiotic association with leguminous plants living within root nodules formed in response to the invasion of plant roots by these bacteria (Vincent 1981). *Agrobacterium* species, which do not fix nitrogen, produce tumorous growths on infected plants, known as galls (Lippincott et al. 1981). *Agrobacterium tumefaciens* causes galls of many different plants and is a plant pathogen of economic significance in agriculture.

The family Methylomonadaceae includes bacteria that can utilize carbon monoxide, methane, or methanol as a sole source of carbon (Whittenbury and Dalton 1981). The ability to use one-carbon-containing (C_1) organic compound as the sole source of carbon and energy requires a special metabolic capability. These bacteria are ecologically important because they maintain the availability of carbon that otherwise would be lost from biological cycling.

Several genera of uncertain affiliation are Gram-negative aerobic rods. These include the genera *Brucella, Bordetella, Francisella, Alcaligenes, Acetobacter,* and *Thermus. Thermus* is an ecologically interesting genus that grows well at temperatures over 70°C. Strains of this organism have been isolated from hot springs and the hot-water tanks of homes and laundromats (Brock 1981).

Gram-negative Facultatively Anaerobic Rods There are two major families of Gram-negative facul-

Table A2.7

Some characteristics of Gram-negative aerobic rods and cocci

Taxonomic group	Flagella	Carbon source	Denitrification	N_2 fixation
Pseudomonadaceae	Polar	Diverse	Some species	No
Azotobacteraceae	Peritrichous or polar	Diverse	No	Yes
Rhizobiaceae	Peritrichous or polar	Diverse	No	Yes
Methylomonadaceae	Polar or none	One-carbon compounds only	No	No

tatively anaerobic rods: the Enterobacteriaceae (non-motile or motile by means of peritrichous flagella) and the Vibrionaceae (nonmotile or motile by means of polar flagella) (Buchanan and Gibbons 1974).

The family Enterobacteriaceae includes the genera *Escherichia, Edwardsiella, Citrobacter, Salmonella, Shigella, Klebsiella, Enterobacter, Hafnia, Serratia, Proteus, Yersinia,* and *Erwinia* (Sanderson 1976; Brenner 1981). Members of the genus *Escherichia* occur in the human intestinal tract; *E. coli* is employed as an indicator of fecal contamination in environmental microbiology.

The family Vibrionaceae includes the genera *Vibrio, Aeromonas, Plesiomonas, Photobacterium,* and *Lucibacterium.* Many of the *Vibrio* have curved, rod-shaped cells. The habitat of *Vibrio* species is generally aquatic. *Vibrio cholerae,* which causes cholera, appears to be a widespread inhabitant of estuaries. *Photobacterium* and *Lucibacterium* are interesting because of their ability to luminesce (Hastings and Nealson 1977). Some species of luminescent bacteria occur in association with fish; some of these fish are known as "flashlight fish" because of the luminescent organs housing these bacteria. The association of luminescent bacteria is important in various behavioral aspects of these fish, including schooling, feeding, and mating activities.

Gram-Negative Anaerobic Bacteria   There is only one family, Bacteroidaceae, and relatively few genera in the group of Gram-negative, anaerobic, rod-shaped bacteria (Macy and Probst 1979; Gottschalk 1981). The genera included in the Bacteroidaceae are *Bacteroides, Fusobacterium,* and *Leptotrichia. Bacteroides* are dominant members of the normal intestinal microbiota of humans and other animals.

Desulfovibrio is considered a genus of uncertain affiliation within the group of Gram-negative anaerobic rods (Pfennig et al. 1981). Members of the genus *Desulfovibrio* are curved rods capable of reducing sulfates, or other reducible sulfur compounds, to hydrogen sulfide. *Desulfovibrio desulfuricans,* normally found in anaerobic sediments, plays an important role in the biogeochemical cycling of sulfur. Other genera affiliated with the Gram-negative anaer-

obic rods include *Butyrovibrio, Lachnospira, Succinovibrio, Succinimonas,* and *Selenomonas.* Species of these genera occur in the rumen (a compartment in the stomachs of cows and other ruminant animals), where they play a critical metabolic role in digesting cellulose and producing low–molecular-weight fermentation products that the animal can absorb.

The Gram-negative anaerobic cocci include only four genera: *Veillonella, Acidaminococcus, Megasphaera,* and *Gemmiger.* Each of these genera contains few species. *Veillonella* species have complex nutritional requirements and are unable to grow on individual organic substrates. Although these organisms are fastidious in their nutritional requirements, they comprise part of the normal intestinal microbiota of humans.

Gram-negative Cocci and Coccobacilli   The Gram-negative cocci and coccobacilli occur in the family Neisseriaceae (Bøvre and Hagen 1981). The family includes the genera *Neisseria, Branhamella, Moraxella,* and *Acinetobacter* (Henricksen 1976). Some members of the genera *Neisseria, Branhamella,* and *Moraxella* are parasitic. *Acinetobacter* species are saprophytic. They are nutritionally versatile and can utilize a variety of organic compounds as a sole source of carbon and energy.

Gram-negative Chemolithotrophic Bacteria   The metabolic activities of the Gram-negative chemolithotrophic bacteria are extremely important in biogeochemical cycling processes (Kelly 1981) (Table A2.8). These bacteria oxidize inorganic compounds in order to generate ATP, and because the energy yield of some of these reactions is low, they must metabolize large amounts of substrate in order to meet their energy requirements. The metabolic transformations of inorganic compounds mediated by these organisms cause global-scale cycling of various elements between the air, water, and soil.

Members of the family Nitrobacteraceae oxidize ammonia or nitrite and are referred to as nitrifying bacteria; they are commonly found in soil, fresh water, and seawater (Watson et al. 1981). One group of nitrifying bacteria oxidizes ammonia to nitrite, and

Table A2.8

Some characteristics of Gram-negative chemolithotrophic bacteria

Genus	Description
Nitrobacter	Oxidizes nitrite to nitrate; cells short rods; cells possess a polar cap of cytomembranes; distributed in soil, fresh water, and seawater
Nitrospina	Oxidizes nitrite to nitrate; cells straight, slender rods; no extensive cytomembrane system; found in Atlantic Ocean
Nitrococcus	Oxidizes nitrite to nitrate; cells spherical; motile by subterminal flagella; Gram-negative; found in Pacific Ocean
Nitrosomonas	Oxidizes ammonia to nitrite; cells ellipsoidal or short rods; cells occur in pairs or short chains; possess cytomembranes that occur in flattened vesicles in the peripheral regions of the cytoplasm; widely distributed in soils
Nitrosospira	Oxidizes ammonia to nitrite; cells spiral shaped; cells lack cytomembranes; widely distributed in soils
Nitrosococcus	Oxidizes ammonia to nitrite; cells spherical; found in soils and seawater
Nitrosolobus	Oxidizes ammonia to nitrite; cells pleomorphic and lobate; cells partly compartmentalized by invagination of the cytoplasmic membrane; motile by peritrichous flagella; found in some soils
Thiobacillus	Oxidizes sulfur and sulfur compounds; sulfate is final oxidation product; cells rod shaped; motile by polar flagella; widely distributed in soils, fresh water, and seawater
Sulfolobus	Oxidizes sulfur and sulfur compounds; cells spherical with lobes; cell wall lacks murein; optimal growth 70°C–75°C; found in hot, acidic, sulfur-rich soils and water
Macromonas	Oxidizes sulfur and sulfur compounds; cells cylindrical to bean shaped; inclusions of calcium carbonate sometimes accompanied by sulfur globules; found in fresh water
Thiovulum	Oxidizes sulfur and sulfur compounds; cells round to ovoid; cytoplasm normally contains sulfur inclusions; motile by peritrichous flagella; found in fresh water and also in seawater
Thiospira	Oxidizes sulfur and sulfur compounds; cells spiral shaped, usually with pointed ends; motile by peritrichous flagella; found in water overlying sulfur-rich muds
Siderocapsa	Oxidizes iron or manganese; cells spherical and embedded in capsule encrusted with iron or manganese oxides; common in fresh water
Naumanniella	Oxidizes iron; cells rod shaped; cells surrounded by capsule containing iron oxides; widely distributed in iron-bearing fresh water
Ochrobium	Oxidizes iron; cells ellipsoidal to rod shaped; cells surrounded by capsule containing iron oxides; widely distributed in iron-bearing fresh water
Siderococcus	Iron but not manganese deposited; cells spherical; cells not coated by iron oxides; widely distributed in fresh water and sediments

a second group oxidizes nitrite to nitrate. Most members of the family Nitrobacteraceae are obligate chemolithotrophs. *Nitrosomonas* species are extremely important nitrifiers in soil, oxidizing ammonia to nitrite. Likewise, *Nitrobacter* species are important nitrifiers in soil, oxidizing nitrite to nitrate. The combined actions of the members of the genera *Nitrosomonas* and *Nitrobacter* permit the conversion of ammonia to nitrate. The reversal of electronic charge between NH_4^+ and NO_3^- alters the mobility of these nitrogenous ions in soil and has a major influence on soil fertility.

Several different genera of chemolithotrophic bacteria metabolize sulfur and sulfur-containing inorganic compounds (Kuenen and Tuovinen 1981). The genus *Thiobacillus* derives energy from the oxidation of reduced sulfur compounds; some members of the genus *Thiobacillus* oxidize only sulfur compounds, whereas others, such as *Thiobacillus ferrooxidans,* can also oxidize ferrous iron to ferric iron in order to generate ATP. *Thiobacillus thiooxidans* is often found in association with waste coal heaps and strip-mined areas. The metabolic activities of this organism produce acid mine drainage that constitutes a serious ecological problem.

Members in the family Siderocapsaceae are able to oxidize iron or manganese, depositing iron and/or manganese oxides in capsules or in extracellular material. *Siderocapsa* species are widely distributed in nature, and their metabolic activities are of geological importance. Members of this family are found in iron-bearing waters, forming high concentrations in the lower portions of some lakes.

Gram-positive Cocci   The Gram-positive cocci include three families: the Micrococcaceae, the Streptococcaceae, and the Peptococcaceae (Buchanan and Gibbons 1974). The coccoid cells of the Micrococcaceae may occur singly or as irregular clusters; for example, the genus *Staphylococcus* typically forms grapelike clusters. In the family Streptococcaceae, Gram-positive cocci occur as pairs or chains. The Peptococcaceae have complex nutritional requirements; they are obligately anaerobic and produce low–molecular-weight volatile fatty acids, carbon dioxide, hydrogen, and ammonia as the main products of amino acid metabolism.

Endospore-forming Rods and Cocci   The endospore-forming rods and cocci are extremely important because of the heat resistance of the endospore structure (Buchanan and Gibbons 1974). The genera *Bacillus, Sporolactobacillus, Clostridium, Desulfotomaculum, Sporosarcina,* and *Thermoactinomyces* are all characterized by the formation of endospores.

Gram-positive, Asporogenous, Rod-shaped Bacteria   The Gram-positive, asporogenous (non-sporulating), rod-shaped bacteria include the family Lactobacillaceae (Buchanan and Gibbons 1974). These are Gram-positive rods with lactic acid as their major fermentation product; they occur in fermenting plant and animal products that contain available carbohydrate substrates; they also comprise part of the normal intestinal microbiota of humans.

There are several genera of uncertain affiliation that are Gram-positive, non-spore-forming, rod-shaped bacteria. These include the genera *Listeria, Erysipelothrix,* and *Caryophanon.* The *Listeria* are Gram-positive rods with a tendency to produce chains. Several species of *Listeria* are animal pathogens. *Caryophanon latum* forms rods or filaments up to 3 μm in diameter; this organism is normally found on animal fecal matter.

Actinomycetes and Related Organisms   The coryneform group of bacteria is a heterogeneous group defined by the characteristic irregular morphology of the cells and the tendency of the cells to show incomplete separation following cell division (Goodfellow and Minnikin 1981). Although they do not form true filaments, the irregular morphology and the association of the cells after division indicates a relationship between the coryneforms and the filament-forming actinomycetes. This group includes the genera *Corynebacterium, Arthrobacter, Brevibacterium, Cellulomonas,* and *Kurthia.* The genus *Arthrobacter,* which is widely distributed in soils, is interesting because it exhibits a simple life cycle, in which there is a change from rod-shaped cells to coccoid cells. The sequence of morphological changes in the growth cycle distinguishes *Arthrobacter* from other genera. The coccoid cells present during the stationary growth phase are sometimes referred to as arthrospores and cystites. The formation of arthrospores represents the beginning of a regular life cycle that is characteristic of eukaryotic microorganisms but is rare among the prokaryotes.

The order Actinomycetales contains bacteria characterized by the formation of branching filaments (Cross and Goodfellow 1973; Lechevalier and Lechevalier 1981). Many of the more evolved actinomycetes resemble the fungi in appearance, but their cells are prokaryotic, and they are clearly bacteria. The various families of the order Actinomycetales are distinguished from one another by the nature of their mycelia and spores (Table A2.9). The actinomycetes are widely distributed in nature. The main ecological role of actinomycetes appears to be the decomposition of organic matter in soil. Many actinomycetes produce antibiotics, which, in addition to their medical importance, may play a role in ecological relationships among differing microbial populations.

Rickettsias and Chlamydias The rickettsias are obligate intracellular parasites with insufficient capacity for ATP synthesis to support independent reproduction. Some are adapted to existence in arthropods but are capable of infecting vertebrate hosts, including humans. For example, *Rickettsia rickettsii* is transmitted by ticks and causes Rocky Mountain spotted fever.

The chlamydias are also obligate intracellular parasites; they have sometimes been referred to as large viruses, but they are true Gram-negative bacteria (Becker 1978). Chlamydias cause human respiratory and urinary-genital tract diseases. In birds and sometimes humans they cause respiratory and generalized infections; for example, *Chlamydia psittaci* (causes parrot fever) and when infected tropical birds are detected, import into the United States is restricted to prevent spread to humans.

Mycoplasmas The mycoplasmas are bacteria that lack a cell wall (Razin 1978; Whitcomb 1980). They are the smallest organisms capable of self-reproduction. When growing on artificial media, mycoplasmas form small colonies that have a characteristic "fried egg" appearance. Members of the genera *Mycoplasma* and *Spiroplasma* cause diseases in plants and animals.

Table A2.9
Properties of the actinomycetes

Family	Description
Actinomyceataceae	Mycelium not formed; no spores formed; not acid fast; representative genera are *Actinomyces, Arachnia, Bifidobacterium, Bacterionema,* and *Rothia*
Mycobacteriaceae	Mycelium not formed; no spores formed; acid fast; representative genus is *Mycobacterium*
Frankiaceae	Mycelium formed; symbionts in plant nodules with free stage in soil; representative genus is *Frankia*
Actinoplanaceae	Mycelium formed; saprophytes or facultative parasites; spores borne inside sporangia; representative genera are *Actinoplanes, Spirillospora, Streptosporangium, Amorphosporangium, Ampullariella, Pilimelia, Planomonospora, Planobispora, Dactylosporangium,* and *Kitasatoa*
Dermatophilaceae	Mycelium divides transversely to form motile cocci; saprophytes or facultative parasites; spores not borne in sporangia; representative genera are *Dermatophilus* and *Geodermatophilus*
Nocardiaceae	Mycelium fragments to form nonmotile cells; saprophytes or facultative parasites; spores not borne in sporangia; aerial spores usually absent; representative genera are *Nocardia* and *Pseudonocardia*
Streptomycetaceae	Mycelium tends to remain intact; saprophytes or facultative parasites; spores not borne in sporangia; usually abundant aerial mycelium and long spore chains; representative genera are *Streptomyces, Streptoverticillium, Sporichthya,* and *Microellobosporia*
Micromonosporaceae	Mycelium remains intact; saprophytes or facultative parasites; spores not borne in sporangia; spores formed singly or in short chains; representative genera are *Micromonospora, Thermoactinomyces, Actinobifida, Thermomonospora, Microbispora,* and *Micropolyspora.*

Endosymbionts Several bacterial genera have been recognized that are obligate endosymbiotic of invertebrates; that is, they live within the cells of invertebrate animals without adversely affecting the animal. For example, the protozoan *Paramecium aurelia* can harbor a variety of endosymbiotic bacteria (Preer 1981) (Table A2.10). A number of new genera of endosymbiotic bacteria have recently been described for other protozoa, insects, and various other invertebrates. An understanding of the nutritional requirements of these bacteria has permitted the creation of complex media for their culture and identification.

EUKARYOTES (EUCARYA)

Fungi

The fungi are eukaryotic, heterotrophic microorganisms that typically form reproductive spores; some fungi are unicellular, but many form filaments of vege-

tative cells known as mycelia (Ainsworth and Sussman 1965–1973; Alexopoulos and Mims 1979; Moore-Landecker 1982). Mycelia, which usually exhibit branching and are typically surrounded by cell walls containing chitin and/or cellulose, are integrated masses of individual, tubelike filaments of hyphae. By definition the fungi are achlorophyllous, saprophytic, or parasitic, with unicellular or, more typically, filamentous vegetative structures, usually surrounded by cell walls composed of chitin or other polysaccharides, propagating with spores and normally exhibiting both asexual and sexual reproduction. The broadness of this definition reflects the great morphological and physiological diversity of the fungi.

The primary taxonomic groupings of the fungi are based on their sexual spores (Alexopoulos and Mims 1979). To a lesser extent, fungal systematics relies on the morphological characteristics of the vegetative cells. Physiological features are particularly important in the classification of yeasts, which are primarily unicellular fungi. In a formal systematic sense, yeasts are not recognized as being separate from the rest of the fungi and are classified along with their fil-

Table A2.10

Symbionts of *Paramecium aurelia*

Genus	Common names	Description
Caedibacter	Kappa	Varying in size and distinguished by the presence of a 0.5-μm-diameter inclusion within the host cell; exhibits killing of sensitive strains
Pseudocaedibacter	Pi	Slender rod; until recently, considered as a mutant of kappa; nonkilling symbiont
Pseudocaedibacter	Mu	Slender rod, often elongated; distinguished because its killing action is wholly dependent on cell-to-cell contact between mating paramecia
Pseudocaedibacter	Nu	Nonkilling symbiont similar in appearance to pi and mu
Pseudocaedibacter	Gamma	Diminutive bacterium, frequently appearing as doublets; strong killing of other strains is shown by gamma bearers
Tectobacter	Delta	Rod distinguished by an electron-dense material surrounding the outer of its two membranes
Lyticum	Lambda	Appears as a typical motile bacterium with peritrichous flagella, although its movement within the cytoplasm has not been observed
Lyticum	Sigma	Largest of all endosymbionts of *Paramecium aurelia;* curved flagellated rod resembles lambda

amentous counterparts. In practice, however, the yeasts are typically treated separately from the filamentous fungi in both classification and identification systems.

The slime molds, which are placed in the division Gymnomycota, represent a borderline case between the fungi and the protozoa (Gray and Alexopoulos 1968). They could just as well be classified with the protozoa, but mycologists have traditionally studied them. The vegetative cells of the slime molds are amoeboid and generally lack a cell wall, making them similar to protozoa. However, the slime molds also resemble true fungi because they produce spores that are surrounded by wall structures.

Gymnomycota   All slime molds exhibit characteristic life cycles, a feature used in subdividing this division (Gray and Alexopoulos 1968; Alexopoulos and Mims 1979). The Acrasiales, or cellular slime molds, form fruiting (spore-bearing) bodies known as sporocarps, which are generally stalked structures. The stalks normally consist of walled cells, and this characteristic forms the basis for designating these organisms as the cellular slime molds. The sporocarp releases spores that germinate, forming myxamebae (amoeboid cells that form pseudopodia), and the myxamebae swarm together or aggregate to form a pseudoplasmodium. Within the pseudoplasmodium, the cells of the cellular slime molds do not lose their integrity. The pseudoplasmodium undergoes a developmental sequence (differentiation), culminating in the formation of a special type of fruiting body that bears a mucoid droplet at the tip of each branch, containing spores with cell walls.

The pseudoplasmodium formation of the cellular slime molds is interesting because of the biochemical communication involved in initiating swarming activity. When food sources become limiting, the myxamebae of *Dictyostelium discoideum,* a well-studied cellular slime mold, cease their feeding activity and swarm to an aggregation center. The swarming activity is initiated when one or more cells at the aggregation center release cyclic AMP (acrasin); the myxamebae move along the concentration gradient of cyclic AMP until they reach the center of aggregation.

When the myxamebae reach the center of aggregation, they mass together to form a pseudoplasmodium.

The Myxomycetes are known as the true slime molds because their myxamebae or swarm cells fuse together to form a true plasmodium. The plasmodium of the Myxomycetes is a multinucleate protoplasmic mass that is devoid of cell walls and is enveloped in a gelatinous slime sheath. The plasmodium gives rise to brilliantly colored fruiting bodies that are often seen on decaying logs or other moist areas of decaying organic matter.

Mastigomycota   As opposed to the phagotrophic mode of nutrition exhibited by members of the Gymnomycota, Mastigomycota accomplish nutrition by absorbing nutrients (Alexopoulos and Mims 1979). Mastigomycota typically produce motile cells with flagella during part of their life cycle. The division Mastigomycota includes four classes: Chytridiomycetes, Hyphochytridiomycetes, Plasmodiophoromycetes, and Oomycetes (Table A2.11).

The chytrids are differentiated from all other fungi by the production of zoospores, which are motile with a single posterior flagellum of the whiplash type. Many chytrids are parasitic on other fungi, algae, and plants.

The oomycetes, known as the water molds, reproduce by using flagellated zoospores. The zoospores typically have two flagella, one of the tinsel type and the other of the whiplash type. Sexual reproduction in the oomycetes typically involves the formation of oospores, which are thick-walled spores that develop by contact with specialized gametangia (structures containing differentiated cells involved in sexual reproduction). Members of the Saprolegniales, an order of the Oomycetes, are abundant in aquatic ecosystems. Several species in this order are important animal and plant pathogens. For example, *Phytophthora infestans* causes potato blight and was responsible for the great Irish potato famine of 1845 and 1846, which resulted in the great wave of immigration from Ireland to the United States.

Amastigomycota   Unlike the Gymnomycota and Mastigomycota, the Amastigomycota do not pro-

Table A2.11

Descriptions of the mastigomycota

Class	Description
Chytridiomycetes	Vegetative form varied; produces posteriorly uniflagellate motile cells with whiplash flagella
Hyphochytridiomycetes	Small group of fungi, produces motile anteriorly uniflagellate cells with tinsel flagella
Plasmodiophoromycetes	Parasitic fungi with multinucleate thalli (plasmodia) within the cells of their hosts; resting cells (cysts) produced in masses but not in distinct sporophores; motile cells with two anterior whiplash flagella
Oomycetes	Vegetative form varied, usually filamentous, with a coenocytic, walled mycelium; produces zoospores, each with one whiplash and one tinsel flagellum; sexual reproduction oogamous, resulting in the formation of oospores

duce motile cells. There are four subdivisions in the Amastigomycota: Zygomycotina, Ascomycotina, Basidiomycotina, and Deuteromycotina (Alexopoulus and Mims 1979) (Table A2.12).

The Zygomycotina typically have coenocytic mycelia and are characterized by the formation of a zygospore, a sexual spore that results from the fusion of gametangia. Species of the Trichomycetes, a subdivision of the Zygomycotina, are obligately associated with arthropods and normally grow within the guts of these animals, where they attach to the chitinous lining by means of a specialized structure known as a holdfast.

Several species of Mucorales, an order of the Zygomycotina, exhibit an interesting morphological change known as dimorphism, occurring under some conditions as filamentous mycelia and under other conditions as yeastlike unicellular forms. *Pilobolus* species, which occur in this order, are interesting because of their forceful mode of spore discharge. *Pilobolus* species can shoot their spores several meters into the air with the entire spore cluster ejected in the direction of highest light intensity. In this way, the spores of *Pilobolus* are released to where air currents are likely to further disperse them.

Table A2.12

Descrptions of the amastigomycota

Subdivision	Description
Zygomycotina	Saprophytic, parasitic, or predatory fungi; coenocytic mycelium; asexual reproduction, usually by sporangiospores; sexual reproduction, where known, by fusion of equal or unequal gametangia, resulting in the formation of zygosporangia containing zygospores
Ascomycotina	Saprophytic, symbiotic, or parasitic fungi; unicellular or with a septate mycelium, producing ascospores in saclike cells (asci)
Basidiomycotina	Saprophytic, symbiotic, or parasitic fungi; unicellular or, more typically, with a septate mycelium, producing basidiospores on the surface of various types of basidia
Deuteromycotina	Saprophytic, symbiotic, parasitic, or predatory fungi; unicellular or, more typically, with a septate mycelium, usually producing conidia from various types of conidiogenous cells; sexual reproduction unknown

Members of the subdivision Ascomycotina produce sexual spores within a specialized saclike structure known as the ascus. The members of the subclass Hemiascomycetidae, in which ascomycetous yeasts occur, are morphologically simple ascomycetes that generally lack hyphae; many yeasts are ascomycetes, and the morphology of the ascospore is a critical taxonomic feature for classifying yeasts to the genus level (Rose and Harrison 1969; Lodder and Kreger-van Rij 1970; Phaff et al. 1978; Barnett et al. 1979). In addition to the ascosporogenous yeasts, the subclass Hemiascomycetidae includes the order Taphrinales; these fungi resemble yeasts in that they reproduce asexually by budding and sexually by the production of ascospores, but they differ from the yeasts in that they produce a true mycelium. Members of the Taphrinales are parasitic on plants.

The Euascomycetidae (true ascomycetes) produce asci that normally develop from dikaryotic hyphae (Alexopoulos and Mims 1979). The asci are produced in or on a specific structure known as the ascocarp. The euascomycetes are divided according to the structure of the ascocarp into the plectomycetes, in which the ascocarp has no special opening; the pyrenomycetes, in which the ascocarp is shaped like a flask; and the discomycetes, in which the ascocarp is cup shaped. The pyrenomycete *Endothia parasitica* is the causative agent of chestnut blight. This organism was introduced into North America from eastern Asia in the early twentieth century and quickly devastated the chestnut trees of the United States and Canada. The plectomycetes also include some important plant and animal pathogens. This group includes the black molds, blue molds, and ringworms. *Ceratocystis ulmi,* a member of the plectomycete, is the causative agent of Dutch elm disease, a great threat to elm trees in North America. The Discomycetes include cup fungi, morels, and truffles.

The basidiomycetes (subdivision Basidiomycotina) are distinguished from other classes of fungi by the fact that they produce sexual spores, known as basidiospores, on the surfaces of specialized spore-producing structures, known as basidia (Alexopoulos and Mims 1979). Smuts, rust, jelly fungi, shelf fungi, stinkhorns, bird's-nest fungi, puffballs, and mushrooms are all basidiomycetes. The Basidiomycete order Aphyllophorales contains the shelf, or bracket, fungi. These are some of the most conspicuous fungi, often seen growing on trees. The growth of fungi in this group on wood results in two characteristic types of decay, called brown rots and white rots because of the color of the rotted wood. In brown rot, only the cellulose component of wood is decomposed, leaving the brown, powdery lignins. In white rot, primarily lignin is degraded, leaving white-colored cellulose fibers.

The Basidiomycete order Agaricales includes the mushrooms; mushrooms are the fruiting bodies (basidiocarps) of basidiomycetes (Miller 1979). Some mushrooms are edible, but others are extremely poisonous; therefore, their proper identification is critical as it is sometimes easy to confuse an edible species with one that is poisonous. Several species of the genus *Amanita,* characterized by free gills and the presence of both an annulus and a volva, are deadly; *Amanita phalloides* is known as the death cap because most deaths due to mushroom poisoning have been attributed to the ingestion of this species.

Several orders comprise the gastromycetes, another Basidiomycete group, including the puffballs, earthstars, stinkhorns, and bird's-nest fungi. Members of this taxonomic group are commonly seen in a number of habitats. The order Phallales (stinkhorns) produce a green gelatinous ooze and a foul smell when the basdiocarp undergoes autodigestion, releasing the basidiospores. Although humans find the odor offensive, flies are attracted by the smell; some of the ooze containing the basidiospores adheres to the flies, providing a mechanism for the disseminaton of the basidiospores of these fungi.

Only a few yeasts are Basidiomycetes. These are placed into the subclass Teliomycetidae, along with the rusts and smuts. *Rhodosporidium* and *Leucosporidium,* which have been placed in this subclass, are basidiomycetous yeasts found in marine environments. The many species of rust and smut fungi, the most serious fungal plant pathogens, occur in the subclass Teliomyctidae. There are more than 20,000 species of rust fungi and more than 1,000 species of smut fungi. Rusts and smuts are characterized by the production of a resting spore known as a teliospore,

Table A2.13

Descriptions of some representative deuteromycetes

Genus	Description
Alternaria	Soil saprophytes and plant pathogens, muriform spores fit together like bricks of a wall
Arthrobotrys	Soil saprophytes; some form organelles for capture of nematodes
Aspergillus	Common molds; radially arranged; colored, often black; conidiospores
Aureobasidium (Pulullaria)	Short mycelial filaments; lateral blastospores; often damage painted surfaces
Candida	Common yeast; some cause mycoses; some species able to grow in concentrated sugar solutions; some species able to grow on hydrocarbons
Cryptococcus	Yeasts; saprophytic in soil, but some may cause mycoses in animals and humans
Geotrichum	Common soil fungus; older mycelial filaments break up into arthrospores
Helminthosporium	Cylindrical, multiseptate spores; many are economically significant plant pathogens
Penicillium	Common mold with colored, often green, conidiospores arranged in brush shape
Trichoderma	Common soil saprophyte with highly branched conidiospores

which is thick walled and binucleate. The rusts, all of which are plant pathogens, require two unrelated hosts for the completion of their normal life cycle. For example, white pine blister rust uses gooseberry bushes as its alternate host.

The final subdivision of the fungi is the Deuteromycotina; members of this subdivision are known as the deuteromycetes, or fungi imperfecti (Alexopoulos and Mims 1979; Cole and Kendrick 1981). There are about 15,000 species in the fungi imperfecti; Table A2.13 lists some representative genera of deuteromycetes. The vegetative structures of most of the fungi in this class resemble ascomycetes, although the vegetative structures of a few members of the fungi imperfecti resemble basidiomycetes. Sexual forms of reproduction in the deuteromycetes, however, do not occur or have not been detected, and it is the lack of observed sexual spores that places fungal species into this group. The fungi imperfecti are classified largely on the basis of the morphological structure of the vegetative phase and on the types of asexual spores produced. Without observation of the sexual reproductive stage, a definite placement of the deuteromycetes into either the ascomycetes or basidiomycetes is not possible, but many fungi imperfecti are closely related to either the ascomycetes or the basidiomycetes.

Algae

The algae are eukaryotic photosynthetic organisms that are separated from plants by their lack of tissue differentiation (Chapman and Chapman 1975; Trainor 1978; Lee 1980; Bold and Wynne 1985). In R. H. Whittaker's five-kingdom classification system, some of the algae are placed in the kingdom Protista along with the protozoa, and other algae, exhibiting more extensive organizational development, are placed in the kingdom Plantae (Whittaker 1969). Indeed, some organisms that are classified as algae are on the borderline of higher plants, whereas others are on the borderline of the protozoa.

The algae are classified into divisions, based largely on the types of photosynthetic pigments that are produced, the types of reserve materials that are stored intracellularly, and the morphological characteristics of the cell (Table A2.14). The relative concentrations of the various photosynthetic pigments give the algae characteristic colors. Many of the major algal divisions, such as the green algae, red algae, and brown algae, have common names based on these characteristic colors.

Chlorophycophyta The Chlorophycophyta, or green algae, are widely distributed in aquatic ecosystems (Pickett-Heaps 1975). The cellular organization

Table A2.14

Descriptions of the algae

Division	Description
Chlorophycophyta	Green algae; photosynthetic pigments—chlorophylls and b, carotenes, several xanthophylls; storage product—starch; cell wall—cellulose, xylans, mannans, absent in some, calcified in some; flagella if present—1, 2–8, many, equal, apical
Chrysophycophyta	Golden and yellow-green algae. Includes the diatoms; photosynthetic pigments—chlorophylls a and c, carotenes, fucoxanthin and several other xanthophylls; storage product—chrysolaminarin; cell wall—absent; flagella—2, unequal, subapical
Euglenophycophyta	Euglenoids; photosynthetic pigments—chlorophylls a and b, carotenes, several xanthophylls; storage product—paramylon; cell wall—absent; flagella—1–3, apical, subapical
Phaeophycophyta	Brown algae; photosynthetic pigments—chlorophylls a and c, carotenes, fucoxanthin, and several other xanthophylls; storage product—laminarin; cell wall—cellulose, alginic acid, sulfated mucopolysaccharides; flagella if present—2, unequal, lateral
Pyrrophycophyta	Dinoflagellates; photosynthetic pigments—chlorophylls a and c, carotenes, several xanthophylls; storage product—starch; cell wall—cellulose or absent; flagella—2, one trailing, one girdling
Rhodophycophyta	Red algae; photosynthetic pigments—chlorophyll a (also d in some), phycocyanin, phycoerythrin, carotenes, several xanthophylls; storage product—Floridean starch; cell wall—cellulose, xylans, galactans; flagella—absent

in different species of the Chlorophycophyta may be unicellular, colonial, or filamentous. The order Volvocales includes the unicellular green algae, which are normally motile by means of flagella. Several well-known genera of algae occur in this order, including *Chlamydomonas* and *Volvox*. There are several filamentous types of green algae, including members of the genera *Ulothrix* and *Spirogyra*. *Ulva,* commonly known as sea lettuce, is restricted to marine habitats, growing attached to rocks and other surfaces.

Euglenophycophyta   The Euglenophycophyta are similar to the Chlorophycophyta in that they contain chlorophylls a and b and typically appear green in color, but they differ with respect to their cellular organization and their intracellular reserve storage products. The Euglenophycophyta are unicellular and do not store starch but do store paramylon, a β-1,3-glucose polymer. These algae are widely distributed in aquatic and soil habitats.

Chrysophycophyta   The division Chrysophycophyta includes the classes Xanthophyceae (yellow-green algae), Chrysophyceae (golden algae), and Bacillariophyceae (the diatoms). Chrysophycophyta produce a diversity of pigments, cell-wall constituents, and cell types. The Chrysophycophyta are unified by the production of the same reserve storage material, chrysolaminarin, a β-linked polymer of glucose. Chrysophycophyta species produce carotenoid and xanthophyll pigments that tend to dominate over the chlorophyll pigments; this confers the golden-brown-colored hues on members of this division. *Vaucheria,* a yellow-green algal genus, is known as the water felt and is widely distributed in moist soils and aquatic habitats.

The Bacillariophyceae, or diatoms, produce distinctive cell walls known as frustules. The frustules of diatoms have two overlapping halves that fit together like a petri dish. Some diatoms are benthic, living at the bottom of aquatic ecosystems at the sediment layer, and other diatoms are planktonic, living suspended in open water bodies. The growth of diatoms is dependent on the concentrations of available silica, because the cell walls of the diatom are impregnated with this mineral. The frustules of diatoms are relatively resistant to natural degradation and accumulate over geologic periods; as a result, diatoms are preserved in fossil records dating back to the Cretaceous period 65 million years ago.

Pyrrophycophyta The Pyrrophycophyta, or fire algae, are generally brown or red in color because of the presence of xanthophyll pigments. The Pyrrophycophyta are unicellular, are biflagellate, and store starch or oils as their reserve material. In the dinoflagellates, members of the Pyrrophycophyta, the two flagella emerge from an opening in a transverse groove. The cell walls of Pyrrophycophyta contain cellulose and sometimes form structured plates, called theca. Several dinoflagellates exhibit bioluminescence. Some species also exhibit regular 24-hour behavioral patterns, known as circadian rhythms. For example, *Gonyaulax polyedra* exhibits a cyclic expression of luminescence with peak luminescence occurring in the middle of the dark period. The glow rhythm is associated with a nightly increase in the level of luciferin and luciferase, the same enzyme substrate system that is operative in fireflies. Species of *Gonyaulax* and other dinoflagellates produce the toxic blooms known as red tides that tend to color water red or red-brown in the vicinity of the bloom; the toxins of dinoflagellates during such blooms may kill other marine organisms. They also render shellfish toxic for human consumption.

Some dinoflagellates tend to enter into mutually beneficial (symbiotic) relationships with a variety of marine invertebrates. Such associations are termed zooxanthellae. Within such associations, the animal cell provides protection and carbon dioxide for photosynthesis for the dinoflagellates. The algae provide the animal with oxygen and organic carbon for its nutritional needs. Often dinoflagellates grow on ingested bacteria and other algal species. As such, dinoflagellates are mixotrophic, capable of both heterotrophic and photoautotrophic metabolism.

Rhodophycophyta The Rhodophycophyta, or red algae, contain phycocyanin and phycoerythrin in addition to chlorophyll pigments; their red color is due to phycoerythrin. The primary reserve material in the Rhodophycophyta is Floridean starch, a polysaccharide similar to the amylopectin found in higher plants. Most red algae occur in marine habitats. Red algae typically have a bilayered cell wall with an inner microfibrillar rigid layer and an outer mucilaginous layer. Various biochemicals, including agar and carrageenan, occur in the cell walls of red algae.

Phaeophycophyta The division Phaeophycophyta, or brown algae, includes more than 200 genera and 1,500 species. The Phaeophycophyta produce xanthophylls that dominate over the carotenoid and chlorophyll pigments and impart a brown color to these organisms. The main reserve materials for the brown algae are laminarin and mannitol. The Phaeophycophyta, most of which are marine organisms, are found primarily in coastal zones. The kelps, which are brown algae, can form macroscopic structures up to 50 m in length. The genera *Fucus* and *Sargassum* are important representatives of Phaeophycophyta; large populations of *S. natans* occur in the Atlantic Ocean in the region known as the Sargasso Sea, and species of *Fucus* commonly occur along rocky shores, attached to the rocks by disclike holdfasts. These brown algae clearly are the most complex organisms classified as algae, or for that matter, as microorganisms, representing a borderline case between algae and plants.

Protozoa

According to the revised 1980 classification of the protozoa reported by the Committee on Systematics and Evolution of the Society of Protozoologists, the protozoa, which are essentially unicellular organisms, are divided into seven phyla (Levine et al. 1980) (Table A2.15). This classification system departs from earlier classification systems that recognized four phyla primarily based on their means of motility: the Mastigophora (flagellates); Sarcodina (pseudopodia formers); Ciliophora (ciliates); and Sporozoa (spore formers) (Westphal 1976; Kudo 1977; Jahn et al. 1979; Farmer 1980). The current classification system of protozoa

Table A2.15

Descriptions of the protozoa

Group	Description
Sarcomastigophora	Locomotion—flagella, pseudopodia, or both; sexual reproduction, when present, essentially syngamy; representative genera—*Monosiga, Bodo, Leishmania, Trypanosoma, Giardia, Opalina, Amoeba, Entamoeba, Difflugia*; in old classification system, these were separated into the Sarcodina (locomotion: pseudopodia—false feet; reproduction: asexual, binary fission; nutrition: phagocytic) and the Mastigophora (locomotion: usually paired flagella; reproduction: asexual, longitudinal fission; nutrition: heterotrophic, absorptive)
Ciliophora	Cilia produced at some stage in life cycle; reproduction by binary transverse fission, budding and multiple fission also occur; sexual reproduction involving conjugation, autogamy, and cytogamy; most are free-living heterotrophs; representative genera—*Didinium, Tetrahymena, Paramecium, Stentor*
Acetospora	Spore multicellular; no polar capsules or polar filaments; all species parasitic; representative genus—*Paramyxa*; previously included in the Sporozoa
Apicomplexa	Produce apical complex visible with electron microscope; all species parasitic; representative genera—*Eimeria, Toxoplasma, Babesia, Theileria*; previously included in the Sporozoa
Microspora	Unicellular spores, each with imperforate wall; obligate intracellular parasites; representative genus—*Metchnikovella*; previously included in the Sporozoa
Myxospora	Spores of multicellular origin with one or more polar capsules; all species parasitic; representative genera—*Myxidium, Kudoa*; previously included in the Sporozoa
Labyrinthomorpha	Synonymous with the net slime molds; produce ectoplasmic network with spindle-shaped or spherical nonamoeboid cells; in some genera, amoeboid cells move within network by gliding

encompasses several groups claimed by other disciplines, including many algae and the lower fungi.

Sarcomastigophora   The Sarcomastigophora includes the Sarcodina (protozoa motile by pseudopod or false foot formation) and the Mastigophora (protozoa motile by flagella). The Sarcodina are motile by means of pseudopodia, which are cytoplasmic extensions that are sometimes referred to as false feet, or rhizopods. The pseudopodia are used for engulfing and ingesting food as well as for locomotion.

Members of the genus *Amoeba* have no distinct shape because the flow of cytoplasm continuously changes the shape of true amoeba. *Amoeba* species are readily found in pond-water samples. *Amoeba*

feed on a number of smaller organisms, including bacteria and other protozoa.

The Foraminiferida are marine members of the Sarcodina. Foraminiferans form one or many chambers composed of siliceous or calcareous tests. The tests accumulate in marine sediments and are preserved in the geological record; for example, the white cliffs of Dover are composed largely of foraminiferan deposits.

The Mastigophora are the flagellate protozoa. Because some members of the Mastigophora are able to produce pseudopodia in addition to possessing flagella, these organisms are now classified together with the Sarcodina. In this subphylum, protozoologists place the dinoflagellates, euglenoids, and various other algae with flagellate stages in their life cycles. Many members of the Mastigophora are plant and animal parasites. Human diseases, such as sleeping sickness (trypanosomiasis), caused by flagellate protozoa, are normally transmitted by arthropods, and control of many of these diseases rests with controlling the carrier rather than eliminating the disease-causing protozoa.

Ciliophora Ciliophora are protozoa that are motile by means of cilia (Corliss 1979). In addition to their role in locomotion, the cilia of many of these protozoa move food particles into a mouthlike region known as the cytostome. The ciliate protozoa reproduce by various asexual and sexual means. *Paramecium* is perhaps the best-known genus of ciliate protozoa. Some ciliate protozoa consume smaller microorganisms, such as bacteria, yeasts, and algae,

as their food source. Others, such as *Didinium*, can engulf ciliates equal to their own size; *Paramecium* is a common prey.

Sporozoa (Acetospora, Apicomplexa, Microspora, and Myxospora) The Sporozoa—a group comprising four phyla—are parasites, exhibiting complex life cycles. The adult forms are nonmotile, but immature forms and gametes may be motile. The sporozoans derive nutrition by absorption of nutrients from the host cells they inhabit. The immature stages are referred to as sporozoites. Reproduction of the trophozoite, the adult stage of a sporozoan, occurs asexually by multiple fission; the multiple fission process can result in the production of thousands of spores. *Plasmodium vivax* and various other species of *Plasmodium* cause malaria, one of the most widespread infectious diseases of humans.

REFERENCES & SUGGESTED READINGS

Ainsworth, G. C., and A. S. Sussman (eds.). 1965–1973. *The Fungi: An Advanced Treatise.* 4 vol. Academic Press, New York.

Alexopoulos, C. J., and C. W. Mims. 1979. *Introductory Mycology.* John Wiley and Sons, New York.

Balch, W. E., G. E. Fox, L. J. Magrum, C. R. Woese, and R. S. Wolfe. 1979. Methanogens: Reevaluation of a unique biological group. *Microbiological Reviews* 43:260–296.

Barksdale, L., and K. S. Kim. 1977. Mycobacterium. *Bacteriological Reviews* 41:217–372.

Barnett, J. A., R. W. Payne, and D. Yarrow. 1979. A *Guide to Identifying and Classifying Yeasts.* Cambridge University Press, New York.

Becker, Y. 1978. The chlamydia: Molecular biology of procaryotic obligate parasites of eucaryocytes. *Microbiological Reviews* 42:274–306.

Becking, J. H. 1981. The family Azotobacteraceae. In M. P. Starr, H. Stolp, H. G. Trüper, A. Ballows, and H. G. Schlegel (eds.). *The Prokaryotes: A Handbook on Habitats, Isolation, and Identification of Bacteria.* Springer-Verlag, Berlin, pp. 795–817.

Biebl, H., and N. Pfennig. 1981. Isolation of members of the family Rhodospirillaceae. In M. P. Starr, H. Stolp, H. G. Trüper, A. Ballows, and H. G. Schlegel (eds.). *The Prokaryotes: A Handbook on Habitats, Isolation, and Identification of Bacteria.* Springer-Verlag, Berlin, pp. 267–273.

Bold, H. C., and M. Wynne. 1985. *Introduction to the Algae.* Prentice-Hall, Englewood Cliffs, N.J.

Bøvre, K., and N. Hagen. 1981. The family Neisseriaceae: Rod-shaped species of the genera *Moraxella, Acinetobacter, Kingella,* and *Neisseria,* and the *Branhamella* group of cocci. In M. P. Starr, H. Stolp, H. G. Trüper, A. Ballows, and H. G. Schlegel (eds.). *The Prokaryotes: A Handbook on Habitats, Isolation, and Identification of Bacteria.* Springer-Verlag, Berlin, pp. 1506–1529.

Brenner, D. J. 1981. Introduction to the family Enterobacteriaceae. In M. P. Starr, H. Stolp, H. G. Trüper, A. Ballows, and H. G. Schlegel (eds.). *The Prokaryotes: A Handbook on Habitats, Isolation, and Identification of Bacteria.* Springer-Verlag, Berlin, pp. 1105–1127.

Brock, T. D. 1981. Extreme thermophiles of the genera *Thermus.* In M. P. Starr, H. Stolp, H. G. Trüper, A. Ballows, and H. G. Schlegel (eds.). *The Prokaryotes: A Handbook on*

Habitats, Isolation, and Identification of Bacteria. Springer-Verlag, Berlin, pp. 978–984.

Buchanan, R. E., and N. E. Gibbons (eds.). 1974. *Bergey's Manual of Determinative Bacteriology,* 8th ed. Williams and Wilkins, Baltimore.

Castenholz, R. W., and B. K. Pierson. 1981. Isolation of members of the family Chloroflexaceae. In M. P. Starr, H. Stolp, H. G. Trüper, A. Ballows, and H. G. Schlegel (eds.). *The Prokaryotes: A Handbook on Habitats, Isolation, and Identification of Bacteria.* Springer-Verlag, Berlin, pp. 290–298.

Chapman, V. J., and D. J. Chapman. 1975. *The Algae.* St. Martin's Press, New York.

Cole, G. T., and B. Kendrick (eds.). 1981. *Biology of Conidial Fungi.* Academic Press, New York.

Colwell, R. R. 1973. Genetic and phenetic classification of bacteria. *Advances in Applied Microbiology* 16:137–176.

Corliss, J. O. 1979. *The Ciliated Protozoa: Characterization, Classification and Guide to the Literature.* Pergamon Press, New York.

Cowan, S. T., and L. R. Hill (eds.). 1978. *A Dictionary of Microbial Taxonomy.* Cambridge University Press, New York.

Cross, T., and M. Goodfellow. 1973. Taxonomy and classification of the actinomycetes. In G. Sykes and F. A. Skinner (eds.). *Actinomycetales: Characteristics and Practical Importance.* Academic Press, London, pp. 11–112.

Dalton, A. J., and F. Haguenau (eds.). 1973. *Ultrastructure of Animal Viruses and Bacteriophages: An Atlas.* Academic Press, New York.

Dickerson, R. E. 1978. Chemical evolution and the origin of life. *Scientific American* 239(3):70–76.

Diener, T. O. 1981. Viroids. *Scientific American* 224(1):66–73.

Farmer, J. N. 1980. *The Protozoa: Introduction to Protozoology.* C. V. Mosby Co., St. Louis, Mo.

Fenner, F. 1976. The classification and nomenclature of viruses. *Journal of General Virology* 31:463–470.

Fenner, F., B. R. McAuslan, C. A. Mims, J. Sambrook, and D. O. White. 1974. *The Biology of Animal Viruses.* Academic Press, New York.

Fox, G. E., E. Stackebrandt, R. B. Hespell, J. Gibson, J. Maniloff, T. A. Dyer, R. S. Wolfe, W. E. Balch, R. S. Tanner, L. J. Magrum, L. B. Zablen, R. Blakemore, R. Gupta, L. Bonen, B. J. Lewis, D. A. Stahl, K. R. Luehrsen, K. N. Chen, and C. R. Woese. 1980. The phylogeny of prokaryotes. *Science* 209:457–463.

Fraenkel-Conrat, H., and R. R. Wagner (eds.). 1977. *Comprehensive Virology: Regulation and Genetics— Plant Viruses.* Plenum Press, New York.

Fraenkel-Conrat, H., and R. R. Wagner (eds.). 1978. *Comprehensive Virology: Newly Characterized Protist and Invertebrate Viruses.* Plenum Press, New York.

Fraenkel-Conrat, H., and P. C. Kimball. 1982. *Virology.* Prentice-Hall, Englewood Cliffs, N.J.

Gibbons, N. E., and R. G. E. Murray. 1978. Proposals concerning the higher taxa of bacteria. *International Journal of Systematic Bacteriology* 28:1–6.

Goodfellow, M., and D. E. Minnikin. 1977. Nocardioform bacteria. *Annual Review of Microbiology* 31:159–180.

Goodfellow, M., and D. E. Minnikin. 1981. Introduction to the coryneform bacteria. In M. P. Starr, H. Stolp, H. G. Trüper, A. Ballows, and H. G. Schlegel (eds.). *The Prokaryotes: A Handbook on Habitats, Isolation, and Identification of Bacteria.* Springer-Verlag, Berlin, pp. 1811–1826.

Gordon, J. K. 1981. Introduction to the nitrogen-fixing prokaryotes. In M. P. Starr, H. Stolp, H. G. Trüper, A. Ballows, and H. G. Schlegel (eds.). *The Prokaryotes: A Handbook on Habitats, Isolation, and Identification of Bacteria.* Springer-Verlag, Berlin, pp. 781–794.

Gottschalk, G. 1981. The anaerobic way of life of prokaryotes. In M. P. Starr, H. Stolp, H. G. Trüper, A. Ballows, and H. G. Schlegel (eds.). *The Prokaryotes: A Handbook on Habitats, Isolation, and Identification of Bacteria.* Springer-Verlag, Berlin, pp. 1415–1424.

Gray, W. D., and C. J. Alexopoulos. 1968. *Biology of the Myxomycetes.* The Ronald Press, New York.

Häeckel, E. 1866. *Generelle Morphologie der Organismen,* II. Georg Reiner, Berlin.

Halvorson, H. O., and K. E. Van Holde (eds.). 1980. *The Origins of Life and Evolution.* MBL Lectures in Biology. Vol. 1. Marine Biological Laboratory, Woods Hole, Mass.

Hastings, J. W., and K. H. Nealson. 1977. Bacterial bioluminescence. *Annual Review of Microbiology* 31:549–595.

Henriksen, S. D. 1976. *Moraxella, Branhamella,* and *Acinetobacter. Annual Review of Microbiology* 30:63–83.

Hollings, M. 1978. Mycoviruses: Viruses that infect fungi. *Advances in Virus Research* 22:2–54.

Holt, J. G., and N. R. Krieg (eds). 1984. *Bergey's Manual of Systematic Bacteriology.* 9th ed. Vol. I. Williams and Wilkins, Baltimore.

Holt, S. C. 1978. Anatomy and chemistry of spirochetes. *Microbiological Reviews* 42:114–160.

Jahn, T. L., E. C. Bovee, and F. F. Jahn. 1979. *How to Know the Protozoa.* William C. Brown Co., Dubuque, Iowa.

Johnson, R. C. 1976. The spirochetes. *Annual Review of Microbiology* 31:39–61.

Johnson, R. C. 1981. Introduction to the spirochetes. In M. P. Starr, H. Stolp, H. G. Trüper, A. Ballows, and H. G. Schlegel (eds.). *The Prokaryotes: A Handbook on Habitats, Isolation, and Identification of Bacteria.* Springer-Verlag, Berlin, pp. 533–537.

Kelly, D. P. 1981. Introduction to the chemolithotrophic bacteria. In M. P. Starr, H. Stolp, H. G. Trüper, A. Ballows, and H. G. Schlegel (eds.). *The Prokaryotes: A Handbook on Habitats, Isolation, and Identification of Bacteria.* Springer-Verlag, Berlin, pp. 997–1004.

Krieg, N. R. 1976. Biology of the chemoheterotrophic spirilla. *Bacteriological Reviews* 40:55–115.

Kudo, R. R. 1977. *Protozoology.* Charles C. Thomas, Springfield, Ill.

Kuenen, J. G., and O. Tuovinen. 1981. The genera Thiobacillus and Thiomicrospira. In M. P. Starr, H. Stolp, H. G. Trüper, A. Ballows, and H. G. Schlegel (eds.). *The Prokaryotes: A Handbook on Habitats, Isolation, and Identification of Bacteria.* Springer-Verlag, Berlin, pp. 1023–1036.

Lapage, S. P., P. H. A. Sneath, E. F. Lessel, V. B. D. Skerman, H. P. R. Seeliger, and W. A. Clark (eds.). 1975. *International Code of Nomenclature of Bacteria.* American Society for Microbiology, Washington, D.C.

Larsen, H. 1981. The family Halobacteriaceae. In M. P. Starr, H. Stolp, H. G. Trüper, A. Ballows, and H. G. Schlegel (eds.). *The Prokaryotes: A Handbook on Habitats, Isolation, and Identification of Bacteria.* Springer-Verlag, Berlin, pp. 985–994.

Laskin, A. I., and H. A. Lechevalier. 1977. *Handbook of Microbiology: Bacteria.* CRC Press, Cleveland, Ohio.

Laskin, A., and H. A. Lechevalier. 1979. *Handbook of Microbiology: Fungi, Algae, Protozoa, and Viruses.* CRC Press, Cleveland, Ohio.

Lechevalier, H. A., and M. P. Lechevalier. 1981. Introduction to the order Actinomycetales. In M. P. Starr, H. Stolp, H. G. Trüper, A. Ballows, and H. G. Schlegel (eds.). *The Prokaryotes: A Handbook on Habitats, Isolation, and Identification of Bacteria.* Springer-Verlag, Berlin, pp. 1915–1922.

Lee, R. E. 1980. *Phycology.* Cambridge University Press, Cambridge, England.

Lemke, P. A., and C. H. Nash. 1974. Fungal viruses. *Bacteriological Reviews* 38:29–56.

Levine, N. D., J. O. Corliss, F. E. G. Cox, G. Deroux, J. Grain, B. M. Honigberg, G. F. Leedale, A. R. Loeblich, J. Lom, D. Lynn, E. G. Meringeld, F. C. Page, G. Poljansky, V. Sprague, J. Vavra, and F. G. Wallace. 1980. A newly revised classification of the Protozoa. *Journal of Protozoology* 27:37–58.

Lewin, R. A. The Prochlorophytes. 1981. In M. P. Starr, H. Stolp, H. G. Trüper, A. Ballows, and H. G. Schlegel (eds.). *The Prokaryotes: A Handbook on Habitats, Isolation, and Identification of Bacteria.* Springer-Verlag, Berlin, pp. 257–266.

Linnaeus, C. 1759. *Systema Naturae per regna tria naturae, secundum classes, ordines, genera, species cum characteribus, differentiis, synonymis, locis.* Editio decima, reformata, Tom. I. Laurentii Salvii, Holmiae. 1964. Lubrect and Kramer, Forestburgh, N.Y.

Lippincott, J. A., B. Lippincott, and M. Starr. 1981. The genus *Agrobacterium.* In M. P. Starr, H. Stolp, H. G. Trüper, A. Ballows, and H. G. Schlegel (eds.). *The Prokaryotes: A Handbook on Habitats, Isolation, and Identification of Bacteria.* Springer-Verlag, Berlin, pp. 842–855.

Lodder, J., and N. Kreger-van Rij. 1970. *The Yeasts: A Taxonomic Study.* North Holland Publications, Amsterdam, The Netherlands.

Luria, S. E., J. E. Darnell, Jr., D. Baltimore, and A. Campbell. 1978. *General Virology.* John Wiley and Sons, New York.

Macy, J. M., and L. Probst. 1979. The biology of gastrointestinal bacteriodes. *Annual Review of Microbiology* 33:561–594.

Mah, R. A., and M. Smith. 1981. The methanogenic bacteria. In M. P. Starr, H. Stolp, H. G. Trüper, A. Ballows, and H. G. Schlegel (eds.). *The Prokaryotes: A Handbook on Habitats, Isolation, and Identification of Bacteria.* Springer-Verlag, Berlin, pp. 948–977.

Maramorosch, K. (ed.). 1977. *Insect and Plant Viruses: An Atlas.* Academic Press, New York.

Miller, O. K. 1979. *Mushrooms of North America.* E. P. Dutton Co., New York.

Moore-Landecker, E. 1982. *Fundamentals of the Fungi.* Prentice-Hall, Englewood Cliffs, N.J.

Mulder, E. G., and M. H. Deinema. 1981. The sheathed bacteria. In M. P. Starr, H. Stolp, H. G. Trüper, A. Ballows, and H. G. Schlegel (eds.). *The Prokaryotes: A Handbook on Habitats, Isolation, and Identification of Bacteria.* Springer-Verlag, Berlin, pp. 425–440.

Padau, E. 1979. Impact of facultatively anaerobic photoautotrophic metabolism on ecology of cyanobacteria (blue-green algae). *Advances in Microbial Ecology* 3:1–48.

Palleroni, N. J. 1981. Introduction to the family *Pseudomonadaceae*. In M. P. Starr, H. Stolp, H. G. Trüper, A. Ballows, and H. G. Schlegel (eds.). The *Prokaryotes: A Handbook on Habitats, Isolation, and Identification of Bacteria*. Springer-Verlag, Berlin, pp. 655–665.

Palmer, E. L., and M. L. Martin. 1982. *An Atlas of Mammalian Viruses*. CRC Press, Cleveland, Ohio.

Pfennig, N. 1977. Phototrophic green and purple bacteria: A comparative systematic survey. *Annual Review of Microbiology* 31:275–290.

Pfennig, N., F. Widdel, and H. Trüper. 1981. The dissimilatory sulfate-reducing bacteria. In M. P. Starr, H. Stolp, H. G. Trüper, A. Ballows, and H. G. Schlegel (eds.). The *Prokaryotes: A Handbook on Habitats, Isolation, and Identification of Bacteria*. Springer-Verlag, Berlin, pp. 926–940.

Pfennig, N., and H. G. Trüper. 1981. Isolation of members of the families Chromatiaceae and Chlorobiaceae. In M. P. Starr, H. Stolp, H. G. Trüper, A. Ballows, and H. G. Schlegel (eds.). *The Prokaryotes: A Handbook on Habitats, Isolation, and Identification of Bacteria*. Springer-Verlag, Berlin, pp. 279–289.

Phaff, H. J., M. W. Miller, and E. M. Mrak. 1978. *The Life of Yeasts*. Harvard University Press, Cambridge, Mass.

Pickett-Heaps, J. D. 1975. *Green Algae: Structure, Reproduction and Evolution in Selected Genera*. Sinauer Associates, Sunderland, Mass.

Poindexter, J. S. 1981. The caulobacters: Ubiquitous unusual bacteria. *Microbiological Reviews* 45:123–179.

Preer, L. B. 1981. Prokaryotic symbionts of *Paramecium*. In M. P. Starr, H. Stolp, H. G. Trüper, A. Ballows, and H. G. Schlegel (eds.). *The Prokaryotes: A Handbook on Habitats, Isolation, and Identification of Bacteria*. Springer-Verlag, Berlin, pp. 2127–2136.

Prusiner, S. B. 1982. Novel proteinaceous infectious particles cause scrapie. *Science* 216:136–144.

Prusiner, S. B. 1984. Prions. *Scientific American* 251(4):50–59.

Raj, H. D. 1981. The genus *Microcyclus* and related bacteria. In M. P. Starr, H. Stolp, H. G. Trüper, A. Ballows, and H. G. Schlegel (eds.). *The Prokaryotes: A Handbook on Habitats, Isolation, and Identification of Bacteria*. Springer-Verlag, Berlin, pp. 630–644.

Razin, S. 1978. The mycoplasmas. *Microbiological Reviews* 42:414–470.

Reichenbach, H., and M. Dworkin. 1981. Introduction to the gliding bacteria. In M. P. Starr, H. Stolp, H. G. Trüper, A. Ballows, and H. G. Schlegel (eds.). *The Prokaryotes: A Handbook on Habitats, Isolation, and Identification of Bacteria*. Springer-Verlag, Berlin, pp. 315–327.

Rose, A. H., and J. S. Harrison (eds.). 1969. *The Yeasts*. Academic Press, New York.

Sanderson, K. E. 1976. Genetic relatedness in the family Enterobacteriaceae. *Annual Review of Microbiology* 30:327–349.

Shapiro, L. 1976. Differentiation in the Caulobacter cell cycle. *Annual Review of Microbiology* 30:377–408.

Skerman, V. B. D. 1967. *A Guide to the Identification of the Genera of Bacteria*. Williams and Wilkins, Baltimore.

Skerman, V. B. D., V. McGowan, and P. H. A. Sneath (eds.). 1980. *Approved Lists of Bacterial Names*. American Society for Microbiology, Washington, D.C.

Skinner, F. A., and D. W. Lovelock. 1980. *Identification Methods for Microbiologists*. Academic Press, New York.

Sneath, P. H. A. 1978a. Classification of microorganisms. In J. R. Norris (ed.). *Essays in Microbiology*. John Wiley and Sons, Chichester, England, chapter 9.

Sneath, P. H. A. 1978b. Identification of microorganisms. In J. R. Norris (ed.). *Essays in Microbiology*. John Wiley and Sons, Chichester, England, chapter 10.

Sneath, P. H. A., and R. R. Sokal. 1973. *Numerical Taxonomy: The Principles and Practice of Numerical Classification*. W. H. Freeman, San Francisco.

Staley, J. T., P. Hirsch, and J. M. Schmidt. 1981. Introduction to the budding and/or appendaged bacteria. In M. P. Starr, H. Stolp, H. G. Trüper, A. Ballows, and H. G. Schlegel (eds.). *The Prokaryotes: A Handbook on Habitats, Isolation, and Identification of Bacteria*. Springer-Verlag, Berlin, pp. 449–450.

Stanier, R. Y., and G. Cohen-Bazire. 1977. Phototrophic prokaryotes: The cyanobacteria. *Annual Review of Microbiology* 31:225–274.

Stanier, R. Y., N. Pfennig, and H. G. Trüper. 1981. Introduction to the phototrophic prokaryotes. In M. P. Starr, H. Stolp, H. G. Trüper, A. Ballows, and H. G. Schlegel (eds.). *The Prokaryotes: A Handbook on Habitats, Isolation, and Identification of Bacteria*. Springer-Verlag, Berlin, pp. 197–211.

Starr, M. P., H. Stolp, H. G. Trüper, A. Ballows, and H. G. Schlegel (eds.). 1981. *The Prokaryotes: A Handbook on Habitats, Isolation, and Identification of Bacteria*. Springer-Verlag, Berlin.

Stolp, H. 1981. The genus *Bdellovibrio*. In M. P. Starr, H. Stolp, H. G. Trüper, A. Ballows, and H. G. Schlegel (eds.). *The Prokaryotes: A Handbook on Habitats, Isolation, and Identification of Bacteria*. Springer-Verlag, Berlin, pp. 618–629.

Trainor, F. R. 1978. *Introductory Phycology*. John Wiley and Sons, New York.

Trüper, H. G., and N. Pfennig. 1981. Characterization and identification of the anoxygenic phototrophic bacteria. In M. P. Starr, H. Stolp, H. G. Trüper, A. Ballows, and H. G. Schlegel (eds.). *The Prokaryotes: A Handbook on Habitats, Isolation, and Identification of Bacteria*. Springer-Verlag, Berlin, pp. 299–312.

Valentine, J. W. 1978. The evolution of multicellular plants and animals. *Scientific American* 239(3):140–158.

van Veen, W. L., E. G. Mulder, and M. H. Deinema. 1978. The *Sphaerotilus-Leptothrix* group of bacteria. *Microbiological Reviews* 42:329–356.

Vincent, J. M. 1981. The genus *Rhizobium*. In M. P. Starr, H. Stolp, H. G. Trüper, A. Ballows, and H. G. Schlegel (eds.). *The Prokaryotes: A Handbook on Habitats, Isolation, and Identification of Bacteria*. Springer-Verlag, Berlin, pp. 818–841.

Watson, S. W., F. Valois, and J. Waterbury. 1981. The family Nitrobacteraceae. In M. P. Starr, H. Stolp, H. G.

Trüper, A. Ballows, and H. G. Schlegel (eds.). *The Prokaryotes: A Handbook on Habitats, Isolation, and Identification of Bacteria*. Springer-Verlag, Berlin, pp. 1005–1022.

Westphal, A. 1976. *Protozoa*. Blackie and Son, Glasgow, Scotland.

Whitcomb, R. F. 1980. The genus *Spiroplasma*. *Annual Review of Microbiology* 34:677–709.

Whittaker, R. H. 1969. New concepts in kingdoms of organisms. *Science* 163:150–160.

Whittenbury, R., and H. Dalton. 1981. The methylotrophic bacteria. In M. P. Starr, H. Stolp, H. G. Trüper, A. Ballows, and H. G. Schlegel (eds.). *The Prokaryotes: A Handbook on Habitats, Isolation, and Identification of Bacteria*. Springer-Verlag, Berlin, pp. 894–902.

Wildy, P. 1971. Classification and nomenclature of viruses. *Monographs in Virology* 5:1–81.

Woese, C. R. 1981. Archaebacteria. *Scientific American* 244(6):98–122.

Woese, C. R., O. Kandler, and M. L. Wheelis. 1990. Towards a natural system of organisms: Proposal for the domains Archea, Bacteria, and Eucarya. *Proceedings of the National Academy of Science USA* 87:4576–4579.

Zeikus, J. G. 1977. The biology of methanogenic bacteria. *Bacteriological Reviews* 41:514–541.

Glossary

A horizon   the uppermost layer of soil; topsoil.

Abiotic   referring to the absence of living organisms.

Absorption   the uptaking, drinking in, or imbibing of a substance; the movement of substances into a cell; the transfer of substances from one medium to another, e.g., the dissolution of a gas in a liquid; the transfer of energy from electromagnetic waves to chemical bond and/or kinetic energy, e.g., the transfer of light energy to chlorophyll.

Accessory pigments   pigments including carotenoids and chlorophylls that harvest light energy and transfer it to the primary photosynthetic reaction centers; these pigments allow capture of light at different wavelengths.

Acellular   lacking cellular organization; not having a delimiting cytoplasmic membrane; organizational description of viruses, viroids, and prions.

Acetylene reduction assay   method for measuring rates of nitrogen fixation based upon the conversion of acetylene to ethylene by nitrogenase, the enzyme responsible for nitrogen fixation.

Acid mine drainage   consequence of the metabolism of sulfur- and iron-oxidizing bacteria when coal mining exposes pyrite to atmospheric oxygen and the combination of autoxidation and microbial sulfur and iron oxidation produces large amounts of sulfuric acid, which kills aquatic life and contaminates water.

Acidic   a compound that releases hydrogen (H^+) ions when dissolved in water; a compound that yields positive ions upon dissolution; a solution with a pH value less than 7.0.

Acidophiles   microorganisms that show a preference for growth at low pH, e.g., bacteria that grow only at very low pH values, ca. 2.0.

Acrasin   a substance (3´5´ cyclic AMP) secreted by a slime mold that initiates aggregation to form a fruiting body .

Acridine orange direct count (AODC)   a direct count method using the fluorescent dye acridine orange to stain bacterial cells; it detects total numbers of cells, both living and dead.

Actinomycetes   members of an order of bacteria in which species are characterized by the formation of branching and/or true filaments.

Activated sludge process   an aerobic secondary sewage treatment process using sewage sludge containing active complex populations of aerobic microorganisms to break down organic matter in sewage.

Adaptation   a structure or behavior that enhances survival and reproductive potential of an organism in an environment.

Adhesins   substances involved in the attachment or adherence of microorganisms to solid surfaces; factors that increase adsorption.

Adhesion factors   substances involved in the attachment of microorganisms to solid surfaces; factors that increase adsorption.

Adsorption   a surface phenomenon involving the retention of solid, liquid, or gaseous molecules at an interface.

Aerated pile method   method of composting for the decomposition of organic waste material where the wastes are heaped in separate piles and forced aeration provides oxygen.

Aerobes   microorganisms whose growth requires the presence of air or free oxygen.

Aerobic   having molecular oxygen present; growing in the presence of air.

Aerosol   a fine suspension of particles or liquid droplets sprayed into the air.

Algae   a heterogeneous group of eukaryotic, photosynthetic, unicellular and multicellular organisms lacking true tissue differentiation.

Algicides   chemical agents that kill algae.

Alkaline a condition in which hydroxyl (OH⁻) ions are in abundance; solutions with a pH of greater than 7.0; basic.

Allochthonous an organism or substance foreign to a given ecosystem.

Amensalism an interactive association between two populations that is detrimental to one while not adversely affecting the other.

Ammonification the release of ammonia from nitrogenous organic matter by microbial action.

Anaerobes organisms that grow in the absence of air or oxygen; organisms that do not use molecular oxygen in respiration.

Anaerobic the absence of oxygen; able to live or grow in the absence of free oxygen.

Anaerobic digestor a secondary sewage treatment facility used for the degradation of sludge and solid waste.

Anions negatively charged ions.

Anoxic absence of oxygen; anaerobic.

Anoxygenic photosynthesis photosynthesis that takes place in the absence of oxygen and during which oxygen is not produced; photosynthesis that does not split water and evolve oxygen.

Anoxyphotobacteria (anaerobic photosynthetic bacteria) bacteria that have only photosystem I and do not evolve oxygen in the course of their photosynthesis. They live in anaerobic aquatic habitats that receive some light.

Antagonism the inhibition, injury, or killing of one species of microorganism by another; an interpopulation relationship in which one population has a deleterious (negative) effect on another.

Aquatic growing, living in, or frequenting water; a habitat composed primarily of water.

Aquifer a geological formation containing water, such as subsurface water bodies that supply the water for wells and springs; a permeable layer of rock or soil that holds and transmits water.

Arbuscules specialized inclusions in root cortex in the vesicular-arbuscular type of mycorrhizal association.

Archaea (archaebacteria) prokaryotes with cell walls that lack murein, having ether bonds in their membrane phospholipids; analysis of rRNA indicates that the archaebacteria represent a primary biological kingdom distinct from both eubacteria and eukaryotes.

Ascomycetes members of a class of fungi distinguished by the presence of an ascus, a saclike structure containing sexually produced ascospores.

Asexual reproduction reproduction without union of gametes; formation of new individuals from a single individual.

Assimilation the incorporation of nutrients into the biomass of an organism.

Atmoecosphere that portion of the atmosphere in which living organisms are found and which is chemically transformed through the metabolism of organisms.

Atmosphere (atm) the whole mass of air surrounding the Earth; a unit of pressure approximating 1×10^6 dynes/cm².

Autecology branch of ecology that examines individual organisms in relation to their environment, emphasizing the "self-properties" of an organism's physiological attributes.

Autochthonous microorganisms and/or substances indigenous to a given ecosystem; the true inhabitants of an ecosystem; refering to the common microbiota of the body or soil microorganisms that tend to remain constant despite fluctuations in the quantity of fermentable organic matter.

Autotrophs organisms whose growth and reproduction are independent of external sources of organic compounds, the required cellular carbon being supplied by the reduction of CO_2 and the needed cellular energy being supplied by the conversion of light energy to ATP or the oxidation of inorganic compounds to provide the free energy for the formation of ATP.

Autoxidation the oxidation of a substance upon exposure to air.

B horizon the soil layer beneath the A horizon, consisting of weathered material and minerals leached from the overlying soil.

Bacteria members of a group of diverse and ubiquitous prokaryotic, single-celled organisms; organisms with prokaryotic cells, i.e., cells lacking a nucleus.

Bactericidal any physical or chemical agent able to kill some types of bacteria.

Bacteriochlorophyll (bacterialchlorophyll) photosynthetic pigment of green and purple anaerobic photosynthetic bacteria.

Bacteriophage a virus whose host is a bacterium; a virus that replicates within bacterial cells.

Bacteriostatic an agent that inhibits the growth and reproduction of some types of bacteria but need not kill the bacteria

Bacteroids irregularly shaped (pleomorphic) forms that some bacteria can assume under certain conditions, e.g., Rhizobium in root nodules.

Barophiles organisms that grow best and/or only under conditions of high pressure, e.g., in the ocean's depths.

Barotolerant organisms that can grow under conditions of high pressure but do not exhibit a preference for growth under such conditions.

Basidiomycetes a group of fungi distinguished by the formation of sexual basidiospores on a basidium.

Benthos the bottom region of aquatic habitats; collective term for the organisms living at the bottom of oceans and lakes.

Beta-oxidation (β-oxidation) metabolic pathway for the oxidation of fatty acids resulting in the formation of acetate and a new fatty acid that is two carbon atoms shorter than the parent fatty acid.

Biocide an agent that kills microorganisms.

Biodegradable a substance that can be broken down into smaller molecules by microorganisms.

Biodegradation the microbially mediated process of chemical breakdown of a substance to smaller products caused by microorganisms or their enzymes.

Biodeterioration the chemical or physical alteration of a product that decreases the usefulness of that product for its intended purpose.

Biodisc system a secondary sewage treatment system employing a film of active microorganisms rotated on a disc through sewage.

Biofilm a microbial community occurring on a surface as a microlayer.

Biofilter a device used for the bioremediation of air pollutants consising of an immobilized microbial community as a biofilm through which air is passed to detoxify contaminants.

Biogas gas produced by anaerobic microorganisms, primarily methane.

Biogenic element an element that is incorporated into the biomass of living organisms.

Biogeochemical cycling the biologically mediated transformations of elements that result in their global cycling, including transfer between the atmosphere, hydrosphere, and lithosphere.

Bioleaching the use of microorganisms to transform elements so that the elements can be extracted from a material when water is filtered through it.

Biological control the deliberate use of one species of organism to control or eliminate populations of other organisms; used in the control of pest populations.

Biological oxygen demand (BOD) the amount of dissolved oxygen required by aerobic and facultative microorganisms to stabilize organic matter in sewage or water; also known as biochemical oxygen demand.

Bioluminescence the generation of light by certain microorganisms; proteins called luciferins are converted in the presence of oxygen to oxyluciferins by luciferase with the liberation of light.

Biomagnification an increase in the concentration of a chemical substance, such as a pesticide, as the substance is passed to higher members of a food chain.

Biomass the dry weight, volume, or other quantitative estimation of organisms; the total mass of living organisms in an ecosystem.

Bioremediation the use of biological agents to reclaim soils and waters polluted by substances hazardous to human health and/or the environment; it is an extension of biological treatment processes that have traditionally been used to treat wastes in which microorganisms typically are used to biodegrade environmental pollutants.

Bioscrubber a device in which air moves through a fine spray or a microbial suspension in order to remove pollutants from the air.

Biosensor an immunological or genetic method of detecting chemicals or microbial activities, e.g., based upon the emission of light or electrical detection.

Biosphere the part of the Earth in which life can exist; all living things together with their environment.

Biosurfactant a surface active agent produced by microorganisms.

Biotic of or relating to living organisms, caused by living things.

Biotower a device similar to a trickling filter that employs a biofilm through which air is passed to remove pollutants.

Bioturbation mixing caused by movement of living organisms, responsible in part for the aeration of soils and sediments.

Black smoker a volcanic eruption occurring in deep sea regions that emits metal- and sulfide-rich waters that form black metal precipitates.

Blight any plant disease or injury that results in general withering and death of the plant without rotting.

Bloom a visible abundance of microorganisms, generally referring to the excessive growth of algae or cyanobacteria at the surface of a body of water.

Bottle effect growth of microorganisms within a collection vessel that results in artificially elevated microbial counts.

Breakpoint chlorination procedure for the removal and oxidation of ammonia from sewage to molecular nitrogen by the addition of hypochlorous acid.

C horizon the soil layer beneath the B horizon consisting of the broken or partially decomposed underlying bedrock.

Calvin cycle the primary pathway for carbon dioxide fixation (conversion of carbon dioxide to organic matter) in photoautotrophs and chemolithotrophs.

Cankers plant diseases, or conditions of those diseases, that interfere with the translocation of water and minerals to the crown of the plant.

Carbon cycle the biogeochemical cycling of carbon through oxidized and reduced forms, primarily between organic compounds and inorganic carbon dioxide.

Carrying capacity the largest population that a habitat can support.

Catabolic pathway a degradative metabolic pathway; a metabolic pathway in which large molecules are broken down into smaller ones.

Cations positively charged ions.

Cellulase an extracellular enzyme that hydrolyzes cellulose.

Cellulose a linear polysaccharide of β-D-glucose.

Chelator a substance that binds metallic ions.

Chemical oxygen demand (COD) the amount of oxygen required to oxidize completely the organic matter in a water sample.

Chemoautotrophs microorganisms that obtain energy from the oxidation of inorganic compounds and carbon from inorganic carbon dioxide; organisms that obtain energy through chemical oxidation and use inorganic compounds as electron donors; also known as chemolithotrophs.

Chemocline a boundary layer in an aquatic habitat formed by a difference in chemical composition, such as a halocline formed in the oceans by differing salt concentrations.

Chemolithotrophs microorganisms that obtain energy through chemical oxidation and use inorganic compounds as electron donors and cellular carbon through the reduction of carbon dioxide; also known as chemoautotrophs.

Chemoorganotrophs organisms that obtain energy from the oxidation of organic compounds and cellular carbon from preformed organic compounds.

Chemostat an apparatus used for continuous-flow culture to maintain bacterial cultures in a selected phase of growth, based on maintaining a continuous supply of a solution containing a nutrient in limiting quantities that controls the growth rate of the culture.

Chemotaxis a locomotive response in which the stimulus is a chemical concentration gradient; movement of microorganisms toward or away from a chemical stimulus.

Chitin a polysaccharide composed of repeating N-acetylglucosamine residues that is abundant in arthropod exoskeletons and fungal cell walls.

Chloramination the use of chloramines to disinfect water.

Chlorination the process of treating with chlorine, as in disinfecting drinking water or sewage.

Chlorophycophyta green algae; may be unicellular, colonial, or filamentous; most cells are uninucleate; some form coenocytic filaments, contain contractile vacuoles, or store starch as reserve material; their cell walls are composed of cellulose, mannans, xylans, or protein.

Chlorophyll the green pigment responsible for photosynthesis in plants; the primary photosynthetic pigment of algae and cyanobacteria.

Chlorosis the yellowing of leaves and/or plant components due to bleaching of chlorophyll, often symptomatic of microbial disease.

Chytrids members of the Chytridiales, which are mainly aquatic fungi that produce zoospores with a single posterior flagellum.

Ciliophora members of one subphylum of protozoa that possess simple to compound ciliary organelles in at least one stage of their life cycle; these protozoa are motile by means of cilia.

Circadian rhythms daily cyclical changes that occur in an organism even when it is isolated from the natural daily fluctuations of the environment.

Climax community the organisms present at the endpoint of an ecological succession series.

Coliforms Gram-negative, lactose-fermenting, enteric rods, e.g., *Escherichia coli.*

Colonization   the establishment of a site of microbial reproduction on a material, animal, or person without necessarily resulting in tissue invasion or damage.

Colony   the macroscopically visible growth of microorganisms on a solid culture medium.

Colony forming units (CFUs)   number of microbes that can replicate to form colonies, as determined by the number of colonies that develop.

Cometabolism   the gratuitous metabolic transformation of a substance by a microorganism growing on another substrate; the cometabolized substance is not incorporated into an organism's biomass, and the organism does not derive energy from the transformation of that substance.

Commensalism   an interactive association between two populations of different species living together in which one population benefits from the association, while the other is not affected.

Community   highest biological unit in an ecological hierarchy composed of interacting populations.

Compensation depth   the depth of an aquatic habitat at which photosynthetic activity balances respiratory activity; in lakes, the depth of effective light penetration, separating the limnetic and profundal zones.

Competition   an interactive association between two species, both of which need some limited environmental factor for growth and thus grow at suboptimal rates because they must share the growth-limiting resource.

Competitive exclusion principle   the statement that competitive interactions tend to bring about the ecological separation of closely related populations and precludes two populations from occupying the same ecological niche.

Competitive inhibition   the inhibition of enzyme activity caused by the competition of an inhibitor with a substrate for the active (catalytic) site on the enzyme; impairment of the function of an enzyme due to its reaction with a substance chemically related to its normal substrate.

Composting   the decomposition of organic matter in a heap by microorganisms; a method of solid waste disposal.

Concentration gradient   condition established by the difference in concentration on opposite sides of a membrane.

Consortium   an interactive association between microorganisms that generally results in combined metabolic activities.

Coprophagous   capable of growth on fecal matter; feeding on dung or excrement.

Corrosion   the eating away of a metal resulting from changes in the oxidative state.

$C_0t_{1/2}$ a measure of genetic diversity given as the initial concentration of single-stranded DNA multiplied by the time that it takes for half of the single-stranded DNA to reanneal to form double-stranded DNA.

Crop rotation   the alternation of the types of crops planted in a field.

Cross-feeding   the phenomenon that occurs when two organisms mutually complement each other in terms of nutritional factors or catabolic enzymes related to substrate utilizations; also termed syntrophism.

Crown gall   plant disease caused by *Agrobacterium tumefaciens*, which infects fruit trees, sugar beets, and other broad-leafed plants, manifested by the formation of a tumor growth.

Culture   to encourage the growth of particular microorganisms under controlled conditions; the growth of particular types of microorganisms on or within a medium as a result of inoculation and incubation.

Cyanobacteria   prokaryotic, photosynthetic organisms containing chlorophyll a, capable of producing oxygen by splitting water; formerly known as blue-green algae.

Cyst   a dormant form assumed by some microorganisms during specific stages in their life cycles, or assumed as a response to particular environmental conditions in which the organism becomes enclosed in a thin- or thick-walled membranous structure, the function of which is either protective or reproductive.

Decomposers   organisms, often bacteria or fungi, in a community that convert dead organic matter into inorganic nutrients.

Dendrograms   graphic representations of taxonomic analyses, showing the relationships between the organisms examined.

Denitrification   the formation of gaseous nitrogen or gaseous nitrogen oxides from nitrate or nitrite by microorganisms.

Desert   a region of low rainfall; a dry region; a region of low biological productivity.

Desiccation   removal of water; drying.

Desulfurization   removal of sulfur from organic compounds.

Detergent   a synthetic cleaning agent containing surface-active agents that do not precipitate in hard water; a surface-active agent having a hydrophilic and a hydrophobic portion.

Detrital food chain   a food chain based on the biomass of decomposers rather than on that of primary producers.

Detritivore   an organism that feeds on detritus; an organism that feeds on organic wastes and dead organisms.

Detritus   waste matter and biomass produced from decompositional processes.

Diatomaceous earth   a silicaceous material composed largely of fossil diatoms, used in microbiological filters and industrial processes.

Diatoms   unicellular algae having a cell wall composed of silica, the skeleton of which persists after the death of the organism.

Dinoflagellates   algae of the class Pyrrhophyta, primarily unicellular marine organisms, possessing two unequal flagella.

Direct counting procedures   methods for the enumeration of bacteria and other microbes that do not require the growth of cells in culture but rather rely upon direct observation or other detection methods by which the undivided microbial cells can be counted.

Direct viability count   a direct microscopic assay that determines whether or not microorganisms are metabolically active, i.e., viable.

Diversity   the heterogeneity of a system; the variety of different types of organisms occurring together in a biological community.

Diversity index   a mathematical measure which describes the species richness and apportionment of species within the community.

Domestic sewage   household liquid wastes.

Dormant   an organism or spore that exhibits minimal physical and chemical change over an extended period of time but remains alive.

Dwarfism   plant condition resulting from degradation or inactivation of plant growth substances by microorganisms.

Ecological balance   the totality of the interactions of organisms within an ecosystem that describes the stable relationships among populations and environmental quality.

Ecological niche   the functional role of an organism within an ecosystem; the combined description of the physical habitat, functional role, and interactions of the microorganisms occurring at a given location.

Ecological succession   a sequence in which one ecosystem is replaced by another within a habitat until an ecosystem that is best adapted is established.

Ecology   the study of the interrelationships between organisms and their environments.

Ecosystem   a functional self-supporting system that includes the organisms in a natural community and their environment.

Ectomycorrhizae   a stable, mutually beneficial (symbiotic) association between a fungus and the root of a plant where the fungal hyphae occur outside the root and between the cortical cells of the root.

Effluent   the liquid discharge from sewage treatment and industrial plants.

Endomycorrhizae   mycorrhizal association in which there is fungal penetration of plant root cells.

Endophytic   a photosynthetic organism living within another organism.

Endospores   thick-walled spores formed within a parent cell; in bacteria, heat-resistant spores.

Endosymbiotic   a symbiotic (mutually dependent) association in which one organism penetrates and lives within the cells or tissues of another organism.

Endosymbiotic evolution   theory that bacteria living as endosymbionts within eukaryotic cells gradually evolved into organelle structures.

Enrichment culture   any form of culture in a liquid medium that results in an increase in a given type of organism while minimizing the growth of any other organism present.

Enteric bacteria   bacteria that live within the intestinal tract of mammals.

Entomogenous fungi   fungi living on insects; fungal pathogens of insects.

Enumeration   determination of the number of microorganisms.

Enzyme-linked immunosorbent assay (ELISA)   a technique used for detecting and quantifying specific serum antibodies based upon tagging the antigen–antibody complex with a substrate that can be enzymatically converted to a readily quantifiable product by a specific enzyme.

Epilimnion   the warm layer of an aquatic environment above the thermocline.

Epiphytes   organisms growing on the surface of a photosynthetic organism, e.g., bacteria growing on the surface of an algal cell.

Epizootic an epidemic outbreak of infectious disease among animals other than humans.

Equitability the measure of the proportion of individuals among the species present.

Erosion breakdown of material from the earth's crust by various physical and chemical processes.

Estuary a water passage where the ocean tide meets a river current; an arm of the sea at the lower end of a river.

Eubacteria prokaryotes other than archaebacteria.

Euglenophycophyta unicellular division of algae that contain chlorophylls a and b and appear green, lack a cell wall, and are surrounded by a pellicle; they store paramylon as reserve material and reproduce by longitudinal division; they are widely distributed in aquatic and soil habitats.

Eukaryotes cellular organisms having a membrane-bound nucleus within which the genome of the cell is stored as chromosomes composed of DNA; eukaryotic organisms include algae, fungi, protozoa, plants, and animals.

Euphotic the top layer of water, through which sufficient light penetrates to support the growth of photosynthetic organisms.

Eurythermal microorganisms that grow over a wide range of temperatures.

Eutrophic containing high-nutrient concentrations, such as a eutrophic lake with high phosphate concentration that will support excessive algal blooms.

Eutrophication the enrichment of natural waters with inorganic materials, especially nitrogen and phosphorus compounds, that support the excessive growth of photosynthetic organisms.

Evenness a description of the distribution of microbial populations within a community based on the apportionment of individuals and species.

Evolution the directional process of change of organisms by which descendants become distinct in form and/or function from their ancestors.

Extreme environments environments characterized by extremes in growth conditions, including temperature, salinity, pH, and water availability, among others.

Floc a mass of microorganisms cemented together in a slime produced by certain bacteria, usually found in waste treatment plants.

Food web an interrelationship among organisms in which energy is transferred from one organism to another; each organism consumes the preceding one and in turn is eaten by the next higher member in the sequence.

Freshwater habitats lakes, ponds, swamps, springs, streams, and rivers.

Fruiting bodies specialized microbial structures that bear sexually or asexually derived spores.

Frustules the siliceous cell walls of a diatom.

Fungal gardens fungi grown in pure culture by insects.

Fungi a group of diverse, unicellular and multicellular eukaryotic organisms, lacking chlorophyll, often filamentous and spore-producing.

Fungicides agents that kill fungi.

Fungistasis the active prevention or hindrance of fungal growth by a chemical or physical agent.

Galls abnormal plant structures formed in response to parasitic attack by certain insects or microorganisms; tumorlike growths on plants in response to an infection.

Genetic engineering the deliberate modification of the genetic properties of an organism by the application of recombinant DNA technology.

Germ-free animal an animal without microbiota; all of its surfaces and tissues are sterile, and it is maintained in that condition by being housed and fed in a sterile environment.

Gnotobiotic culture or environment containing only defined forms of life.

Grazers organisms that prey upon primary producers; protozoan predators that consume bacteria indiscriminately; filter-feeding zooplankton.

Greenhouse effect rise in the concentration of atmospheric CO_2 and a resulting warming of global temperatures.

Gross primary production total amount of organic matter produced in an ecosystem.

Groundwater subsurface water in terrestrial environments.

Growth rate increase in the number of microorganisms per unit time.

Guild structure populations within a community which use the same resources.

Habitat a location where living organisms occur.

Halophiles organisms requiring NaCl for growth; extreme halophiles grow in concentrated brines.

Herbicides chemicals used to kill weeds.

Heterocysts   cells that occur in the trichomes of some fila-mentous cyanobacteria that are the sites of nitrogen fixation.

Heterotrophs   organisms requiring organic compounds for growth and reproduction, the organic compounds serve as sources of carbon and energy.

Holdfast   a structure that allows certain algae and bacteria to remain attached to the substratum.

Host   a cell or organism that acts as the habitat for the growth of another organism; the cell or organism upon or in which parasitic organisms live.

Hot springs   thermal springs with a temperature greater than 37°C.

Humic acids   high molecular weight irregular organic polymers with acidic character; the portion of soil organic matter soluble in alkali but not in acid.

Humus   the organic portion of the soil remaining after microbial decomposition.

Hydrosphere   the aqueous envelope of the Earth, including bodies of water and aqueous vapor in the atmosphere.

Hydrostatic pressure   pressure exerted by the weight of a water column; it increases approximately 1 atm with every 10 m/in depth.

Hypolimnion   the deeper, colder layer of an aquatic environment; the water layer below the thermocline.

Immobilization   the binding of a substance so that it is no longer reactive or able to circulate freely.

In situ   in the natural location or environment.

In vitro   in glass; a process or reaction carried out in a culture dish or test tube.

In vivo   within the living organism.

Indicator organism   an organism used to identify a particular condition, such as *Escherichia coli* as an indicator of fecal contamination.

Indigenous   native to a particular habitat.

Insecticides   substances destructive to insects; chemicals used to control insect populations.

Ionizing radiation   radiation, such as gamma and X-radiation, that forms toxic-free radicals, disruptive to the biochemical organization of cells.

Kelp   brown algae with vegetative structures consisting of a holdfast, stem, and blade, that form large macroscopic structures.

Landfarming (landtreatment)   the application of toxic organic wastes to soils for the purpose of biodegradation.

Landfill   a site where solid waste is dumped and allowed to decompose; a process in which solid waste containing both organic and inorganic material is deposited and covered with soil.

Landtreatment (landfarming)   the application of toxic organic wastes to soils for the purpose of biodegradation.

Leach   to wash or extract soluble constituents from insoluble materials.

Leguminous crop   plants belonging to the *Leguminosae* which have a seed pod divided into two parts or valves.

Lichens   a large group of composite organisms consisting of a fungus in symbiotic association with an alga or cyanobacterium.

Lignin   a class of complex polymers in the woody material of higher plants, second in abundance to cellulose only.

Limnetic zone   in lakes, the portion of the water column excluding the littoral zonewhere primary productivity exceeds respiration.

Liquid wastes   waste material in liquid form, the result of agricultural, industrial, and all other human activities.

Lithosphere   the solid part of the Earth.

Lithotrophs   microorganisms that live in and obtain energy from the oxidation of inorganic matter; chemoautotrophs.

Littoral   situated or growing on or near the shore; the region between the high and low tide marks.

Magnetotaxis   motility directed by a geomagnetic field.

Manganese nodules   nodules (round, irregular mineral masses) produced in part by microbial oxidation of bivalent manganese oxides.

Marine   of or relating to the oceans.

Mesophiles   organisms whose optimum growth is in the temperature range of 20°C–45°C.

Methanogens   methane-producing prokaryotes; a group of archaebacteria capable of reducing carbon dioxide or low–molecular-weight fatty acids to produce methane.

Methylation   the process of substituting a methyl group for a hydrogen atom.

Microbial ecology   the field of study that examines the interactions of microorganisms with their biotic and abiotic surroundings.

Microbial mining   a mineral recovery method that uses

bioleaching to recover metals from ores not suitable for direct smelting.

Microbial pesticides preparations of pathogenic or predatory microorganisms that are antagonistic toward a particular pest population.

Microbiology the study of microorganisms and their activities.

Microbiota the totality of microorganisms associated with a given environment.

Microorganisms microscopic organisms, including algae, bacteria, fungi, protozoa, and viruses.

Mildew any of a variety of plant diseases in which the mycelium of the parasitic fungus is visible on the affected plant; biodeterioration of food or fabric due to fungal growth.

Mineralization the microbial breakdown of organic materials into inorganic materials brought about mainly by microorganisms.

Mixotrophs organisms capable of utilizing both autotrophic and heterotrophic metabolic processes, e.g., the concomitant use of organic compounds as sources of carbon and light as a source of energy.

Most probable number (MPN) a method for determination of viable organisms using statistical analyses and successive dilution of the sample to reach a point of extinction.

Mutualism a stable condition in which two organisms of different species live in close physical association, each organism deriving some benefit from the association; symbiosis.

Mycangia see mycetangia.

Mycelia the interwoven mass of discrete fungal hyphae.

Mycetangia specialized pocketlike invaginations of fungus-cultivating insects for storage of mycelial inoculum.

Mycobiont the fungal partner in a lichen.

Mycorrhizae a stable, symbiotic association between a fungus and the root of a plant; the term also refers to the root–fungus structure itself.

Necrosis the pathological death of a cell or group of cells in contact with living cells.

Net primary production amount of organic carbon in the form of biomass and soluble metabolites available for heterotrophic consumers in terrestrial and aquatic habitats.

Neuston the layer of organisms growing at the interface between air and water.

Neutralism the relationship between two different microbial populations characterized by the lack of any recognizable interaction.

Niche the functional role of an organism within an ecosystem; the combined description of the physical habitat, functional role, and interactions of the microorganisms occurring at a given location.

Nitrification the process in which ammonia is oxidized to nitrite and nitrite to nitrate; a process primarily carried out by the strictly aerobic, chemolithotrophic bacteria of the family Nitrobacteraceae.

Nitrifying bacteria Nitrobacteraceae; Gram-negative, obligately aerobic, chemolithotrophic bacteria occurring in aquatic environments and in soil that oxidize ammonia to nitrite or nitrite to nitrate.

Nitrite ammonification reduction of nitrite to ammonium ions by bacteria; does not remove nitrogen from the soil.

Nitrogen fixation the reduction of gaseous nitrogen to ammonia, carried out by certain prokaryotes.

Nitrogenase the enzyme that catalyzes biological nitrogen fixation.

Nodules tumorlike growths formed by plants in response to infections with specific bacteria within which the infecting bacteria fix atmospheric nitrogen; a rounded, irregularly shaped mineral mass.

No tillage farming a type of soil management involving elimination of plowing, retention of crop residues as mulch cover, and planting done with drill seeding machines.

Numerical taxonomy a system that uses overall degrees of similarity and large numbers of characteristics to determine the taxonomic position of an organism; allows organisms of unknown affiliation to be identified as members of established taxa.

O horizon the organic layer of soil, consisting of humic substances.

Obligate aerobes organisms that grow only under aerobic conditions, i.e., in the presence of air or oxygen.

Obligate anaerobes organisms that cannot use molecular oxygen; organisms that grow only under anaerobic conditions, i.e., in the absence of air or oxygen; organisms that cannot carry out respiratory metabolism.

Obligate intracellular parasites organisms that can live and reproduce only within the cells of other organisms, such as viruses, all of which must find suitable host cells for their replication.

Obligate thermophiles organisms restricted to growth at high temperatures.

Oligotrophic lakes and other bodies of water that are poor in those nutrients that support the growth of aerobic, photosynthetic organisms; microorganisms that grow at very low nutrient concentrations.

Optimal growth temperature the temperature at which microbes exhibit the maximal growth rate.

Optimal oxygen concentration the oxygen concentration at which microbes exhibit the maximal growth rate with maximal product yield.

Osmophiles organisms that grow best or only in or on media of relatively high osmotic pressure.

Osmotic pressure the force resulting from differences in solute concentrations on opposite sides of a semipermeable membrane.

Osmotolerant organisms that can withstand high osmotic pressures and grow in solutions of high solute concentrations.

Oxidation pond a method of aerobic waste disposal employing biodegradation by aerobic and facultative microorganisms growing in a standing water body.

Oxidation-reduction potential a measure of the tendency of a given oxidation-reduction system to donate electrons, i.e., to behave as a reducing agent, or to accept electrons, i.e., to act as an oxidizing agent; determined by measuring the electrical potential difference between the given system and a standard reference system.

Ozonation the killing of microorganisms by exposure to ozone.

P/R ratio the relationship between gross photosynthesis and rate of community respiration.

Paralytic shellfish poisoning caused by toxins produced by the dinoflagellate *Gonyaulax*, which concentrates in shellfish such as oysters and clams.

Parasites organisms that live on or in the tissues of another living organism, the host, from which they derive their nutrients.

Parasitism an interactive relationship between two organisms or populations in which one is harmed and the other benefits; generally, the population that benefits, the parasite, is smaller than the population that is harmed.

Pathogens organisms capable of causing disease in animals, plants, or microorganisms.

Pelagic zone the portion of the marine environment beyond the edge of the continental shelf, comprising the entire water column but excluding the sea floor.

Pest a population that is an annoyance for economic, health, or aesthetic reasons.

Pesticides substances destructive to pests, especially insects.

pH the symbol used to express the hydrogen ion concentration, signifying the logarithm to the base 10 of the reciprocal of the hydrogen ion concentration.

Photoautotrophs organisms whose source of energy is light and whose source of carbon is carbon dioxide; characteristic of plants, algae and some prokaryotes.

Photoheterotrophs organisms that obtain energy from light but require exogenous organic compounds for growth.

Photolysis liberation of oxygen by splitting of water during photosynthesis.

Photosynthesis the process in which radiant (light) energy is absorbed by specialized pigments of a cell and is subsequently converted to chemical energy; the ATP formed in the light reactions is used to drive the fixation of carbon dioxide, with the production of organic matter.

Photosynthetic membranes specialized membranes in photosynthetic bacteria that are the anatomical sites where light energy is converted to chemical energy in the form of ATP during photosynthesis.

Phototaxis the ability of bacteria to detect and respond to differences in light intensity, moving toward or away from light.

Phototrophs organisms whose sole or principal primary source of energy is light; organisms capable of photophosphorylation.

Phycobiont the algal partner of a lichen.

Phytoplankton passively floating or weakly motile photosynthetic aquatic organisms, primarily cyanobacteria and algae.

Phytoplankton food chain a food chain in aquatic habitats based on the consumption of primary producers.

Plankton collectively, all microorganisms and invertebrates that passively drift in lakes and oceans.

Plate counting method of estimating numbers of microorganisms by diluting samples, culturing on solid media, and counting the colonies that develop to estimate the number of viable microorganisms in the sample.

Pollutants materials that contaminate air, soil, or water; substances—often harmful—that foul water or soil, reducing their purity and usefulness.

Predation   a mode of life in which food is primarily obtained by killing and consuming animals; an interaction between organisms in which one benefits and one is harmed, based on the ingestion of the smaller organism, the prey, by the larger organism, the predator.

Predators   organisms that practice predation.

Preemptive colonization   alteration of environmental conditions by pioneer organisms in a way that discourages further succession.

Prey   an animal taken by a predator for food.

Primary producers   organisms capable of converting carbon dioxide to organic carbon, including photoautotrophs and chemoautotrophs.

Primary sewage treatment   the removal of suspended solids from sewage by physical settling in tanks or basins.

Prions   infectious proteins; substances that are infectious and reproduce within living systems but appear to be proteinaceous, based on degradation by proteases, and to lack nucleic acids based on resistance to digestion by nucleases.

Profundal zone   in lakes, the portion of the water column where respiration exceeds primary productivity.

Proto-cooperation   synergism; a nonobligatory relationship between two microbial populations in which both populations benefit.

Protozoa   diverse eukaryotic, typically unicellular, non-photosynthetic microorganisms generally lacking a rigid cell wall.

Psychrophile   an organism that has an optimum growth temperature below 20°C.

Psychrotroph   a mesophile that can grow at low temperatures.

Pure culture   a culture that contains cells of one kind; the progeny of a single cell.

Putrefaction   the microbial breakdown of protein under anaerobic conditions.

Pyrite   a common mineral containing iron disulfite.

Recalcitrant   a chemical that is totally resistant to microbial attack.

Red tides   aquatic phenomenon caused by toxic blooms of *Gonyaulax* and other dinoflagellates that color the water and kill invertebrate organisms; the toxins concentrate in the tissues of filter-feeding molluscs, causing food poisoning.

Reporter gene   a gene whose expression is easily detected and that can be used to track the transcription of other genes, e.g., the *lux* gene which codes for light production and can be used to detect the expression of other genes.

Resistant crop varieties   species of agricultural plants that are not susceptible to particular plant pathogens.

Rhizosphere   an ecological niche that comprises the surfaces of plant roots and the region of the surrounding soil in which the microbial populations are affected by the presence of the roots.

Rhizosphere effect   evidence of the direct influence of plant roots on bacteria, demonstrated by the fact that microbial populations usually are higher within the rhizosphere (the region directly influenced by plant roots) than in root-free soil.

Rotating biological contactor   see biodisc system.

Rots   plant diseases characterized by the breakdown of tissue caused by any of a variety of fungi or bacteria.

Rusts   plant diseases caused by fungi of the order Uredinales, so called because of the rust-colored spores formed by many of the causal agents on the surfaces of the infected plants.

Salt lake   a landlocked water body with a high salt concentration often approaching saturation.

Sanitary landfill   a method for disposal of solid wastes in low-lying areas; the deposited wastes are covered with a layer of soil each day.

Sarcodina   a major taxonomic group of protozoa characterized by the formation of pseudopodia.

Secondary sewage treatment   the treatment of the liquid portion of sewage containing dissolved organic matter, using microorganisms to degrade the organic matter that is mineralized or converted to removable solids.

Self-purification   inherent capability of natural waters to cleanse themselves of pollutants based on biogeochemical cycling activities and interpopulation relationships of indigenous microbial populations.

Septic tank   a simple anaerobic treatment system for waste water where residual solids settle to the bottom of the tank and the clarified effluent is distributed over a leaching field.

Sewage   the refuse liquids or waste matter carried by sewers.

Sewage treatment   the treatment of sewage to reduce its biological oxygen demand and to inactivate the pathogenic microorganisms present.

Sludge   the solid portion of sewage.

Smuts   plant diseases caused by fungi of the order Ustilaginales; typically involve the formation of masses of dark-colored teliospores on or within the tissues of the host plant.

Soil horizon   a layer of soil distinguished from layers above and below by characteristic physical and chemical properties.

Solid waste   refuse, waste material composed both of inert materials—glass, plastic, and metal—and decomposable organic wastes, including paper and kitchen scraps.

Stenothermophiles   microorganisms that grow only at temperatures near their optimal growth temperature.

Succession   the replacement of populations by other populations better adapted to fill the ecological niche.

Sulfur cycle   biogeochemical cycle mediated by microorganisms that changes the oxidation state of sulfur within various compounds.

Symbiosis   an obligatory interactive association between members of two populations, producing a stable condition in which the two organisms live together in close physical proximity to their mutual advantage.

Symbiotic nitrogen fixation   fixation of atmospheric nitrogen by bacteria living in mutually dependent associations with plants.

Synergism   in antibiotic action, when two or more antibiotics are acting together, the production of inhibitory effects on a given organism that are greater than the additive effects of those antibiotics acting independently; an interactive but nonobligatory association between two populations in which each population benefits.

Syntrophism   the phenomenon that occurs when two organisms mutually complement each other in terms of nutritional factors or catabolic enzymes related to substrate utilization, also termed cross-feeding.

Temperature growth range   the range between the maximum and minimum temperatures at which a microorganism can grow.

Tertiary recovery of petroleum   the use of biological and chemical means to enhance oil recovery.

Tertiary sewage treatment   a sewage treatment process that follows a secondary process, aimed at removing nonbiodegradable organic pollutants and mineral nutrients.

Thermal stratification   division of temperate lakes into an epilimnion, thermocline, and hypolimnion, subject to seasonal change; zonation of lakes based on temperature where warm and cold water masses do not mix.

Thermal vents   hot areas located at depths of 800–1,000 m on the sea floor, where spreading allows seawater to percolate deeply into the crust and react with hot core materials; life around the vents is supported energetically by the chemoautotrophic oxidation of reduced sulfur.

Thermocline   zone of water characterized by a rapid decrease in temperature, with little mixing of water across it.

Thermophiles   organisms having an optimum growth temperature above 45°C.

Tolerance range   the range of a parameter, such as temperature, over which microorganisms survive.

Trickling filter system   a simple, film-flow aerobic sewage treatment system; the sewage is distributed over a porous bed coated with bacterial growth that mineralizes the dissolved organic nutrients.

Trophic level   the position of an organism or population within a food web: primary producer, grazer, predator, etc.

Trophic structure   the collection of steps in the transfer of energy stored in organic compounds from one to another.

Trophozoite   a vegetative or feeding stage in the life cycle of certain protozoa.

Turbidity   cloudiness or opacity of a suspension .

Turbidostat   a system in which an optical sensing device measures the turbidity of the culture in a growth vessel and generates an electrical signal that regulates the flow of fresh medium into the vessel and the release of spent medium and cells.

Ultraviolet light (UV)   short wavelength electromagnetic radiation in the range 100–400 nm.

Vesicular arbuscular mycorrhizae   a common type of mycorrhizae characterized by the formation of vesicles and arbuscules.

Viable nonculturable microorganism   microorganisms that do not grow in viable culture methods, but which are still metabolically active and capable of causing infections in animals and plants.

Viable plate count   method for the enumeration of bacteria whereby serial dilutions of a suspension of bacteria are plated onto a suitable solid grown medium, the plates are incubated, and the number of colony-forming units is counted.

Viroids   the causal agents of certain diseases, resembling viruses in many ways but differing in their apparent lack of a viruslike structural organization and their resistance to a wide variety of treatments to which viruses are sensitive; naked infective RNA.

Virus   a noncellular entity that consists minimally of protein and nucleic acid and that can replicate only after entry into specific types of living cells; it has no intrinsic metabolism, and its replication is dependent on the direction of cellular metabolism by the viral genome; within the host cell, viral components are synthesized separately and are assembled intracellularly to form mature, infectious viruses.

Visible light   radiation in the wavelength range of 400–800 nm that is required for photosynthesis but can be lethal to some nonphotosynthetic microorganisms.

Volatile organic chemical (VOC)   organic compound that vaporizes into the atmosphere.

Water activity (A_w)   a measure of the amount of reactive water available, equivalent to the relative humidity; the percentage of water saturation of the atmosphere.

Wilts   plant diseases characterized by a reduction in host tissue turgidity, commonly affecting the vascular system.

Windrow method   a slow composting process that requires turning and covering with soil or compost.

Xenobiotic   a synthetic product not formed by natural biosynthetic processes; a foreign substance or poison.

Xerotolerant   able to withstand dryness; an organism capable of growth at low water activity.

Yeasts   a category of fungi defined in terms of morphological and physiological criteria; typically, unicellular, saprophytic organism that characteristically ferment a range of carbohydrates and in which asexual reproduction occurs by budding.

Zymogenous   term used to describe soil microorganisms that grow rapidly on exogenous substrates.

Index

Abyssal plain, 266
Acantharia, 271
Acclimation, 418
Acetic acid, 352
Acetobacter, 515
Acetobacter aceti, 352
Acetogenic bacteria, 109
Acetospora, 527, 528
Acetylene blockage assay, 323
Acetylene reduction assay, 317
Achromobacter, 255, 260, 365
Acid mine drainage, 383, 404
Acidaminococcus, 516
Acidianus, 510
Acidophiles, 325, 335, 404
Acineta, 256
Acinetobacter, 278, 373, 389, 516
Acrasiales, 521
Acrasin, 521
Acremonium, 85
Acridine orange direct count, 180
Actinomyces, 278
Actinomycetes, 278, 518, 519
 nitrogen fixation, 82
Actinoplanaceae, 519
Activated sludge process, 365, 367
Acylanilide herbicides, 403
Adaptation, 418
Adenylate pool, 191
Aeromonas, 322, 516
Aerosols, 250
Aflatoxin, 351
African sleeping sickness, 123
Agar film technique, 179
Agar plate count method, 186
Agaricales, 523

Agaricus, 60
Agaricus bisporus, 454
Agricultural management, 461
Agricultural soil, 354
Agrobacterium, 73, 90, 131, 278, 466, 515
Agrobacterium radiobacter, 466
Agrobacterium tumefaciens, 91, 92, 93, 466, 515
Agrocins, 466
AIDS, 124
Air emissions, 431
Air pollutants
 effects on lichens, 51
Air sampler, 169
Alcaligenes, 255, 270, 278, 308, 322, 335, 387, 389, 515
Alcaligenes eutrophus, 392
Alcohols, basis for antagonism, 56
Aldrin, 401
Alexander, Martin, 11
Alfalfa, nitrogen fixation, 82
Algae, 524
 brown, 516
 chlorophyll determination, 192
 coral reef, 268
 freshwater, 261
 marine, 270
 production as food, 453
 soil, 280
Algal biomass
 chlorophyll determination, 192
Algal bloom, 136
Alicyclic hydrocarbons, 395
Alkyl benzyl sulfonates, 392
Allee's principle, 37
Allelopathy, 56, 73

Allochthonous microorganisms, 137, 246, 247
Alternaria, 84, 524
Alteromonas, 178
"Alvin" deep sea submersible, 169
Amanita, 75
Amanita phalloides, 523
Amastigomycota, 521
Ambrosia beetles, 108
 cultivation of fungi, 107, 108
Amensalism, 56–57, 466
Amidase inhibitors, 401
Ammonia, 319
 removal, 373
 stripping, 372, 374
Ammonia monooxygenase, 320
Ammonification, 319
Ammonium ions, 319
Amoeba, 261, 527
Amylase, 198
Anabaena, 84, 154, 255, 260, 278, 318, 319, 512
 nitrogen fixation, 83
Anabaena flos-aquae, 475
Anabaena oscillatorioides, 47
Anabaena-Azolla association, 85
Anaerobic digestor, 369, 370, 371
Anaerobic lake, 155
Anaerobic sewage treatment, 371
Anaerobic wastewater treatment, 369
Analysis of variance, 492, 493, 494
Andersen air sampler, 169
Animal
 coprophagous, 104
 diseases, 120
 viruses, 507

Anomalops, 119
ANOVA, 492
Anoxyphotobacteria, 308, 511
Antagonism, 56–57, 464
Antarctic ozone hole, 248
Antibiosis, 56
Antibiotic resistance, 138
Antibiotic resistance genes, 139
Antibiotics, 57
Ants
  cultivation of fungi, 106
  fungal gardens, 105
AODC, 180
Apatite, 332
Aphanizomenon, 255, 260, 318
Apicomplexa, 527, 528
Appressoria, 50
Aquatic habitat
  biodegradation, 306
  carbon flow, 306
Aquifer, 357
Arbovirus, 123
Arbuscules, 76
Arcahaeoglobus, 510
Arcella, 256
Archaea, 25, 508
  characteristics, 26
  cofactors, 26
  evolution, 25
Archaebacteria, 25, 215, 328, 508, 509,
  511
Archangium, 278
Armillaria mellea, 76
Arochlor, 390
Aromatic
  hydrocarbons, 396
  solvents, 429
Arsenic, 410
Arthrobacter, 48, 73, 278, 304, 321,
  336, 365, 518
Arthrobotrys, 117, 524
Ascocarp, 523
Ascochytula, 84
Ascomycetous yeasts, 523
Ascomycotina, 522, 523
Aspergillus, 120, 131, 277, 279, 321, 524
Aspergillus flavus, 351
Aspergillus fumigatus, 359

Asphaltenes, 398
Assimilatory nitrate reduction, 322
Assimilatory sulfate reduction, 324
Asterionella formosa, 55
Atlas, Ronald, 11
Atmosphere
  carbon dioxide in, 292
  dispersal of microorganisms, 248
  methane in, 293
  microorganisms in, 253
  properties, 247
ATP, luciferin-luciferase assay, 190
Attine ants, 105, 106, 107
Aureobasidium, 524
Aureobasidium pullulans, 83
Autecology, 130, 213
Autochthonous microorganisms, 8,
  246, 247
Autotrophic succession, 132
Auxins, 72
Average linkage clustering techniques,
  500
Axenic cultures, 145
Azolla, 84, 85, 319
Azorhizobium, 79
Azospirillum, 318, 319, 322
Azotobacter, 317, 318, 319, 420, 515

Bacillus, 63, 131, 260, 270, 277, 278,
  318, 322, 335, 336, 366, 465, 518
Bacillus cereus, 392
Bacillus lentimorbus, 468, 471
Bacillus popilliae, 468, 471
Bacillus pumilus, 350
Bacillus sphaericus, 468
Bacillus stearothermophilus, 218, 359
Bacillus thuringiensis, 468
  genetics, 471
  recombinant strains, 477
  targets, 470
  toxins, 469
Bacteria
  evolution, 27
  iron-oxidizing, 335
  marine, 270
  soil, 278
Bacterial pesticides, 468
Bacteriophage, 58, 508

Bacteroidaceae, 516
Bacteroides, 111, 113, 134, 135, 516
Bacteroides amylophilus, 114
Bacteroides ruminicola, 114
Bacteroides succinogenes, 114
Bacteroids, 80
Baculoviruses, 468
Baeocyte, 512
Bartha, Richard, 11
Basidiomycetes, 523
Basidiomycotina, 522
Batch systems, 145
Bathurst burr, 475
Bathyl region, 266
Bathypelagic zone, 267
Bdelloplast, 59
Bdellovibrio, 58, 59, 60, 367, 515
Beauveria, 472
Beauveria bassiana, 472
Beeswax, 110
Beetles, bark feeding, 108
Beggiatoa, 8, 45, 73, 324, 325, 326,
  366, 514
Beijerinck, Martinus, 8, 154, 328
Beijerinckia, 318, 319, 515
Beneckea, 119
Benthos, 266
Benzene, 429
Benzoate, 352
Beta-oxidation, 395, 400
Bias, 486
Bicarbonate, 336
Bifidobacterium, 111, 134
Bioaccumulation
  heavy metals, 410, 442
  radionuclides, 411
Bioassays, 423
Biocides, 352
Biodecontamination, 426
Biodegradability, 418
  plastics, 392
  synthetic compounds, 384
  xenobiotic compounds, 384
Biodegradation
  asphaltic compounds, 399
  DDT, 385
  herbicides, 402
  hydrocarbons, 394

measurement, 418
oil, 398
pesticides, 400
petroleum, 398
plastics, 391
polymers, 391
resistance to, 384
solid hydrocarbons, 399
Biodegradative capacity, 384
Biodeterioration, 349
leather, 351
wood, 351
Biodisc system, 364
Biofilm, 353
Biofilters, 432
Biogas, 359
Biogenic elements, 290
Biogeochemical cycling, 289–310,
314–341
human impact on, 292
interrelationships, 339
Biohydrometallurgy, 439
Bioleaching, 440
copper, 442–444
influences of ecological factors, 441
oil shale, 447
uranium, 444
Biological control, 460, 465, 467
practical considerations, 477
Biological incinerators, 384
Biological oxygen demand, 360
Bioluminescence, 178, 526
reporter strain, 178
Biomagnification, 385, 386, 390, 407,
419, 420
chlorinated hydrocarbons, 386
Biomarker, 177
Biomass, 189
Biometer flask, 197
Bioremediation, 421–433
Alaskan oil spill, 422
efficacy testing, 422
fertilizers, 424
groundwater, 425
hydrocarbons, 425
in situ, 430
microbial seeding, 426
oxygenation, 425

seed cultures, 426
side-effects testing, 423
Bioscrubbers, 432
Biosensors, 420, 421
immunological, 421
lux gene, 421
Biosphere, 246
nitrogen transfer, 315
Biosurfactants, 446
Biotower, 432
Bioturbation, 309
Biphenyl, 389
Bird's-nest fungi, 250, 523
Birds, associated microorganisms, 110
Bitumen, 446
Black molds, 523
Black root rot of tobacco, 95
Black smoker, 326
Blastocrithidia, 51
Blight, 88
cotton, 91
Blue baby syndrome, 321
Blue molds, 523
BOD, 360
Bodo, 261
Bogs, 256
iron deposits, 335
Boletus, 75
Bordetella, 515
Borrelia burgdorferi, 123, 124
Borreliosis, 123
Botrydiopsis, 256
Botrytis cinerea, 87, 465
Bottle effect, 171
Box corer, 169
Bracket fungi, 523
Bradyrhizobium, 79, 80, 278, 515
Branhamella, 516
Breakpoint chlorination, 373
Brevibacterium, 255, 260, 278, 518
Bromoaniline, 402
Bromuron, 402
Brown algae, 526
Brucella, 515
Brush skeleton weed, 475
BTX, 429
Bulk density, 275
Buried slide technique, 166

Butyrivibrio, 113, 516
Butyrivibrio fibrisolvens, 114

Cadmium, 411
Caedibacter, 52, 520
Calcium cycling, 336
Calothrix, 155, 272, 278, 318
Calothrix crustacea, 268
Calpytogena magnifica, 112
Calvin cycle, 308
Candida, 261, 270, 279, 524
Candida albicans, 56, 110, 122
Candida lipolytica, 452
Candida pulcherrima, 84
Candida reukaufii, 84
Candida utilis, 453
Canker, 88
apple trees, 466
Canning, 350
Capillary techniques, 167
Carbaryl, 401
Carbon cycle, 32, 292–306
human impact, 293
oceans, 293
Carbon dioxide, atmospheric, 292
Carbon reservoirs, 292
Carbonaceous sedimentary rock, 292
Carbonate, 336
precipitation, 336, 337
Carcinogen, 351
Caryophanon, 518
Caryophanon latum, 518
Catechol, 396
Cation exchange capacity, 274
Caulerpa, 453
Caulobacter, 167, 255, 260, 278, 514
Causal loop diagram, 149
Cellular evolution, 24–28
Cellular slime molds, 521
Cellulases, 59, 107, 109, 198
Cellulomonas, 278, 518
Cephalosporium, 107, 366
Cephalosporium gramineum, 57
Ceratocystis, 107
Ceratocystis ulmi, 523
Chaetomium thermophile, 359
Chagas' disease, 123
Chamaesiphon, 512

Chelated iron, 334
Chelators, 410
Chemical elements
  biogeochemical cycling, 290
  evolution, 23
  trace, 290
Chemical oxygen demand, 361
Chemiluminescence, 192
Chemocline, 258
Chemolithotrophs, 111
  hydrothermal vents, 327
  nitrification, 320
  sulfur metabolism, 325, 326
Chemostat, 53, 146, 147, 426
Chemosterilization, 350
Chemotaxis, 48, 354
  plant pathogens, 86
Chestnut blight, 86, 523
Chi square, 495
Chitinase, 59, 110, 198
Chlamydia psittaci, 519
Chlamydias, 519
Chlamydomonas, 525
Chloramination, 376
Chlorella, 51, 453
Chlorinated hydrocarbons, 430
  biomagnification, 386
Chlorinated pesticides, 387
  biodegradation, 388
Chlorination, 367, 373, 375, 376
  breakpoint, 373
Chlorine, 375
Chlorobiaceae, 260, 326, 512, 513
Chlorobium, 46, 47, 155, 318, 319, 513
Chlorodibenzodioxins, 389
Chlorodibenzofurans, 389
Chloroflexaceae, 513, 514
Chloroflexus, 514
Chlorophycophyta, 524, 525
Chlorophyll a, 511
Chlorophyll b, 511
Chloroplasts, 28
Chloropseudomonas, 318, 319, 513
Chlorosis, 88
Choke disease, 85
Cholodny-Rossi buried slide technique,
  166

Chondrococcus, 278
Choroflexus aurantiacus, 514
Chromatiaceae, 260, 326, 512, 513
Chromatium, 155, 318, 319
Chromatium vinosum, 54
Chromatium weissei, 54
Chromulina, 255
Chronic toxicities, 423
Chroococcacean cyanobacteria, 512
Chroococcus, 278
Chrysanthemum stunt, 89, 506
Chrysophycophyta, 525
Chytridiomycetes, 521, 522
Chytrids, 521
  parasitism, 60
Ciliates, 367
  protozoa, 528
Ciliophora, 526, 527, 528
Citrobacter, 516
Citrus exocortis, 89, 506
Cladosporium, 107, 253, 255, 366
Clathrulina, 256
Clay particles, 274
  effect on parasitism, 58
Climax community, 132, 134
Cloroflexaceae, 512
Clostridium, 218, 261, 278, 318, 319, 518
Clostridium botulinum, 350, 352, 406
Clostridium malacosome, 468
Clostridium tetani, 462
Clostridium thermocellum, 359
Clouds, 248
Clover, nitrogen fixation, 82
Club root of crucifers, 96
Cluster analysis, 498, 499, 500
Coacervates, 24
Coal, 404
Coal tar, 352
Coccolithophoridae, 271
Cockroaches, 109
Codling moth, 472
Codonosigna, 256
Coelomomyces, 472, 473
Coexistence
  of competing populations, 54
Coliform bacteria, 135, 187, 365
  detection, 377

Coliphage, 508
Colletotrichum, 84
Colletotrichum xanthii, 475
Collybia, 304
Colonization, 353
Colony formation, 39
Colony hybridization, 187, 188
Colorado beetle, 472
Cometabolism, 44, 396, 400, 430
Commensalism, 43–45
Communities, 130
  biologically accommodated, 143
  climax, 132
  diversity, 140
  freshwater, 260
  gene transfer, 138
  limnetic, 151
  pelagic, 151
  pioneer, 132
  population diversity, 141
  recovery of microbial DNA, 144
  respiration, 132
  soil, 276
  stability, 140
  stressed, 143
  succession, 131, 132
Compensation depth, 256
Competition, 40, 53–56
  intrapopulation, 40
  resource based, 54
Competitive exclusion, 53, 55
Complete linkage clustering, 500
Composting, 357
  aerated pile process, 358
  optimal conditions, 359
  thermophiles, 359
Computers, 503
Conjugation, 139
Conservation till farming, 355
Contact slide, 11
Continental shelf, 266
Continental slope, 266
Continuous feed reactor, 357
Convoluta roscoffensis, 115
Cooperation
  Dictyostelium, 39
  genetic basis, 40

intrapopulation, 38
metabolic basis, 39
Copper
bioleaching, 440, 442
Coprophagous animals, 104
Corals, 115
polyp, 115
reefs, 268, 337
Correlation, 496
Corynebacterium, 90, 122, 131, 278, 518
Coulter counter, 185
CPE, 189
Crenarchaeota, 25
Creosote, 352
bioremediation, 422
Crithidia, 51
Crofton weed, 475
Cross-feeding, 45
Crown gall, 91, 92
Crude oil biodegradation, 398
Crustose lichens, 50
Cryptococcus, 261, 279, 524
Cryptococcus laurentii, 83
Cultural eutrophication, 372
Cup fungi, 523
Cyanellae, 51, 115
Cyanobacteria, 131, 260, 278, 318, 511, 512
blooms, 333
evolution, 26
heterocysts, 31
Cyanobacteriales, 511
Cyanophage, 508
Cyclohexane, 396
Cyclotella meneghiniana, 55
Cylindrospermum, 278, 318
Cyrptococcus, 270
Cytopathic effect, 189
Cytophaga, 59, 260, 270, 514
Cytophagales, 514
Cytoplasmic polyhedrosis viruses, 468
C_0t plots, 144
$C_0t_{1/2}$, 144

2,4-D, 388, 401
Dactylaria, 117

Dactylella, 117
Daphnia, 104, 423
DAPI, 180
Daptobacter, 155
DDT, 384, 385, 386, 387, 389, 401, 417, 419, 420, 426
Dealkylation, 401
Deep-diving submersibles, 169
Deep-sea hydrothermal vents, 111
Defined substrate technology, 377
Dehalogenation, 386, 388
Dehydrogenase, 198
Dendrogram, 501
Denitrification, 322, 323, 405, 407
acetylene blockage assay, 323
products, 322
Dental caries, 148
Dependent variable, 485
Dermatophilaceae, 519
Desert crusts, 154, 155
Desert soils, 277
Desertification, 355
Desulfotomaculum, 318, 319, 405, 518
Desulfovibrio, 44, 47, 155, 261, 318, 319, 405, 445, 447, 516
Desulfovibrio desulfuricans, 409, 516
Desulfovibrio vulgaris, 52
Desulfuration, 324
Desulfurococcus, 218, 510
Desulfuromonas acetoxidans, 328
Detection
gene probes, 174
immunological, 173
Detergents, 392
Detrital particles, 104, 105, 133
microbial succession, 133
Detritivores, 306
Detritus feeders, 104
Deuteromycotina, 522, 524
Diatoms, 54, 271, 338, 525
communities, 141
Dichloroaniline, 49, 402, 424
Dichloroethylene, 430
Dictyostelium, 39
Dictyostelium discoideum, 521
Didinium, 61, 261
Didinium nasutum, 11, 61

Differential media, 187
Difflugia, 256
Digestor, anaerobic, 369, 370, 371
Dimethylmercury, 408
Dimethylselenide, 410
Dimethylsulfide, 324
Dimorphism, 522
Dinoflagellates, 115, 271, 526
in corals, 115
Diols, 397
Dioxins, 389, 390
Dioxygenases, 394, 396
Direct count procedures, 179–185
Direct microscopic observations, 179
Direct viability count, 184
Disinfection, 367, 375, 376, 463
Dissimilatory nitrate reduction, 322, 323
Dissimilatory nitrite reductase, 323
Dissolved oxygen, 361
Disturbance
effect on succession, 132, 137
Diversity, 140–145
effect of pollution, 143
effect of stress, 142
genetic, 144
Shannon-Weaver index, 141
DNA
recovery from environmental samples, 174
reannealing
index of diversity, 144
Dolomite, 292
Downy mildew, 96
Drinking water, 376
standards, 378
Dum-dum fever, 123
Duncan procedure, 494
Dunnett procedure, 494
Durham tube, 377
Dutch elm disease, 523

Earth
crust, 271
early conditions, 23
Earthstars, 523
EC ratio, 191
Eco-core, 420

Ecological balance, 37
Ecological control, 460
Ecological hierarchy, 130
Ecological niche, 131
Ecological separation, 53
Ecology, definition, 3
Ecosphere, 246, 289
Ecosystem, 145
  energy flow, 289
  experimental models, 145
Ectomycorrhizae, 74, 75–76
Ectomycorrhizal fungi, 75
Ectoparasite, 58
Ectothiorhodospiraceae, 326
Edwardsiella, 516
Eigenvalue, 501
Eijkman test, 377
Elements, biogenic, 290
ELISA, 174, 421
EMB agar, 187
Endogone, 76
Endomyces, 107
Endomycopsis, 107
Endomycopsis fibuliger, 453
Endomycorrhizae, 74, 76–77
Endoparasite, 58
Endospores, 6, 350
Endosymbionts, 51, 520
Endothia parasitica, 523
Endozoic algae, 116
Energy charge, 191
Enrichment culture, 9, 173, 418
Entamoeba histolytica, 375
Enterobacter, 322, 365, 516
Enterobacter aerogenes, 63, 189, 377, 385
Enterobacter agglomerans, 109
Enterobacteriaceae, 516
Enterobactin, 334
Enterochelin, 334, 410
Enterococci, 134
Enteromorpha, 272
Enteroviruses, 378
  detection, 379
Entodinium caudatum, 63
Entodinium vorax, 63
Entomogenous fungi, 472

Entomophthora, 472, 473
Enumeration
  acridine orange direct count, 180
  agar film technique, 179
  agar plate count, 186
  autoradiography, 184
  colony hybridization, 187
  DAPI direct count, 180
  direct count procedures, 179–185
  direct viability count, 184
  DVC procedure, 184
  effects of diluents, 171
  effects of mixing time, 171
  electron microscopic, 184
  epifluorescence microscopy, 181
  FDC method, 182
  fluorescent antibody technique, 182
  INT method, 183
  microorganisms, 178–189
  most probable number method, 189
  particle counter, 185
  plate count media, 186
  pour plate method, 186
  psychrophiles, 172
  respiring cells, 184
  roll-tube method, 186
  surface spread method, 186
  viable count procedures, 186–189
  viruses, 172
Environmental determinants, 213
Enzyme-linked immunosorbant assays, 421
Enzymes
  assays, 198
  from thermophiles, 214
Eosin methylene blue agar, 187
Epichloe typhina, 85
Epicoccum, 84
Epifluorescence microscopy, 181
Epilimnion, 258
Epiphytic microorganisms
  bacteria, 47
  on flowers, 84
  plant surfaces, 83
Epizootics, 468
Epoxidation, 401
Equitability, 141

Ericales, 76
Erosion, 355
Erwinia, 90, 278, 516
Erwinia amylovora, 88, 90, 91
Erwinia herbicola, 85
Erysipelothrix, 518
Escherichia, 322, 365, 516
Escherichia coli, 46, 55, 56, 62, 134, 177, 179, 247, 367, 377
Estuaries, 264
Ethanol, 352
  production as fuel, 447, 448
Euascomycetidae, 523
Eubacteria, 25
Eubacterium, 111, 113
Eubacterium ruminantium, 114
Eucarya, 25
Euglena, 117
Euglenophycophyta, 525
Eukaryotes
  characteristics, 26
  evolution, 25
  flagella evolution, 29
Euphotic zone, 256, 266
Euplotes octocarinatus, 63
Euryhaline microorganism, 214
Eutrophic lakes, 258
Eutrophication, 333, 372
Eutypa armeniaceae, 466
Evenness, 141
Evolution
  abiotic, 23–24
  archaea, 26
  archaebacteria, 26
  bacteria, 27
  biochemical pathways, 30
  cellular, 24–28
  chemical, 23
  events during, 22
  genetic basis, 29–30
  metabolism, 31
  microorganisms, 21–32
  photosynthesis, 31
  physiological diversity, 31–32
  ribosomal RNA, 25
  sulfur metabolism, 31
exo genes, 81

Exoenzymes, 428
Exoskeleta, 336
Extreme thermophiles, 215
Exxon Valdez oil spill, 422, 430

F statistic, 492
Factor analysis, 501
Fatty acids, 395
  basis for antagonism, 56
fdA gene, 178
Fecal coliforms, 378
  detection, 377
Fecal contamination, 377
Fermentation
  in rumen, 113
Ferric hydroxide, 334
Ferrioxamines, 410
Ferromanganese nodules, 336
Fertilization, 425
Fertilizer, 407, 429
FIAs, 421
Film-flow aerobic sewage treatment, 363
Film-mulching technique, 391
Filter feeding, 104–105
Filtration, 351, 376
Fire blight, 88, 90, 91
FITC method, 180, 182
fix genes, 81
Flashlight fish, 119
Flavobacterium, 134, 135, 136, 255,
  260, 270, 278, 304, 322, 365
Flavobacterium brevis, 43
Floridean starch, 526
Flowers, epiphytic microorganisms, 84
Flow-through systems, 146
Fluorescein diacetate stain, 179
Fluorescent antibody technique, 182
Fluorocarbons, 407
Foliose lichens, 50
Food preservation, 349
Food web, 294
  carbon transfer, 294
Foraminiferans, 51, 337, 527
Fossil fuels, 290, 292
Fossils, 21
Fouling, 352, 353
Fox, Sidney, 24

Francisella, 515
Frankia nitrogen fixation, 82
Frankiaceae, 519
Freeze-drying, 351
Freezing, 351
Frequency of dividing cells, 182
Freshwater
  algae, 261
  allochthonous microorganisms, 261
  autochthonous microorganisms, 261
  decomposition, 263
  habitats, 255
  microorganisms, 260
  organic matter turnover times, 262
  phosphates, 333
Frog hoppers, 473
Frost damage, 86
  control, 476
Fruiting myxobacteria, 514
Frustule, 525
Fruticose lichens, 50
Fucus, 526
Fuels
  bioremediation, 429
  production, 446
  spills, 424
Fuligo, 84
Fuller's Earth, 338
Fulvic acid, 274
Fungal diseases of plants, 93, 94–96
Fungal pesticides, 472
Fungal plant pathogens, 93
  spores, 86
Fungi, 279, 520
  abundance in atmosphere, 253
  associated with bark beetles, 108
  associated with ship timber worms, 108
  cultivation by ants, 105, 106
  cultivation by wood-inhabiting in-
    sects, 107
  entomogenous, 472
  freshwater, 261
  in atmosphere, 249
  lignin degradation, 304
  marine, 270
  metal mobilization, 411
  nematode-trapping, 117, 474

  rotifer-trapping, 118
  spores, 279
Fungistasis, 56, 279, 465
Fusarium, 88, 95, 304, 465
Fusarium lateritium, 466
Fusarium solani, 87, 465
Fusobacterium, 111, 516

Gaia hypothesis, 21
Gall, 88
Gall midges, 108
Gallionella, 167, 335, 336
Gamma irradiation, 350
Gas vacuoles, 271
Gasohol, 447
Gastromycetes, 523
Gause, George, 11, 53, 61
Gauze pad method, 379
Gemmiger, 516
Gene probe detection, 174
Gene probes, 188, 378
Genetic, 40
Genetic diversity, 144
Genetic engineering
  for biological control, 476
  ice-minus bacteria, 86
Genetic fitness, 138
Genetic transfer mechanisms, 139
Genetically engineered microorganisms
  detection by PCR, 177
  hydrocarbon-degrading, 427
  regulations, 427
Geotrichum, 279, 366, 524
Geotrichum candidum, 49, 359
Germfree animals, 148, 464
Giardia lamblia, 379
Gibberella zeae, 96
Gibberellins, 72, 73
Gliding motility, 514
Global warming, 293
Gloeocapsa, 155
Gloeotrichia, 318
Glutamate synthase, 320
Glutamine synthetase, 320
Gnotobiotic, 145, 148
Gonyaulax polyedra, 526
Grab samplers, 169

Gram-negative bacteria, 191
Granulosis viruses, 468
Grazing, 63, 103–104
Great Lakes, 408
Great Salt Lake, 259
Green algae, 524
Green bacteria, 511
Green flexibacteria, 512, 513
Green sulfur bacteria, 257, 512, 513
Greenhouse effect, 293
Grey speck disease, 73
Gross photosynthesis, 132
Gross primary productivity
    measurement, 196
Groundwater, 321, 425
Growth factors, 43, 111
Growth kinetics in chemostat, 147
Guano, 315
Guild structure, 130
Gun cell, 118
Gymnomycota, 521
Gypsum, 155

Haber-Bosch process, 315
Habitat, 37, 246
Hadal region, 266
Hafnia, 516
Haldane, John, 23
Halo blight of beans, 91
Halobacterium, 154, 509, 510, 511
Halobacterium salinarium, 253
Halocarbons, 428
    propellants, 386
    resistance to biodegradation, 386
Halococcus, 510
Halophiles, 351, 509
Hansenula, 279
Haptoglossa mirabilis, 118
Haustorium, 118
Heavy metals, 407, 410
    bioaccumulation, 442
Heavy-metal pollution, 383
Helminthosporium, 524
Herbicide
    acylanilide, 403
    biodegradation, 402
Heterobasidion annosum, 465

Heterocysts, 31, 512
Heterotrophic potential, 194–196
Heterotrophic succession, 132, 133
Hirsutella, 473
Hoechst 33258, 192
Hohenbuehelia, 118
Homeostasis, 135, 137
Hooke, Robert, 4
Host-parasite relationship, 57
Hot acidic habitats, 326
Hot springs, 217, 338
    microbial populations, 140
Humic acid, 274
    bound herbicides, 403
    materials, 274
Humification, 403
Humin, 274
Humotex, 360
Humus, 306, 355
    degrading bacteria, 131
Hungate, Robert, 10
hup genes, 82
Hydrocarbons, 396, 427
    biodegradation, 393
    biosynthesis, 450
    landfarming, 428
    rates, 398
    susceptibility to biodegradation, 394
Hydrogen
    bacteria, 307
    consumer, 340
    cycle, 306
    interspecies transfer, 340
    production, 451
    production as fuel, 450
Hydrogen peroxide, 425
Hydrogen sulfide, 73, 324, 326
    oxidation, 325
Hydrogenase, 308
    hup genes, 82
Hydrogentrophs, 340
Hydrosphere, 255
Hydrothermal fluid, 326
Hydrothermal vents, 111, 217, 326, 327
Hydroxylamine, 320
Hydroxylamine oxidoreductase, 320
Hyperparasitism, 60, 119, 467

Hyperplasia, 88
Hyphochytridiomycetes, 521, 522
Hyphomicrobium, 255, 260, 270, 335
Hypochlorite, 375
Hypolimnion, 258
Hypoplasia, 88
Hypothesis testing, 489

Ice-minus bacteria, 86, 172, 476
Image analysis, 183
Immunity, 461
    animals, 462
    plants, 462
Immunoassays, 421
    biosensors, 421
Immunofluorescence, 173
inaZ gene, 178
Independent variable, 485
Indicator organisms, 350, 377, 384
Indoleacetic acid, 73
Infection thread, 80
Inipol, 431
Inositol hexaphosphate, 333
In-plant nitrification, 373
Insects
    bacterial pathogens, 468
    bloodsucking, 150
INT method of enumeration, 184
Interactions
    interpopulation, 41–65
    microorganisms and animals, 103–125
    negative, 40–41
    positive, 38–40, 41
Interspecies
    electron transfer, 52
    hydrogen transfer, 340
Iron
    bacteria, 335
    chelated, 334
    cycle, 334
    reduction, 335
Itai-itai disease, 411

Jaccard coefficient, 498
Jannasch, Holger, 12
Japanese beetles, 471
J–Z sample, 168

K strategist, 131
Kala-azar disease, 123
Kappa particle, 52
Kelps, 526
Kepone, 387
Klebsiella, 318, 516
Klebsiella pneumoniae, 62
Kloeckera, 84
Kluyver, Albert Jan, 9
Kluyveromyces, 279
Koch, Robert, 6, 7
Kurthia, 518
Kyasanur forest disease, 123

Labyrinthomorpha, 527
Labyrinthula, 270
Laccases, 305, 426
Lachnospira, 516
Lactic acid, 352
Lactobacillus, 40, 113, 134, 148
Lactobacillus arabinosus, 46
Lactobacillus vitulinus, 114
lacZ gene, 177
Laetisaria arvalis, 465
Lakes, 256
    anaerobic, 155
    composition of microbial community,
        260
    light penetration, 256
    nutrient concentrations, 258
    oligotrophic, 333, 372
    oxygen concentrations, 258
    pH, 259
    phosphate-limited, 333
    photosynthetic sulfur bacteria, 257
    profundal zone microorganisms, 257
    thermal stratification, 258
Laminaria, 453
Laminarinase, 59
Landfarming, 428
Landfills, 356
    anaerobic decomposition, 356
    methane, 357
    methane seepage, 356
Landtreatment, 428, 429
Lantern of Aristotle, 104
Leaching field, 369

Leaf blight of maize, 86
Leaf cutting ants, 106
Leaf hoppers, 473
Leaf-nodule, 85
Leaking underground storage tanks, 425
Lecithinase, 48
Leeuwenhoek, Antonie van, 4
    microscopic observations, 5
Leghemoglobin, 80
Legionella, 124
Legionella pneumophila, 43, 59, 177
    detection by PCR, 177
Legionnaires disease, 124
Leguminous crops
    interactions with rhizobia, 78
    nitrogen fixation, 82
Leishmania donovani, 123
Lembladion lucens, 63
Lentic habitats, 255
Lentinus edodes, 454
Leptosphaeria, 84
Leptospirillum ferrooxidans, 335
Leptothrix, 45, 255, 335, 336, 514
Leptotrichia, 516
Leucosporidium, 523
Licea, 84
Lichens, 50, 84, 104
    sensitivity to air pollutants, 51
Liebig, Justus, 212
Liebig's Law of the Minimum, 212,
    350, 372
Lignin biodegradation, 306
Limestone, 292
    deposits, 290, 337
Limnetic zone, 256
Limnoria, 110
Limulus amoebocyte lysate, 191
Linear ABS, 393
Lipomyces, 279
Liquid waste treatment
    tertiary, 372
Liquid wastes, 360
Listeria, 518
Lithosphere, 271–280
Little leaf disease, 73
Littoral zone, 256, 265
Lotic habitats, 255

Lotka-Volterra model, 61
LPS, 191
lps genes, 81
Lucibacterium, 516
Luciferin-luciferase assay, 190
Luminescent bacteria, 119
lux genes, 178, 421
Lyme disease, 123, 124
Lyngbya, 278, 318
Lyticum, 520

MacConkey's agar, 187
Macromonas, 517
Macrophytes, 153
Magnaporthe grisea, 86
Manganese
    cycle, 335
    nodules, 335, 336
    oxidation, 336
Mann Whitney U test, 495
Margulis, Lynn, 28
Marine
    algae, 270
    bacteria, 270
    ecosphere, 266
    fungi, 270
    habitats, 255, 263–271
    microorganisms, 269
    protozoa, 271
    sediments, 268
Mastigomycota, 521
Mastigophora, 526
Mathematical models, 148–151
    causal loop diagram, 150
    components, 149
    flow diagram, 149
    population descriptions, 151
    predator-prey relationship, 150
Mats, 154
Mean, 487
Media, differential, 187
Median, 487
Megasphaera, 516
Meiofauna, 104
Membrane filtration test, 377
Mercaptans, 324
Mercury, 407, 408

Mesophiles, 215
Metabolic pathway evolution, 31
Metabolism,unity, 9
Metallogenium, 167, 335, 336, 404
Metal, recovery, 439
Metamorphic rocks, 272
Metarrhizium, 472, 473
Methane, 356, 370
  atmospheric, 293
  monooxygenase, 387, 430
  production as fuel, 449
Methanobacillus, 318
Methanobacterium, 44, 113, 135, 510
Methanobacterium omelianskii, 52
Methanobacterium ruminantium, 114
Methanobrevibacter, 510
Methanococcoides, 510
Methanococcus, 326, 510
Methanogenesis, 409
Methanogenium, 510
Methanogens, 52, 340, 370, 450, 509
Methanomicrobium, 510
Methanomonas fervidus, 218
Methanosarcina, 510
Methanospirillum, 510
Methanothermus, 510
Methanothrix, 510
Methanotrophic bacteria, 387
Methoxychlor, 401
Methylation
  mercury, 408
  selenium, 410
Methylcobalamine, 408, 410
Methylmercury, 408, 409
Methylococcus capsulatus, 387
Methylomonadaceae, 515
Methylomonas, 399
Methylparaben, 352
Michaelis-Menten enzyme kinetics, 194
Microbial biomass, 189
  ATP determinations, 190
  based on respiration, 194
  biochemical assays, 190
  chloroform fumigation method, 193
  DNA determination, 192
  lipopolysaccharide determination, 191
  muramic acid determination, 191

  physiological measures, 193
  protein determination, 192
Microbial biomass production, 451
Microbial communities, 37, 130, 132,
  140, 141, 151, 276
  freshwater, 260
Microbial control, 467
Microbial ecology
  books, 14
  career opportunities, 17–18
  definition, 4
  history, 4–11
  international commission, 10
  international symposia, 10
  modern history, 8–11
  periodicals, 13
  relation to ecology, 11–12
  scope of, 3–4
  sources of information, 12–17
Microbial evolution, 21–32
Microbial growth
  at high temperature, 216
  effects of temperature, 215–218
  limiting factors, 212
  rates, 196
Microbial herbicides, 475
Microbial infallibility, 384
Microbial intrapopulation interactions,
  37–41
Microbial mats, 154, 337, 338
Microbial mining, 439
Microbially enhanced oil recovery, 444
Microbiology
  early history, 4–6
  pure culture period, 6–7
Micrococcus, 255, 260, 278, 304, 365
Micrococcus cerolyticus, 110
Micrococcus luteus, 252, 450
Microcoleus, 278, 318
Microcoleus chthonoplastes, 154
Microcosms, 145, 147, 420
  flow-through, 148
Microcyclus, 136, 270
Microcystis, 255, 260
Micromonospora, 260, 278
Micromonosporaceae, 519
Microorganisms

  autotochthonous, 246
  allochthonous, 246
  denitrifying, 322
  diluents in count procedures, 171
  freshwater, 260
  hydrocarbon-utilizing, 430
  in atmosphere, 248
  marine, 269
  phenotypic detection, 172–173
  recovery, 171
  sampling in soil, 166
  soil, 276
Microspora, 527, 528
Microtox assay, 423
Miller, Stanley, 23
Mima, 373
Mirex, 387
Mitochondria, 28
Mixed function oxidases, 394
Mixotricha paradoxa, 29
Mixotrophy, 526
Mobilization, heavy metals, 411
Mode, 487
Model ecosystem, 419
Molecular breeding, 427
Monilia, 107
Monochloramine, 376
Monooxygenases, 394
Moraxella, 373, 516
Morels, 523
Mosquito control, 464
Most probable number (MPN), 189
Mucor, 277
Mucor pusillus, 359
Mucorales, 522
Multiple fission, 528
Multiple range tests, 494
Multivariate analysis of variance, 494
Mu-*lux* vector, 178
Municipal water disinfection, 376
Muramic acid, 191
Mushrooms, 454
  parasitism, 60
Mutagenic residues, 424
Mutualism, 49–52
  lichens, 50
Mycangia, 108

Mycetangia, 108
Mycobacteriaceae, 519
Mycobacterium, 278
Mycobiont, 50
Mycoparasitism, 465
Mycoplasma, 90, 519
Mycorrhizae, 74–77
  habitat restoration, 77
Mycotoxin, 351
Mycoviruses, 508
Myxobacterales, 514
Myxobacteria, 59, 278, 514
Myxococcus, 278
Myxococcus xanthus, 39
Myxoma virus, 474
Myxomycetes, 84, 521
Myxospora, 527, 528

nah gene, 178
Nalidixic acid, 184
Nansen sampling bottle, 168
National Environmental Technology
  Applications Corporation, 423
Natronobacterium, 510
Natronococcus, 510
Natural gas seepages, 399
Naumanniella, 335, 517
Nautococcus, 256, 266
Navicula, 256
ndv genes, 81
Nearshore zone, 266
Necrosis, 88
Nectria galligena, 466
Negleria fowleri, 379
Neisseria, 516
Nematode-trapping fungi, 117, 474
Neritic zone, 266
Neurospora, 249
Neuston, 255–256
Neutralism, 41–43
Nevskia, 255
New Zealand hedgehog, 57
Niche, 130, 131
  competition for, 53
nif genes, 81, 82
Niskin sterile bag water sampler, 168, 169
Nitragin, 319

Nitrapyrin, 407
Nitrate, 320, 405
  ammonification, 322
  deposits, 315
  in groundwater, 321
  leaching to groundwater, 322
  reductase, 198, 322
  reduction, 322, 405
  respiration, 322
Nitrification, 8, 320–322, 323, 373,
  405, 407
  inhibition in climax community, 134
  inhibitors, 321
Nitrifying bacteria, 134, 320, 516
Nitrite, 320, 405, 406
  groundwater, 321
  oxidation, 320
  reductases, 322
Nitroaromatics, 401
Nitrobacter, 56, 260, 320, 321, 420, 517
Nitrobacteraceae, 516
Nitrococcus, 320, 517
Nitrogen, 314, 322
  anthropogenic inputs, 316
  cycle, 314–323, 405
  fertilizer, 316, 407
Nitrogen fixation, 80, 308, 315, 316–319
  acetylene reduction assay, 317
  Anabaena–Azolla, 84
  effect of oxygen, 318
  energy requirements, 319
  exo genes, 82
  ammonification, 322
  Frankia, 82
  free-living bacteria, 319
  in phyllosphere, 85
  in termite gut, 109
  lichens, 50
  lps genes, 82
  ndv genes, 82
  nonleguminous plants, 82–83
  soybeans, 82
  Sym plasmids, 82
  symbiotic, 77–83
Nitrogenase, 80, 198, 316, 317
Nitrogenous wastes, 320
Nitrosamines, 321, 406

Nitrosococcus, 320, 517
Nitrosolobus, 320, 517
Nitrosomonas, 260, 320, 321, 333,
  407, 517
Nitrosospira, 320, 517
Nitrosovibrio, 320
Nitrospina, 320, 517
Nitrospira, 320
Nitrous oxide, 322, 323, 406
Nocardia, 46, 260, 278, 322
Nocardiaceae, 519
nod genes, 81, 82
nodABC genes, 81
nodD gene, 81
NodD protein, 82
Nodularia, 278
Nodulation, 79, 80
  genetics, 81
  role of flavonoids, 81, 82
Nodulin gene expression, 81
Nonlinear ABS, 393
Nonparametric statistical tests, 495
Nostoc, 154, 155, 278, 318, 512
  nitrogen fixation, 83
No-tillage farming, 355
Nuclear polyhedrosis viruses, 468
Nucleic acid hybridization, 174, 187
Nucleus, evolution, 29
Null hypothesis, 489
Numerical taxonomy, 141, 499
Nutrient enrichment, 372
N_2O, 248

Ocean dumping, 366
Ocean trench, 266
Oceanic province, 266
Oceans, 265
  carbon cycling, 293
Ochrobium, 335, 517
Octanol partitioning, 419
Octopine, 92
Odors, 432
Oidium, 86
Oil biodegradation, 393, 398
  bioremediation, 425
  limiting factors, 398
  pollution, 383

rates, 398
shale, 447
spills, 399
Oily sludges
landfarming, 424, 428
Oleophilic fertilizers, 425, 431
Oligoheterotrophs, 306
Oligotrophic lakes, 258
Oomycetes, 521, 522
Oparin, Alexander, 23
Opines, 93
Optimal growth temperature, 215
Orcinol, 48
Organelles, 28
evolution, 28–29
Organic
compound persistence, 384
matter decomposition, 263
molecules, abiotic formation, 23
Organomercury fungicides, 408
Organophosphate insecticide, 49
Organosulfur decomposition, 324
Oscillatoria, 154, 278, 318
Oscillatorian cyanobacteria, 512
Oxidation ditches, 368
Oxidation lagoons, 368
Oxygen, 309
basis for antagonism, 56
cycle, 308
impact on evolution, 32
in soil, 275
Oxygenic photosynthesis, 31
Oxyphotobacteria, 511
Ozone, 32, 248, 309, 376, 406
depletion, 407

P/R ratio, 132
Paracoccus denitrificans, 322
Paralytic shellfish poisoning, 121
Paramecium, 51, 63, 141, 261, 528
Paramecium aurelia, 52, 53, 520
Paramecium bursaria, 11, 53, 61
Paramecium caudatum, 11, 53, 61
Parametric statistical tests, 490
Parasitism, 57–60
Parathion, 49
Pasteur, Louis, 6, 7

Pasteurella multocida, 474
Pasteurization, 350
Pathogens
opportunistic, 122
PBBs, 389, 390
PCBs, 388, 389, 390, 424
PCR, 175
detection of genetically engineered
microorganisms, 177
Pearson product moment correlation
coefficient, 496
Pectinases, 87
Pedomicrobium, 335
Pedoscope, 167, 168
Pelagic, 266
Pelagic zone, 266
Pelodictyon, 513
Peloscope, 168
Peltigera, 50
Penicillin, 7
Penicillium, 7, 107, 131, 277, 279,
366, 524
Penicillium piscarium, 40, 49
Peniophora gigantea, 466
Percent respiration, 195
Peroxidase, 49, 304, 305, 426
Pesticides, 387, 426
biodegradation, 400
biodegradability, 401
Petroff-Hauser chamber, 179
Petroleum, 393
biodegradation, 393
bioremediation, 422
tertiary recovery, 444
pH, lakes, 259
Phaeophycophyta, 525, 526
Phagocytosis, 63
Phanerochaete, 389
Phanerochaete chrysosporium, 389,
426, 428
PHB, 373
Phenylmercury, 408, 409
Phenylurea herbicide, 402
Phoma, 84
Phormidium, 155, 278
Phosphatase, 198, 333
Phosphate

availability in rhizosphere, 73
in soils, 333
organic mineralization, 333
precipitation, 372
Phosphate fertilizer, 332
Phosphate filler, 372
Phosphate solubilization, 333
Phosphates, 332
Phosphine, 332
Phosphorus
immobilization, 333
transformations, 332
Photobacterium, 119, 178, 516
Photobacterium phosphoreum, 423
Photoblepharon, 119
Photoheterotrophs, 511
Photosynthesis
evolution, 31
gross, 132
measurement, 196
oxygen production, 308
sulfur bacteria, 326
Phototrophic bacteria, 511
Phycobiont, 50
Phycocyanin, 526
Phycoerythrin, 526
Phycoviruses, 508
Phylloplane, 83
Phyllosphere, 83
allochthonous fungal populations, 84
ice crystal formation, 85
Phylogenic classification, 25
Phytase, 333
Phytophthora infestans, 521
Phytophthora palmivora, 464
Pilobolus, 251, 522
Pioneer communities, 132
Pisolithus tinctorius, 77
Pityrosporum ovale, 122
Planctomyces, 335
Planktonic diatoms, 271
Plants
bacterial diseases, 90
density and disease, 461
disease symptoms, 88
diseases, 86
dispersal of pathogens, 86

epiphytic microorganisms, 83
fungal diseases, 464
immunization, 462
pathogens, 86
pathology, 86–96
pests, 86
resistance to disease, 461
viral diseases, 89
viruses, 507
Plasmids, 139
assisted molecular breeding, 427
survival, 41
Plasmodiophora brassicae, 96
Plasmodiophoromycetes, 521, 522
Plasmodium vivax, 528
Plasticizers, 389, 391
Plastics
biodegradable, 392
biodegradation, 391
Plating procedures, 173
Platychrisis, 256
Platymonas convolutae, 115
Plectonema, 278
Pleospora, 84
Plesiomonas, 516
Pleurocapsalean cyanobacteria, 512
Pleuston, 266
Pollutants, 383
biodegradation, 421
detection, 421
Polyangium, 278
Poly-β-hydroxyalkanoates, 392
Poly-β-hydroxybutyrate, 373
Polybrominated biphenyls, 389
Polychlorinated biphenyls, 388
Polychlorinated terphenyls, 390
Polyethylene, 391
Polygalacturonase in nodule formation, 80
Polyketols, 88
Polymerase chain reaction (PCR), 175, 176, 378
Polymers, 391
Polypeptides
abiotic formation, 24
Polyphenoloxidases, 305
Polyphosphates, 373

Polystyrene, 391
Polyvinyl chloride, 391
Populations, 37, 130
mathematical expressions for growth, 152
Porphyra, 272, 453
Potatoes, 476
scab disease, 354
spindle disease, 89
spindle tuber disease, 506
Pour plate method, 186
Predation, 11, 60–63, 367
animals on microbes, 103–105
grazing, 104
Predator-prey relationships, 60
Lotka-Volterra model, 61
Preemptive colonization, 56, 111, 132
Presence-absence test, 377
Preservation of food, 218
Primary production
measurement, 196
phosphate limitation, 333
Primary succession, 132
Principal component analysis, 501, 502
Principal factor analysis, 503
Principle of microbial infallibility, 384
Prions, 506
Prochlorales, 511, 512
Prochlorobacteria, 511
Prochloron, 512
Prochlorophytes, 511
Productivity, phosphate limitation, 333
Profundal zone, 256
Progenotes, 25
Propachlor, 401
Propanil, 49, 402
Propham, 401
Propionate, 352
Propionibacterium, 322
Propylparaben, 352
Prosthecae, 514
Prosthecate bacteria, 255
Protease, 198
Proterospongia, 256
Proteus, 335, 516
Protobionts, 25
Protocooperation, 45–49

Protozoa, 366, 526
endosymbionts, 51
in rumen, 113
marine, 271
parasitism by *Legionella,* 59
pesticides, 472
soil, 280
Pseudanabaena, 512
Pseudocaedibacter, 520
Pseudomonads, 392
Pseudomonas, 46, 48, 73, 86, 90, 131, 177, 178, 255, 260, 261, 266, 270, 277, 278, 304, 318, 322, 335, 336, 399, 426, 427, 515
in phyllosphere, 83
metabolic diversity, 10
soft rots, 90
Pseudomonas aeruginosa, 49, 468, 471
Pseudomonas cepacia, 177, 388, 427
Pseudomonas denitrificans, 322
Pseudomonas fluorescens, 188
Pseudomonas phaseolicola, 91
Pseudomonas putida, 188
Pseudomonas stutzeri, 49
Pseudomonas syringae, 85, 476
ice-minus, 86
Pseudoparenchymatous sheath, 74
Psychrophiles, 54, 213, 215
enumeration, 172
membranes, 216
Psychrotroph, 54
Puccinia chondrillina, 475
Puccinia graminis, 253, 461, 463
Puffballs, 523
Pure culture methods, 6, 173
Purple bacteria, 511
Purple nonsulfur bacteria, 512, 513
Purple sulfur bacteria, 154, 257, 512, 513
Pyrite, 404, 405
Pyrococcus, 510
Pyrodictium, 218, 510
Pyrrophycophyta, 525, 526
Pythium, 461
Pythium ultimum, 465

Q_{10}, 219
R body, 52

r strategist, 131
Rabbits
  biological control, 474
Rabies, 123
Radiation sterilization, 350
Radioimmunoassays, 421
Radiolaria, 271
Radionuclides, 411
Rain, liberation of spores, 250
Rainforest, 77
Range, 488
Recalcitrant
  compounds, 384
  pesticides, 401
Recombinant DNA technology, 427, 471
Recombination, 30, 139
Red tides, 271, 526
Redox potential, 339
Reductive sulfur transformations, 328
Refrigeration, 351
Regolith, 272
Regression analysis, 497
Reporter genes, 177, 421
Residual chlorine, 375
Respiration, 309
  biometer flasks, 197
  effects of temperature, 219
  measurement, 196
  oxygen consumption, 198
  radiolabeled substrates, 197
Resupinatus, 118
Rhabdovirus, 123
Rhizobia, 317
  interactions with leguminous plants, 78
Rhizobiaceae, 515
Rhizobium, 79, 80, 81, 82, 278, 306,
  322, 355, 515
  effect of soil conditions, 80
Rhizobium meliiloti, 82
Rhizoctonia, 461
Rhizoctonia solani, 76, 465
Rhizosolenia, 266
Rhizosphere
  microbial effects on plants, 72–74
  mineral immobilization, 73
  phosphate availability, 73
Rhodomicrobium, 318, 319

Rhodophycophyta, 525, 526
Rhodopseudomonas, 318, 319, 322, 451
Rhodospirillaceae, 261, 512, 513
Rhodospirillum, 318, 319, 451
Rhodosporidium, 270, 523
Rhodotorula, 84, 261, 270, 279
Rhodotorula glutinis, 83
Rhodotorula mucilaginosa, 83
RIAs, 421
Ribosomal RNA
  16S, 25
Rice blast disease, 86
Rickettsia prowazekii, 123
Rickettsia rickettsii, 123, 519
Rickettsia typhi, 123
Rickettsias, 519
Rickettsiella popillae, 468
Riftia pachyptila, 112
Ringworms, 523
Rivers, 259
  nutrients, 259
r–K gradient, 131
RNA library, 25
Rocks, 271, 272
Rocky Mountain spotted fever, 123
Roll-tube method, 186
Root rot, 464
Rotating biological contactor, 364
Rumen, 113
  bacterial populations, 113
  digestion within, 112–115
  fermentation, 113
  protozoan populations, 113
Ruminant animals, 112, 114
Ruminococcus, 113, 135
Ruminococcus albus, 114
Ruminococcus flavefaciens, 114
Rust fungi, 87, 94, 523

Sabouraud dextrose agar, 187
Saccharomyces, 40, 84, 270
Saccharomyces cerevisiae, 452
Salinity, marine habitats, 265
Salmonella, 278, 367, 376, 377, 516
Salmonella typhi, 375
Salt lake, microbial populations, 140
Salt marsh, 151, 264

Salting, 351
Sampling
  air samples, 169
  biological specimens, 170
  bottles for water, 168
  procedures, 166–172
  processing, 171
  sediment samples, 169
  soils, 166
  water, 167
Sanitary landfill, 356
Sanitation, 376
Sarcodina, 526
Sarcomastigophora, 527
Sargassum, 266, 526
Scale insects
  fungal association, 118
  fungal determination of sex, 119
Schizoblastosporion, 279
Schizophyllum, 304
Schizosaccharomyces pombe, 11, 61
Schizothrix, 155, 278
Schwanniomyces, 279
Sclerotium rolfsii, 465
Scopulariopsis brevicaulis, 410
Scytonema, 155, 278, 318
Sea ice algae, 136
Secondary succession, 132
Sedimentary rocks, 272
Seed cultures, 426
Selenium, 409
  methylation, 410
Selenomonas, 113, 135, 516
Selenomonas ruminantium, 114
Seliberia, 335
Septic tank, 369
Septobasidium, 119
Serial symbiosis, 28, 29
Serratia, 516
Sewage, 259, 361
  aerobic treatment, 363
  anaerobic treatment, 371
  bulking, 366
  disinfection, 375
  fungus, 514
  pathogens, 367
  protozoa, 366

primary treatment, 362
secondary treatment, 362
settling, 366
sludge floc, 366
stages of treatment, 362
tertiary treatment, 373
Shannon-Weaver index, 141, 142
Sheathed bacteria, 514
Shelf fungi, 523
Shelf life, 351
Shelford's Law of Tolerance, 212, 350
Shellfish, 378
Shigella, 367, 375, 376, 377, 516
Ship timber worms, 108
Siderocapsa, 335, 517, 518
Siderococcus, 335, 517
Siderophores, 334
Significance levels, 489
Silica, 338
Silicic acid, 337
Silicon cycling, 337
Siloxanes, 337
Simple matching coefficient, 498
Simulation models, 148
Single linkage clustering, 500
Single-cell protein, 451, 452, 453
Skin, microbial populations, 122
Slides
   Cholodny-Rossi technique, 166
   contact, 11
Slime molds, 521
Slime trails, 104
Sludge, 365, 367, 371
Smut fungi, 94, 523
Snow mold, 95
Soil, 272
   agricultural, 354
   algae, 280
   allochthonous microorganisms, 278
   atmosphere, 275
   bacteria, 278
   chemical properties, 274
   classification, 273
   conditioner, 359
   conservation, 355
   corer, 166
   desert, 277

fungi, 279
fungistasis, 279, 465
horizons, 272
humus, 306
layers, 275
microorganisms, 276
organic matter, 274, 355
pH, 354
protozoa, 280
samples, 166
texture, 273
texture triangle, 273
Solid waste, 355
Soredia, 50
Spallanzani, Lazzaro, 5
Spartina, 151, 264
Species assemblage, 130
Species diversity, 140
   indices, 141
   response to thermal shock, 143
Species richness, 140, 141
Sphaerobolus, 251
Sphaerocystis, 104
Sphaeroeca, 256
Sphaerotilus, 335, 336, 366
Sphaerotilus natans, 514
Sphagnum moss, 256
Spirillum, 47, 55, 260, 270, 322
Spirillum serpens, 59
Spirochetes, 514
Spirogyra, 525
Spiroplasma, 90, 519
Spirulina, 154, 512
Spirulina maxima, 453
Splash cups, 250
Spontaneous generation, 5, 6
Spores, 249
   dispersal through air, 249
   survival in atmosphere, 248
Sporobolomyces, 83
Sporobolomyces roseus, 83
Sporolactobacillus, 518
Sporosarcina, 518
Sporozoa, 526, 528
Sporozoites, 528
Stability, relation to diversity, 140
Standard deviation, 488

Standard error, 488
Stanier, Roger, 9
Staphylococcus, 7, 57, 122, 278, 322, 518
Staphylothermus, 510
Static piles, 357
Statistical computer programs, 503
Statistics, 485–504
Stemphylium, 84
Stenohaline microorganism, 214
Stenotolerant, 140
Stentor, 63, 261
Sterilization, 350
Stinkhorns, 523
Stratosphere, 248
Streptococcus, 113, 278
Streptococcus bovis, 114
Streptococcus faecalis, 46, 134
Streptomyces, 48, 90, 178, 260, 278, 465
Streptomyces scabies, 354
Streptomycetaceae, 519
Streptomycetes, 334
   soil, 131
Strip mining, 323, 404
Stromatolites, 154, 337, 338
Stuck digestor, 370
Student Newman-Keuls test, 494
Student *t* test, 491
Stylonychia, 256
Sublittoral zone, 265
Submarine ridge, 266
Subterminal oxidation, 396
Succession, 353
   autotrophic, 132
   gastrointestinal tract, 134
   heterotrophic, 132
   of populations, 131
   on detrital particles, 133
   primary, 132
   secondary, 132
Succinimonas, 113, 516
Succinimonas amylolytica, 114
Succinivibrio, 113, 516
Succinivibrio dextrinosolvens, 114
Sulfate, 323
   preferential utilization of sulfur iso-
      topes, 329
   reducing bacteria, 219

reduction, 409
Sulfide, 324
    biologically generated, 329
Sulfolobus, 25, 326, 442, 510, 511, 517
Sulfolobus acidocaldarius, 218, 335
Sulfur, 323
    cycle, 323–332
    deposition, 326
    elemental, 326
    evolution of metabolism, 31
    globules, 326
    isotope fractionation, 329
    isotopic enrichment, 31
    metabolism, 25
    oxidation, 404
    oxidizing archaebacteria, 511
    oxidizing bacteria
    reductive transformations, 328
    thermal vents, 112
Surface fouling, 352, 353
Surface spread method, 186
Swamps, 256
Sym plasmids, 82
Symbiosis, 49–52
    associations of animals with
        photosynthetic microorganisms, 115
    nitrogen fixation, 77
Synechococcus, 512
Synechocystis, 512
Synecology, 130
Synergism, 45–49
    based on detoxification, 47
    Lactobacillus arabinosus and
        Streptococcus faecalis, 46
    Streptococcus faecalis and
        Escherichia coli, 46
Synthetic polymers, 391
Syntrophism, 45, 46

2,4,5-T, 388, 401, 427
Tanning, 352
Taphrina, 84
Taphrinales, 523
Taq DNA polymerase, 175
Tar globules, 399
TCDD, 390
TCE, 387

TDT, 218
Tectobacter, 520
Temperature
    effect on microbial growth, 215
    growth at high, 216
    growth range, 215
    killing of microorganisms, 218
    optimal, 215
Termites, 107
    associations with microorganisms, 109
    gut microbiota, 109
    higher, 109
    lower, 109
Termitomyces, 109
Tertiary oil recovery, 445
Tertiary sewage treatment, 372
Tetrachloroazobenzene, 49, 402
Tetrachloroethene, 387
Tetrachloroethylene degradation, 387
Tetrahymena pyriformis, 62
Theoretical models, 148
Thermal death time, 218
Thermal proteinoids, 24
Thermal vents
    communities, 326
    sulfur-oxidizing bacteria, 112
Thermoactinomyces, 359, 518
Thermoanaerobacter, 218
Thermoascus auranticus, 359
Thermobacteroides, 218
Thermocline, 258
Thermococcus, 218, 510
Thermodesulfobacterium, 218
Thermodiscus, 218, 510
Thermofilum, 510
Thermomonospora, 359
Thermophiles, 25, 215
    archaebacteria, 217
    bacteria, 359
    bioleaching, 442
    eubacteria, 217
    microorganisms, 216
Thermoproteus, 218, 510
Thermus, 515
Thermus aquaticus, 175, 218
Thielaviopsis basicola, 95
Thiobacillus, 260, 318, 325, 333, 442,

517, 518
Thiobacillus denitrificans, 325, 339
Thiobacillus ferrooxidans, 335, 439,
    441, 443, 444, 518
Thiobacillus novellus, 325
Thiobacillus thiooxidans, 56, 355,
    404, 518
Thiobacillus thioparus, 325
Thiodictyon, 513
Thiomicrospira, 326
Thiopedia, 513
Thioploca, 324
Thiospira, 517
Thiothrix, 324, 366
Thiovulum, 517
Thymidine uptake, 196
Till farming, 355
Times Beach, 389
Tintinnidium, 271
Ti-plasmid, 92
Tissue culture infectious dose, 189
TNT, 401
Tobacco mosaic virus, 462
Togavirus, 123
TOL plasmid, 177
Tolerance range, 213
Tolypothrix, 278, 318
Torula thermophila, 359
Torulopsis, 84, 261, 270
Torulopsis ingeniosa, 83
TOSCA, 418
Total adenylate pool, 191
Toxic Substances Control Act, 418
Toxicity tests, 423
Trace elements, 290
Transduction, 139
Transformation, 139
Trebouxia, 50
Trichia, 84
Trichloroethylene, 430, 432
Trichlorophenoxyacetic acid, 388
Trichoderma, 60, 279, 465, 524
Trichoderma harzianum, 465
Trichoderma viride, 466
Trichodesmium, 266, 271, 318
Trichophyton mentagrophytes, 57
Trichosporon, 270

Trichothecium, 117
Trickling filters, 363, 432
Trihalomethane, 376
Trimethylarsines, 410
Tristeza virus, 462
Trophosome, 112
Trophozoites, 528
Tropical oceans, 268
Troposphere, 247
Truffles, 523
Trypanosoma, 464
Trypanosoma cruzi, 123
Trypanosoma gambiense, 123
Tukey test, 494
Tyndall, John, 6
Typhula, 95
Typhus, 123

Ulothrix, 525
Ultraviolet light, 32
Ulva, 525
Underground storage tanks, 425
Unit community, 130
UNOX process, 368
Upwelling, 267, 268
Uranium
    bioleaching, 444
Urey, Harold, 23
Urkaryotes, 25
Uromyces appendiculatus, 87
Uromyces rumicis, 475
UV light radiation, 248

Vaccination, 462
Van Niel, C. B., 9
Van Dorn sampling bottle, 168
Variance, 488
Vaucheria, 525
Vectors, 464
Veillonella, 135, 516
Veillonella alcalescens, 114

Veillonella alcaligenes, 409
Venturia, 95
Verticillium, 461
Vesicular-arbuscular (VA) mycor-
    rhizae, 76
Vexillifera, 62
Viable count procedures, 173, 186–189
Viable plate count
    differential media, 187
    selective media, 187
Vibrio, 110, 178, 260, 270, 318, 322, 516
Vibrio cholerae, 173, 375, 516
Vibrio fischeri, 178
Vibrio parahaemolyticus, 136
Vibrionaceae, 516
Vinyl chloride, 430
Viral pesticides, 468
Viroids, 89, 506
Viruses, 378, 506
    detection in water, 380
    enumeration, 172
    obligate intracellular parasites, 58
    of microorganisms, 508
    pathogenic to plants, 89
    recovery, 172
Vitamin K, 111
Vitamins, 48
Vmax, 194
Volatile organic chemicals (VOCs), 432
Volvariella volvacea, 454
Volvox, 525
Vorticella, 63, 256, 261, 366

Wall effect, 54
Wastes
    anaerobic digestion, 370
    minimization, 357
    separation, 357
Wastewater treatment, anaerobic, 369
Water molds, 521
Water quality testing, 377

White Cliffs of Dover, 337, 527
White rot fungus, 304, 426
White smokers, 326
Wilt, 88
Wilt of tomatoes, 88
Windrows, 357
Winogradsky, Sergei, 8, 276, 325
Wood wasps, 108
Wood-boring shipworms, 110

Xanthan gums, 445
Xanthomonas, 90, 278
Xanthomonas campestris, 445
Xanthomonas malvacearum, 91
Xenobiotic chemicals, 428
    interactions, 420
Xenobiotic compounds, 384
Xenobiotic pollutants, 386
    biodegradation, 422
    detection, 421
    lipophilic character, 419
Xenorhabdus, 178
Xenorhabdus nematophilus, 468
Xenospores, 248
xylE reporter gene, 178

Yeasts, 279, 523
    phyllosphere, 83
Yellow fever, 123
Yersinia, 516
Yersinia pestis, 123

ZoBell, Claude, 12
Zoochlorellae, 51, 115
Zooflagellates, 133
Zooglea, 365
Zooglea ramigera, 363, 366
Zooxanthellae, 51, 115
Zygomycotina, 522
Zymogenous microorganisms, 8, 131,
    277